The following supplements are compatible with the following textbooks from Addison-Wesley Publishing Co.:

Drafting Technology,

Engineering Design Graphics,

Design Drafting,

Graphics for Engineers,

Geometry for Engineers

DRAFTING TECHNOLOGY PROBLEMS (With Computer Graphics) is a problem book designed to cover basic graphics, descriptive geometry, and specialty drafting areas. It is designed to accompany *Drafting Technology,* that is available from Addison-Wesley Publ. Co., Reading, Mass. 01867. 131 pages.

BASIC DRAFTING (With Computer Graphics) is a problem book that covers the basics for a one-semester high school drafting course. 67 pages.

CREATIVE DRAFTING (With Computer Graphics) is a problem book that covers mechanical drawing and architectural drafting for the high school. 106 pages.

DRAFTING & DESIGN (With Computer Grahpics) is a problem book for mechanical drawing for a high school or college course. 106 pages.

DRAFTING FUNDAMENTALS 1 is a problem book for a one-year high school course in mechanical drawing. 94 pages.

DRAFTING FUNDAMENTALS 2 is a second version of **DRAFTING FUNDAMENTALS 1** for the same level and content. 94 pages.

TECHNICAL ILLUSTRATION is a problem book for a course in pictorial drawing for the college or high school. 70 pages.

ARCHITECTURAL DRAFTING is a problem book for a first course in architectural drafting. 71 pages.

GRAPHICS FOR ** ** **Computer Graphic** first course in enginee student. 100 pages.

GRAPHICS FOR ENGINEERS 2 (With Computer Graphics) is a problem book for a first course in engineering graphics for the college student. 100 pages.

GRAPHICS FOR ENGINEERS 3 (With Computer Graphics) is a problem book for a first course in engineering graphics for the college student. 108 pages.

GEOMETRY FOR ENGINEERS 1 (With Computer Graphics) is a problem book for a college-level descriptive geometry course. 100 pages.

GEOMETRY FOR ENGINEERS 2 (With Computer Graphics) is a problem book for a college-level descriptive geometry course. 100 pages.

GEOMETRY FOR ENGINEERS 3 (With Computer Graphics) is a problem book for a college-level descriptive geometry course. 100 pages.

GRAPHICS & GEOMETRY 1 (With Computer Graphics) is a problem book for a college course in graphics and descriptive geometry. 119 pages.

GRAPHICS & GEOMETRY 2 (With Computer Graphics) is a problem book for a college course in graphics and descriptive geometry. 121 pages.

GRAPHICS & GEOMETRY 3 (With Computer Graphics) is a problem book for a college course in graphics and descriptive geometry. 138 pages.

CREATIVE PUBLISHING CO.
Box 9292   College Station, Texas   77840
Phone 409-775-6047

Engineering
Design
Graphics

SIXTH EDITION

James H. Earle

Texas A&M University

Engineering Design Graphics

SIXTH EDITION

Addison-Wesley Publishing Company

Reading, Massachusetts · Menlo Park, California · New York
Don Mills, Ontario · Wokingham, England · Amsterdam
Bonn · Sydney · Singapore · Tokyo · Madrid · San Juan

Tom Robbins, David L. Rogelberg, Katherine Harutunian Sponsoring Team
Karen Myer Production Supervisor
Patsy DuMoulin Production Coordinator
Joseph K. Vetere Technical Art Consultant
Lorraine Hodsdon Layout Artist
Hugh Crawford Manufacturing Supervisor
Pat Steele Copyeditor
Melinda Grosser for *silk* Interior Designer
Robert P. McCormack Cover Designer

Many of the designations used by manufacturers and sellers to distinguish their products are claimed as trademarks. Where those designations appear in this book, and Addison-Wesley was aware of a trademark claim, the designations have been printed in initial caps or all caps.

The program and applications presented in this book have been included for their instructional value. They have been tested with care, but are not guaranteed for any particular purpose. The publisher does not offer any warranties or representations, nor does it accept any liabilities with respect to the programs or applications.

Library of Congress Cataloging-in-Publication Data

Earle, James H.
  Engineering design graphics/James H. Earle.—6th ed.
    p.   cm.
  Includes index.
  ISBN 0-201-16893-6
  1. Engineering design. 2. Engineering graphics. I. Title.
  TA174.E23   1990
  620′.0042′0222—dc20
                                                                    89-34324
                                                                    CIP

Reprinted with corrections June, 1990

CDEFGHIJ-HA-943210

Preface

Engineering Design Graphics covers the principles of engineering drawing, computer graphics, descriptive geometry, design, and problem solving for a college-level course. The sixth edition is a major revision of the previous edition, but it retains the same class-tested format and general sequence of topics.

Objectives

The objective of this book is to support a course in which the student learns:

- ANSI standards and techniques of preparing engineering drawings.
- How to solve three-dimensional problems by descriptive geometry.
- The principles and applications of the design process.
- How to use drafting instruments to prepare drawings.
- How to use computer graphics to prepare engineering drawings.

Above all, this textbook is designed to help students *expand their creative talents* and *communicate their ideas* in an effective manner.

Format

Engineering Design Graphics provides self-instructional examples that enable the student to work independently. Many examples are presented in a step-by-step sequence to illustrate how problems are solved. A second color is used to emphasize sequential points of problem solution. The problems at the ends of chapters are graduated from simple to difficult to offer a range of assignments.

The computer graphics examples and illustrations have been tinted in a second color to highlight them for easy reference. Key points, not related to computer methods, have been tinted in gray to assist the student in browsing through the text.

Computer Graphics

Chapters 10 and 38 are devoted to computer graphics. Chapter 10 gives a general overview of hardware and software. Chapter 38 covers the use of AutoCAD Release 10 software for the microcomputer.

In addition, AutoCAD applications have been integrated throughout the other chapters to give a dual approach (by pencil and computer) to the solution of most types of problems.

AutoCAD software was selected as the software to feature because it is the most widely sold computer graphics software for the microcomputer. Therefore, students are most likely to encounter AutoCAD in industry.

All computer graphics principles have been presented in illustrations that use a two-step, three-step, or four-step format. Within each step are prompts from AutoCAD as the user would see them on the screen and the responses the user would type into the system. The prompts and responses are in a different typeface to distinguish them from the main text. With this type of presentation, the student will be better able to progress on his own.

Institutions not equipped to cover computer graphics, or AutoCAD, will find the coverage of computer graphics and AutoCAD in this book will provide a general introduction to this method of drawing. *Engineering Design Graphics* has been designed for use with or without computer graphics: **Learning graphics is the major theme.** Therefore, the student will benefit from reading about computer graphics, even if he does not use computer graphics techniques.

As an Aid to Learning

This sixth edition of *Engineering Design Graphics* has been revised to keep abreast of the needs of education and industry. Above all, the book has been made as **teachable** as possible. That is, the examples, illustrations, applications, format, text, and problems have been modified or newly done to assist the teacher in transmitting these important principles to the student.

Additional Revision Features

Some of the major revision features of the sixth edition of *Engineering Design Graphics* are:

- The revision of several hundred illustrations to enhance clarity and student learning.

- The inclusion of several hundred new drawings.
- The inclusion of new working drawing problems in Chapter 24.
- The inclusion of computer graphics examples based on the latest version of AutoCAD.
- The inclusion of computer graphics techniques in all appropriate chapters.
- The addition of many three-dimensional pictorial drawings to improve the clarity of complex examples.

Problems and Supplementary Problems

This text contains over 500 problems; more than 100 are new. These problems offer a range of assignments to aid the student in grasping the principles covered in each chapter.

Seventeen problem books and teachers' guides (with outlines, problem solutions, tests, and test solutions) are available for use with this textbook, and other titles will be introduced in the future. A listing of these books and their source is given in the front endpapers. Thirteen of the manuals have computer graphics problems on the backs of the problem sheets, which allows solution of the problems by both computer and pencil.

Sixteen modules of *SoftVisuals* are available on disks from which multi-colored overhead transparencies can be plotted on transparency film for classroom presentations. Transparency selection can be made from over 500 *SoftVisuals* keyed to this textbook that can be plotted with AutoCAD 2.52 and later versions.

An Educational System

This textbook, used in combination with a correlated problem book, teachers' guide, and *SoftVisuals* transparencies, comprises a complete **educational system** for achieving the maximum in efficiency and effectiveness.

Acknowledgments

We are grateful for the assistance of many who have influenced the development of this volume. Many industries have furnished photographs, drawings, and applications that have been acknowledged in the corresponding legends. The Engineering Design Graphics staff of Texas A&M University has been helpful in making suggestions for the revision of this book. Professor Tom Pollock provided valuable information on various metals for Chapter 20. Professor Leendert Kersten of the University of Nebraska, Lincoln, kindly provided his descriptive geometry programs for inclusion, and his cooperation is appreciated.

We are indebted to Mary Ann Zadfar, Josef Woodman, and Joseph Oakey, of Autodesk, Inc., for their assistance with AutoCAD. We appreciate the assistance and cooperation of Karen Kershaw of MEGACAD, Inc. After our association with these individuals and their companies, it is easy to understand why they are leaders in their respective fields.

Our thanks go to Professor Gerald Vinson, and Russell Echols of Texas A&M for their assistance in checking the manuscript for accuracy.

We are appreciative of the many institutions that have thought enough of our publications to adopt them for classroom use. It is an honor for one's work to be accepted by his colleagues. We are hopeful that this textbook will fill the needs of engineering and technology programs. As always, comments and suggestions for improvement and revision of this book will be appreciated.

College Station, Texas   Jim Earle

Brief Contents

xii

Contents

18   Screws, Fasteners, and Springs **253**

19   Gears and Cams **283**

20   Materials and Processes **298**

21 Dimensioning 312

22 Tolerances 340

27 Points, Lines, and Planes 469

28 Primary Auxiliary Views in Descriptive Geometry 482

Introduction to Engineering and Technology

1.1 Introduction

This book introduces engineering design concepts and applies engineering graphics to the design process. We give examples that have an engineering problem at the core and that require organization, analysis, problem-solving principles, communication, and skill (Fig. 1.1).

Creativity and imagination are essential to the engineering profession. Albert Einstein said, "Imagination is more important than knowledge, for knowledge is limited, whereas imagination

Figure 1.2 "Imagination is more important than knowledge."

embraces the entire world . . . stimulating progress, or, giving birth to evolution" (Fig. 1.2).

1.2 Engineering Graphics

Engineering graphics is the total field of graphical problem solving and includes two areas of specialization: descriptive geometry and working

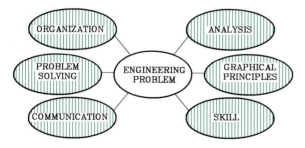

Figure 1.1 Problems in this text require a total engineering approach with the engineering problem as the central theme.

1

drawings. Other areas that can be used for a wide variety of applications are nomography, graphical mathematics, empirical equations, technical illustration, vector analysis, data analysis, and computer graphics. Graphics is one of the designer's main methods of thinking, solving problems, and communicating ideas.

Descriptive Geometry

Gaspard Monge (1746–1818) is the "father of descriptive geometry" (Fig. 1.3). While a military student in France, young Monge used this graphical method to solve design problems related to fortifications and battlements. For not solving problems of this type by the usual (long and tedious) mathematical process, he was scolded by his headmaster. Only after long explanations and comparisons of both methods' solutions was he able to convince the faculty that his graphical method solved problems in considerably less time. Descriptive geometry was such an improvement over the mathematical method that it was kept a military secret for 15 years before it was allowed to be taught as part of the technical curriculum. During Napoleon's reign, Monge became a scientific and mathematical aide to the emperor.

Descriptive Geometry is the projection of three-dimensional figures onto a two-dimensional plane of paper in a manner that allows geo-

metric manipulations to determine lengths, angles, shapes, and other descriptive information about the figures.

1.3   The Technological Team

So rapidly has the scope of technology broadened that it has become necessary for professional responsibilities to be performed by people with specialized training. Technology has become a team effort involving the scientist, engineer, technologist, technician, and craftsman (Fig. 1.4).

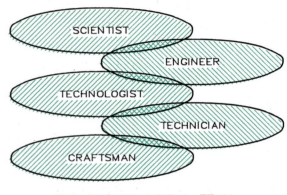
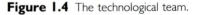

THE TECHNOLOGICAL TEAM

Figure 1.4 The technological team.

Practically all of today's projects require the interaction among team members with a variety of backgrounds (Fig. 1.5).

Scientist

The scientist is a researcher who seeks to establish new theories and principles through experimentation and testing (Figs. 1.6 and 1.7). Scientists are more concerned with the discovery of scientific principles than the application of the principles to products and systems. Scientific discoveries are the basis for the development of practical applications that may not exist until years after discovery.

GASPARD MONGE

Figure 1.3 Gaspard Monge, the "father of descriptive geometry."

Figure 1.5 The technological team must communicate and interact with team members with varying background and areas of expertise. (Courtesy of Honeywell, Inc.)

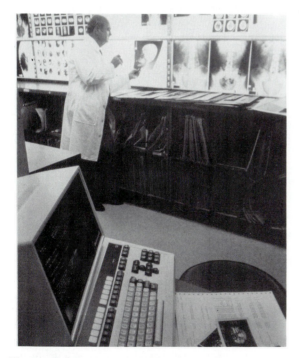

Figure 1.6 Science and technology are merged together to provide a comprehensive information system for radiology through computerization. (Courtesy of Honeywell Inc.)

Figure 1.7 This geologist is studying seismological charts to determine the likelihood of petroleum deposits. (Courtesy of Texas Eastern; photo by Bob Thigpen.)

Engineer

Engineers are trained in science, mathematics, and industrial processes to prepare them to apply the principles discovered by scientists (Fig. 1.8). They are concerned with converting raw materials and power sources into needed products and services. Creatively applying these principles to new products or systems is the design process, the engineer's most unique function. Generally, the engineer uses known principles and available resources to achieve a practical end at a reasonable cost.

Figure 1.8 These engineers are discussing geological data and area surveys to arrive at the best placement of exploratory oil wells. (Courtesy of Texas Eastern; photo by Bob Thigpen.)

Figure 1.9 Engineering technologists combine their knowledge of production techniques with the design talents of the engineers to produce a product. (Courtesy of Omark Industries, Inc.)

Technologist

Technologists are usually graduates of four-year programs in engineering technology where they are trained in science, mathematics, and industrial processes. Whereas the engineer is responsible for research and design, the technologist is concerned with the application of engineering principles to planning, detail design, and production (Fig. 1.9).

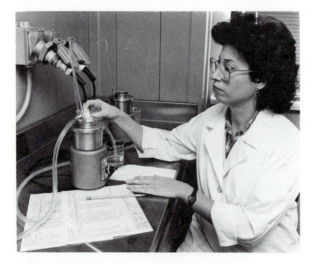

Figure 1.10 This engineering technician performs laboratory tests as a member of the technological team. (Courtesy of Texas Eastern; photo by Bob Thigpen.)

Technologists apply their knowledge of engineering principles, manufacturing, and testing to assist in the implementation of projects and products. They also provide support and act as a liaison to the consumer once the products are in service.

Technician

Technicians assist the engineer and technologist at a theoretical level below the technologist's. In general, they work as liaisons between the technologist and the craftsman (Fig. 1.10). Their work may vary from conducting routine laboratory experiments to supervising craftsmen in manufacturing or construction. Usually the technician is required to have two years of technical training beyond high school.

Figure 1.11 This craftsman is a welder skilled in joining metal parts according to prescribed specifications. (Courtesy of Texas Eastern: *TE Today;* photo by Bob Thigpen.)

Craftsman

Craftsmen are responsible for implementing the engineering design by producing it according to the engineer's specifications. Craftsmen may be machinists who fabricate the product's components or electricians who assemble electrical components. The craftsman's ability to produce a part according to design specifications is as necessary as the engineer's ability to design the part. Craftsmen include electricians, welders, machinists, fabricators, drafters, and members of many other occupational groups (Fig. 1.11).

Designer

The designer may be an engineer, an inventor, or a person who has special talents for devising creative solutions. Designers often do not have an engineering background especially in newer technologies where there may be little precedent for the work involved. Thomas A. Edison (Fig. 1.12) had little formal education, but he had an exceptional ability to design and implement some of the world's most useful inventions.

Figure 1.12 Thomas A. Edison had essentially no formal education, but he gave the world some of its most creative designs.

Stylist

Concerned with the outward appearance of a product rather than the development of a functional design (Fig. 1.13), stylists may design an automobile body or the configuration of an electric iron. An automobile stylist considers the functional requirements of the body, driver vision, enclosure of passengers, space for a power unit, and so forth. But the stylist is not involved with the design of internal details of the product such as the engine or steerage linkage. The stylist must have a high degree of aesthetic awareness and a feel for the consumer's acceptance of designs.

Figure 1.13 The stylist is more concerned with the outward appearance of a product than the functional aspects of its design. (Courtesy of Ford Motor Corp.)

1.4   Engineering Fields

Recent changes in engineering include the emergence of the technologist and technician and the growing number of women pursuing engineering careers. Indeed, more than 15% of today's freshman engineering students are women.

Aerospace Engineering

Aerospace engineering deals with all aspects (all speeds and altitudes) of flight. Aerospace engineering assignments range from complex vehicles traveling 350 million miles to Mars to hovering aircraft used in deep-sea exploration. In the space exploration branch of this profession, aerospace engineers work on all types of aircraft and spacecraft—state-of-the-art missiles and rockets, as well as conventional propeller-driven and jet-powered planes (Fig. 1.14).

Second only to the auto industry in sales, the aerospace industry contributes immeasurably to national defense and the economy. Specialized areas include (1) aerodynamics, (2) structural design, (3) instrumentation, (4) propulsion systems, (5) materials, (6) reliability testing, and (7) pro-

Figure I.14 The aerospace engineer will design and develop systems for space outposts, such as this attitude control system. (Courtesy of Honeywell, Inc.)

duction methods. Aerospace engineers may also specialize in a particular product, such as conventional-powered planes, jet-powered military aircraft, rockets, satellites, or manned space capsules.

Aerospace engineers can be divided into two major areas: research engineering and design engineering. The research engineer investigates known principles in search of new ideas and concepts. The design engineer translates these new concepts into workable applications for improving the state of the art. This approach has elevated the field of aerospace engineering from the Wright brothers' first flight at Kittyhawk, North Carolina, to the penetration of outer space.

The professional society for aerospace engineers is the American Institute of Aeronautics and Astronautics (AIAA).

Agricultural Engineering

Agricultural engineers are trained to serve the world's largest industry, agriculture. Agricultural engineering problems deal with the production, processing, and handling of food and fiber.

Mechanical Power The agricultural engineer who works with manufacturers of farm equipment is concerned with gasoline and diesel engine equipment such as pumps, irrigation machinery, and tractors. Machinery must be designed for the electrical curing of hay, milk processing and pasteurizing, fruit processing, and heating environments in which to raise livestock and poultry. Farm machinery designed by agricultural engineers has been largely responsible for the increased productivity in agriculture.

Farm Structures The construction of barns, shelters, silos, granaries, processing centers, and other agricultural buildings requires specialists in agricultural engineering. Engineers must understand heating, ventilation, and chemical changes that might affect the storage of crops.

Electrical Power Agricultural engineers design electrical systems and select equipment that will provide efficient operation to meet the requirements of a situation. They may serve as consultants or designers for manufacturers or processors of agricultural products.

Soil and Water Control The agricultural engineer is responsible for devising systems for improving drainage and irrigation systems, resurfacing fields, and constructing water reservoirs (Fig. 1.15). These activities may be performed in association with the U.S. Department of Agriculture, the U.S. Department of the Interior, state agricultural universities, consulting engineering firms, or irrigation companies.

Most agricultural engineers are employed in private industry especially by manufacturers of heavy farm equipment and specialized lines of field, barnyard, and household equipment; electrical service companies; and distributors of farm equipment and supplies. Although few agricultural engineers live on farms, it is helpful if they have a clear understanding of agricultural problems, farming, crops, animals, and farmers themselves. Thus agricultural engineering has helped increase the farmer's efficiency: Today, a farmer

Figure 1.15 The agricultural engineer of the future will design food production environments and systems for space settlements. (Courtesy of NASA.)

produces enough food for 80 people, whereas one hundred years ago, a farmer was able to feed only 4 people.

The professional society for agricultural engineers is the American Society for Agricultural Engineers.

Chemical Engineering

Chemical engineering involves the design and selection of equipment that facilitates the processing and manufacturing of large quantities of chemicals. Chemical engineers design unit operations, including fluid transportation through ducts or pipelines, solid material transportation through pipes or conveyors, heat transfer from

one fluid or substance to another through plate or tube walls, absorption of gases by bubbling them through liquids, evaporation of liquids to increase concentration of solutions, distillation under carefully controlled temperatures to separate mixed liquids, and many other similar chemical processes. Chemical engineers may employ chemical reactions of raw products such as oxidation, hydrogenation, reduction, chlorination, nitration, sulfonation, pyrolysis, and polymerization (Fig. 1.16).

Process control and instrumentation are important specialities in chemical engineering. With the measurement of quality and quantity by instrumentation, process control is fully automatic.

Chemical engineers develop and process chemicals such as acids, alkalies, salts, coal-tar products, dyes, synthetic chemicals, plastics, insecticides, and fungicides for industrial and domestic uses. They help develop drugs and medicine, cosmetics, explosives, ceramics, cements, paints, petroleum products, lubricants, synthetic fibers, rubber, and detergents. They also design equipment for food preparation and canning plants.

Approximately 80% of chemical engineers work in the manufacturing industries, primarily

Figure 1.16 The efforts of scientists, chemical engineers, and others are realized in the construction and operation of a refinery that produces vital products for our economy. (Courtesy of Houston Oil and Refining Co.)

the chemical industry. The other 20% or so work for government agencies, independent research institutes, and as independent consulting engineers. New fields requiring chemical engineers are nuclear sciences, rocket fuels, and environmental pollution.

The professional society for chemical engineers is the American Institute of Chemical Engineers (AIChE).

Civil Engineering

Civil engineering, the oldest branch of engineering, is closely related to practically all of our daily activities. The buildings we live and work in, the transportation we use, the water we drink, and the drainage and sewage systems we rely on are all the results of civil engineering. Civil engineers design and supervise the construction of roads, harbors, airfields, tunnels, bridges, water supply and sewage systems, and many other types of structures.

Construction Engineers manage the resources, workers, finances, and materials needed for construction projects, which vary from erecting skyscrapers to moving concrete and earth.

City Planners develop plans for the future growth of cities and the systems related to their operation. Street planning, zoning, and industrial site development are problems that city planners solve.

Structural Engineers design and supervise the erection of buildings, dams, powerhouses, stadiums, bridges, and other structures (Fig. 1.17).

Hydraulic Engineers work with the behavior of water from its conservation to its transportation. They design wells, canals, dams, pipelines, drainage systems, and other methods of controlling and using water and petroleum products.

Figure 1.17 The civil engineer may work on such projects as this massive offshore platform jacket that will be installed in water 1200 feet deep. (Courtesy of Exxon Company; photo by Doug Hoffman.)

Transportation Engineers develop and improve railroads and airlines in all phases of their operations. Railroads are built, modified, and maintained under the supervision of transportation engineers. Design and construction of airport runways, control towers, passenger and freight stations, and aircraft hangars are also done by transportation engineers.

Highway Engineers develop the complex network of highways and interchanges for moving automobile traffic. These systems require the design of tunnels, culverts, and traffic control systems.

Sanitary Engineers help maintain public health through the purification of water and control of water pollution and sewage. These systems involve the design of pipelines, treatment plants, dams, and so on.

Many civil engineers find positions in administration and municipal management. Most civil engineers are associated with federal, state, and local government agencies and the construction industry. Many work as consulting engineers for architectural firms and independent consulting engineering firms. The remainder work for public utilities, railroads, educational institutions, steel industries, and other manufacturing industries.

The professional society for civil engineering is the American Society of Civil Engineers (ASCE). Founded in 1852, it is the oldest engineering society in the United States.

Electrical Engineering

Electrical engineers are concerned with the use and distribution of electrical energy. The two main divisions of electrical engineering are (1) power, which deals with the control of large amounts of energy used by cities and large industries, and (2) electronics, which deals with the small amounts of power used for communications and automated operations that have become part of our everyday lives. These two divisions of electrical engineering have many areas of specialization.

Power Generation poses many electrical engineering problems from the development of transmission equipment to the design of generators for producing electricity (Fig. 1.18).

Power Applications are quite numerous in homes, where toasters, washers, dryers, vacuum cleaners, and lights are used. Of total energy consumption, only about one quarter is used in the home. About half of all energy is used by industry for metal refining, heating, motor drives, welding, machinery controls, chemical processes, plating, electrolysis, and so forth.

Illumination is required in nearly every area of modern life. Improving the efficiency and economy of illumination systems is a challenging area for the electrical engineer.

Computers are a gigantic industry and the domain of electrical engineers. Used with industrial electronics, computers have changed industry's

Figure 1.18 Electrical engineers work with engineers from other disciplines to design power plants such as this one. (Courtesy of Kaiser Engineers.)

manufacturing and production processes, resulting in greater precision and fewer employees.

Communications is devoted to the improvement of radio, telephone, telegraph, and television systems, the nerve centers of most industrial operations.

Instrumentation is the study of systems of electronic instruments used in industrial processes. Extensive use has been made of the cathode-ray tube and the electronic amplifier in industry and atomic power reactors. Increasingly, instrumentation is applied to medicine for diagnosis and therapy.

Military Electronics is used in practically all areas of military weapons and tactical systems from the walkie-talkie to the distant radar networks for detecting enemy aircraft. Remote-controlled electronic systems are used for navigation and interception of guided missiles.

More electrical engineers are employed than any other type of engineer. The increasing need for electrical equipment, automation, and computerized systems is expected to contribute to the growth of this field.

The professional society for electrical engineers is the Institute of Electrical and Electronic Engineers (IEEE). Founded in 1884, the IEEE is the world's largest technical society.

Industrial Engineering

Industrial engineering, one of the newer engineering professions, is defined by the National Professional Society of Industrial Engineers as follows:

> Industrial engineering is concerned with the design, improvement, and installation of integrated systems of men, materials, and equipment. It draws upon specialized knowledge and skill in the mathematical, physical, and social sciences together with the principles and methods of engineering analysis and design to specify, predict, and evaluate the results to be obtained from such systems.

Industrial engineering differs from other branches of engineering in that it is more closely related to people and their performance and working conditions (Fig. 1.19). Often the industrial engineer is a manager of people, machines, materials, methods, money, and the markets involved.

Figure 1.19 Industrial engineers design and lay out automatic conveyor systems such as are used at the Sara Lee plant. (Courtesy of Honeywell, Inc.)

The industrial engineer may be responsible for plant layout, the development of plant processes, or the determination of operating standards that will improve the efficiency of a plant operation. They also design and supervise systems for improved safety and production.

Specific areas of industrial engineering include management, plant design and engineering, electronic data processing, systems analysis and design, control of production and quality, performance standards and measurements, and research. Industrial engineers are also increasingly involved in implementing automated production systems.

People-oriented areas include the development of wage incentive systems, job evaluation, work measurement, and the design of environmental systems. Industrial engineers are often involved in management–labor agreements that affect the operation and production of an industry.

More than two thirds of all industrial engineers are employed in manufacturing industries. Others work for insurance companies, construction and mining firms, public utilities, large businesses, and governmental agencies.

The professional society for industrial engineers is the American Institute of Industrial Engineers (AIIE), which was organized in 1948.

Mechanical Engineering

The mechanical engineer's major areas of specialization are power generation, transportation, aeronautics, marine vessels, manufacturing, power services, and atomic energy.

Power Generation requires that prime movers (machines that convert natural energy into work) be developed to power electrical generators that will produce electrical energy. Mechanical engineers design and supervise the operation of steam engines, turbines, internal combustion engines, and other prime movers (Fig. 1.20).

Figure 1.20 Mechanical engineers designed and supervised the building of this 1.7 liter 4-cylinder automobile engine. (Courtesy of Chrysler Corp.)

Transportation including trucks, buses, automobiles, locomotives, marine vessels, and aircraft are designed and manufactured by mechanical engineers.

Aeronautics requires mechanical engineers to develop aircraft engines as well as aircraft controls and environmental systems.

Marine Vessels powered by steam, diesel, or gas-generated engines are designed by mechanical engineers, as are power services throughout the vessel such as lighting, water, refrigeration, and ventilation.

Manufacturing requires mechanical engineers to design new products and factories to build them in. Economy of manufacturing and uniform quality of products are major functions of manufacturing engineers. The professional society for manufacturing engineers is the Society of Manufacturing Engineers (SME).

Power Services include the movement of liquids and gases through pipelines, refrigeration systems, elevators, and escalators. These mechanical engineers must have a knowledge of pumps, ventilation equipment, fans, and compressors.

Atomic Energy needs mechanical engineers for the development and handling of protective equipment and materials. The mechanical engineer plays an important role in the construction of nuclear reactors.

> The professional society for mechanical engineers is the American Society of Mechanical Engineers (ASME).

Mining and Metallurgical Engineering

Mining engineers are responsible for extracting minerals from the earth and preparing them for use by manufacturing industries. Working with geologists to locate ore deposits, which are exploited through the construction of tunnels and underground operations, mining engineers must have an understanding of safety, ventilation, water supply, and communications. Two main areas of metallurgical engineering are (1) extractive metallurgy, the extraction of metal from raw ores to form pure metals, and (2) physical metallurgy, the development of new products and alloys.

Many metallurgical engineers work on the development of machinery for electrical equipment and in the aircraft and aircraft parts industries. The development of new lightweight, high-strength materials for space flight vehicles, jet aircraft, missiles, and satellites will increase the need for metallurgical engineers (Fig. 1.21). Mining engineers who work at mining sites are usually employed near small, out-of-the-way communities, whereas those in research and consulting often work in metropolitan areas.

> The professional society for mining and metallurgical engineers is the American Institute of Mining, Metallurgical, and Petroleum Engineering (AIME).

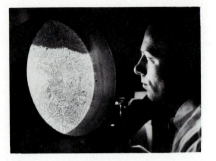

Figure 1.21 A metallograph shows the structure of an alloy that may be used in the construction of a refinery unit. Materials are specially developed for specific applications by metallurgical engineers. (Courtesy of Exxon Corp.)

Nuclear Engineering

The earliest work in the nuclear field has been for military and defense applications. However, nuclear power for domestic needs is being developed for the medical profession and other areas.

Peaceful applications of nuclear engineering are divided into two major areas: radiation and nuclear power reactors. Radiation is the propagation of energy through matter or space in the form of waves. In atomic physics, radiation includes fast-moving particles (alpha and beta rays, free neutrons, and so on), gamma rays, and x rays. Nuclear science is closely allied with botany, chemistry, medicine, and biology.

Figure 1.22 Much of the energy of the future will be economically produced by nuclear reactor plants designed by the nuclear engineer.

The production of nuclear power in the form of mechanical or electrical power is a major peaceful use of nuclear energy (Fig. 1.22). For the production of electrical power, nuclear energy is used as the fuel for producing steam that will drive a turbine generator in the conventional manner.

Although nuclear engineering degrees are offered at the bachelor's level, advanced degrees are recommended for this area. Most of the nuclear engineer's training centers on the design, construction, and operation of nuclear reactors. Other areas of study include the processing of nuclear fuels, thermonuclear engineering, and the use of various nuclear by-products.

> The professional society for nuclear engineers is the American Nuclear Society.

Petroleum Engineering

The recovery of petroleum and gases is the primary concern of petroleum engineers, but they also develop methods for the transportation and separation of various products. Moreover, they are responsible for the improvement of drilling equipment and its economy of operation (Fig. 1.23). In exploring for petroleum, the petroleum engineer is assisted by the geologist and by instruments like the airborne magnetometer, which indicates uplifts on the earth's subsurfaces that could hold oil or gas.

Oil well drilling is supervised by the petroleum engineer, who also develops the equipment to most efficiently remove the oil. When oil is found, the petroleum engineer must design piping systems to remove and transport the oil to its next point of processing. Processing itself is a joint project with chemical engineers.

> The Society of Petroleum Engineers is a branch of AIME, which includes mining and metallurgical engineers and geologists.

Figure 1.23 The petroleum engineer supervises the operation of offshore platforms used in the exploration for oil. (Courtesy of Texas Eastern; photo by Bob Thigpen.)

1.5 Technologists and Technicians

As technology has become more complex, two new members of the technological team have emerged: the technologist and the technician. The primary mission of both is to assist the engineer at a technical level below that of the engineer and above that of the craftsman.

The Technologist works under the supervision of an engineer but has considerable responsibility, thereby freeing the engineer for more advanced applications. Most technologists have a four-year college background in a specialty area of engineering technology that enables them to perform semiprofessional jobs with a high degree of skill. Their interest in the practical aspects of engineering qualifies them to offer advice to the engineer about production specifications and on-the-site procedures when new projects are being designed.

The Technician performs tasks less technical than those performed by the technologist. Most technicians have graduated from a two-year technical program that enables them to be repairers, inspectors, production specialists, or survey-

ors (Fig. 1.24). They may work under the supervision of an engineer, but ideally they are supervised by a technologist. In turn, the technician coordinates the activities between the technologist and the craftsman (Fig. 1.25). These levels of responsibility ensure that the proper skills and qualifications are available throughout the chain of command.

Figure 1.24 These technicians are making a cartographic survey. Data from the tellurometer are being recorded in a notebook. (Courtesy of the U.S. Forest Service.)

Figure 1.25 Technicians monitor quality control in the assembly and testing of microcomputers. (Courtesy of Tandy Corporation.)

1.6 Drafters

Today's drafters carry a great responsibility in assisting the engineer and designer. The experienced drafter may be involved in preparing complex drawings, selecting materials, detailing designs, and writing specifications.

Design and Construction Drawings are made by drafters to explain how to fabricate, build, or erect a project or product in fields such as aerospace engineering, architecture, machine design, mechanical engineering, and electrical and electronic engineering.

Techincal Illustration is a type of graphics that is usually prepared as three-dimensional pictorials to illustrate a project or product as realistically as possible. Technical illustration is the most artistic area of engineering design graphics.

Maps, Geological Sections, and Highway Plats are used for locating property lines, physical features, strata, right-of-way, building sites, bridges, dams, mines, utility lines, and so forth. Drawings of this type are usually prepared as permanent ink drawings.

There are three levels of certification for drafters: drafters, design drafters, and engineering designers.

Drafters are graduates of a two-year, post-high-school curriculum in engineering design graphics.

Figure 1.26 This drafter is aided by a PC computer, graphics software, graphics display, input devices, and a drafting plotter. (Courtesy of Hewlett Packard.)

Design Drafters complete two-year programs in an approved junior college or technical institute.

Engineering Designers are drafters who have completed a four-year college course in engineering design graphics. Graduates of these programs can become certified as technologists.

Computerized systems are being adopted by industry to improve the drafter's productivity. Computer graphics systems do not lessen the need for drafters or engineers to know graphical principles; rather, they offer a different medium of expression (Fig. 1.26).

Problems

1. Write a report that outlines the specific duties and relationships between the scientist, engineer, craftsman, designer, and stylist in an engineering field of your choice. For example, explain this relationship for an engineering team involved in an aspect of civil engineering. Your report should be supported by factual information obtained from interviews, brochures, or library references.

2. Write a report that investigates the employment op-

portunities, job requirements, professional challenges, and activities of your chosen branch of engineering or technology. Illustrate this report with charts and graphs where possible for easy interpretation. Compare your personal abilities and interests with those required by the profession.

3. Arrange a personal interview with a practicing engineer, technologist, or technician in the field of your interest. Discuss with him or her the general duties and responsibilities of the position to gain a better understanding of this field. Summarize your interview in a written report.

4. Write to the professional society of your field of study for information about this area. Prepare a notebook of these materials for easy reference. Include in the notebook a list of books that would provide career information for the engineering student.

Addresses of Professional Societies

Publications and information from these societies were used in preparing this chapter.

American Ceramic Society
65 Ceramic Drive, Columbus, Ohio 43214

The American Institute of Aeronautics and Astronautics
1290 Avenue of the Americas,
New York, N.Y. 10019

American Institute of Chemical Engineers
345 East 47th Street, New York, N.Y. 10017

American Institute for Design and Drafting
3119 Price Road, Bartlesville, Okla. 74003

The American Institute of Industrial Engineers
345 East 47th Street, New York, N.Y. 10017

American Institute of Mining, Metallurgical, and Petroleum Engineering
345 East 47th Street, New York, N.Y. 10017

American Nuclear Society
244A East Ogden Avenue,
Hinsdale, Ill. 60521

American Society of Agricultural Engineers
2950 Niles Road, St. Joseph, Mich. 49085

American Society of Civil Engineers
345 East 47th Street, New York, N.Y. 10017

American Society for Engineering Education
11 DuPont Circle, Suite 200, Washington, D.C. 20036

American Society of Mechanical Engineers
345 East 47th Street, New York, N.Y. 10017

The Institute of Electrical and Electronic Engineers
345 East 47th Street, New York, N.Y. 10017

National Society of Professional Engineers
2029 K Street, N.W.,
Washington, D.C. 20006

Society of Petroleum Engineers (AIME)
6300 North Central Expressway,
Dallas, Texas 75206

Society of Women Engineers
United Engineering Center, Room 305,
345 East 47th Street, New York, N.Y. 10017

2 The Design Process

2.1 Introduction

Engineering graphics and descriptive geometry provide methods of solving technical problems. An engineer working on a design solution must make many sketches and drawings to develop preliminary ideas before effective communication with associates is possible.

2.2 Types of Design Problems

Most design problems fall into one of two categories: **systems design** and **product design.**

Systems Design

Systems design deals with the arrangement of available products and components into a unique combination that yields the desired result. A residential building is a complex system made up of components and products: a heating-cooling system, a utility system, a plumbing system, a gas system, an electrical system, and other systems that form the overall system (Fig. 2.1).

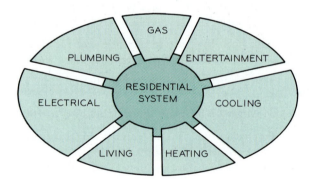

Figure 2.1 The typical residence is a system composed of many component systems.

The engineer who develops a traffic system will need assistance from other professionals, for the project will also involve legal problems, economics, historical data, human factors, social considerations, scientific principles, and politics (Fig. 2.2). By applying engineering principles, the engineer can design a driving surface, drainage systems, overpasses, and other components of the traffic system; however, adhering to budgetary constraints is equally necessary, and the budget is closely related to legal and political requirements.

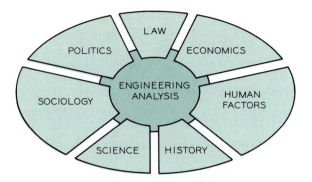

Figure 2.2 An engineering project may involve the interaction of many professions, with engineering analysis the central function.

Traffic laws, zoning ordinances, right-of-way possession, and liability clearances are other legal areas that must be considered. Past trends and historical data also should be reviewed, as should human factors including driver characteristics and safety features that would affect the function of the traffic system. Social problems may arise if heavily traveled highways attract commercial establishments, shopping centers, and filling stations. Finally, scientific principles developed through research can be applied in building more durable roads, cheaper bridges, and a more functional system.

Product Design

Product design is concerned with the design, testing, manufacture, and sale of an item that will be mass-produced such as an appliance, a tool, or a toy (Fig. 2.3). The primary function of an automotive system is to provide transportation.

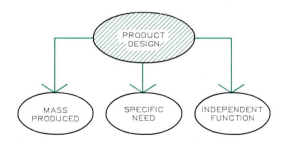

Figure. 2.3 Product design is more limited in scope and specific in application than systems design.

However, the automobile must also provide communication, illumination, comfort, and safety, which seemingly would classify it as a system. Nevertheless, since it is mass-produced for a large consumer market the automobile is regarded as a product.

Product design is related to current market needs, production costs, function, sales, distribution methods, and profit predictions (Fig. 2.4). This concept may encompass a complex system; for example the automobile has expanded to a system of highways, service stations, repair shops, parking lots, drive-in businesses, residential garages, traffic enforcement, and endless other related components.

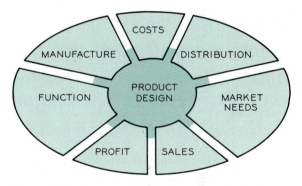

Figure 2.4 Areas associated with product design are related to the manufacture and sale of completed products.

2.3 The Design Process

Design, the responsibility that most distinguishes the engineer from the scientist and technologist, is the act of devising an original solution to a problem by a combination of principles, resources, and products.

This book emphasizes a six-step design process: (1) problem identification, (2) preliminary ideas, (3) problem refinement, (4) analysis, (5) decision, and (6) implementation (Fig. 2.5). Designers work sequentially from step to step, but they may recycle to previous steps as they progress.

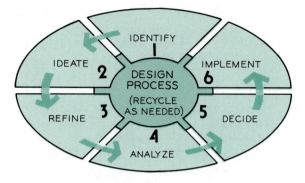

Figure 2.5 The steps of the design process. Each step can be recycled when needed.

Problem Identification

Most engineering problems are not clearly defined at the outset, so they must be identified before an attempt is made to solve them (Fig. 2.6).

A prominent concern in our society is air pollution, but before this problem can be solved, we must identify what air pollution is and what causes it. Is pollution caused by automobiles, factories, atmospheric conditions that harbor impurities, or geographic features that contain impure atmospheres? When you enter a street intersection where traffic is unusually congested, do you identify the reasons for the congestion? Are there too many cars? Are the signals poorly synchronized? Are there visual obstructions?

Problem identification requires a good deal more study than just a simple statement like "solve air pollution." You will need to gather data of several types: field data, opinion surveys,

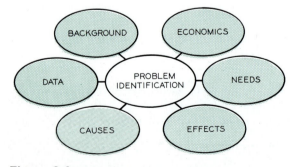

Figure 2.6 Problem identification requires accumulating as much information about the problem as possible before attempting a solution.

historical records, personal observations, experimental data, and physical measurements and characteristics (Fig. 2.6).

Preliminary Ideas

The second step is to accumulate as many ideas for solving the problem as possible (Fig. 2.7). Preliminary ideas should be broad enough to allow for unique solutions that could revolutionize present methods. Many rough sketches of preliminary ideas should be made and retained to generate original ideas and stimulate the design process. Ideas and comments should be noted on the sketches. The more ideas, the better.

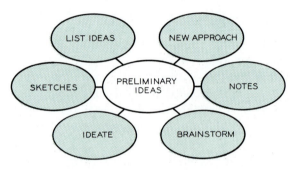

Figure 2.7 Preliminary ideas are developed after the identification process has been completed. All ideas should be listed and sketched to give the designer a broad selection to work from.

Problem Refinement

Next, several of the better preliminary ideas are selected for refinement to determine their true merits. Rough sketches are converted to scale drawings that will permit space analysis, critical measurements, and the calculation of areas and volumes affecting the design (Fig. 2.8). Consideration is given to spatial relationships, angles between planes, lengths of structural members, intersections of surfaces and planes.

The design of the offshore platform structure (Fig. 2.9) required the application of descriptive geometry principles at a massive scale. Lengths, angles, connections, and other geometry was de-

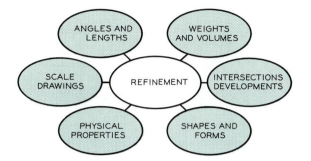

Figure 2.8 Refinement begins with the construction of scale drawings of the better preliminary ideas. Descriptive geometry and graphical methods are used to find necessary geometric characteristics.

Figure 2.9 The refinement of this offshore platform structure required the use of descriptive geometry and many other graphical techniques. (Courtesy of Exxon Company; photo by Doug Hoffman.)

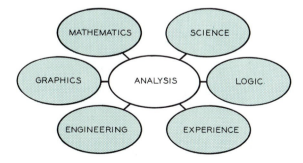

Figure 2.10 In the analysis phase of the design process all available technological methods, from science to graphics, are used to evaluate the refined designs.

termined during the design process before construction began.

Analysis

The step of the design process where engineering and scientific principles are most often used is analysis (Fig. 2.10)—the evaluation of the best designs to determine the comparative merits of each with respect to cost, strength, function, and market appeal. Graphical methods of analysis are means of checking a solution. Data that are difficult to mathematically interpret can be graphically analyzed. Models constructed at reduced scales are also valuable analytical tools. They help to establish relationships of moving parts and outward appearances and to evaluate other design characteristics.

Decision

At this stage a decision must be made. A single design must be selected as the solution of the design problem (Fig. 2.11). Often, the final design is a compromise that offers many of the best features of several designs. The decision may be made by the designer alone, or it may be made by several associates. Regardless of who decides, graphics is a primary means of presenting the proposed designs for a decision. The outstanding aspects of each design usually lend themselves to graphical presentations that compare manufacturing costs, weights, operational characteristics, and other data that would be considered before arriving at the final decision.

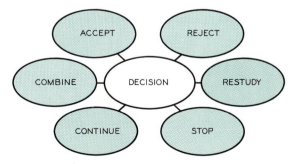

Figure 2.11 Decision is the selection of the best design or design features to be implemented.

Implementation

The final design concept must be presented in a workable form. Working drawings and specifications are usually used as the instruments for fabrication of a product, whether it is a small piece of hardware or a huge bridge (Fig. 2.12). Workers must have detailed instructions for the manufacture of each part, measured to a thousandth of an inch to ensure its proper manufacture. Working drawings must be sufficiently explicit to provide a legal contractual basis for the contractor's bid on the job.

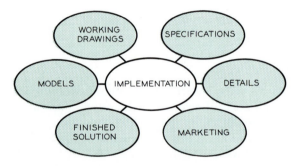

Figure 2.12 During implementation, the final step of the design process, drawings and specifications are prepared from which the final product can be constructed.

2.4 Application of the Design Process to a Simple Problem

To illustrate the steps of the design process as they would be applied to a simple design problem, the following example is given.

Swing-Set Anchor Problem

A child's swing set is unstable during the peak of the swing. The momentum of the swing causes the A-frame to tilt with a possibility of overturning and causing injury. The swing set can accommodate three children at a time. Design a device that eliminates this hazard and has market appeal for owners of swing sets of this type.

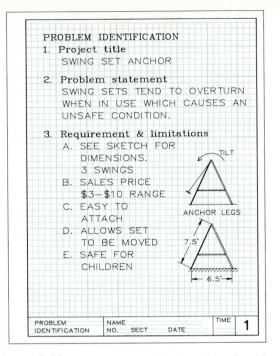

Figure 2.13 A worksheet showing a portion of problem identification.

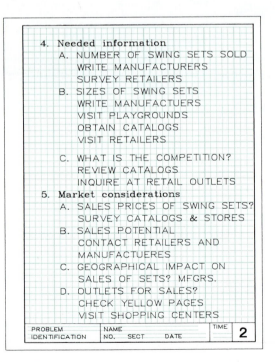

Figure 2.14 The information needed to identify the problem is listed. This information needs to be gathered to complete this part of the design process.

Problem Identification First, the designer writes down the problem statement (Fig. 2.13) and a statement of need. The limitations and desirable features are listed, along with necessary sketches, to enable the designer to better understand the problem requirements (Figs. 2.13 and 2.14). Much of the information in the problem identification step may be obvious to the designer, but making sketches and writing statements about the problem help the designer get off "dead center," a common difficulty at the beginning of the creative process.

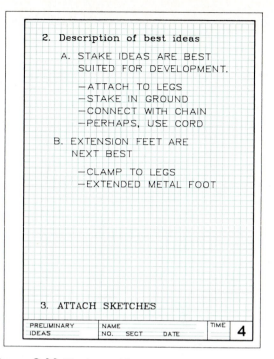

Figure 2.16 The better ideas are selected from the brainstorming list to be developed as preliminary ideas.

PRELIMINARY IDEAS
1. Brainstorming ideas
   A. SANDBAGS
   B. STAKE IN GROUND
   C. SET IN CONCRETE
   D. WIDEN BASE
   E. FOOT ATTACHED TO LEGS
   F. WEIGHTS ON LEGS
   G. REDESIGN A—FRAME
   H. STAKE WITH NYLON CORD
   I. STAKE WITH ROPE
   J. SUCTION CUPS ON PATIO
   K. WATER—FILLED WEIGHTS
   L. BRICKS AT BASE
   M. LIMIT SWING OF SEATS
   N. A—FRAME ON PLATFORM
   O. PROTECT WITH AIR BAGS
   P. WEIGHT LEGS OF A—FRAME
   Q. ROLL BARS ON A—FRAME
   R. PADDED CRASH AREA
   S. FLANGE ON LEGS
   T. SET LEGS IN GROUND
   U. SUPPORT WITH BALLOONS

| PRELIMINARY IDEAS | NAME NO. SECT DATE | TIME | 3 |

Figure 2.15 After holding a brainstorming session, all ideas are listed.

Preliminary Ideas A worksheet is used to list brainstorming ideas (Fig. 2.15). The better ideas and design features are then summarized on the worksheet (Fig. 2.16). These verbal ideas are then translated into rapidly drawn freehand sketches (Fig. 2.17). The designer should attempt to develop as **many ideas** as possible.

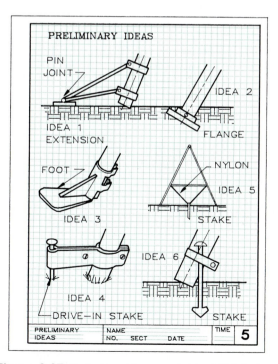

Figure 2.17 Preliminary ideas are sketched. This is the most creative step of the design process.

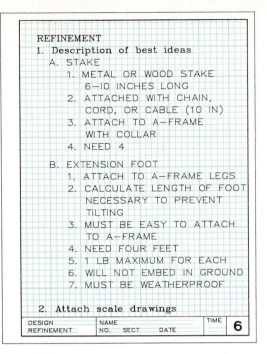

```
REFINEMENT
1. Description of best ideas
   A. STAKE
      1. METAL OR WOOD STAKE
         6-10 INCHES LONG
      2. ATTACHED WITH CHAIN,
         CORD, OR CABLE (10 IN)
      3. ATTACH TO A-FRAME
         WITH COLLAR
      4. NEED 4
   B. EXTENSION FOOT
      1. ATTACH TO A-FRAME LEGS
      2. CALCULATE LENGTH OF FOOT
         NECESSARY TO PREVENT
         TILTING
      3. MUST BE EASY TO ATTACH
         TO A-FRAME
      4. NEED FOUR FEET
      5. 1 LB MAXIMUM FOR EACH
      6. WILL NOT EMBED IN GROUND
      7. MUST BE WEATHERPROOF

2. Attach scale drawings
```
| DESIGN REFINEMENT | NAME NO. SECT DATE | TIME 6 |

Figure 2.18 Refining preliminary ideas begins by giving verbal descriptions of the selected designs.

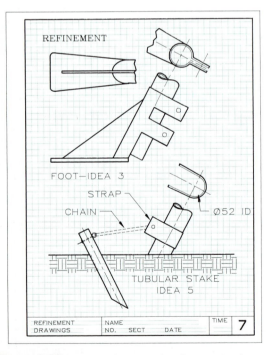

REFINEMENT

FOOT—IDEA 3
STRAP
CHAIN
Ø52 ID
TUBULAR STAKE
IDEA 5

| REFINEMENT DRAWINGS | NAME NO. SECT DATE | TIME 7 |

Figure 2.19 Scaled refinement drawings are used to develop two or more of the designs. Only a few dimensions are needed.

Problem Refinement A verbal description of the design features of one or more of the preliminary ideas is listed on a worksheet (Fig. 2.18) for comparison. The better designs are drawn to scale in preparation for their analysis (Fig. 2.19). Only a few dimensions need to be given at this stage.

Orthographic projection, working drawings, and descriptive geometry may be used. In this example (Fig. 2.19), orthographic views with auxiliary views depict the two designs.

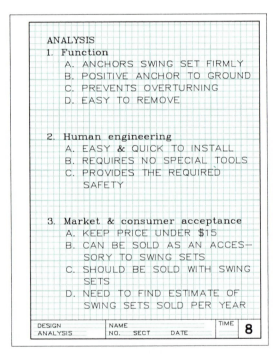

```
ANALYSIS
1. Function
   A. ANCHORS SWING SET FIRMLY
   B. POSITIVE ANCHOR TO GROUND
   C. PREVENTS OVERTURNING
   D. EASY TO REMOVE

2. Human engineering
   A. EASY & QUICK TO INSTALL
   B. REQUIRES NO SPECIAL TOOLS
   C. PROVIDES THE REQUIRED
      SAFETY

3. Market & consumer acceptance
   A. KEEP PRICE UNDER $15
   B. CAN BE SOLD AS AN ACCES-
      SORY TO SWING SETS
   C. SHOULD BE SOLD WITH SWING
      SETS
   D. NEED TO FIND ESTIMATE OF
      SWING SETS SOLD PER YEAR
```
| DESIGN ANALYSIS | NAME NO. SECT DATE | TIME 8 |

Figure 2.20 An analysis worksheet.

Analysis An analysis worksheet is used to analyze the tubular stake design (Figs. 2.20, 2.21, and 2.22). If several solutions are being considered each design should be analyzed.

The force F at the critical angle can be measured or estimated (Fig. 2.23) and used as the basis of a vector polygon to determine the magnitude of R at the base of the swing that must be resisted by the anchor. Again, graphics is used as a design tool.

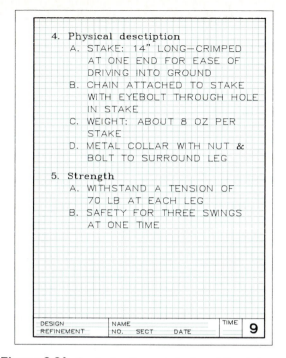

Figure 2.21 Continuing the analysis step of the design process.

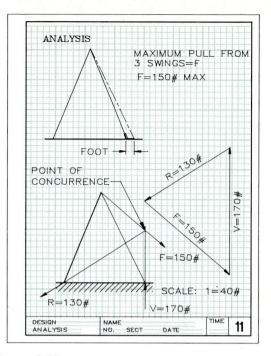

Figure 2.23 Descriptive geometry and graphics can be used to analyze a solution. Here, the force (*F*) is calculated graphically.

Decision Two designs, the stake and chain and the extension foot, are compared in the decision table (Fig. 2.24). Factors for analysis are assigned points for a total value of ten points. The factors for each design are evaluated to determine which has the best overall score.

Conclusions are then given: a summary of the recommended design and the product's profit outlook in the marketplace (Fig. 2.25). A decision is made to implement the stake and chain solution.

Implementation The tubular stake design is presented in a working drawing, where each individual part is detailed and dimensioned. All principles of graphical presentation are used, including a freehand sketch illustrating how the parts will be assembled (Fig. 2.26).

Standard parts—the nuts, bolts, and chain—should be noted but not drawn, since they will not be specially fabricated. With this working drawing, the design has been implemented as far as can be done without actually building a prototype or model.

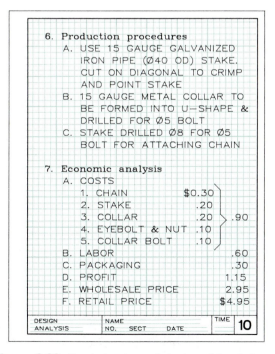

Figure 2.22 Continuing the analysis step of the design process.

DECISION
1. Decision table for evaluation
DESIGN 1: STAKE AND CHAIN
DESIGN 2: EXTENSION FOOT
DEISGN 3:
DESIGN 4:
DESIGN 5:

MAX	FACTORS	1	2	3	4	5
2	FUNCTION	2	1			
2	HUMAN FACT.	1.5	1.5			
1	MARKET ANAL	.5	.2			
1	STRENGTH	1	1			
1	PRODUCTION	.5	.5			
1	COST	.5	.2			
2	PROFITABILITY	1.5	1			
0	APPEARANCE	0	0			
10	TOTALS	8	5.9			

DESIGN DECISION | NAME NO. SECT DATE | TIME | 12

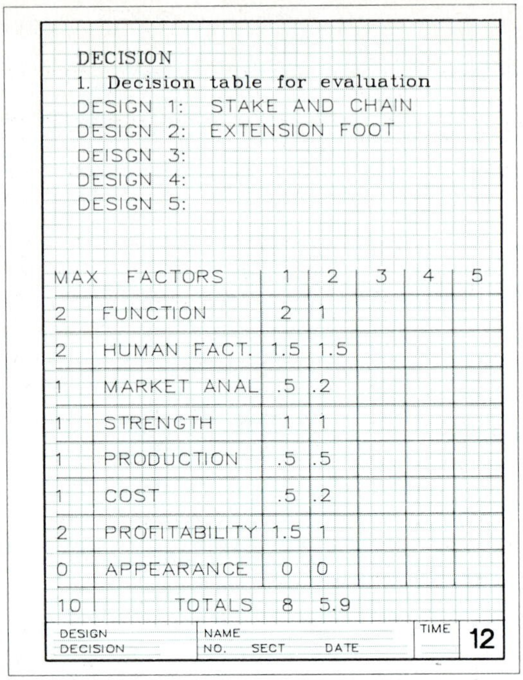

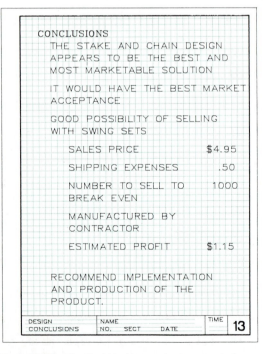

Figure 2.24 A worksheet with a decision table that is used to evaluate the final designs.

CONCLUSIONS
THE STAKE AND CHAIN DESIGN APPEARS TO BE THE BEST AND MOST MARKETABLE SOLUTION

IT WOULD HAVE THE BEST MARKET ACCEPTANCE

GOOD POSSIBILITY OF SELLING WITH SWING SETS

SALES PRICE	$4.95
SHIPPING EXPENSES	.50
NUMBER TO SELL TO BREAK EVEN	1000
MANUFACTURED BY CONTRACTOR	
ESTIMATED PROFIT	$1.15

RECOMMEND IMPLEMENTATION AND PRODUCTION OF THE PRODUCT.

DESIGN CONCLUSIONS | NAME NO. SECT DATE | TIME | 13

Figure 2.25 The final decision and other conclusions are summarized on this worksheet.

The actual part as it would be available for the market is shown in Fig. 2.27. A market analysis or an evaluation of the commercial prospects of the item would be another requirement for implementing any device produced for general market consumption.

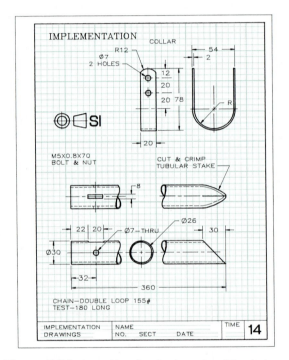

Figure 2.26 A working drawing is made of the design that is to be implemented.

Figure 2.27 The completed swing-set anchor.

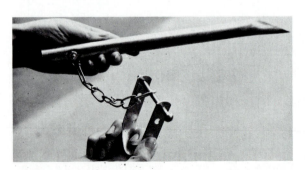

Problems

Most problems are to be solved on 8½-by-11-inch paper, using instruments or drawing freehand as specified. Paper, with a 0.20 inch printed grid, or plain paper can be used for laying out the problems with an engineers' scale. The grid of the given problems and problems in later chapters represents 0.20-in. intervals that can be counted and transferred to a like grid paper or scaled on plain paper.

Each problem sheet should be endorsed and should include the student's name and seat number, the date, and the problem number. All points, lines, and planes should be lettered using ⅛-in. letters with guidelines in all cases.

Essay problems should have their answers lettered, using approved, single-stroke Gothic lettering, as introduced in Chapter 12.

1. Outline your plan of activities for the weekend. Indicate areas in your plans that you feel display a degree of creativity or imagination. Explain why.

2. Write a short report on the engineering achievement or person that you feel has exhibited the highest degree of creativity. Justify your selection by outlining the creative aspects of your choice. Your report should not exceed three typewritten pages.

3. Test your creativity in recognizing needs for new designs. List as many improvements for the typical automobile as possible. Make suggestions for implementing these improvements. Follow this same procedure in another area of your choice.

4. List as many systems as possible that affect your daily life. Separate several of these systems into component parts or subsystems.

5. Subdivide the following systems into components: (a) a classroom, (b) a wrist watch, (c) a movie theater, (d) an electric motor, (e) a coffee percolator, (f) a golf course, (g) a service station, (h) a bridge.

6. Indicate which of the items in Problem 5 are systems and which are products. Explain your answers.

7. Make a list of products and systems that you would anticipate for life on the moon.

8. Assume you have been assigned the responsibility for organizing and designing a skate board installation on your campus. This must be a self-supporting enterprise. Write a paragraph on each of the six steps of the design process explaining how the steps would be applied to the problem. For example, what action would you take to identify the problem?

9. You are responsible for designing a motorized wheelbarrow to be marketed for home use. Write a paragraph on each of the six steps of the design process explaining how the steps would be applied to the problem. For example, what action would you take to identify the problem?

10. List and explain a sequence of steps that you feel would be adequate for, yet different from, the design process given in this chapter. Your version of the design process may contain as many of the steps discussed here as you desire.

11. Can you design a simple device for holding a fishing pole in a fishing position while you are rowing a boat? Make sketches and notes to describe your design.

12. Design a doorstop that could be used to prevent a door from slamming into a wall. Make sketches and notes as necessary to give tangible evidence that you have proceeded through the six steps, and label each step. Your work should be entirely freehand and rapid. Do not spend more than thirty minutes on this problem. Indicate any information you would need in a final design that may not be accessible to you now.

13. List areas that you must consider during the problem identification phase of a design project for the following products: a new skillet design, a bicycle lock, a handle for a piece of luggage, an escape from prison, a child's toy, a stadium seat, a desk lamp, an improved umbrella, a hot-dog stand.

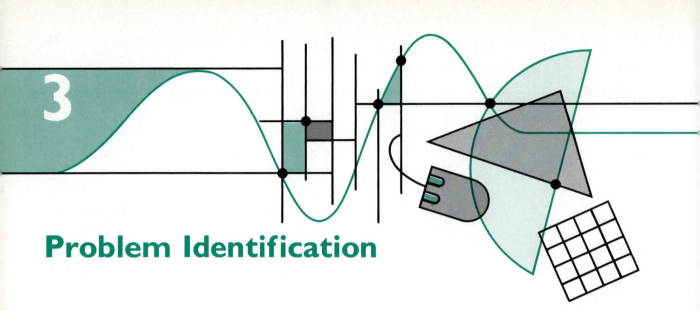

Problem Identification

3.1 Introduction

Problem identification, the initial step that a designer takes to solve a problem, can be either of two general types: (1) identification of a need or (2) identification of design criteria (Fig. 3.1).

Identification of a Need is the beginning of the process. Recognizing a problem, a defect, or a shortcoming in an existing product or situation, the designer investigates the causes and effects, which may result in the development of a new product or solution. The need may be for an improved automobile safety belt, a solution to air pollution, or a special hunting seat.

Identification of Design Criteria is that part of the problem where the designer identifies the specifications that must be met by a new design.

3.2 The Problem Identification Process

Problem identification requires the designer to analyze requirements, limitations, and other background information without becoming involved with the solution to the problem. The following steps should be used in problem identification (Fig. 3.2).

1. **Problem statement.** Write down the problem statement to begin the thinking process.
2. **Problem requirements.** List factors that the design must meet. Some of these statements may be questions that will be answered later when data have been gathered.

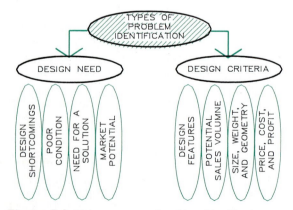

Figure 3.1 The two types of problem identification.

3. **Problem limitations.** List factors that will help determine the design specifications. For example, cannot weigh more than twenty-five pounds; must fit in the trunk of a car.
4. **Sketches.** Make sketches, with notes and dimensions, of the problem's physical characteristics.
5. **Data collection.** Gather data on population trends, related designs, physical characteristics, sale records, market studies, and so on. Once collected, the data should be graphed for easy interpretation.

The designer should review the identification of the problem throughout the design process. Any new information about the problem should be added to the notes.

Figure 3.3 The design of a new automobile demonstrates the problem identification step of the design process. (Courtesy of Ford Motor Co.)

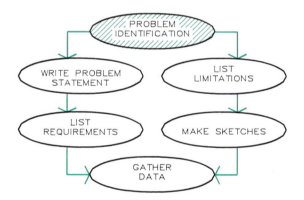

Figure 3.2 The steps suggested for identifying a problem.

3.3 Automobile Design—Problem Identification

Suppose you were involved in the development of a new model automobile that would have a broad market appeal (Fig. 3.3). It would be necessary to consider the prevailing climate of the marketplace: need for fuel economy, rigorous emission controls, expensive safety standards, and everchanging lifestyles. Knowing as much as possible about the consumer is also part of problem identification.

The data graphed in Fig. 3.4 were gathered in a national study about the backgrounds of new-car buyers. As the graph shows, most car buyers are under the age of thirty-six. The most rapid growth has been among female car buyers. Single people buy an increasingly larger portion of new cars. The data in Fig. 3.5 reveal that 84% of all "person trips" are made in private cars, and the remaining 16% is divided among other forms of transportation. Other data show that Americans drive more than a trillion miles per year: 42% for business; 33% for recreation and vacations; 19% for errands; and 5% for education and

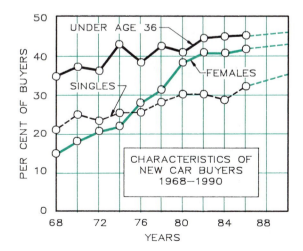

Figure 3.4 The data plotted here were gathered to identify the changing characteristics of new-car buyers.

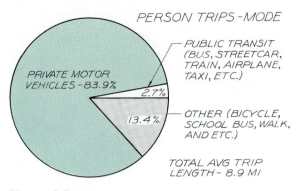

Figure 3.5 When a car is to be designed, it is important to know what kinds of trips it will be used for. This pie chart shows that 84 percent of person trips are made in private automobiles.

civics. Figure 3.6 reflects the boom in the number of men, women, and children participating in outdoor sports. How are the data interpreted? What does it mean? Experience as well as instinct are used to answer these questions.

Ford Motor Company concluded that a car was needed that was economical, fun to drive, lively but safe, and sporty but functional. Since younger people and singles were a greater part of the market, there were things the car did not have to be. It did not need a lot of room for pas-

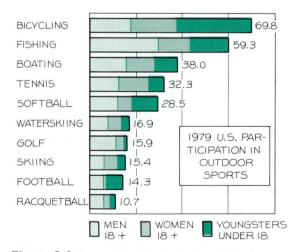

Figure 3.6 Survey data showing the recreational activities the public participates in are presented in this bar graph. These findings suggest that a car should be designed to meet the needs of consumers active in outdoor sports.

sengers; a one- or two-seater with room for luggage was adequate. Likewise, the car did not have to be "hot" to be sporty. Instead, the car needed to be lively and attractive and give good overall performance. The collection of this data, and the knowledge gained from it, resulted in the production of the Ford 1986½ EXP (Fig. 3.7).

Figure 3.7 Ford introduced the 1986½ EXP with features and characteristics identified by the problem identification phase of their design process. (Courtesy of Ford Motor Co.)

3.4 Design Worksheets

Throughout the design process, designers must make numerous notes and sketches as they search for a solution. Periodically reviewing earlier ideas and notes is important to avoid overlooking previously identified concepts. Moreover, a written record of a designer's work will help establish ownership to patentable ideas.

Materials The following materials will aid designers in maintaining permanent records for their design activities.

1. **Sketchpad (8½ by 11 inches).** A sketchpad can be either grid-lined or plain, depending on the person's preference. Sheets should be punched for insertion in a notebook or file.
2. **Pencils.** A medium-grade pencil is adequate for most papers.

3. **Binder or envelope.** All worksheets should be kept in a binder or envelope for reference.

3.5 Hunting Seat—Problem Identification

The following example is used to illustrate the problem identification step of designing a hunting seat.

Hunting Seat

Many hunters hunt from trees to obtain a better vantage point. Design a seat that provides hunters with comfort and safety while hunting from a tree and that meets the requirements of economy and the limitations of hunting.

Worksheet Completion The worksheets in Fig. 3.8 and Fig. 3.9 are typical of the information the designer needs to understand the background of the problem.

PROBLEM IDENTIFICATION

1. **Project title**
   SEAT FOR HUNTING FROM A TREE

2. **Problem statement**
   MANY HUNTERS SIT ON TREE LIMBS WHILE HUNTING, WHICH IS UNSAFE AND UNCOMFORTABLE. A TREE PROVIDES A DESIRABLE VANTAGE POINT FROM WHICH TO HUNT. A SEAT IS NEEDED TO TAKE ADVANTAGE OF THE HEIGHT AND CONCEALMENT OF TREES.

3. **Requirements and limitations**
   A. MUST BE CARRIED TO SITE BY THE HUNTER
   B. MUST PROVIDE SAFETY AND COMFORT
   C. 7'–12' ABOVE GROUND
   D. PROTECTION FROM BAD WEATHER
   E. CONSIDER METHOD OF ASCENDING
   F. COULD DOUBLE AS A BACKPACK
   G. PRICE: $75–$200
   H. WEIGHT: 20 LB MAX
   I. FIT IN TRUNK OF CAR
   J. COLORS TO BE ACCEPTABLE TO HUNTERS

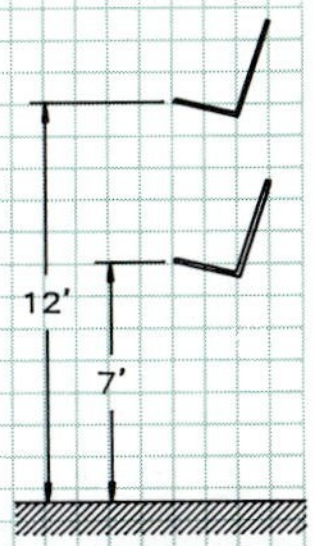

PROBLEM IDENTIFICATION | NAME | FILE | SEC | DATE | GRADE **1**

Figure 3.8 A worksheet for the problem identification step for the hunting seat design.

4. **Needed information**
   A. NUMBER OF HUNTERS WHO HUNT FROM TREES? —CONTACT STATE & NATIONAL GAME OFFICES
   B. WHAT ARE METHODS OF HUNTING? —SURVEY HUNTERS
   C. WHAT ARE THE TECHNIQUES OF HUNTING FROM TREES WITHOUT SEATS? —SURVEY HUNTERS
   D. LAWS CONCERNING HUNTING FROM TREES? CONTACT GAME COMMISSIONS
   E. NUMBER OF BOW & ARROW HUNTERS? —CONTACT ARCHERY ASSOCIATIONS
   F. WHEN AND HOW LONG ARE VARIOUS HUNTING SEASONS? CALL GAME OFFICE
   G. WHAT EQUIPMENT IS CARRIED BY THE HUNTER? —SURVEY HUNTERS, INTERVIEW SPORTING GOODS CLERKS

5. **Market considerations**
   A. WHAT WOULD RETAIL OUTLETS THINK OF A HUNTING SEAT? —INTERVIEW SPORTING GOODS DEALERS
   B. ARE THERE COMPETING PRODUCTS? —LOOK AT ADS IN HUNTING MAGAZINES
   C. WHAT IS BEST PRICE RANGE? —SURVEY DEALERS
   D. HOW MUCH MONEY IS SPENT PER YEAR BY A TYPICAL HUNTER? —SURVEY DEALERS AND HUNTERS
   E. FEATURES TO INCLUDE IN DESIGN? —SURVEY DEALERS & HUNTERS
   F. WHAT WOULD BE POSSIBLE MARKET OUTLETS WHERE SEATS COULD BE SOLD? —VISIT RETAILERS, CHECK CATALOGS

PROBLEM IDENTIFICATION | NAME | FILE | SEC | DATE | GRADE **2**

Figure 3.9 A worksheet that is a continuation of problem identification.

Title and Problem Statement The title of the project is recorded along with a brief problem statement.

Requirements and Limitations The requirements and limitations are listed along with any sketches that would aid in understanding the problem. You might have to list some requirements as questions for the time being.

After further investigation, you should list the limits, for example, must cost between $60 and $100. It is better to give a range of prices or weights rather than to attempt to be exact. Catalogs offering similar products is one source of information on sales prices, weights, and sizes.

Needed Information How many hunters are there? How many hunt from trees or elevated blinds? This and similar information is available from your state game office.

What is the average income of the hunter? How much do they spend on their hobby per year? Sporting-goods dealers could help you by sharing their experiences, and perhaps they could direct you to other sources for answers to these questions.

Market Considerations The designer must think about cost control even in the problem identification step (Figs. 3.8, 3.9, and 3.10).

Figure 3.11 shows how an item is priced from wholesale to retail. Percentages vary by product: There is less profit in retail food sales than in furniture sales. If an item retails for $50, the production cost cannot be more than about $20 to maintain the necessary margins.

Another method of collecting information is surveying hunters about the merits of introducing a hunting seat to the market (Fig. 3.10). The data show that twenty out of fifty people gave the market potential of the seat a high ranking.

Graphs Data is easier to interpret if presented graphically. For example, the number of hunters is compared with those who hunt from trees in

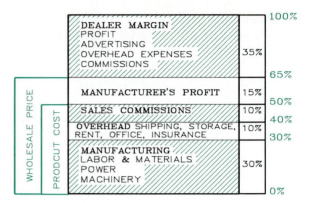

Figure 3.11 A model showing the breakdown of expenses and costs involved in arriving at the retail price of a product.

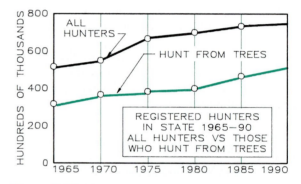

Figure 3.12 The survey data plotted in this graph describes the population trends of potential customers for the hunting seat.

Fig. 3.12. Thorough problem identification includes graphs, sketches, and schematics that improve the communication of the findings of the designer.

The problem identification of this problem is not yet complete. By following the method in this example, you should be able to incorporate your own innovations to arrive at a more thorough problem identification.

3.6 Organization of Effort

A schedule of the activities that must be performed in achieving a solution should be prepared immediately after the identification of a

```
PROBLEM  IDENTIFICATION

NUMBER OF HUNTING LICENSES SOLD IN STATE
AND ESTIMATES OF NUMBER OF THOSE WHO
HUNT FROM TREES:
     YEAR     TOTAL        FROM TREES
     1970     467,000      305,000
     1975     481,000      370,000
     1980     520,000      380,000
     1985     542,000      392,000
     1990     544,000      393,000

CONSUMER SURVEY: QUESTIONNAIRE GIVEN TO
50 HUNTERS. RANKING: 1-HIGH; 4-LOW

OPINION OF THE SEAT AS A GOOD PRODUCT:
     RANK      NO.      %
      1        20       40
      2        15       30
      3         8       16
      4         7       14
               50      100  TOTAL

RETAILER SURVEY

  4 DEALERS WERE POSITIVE ABOUT THE
  PROSPECTS FOR THE HUNTING SEAT.
  SUGGESTED A PRICE OF $100.

ANNUAL EXPENDITURES OF HUNTERS:

     AMMUNITION      $ 50
     CLOTHING          70
     LEASES           100
     GUN              100
     TRAVEL            70
                     $390  TOTAL

PROBLEM          NAME                          GRADE
IDENTIFICATION   FILE      SEC      DATE          3
```

Figure 3.10 A worksheet for collecting data.

need has been sufficiently established. One technique of scheduling project work is Project Evaluation and Review Technique (PERT), which was developed by industries participating in certain governmental projects where coordination of many activities was essential to the successful completion of the project. PERT provides a means of scheduling the activities in their appropriate sequence and reviewing the progress being made toward their completion.

The **critical-path method** of scheduling project activities evolved from, and is used with, PERT. Since some jobs cannot be performed until others have been completed, the critical-path method determines the chain of events that depend on other activities. The critical path is the sequence of tasks requiring the **longest** time and the least flexibility before the project can be completed. Other activities, not in the critical path, can be scheduled to receive secondary emphasis.

3.7 Planning Design Activities

The flowchart in Fig. 3.13 suggests steps for achieving the best results in planning your project. These steps are (1) list the jobs that must be performed on a form called a Design Schedule and Progress Record; (2) prepare an Activities Network to place the jobs in sequence, and (3) prepare an Activity Sequence Chart that graphs the jobs in the same sequence shown in the network.

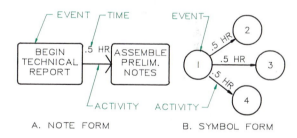

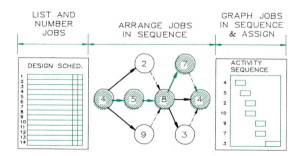

Figure 3.13 The three steps of planning and scheduling a project.

Design Schedule & Progress Record

TEAM 7 PROJECT PRODUCT DESIGN
WORK PERIODS 11 MAN-HOURS 70 FINISHING DATE 4-11 SECTION 501

JOB	ASSIGNMENT	ESTIMD HOURS	ACTUAL HOURS	PER CENT COMPLETE 0 50 100
1	WRITE REPORT-ALL	15		
2	BRAINSTORM-ALL	.5	1	
3	WRITE MFGRS-JHE	2		
4	GRAPH DATA-HLE	2		
5	MARKET ANAL-DL	3	4	
6	COLLECT DATA-TT	2		

Figure 3.14 A Design Schedule and Progress Record showing typical entries to identify example design tasks.

Design Schedule and Progress Record Figure 3.14 suggests a layout for this form. Each job is broken into reasonably small tasks, which are listed on the Design Schedule and Progress Record (DS&PR) form with no concern for their sequence. When these are all listed, estimate the time required for each job in the second column. The sum of the times for each job should be adjusted to approximate the total time allotted for the project. Extra time may be left unassigned for emergencies. Each job is given a number to identify it.

Figure 3.15 Two ways of preparing an Activities Network, note form (A) and symbol form (B), which graphically places the jobs in sequence.

Activities Network The jobs listed on the DS&PR must now be arranged in proper sequence (Fig. 3.15). The note form (Fig. 3.15A) lists the events by name, and the symbol form (Fig. 3.15B) uses the job numbers as the events. An **event** is some specific point in the Activities Network. Events do not require time but are milestones along the sequence of activities.

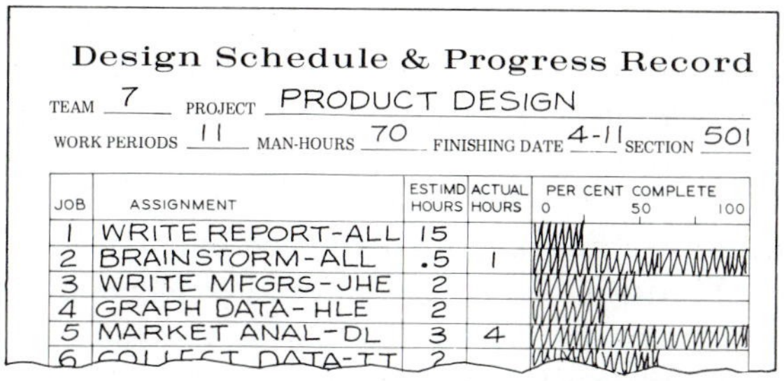

Events are connected by arrows that represent the **activities** that consume time from one event to the next. For example, in Fig. 3.15A, it requires one-half hour to assemble preliminary notes after beginning the technical report. The time is marked on the activity arrow.

Dummy activities indicate a connection between activities even though they require no time. All events must be connected by activity arrows coming to and going from the event. Two dummy activities are used in the Activities Network in Fig. 3.16 to show a sequential connection but no expenditure of time. Dashed lines are used for dummy activities.

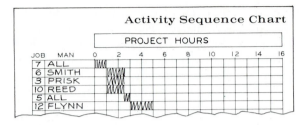

Figure 3.17 The jobs listed on the DS&PR are converted into an Activities Network and then graphed on an Activity Sequence Chart (ASC), which assigns each job to a fixed time schedule in the proper order.

Activity Sequence Chart The jobs of the Activities Network are listed on the Activity Sequence Chart (ASC) (Fig. 3.17). The job numbers are taken from the Network, listed in sequence, and assigned to members of the team. A bar graph shows the time for each job. For example, one hour of the project will be used to perform job 2, which is "brainstorming" on the DS&PR. If your team has four members, this is equivalent to four man-hours of work. Continue until all the jobs are listed. The project hours across the top of the ASC can be used to schedule the events and to estimate when each event should be reached during the project.

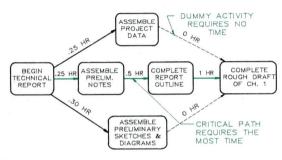

Figure 3.16 The critical path is the path that requires the most time to arrive at the last event of a project.

The critical path marked in the portion of the Activity Network shown in Fig. 3.16 is the longest time path from the first event to the completion of the project; one and three-quarters hours are required to complete a draft of the first chapter of the report.

As progress is made, the status of each assignment can be graphed on the DS&PR (Fig. 3.14). When a job is completed, the hours it required and the completion date are listed in the last two columns. If more time was required than was scheduled, adjustments can be made in subsequent jobs to compensate for this loss.

 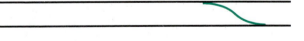

Problems

Problems should be presented on 8½-by-11-inch paper, grid or plain. All notes, sketches, drawings, and graphical work should be neatly prepared in keeping with good practices. Written matter should be typed or lettered using ⅛-in. guidelines.

General

1. Identify a need for a design solution that could be used as a short design problem for a class assignment (less than three man-hours for complete solution). Submit a proposal outlining this need and your general plan for solution. Limit the proposal to two typewritten pages.

2. Assume you were marooned on a deserted island with no tools, supplies, or anything. Identify major problems that you would be required to solve. List factors that would identify the problem in detail. For example, need for food: Determine (a) available sources of food on island, (b) method of storing food supply, (c) method of cooking, (d) method of hunting and trapping. Although you are not in a position to gather data or supply answers to these questions, list factors of this type that would need to be answered before a solution could be attempted.

3. While walking to class, what irritations or discomforts did you recognize? Using worksheets, identify the problem causing these irritations; write down the problem statement, the recognition of a need, the requirements, and the limitations.

4. Apply the criteria given in Problem 3 to your living quarters, classroom, recreation facilities, dining facilities, and other environments with which you are familiar.

Product Design Problems

5. Assume you are attempting to reconstruct the designer's approach to the development of the self-opening can. Even though the problem has been solved and completed, follow the identification steps with which the designer was concerned. Using the procedure outlined in Section 3.2, list these on worksheets. After identifying the problem, do you feel that the solution is the most appropriate one, or does your identification suggest other designs?

6. Follow the same procedure outlined in Problem 5 in identifying the problem of designing a travel iron for pressing clothes.

7. Identify the problems of designing a motorized wheelbarrow. List your ideas on worksheets.

8. You have noticed the need for a device that, attached to a bicycle, would allow the bicycle to ride over street curbs to sidewalk level. Identify the problem to determine its application to the general market.

9. Identify the problem and need for the development of a portable engineering travel kit that would provide the engineer with on-the-road facilities to make engineering calculations, notes, sketches, and drawings. This may take the form of a case that includes a calculator, drawing instruments, paper, reference material, and so on.

4

Preliminary Ideas

4.1 Introduction

The relationship between creativity and accumulating information during the design process is shown in Fig. 4.1. You have no limitations on generating preliminary ideas; be as creative as possible, as wild as you like. During the later stages of the design process, the need for creativity diminishes while the need for information increases.

Figure 4.2 The Xerox Conference Copier is a unique design that duplicates anything drawn on the easel board into multiple paper copies. (Courtesy of the Xerox Corporation.)

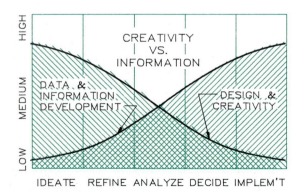

Figure 4.1 Engineering creativity is highest during the initial stages of the design process, whereas data and information development increase during the final stages.

The Xerox Conference Copier (Fig. 4.2) is an example of an advanced design idea to convert drawings and notes on an easel board to paper copies. Anything drawn on the board can be immediately duplicated on paper by the press of a button.

4.2 Individual Versus Team

Designers work as individuals and as members of design teams. Each approach is discussed below.

Individual Approach

Designers who work alone must keep notes and make sketches to communicate not with others but with themselves. Obtaining as many ideas as possible is their primary goal for the better ideas will more likely come from a long list than from a short list.

Possible solutions are sketched and accompanied by notes and schematics. These sketches are not working drawings, but rapid freehand sketches used to retain ideas that might otherwise be lost.

Team Approach

The more people involved in a project, the more problems of management and human relations exist. Teams perform best if the members select a leader to be responsible for moderating and guiding the team's activities.

Design teams should alternate between individual and group work. For example, members could individually develop preliminary ideas, which would then be discussed with the group.

4.3 Plan of Action

In gathering preliminary ideas for a design problem, the following sequence of steps is suggested: (1) hold brainstorming sessions, (2) prepare sketches and notes, (3) research existing designs, and (4) conduct surveys (Fig. 4.3). Periodically reviewing your notes and worksheets from the previous step, problem identification, will ensure that your efforts are directed toward the "target."

4.4 Brainstorming

Brainstorming is a problem-solving technique in which all members of a group spontaneously contribute ideas.

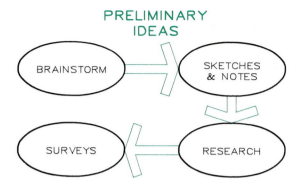

Figure 4.3 A suggested plan of action for gathering preliminary ideas for a design solution.

Rules of Brainstorming

The fundamental guidelines of a brainstorming session are as follows:★

1. **Criticism is ruled out.** Adverse judgment of ideas must be withheld until later.
2. **"Free-wheeling" is welcomed.** The wilder the idea, the better; it is easier to tame down than to think up.
3. **Quantity is wanted.** The greater the number of ideas, the more likelihood of useful ideas.
4. **Combination and improvement are sought.** Participants should seek ways of improving the ideas of others.

Organization of a Brainstorming Session

The organizational steps of a brainstorming session are (1) selection of the panel, (2) preliminary group work (become familiar with the problem), (3) selection of a moderator and a recorder, (4) the session, and (5) the follow-up.

Panel Selection The optimum number of participants in a brainstorming session is twelve. To encourage a variety of ideas, the group should include people with and people without knowledge of the subject. Supervisors often restrict a flow

---

★From Alex F. Osborn, *Applied Imagination.* New York: Scribner, 1963, p. 156.

of ideas, so it is desirable that panels be composed of people with a similar professional status.

Preliminary Work   A one-page outline of information about the session should be given to panel members a couple of days beforehand the session to allow ideas to incubate.

The Problem   The problem to be brainstormed should be concisely defined. For example, instead of presenting the problem as "how to improve our campus," present it as "how to improve student parking."

The Moderator and Recorder   A moderator is selected to be in charge of the session. The recorder is assigned the job of recording ideas as they are verbalized by the panel.

The Session   The moderator tosses out the problem and recognizes the first member holding up a hand. That person responds with an idea in as few words as possible. The moderator then recognizes the next person who holds up a hand, and the process continues in this manner.

Ideally, a suggestion made by one member stimulates ideas in other members, who would then hold up their hands and snap their fingers signifying they want to "hitchhike" on the previous idea. This kind of idea development is central to the brainstorming session.

Length   A session should move at a brisk pace and should be called to a halt when ideas slow to an unproductive rate. An effective session can last from a few minutes to an hour, but twenty-minutes is considered about the best length.

Follow-up   The recorder should reproduce the list of ideas gathered during the session for distribution to the participants. Approximately 100 ideas are usually gathered during a twenty-five-minute session. The list should be pared down to ten or twelve ideas believed to have the most merit.

Figure 4.4 Designers communicate with themselves, and others through sketches and notes to develop their preliminary ideas. (Courtesy of Ford Motor Co.)

4.5   Sketching and Notes

Sketching is the designer's most useful medium for developing preliminary ideas (Fig. 4.4).

By sketching, the designers's ideas take form as three-dimensional pictorials (Fig. 4.5) or as two-dimensional views. When properly used, sketching is a rapid, visual extension of the thinking process. The sketch shown in Fig. 4.5 was used in completing the styling features of an automobile. Additional sketches were used to develop

Figure 4.5 Sketching is used by the designer and stylist to develop the exterior design of an automobile. (Courtesy of Ford Motor Co.)

preliminary designs for a rearview mirror for an automobile (Fig. 4.6).

The Transportable Uni-Lodge (Fig. 4.7) illustrates the value of sketches in developing and communicating ideas. The Uni-Lodge could be transported by helicopter to previously unreachable areas. Retractable legs with pontoons permit it to float on water. Most ideas are indicated on the sketches by notes.

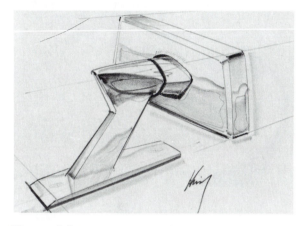

Figure 4.6 Preliminary sketch of a rearview mirror design. (Courtesy of Ford Motor Co.)

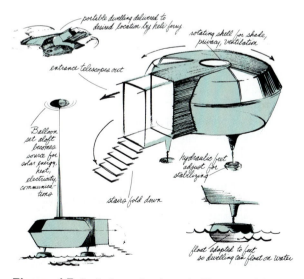

Figure 4.7 Preliminary sketches of a Transportable Uni-Lodge, a mobil dwelling of the future. (Courtesy of Lippincott and Margulies, Inc., and Charles Bruning Co.)

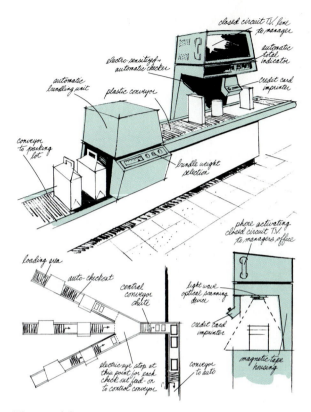

Figure 4.8 The conceptualization of an automatic checkout/packaging unit for a supermarket is shown in a sequence of sketches. (Courtesy of Lester Beall, Inc., and Charles Bruning Co.)

Sketching was also used to develop the concept shown in Fig. 4.8, a check-out system for a grocery store. The shopper sets the machine in operation by inserting a credit card in a slot. After the card is scanned, the customer receives an order number tag, and the conveyor moves the items under a scanner that totals the prices. If the customer has any questions, the machine can be stopped for communication with the store's employees by lifting the phone. The items are then conveyed to a unit where they are packaged in plastic containers marked with the customer's number. Large orders are transported by a central conveyor to an exterior pick-up near where the customer is parked.

Another concept is a self-contained pipelayer (Fig. 4.9) that can be used to lay pipe to provide

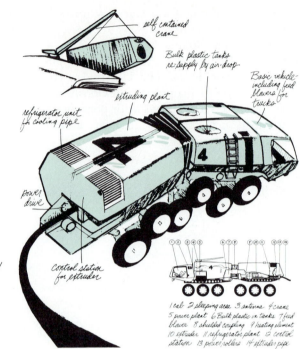

Figure 4.9 A designer's preliminary design of a self-contained pipelayer, which is intended to lay pipe for the irrigation of large tracts of desert. (Courtesy of Donald Desky Associates, Inc., and Charles Bruning Co.)

irrigation of desert areas. The first unit is the tractor, which includes a cab, sleeping accommodations, radio equipment, power plant, and bulk storage tanks for plastic. The second unit, the van, consists of an extrusion machine, a refrigeration unit, and a control station. The pipelayer is capable of transporting bulk plastic and machinery to extrude and lay approximately two miles of plastic pipe from each pair of storage tanks. The tanks are discarded when empty and are replaced by new tanks that are air-dropped into the area.

4.6   Research Methods

One way of gathering preliminary ideas is to research similar products and designs in use. Applying known principles to new applications and designs is called **synthesis.** Some reference sources available for this research are suggested below.

Technical Magazines Articles in technical magazines often give detailed explanations of unique designs, complete with sketches and photographs. Advertisements in these magazines can furnish information on materials and innovations that may be helpful.

General Magazines Significant design developments are often reported in popular periodicals. Magazines that are several years old can also be helpful in finding ideas.

Patents Patents from the U.S. Patent and Trademark Office can be used to good advantage by the designer. Patents are available to anyone for $1.50 each.

Consultants A comprehensive design may require a team of specialists in structures, electronics, power systems, and instrumentation. Manufacturers' representatives are also available to assist with problems related to their products.

4.7 Survey Methods

Survey methods are used to gather opinions and reactions to a preliminary or completed design. This is especially important when a product is being designed for the general market.

Opinions

Designers need to know consumers' attitudes toward their preliminary designs. Are potential buyers excited about the design, or do they show little interest in it? Before conducting the survey, the particular group whose opinions would be most valuable should be targeted for survey by (1) personal interview, (2) telephone interview, or (3) mail questionnaire.

The Personal Interview should be organized to provide reliable, unbiased opinions. Questions should be true-false or multiple choice to make conclusions easier to determine.

Interviewers should introduce themselves, mention the purpose of the interview, and ask for permission to proceed. They should then tabulate the responses and thank the interviewee for participating.

The Telephone Interview can be conducted by randomly picking names from the directory if the opinions of a cross section of the public are desired. If the opinions of a select group—sporting goods retailers—would be more valuable, a Business-to-Business Yellow Pages would supply a list of prospects.

The Mail Questionnaire is an economical method of contacting large numbers of people in a wide range of locations. To identify questions that need to be revised, the questionnaire should first be sent to a small group as a pilot test. Send at least three times as many questionnaires as the number of responses that you wish to receive. A self-addressed, stamped envelope should be included to improve response to the inquiry.

What types of opinions would be helpful to you? First, is there a need for the product? What features do consumers like or dislike? What price would they willingly pay for it? What do retailers think it will sell for? Does size matter? Color?

Where appropriate, findings and data should be graphed for study and analysis.

4.8 Hunting Seat—Preliminary Ideas

The design of the tree-borne hunting seat is used to illustrate the preliminary ideas step of the design process. The problem is restated below.

Hunting Seat Many hunters hunt from a sitting position in trees to obtain a better vantage point. Design a seat that provides comfort and safety and that meets the requirements of economy and the limitations of hunting.

Brainstorming Ideas are gathered from a brainstorming session with classmates. All ideas are listed on a worksheet (Fig. 4.10). Remember, wild ideas are encouraged. The better ideas are then selected and their features described on the worksheet (Fig. 4.11). You may list more features than would be possible to include in a single design, and be sure that no ideas are forgotten or lost at this state.

Sketches of Preliminary Ideas are drawn on worksheets, using rapid freehand techniques. Orthographic views and pictorial methods are both used. Thoughts or questions that come to mind while sketching should be noted on the drawings. Lettering and sketching techniques need not be highly detailed or precisely executed.

In Fig. 4.12, notice how ideas have been adapted from various types of chairs, lawn chairs in particular. Each idea is numbered for identification. Another worksheet (Fig. 4.13) shows a fourth idea, and idea #2 is modified to include a footrest and to suggest a method of guying the seat while suspended.

As many ideas of this type as possible need to be developed and sketched during the preliminary ideas step of the design process.

```
PRELIMINARY IDEAS
1. Brainstorming ideas
    A. USE LAWN CHAIR
    B. CHAIR ON STILTS
    C. INFLATABLE CHAIR
    D. PROVIDE A FOOTREST
    E. ROOF FOR RAIN PROTECTION
    F. PADDED SEAT
    G. HEADREST
    H. RIFLE REST
    I. AMMUNITION COMPARTMENT
    J. REFRESHMENTS COMPARTMENT
    K. RADIO COMPARTMENT
    L. TV COMPARTMENT
    M. CB RADIO
    N. ENTERTAINMENT COMPARTMENT
    O. HOIST SYSTEM
        1. PULLEYS & CABLES
        2. TREE CLIMBER
        3. LADDER
        4. STEPS
        5. BALLOON
    P. PLATFORM FOR STANDING
    Q. PLATFORM FOR SLEEPING
    R. LIGHTS
    S. SAFETY BELT
    T. SAFETY BELT FOR ARCHER
    U. SAFTEY BELT FOR RIFLEMAN
    V. DOUBLES AS BACKPACK
    W. DOUBLES AS TENT
    X. CARRYING CASE FOR SEAT
    Y. SEAT ON WHEELS
    Z. MOTORIZED SEAT
    A1. TELESCOPE MOUNT
    B1. HEATING SYSTEM
    C1. ALARM SYSTEM
    D1. SUN SHADE
```

PRELIMINARY IDEAS	NAME FILE SEC DATE	GRADE **4**

Figure 4.10 A worksheet listing the brainstorming ideas recorded by a design team.

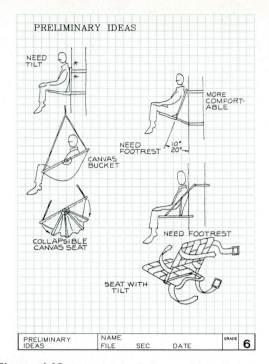

PRELIMINARY IDEAS

PRELIMINARY IDEAS	NAME FILE SEC DATE	GRADE **6**

Figure 4.12 A worksheet for presenting preliminary ideas on the development of a hunting seat.

```
2. Description of best ideas

    A. INCLUDE A FOOTREST

    B. INCLUDE METHOD OF ASCENDING—CABLE
       AND PULLEY SYSTEM

    C. ACCESSORIES

        1. STANDING PLATFORM

        2. RIFLE RACK

        3. BOW RACK

        4. CLIMBING ACCESSORY FOR SCALING
           TREE

        5. SUN SHADE

    D. SAFETY BELT

    E. SEAT SHOULD FOLD FOR EASY
       CARRYING

    F. SHOULD WEIGH UNDER 10 POUNDS

    G. PLASTIC COVER FOR CARRYING AND
       FOR RAIN PROTECTION

    H. SHOULD FOLD UP IN AN EASY-TO-SHIP
       CARTON

    I. MUST BE STABLE AND STURDY

2. Attach sketches
```

PROBLEM IDENTIFICATION	NAME FILE SEC DATE	GRADE **5**

Figure 4.11 A description of the better ideas selected from the original brainstorming ideas.

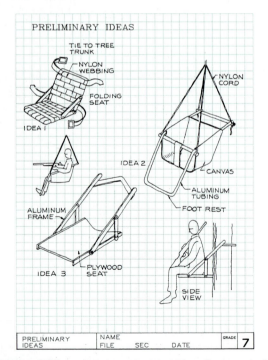

PRELIMINARY IDEAS

PRELIMINARY IDEAS	NAME FILE SEC DATE	GRADE **7**

Figure 4.13 Additional preliminary ideas illustrate design concepts for the hunting seat problem.

40

Problems

Problems should be presented on 8½-by-11-inch paper, grid or plain. All notes, sketches, drawings, and graphs should be neatly prepared in keeping with the practices introduced in this volume. Written matter should be legibly lettered using ⅛-in. guidelines.

1. Select one or several of these items and list as many uses as possible: empty vegetable cans (3 in. diameter (or DIA) by 5 in.), two thousand sheets of 8½-by-11-inch bond paper, one cubic yard of dirt, three empty oil drums (24 in. DIA by 36 in.), a load of egg cartons, twenty-five bamboo poles (10 ft. long), ten old tires, or old newspapers.

2. If you were going to select an ideal team to develop an engineering problem, what would you look for? List the characteristics with your explanations.

3. What are the advantages and disadvantages of working independently on a project? What are the advantages and disadvantages of working as a member of a design team? List your reasons and give examples of the types of problems where each approach would have the greater advantage.

4. Research the materials available to you to accumulate information on one of the following design problems or on one of your own selection. You are concerned with costs, methods of construction, dimensions, existing models, estimates of need, and other information of a general nature that will assist you in understanding the problem and deciding whether it is a feasible project. List the references you used. Prepare your research in a presentable form that could be reviewed by your instructor. The design problems are: a one-person canoe, a built-in car jack, an automatic blackboard eraser, a built-in coffee maker for an automobile, a self-opening door to permit a pet to leave or enter the house, an emergency fire escape for a two-story building, a rain protector for persons attending outdoor spectator activities, a new household appliance, a home exerciser.

5. List and describe the type of engineering or professional consulting services that would be required in the following design projects: a zoning system for a city of twenty thousand people, a shopping center, a go-cart, a water-purification facility, a hydroelectrical system, a nuclear fallout disaster plan, a processing plant for refining petroleum products, a drainage system for residential and rural areas.

6. Develop a questionnaire that could be given to the general public to determine its attitude toward a particular product. Select a product, and prepare the questionnaire to measure response to its unique features. Indicate how you would tabulate the information received from your questionnaire.

7. Review the brainstorming techniques, and then organize a group of associates to brainstorm a selected problem or to determine a problem in need of solution. List all ideas as they are suggested.

5

Design Refinement

5.1 Introduction

Descriptive geometry has its greatest application in the refinement step of the design process, for in this step it is necessary to make scale drawings with instruments to check the critical dimensions and geometry that cannot accurately be shown in sketches (Fig. 5.1).

Figure 5.1 The designer's first step in the refinement step of the design process is to draw preliminary ideas as scale drawings with instruments. (Courtesy of Keuffel & Esser Company.)

Refinement is the first departure from unrestricted creativity and imagination. Practicality and function must now be given primary consideration.

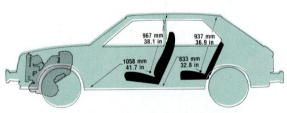

HEAD ROOM AND LEG ROOM DIMENSIONS

Figure 5.2 This scale drawing gives the dimensions and clearances necessary for a comfortable automobile interior. (Courtesy of Chrysler Corp.)

5.2 Physical Properties

One of the important concerns of the design's refinement is determining the physical properties of the proposed solutions. An example of a refinement drawing of the profile of an automobile with the appropriate overall dimensions is shown in Fig. 5.2.

5.3 Application of Descriptive Geometry

Descriptive geometry is the study of points, lines, and surfaces in three-dimensional space. The calculation of practically any given properties begins with these geometric elements.

Before descriptive geometry can be applied, orthographic views must be drawn to scale from which auxiliary views can be projected. An ex-

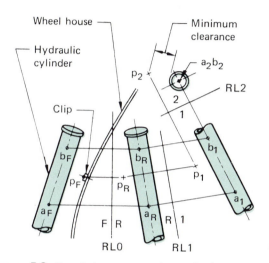

Figure 5.3 Descriptive geometry is an effective means of determining clearances between components. Shown here is the clearance between a hydraulic cylinder and a fender. (Courtesy of General Motors Corp.)

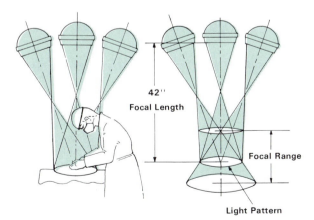

Figure 5.4 A well-adapted surgical lamp emits light that passes around the surgeon's shoulders with the minimum of shadow. The focal range of this surgical lamp is between 30 and 60 inches. (Courtesy of Sybron Corp.)

ample problem has been solved in Fig. 5.3, where descriptive geometry determined the clearance between a hydraulic cylinder and the fender of an automobile.

The design of a surgical light (Fig. 5.4) involves the application of geometry. The light fixture had to be designed to provide the maximum of light on the operating area. A scaled refinement drawing shows the converging beams of light that are emitted from the reflectors. The beams are very narrow at their centers and are positioned at shoulder level to minimize shadows cast by the surgeon's shoulders, arms, and hands.

From these scale drawings, measurements, angles, areas, and other geometry can be determined (Fig. 5.5). Figure 5.6 is a refinement drawing that shows critical dimensions of the surgical lamp.

Figure 5.5 By using scale drawings developed in the refinement step of the design process, the geometry of a surgical lamp can be studied. (Courtesy of Sybron Corp.; photo by Brad Bliss.)

5.4 Refinement Considerations

Advanced designs, such as the development of a new model automobile, have numerous features that must be improved and optimized at the refinement stage. In Fig. 5.7 the arrangement of the exhaust system is shown pictorially. Thus, many refinement drawings were required before

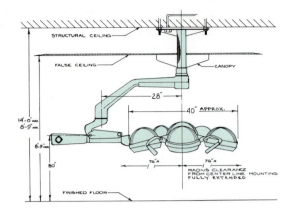

Figure 5.6 The overall dimensions of the final design of the surgical lamp are shown in this refinement drawing. (Courtesy of Sybron Corp.)

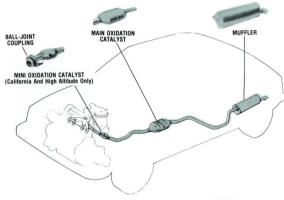

CALIFORNIA OXIDATION CATALYST SYSTEM

Figure 5.7 An automobile's exhaust system must be developed by using descriptive geometry to determine the lengths and angles necessary to clear the structural members of the chassis. (Courtesy of Chrysler Corp.)

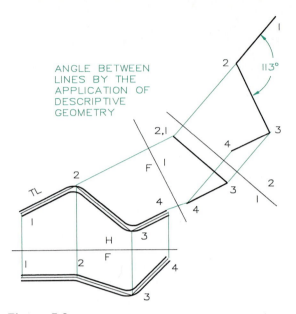

ANGLE BETWEEN LINES BY THE APPLICATION OF DESCRIPTIVE GEOMETRY

Figure 5.8 An application of descriptive geometry is shown in this example, which finds the lengths and angles between exhaust pipe segments.

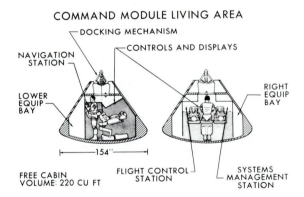

COMMAND MODULE LIVING AREA

Figure 5.9 Information and specifications for the orbiter vehicle are given in this scaled refinement drawing. (Courtesy of NASA.)

the general pictorials were possible; for example, descriptive geometry had to be used to determine the bend angles and clearances in the exhaust pipe to fit a particular chassis (Fig. 5.8). Computer graphics is a powerful tool that can be used to refine a preliminary idea.

The Command Module was also refined using a combination of orthographic drawings and notes. As orthographic views of the Module, (Fig. 5.9) gives the overall size of the craft, and a series of notes show its geometry and specifications.

5.5   Hunting Seat—Refinement

The hunting seat problem illustrates the method of refining a preliminary product design.

Hunting Seat   Many hunters hunt from trees to obtain a better vantage point. Design a seat that provides the hunter with comfort and safety

Figure 5.10 A design's specifications and desirable features are listed on a worksheet of this type. In this example, the hunting seat is refined.

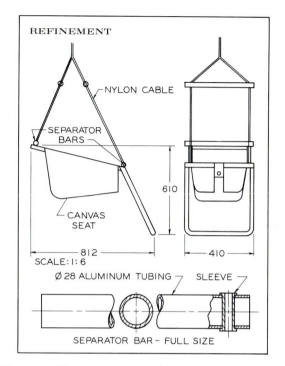

Figure 5.11 A refinement drawing of idea #2 for a hunting seat. Note that only general dimensions are given on the scale drawing.

while hunting from a tree and that meets the requirements of economy and the limitations of hunting.

Refinement Preliminary ideas for this design were developed in Chapter 4. The features to be incorporated into the design are listed on a worksheet (Fig. 5.10).

Idea #2 is refined in Fig. 5.11, where a scale drawing of the seat is shown orthographically. Tubular parts, like the separator bars, are blocked in to expedite the drawing process; also, some hidden lines are omitted.

It is important that refinement drawings be made to scale to give an accurate proportion of the design and to serve as a basis for finding angles, lengths, shapes, and other geometric specifications. Only overall dimensions are given on the drawing, but specific details are shown for the separator bar to explain an idea for a sleeve to protect the nylon cord from being cut. Another

design concept is refined in Fig. 5.12. Once again, only the major dimensions are given on these scaled refinement drawings made with instruments. These worksheets do not represent a complete refinement of the design; they are merely examples of the type of drawings required in this step of the design process.

5.6 Standard Parts

When preparing refinement drawings, attempt to use standard, commercially available parts because they are cheaper and more readily available. Information on standard parts can be obtained from merchandise catalogs, descriptive sales brochures, magazine advertisements, newspapers, and many other handy sources.

Engineers and designers make a practice of keeping a file on stock items that they use in developing products. Examples of wheels that

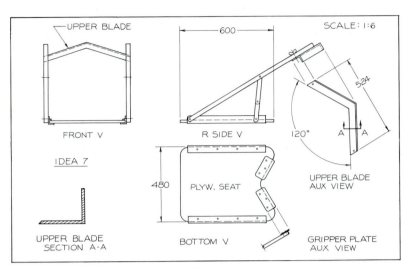

Figure 5.12 Another design concept for a hunting seat is shown as a refinement drawing.

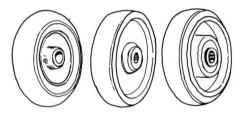

Figure 5.13 Examples of types of wheels that are available from commercial sources for light applications.

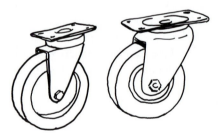

Figure 5.14 Typical casters for light applications.

might be used in refinement drawing are shown in Fig. 5.13. Typical casters are shown in Fig. 5.14. Parts of this type can be specified and drawn in refinement drawings by referring to literature published by manufacturers and vendors.

To become a better designer, learn to observe how components and parts are made and assembled.

Problems

Problems should be presented on 8½-by-11-inch paper, grid or plain. All notes, sketches, drawings, and graphical work should be neatly prepared in keeping with the practices covered in this book. Written matter should be legibly lettered using ⅛ in.-guidelines.

1. When refining a design for a folding lawn chair, what physical properties would a designer need to determine? What physical properties would be needed for the following items: a TV-set base, a golf cart, a child's swing set, a portable typewriter, an earthen dam, a shortwave radio, a portable camping tent, a warehouse dolly used for moving heavy boxes?

2. Why should scale drawings rather than freehand sketches be used in the refinement of a design? Explain.

3. List five examples of problems involving spatial relationships that could be solved by the application of descriptive geometry. Explain your answers.

4. Make a freehand sketch of two oblique planes that intersect. Indicate by notes and algebraic equations how you would mathematically determine the angle between these planes.

5. What is the difference between a working drawing and a refinement drawing? Explain your answer and give examples.

6. In the refinement step, how many preliminary designs should be refined? Explain.

7. Prepare refinement drawings of Problems 5–9 of Chapter 3. It may be necessary to postpone preparing these refinement drawings until you read some of the succeeding chapters and understand sufficient theory. Keep all these drawings together as they accumulate throughout the design process.

8. Make a list of refinement drawings that would be needed to develop the installation and design of a 100-foot radio antenna. Make rough sketches indicating the type of drawings needed with notes to explain their purposes.

9. Make a list of refinement drawings that would be needed to refine a preliminary design for a rearview mirror that will attach to the outside of an automobile. Refer to Fig. 4.6.

10. After a refinement drawing has been made, the design is found to be lacking in some respects so that it is eliminated as a possible solution. What should be the designer's next step? Explain.

11. Would a pictorial be helpful as a refinement drawing? Explain your answer.

12. List several design projects that an engineer or technician in your particular field would probably be responsible for. Outline the type of refinement drawings that would be necessary.

13. For the hunting seat design discussed in Section 4.8, what refinement drawings are necessary that were not given on the worksheets? Make freehand sketches of the necessary drawings with notes to explain what they would reveal.

Refinement Problems

The following sketches show products that a designer has developed as preliminary ideas. Prepare scaled refinement drawings are made with a instruments of each to better understand how the products are to be made. Types of refinement drawings you can make are

a) Orthographic views of each individual part of the design.

b) Orthographic views of the products' assemblies and subassemblies.

c) Pictorials to explain relationships between parts.

You must develop and design the details as they are drawn since the preliminary sketches are just that—preliminary. Devise solutions that will work by adding your own inventiveness to the refinement drawings.

14. (Fig. 5.15) Make scale drawings of one of the pipe hangers; the base plates are about 6 inches in size. The hangers are used to support overhead plumbing pipes. Modify the given designs, if you like.

Figure 5.15 Problem 14. Pipe hangers with six-inch plates.

15. (Fig. 5.16) Sit-up device. A device that clamps to an interior door to hold one's feet while doing sit-ups for fitness.

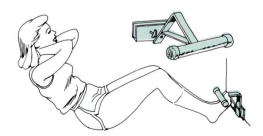

Figure 5.16 Problem 15. Sit-up device refinement.

16. (Fig. 5.17) Fireplace caddy. A small handcart for carrying firewood decorative enough to be kept indoors as a holder for the wood.

17. (Fig. 5.18) Woodworking clamp. A clamp permanently attached to a workbench for holding wood up to 6 inches (150 mm) thick. The design involves refining the mechanism and the collar that attaches to the workbench to permit height adjustment and easy removal.

18. (Fig. 5.19) Hold-down clamp. A clamp designed to attach to a workbench by drilling and counterboring a hole for a mounting bolt that fits in the T-slot of the clamp. When the clamp is removed, the bolt will drop below the surface of the workbench.

19. (Fig. 5.20) Pointer mount. A mount that connects on the top of a drawing table to hold a rotational pencil pointer.

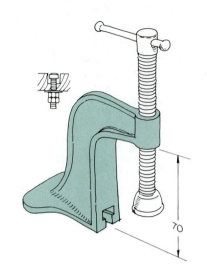

Figure 5.19 Problem 18. Hold-down clamp refinement.

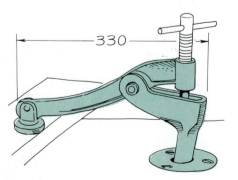

Figure 5.17 Problem 16. Fireplace caddy refinement.

Figure 5.18 Problem 17. Woodworking clamp refinement.

Figure 5.20 Problem 19. Pointer mount refinement.

20. (Fig. 5.21) Sharpener guide. A device for holding a chisel cutting edge at a constant angle while sharpening it on a whetstone.

21. (Fig. 5.22) Luggage carrier. A portable cart for carrying luggage that will fold up to as small a size as possible.

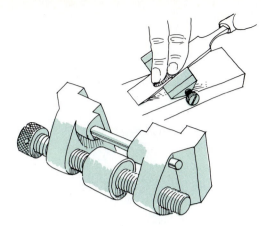

Figure 5.21 Problem 20. Sharpener guide refinement.

Figure 5.22 Problem 21. Luggage carrier refinement.

22. (Fig. 5.23) Sideview mirror. A fully adjustable mirror to be mounted on the side of an automobile.

23. (Fig. 5.24) Rotary pump. A rotary pump that operates by squeezing liquid through a flexible tube.

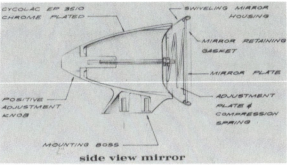

Figure 5.23 Problem 22. Sideview mirror refinement.

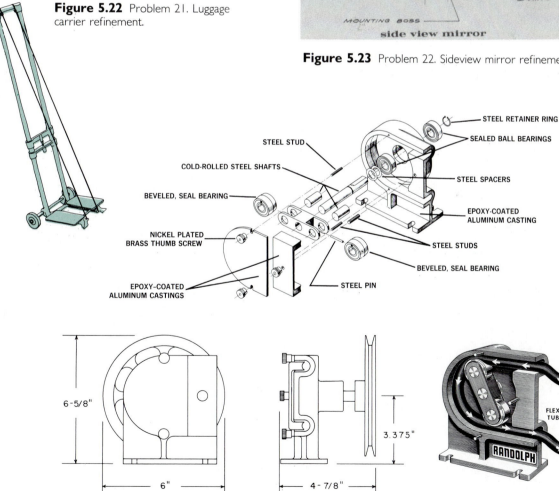

Figure 5.24 Problem 23. Rotary pump refinement.

24. (Fig. 5.25) Safety hook. A self-locking lifting hook.

25. (Fig. 5.26) Make refinement drawings of the portable crane that has a maximum height reach of 8.5 feet.

26. (Fig. 5.27) Make refinement drawings of the parts of the overhead conveyor mechanism that rolls on the flanges of an I-beam. The wheels are 3 inches in diameter.

Figure 5.25 Problem 24. Safety hook refinement.

Figure 5.27 Problem 26. Overhead conveyor wheels. (Courtesy of Mechanical Handling Systems, Inc.)

Figure 5.26 Problem 25. Portable crane. (Courtesy of Air Technical Industries.)

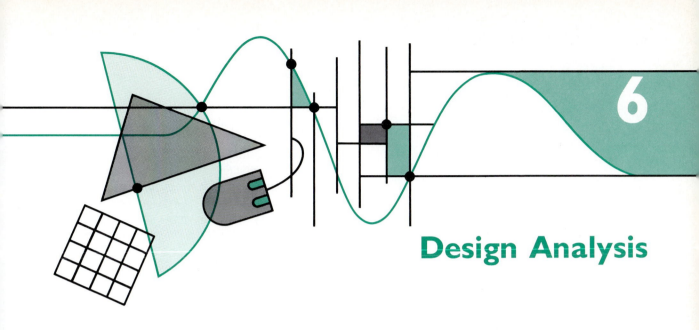

Design Analysis

6.1 Introduction

Design analysis is the process most commonly associated with traditional engineering courses. For example, a bridge design must be analyzed to select the proper structural materials and components for it to be strong, economical, and functional. Analysis is the evaluation of a proposed design. This stage is characterized by objective thinking and the application of technical knowledge. Less creativity is employed during this stage than during the previous stages of the design process.

6.2 Types of Analysis

The general areas of analysis are

1. Functional Analysis.
2. Human Engineering Analysis.
3. Market and Product Analysis.
4. Specifications Analysis.
5. Strength Analysis.
6. Economic Analysis.
7. Model Analysis.

Functional Analysis Functional analysis is the most important characteristic of a design; if a design does not function, it is a failure regardless of its other desirable features (Fig. 6.1). A doorknob that will not open a door is an unacceptable design even if it is attractive, strong, and economic.

Human Engineering All designs must serve the humans who use the product, travel on or in

Figure 6.1 Experimental automobile designs are analyzed and tested to arrive at the most functional and efficient performance. (Courtesy of Ford Motor Company.)

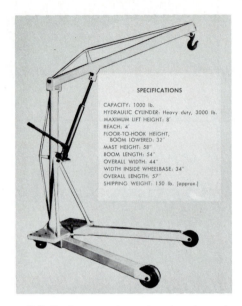

Figure 6.2 Designs must be analyzed to determine their physical properties, including dimensions, weights, and capacities. (Courtesy of Air Technical Industries.)

it, or profit from its existence. Therefore the designer must consider the physical, mental, and emotional characteristics of the user of the product.

Market and Product Analysis   The market for which a product is designed is studied during the early stages of the product's development (Chapters 3 and 4) and before its production. To determine consumer attitude toward a proposed concept, an initial market survey is done.

Physical Specifications Analysis   In the refinement stage, various specifications, such as lengths, areas, shapes, and angles were obtained. In the analysis stage, other specifications, such as weights, volumes, capacities, velocities, and ranges of operation, must also be obtained (Fig. 6.2).

Strength Analysis   A proposed design must be strong enough to support the maximum design load that can be anticipated. Strength is closely

associated with function, since a weak design is not a functional design.

Economic Analysis   Designs that are unduly expensive have little chance of being profitable in a competitive marketplace. As the design nears the completion stage, the designer must consider economy and type of fabrication.

Model Analysis   A proposed design is seldom produced before a model or prototype has been

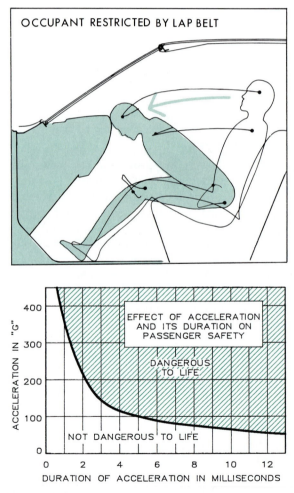

Figure 6.3 Empirical data obtained from laboratory experiments can be analyzed more efficiently by graphical techniques than by mathematical methods. (Courtesy of General Motors Corp.)

constructed for analysis and evaluation. Extensive tests may be run on a functional design to gather data to support its acceptance or rejection.

6.3 Graphics and Analysis

Engineering graphics and descriptive geometry are valuable in the analysis step of the design process. Empirical data obtained from laboratory experiments and field data can be transformed into algebraic equations by graphical techniques (Fig. 6.3). Mathematical evaluation can also be handled graphically.

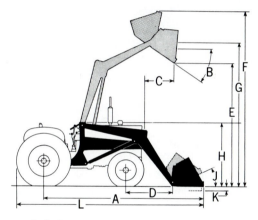

Figure 6.4 Clearance between functional parts and linkage systems can be analyzed efficiently by using graphical methods. (Courtesy of Navistar International.)

When the data does not fit an algebraic equation, but instead plots as an irregular curve, graphical calculus must be used. Examples of graphical analysis include analyzing clearances between parts and linkage systems (Fig. 6.4) and presenting and analyzing market surveys, populations, trends, and technical data (Fig. 6.5).

6.4 Functional Analysis

Functional analysis requires applying judgment and experience in order to balance the economy, durability, appearance, and marketability of a product's design. For example, in times of high

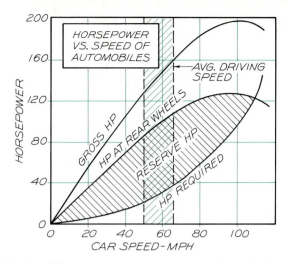

Figure 6.5 An automobile's power system is analyzed by plotting experimental data on a graph of this type.

fuel costs, a need for an economical automobile exists, but simply designing a car that gets good mileage is not enough. Most buyers are willing to give up some luxuries, but few buyers would give them all up just for better gasoline economy.

Certain functional characteristics can be analytically evaluated when function is the only consideration. The function of a hand-operated clamp is analyzed for its function of clamping a part in position while the part is machined or drilled (Fig. 6.6). In cases like these, function is limited to a narrow range of requirements.

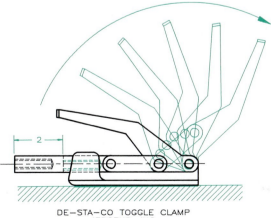
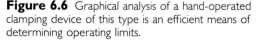

Figure 6.6 Graphical analysis of a hand-operated clamping device of this type is an efficient means of determining operating limits.

53

6.5 Human Engineering

Human engineering is defined by Woodson* as follows:

> The design of human tasks, man-machine systems, and specific items of man-operated equipment for the most effective accomplishments of the job, including displays for presenting information to the human senses, controls for human operation, and complex man-machine systems. In the design of equipment, human engineering places major emphasis upon efficiency, as measured by speed and accuracy, of human performance, in the use and operation of equipment. Allied with efficiency are safety and comfort of the operator.

Figure 6.7 Leonardo da Vinci analyzed body dimensions and proportions more than four hundred years ago.

Leonardo da Vinci analyzed body dimensions and proportions in about 1473 (Fig. 6.7). A close similarity can be seen in Fig. 6.8, where human factors are measured to determine the restriction of human mobility by a radiation protection garment used by astronauts.

Human factors are very important to a successful space program. In a weightless atmosphere even the simplest, most familiar tasks re-

*Woodson, W. E., *Human Engineering Guide for Equipment Designer,* Berkeley, CA.: University of California Press, 1954.

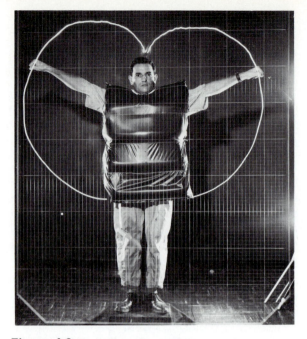

Figure 6.8 Body dimensions and movements are measured to analyze human mobility that is restricted by a radiation protection vest used by astronauts in space travel. (Courtesy of General Dynamics Corp.)

quire training and getting used to. What's more, the spacecraft's cabin must be designed to provide the proper life systems as well as a work area in which the on-orbit assignments can be performed (Fig. 6.9).

Dimensions A design must take into account the dimensions, ranges of manipulations, and senses of the person who will be using the fin-

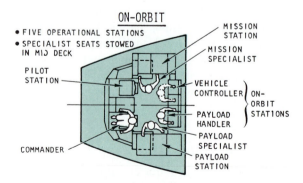

Figure 6.9 Human factors and living environments had to be analyzed when the crew cabin was designed for the space shuttle. (Courtesy of Rockwell International.)

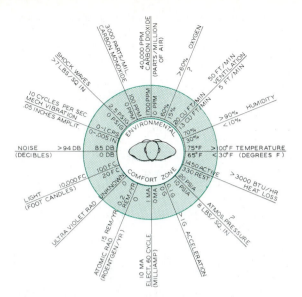

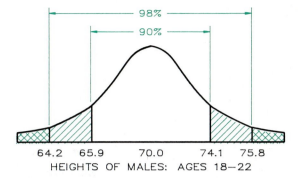

Figure 6.10 The inner circle is the environmental comfort zone, and the outer circle is the bearable limit zone of human environment. (Courtesy of Henry Dreyfuss, *The Measure of Man*, New York: Whitney Library of Design, 1967.)

HEIGHTS OF MALES: AGES 18—22

Figure 6.11 The distribution of the average heights in inches of American males from 18 to 22 years of age. (Based on data gathered by B. D. Corpinos, *Human Biology* 30:292.) Fifty percent of American males in this age range are taller than 70 inches, and 50 percent are shorter.

ished product (Fig. 6.10). Variations in physical characteristics tend to conform to the normal distribution curve shown in Fig. 6.11.

Body dimensions of the average American male are shown in Fig. 6.12. Collected by Henry Dreyfuss, the measurements are used for industrial designs. The average female body dimensions are given in Fig. 6.13.

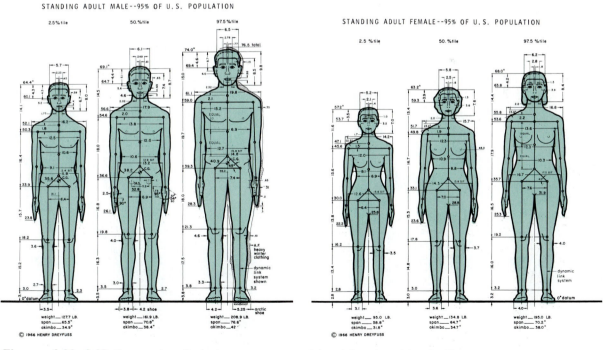

Figures 6.12—6.13 Front and profile body measurements of the adult male and female. These measurements describe 95 percent of the U.S. adult population. (Courtesy of Henry Dreyfuss, *The Measure of Man*, New York: Whitney Library of Design, 1967.)

Figure 6.14 The safety bar of the child's seat (Left) is more difficult for the child to remove since the seat is unattended in the car. The other seat (Right) has a safety bar that is easier to lift above the child since the seat will be placed in the front seat where there will be more supervision.

Two car seats are shown in Fig. 6.14. The safety bar of one seat can easily be raised to clear the child's head and thus permits her to dismount from the seat with a minimum of assistance. The other car seat has a safety bar that cannot be lifted easily, which makes it better to protect a child riding in the back seat.

Motion   The study of body motion includes the amount of space required to function comfortably, safely, and efficiently. A great deal of anal-

ysis is devoted to the interior of an automobile (Fig. 6.15) in order to adapt it to the human body.

Vision   Designs with gauges and controls are developed to take advantage of the most visually effective means of aiding the operator. The automobile dashboard configuration shown in Fig. 6.16 is very similar to those of other models. Lights make it easy for the driver to manipulate the controls and to obtain information from the instruments.

Figure 6.15 The design of an automobile is based on the dimensions of the consumers who will use it. This is an application of human engineering. (Courtesy of General Motors Corp.)

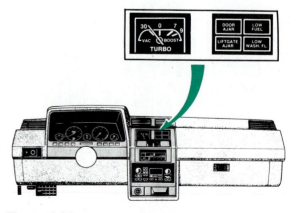

Figure 6.16 The design of efficient, safe automobile instrument panels is a problem involving human engineering. (Courtesy of Chrysler Corp.)

Sound To be audible, sound must be within specified frequencies. Studies indicate that sound affects a person's stress level and, consequently, productivity.

Environment Working environments may include the entire layout of an industrial plant, the conditions in a particular workstation, or a specialized location, such as the cockpit of an airplane. Environment comprises temperature, lighting, color scheme, sound control, and comfort of operation.

6.6 Market and Product Analysis

Areas of product analysis are

1. potential market evaluation.
2. market outlets.
3. sales features.
4. advertising methods.

Potential Market Analysis General information about the market should be determined by including predicted age groups, income brackets, and geographical locations of prospective purchasers of the product. This kind of market information will also be helpful in planning advertising campaigns to reach the customer.

Market Outlets A product may be introduced to the market through existing distribution channels, such as retail outlets, or through newly established dealerships. Large mainframe computers are not suitable for distribution through retail outlets, so technical representatives work with clients individually. On the other hand, a sportsman's camping seat can be sold effectively through department stores and sporting goods outlets.

Sales Features A list of unique sales features used in a new design should be kept to support the feasibility of the proposed design. Unique features are likely to stimulate interest in a product and attract consumers.

Advertising Methods Advertising a product can be done through several media including personal contact, direct mail, radio, television, newspapers, and periodicals. Advertising costs vary widely and should be considered before selecting a medium. For products used seasonally, timing is also important.

6.7 Physical Specifications Analysis

To complete the design, the product's physical specifications must be analyzed.

Sizes A product's overall size and dimensions must be evaluated to ensure it meets the standards of size (if any), such as permissible widths in an automobile design. Is its weight within the prescribed range? And what is the size of the product when extended, contracted, or positioned differently?

Ranges Many products have ranges of operation, capacities, and speeds that need to be analyzed before a design can be made final. The designer must calculate or estimate ranges such as seating capacity, miles per gallon, pounds of laundry per cycle, flow in gallons per minute, or electrical power required for operation.

Shipping Specifications Finally, the designer must be concerned with the packaging and shipping of the product. Will it be shipped by air, mail, rail, or truck? Will it be shipped one at a time or in quantities? Will it be shipped assembled, partially assembled, or disassembled? How much will the shipping crate cost, and how much will shipping cost? What will be the size of the product when crated for shipment?

6.8 Strength Analysis

Much of engineering is devoted to analyzing the design's strength to support dead loads, withstand shocks, and endure repetitive usage at motions ranging from slow to fast. Data are also

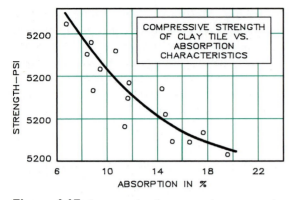

Figure 6.17 An example of an approximate curve that represents the strength of a structural clay material related to its percent of absorption.

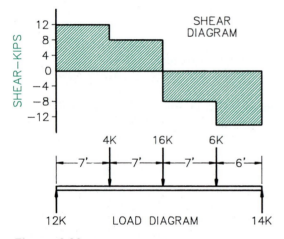

Figure 6.18 Graphical calculus can be used as an analysis technique to determine shear in a structural beam.

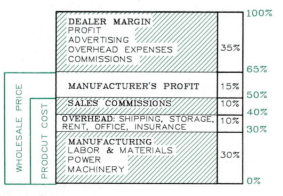

Figure 6.19 A price structure model that can be used as a guide in the economic analysis of a product.

gathered from experiments and plotted graphically to establish strength characteristics; for example, the strength of clay tile as related to its absorption characteristics (Fig. 6.17).

In Fig. 6.18 the shear for a beam has been determined by graphical calculus. The resulting load calculations are necessary to select the materials and their sizes required by the design.

6.9   Economic Analysis

Before a product is released for production, a cost analysis must be done to determine the item's production cost and the margin of profit that can be realized from it. For the average student assignment, two methods of pricing a product or project are (1) itemizing and (2) comparative pricing.

Itemizing   Industrial firms staffed with experienced estimators can determine a product's production cost by itemizing each expense. A number of costs that must be itemized are shown in Fig. 6.19; obviously, the percentages vary for different areas of manufacturing and retailing.

By studying the project's working drawings, the cost analyst can estimate the costs of materials, manufacturing, labor, overhead, and so forth. Finally, the production cost for the product, or the construction cost for the project, is determined. The profit is then added to arrive at the wholesale price.

Comparative Pricing   By comparing the proposed product with the cost of similar products on the market, the product's cost can be approximated.

An example of comparative pricing is shown in Fig. 6.20; each power tool is priced at $99.

These tools are very similar; all have the same power sources, the same materials, and the same styling. More important, each has identical market potential.

Figure 6.20 These three products retail for $99 each. You can see that each is similar in design and will have essentially identical market volume. This is an example of comparative pricing. (Courtesy of Sears, Roebuck and Co.)

Approximately the same numbers of drills are sold as sanders and saws; consequently, production costs and retail price will be similar for each.

Figure 6.21 This hunting seat can be compared with the exercise apparatus shown in Fig. 6.22. Each is comparatively priced from $90 to $100 since both have similar manufacturing problems and market potentials. (Courtesy of Baker Manufacturing Co., Valdosta, Ga.)

Another example of comparative pricing is the price of the hunting seat (Fig. 6.21) compared with that of the exercise bench sold by Sears Company (Fig. 6.22). Both products' manufacturing requirements and market volumes are similar, both sell for about $100. On the other hand, the baby stroller sells in the $50 range because it has a larger market volume even though it is similar to the previous products (Fig. 6.23).

Comparative pricing is used by manufacturers by determining factors like cost per square

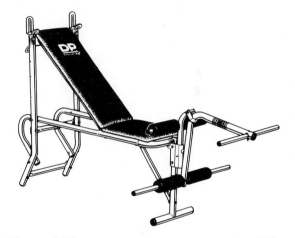

Figure 6.22 This exercise apparatus is priced at $100, which is similar in manufacture and market appeal to the hunting seat shown in the previous figure. (Courtesy of Diversified Products Corp.)

Figure 6.23 These baby strollers are priced from $55 to $75, less than the hunting seat and the exercise apparatus. The market potential for the strollers is greater than the other two products, which makes them cheaper because they can be mass marketed. (Courtesy of Strolee of California.)

foot, cost per mile, cost per cubic foot, or cost per day. These factors enable them to formulate rough cost estimates before doing more detailed studies.

Miscellaneous Expenses   It is easy to overlook some expenses that are usually incurred in developing a new product. Shipping and packaging costs must be considered in pricing a product, and this may be a sizable expense.

   Warehousing and storage of finished products must be absorbed in the price of a product. Coupled with warehousing is the cost of insurance, temperature control, shelving, forklifts, and of course, employees. Failing to recognize expenses like these can affect a product's profitability.

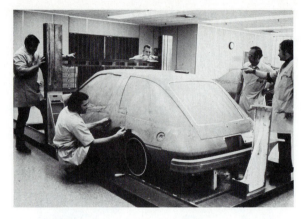

Figure 6.24 A mock-up is a full-size dummy of a product made to represent the final appearance but not necessarily made to operate. (Courtesy of Chrysler Corp.)

6.10   Model Analysis

Models are effective aids in analyzing a design in the final stages of its development. A three-dimensional model can be used to study the design's proportion, operation, size, function, and efficiency. The types of models are

1. Conceptual models.
2. Mock-ups.
3. Prototypes.
4. System layout models.

Conceptual Models   A conceptual model is a rough model made for the designers' own use to help them analyze a design or feature.

Mock-ups   Mock-ups are full-size dummies of the finished design that give the proper appearance of the product (Fig. 6.24). Mock-ups are constructed more for the presentation of size, shape, appearance, and component relationships than for operational movements.

Prototypes   A prototype is a full-size working model that follows the final specifications in all respects. The only exceptions may be in the use of materials. Since a prototype is made mostly by

hand, materials that are easier to fabricate by hand may be used instead of those used in final production.

System Layout Models   System layout models are used to show the relationships between large layouts such as manufacturing systems, architectural developments, and traffic systems. System layout models of refineries are often constructed to supplement working drawings during the construction of the facility (Fig. 6.25).

Figure 6.25 A system layout model is used to analyze the details of construction of refinery design. (Courtesy of E. I. du Pont de Nemours and Co.)

Figure 6.26 A student model of a portable home caddy was designed to demonstrate how the device folds flat for ease of storage.

Model Construction

Model Materials Balsawood is commonly used in model construction because it is easy to shape with few tools. Standard parts such as wheels, tubing, figures, dowels, and other structural shapes can be purchased, rather than made, to save time and effort.

Plexiglas can be used to construct models that illustrate inside and outside design features.

Models should be finished to give a realistic impression of the completed design, especially when they are used for presentation to others.

Model Scale A model that is used to analyze moving parts of a functional product should be scaled so the smallest moving part will operate.

For example, a student model of a portable home caddy is shown in Fig. 6.26. This balsawood model is constructed to demonstrate the function of a linkage system that permits the wheels to be collapsed for convenience of storage. Although the model is small, the linkage system operates the same way as in the completed product.

Model Testing The analysis process can be aided by testing models built to the specifications of the final design. The aerodynamic characteristics of the rear styling of an automobile can be evaluated by wind tunnel tests (Fig. 6.27).

Physical relationships and the functional workings of movable components, like the hatch and storage area of a car, can be tested in a pro-

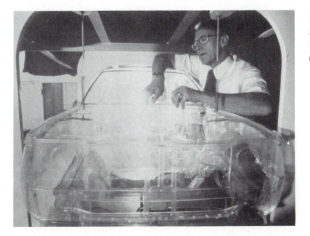

Figure 6.27 A plastic model of this automobile body to study the stresses in it with simulated endurance drives. (Courtesy of Chrysler Corporation.)

Figure 6.28 The physical relationships of design components can be observed and tested with the use of full-size models. (Courtesy of Chrysler Corp.)

totype (Fig. 6.28). Models also are used to test consumer reactions to new products before proceeding with the production.

6.11 Hunting Seat Analysis

To illustrate a method of analyzing a product design, we return to the hunting seat problem.

Hunting Seat Many hunters hunt from trees to obtain a better vantage point. Design a seat that provides comfort and safety and that meets the requirements of economy and the limitations of hunting.

Analysis

The major areas of analysis are listed on the worksheets in Fig. 6.28 through Fig. 6.31. Additional worksheets should be used to elaborate on each of these areas as required.

The strength of the hunting seat's support system of one design is determined by using graphical vectors (Fig. 6.32). Once the loads in

```
ANALYSIS
1.  Function

    A.  PROVIDES A METHOD OF CLIMBING
        TREES

    B.  PROVIDES COMFORTABLE SEATING

    C.  PROVIDES DECK FOR STANDING

2.  Human engineering

    A.  SAFETY BELT INCLUDED

    B.  FOOTREST FOR COMFORT

    C.  PORTABLE: 10-15 LBS IN WEIGHT

    D.  SHOULDER STRAPS FOR CARRYING

    E.  360 DEGREES OF VISION

3.  Market & consumer acceptance

    A.  POTENTIAL MARKET
        1.  STATE: 40,000
        2.  NATION: 1,800,000

    B.  CHEAPER THAN DEER STAND BY 100%-
        500%

    C.  AFFORDABLE AT $100

    D.  ADVERTISE IN HUNTING MAGAZINES

    E.  RETAIL THROUGH SPORTING GOODS
        STORES AND DIRECT MAIL

DESIGN      NAME                        GRADE
ANALYSIS    FILE    SEC     DATE          11
```

Figure 6.29 Worksheet to analyze function, human engineering, and market considerations.

Figure 6.34 The hunting seat used as a platform for standing and the hunting seat used for sitting.

A commercial version of the seat is illustrated in Fig. 6.34, where it is tested to measure its functional features including the method of using it to climb a tree. An analysis drawing of the hunting seat (Fig. 6.35) illustrates the operation of the linkage system that permits the seat to collapse into a single plane for carrying ease. The forces in the members are also found graphically by using vector analysis. The overall features and the physical properties of the Baker Tree Stand are shown in Fig. 6.36. These features are helpful to a consumer in making a purchase.

Figure 6.35 The drawing is used to analyze the linkage system of the hunting seat, and a vector diagram is used to determine the forces in the members when the seat is loaded to maximum.

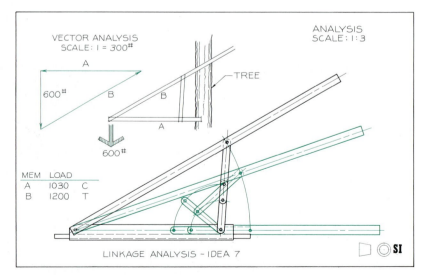

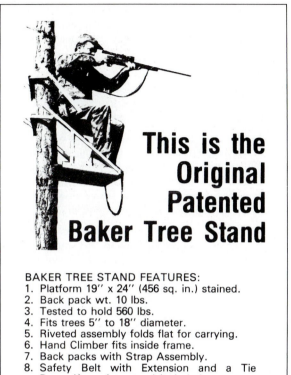

This is the Original Patented Baker Tree Stand

BAKER TREE STAND FEATURES:
1. Platform 19″ x 24″ (456 sq. in.) stained.
2. Back pack wt. 10 lbs.
3. Tested to hold 560 lbs.
4. Fits trees 5″ to 18″ diameter.
5. Riveted assembly folds flat for carrying.
6. Hand Climber fits inside frame.
7. Back packs with Strap Assembly.
8. Safety Belt with Extension and a Tie Down (for safety strap) included.

Figure 6.36 A summary of the features and physical properties of the Baker Tree Stand. (Courtesy of Baker Manufacturing Co.)

 **Problems**

The following problems should be solved on 8½-by-11-inch paper, accompanied by the necessary drawings, notes, and text. Answers to essay problems can be typed or lettered. All sheets should be stapled together or included in a binder or folder.

General

1. Make a list of human factors that must be considered in the design of the following items: a canoe, a hairbrush, a water cooler, an automobile, a wheelbarrow, a drawing table, a study desk, a pair of binoculars, a baby stroller, a golf course, the seating in a stadium.

2. What physical quantities would have to be determined in the designs listed in Problem 1?

3. Select one of the items given in Problem 1 and make an outline of the various steps that should be taken to satisfy the following areas of analysis: (1) human engineering, (2) market analysis, (3) prototype analysis, (4) physical quantities, (5) strength, (6) function, and (7) economy.

Human Engineering

4. Using your body as the average, make a drawing to indicate the optimum working areas for you when in a sitting position at a drawing table. Assume you are to use your measurements as a basis for designing a drawing table to satisfy the needs of your classmates. Your reach, posture, and vision will have considerable effect on its dimensions. Your finished drawing should give three views of the ideal working area for you while drawing; the drawing should also show the most efficient positioning of instruments for working. Experiment with the angle of tilt of the table top to determine the most comfortable position for working.

5. Using the dimensions for the average person given in this chapter, design a stadium seating arrangement that will serve the optimum needs of the spectators. A primary consideration will be the slope of the stadium seating, which should be designed to provide an adequate view of the playing field. The comfort of the average spectator and provision for traffic between seats must also be considered.

6. Compare the dimensions of the students in your class with the standards given in Section 6.5. For example, compare the average height of your class with the national standard height.

7. Design a backpack that will be used on a camping trip. Decide on the minimum belongings a camper should carry, and use their weights and volumes in establishing the design criteria. Make sketches of the pack and the method of attaching it to the body to provide the optimum mobility, comfort, and capacity.

8. Establish the dimensions, facilities, and other provisions that would be needed in a one-person bomb shelter to provide protection for forty-eight hours. Make sketches of the interior in relationship to a person and supplies. How will ventilation, water, food, and other vital resources be provided? Explain your design as it relates to human engineering needs.

9. Design a manhole access to an underground facility. What must the diameter of the manhole be to permit a person to climb a ladder for a distance of ten feet with freedom of movement? Make a sketch of your design and explain your method of solving the problem.

10. Design a one-person facility for temporary observation service in the Arctic. This facility is to be as compact as possible to provide for the needs of a single person during the seventy-two-hour periods that the person is on duty. Determine the facilities and provisions that would be needed, including heat, ventilation, and insulation. Make sketches of your design, and explain items that you consider essential to the human engineering aspects of the problem.

11. Design a configuration for an automobile steering wheel that is different from present-day designs but is just as functional. Base your design on human factors such as arm position, grip, and vision. Make sketches of your design and list items that you considered.

12. Make sketches to indicate safety features that could be built into your automobile to reduce the seriousness of injury caused by accidents. Explain your ideas and the advantages of your designs. Primary consideration should be given to the human aspects of the designs.

13. Assume you prefer to alternate between a sitting and a standing position when working at a drawing table. Determine the ideal height of the table top for working in each position. Indicate how a table could be devised to permit instant conversion from the height for standing to the height for sitting.

14. Identify some human engineering problems that you believe are in need of solving. Present several of these to your instructor for approval. Solve the approved problems. Make a series of sketches and notes to explain your approach.

Market Analysis

15. Assume you are responsible for conducting a market analysis of the drill shown in Fig. 6.20. Include in your analysis all the areas covered in Section 6.6. Assume this product is new and has never been introduced in an electrically powered form before. Outline the steps you would take in conducting a product and market analysis.

16. Make a product and market analysis of the car seat shown in Fig. 6.14, following the steps suggested in Section 6.9. Arrive at a market value that you feel would be satisfactory, and determine the outlets and other information of this type that would be important to your analysis.

17. Assume the cost estimates of producing hunting seats were as follows: 100 seats, $35 each; 200 seats, $20 each; 400 seats, $10 each; 1000 seats, $8.50 each. Using these figures, determine the price at which you could introduce the seats to the market on a trial basis and still have some financial protection. Explain your plan.

18. List as many unique features of the hunting seat as you possibly can that would be important to a sales campaign and to advertising. Make sketches and notes to explain these features.

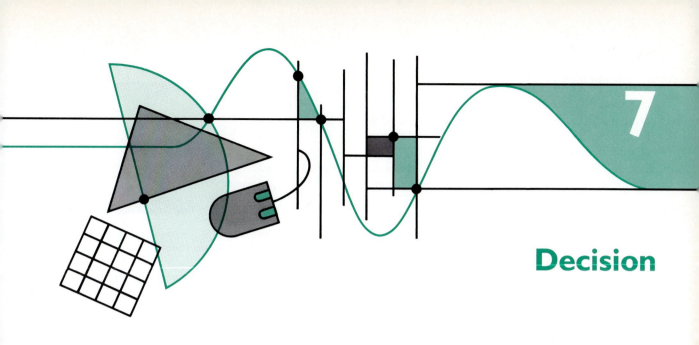

Decision

7

7.1 Introduction

Once the design has been conceived, developed, refined, and analyzed, a decision must be made to determine which design is the most worthy of implementation. The decision step is based on facts and data; still, at best it is subjective and must be made by experienced persons.

7.2 Types of Presentations

Presentations can be made to a few people or to a large group, and the groups can vary from project associates to laymen unfamiliar with the project and its objectives.

Informal Presentation Informal presentations are given to several immediate associates or to a single supervisor. Although specially prepared visual aids rarely are used in these presentations, pictorials, schematics, sketches, and models frequently are.

When only a few people are involved, ideas may be sketched. (Fig. 7.1). In a discussion involving several people, ideas, schematics, and sketches may be drawn on a blackboard.

Figure 7.1 A decision may be the outcome of a presentation, where ideas and designs are discussed and sketched informally. (Courtesy of the MITRE Corporation.)

Formal Presentation Formal presentations receive more emphasis in this chapter because they are usually more crucial than informal presentations.

The group to whom the presentation is made may comprise professional associates, administrators, laymen, or it may be a mixed group. Func-

67

tion and design acceptability are the primary considerations of the engineering associates. Administrators will be concerned with its economic feasibility and estimated profit return. Laymen could be the clients for whom the project is designed, stockholders, or members of the general public who may vote on the approval of a design.

7.3 Organizing a Presentation

One method of planning an oral or written report is by using 3-by-5-inch index cards. Separate ideas that are to be illustrated are first written on separate cards. A rough sketch is made indicating the type of illustration required and the method of reproduction to be used, and brief notes are added outlining the discussion that will accompany the visuals. The sequence of the presentation can be easily reviewed by displaying the cards on a table, bulletin board, or planning board (Fig. 7.2). And the sequence or content of the cards can be easily changed.

The completed cards (Fig. 7.3) should contain the following information:

1. **Number.** The card's position in the sequence of visual aids.
2. **Illustration.** A sketch of the illustration that must be prepared.
3. **Text.** A brief outline of the oral presentation that will accompany the visual aids.

Figure 7.2 Planning cards can be used to prepare the sequence of the presentation by using a planning board (as shown here) or by merely arranging the cards on a table top.

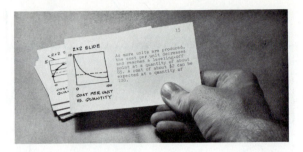

Figure 7.3 An example of the layout of a 3" × 5" card that shows a sketch of the visual and its accompanying text.

If a variety of visuals (flip charts, slides, transparencies, or combinations of these) will be used during the presentation, the material should be arranged to allow smooth transitions from one type of presentation to another.

7.4 Visual Aids for Presentation

In a presentation, the visual aids most commonly used are flip charts, photographic slides, overhead projector transparencies, and models.

The following general rules will improve the effectiveness of any type of visual aid:

1. Convey only one thought on each slide or chart.
2. Reduce lengthy statements to key phrases or words that will communicate the thought.
3. Present tabular data graphically.
4. Use slides and charts that are clearly readable.
5. Use illustrations, color, and attention-getting devices.
6. Prepare a sufficient number of slides or graphs so notes are unneeded.

Flip Chart

Flip charts consist of a series of illustrations prepared on medium-weight paper and mounted on a board or an easel for presentation before a group (Fig. 7.4). Flip charts are used for small conferences or presentations in an area no larger than an average-sized classroom. A 30-by-36-inch flip chart is a common size; anything smaller may be difficult for many people to see.

Figure 7.4 The flip chart can be used to communicate with small groups. (Photo by Hazel Hankin/Stock, Boston.)

Figure 7.5 Felt-tip markers are well-suited to the preparation of flip charts and other types of visual aids.

Figure 7.6 This flip chart was drawn on brown wrapping paper using India ink and a brush. A title is being applied with rubber cement. Colored construction paper can be used to highlight flip charts.

Paper Brown or white wrapping paper is suitable for a series of flip charts that will be used only once. Corrugated cardboard can be sized to serve as a backing board.

Lettering Materials The final lettering can be done with felt-tip markers, ink, tempera or sign paints, or overlay film. Felt-tip markers, used extensively for fast bold lines in a variety of colors, give sophisticated effects if correctly used (Fig. 7.5). India ink is an effective medium for lettering and for adding emphasis to a chart (Fig. 7.6).

Color Color can be added to a chart very easily through the use of construction paper and rubber cement. In general, a chart of this type should be bold and have no tedious or time-consuming details. This technique is especially useful for bar graphs (Fig. 7.7).

Assembly The finished charts are assembled in order, with a title chart covered by a blank sheet of paper. All sheets are stapled or otherwise attached to the backing board. The blank cover sheet conceals the theme of the flip charts and eliminates any anticipation of the presentation.

Presentation Flipped in sequence, the charts are referred to with a pointer to ensure the speaker does not block the audience's view. Since

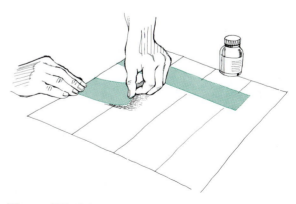

Figure 7.7 Colored construction paper can be cemented onto flip charts and other visuals to give an attractive and colorful appearance with the minimum of effort and expense.

the flip charts serve as notes, written notes can be kept to a minimum, although it may be helpful to write key points lightly (visible only to the presenter) on the charts.

Photographic Slides

Photographic slides are effective when flip charts are too small to be clearly seen, as in large group presentations. They may be necessary to depict actual scenes or examples.

Layout of Artwork for Photographic Slides   Figure 7.8 illustrates a method for sizing the layout in correct proportion for a 35-mm slide. An 8-by-12-inch format is appropriate for most information presented by slides. Combining different colors on each slide will add variety to the sequence and maintain a higher level of interest. Colored construction papers, mat board, and other poster materials can be used in laying out the artwork for a slide.

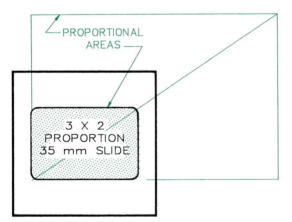

Figure 7.8 This method of proportionately sizing artwork can be used to ensure the artwork will properly fill a photographic slide.

Allow at least an inch margin on all sides of a layout to ensure no edges show when the art is photographed. Selecting the proper size of lettering is critical to ensure readability from any point in the audience.

Figure 7.9 Artwork and charts can be reproduced with a 35-mm reflex camera mounted onto a copy stand, which facilitates the positioning of the camera.

Only capital letters should be used, with space between lines equal to the height of the letters.

Copying the Layouts   To produce slides a camera, copy stand, light meter, and lights are needed. A 35-mm reflex camera is recommended because it has a through-the-lens viewfinder, so the photographer can accurately focus and align each slide before photographing it. The copy stand holds the camera in the proper position during the photographing (Fig. 7.9).

If all layouts are drawn the same size, the camera can be left in the same position. When weather permits, copy work can be photographed in natural light. Book illustrations too small for regular copying can be photographed with a close-up lens that will fill the slide with the area being photographed.

Before the presentation, the mounted slides should be reviewed and sorted to determine their quality (Fig. 7.10). The mounts of all slides should be numbered in their final sequence.

Figure 7.10 Photographic slides should be sorted and arranged in proper sequence prior to loading in the slide tray.

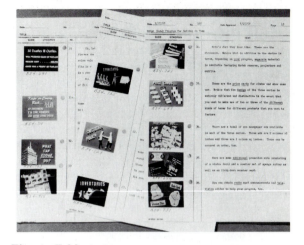

Figure 7.11 A slide manuscript should be prepared for an important slide presentation or for one that will be given on a repetitive basis. (Courtesy of Eastman Kodak.)

The Slide Script A slide presentation that will be used repeatedly needs a script not only for presentations but also for review (Fig. 7.11).

On the left-hand side of the script are black and white photographs of the slides used and on the right-hand side is the narration. The narration is intended as a guide for the presenter.

Overhead Transparencies

Overhead transparencies are reproduced on 8½-by-11-inch acetate by the heat-transfer or diazo process (Fig. 7.12). Tracing paper is most commonly used in the preparation of transparencies because it can be used without special equipment. For best reproduction, line work should be prepared in black India ink. Overlay materials and graphing tapes can be used to give a professional appearance. Lettering should be legible (at least one-quarter inch high). The finished tracing-paper drawing is transferred to the acetate transparencies in much the same way a diazo print is made.

Color Overlays Several color overlays can be hinged to the transparency mount (Fig. 7.12) for a sequential presentation showing the development of an idea or problem. The artwork for overlays is prepared on tracing paper that has been positioned over the basic layout.

Presentation with Transparencies The presenter can stand or sit near the projector in a semilighted room and refer to the transparencies while facing the audience. With a small pointer, the presenter can indicate important points on the stage of the projector, and the image of the pointer will project on the screen. Just as color overlays can be hinged to the mount, so can opaque paper overlays, which are used to conceal parts of the transparency and focus audience attention on a single topic at a time.

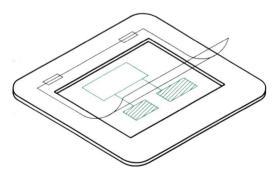

Figure 7.12 A transparency used on an overhead projector is comprised of an 8½″ × 11″ transparency mounted on a 10″ × 12″ frame. The projection area within the frame is about 7½″ × 9½″. To add interest and permit the user to show information in steps, different-colored overlay flips can be used.

Models

A model is an excellent means of communicating the final design concept most realistically. An actual-size prototype of the completed design (Fig. 7.13) gives the most accurate impression of the finished design. Models should be at least twelve inches in size to be visible from more than twelve feet. Photographic slides can effectively supplement the model during a group presentation. A series of close-ups taken from different angles can give each member of the audience a clearer view of a relatively small model.

Figure 7.13 A full-size prototype model of a design can be used as an effective presentation to a small group. (Courtesy of the Chrysler Corp.)

7.5   The Group Presentation

Conference rooms, classrooms, and auditoriums are usually arranged to provide good viewing of visual aids; nevertheless, conditions should be checked by viewing the materials from extreme locations in the audience area before the meeting. All projectors and other visual-aid equipment should be positioned and focused before the audience arrives. The screen or flip charts should be placed where they afford the best view from all locations. Remote controls for slide projectors should be near the speaker's position and ready for use. The speaker should be careful not to block anyone's view. Rehearsing the presentation, with an assistant in the seating area, will help call attention to any blocking movements.

The Presentation

The speaker should give the presentation at a moderate pace, and visual aids should be used as a narration aid (Fig. 7.14). A positive approach in selling ideas should not be confused with possibly deceptive high-pressure salesmanship. The presenter should be the first to point out any weaknesses in the design, but these weaknesses should be offset by alternatives that compensate for them.

Conclusions and recommendations should be given in light of available research and analysis. If after thorough analysis a designer cannot recommend a design, the reasons for that conclusion should be stated. A period for questions and answers is usually provided at the end of the presentation to clarify technical points that may not be fully understood. If possible, the technical report (discussed in the next section) should be available to the audience.

Figure 7.14 A presentation should make full use of graphical aids and models to assist the speaker in communicating ideas. (Courtesy of Bendix Corp.)

Presentation Critique

The evaluation form in Fig. 7.15 aids the student in planning as well as evaluating a presentation. The names of each team member are given at the top of the sheet and their percent contribution is decided by the team as a group. This column must add up to 100 percent. The F-factor is

Oral Report

TEAM NO. _5_ PROJECT _TOY MFGR_

	NAMES	NO. (N = 7)	% CONTRIBU-TION (C)	F = NC	GRADE (G)
1.	BROWN, G.	17	119	91
2.	PRISK, A.	14.3	100	87
3.	SMITH, L.	20	140	95
4.	REED, T.	5.7	40	63
5.	POTTER, M.	14.3	1.00	87
6.	FLYNN, O.	14.3	1.00	87
7.	ROSS, N.	14.4	1.00	87
8.				

EVALUATION BY INSTRUCTOR   100%

		Max. value	Points earned
1.	Introduction of team members	2	2
2.	Statement of purpose of the presentation	5	4
3.	Continuity of presentation	3	3
4.	Use of visuals — point to important points, do not block screen, do not fumble, etc.	10	8
5.	Clear presentation of recommended design	10	8
6.	Presentation of alternative solutions considered	2	2
7.	Coverage of economics — manufacturing, shipping packing, overhead, mark-up, etc.	10	7
8.	Consideration of human factors	5	5
9.	Presentation of an effective conclusion	5	4
10.	Poise and professionalism	2	2
11.	Proper dress for the presentation	2	2
12.	Use of adequate number of visual aids	9	9
13.	Quality of visual aids including model	15	12
14.	Participation of team members (perfect score if all participate)	10	10
15.	Use of allotted time	10	9
		100	87

Additional comments by the instructor on the back of this sheet.

Figure 7.15 An evaluation form for grading a team's oral presentation. Individual grades are found by using the chart in Appendix 51.

Project Uniqueness

TEAM NO. _5_ PROJECT _TOY MFGR_

	NAMES	NO. (N = 7)	% CONTRIBU-TION (C)	F = NC	GRADE (G)
1.	BROWN, G.	14.3	100	83
2.	PRISK, A.	14.3	100	83
3.	SMITH, L.	18.0	126	90
4.	REED, T.	14.3	100	83
5.	POTTER, M.	14.3	100	83
6.	FLYNN, O.	14.3	100	83
7.	ROSS, N.	10.5	74	75
8.				

100%

EVALUATION BY INSTRUCTOR: Degree of uniqueness and originality in solution.

		Max. Value	Points Earned
1.	Serves a needed function	10	8
2.	Functions effectively	10	7.5
3.	Needed by consumers	10	9
4.	Simple, uncomplicated solution	10	8.5
5.	Reasonable, attractive price	10	7
6.	An attractive investment for marketing	10	9
7.	Level of imagination and ingenuity	10	8
8.	Aesthetically pleasing and attractive	10	8
9.	Competition of similar products on the market	10	9
10.	Degree of fulfillment of problem statement	10	9
		100	83

Additional comments by the instructor are on the back of this sheet.

Figure 7.16 An evaluation for grading the uniqueness of a team's project. Individual grades are found by using the chart in Appendix 51.

found for each member by multiplying the number of members on the team by the percent contribution of each. (Use the chart in Appendix 51 to find the grade of each individual.) Another form (Fig. 7.16) is used to evaluate the creativity of the project being presented. This form also allows for the division of the team's grade among the members of the team based on the contribution of each.

7.6 The Technical Report

Engineers, technologists, and technicians at all levels must know how to prepare a written report. The report can be written for a project proposal, a progress report, or a final report (Fig. 7.17).

The Project Proposal

The proposal is written to substantiate the need for a given project that will require the authorization of funds and the use of manpower within the organization. Often, the proposal is a report

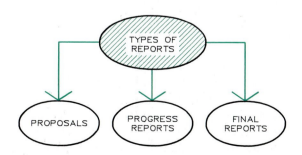

Figure 7.17 The three basic types of technical reports.

submitted to a client, outlining a recommendation that will require an expenditure of funds. The proposal includes data, costs, specifications, time schedules, personnel requirements, completion dates, and any other information that will aid the reader in understanding the project.

> Above all, the purpose and importance of the proposed project must be clearly stated with emphasis on the benefit of the project to the client or organization.

The proposal must also be written in the language of the reader. The businessperson will be interested in profits and benefits that a project promises, whereas the chief engineer will be concerned with its feasibility. Therefore, a typical proposal should contain the following major elements:

- **Statement of the problem.** The problem is clearly identified to present the purpose of the project.
- **Method of approach.** The procedures for attacking the problem are outlined and explained.
- **Personnel needs and facilities.** Requirements for equipment, space, and personnel are itemized.
- **Time schedule.** A time schedule gives an estimate of the completion date for each phase of the project.
- **Budget.** The funds required should be sufficiently detailed.
- **Summary.** The proposal is summarized to emphasize its most important points.

The Progress Report

The progress report is used to periodically review the status of a project. Progress reports which may be in the form of a letter or memo, usually give a projection of an increase or decrease in expenditures or time schedules to permit a revision of project plans. Reference to Project Evaluation and Review Techniques (PERT) introduced in Chapter 3 suggests effective methods of reporting

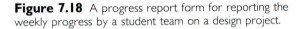

Figure 7.18 A progress report form for reporting the weekly progress by a student team on a design project.

the status of a design project. Figure 7.18 shows a weekly progress report for student projects.

The Final Report

The final report is written at the conclusion of a project. Recommendations in a final report are more conclusive than those in the proposal and progress report, since they are based on the results of the total project rather than on predictions and speculations.

7.7 Organization of a Technical Report

A good technical report comprises the following broad areas:

1. Problem identification.
2. Method.
3. Body.

4. Findings.
5. Conclusions and recommendations.

The order of presenting these areas will vary with the requirements of the governing organization.

For instance, some reports are written with the conclusions and recommendations preceding the body so that the reader immediately sees the results of the report.

Format of the Report

A general sequence of the contents of a technical report is shown in Fig. 7.19.

1. **Cover.** Your finished report should always be bound in an appropriate cover. The title of the report and the name of the person or team that prepared it should be on the cover.
2. **Evaluation sheet.** The first page of the report should be a standard evaluation sheet of the type shown in Fig. 7.20.
3. **Letter of transmittal.** The second page should be a letter of transmittal that briefly describes the contents of the report and the reasons for initiating the project.
4. **Title page.** The title page should contain the title of the report, the name of the person or team that prepared it, and other elements.
5. **Table of contents.** The major headings of the report should be shown.

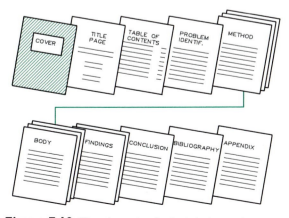

Figure 7.19 The elements of a technical report.

Written Report

TEAM NO. 5   PROJECT   TOY MFGR

NAMES	NO. (N = 7)	% CONTRIBUTION (C)	F = NC	GRADE (G)
1. BROWN, G.		14.3	100	92
2. PRISK, A.		14.3	100	92
3. SMITH, L.		12.0	84	87
4. REED, T.		16.3	114	93
5. POTTER, M.		10.0	70	82
6. FLYNN, O.		8.8	62	79
7. ROSS, N.		14.3	100	92
8.				

EVALUATION BY INSTRUCTOR 100%	Max. Value	Points Earned
1. Use of an appropriate cover	2	2
2. Inclusion of an evaluation sheet	2	2
3. Inculsion of a proper letter of transmittal	2	2
4. Correct title page	2	2
5. Proper table of contents	2	2
6. Sufficient introduction to the the report	5	4
7. Thoroughness in identifying the problem	10	8
8. Continuity and quality of the body the report	10	9
9. Collection and presentation of background data	5	4.5
10. Justification of major decisions	5	5
11. Review of costs, overhead expenses, shipping costs and similar expenses	5	5
12. Arrival at strong conclusion and recommendation	5	4.5
13. Sufficient number of graphs and graphics	10	9
14. Quality of graphics	10	9
15. Bibliography—form and content	5	5
16. Use of footnotes	5	5
17. Appendix—content and form	5	5
18. Form and appearance of report (spelling, punctuation, margins, typing, neatness)	10	9
	100	92

Figure 7.20 A typical evaluation sheet for a student report. Individual grades are taken from the grading chart in Appendix 51.

6. **Table of illustrations.** This listing of illustrations may be omitted in less formal reports.
7. **Problem identification.** (Use a heading appropriate for your report rather than this general term.) This section should establish the importance of and the need for a solution to the problem by reviewing factual information on the problem.

Drawings and illustrations should be included as numbered figures, with legends to describe them and relate them to the text. The text should refer to a figure that follows; for example, "The number of boats sold between 1950 and 1986 is shown in Figure 6."

8. **Method.** (Use a heading appropriate for your report rather than this general term.)

This section should cover the general method used in solving the problem.

9. **Body.** (Use a heading appropriate for your report rather than this general term.) This part of the report elaborates on the solution of the problem. Subheadings should be used to make the parts of the report stand out. Where possible, illustrate your report with sketches, instrument drawings, graphs, or photographs.

Illustrations should be drawn on opaque paper or tracing paper, and a print or photocopy should be made for inclusion in the report. Ink illustrations are preferable. If a drawing must be turned lengthwise on the sheet, the top of the figure should be placed toward the left side of the sheet so it can be read from the right side of the page. Large drawings that must be folded for insertion in a report should be folded to $8\frac{1}{2}$ by 11 inches so they can be conveniently unfolded by the reader.

10. **Findings.** (Use a heading appropriate for your report rather than this general term.) The results of the report should be tabulated and presented graphically with explanatory text. Data are easier to interpret when graphed.

11. **Conclusions.** The conclusions should summarize the entire report and include specific recommendations.

12. **Bibliography.** This is a list of reference material—books, magazines, brochures, conversations—used in the preparation of the report. Bibliographies are usually listed alphabetically by author.

13. **Appendix.** The appendix should include less important information (drawings, sketches, raw data, brochures, letters) not appropriate for inclusion in the main part of the report.

Common Omissions

Topics often omitted from technical reports that should not be omitted include cost estimates, overhead expenses, human and psychological factors, shipping costs, packing specifications, advertising methods, sales considerations, and summarizing recommendations.

7.8 Decision

The purpose of oral and technical reports is to present the basis for a decision on whether or not a recommended design should be implemented.

Acceptance   A design may be accepted in its entirety, which is a compliment to the designer's research and problem solving abilities.

Rejection   A design recommendation may be rejected in its entirety. Changes in the economic climate or moves by competitors may make the design unprofitable.

Compromise   If a design is not approved in its entirety, a compromise might be suggested in one or more areas; for example, the initial production run might be increased or decreased, or several features might be modified to make a design more attractive.

Decision—Hunting Seat

The design of a hunting seat is used to illustrate the decision step of the design process. The problem is restated below.

Hunting Seat   Many hunters hunt from trees to obtain a better vantage point. Design a seat that provides the hunter with comfort and safety while hunting from a tree and that meets the requirements of economy and the limitations of hunting.

Decision Chart   The chart shown in Fig. 7.21 can be used to compare the available designs. Each idea is listed and given a number for identification.

Next, maximum values for the various factors of analysis are assigned so that the total of all factors is ten points. These values will vary from product to product, so use your best judgment to

DECISION
1. Decision for evaluation

DESIGN 1   FOLDING SEAT

DESIGN 2   CANVAS SEAT

DESIGN 3   PLATFORM SEAT

DESIGN 4

DESIGN 5

DESIGN 6

MAX	FACTORS	1	2	3	4
3.0	FUNCTION	2.0	2.3	2.5	
2.0	HUMAN FACTORS	1.6	1.4	1.7	
0.5	MARKET ANALYSIS	0.4	0.4	0.4	
1.0	STRENGTH	1.0	1.0	1.0	
0.5	PRODUCTION	0.3	0.2	0.4	
1.0	COST	0.7	0.6	0.8	
1.5	PROFITABILITY	1.1	1.0	1.3	
0.5	APPEARANCE	0.3	0.4	0.4	
10	TOTALS	7.4	7.3	8.5	

DESIGN DECISION   NAME   FILE   SEC   DATE   GRADE **16**

Figure 7.21 A worksheet with a decision table used to evaluate the developed design alternatives.

determine them. You can now evaluate each factor of the competing designs.

The vertical columns of numbers are summed to determine the design with the highest total, the best design, perhaps. However, your instincts may disagree with your numerical analysis. If so, you should have enough faith in your judgment not to be restricted by your numerical analysis. Decision will always remain the most subjective part of the design process.

Conclusion   Once a decision has been made, it should be clearly stated, along with the reasons for the design's acceptance (Fig. 7.22). Additional information may be given such as number to be produced initially, selling price per unit, expected profit per unit, expected sales during the first year, number that must be sold to break even, and its most marketable features.

It is possible that you would recommend a design not be implemented. The design process should not then be considered a failure; a negative decision could save an investor from large losses.

CONCLUSIONS

THE FLAT FOLDING SEAT IS BELIEVED TO BE THE BEST SOLUTION BECAUSE:

1. GOOD MARKET POTENTIAL

2. SIMPLE DESIGN

3. EASE OF MANUFACTURE

4. FULFILLS PROBLEM REQUIREMENTS

RECOMMEND IMPLEMENTATION AND PRODUCTION OF THIS DESIGN

ECONOMIC FORECAST:

SALES PRICE $100.00

SHIPPING EXPENSES   7.00

ESTIMATED PROFIT PER SEAT $20.00

RECOMMEND THAT SEATS BE PRODUCED BY A QUALIFIED MANUFACTURER ON A CONTRACT BASIS.

BREAK EVEN AFTER 1000 SEATS HAVE BEEN SOLD.

SEAT SHOULD PROVE TO BE AN ATTRACTIVE AND PROFITABLE INVESTMENT.

DESIGN CONCLUSION   NAME   FILE   SEC   DATE   GRADE **17**

Figure 7.22 The decision is summarized on this worksheet to give the designer's conclusion and recommendation concerning the next step, implementation.

Problems

1. Prepare a checklist that could be used to evaluate an oral presentation of one of your classmates. List important items that should be considered, and determine a point system that could be assigned to each. Keep the form simple, yet thorough enough to be of value to the presenter. Devise a means of tabulating the evaluation to arrive at an overall rating.

2. Prepare a series of 3-by-5-inch cards to plan a flip-chart presentation that will last no more than five minutes. The subject of your flip-chart presentation may be of your choosing or assigned by your teacher. Some example topics are: your career plans for the first two years after graduation, the role of this course in your total educational program, the importance of effective communications, the identification of a need for a design project you are proposing, a comparison of the engineering profession with another profession of your selection.

3. Prepare graphical aids for an oral presentation using the methods and materials indicated in this chapter.

4. Using the planning cards developed in Problem 2, prepare a five-minute briefing for a technique that you choose or your instructor assigns. Give this briefing to your class as assigned.

5. Assume you are an engineer responsible for representing your firm in the presentation of a proposal for a sizable contract. Make a list of instructions that you could give to your assistants to coordinate the preparation of your presentation. The presentation will be for a group of 20 persons ranging in background from bankers to engineers. The topic is not so important as the method of presentation. Use a topic of your choice, one assigned by your teacher, or one suggested in Problem 2. Your instructions should outline the materials you need, method of preparation of graphical aids and number required, method of projection or presentation, assistance needed in presenting materials, room seating arrangements, and other factors. Your outline should be complete enough to cover the entire program of an ideal presentation within the time you think most desirable.

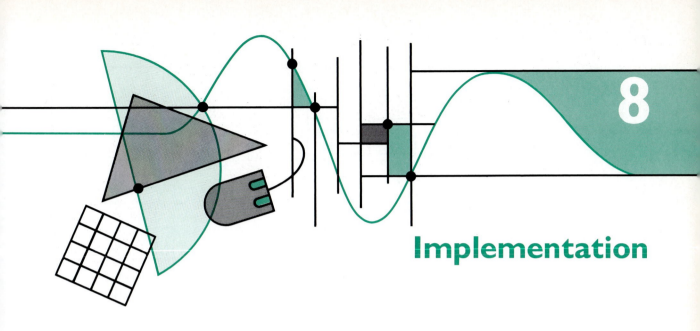

Implementation

8

8.1 Introduction

The final step of the design process is implementation, during which the design becomes a reality. Graphical methods are particularly important during the initial steps of implementation, since all products are constructed from working drawing with specifications.

8.2 Working Drawings

Working drawings are drawn using orthographic views that have been dimensioned and noted to show how the individual parts are to be made. An example of a working drawing generated with computer graphics is shown in Fig. 8.1. A properly drawn working drawing will result in the same product time and again, regardless of the shop in which it is made. When making working drawings, several parts can be drawn on the same sheet without attempting to arrange them in relation to one another. The names of the parts, their identifying numbers, and their materials are given near the views.

8.3 Specifications

Specifications are notes and instructions that supplement working drawings. Specifications may be given in typed documents when graphical representations are unnecessary; for example, instructions like those shown below can be given effectively as written specifications:

METALLURGICAL INSPECTION IS
REQUIRED BEFORE MACHINING.

or

PAINT WITH TWO COATS OF FLAT
BLACK PAINT (NO. 780) AFTER
FINISHING.

However, when space permits, it is more convenient that specifications be given on the working drawing by the drafter.

8.4 Assembly Drawings

Assembly drawings illustrate how parts are put together after they have been made. They can be drawn as pictorial or orthographic views that are

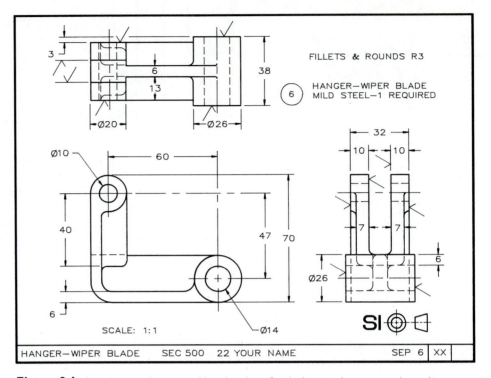

Figure 8.1 A computer-drawn working drawing of a single part that was made as the implementation step of the design process.

fully assembled, fully exploded, or partially exploded. Figure 8.2 is a partially exploded orthographic assembly where sections have been used to clarify internal details.

8.5 Miscellaneous Considerations

In addition to the preparation of drawings and specifications, several other aspects of the implementation step must be considered: packaging, storage, shipping, and marketing of the product.

Packaging In some cases, such as the toy industry, the packaging is very elaborate and may be as expensive as the product. Designers must be aware of the packaging needs as they develop a design, since a product that is difficult to package will cost more. Many products, to make them easier to package and more economical, are shipped partially disassembled.

Storage Most manufacturers maintain a backlog of products for shipment. The cost of keeping an inventory must be figured as part of the product's final selling price.

Shipping Some industries locate their shipping facilities in the middle of their market areas to reduce shipping costs. But companies at the extreme ends of their market's geographical region must raise the cost of their products to cover shipping.

Marketing Designers are concerned with all aspects of a product after it enters the marketplace. They are concerned with its marketability and acceptance by the consumer. Complaints about the product's effectiveness and function are important to designers because they will want to modify any defect in future versions of the product.

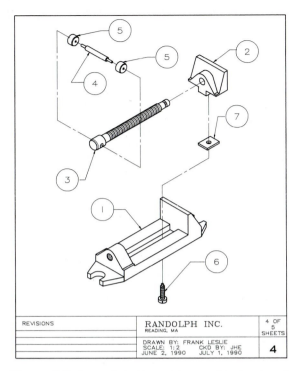

Figure 8.2 An exploded isometric assembly drawing of a vise, which shows how the individual parts of an assembly are put together.

8.6 Implementation—Hunting Seat

The design of a hunting seat illustrates the implementation step of the design process. The problem is restated below.

Hunting Seat Many hunters hunt from trees to obtain a better vantage point. Design a seat that provides the hunter with comfort and safety while hunting from a tree and that meets the requirements of economy and the limitations of hunting.

Four working drawing sheets (Figs. 8.3–8.6) have been prepared to present the details of the hunting seat design. The fifth sheet, Fig. 8.7, is an assembly drawing that illustrates how the parts are assembled once they have been made. (This particular design was developed, patented, and is marketed by Baker Manufacturing Company, Valdosta, Georgia. It is the Baker Favorite Seat, Patent No. 3460649.)

All parts have been dimensioned in millimeters, which are metric units. Standard parts that are purchased from suppliers are not drawn, but are itemized on the drawing, given parts numbers, and listed in the parts list on the assembly drawing. Figure 8.7 is an assembly drawing, which is a pictorial with the different parts iden-

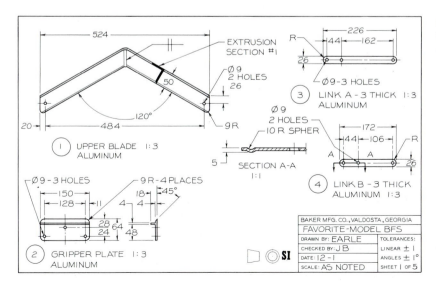

Figure 8.3 A working drawing sheet of parts of a hunting seat design. Sheet 1 of 5.

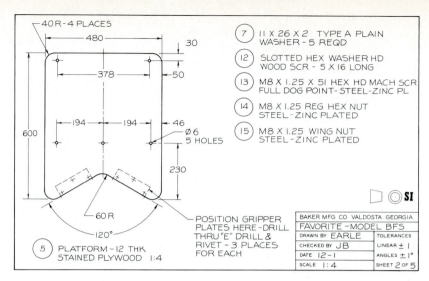

Figure 8.4 A working drawing sheet of parts of a hunting seat design. Sheet 2 of 5.

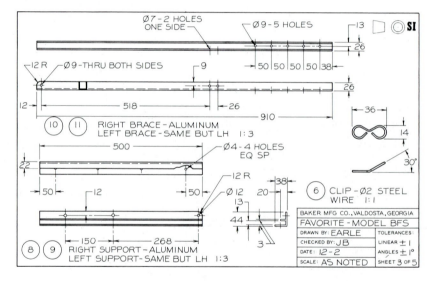

Figure 8.5 A working drawing sheet of parts of a hunting seat design. Sheet 3 of 5.

Figure 8.6 A working drawing sheet of parts of a hunting seat design. Sheet 4 of 5.

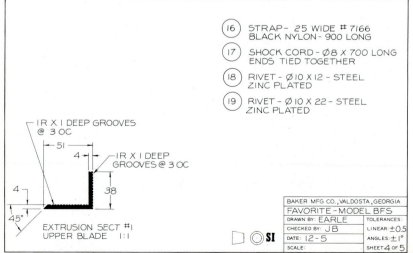

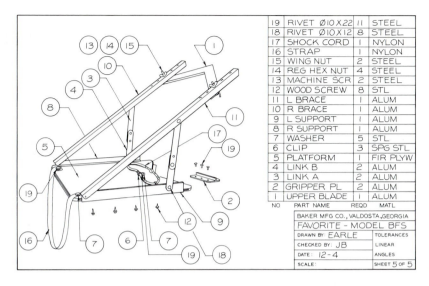

19	RIVET Ø10 X 22	11	STEEL
18	RIVET Ø10 X 12	8	STEEL
17	SHOCK CORD	1	NYLON
16	STRAP	1	NYLON
15	WING NUT	2	STEEL
14	REG HEX NUT	4	STEEL
13	MACHINE SCR	2	STEEL
12	WOOD SCREW	8	STL
11	L BRACE	1	ALUM
10	R BRACE	1	ALUM
9	L SUPPORT	1	ALUM
8	R SUPPORT	1	ALUM
7	WASHER	5	STL
6	CLIP	3	SPG STL
5	PLATFORM	1	FIR PLYW
4	LINK B	2	ALUM
3	LINK A	2	ALUM
2	GRIPPER PL	2	ALUM
1	UPPER BLADE	1	ALUM
NO	PART NAME	REQD	MATL

BAKER MFG CO., VALDOSTA, GEORGIA		
FAVORITE - MODEL BFS		
DRAWN BY: EARLE	TOLERANCES	
CHECKED BY: JB	LINEAR	
DATE: 12-4	ANGLES	
SCALE:	SHEET 5 OF 5	

Figure 8.7 An assembly drawing that demonstrates how the parts of the hunting seat design are assembled. Sheet 5 of 5.

tified by balloons attached to leaders. Each part is listed in the parts list by number with general information to describe it.

Packaging The Baker Favorite Seat is packaged in a corrugated cardboard box and weighs approximately ten pounds (Fig. 8.8). The seat is shipped unassembled so it will fit the shipping carton and be easily handled while in shipment (Fig. 8.9).

Storage The periods before hunting seasons will require more inventory than other times of

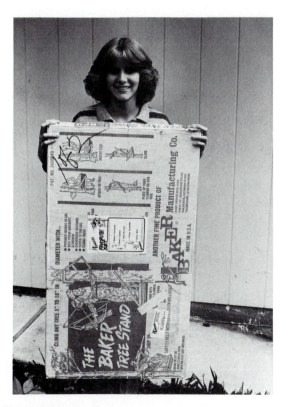

Figure 8.8 The Baker Favorite Seat is shipped in a corrugated cardboard box to the retailer or the consumer.

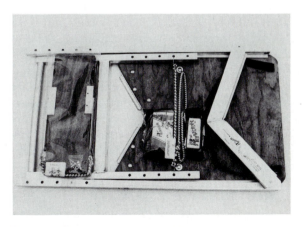

Figure 8.9 The hunting seat is folded into a flat position for ease of packaging before shipment.

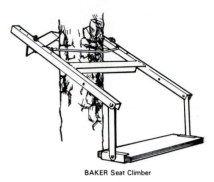

BAKER Seat Climber

Figure 8.10 Examples of accessories designed to accompany the hunting seat. (Courtesy of Baker Manufacturing Co., Valdosta, Ga.)

the year. An inventory of seats waiting to be sold adds to the cost of overhead in the form of interest payments, warehouse rent, warehouse personnel, and loading equipment.

Shipping   Shipping costs for all types of carriers (rail, motor freight, air delivery, mail services) must be evaluated. The shipping cost for a Baker Favorite Seat with its accessories is $5–$10 depending on distance when shipped one at a time by United Parcel Service. The cost per unit is reduced by about fifty percent when they are shipped in bundles of ten to the same destination.

Accessories   Examples of accessories are fold-down seats, hand climbers, and add-on seats (Fig. 8.10). Accessories provide for the special needs of the hunters, increase the marketability of the product, and increase sales volume.

The retail price of the Baker Seat is about $90, five or six times more than the cost of the materials and labor to manufacture it. Retailers

are given approximately a forty percent margin, and distributors earn about ten percent. The remainder of the overhead includes advertising costs and other expenses already mentioned. All expenses must be absorbed by the consumer who ultimately purchases the product.

8.7 Patents

A designer who has developed an original and novel solution to a design problem should investigate the possiblities of obtaining a patent from the U.S. Patent and Trademark Office (PTO) (Fig. 8.11). This must be done before disclosing the invention, since premature disclosure may forfeit the patent right.

A good source for the specific details of the patent procedure is *General Information Concerning Patents*. (This publication, available at no cost from the PTO, was the primary reference for the following text.)

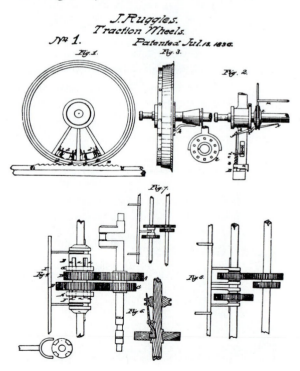

Figure 8.11 The first patent, issued by the U.S. Patent Office in 1836.

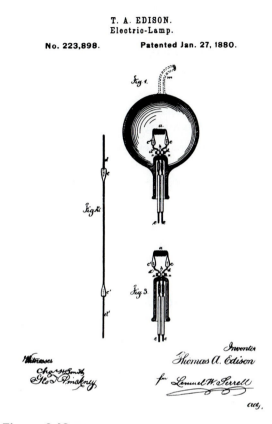

Figure 8.12 Thomas Edison's patent drawing for the electric lamp, 1880.

General Patent Requirements

Designers must have a general understanding of patent requirements, that is, what can be patented and who is eligible to apply for a patent.

What Can Be Patented Any person who "invents or discovers any new and useful process, machine, manufacture, or composition of matter, or any new and useful improvement thereof, may obtain a patent," subject to the conditions and requirements of law. Essentially, these categories include everything made by humans and the processes for making them (Fig. 8.12).

Inventions used solely for the development of nuclear and atomic weapons for warfare are not patentable since they are not considered "useful." Also, a design for a functional mechanism that will not operate in keeping with its intended pur-

pose is not patentable. An **idea** for a new invention or machine is not patentable; the specific design and description of the machine must be available before it can be considered for patent registration.

Who Can Apply For a Patent Only the inventor of a device may apply for a patent. A patent given to a person who was not the inventor would be void, and the person would be subject to prosecution for committing perjury. Application for a patent can, however, be made by the executor of a deceased inventor's estate. Two or more persons may apply for a patent as joint inventors.

> PATENT RIGHTS An inventor granted a patent has the right to exclude others from making, using, or selling the invention throughout the United States for seventeen years.

After expiration of the seventeen-year patent term, the invention may be made, used, or sold by anyone without authorization from the patent holder.

Patented articles must be marked with the word "Patent" and the number of the patent. Markings using the terms "Patent Pending" have no legal effect, since protection does not begin until the actual grant is made.

Application for a Patent

An inventor applying for a patent must include the following:

1. A written document that comprises a petition, a specification (description and claims), and an oath or declaration.
2. A drawing if a drawing is possible.
3. The filing fee.

Petition and Oath The petition and oath are usually on one form. On this form, the inventor petitions or requests to be given a patent on the invention and declares that he or she is the original and first inventor of the device described in the application.

Specification   Written specifications of a patent must be attached to the application describing the invention in detail so that a person skilled in the field to which the invention pertains can produce the item. Drawings should be referred to in the text by figures and part numbers (Fig. 8.13).

Claims are brief descriptions of the features of the invention that distinguish it from already patented material. The claims are the most significant part of the patent, since they will be used to ascertain the novelty and patentability of an invention.

Fee   The application for a patent must be accompanied by the filing fee, $170; after the application has been accepted, notice will be sent to

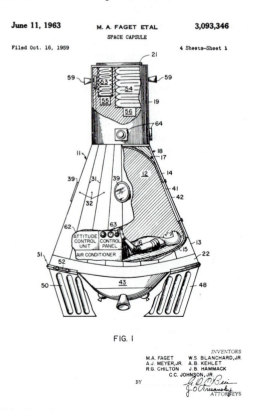

Figure 8.13 The patent drawing of a space capsule developed by the National Aeronautics and Space Administration (NASA). (Courtesy of the U.S. Patent and Trademark Office.)

the applicant giving him or her three months to remit a $280 issue fee. There are other miscellaneous charges for multiple claims within the patent application.

8.8   The Preparation of Patent Drawings

When drawings are necessary to describe an invention, the applicant must submit them as well. A booklet, *Guide for Patent Draftsmen* (available from the U.S. Government Printing Office), outlines the procedures for preparing patent drawings. If the inventor cannot furnish drawings, the Patent Office will recommend a drafter who can prepare them at the inventor's expense.

Patent Drawing Standards

When the patent is issued, the completed drawings are printed and published. To prevent rejection of the application, follow these drawing standards as closely as possible.

Paper and Ink   Drawings must be made on pure white paper of thickness corresponding to a two- or three-ply Bristol board. The surface must be calendered and smooth to permit erasure and correction. Only India ink will secure perfectly black solid lines. The use of white pigment to cover lines is not acceptable.

Sheet Size and Margins   The sheet size must be exactly $8\frac{1}{2}$ by 14 in. (21.6 by 35.6 cm) or exactly 21.0 by 29.7 cm. All sheets in a particular application must be the same size. One of the shorter sides is regarded as the top of the sheet. On $8\frac{1}{2}$-by-14-in. sheets, the top margin must be 2 in. and the side and bottom margins $\frac{1}{4}$ in. Margin border lines cannot be drawn on the sheets, but all work must be included within the margins. The sheets may be punched with two $\frac{1}{4}$-in. holes with their centerlines $\frac{11}{16}$ in. below the top edge and $2\frac{3}{4}$ in. apart and centered from the sides of the sheet. The margins for 21.0-by-29.7-cm sheets are 2.5 cm from the top, 2.5 cm from the left, 1.5 cm from the right, and 1 cm from the bottom.

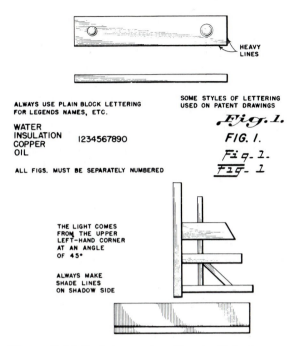

Figure 8.14 Typical examples of lines and lettering recommended for patent drawings.

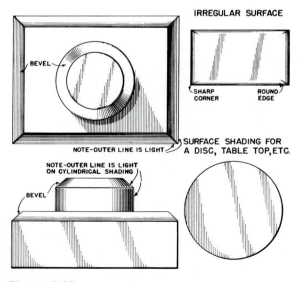

Figure 8.15 Techniques of representing surfaces and beveled planes on patent drawings.

Character of Lines All lines and lettering must be absolutely black regardless of how fine the lines may be. Freehand work should be avoided.

Hatching and Shading Hatching lines, used to shade the surface of an object, should be parallel and not less than $\frac{1}{20}$ in. apart (Fig. 8.14). Heavy lines are used on the shade side of the drawing; however, they should not be used if they are likely to confuse the drawing. The light is assumed to come from the upper left-hand corner at an angle of forty-five degrees. Types of surface delineation are given in Fig. 8.15.

Scale The scale should be large enough to show the mechanism without crowding when the drawing is reduced for reproduction. Certain portions of the mechanism may be drawn at a larger scale to show additional details.

Reference Characters The different views of a mechanism should be identified by consecutive figure numbers. Use plain, legible numerals at least $\frac{1}{8}$ in. high, not encircled, and placed close to the parts to which they apply (Fig. 8.16). Numbers should not be placed on hatched surfaces unless a blank space is provided for them. The same part appearing in more than one view of the drawing should be designated by the same character.

Symbols Symbols used to represent various materials in sections, electrical components, and mechanical devices are suggested by the Patent Office, but these conform to the engineering drawing standards.

Signature and Names The signature of the applicant, or the name of the applicant and the signature of the attorney or agent, may be placed in the lower right-hand corner of each sheet within the marginal lines or below the lower marginal lines.

Views Figures should be numbered consecutively in order of their appearance. Figures may be plan, elevation, section, perspective, or detail

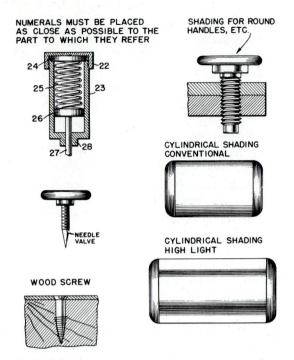

NUMERALS MUST BE PLACED AS CLOSE AS POSSIBLE TO THE PART TO WHICH THEY REFER

SHADING FOR ROUND HANDLES, ETC.

NEEDLE VALVE

CYLINDRICAL SHADING CONVENTIONAL

WOOD SCREW

CYLINDRICAL SHADING HIGH LIGHT

Figure 8.16 Methods of numbering parts and rendering details for patent drawings.

views (Fig. 8.16). Exploded views can be used to advantage to describe the assembly of a number of parts. Large parts may be broken into sections and drawn on several sheets if this does not confuse the matter. Removed sections can be used, provided that the cutting plane is labeled to indicate the section by number. All sheet headings and signatures will be placed in the same position on the sheet whether the drawing is read from the bottom of the sheet or from the right side of the sheet.

No extraneous matter, such as an agent's or attorney's stamp or address, is permitted on the face of the drawing. The completed drawings should be sent flat, protected by heavy board or rolled in a suitable mailing tube.

8.9 Patent Searches

A patent can be granted only after the PTO examiners have searched existing patents to determine whether the invention has been previously patented. With more than 4,500,000 patents on record, a search is the most time-consuming part of obtaining a patent. Many inventors employ patent attorneys or agents to conduct a preliminary search of existing patents to discover whether their invention infringes on another. Patents are filed in the Search Room of the Patent Office by classes and subclasses according to subject matter. Other seemingly unrelated patents may cover the invention being submitted, thereby disallowing the patent.

8.10 Questions and Answers about Patents

For best results, patent applications must be handled with the assistance of a patent attorney or patent agent. The Patent Office has published a pamphlet, *Questions and Answers About Patents.* Most of these questions and answers are listed here.

Nature and Duration of Patents

1. Q. *What is a patent?*
   A. A patent is a grant issued by the U.S. Government giving an inventor the right to exclude all others from making, using, or selling his or her invention within the United States, its territories and possessions.

2. Q. *For how long a term of years is a patent granted?*
   A. Seventeen years from the date on which it is issued; except for patents on ornamental designs, which are granted for terms of $3\frac{1}{2}$, 7, or 14 years.

3. Q. *May the term of a patent be extended?*
   A. Only by special act of Congress, and this occurs very rarely and only in most exceptional circumstances.

4. Q. *Does the patentee continue to have any control over the use of the invention after the patent expires?*
   A. No. Anyone has the free right to use an invention covered in an expired patent, so long as he or she does not use features covered in other unexpired patents in doing so.

5. Q. *On what subject matter may a patent be granted?*
   A. A patent may be granted to the inventor or discoverer of any new and useful process, machine, manufacture, or composition of matter, or any new and useful improvement thereof, or on any distinct and new variety of plant, other than a tuber-propagated plant, which is asexually reproduced, or on any new, original, and ornamental design for an article of manufacture.

6. Q. *On what subject matter may a patent not be granted?*
   A. A patent may not be granted on a useless device, on printed matter, on a method of doing business, on an improvement in a device which would be obvious to a person skilled in the art, or on a machine which will not operate, particularly on an alleged perpetual motion machine.

Meaning of Words "Patent Pending"

7. Q. *What do the terms "patent pending" and "patent applied for" mean?*
   A. They are used by a manufacturer or seller of an article to inform the public that an application for patent on that article is on file in the Patent Office. The law imposes a fine on those who use these terms falsely to deceive the public.

Patent Applications

8. Q. *I have made some changes and improvements in my invention after my patent application was filed in the Patent Office. May I amend my patent application by adding a description or illustration of these features?*
   A. No. The law specifically provides that new matter shall not be introduced into the disclosure of a patent application. However, you should call the attention of your attorney or patent agent promptly to any such changes you may make or plan to make, so that he or she may take or recommend any steps that may be necessary for your protection.

9. Q. *How does one apply for a patent?*

A. By making the proper application to the Commissioner of Patents, Patent and Trademark Office, Washington, DC, 20231.

10. Q. *What are the Patent Office fees in connection with filing of an application for patent and issuance of the patent?*
    A. A filing fee of $170 plus certain additional charges for claims, depending on their number and the manner of their presentation, are required when the application is filed. A final or issue fee of $280 plus certain printing charges are also required if the patent is to be granted. The final fee is not required until your application is allowed by the Patent Office.

11. Q. *Are models required as a part of the application?*
    A. Only in the most exceptional cases. The Patent Office has the power to require that a model be furnished, but rarely exercises it.

12. Q. *Is it necessary to go to the Patent Office in Washington to transact business concerning patent matters?*
    A. No; most business with the Patent Office is conducted by correspondence. Interviews regarding pending applications can be arranged with examiners if necessary, however, and are often helpful.

13. Q. *Can the Patent Office give advice as to whether an inventor should apply for a patent?*
    A. No. It can only consider the patentability of an invention when this question comes regularly before it in the form of a patent application.

14. Q. *Is there any danger that the Patent Office will give others information contained in my application while it is pending?*
    A. No. All patent applications are maintained in the strictest secrecy until the patent is issued. After the patent is issued, however, the Patent Office file containing the application and all correspondence leading up to issuance of the patent is made available in the Patent Office Search Room for inspection by anyone, and copies of these files may be purchased from the Patent Office.

15. Q. *May I write to the Patent Office directly about my application after it is filed?*
    A. The Patent Office will answer an applicant's inquiries as to the status of the appli-

cation and inform him or her whether the application has been rejected, allowed, or is awaiting action by the Patent Office. However, if you have a patent attorney or agent, the Patent Office cannot correspond with both you and the attorney concerning the merits of your application. All comments concerning your invention should be forwarded through your patent attorney or agent.

16. Q. *What happens when two inventors apply separately for a patent on the same invention?*
    A. An "interference" is declared and testimony may be submitted to the Patent Office to determine which inventor is entitled to the patent. Your attorney or agent can give you further information about this if it becomes necessary.

17. Q. *May applications be examined out of their regular order?*
    A. No. All applications are examined in the order in which they are filed, except under certain very special conditions.

When to Apply For a Patent

18. Q. *I have been making and selling my invention for the past 13 months and have not filed any patent application. Is it too late for me to apply for a patent?*
    A. Yes. A valid patent may not be obtained if the invention was in public use or sale in this country for more than one year prior to the filing of your patent application. Your own use and sale of the invention for more than a year before your application is filed will bar your right to a patent just as effectively as though this use and sale had been done by someone else.

19. Q. *I published an article describing my invention in a magazine 13 months ago. Is it too late to apply for a patent?*
    A. Yes. The fact that you are the author of the article will not save your patent application. The law provides that the inventor is not entitled to a patent if the invention has been described in a printed publication any-where in the world more than a year before his patent application is filed.

Who May Obtain a Patent

20. Q. *If two or more persons work together to make an invention, to whom will the patent be granted?*
    A. If each had a share in the ideas forming the invention, they are joint inventors and a patent will be issued to them jointly on the basis of a proper patent application filed by them jointly. If, on the other hand, one of these persons has provided all of the ideas of the invention, and the other has only followed instruction in making it, the person who contributed the ideas is the sole inventor and the patent application and patent should be in his or her name only.

21. Q. *If one person furnishes all of the ideas to make an invention and another employs him or her or furnishes the money for building and testing the invention, should the patent application be filed by them jointly?*
    A. No. The application must be signed, executed, sworn to, and filed in the Patent Office in the name of the inventor. This is the person who furnishes the ideas, not the employer or the person who furnishes the money.

22. Q. *May a patent be granted if an inventor dies before filing his application?*
    A. Yes; the application may be filed by the inventor's executor or administrator.

23. Q. *While in England this summer, I found an article on sale which was very ingenious and has not been introduced into the United States or patented or described. May I obtain a United States patent on this invention?*
    A. No. A United States patent may be obtained only by the true inventor, not by someone who learns of an invention of another.

Ownership and Sale of Patent Rights

24. Q. *May the inventor sell or otherwise transfer his or her right to his patent or patent application to someone else?*

A. Yes. He or she may sell all or part of the interest in the patent application or patent to anyone by a properly worded assignment. The application must be filed in the Patent Office as the invention of the true inventor, however, and not as the invention of the person who has purchased the invention from him.

25. Q. *Is it advisable to conduct a search of patents and other records before applying for a patent?*
A. Yes. If it is found that the device is shown in some prior patent it is useless to make application. By making a search beforehand the expense involved in filing a needless application is often saved.

Technical Knowledge Available from Patents

26. Q. *I have not made an invention but have encountered a problem. Can I obtain knowledge through patents of what has been done by others to solve the problem?*
A. The patents of the Patent Office Search Room in Washington contain a vast wealth of technical information and suggestions, organized in a manner that will enable you to review those most closely related to your field of interest. You may come to Washington and review these patents, or engage a patent practitioner to do this for you and to send you copies of the patents most closely related to your problem.

27. Q. *Can I make a search or obtain technical information from patents at locations other than the Patent Office Search Room in Washington?*
A. Yes. Libraries have sets of patent copies numerically arranged in bound volumes, and these patents may be used for search or other information purposes as discussed in the answer to Question 28.

28. Q. *How can technical information be found in a library collection of patents arranged in bound volumes in numerical order?*
A. You must first find out from the *Manual of Classification* in the library the Patent Office classes and subclasses which cover the field of your invention or interest. You can then, by referring to microfilm reels or volumes of the *Index of Patents* in the library, identify the patents in these subclasses and, thence, look at them in the bound volumes. Further information on this subject may be found in the leaflet *Obtaining Information from Patents,* a copy of which may be requested from the Patent Office.

Infringement of Others' Patents

29. Q. *If I obtain a patent on my invention, will that protect me against the claims of others who assert that I am infringing their patents when I make, use, or sell my own invention?*
A. No. There may be a patent of a more basic nature on which your invention is an improvement. If your invention is a detailed refinement or feature of such a basically protected invention, you may not use it without the consent of the patentee, just as no one will have the right to use your patented improvement without your consent.

Enforcement of Patent Rights

30. Q. *Will the Patent Office help me to prosecute others if they infringe the rights granted to me by my patent?*
A. No. The Patent Office has no jurisdiction over questions relating to the infringement of patent rights. If your patent is infringed, you may sue the infringer in the appropriate United States court at your own expense.

Patent Protection in Foreign Countries

31. Q. *Does a United States patent give protection in foreign countries?*
A. No. The United States patent protects your invention only in this country. If you wish to protect your invention in foreign countries, you must file an application in the Patent Office of each such country within the time permitted by law.

Problems

Working Drawings

1. Prepare working drawings as the implementation step of the design process of one of the problems that was refined at the end of Chapter 5. Draw the working drawings and assembly on 11-by-17-inch sheets of tracing vellum or film.

2. Prepare working drawings of one of the problems assigned or selected from those at the end of Chapter 24. Draw the working drawings and assembly on 11-by-17-inch sheets of tracing vellum or film.

Patents

3. Write for a copy of a patent that would be of interest to you. Make a list of the features that were used as a basis for obtaining the patent.

4. Suggest modifications to the patent mentioned in the previous problem. Make sketches of innovations that would be possible improvements of the patented mechanism.

5. Write the U.S. Patent and Trademark Office for patent application forms. Prepare a patent application for a simple invention that has been previously patented, such as a fountain pen, drafting instrument, or similar item. Determine what drawings and materials are needed to complete your application.

6. Prepare patent drawings in accordance with the standards established in Section 8.8 to depict a simple patented object, such as those mentioned in Problem 5. Strive for a finished technique that would make your drawing acceptable as a patent drawing.

7. Make a list of ideas for products that you believe to be patentable. These may be ideas that you have developed during work on design problems assigned in a class.

8. Write a technical report investigating the history and significance of the patent system and its role in our industrial society. Consult your library and available government publications on patents. Give information and data that will improve your understanding of patents.

Design Problems

9.1 Introduction

This chapter will introduce problems that can be used for class assignments, team projects, and other combinations of approaches to provide experience in applying the principles and techniques covered in this text.

9.2 The Individual Approach

The short problem—requiring one or two hours of work—will probably be assigned for solution by students working individually. Though simple design problems may involve fewer details and less depth, all design steps are applied as in more sophisticated, comprehensive problems.

9.3 The Team Approach

Obviously, an effectively organized team working on a single problem has access to more talent than does the typical individual who devotes the same number of hours to the project. The application of this talent will be a problem if the team is not properly organized. Team management and working effectively with others are invaluable skills that should be developed.

Team Size   A student design team should have from three to seven members. Three is considered the minimum number for a valid team experience. The optimum size, four, lessens the possibility of domination by one or more members.

Team Composition   In practice, the engineering team may be composed of professionals from different firms who may be total strangers. This arrangement can be advantageous in that it reduces the impact of personalities.

Team Leader   A leader is necessary for most teams to function effectively. This person takes the responsibility for making assignments, seeing that deadlines are met, and acting as arbitrator during any disagreements.

9.4 The Selection of a Problem

The best problem for a student design project is one that involves familiar and accessible conditions. A design for a water-ski rack for an automobile is more feasible than one for a support bracket for an airplane.

Student-Proposed Problem The best design problem is one that has been recognized and proposed by the students who will be working toward its solution. When the members of a design team decide on a design project, they should prepare a written proposal identifying the problem and outlining its limits.

Instructor-Assigned Problems Assigned problems are appropriate for classroom situations, and they are analogous to the assignment method used in industry.

9.5 Problem Specifications

An individual or design team must consider the following specifications when preparing a design proposal or outlining an assignment.

The Short Problem The specifications for a short problem could include all or part of the following.

1. Completed worksheets illustrating the development of the design process (Chapter 2).
2. Freehand sketches of the design for implementation (Chapter 14).
3. An instrument drawing of the proposed design (Chapters 8 and 24).
4. A dimensioned instrument drawing of the proposed design (Chapter 24).
5. A pictorial sketch (or one made with instruments) illustrating the design (Chapters 3, 14, and 26).
6. Visual aids, flip charts, or other media to present the design to a group (Chapter 7).

The Comprehensive Problem Comprehensive problems vary in time; typically, one takes an average of 80 to 120 work-hours.

The following specifications apply to a comprehensive problem of the systems- or product-development type.

1. A proposal outlining the problem, the method of approach, and the specifications used in solving the problem (Chapter 7).

2. Completed worksheets illustrating the development of the design process (Chapter 2).
3. Schematic diagrams, flowcharts, or other symbolic methods of illustrating the design and its function (Chapter 3).
4. An opinion survey determining interest for the proposed design (Chapter 3).
5. A market survey evaluating the product's possible acceptance and estimated profit (Chapter 3).
6. A model or prototype for analysis or presentation (Chapter 6).
7. Pictorials explaining features of the final design solution that are not clearly shown in other drawings (Chapter 26).
8. Dimensioned engineering drawings giving details and specifications of the design (Chapter 24).
9. A technical report, fully illustrated with charts and graphs, explaining the activities leading to the solution and ending with conclusions and recommendations (Chapter 7).

9.6 Scheduling Team Activities

The semester schedule in Fig. 9.1 can be used as a guideline to assist team members in working toward the completion of their project.

The product design problems are more comprehensive and require more investigation and background research to fully understand the problem, market needs, and other design factors. These problems are better suited to solution by a team rather than by a single person.

The solution to these problems should follow the steps of the hunting seat example in this chapter.

9.7 Short Design Problems

The following are short problems that can be completed in less than two periods.

1. **Lamp bracket.** Design a simple bracket to attach a desk lamp to a vertical wall for reading in bed. The lamp should be easily remov-

SEMESTER SCHEDULE

	MONDAY	WEDNESDAY	FRIDAY
1			
2			
3		ASSIGN TEAMS	
4		PROB IDENTIF	
5		PROB IDENTIF	
6		BRAINSTORM	
7		PRELIM IDEAS	
8		REFINEMENT	
9		REFINEMENT	
10		ANALYSIS	
11		DECISION	
12		PREPARE	
13		PREPARE	
14		PRESENTATION	
15			
16			

Figure 9.1 This table shows a suggested schedule for using a part of a class period each week for the design project.

able so that it can be used as a conventional desk lamp.

2. **Towel bar.** Design a towel bar for a kitchen or bathroom. Determine optimum size, and consider styling, ease of use, and method of attachment.

3. **Pipe aligner for welding.** The initial problem of joining pipes with a butt weld is the alignment of the pipes in the desired position. Design a device with which to align pipes for on-the-job welding. For ease of operation, a

hand-held device would be desirable. Assume the pipes will vary in diameter from 2 to 4 in.

4. (Fig. 9.2) **Film reel design.** The film reel used on projectors is often difficult to thread because of the limited working space. The figure shows a typical 12-in.-diameter movie reel. Redesign this type of reel to allow more space for threading.

5. **Sideview mirror.** In most cars, rearview mirrors are attached to the side of the automobile to improve the driver's view of the road. Design a sideview mirror that is an improvement over those you are familiar with. Consider the aerodynamics of your design, protection from inclement weather, and other factors that would affect the function of the mirror.

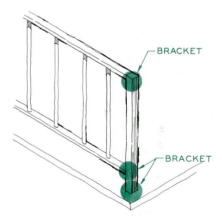

Figure 9.3 Problem 6. A porch railing system.

6. **Railing-post mount.** An ornamental iron railing is to be attached to a wooden porch surface supported by several 1-in.^{2} tubular posts. Design the mounting piece that will attach the posts to the surface. Figure 9.3 shows a typical railing.

A second attachment is needed to assemble the railing with the support post at each end. If two screw holes are to be used, design the part that can secure the two perpendicular members.

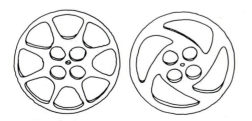

Figure 9.2 Problem 4. A typical movie projector reel.

7. **Nail feeder.** Workers lose time in covering a roof with shingles if they have to fumble for nails. Design a device that can be attached to a worker's chest and will hold nails in such a way that these will be fed in a lined-up position ready for driving.

8. **Cupboard door closer.** Kitchen cupboard and cabinet doors are usually not self-closing and thus are safety hazards and unsightly. Design a device that will close doors left partly open. It would be advantageous to provide a means for disengaging the closer when desired.

9. **Paint-can holder.** Paint cans are designed with a simple wire bail that, when held, makes it difficult to get a paint brush in the can. Wire bails are also painful to hold for any length of time. Design a holding device that can be easily attached and removed from a gallon size paint can ($6\frac{1}{2}$ DIA by $7\frac{1}{2}$ in.). Consider human factors such as comfort, grip, balance, and function.

10. **Self-closing, self-opening hinge.** Interior doors of residences tend to remain partly open. The edge of the door that is ajar in the middle of the hall can be a hazard. Design a hinge that will hold the door in a completely open position.

11. **Tape cartridge storage unit.** Tape players are used frequently in automobiles. Design a storage unit that will hold a number of these cartridges in an orderly fashion, so that the driver can select and insert them with a minimum of motion and distraction. Determine the best location for this unit and the method of attachment to an automobile.

12. **Slide projector elevator.** Most commercial slide and movie projectors have adjustment feet used to raise the projector to the proper position for casting an image on a screen. Study a slide projector to determine the specific needs and limitations of an adjustment. Design a device to serve this purpose as part of the original design or as an accessory that could be used on existing projectors.

13. **Book holder for reading in bed.** As a student, you may often desire to read while lying in bed. Design a holder that can be used for supporting a book in the desired position.

14. **Table leg design.** Do-it-yourselfers build a variety of tables using slab doors, plywood, and commercially available legs. Table tops come in several sizes, but table heights are fairly standard. Determine what the standard table heights are, and design a family of legs that can be attached to table tops with screws. Indicate the method of manufacture, size, cost and method of attachment.

15. **Canoe mounting system.** Canoes and light boats are often transported on the top of automobiles on a luggage rack or similar attachment. Design an accessory that will enable a single person to remove and load a boat on top of an automobile. This attachment should accommodate aluminum boats from 14 to 17 ft. long and weighing from 100 to 200 lb. Give specifications for a method of securing the boat after it is on top of the automobile.

16. **Toothbrush holder.** Design a toothbrush holder that can be attached to a bathroom wall and that can hold a drinking cup and two toothbrushes.

17. (Fig. 9.4) **Wall-mounted stool.** Design a stool that can be attached to a wall that can be swung out of the way when it is not in use.

18. **Book holder.** Design a holder that will support your textbook on your drawing table in a position that will make it more readable and accessible.

Figure 9.4 Problem 17. A wall-mounted stool.

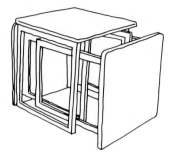

Figure 9.5 Problem 20. Nesting tables.

19. **Clothes hook.** Design a clothes hook that can be attached to a closet door for hanging clothes. It should be easy to manufacture and simple to use.

20. (Fig. 9.5) **Nesting tables.** Design three tables that each have 18 in. square table tops. The tables must nest together to form a cube to save space. The tables should be coffee-table height.

21. **Hammock support.** Design a support for a hammock. It is desirable that the support folds up and requires the minimum of storage space.

22. **Door stop.** Design a door stop that can be attached to a vertical wall or floor to prevent the door knob from bumping the wall.

23. (Fig. 9.6) **Door latch.** Design a door latch that provides security to a home owner or an apartment dweller.

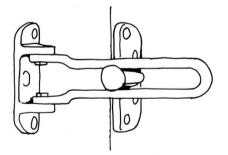

Figure 9.6 Problem 23. A door latch.

24. **Cup dispenser.** Design a paper-cup dispenser that can be attached to a vertical wall. This dispenser should hold a series of cups 2

in. in diameter that measure 6 in. tall when stacked together.

25. **Drawer handle.** Design a handle that would be satisfactory for a standard file cabinet drawer.

26. **Paper dispenser.** Design a dispenser that will hold a 6-×-24-in. roll of wrapping paper. The paper will be used on a table top for wrapping packages.

27. (Fig. 9.7) **Handrail bracket.** Design a bracket that will support a tubular handrail to be used on a staircase. Consider the weight that the handrail must support.

28. (Fig. 9.8) **Latchpole hanger.** Design a hanger that can be used to support a latchpole from a vertical wall. It should be easy to install and use.

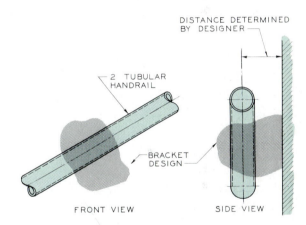

Figure 9.7 Problem 27. A handrail bracket.

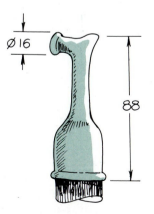

Figure 9.8 Problem 28. A latchpole hanger.

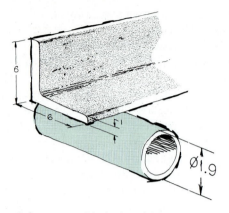

Figure 9.9 Problem 29. A pipe clamp.

29. (Fig. 9.9) **Pipe clamp.** A pipe with a 4-in. diameter must be supported by angles that are spaced 8 ft apart. Design a clamp that will support the pipe without drilling holes in the angles.
30. **TV yoke.** Design a yoke that will support a TV set from the ceiling of a classroom and that will permit it to be adjusted at the best position for viewing.
31. **Flagpole socket.** Design a socket for flags that is to be attached to a vertical wall. Determine the best angle of inclination for the flagpole.
32. **Crutches.** Design a portable crutch that could be used by a person with a temporary leg injury.

33. **Cup holder.** Design a holder that will support a soft-drink can or bottle on the interior of an automobile.
34. **Gate hinge.** Design a hinge that could be attached to a 3-in.-diameter tubular post to support a 3-ft-wide gate.
35. (Fig. 9.10) **Safety lock.** Design a safety lock that will hold a high-voltage power switch in either the "off" or "on" positions to prevent an accident.
36. (Fig. 9.11) **Tubular hinge.** Design a hinge that can be used to hinge 2.5-in. OD high-strength aluminum pipe in the manner shown in this figure. A hinge of this type is needed for portable scaffolding.

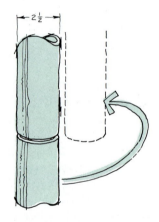

Figure 9.11 Problem 36. A tubular hinge.

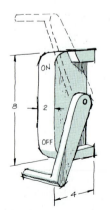

Figure 9.10 Problem 35. A safety lock.

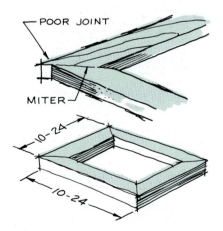

Figure 9.12 Problem 37. A miter jig.

37. (Fig. 9.12) **Miter jig.** Design a jig that can be used for assembling wooden frames at 90° angles. The stock for the frames is to be rectangular in cross sections that vary from 0.75 × 1.5 in. to 1.60 × 3.60 in. Outside dimensions vary from 10 to 24 in.

38. (Fig. 9.13) **Base hardware.** Design the hardware needed at the points indicated for a standard volleyball net. The 7-ft pipes are supported by crossing 2-×-4-in. boards. Design the hardware needed at points a, b, and c.

39. (Fig. 9.14) **Conduit connector hanger.** Design a support that will attach to a 3/4 in. conduit that will support a channel used as an adjustable raceway for electrical wiring. Your design should permit ease of adjustment.

40. (Fig. 9.15) **Fixture design.** Design a fixture that will permit a small-scale manufacturer to saw the corner of the block as shown in this figure.

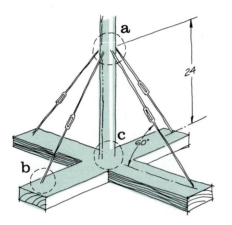

Figure 9.13 Problem 38. Base hardware for a volleyball net.

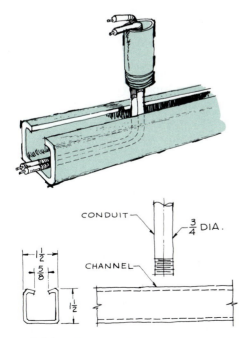

Figure 9.14 Problem 39. A conduit connector hanger.

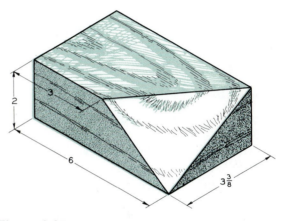

Figure 9.15 Problem 40. A fixture design.

41. (Fig. 9.16) **Drum truck.** Design a truck that can be used for moving a 55-gal drum of turpentine (7.28 lb per gallon). The drum will be kept in a horizontal position, but it would be advantageous to incorporate a feature into the truck permitting the drum to be set in an upright position.

9.8   Systems Design Problems

The systems design problem is a broad engineering problem that requires studying the interrelationships of various components—social, economic, physical, and management considerations. The end result of a systems problem may be a plan, a conclusion, or a recommendation.

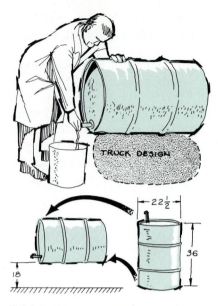

Figure 9.16 Problem 41. A drum truck.

42. **Multipurpose utility meter.** Today's residential units have separate meters for electricity, water, and gas that are checked each month by separate utility companies. Consider the feasibility of combining all meters into a single unit that could be read by a meter-reading service. A unit of this type would reduce the number of meter readers to one third. Outline how such a system could be organized and implemented.

43. **Portable bleachers.** Design a portable bleacher system that can be assembled and disassembled with a minimum of effort and that can be stored. Consider the structure, size, materials and method of construction. Identify a variety of uses for these bleachers that would justify their production for the general market.

44. **Archery range.** Determine the feasibility of providing an archery range for beginning and experienced archers that can be operated at a profit. Your problem is to investigate the need and potential market for such a facility, as well as other factors that would bear upon your decision, including location, equipment needed, method of operation, and costs and fees. Consider site preparation, utilities, concessions, and parking.

45. **Car rental system.** Investigate the possibilities and feasibility of a student-operated rental car system. Data must be collected to determine student interest, cost factors, number of cars needed, rates, personal needs, storage, maintenance, and so forth. Specify the details of your system, including location, cost of operation, and expected income.

46. **Model airplane field.** Model airplane hobbyists who build and fly gasoline engine models often have inadequate facilities for pursuing their hobby. Analyze their needs, including space requirements, type of surfaces, control of sound, safety factors, and method of operation. Come up with a total evaluation of the system. Select a site on or near your campus that you feel is adequate for this facility. Evaluate the equipment, utilities, and site preparation that would be required.

47. **Overnight campsite.** Analyze the feasibility of converting a local lot near a major highway into a complex of campsites for overnight campers. Determine the facilities that would be desired by the campers and the expenditures that would be required for operation. Determine the profit margins for your particular design.

48. **Swimming pool study.** A group of students wishes to construct a swimming pool on your campus that would be self-supporting. Determine the cost of the pool, the area in which it would be located, the students it would serve, and the equipment and labor necessary to operate it. As the conclusion of your problem, determine whether or not it would be feasible.

49. **Outdoor shower.** You are the owner of a weekend cottage, and you do not have a hot and cold water system—you have only cold water. Design a system that would take advantage of the sun's heat in the summer to heat the water used for bathing and kitchen chores. Devise a means of using the same system in the winter with heat provided from some other source (a wood fire, kerosene, or

other method). Can you design a totally portable shower that can be used on camping trips?

50. **Information center.** Visitors to a typical college campus often find it difficult to locate buildings, parking lots, and campus facilities.

   Analyze your college campus to determine the most logical location for a drive-in information center. Determine the informational material that should be included to assist a visitor. Consider using slides, photographs, maps, sound, and other audio-visual aids to accomplish your goal.

   Your design should include the location of the system, its plot plan, equipment, housing and a detailed description of its operation.

51. **Golf driving-range ball return system.** Golf driving ranges are usually designed in such a way that the balls must be retrieved by hand or by means of a specially designed vehicle that collects them from the range. Design a system capable of automatically returning the balls to the driving area.

52. **Instructional system.** Classroom instruction could be improved by providing two-way communication between the teacher and the students. For example, the teacher could proceed at a more efficient rate if he or she had some idea of whether the class understood the points being made. Determine whether a system could be developed that would give the student some means of signaling understanding, or lack of understanding, of the lecture without having to ask questions.

53. (Fig. 9.17) **Instructional console system.** Determine the optimum class size, room layout, lighting requirements, choice of furniture, and other factors of this type that affect the ideal classroom. Study the possibility of developing a visual-aid console that incorporates the latest projectors, recorders, and other devices that would improve instruction and learning.

54. **Pedestrian transport system.** Your campus has considerably more pedestrian traffic in some areas than in others, causing traffic congestion. Consider the possiblity of devel-

Figure 9.17 Problem 53. An instructional console system.

oping a system that would provide a more even flow of pedestrian traffic.

55. **Instant motel.** Many communities have periodic needs for more housing than is available on a regular basis for events like ball games and celebrations. Investigate the different methods of providing an "instant motel" that could fulfill this need for a few days. Consider using tents, vans, trailers, train cars, and so on. Determine the profit margins for your proposed solution.

56. **Helicopter service.** You have been assigned the responsibility of planning a helicopter passenger service that will connect with the local airlines in your community. Analyze the needs of your campus to determine whether such a system could be feasible and self-supporting. Determine the helicopter landing area on your campus and the flight schedule.

57. **Mountain lodge.** Determine what provision should be made for a mountain lodge that was designed for six people to use during the winter months. Assume they may be unexpectedly snowed-in for up to two weeks at a time. What features should be included in the design of the lodge. What food supplies and

other provisions should be stored for their use?

58. **Campus planning.** Assume your college campus is to begin planning for full-capacity use, that is the total use of your present facilities twelve months a year, twenty-four hours a day. Determine how many students could be accommodated in your classrooms, dormitories, and in other facilities without adding new buildings.

   Teaching schedules, faculty, and service personnel should all be evaluated to identify possible problems. Also evaluate the changes that will occur in parking systems, pedestrian traffic, and the management of the total educational system.

59. **Diazo machine operation.** A company's reproduction department will have to reproduce many prints, so a full-time diazo machine operator is required. Establish a system to provide the most efficient use of the equipment and the operator's time to meet the specifications below.

   The machine will accept individual drawings or groups of drawings along its forty-inch belt at a rate of ten feet per minute. The drawing size used most frequently by your company is 11 by 17 in. The diazo paper must be run through the developing chamber of the machine directly above the intake at a speed of 10 ft per minute. The machine is 60-in. wide, 24-in. deep, and 48-in. high. Determine the equipment and tables and their optimum arrangement for most efficient operation.

   What is the cost per drawing of the operator working at peak efficiency? Consider the other duties of the operator, such as stapling sets of drawings together and gathering original drawings to be returned with their prints.

60. **Drive-in theater.** Develop an area for a drive-in motion picture theater. As the chief designer you must determine the optimum size of the drive-in to provide adequate parking and viewing from the audience. Consideration must be given to traffic flow, screen size and position, electrical problems, drainage, utilities, concessions, and other facilities commonly found in a theater of this type. Determine the overall layout of the drive-in, its traffic system, and its major components. Detail a typical parking space for a single car, indicating the contour of the surface to provide the proper viewing angle for the car. List the areas that would require specialists such as electrical engineers and civil engineers.

61. **Boat-launching facility.** Design a boat-launching area at a lake where boat trailers could be positioned. Analyze the requirements of a workable system that will control traffic with a minimum of confusion. Thought must be given to the unloading area, space required, parking area, and type of terrain required. Also determine the charges required to maintain this facility and to pay for any needed help.

62. **Football stadium expansion.** Study the attendance figures of your football stadium to determine what the future holds for it. Will a larger stadium be required? If so, how much extra seating will be required and when? Search historical data for information that would help you make your recommendation.

63. **Shopping checkout system.** An acute problem for shopping centers and grocery markets is checking out goods, payment, and delivering purchases to the customers' automobiles. Develop a system for expediting the transfer of goods selected by the customer, from the store's shelf to the customer's car. Also consider how to expedite packaging and payment.

64. **Injury-proof playground.** Most playgrounds are unsafe for children. Study the activities of children at play. Design a playground system that permits the greatest degree of participation by the children with the least risk of injury.

65. (Fig. 9.18) **Loading dock system.** The trucking industry provides a sizable portion of the transportation of goods and supplies. Unloading is usually performed by a fork lift

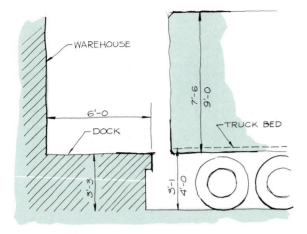

Figure 9.18 Problem 65. The dimensions of an average truck.

at a loading dock that is level with the truck's floor. Because the height of trucks may vary, design a method ensuring that the dock can be adjusted to the level of the truck's floor.

Design a system that will allow the truck to be enclosed or protected from the cold weather during its loading or unloading while taking advantage of the warehouse's heating.

66. **Modification of a drive-in theater.** Assume you are the owner of a drive-in movie theater that has fallen on bad times, and you wish to convert this facility into a different operation that would be profitable. Consider the options available to you having the least expense and the greatest possibility for profit.

67. **Modification of an existing facility.** Isolate an area on your campus or in your community that is inadequate for the demands made on it. This could be a traffic intersection, parking lot, recreational area, or a classroom. Identify the problem and the deficiencies that should be corrected. Propose modifications that would improve the existing facility.

68. **Tape-recording system.** In most classes, students spend much of their time taking notes, which interferes with concentrating on the concepts being presented. Design a system that would provide students with a tape recording of class lectures, which they could take to their rooms and review. Also consider the possibility of providing an automatic system that would record illustrations and diagrams presented by the instructor. Estimate the cost of such a system and determine its value to the educational program.

69. **Service station modification.** Service stations today differ little in arrangement and function from the first filling stations. One major change is the conversion to the credit card system. But the method of servicing automobiles—cleaning windshields, checking oil, and other routine chores—is done in the same manner, with little improvement in technique. Consider the usual operations performed in servicing an automobile to develop a more efficient system.

70. **Recreational facility.** Analyze the various recreational activities on your campus that could be improved with the construction of a multipurpose facility to accommodate them. These activities might include outdoor movies, plays, sports, meetings, and dances. The facility should be analyzed as an outdoor installation with the minimum of conventional structures.

71. (Fig. 9.19) **Educational toy.** Assume you are assigned the responsibility of establishing

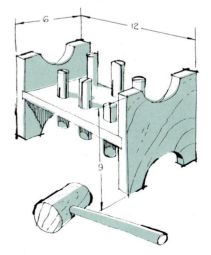

Figure 9.19 Problem 71. An educational toy.

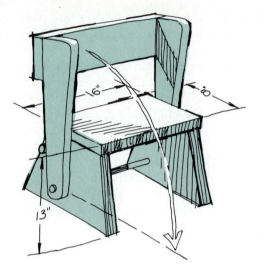

Figure 9.20 Problem 72. Child's furniture.

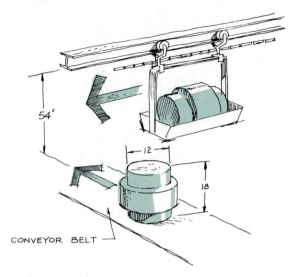

Figure 9.21 Problem 74. A conveyor system.

the production system for manufacturing the educational toy shown in the figure at a rate of five thousand per month. You must determine the square footage needed, the types of machines required, and the space for raw materials, office facilities, and warehousing necessary to sustain this operation. Also determine the number of employees needed, their rate of pay, and the number of work stations required. Compute the expense to produce the item and the selling price necessary to provide the required profit margin.

72. (Fig. 9.20) **Child's furniture.** Proceed as in Problem 71 but assume the child's furniture shown in the figure is the product to be produced. Determine the number to be manufactured per month to break even and establish the selling price. Establish scales for selling prices in quantities in excess of the break-even level.

73. **Simple product.** Select a simple product and establish the production system and requirements for its manufacture proceeding as in Problem 71. Have the product approved by your instructor before going ahead.

74. (Fig. 9.21) **Conveyor system.** You have been assigned the responsibility of designing a conveyor system for delivery of green tire

carcasses to the vulcanizing press. The carcasses from the overhead chain conveyor must be transferred to the belt in an upright position as shown in the figure. Each carcass weighs approximately 30 lb and measures 12 by 18 in. Your solution will probably involve using standard conveyor components.

9.9 Product Design Problems

A product design involves problems developing a device that will perform a specific function and be mass-produced and sold to a broad market.

75. **Hunting blind.** Hunters of geese and ducks must remain concealed while hunting. Design a portable hunting blind to house two hunters. This blind should be completely portable so that it can be carried in separate sections by each of the hunters. Specify its details and how it is to be assembled and used.

76. (Fig. 9.22) **Garden seat.** Design a mobile seat that can be used for projects that require working at low levels such as in the garden.

77. **Writing table for a folding chair.** Design a writing-tablet arm for a folding chair that

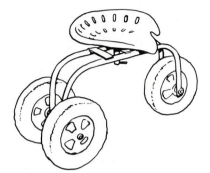

Figure 9.22 Problem 76. A garden seat.

Figure 9.23 Problem 78. A firewood caddy for hauling firewood.

could be used in an emergency or when a class needs more seating. To allow easy storage, the arm must fold with the chair.

78. (Fig. 9.23) **Firewood caddy.** Design a cart that can be used for carrying firewood to a fireplace from the outdoors. This cart, loaded with wood, should be easy to handle when climbing steps. Include other features that would make the caddy attractive as a marketable product.

79. (Fig. 9.24) **Sensor-retaining device.** The Instrumentation Department of the Naval Oceanographic Office uses underwater sensors to learn more about the ocean. These sensors, which each weigh 75 lb, are submerged on cables from a boat on the surface. The winch used to retrieve the sensor frequently overruns (continues pulling when it has been retrieved), causing the cable to break and the sensor to be lost. Design a safety device that will retain the sensor if the cable is broken when the sensor reaches a pulley.

80. (Fig. 9.25) **Portable hauler.** Design a portable, collapsible hauler that can be used of various applications.

81. **Workers' stilts—human engineering.** Workers who apply gypsum board and other types of wallboards to the interiors of buildings must work on scaffolds or wear some type of stilts to be able to reach the ceiling to nail the 4×8-ft boards into position. Design stilts that will provide workers with access to an 8-ft ceiling while permitting them to nail ceiling panels with comfort.

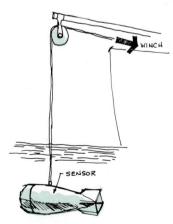

Figure 9.24 Problem 79. An underwater sensor.

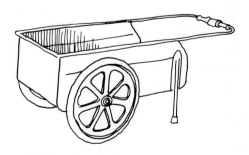

Figure 9.25 Problem 80. A portable hauler.

82. **Pole-vault uprights.** Many pole-vaulters are exceeding the 18-ft height in track meets, which introduces a problem for the officials of this event. The pole-vault uprights must be adjusted for each pole-vaulter by moving them forward or backward plus or minus 18 in. Also, the crossbar must be replaced with great difficulty at these heights by using forked sticks and ladders. Develop a more efficient set of uprights that can be easily repositioned and that will allow the crossbar to be replaced with greater ease.

83. **Sportsman's chair.** Analyze the need for a sportsman's chair that could be used for camping, for fishing from a bank or boat, at sporting events, and for as many other purposes as you can think of. The need is not for a special-purpose chair but for a chair suitable for a variety of uses to fully justify it as a marketable item.

84. (Fig. 9.26) **Bed table.** Modify the design of a hospital table that could be used for study, computing, writing, eating and the like while in bed.

Figure 9.26 Problem 84. A hospital bed table.

85. **Child carrier for a bicycle.** Design a seat that can be used to carry a small child as a passenger on a bicycle. Assume the bicycle will be ridden by an older youth or an adult. Determine the age of the child who would probably be carried as a passenger.

86. (Fig. 9.27) **Sawhorse.** Design a portable sawhorse that can be used for carpentry projects

Figure 9.27 Problem 86. A portable sawhorse.

and that folds up for easy storage. It should be about 36 in. long and 30 in. high.

87. **Power lawn-fertilizer attachment.** The rotary-power lawn mower emits a force through its outlet caused by the air pressure from the rotating blades. This force might be used to distribute fertilizer while the lawn is being mowed. Design an attachment for a power mower that could spread fertilizer while the mower is performing its usual cutting operation.

88. **Car and window washer.** Design an attachment for the typical garden hose that would apply water and agitation (for optimum action) to the surface being cleaned. Consider other applications of the force exerted by water pressure in the performance of yard and household chores.

89. **Projector cabinet.** Design a cabinet that could serve as an end table or some other function while also housing a slide projector ready for use at any time. The cabinet might also serve as storage for slide trays. It should have electrical power for the projector. Evaluate the market potential for a multipurpose cabinet of this type.

90. (Fig. 9.28) **Heavy-appliance mover.** Design a device that can be used for moving large appliances, such as stoves, refrigerators, and washers, about the house. This product would not be used often—only for rearranging, cleaning, and servicing the appliances.

91. **Car jack.** The average car jack does not attach itself adequately to the automobile's

Figure 9.28 Problem 90. A heavy-appliance mover.

frame or bumper, introducing a severe safety problem. Design a jack that would be an improvement over existing jacks and possibly employ a different method of applying a lifting force to a car. Consider the various types of terrain on which the device must serve.

92. **Map holder.** The driver of an automobile traveling alone in an unfamiliar part of the country must frequently refer to a map. Design a system that will give the driver a ready view of the map in a convenient location in the car. Also provide a means of lighting the map during night driving that will not distract the driver.

93. **Bicycle-for-two adapter.** Design the parts and assembly required to convert two bicycles into one bicycle-built-for-two (tandem). Work from an existing bicycle, and consider how each rider can equally share in the pedaling. Determine the cost of your assembly and its method of attachment to the average bicycle.

94. **Automobile unsticker.** Design a kit to be carried in the car trunk in a minimum of space containing the items required to "unstick" a car when no other help is available. This kit can contain one or several items. Investigate the need for such a kit and the main factors that lead to the loss of traction.

95. **Stump remover.** Assume a number of tree stumps must be removed from the ground to clear land for construction. The stumps are dead with partially deteriorated root systems and require a force of approximately two thousand pounds to remove them. Design an apparatus that could be attached to a car bumper that could be used to remove the stumps by either pushing or pulling.

96. **Gate opener.** An annoyance to farmers and ranchers is the necessity of opening and closing gates when driving from one fenced area to the next. A gate that could be opened and closed without getting out of the vehicle would be desirable. Design a manually operated gate that would appeal to this market.

97. **Paint mixer.** Design a product that could be used by the paint store or the paint contractor to quickly mix paint in the store and on the job. Determine the standard-size paint cans for which your mixer will be designed.

98. **Mounting for an outboard motor on a canoe.** Unlike a square-end boat, the pointed-end canoe does not provide a suitable surface for attaching an outboard motor. Design an attachment that will adapt an outboard motor to a canoe.

99. **Automobile coffee maker.** Adequate heat is available in the automobile's power system to prepare coffee in minimum time. Design an attachment as an integral system of an automobile that will serve coffee from the dashboard area. Consider the type of coffee to be used (instant or regular), the method of changing or adding water, the spigot system, and similar details.

100. **Baby seat (cantilever).** Design a child's chair that can be attached to a standard table top and that will support the child at the required height. The chair should be designed to ensure that the child cannot crawl out or detach it from the table top. A possible solution could be a design that would cantilever from the table top, using the child's body as a means of applying the force necessary to grip the table top. The design would be further improved if the chair were collapsible or suitable for other purposes.

101. **Miniature-TV support.** Miniature television sets for close viewing are available with a 5-×-5-in. screen. An attachment is needed that would support sets ranging in size from 6 × 6 in. to 7 × 7 in. for viewing from a bed. Determine the placement of the set for best viewing results. Provide adjustments on the support that will be used to position the set properly.

102. **Panel applicator.** A worker who applies 4-×-8-ft gypsum board or paneling must be assisted by a helper who holds the panel in position while it is being nailed to the ceiling. Design a device that could be used in this capacity; it should be collapsible, economical, and versatile. The average ceiling height is 8 ft, but provide adjustments that would adapt the device to lower or higher ceilings.

103. **Backpack.** Design a backpack that can be used by the outdoorsperson who must carry supplies while hiking. Most of your design effort should be devoted to adapting the backpack to the human body for maximum comfort and the leverage for carrying a load over a long time. Can other uses be made of your design?

104. **Automobile controls.** Design driving controls that can be easily attached to the standard automobile that will permit an injured person to drive a car without the use of his or her legs. This device should be easy to operate with the maximum of safety.

105. **Bathing apparatus.** Design an apparatus that would help a wheelchair-bound person, who does not have use of his or her legs, to get in and out of a bathtub without assistance from others.

106. **Adjustable TV base.** Design a TV base to support full-size TV sets that would allow maximum adjustment: up and down, rotation about a vertical axis and a horizontal axis. Design the base to be as versatile as possible.

107. (Fig. 9.29) **Log splitter.** Design a device to aid in the splitting of logs for firewood.

108. **Projector cabinet.** Design a portable cabinet, which could be left permanently in a classroom, that would house a slide projector and a movie projector. The cabinet should provide both convenience and security from vandalism and theft.

109. (Fig. 9.30) **Boat loader.** Design a rack and a system whereby a single person could load a boat weighing 110 lb on top of a car for transporting from site to site. Use the boat specifications shown in the figure.

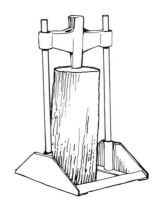

Figure 9.29 Problem 107. A log splitter.

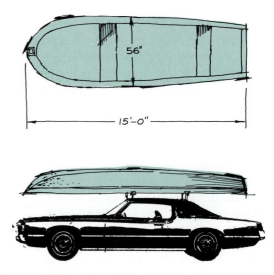

Figure 9.30 Problem 109. Boat specifications.

110. (Fig. 9.31) **Door opener.** Design a method whereby a trucker at a loading dock could open the warehouse door without having to get out of the truck. The doors are dimensioned in the figure, and the dock extends 8 ft from the doors.

111. **Projector eraser.** Design a device that would erase grease pencil markings from the acetate roll of an overhead projector as the acetate is cranked past the stage of the projector.

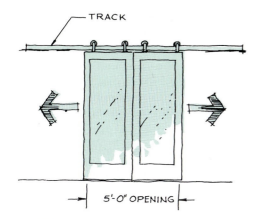

Figure 9.31 Problem 110. Door dimensions.

112. **Cement mixer.** Design a portable and simple cement mixer that can be operated manually by a home owner. The mixer is seldom used, so it should be sufficiently simple and economical to justify its being built.

113. **Boat trailer.** Design a trailer that supports the boat under the trailer rather than on top of it. Develop this design so that a boat can be launched in water more shallow than is required now.

114. **Washing machine.** Design a manually operated washing machine that can be used by people in less-developed countries that may not have power. This could be considered an "undesign" of a powered washing machine.

115. **Pickup truck hoist.** Design a tailgate that can be attached to the tailgate of a pickup truck and that can be used for raising and lowering loads from the ground to the floor of the truck. Design it to be operated without a motor.

116. (Fig. 9.32) **Writing tablet for the handicapped.** Design a writing tablet that will permit a person to write from a prone position using a series of mirrors. This problem involves applying human engineering considerations.

117. (Fig. 9.33) **Portable seat.** Design a portable seat that can be used in as many applications as possible, such as at football games, in fishing boats, and at campsites. It should be easy to carry and collapsible.

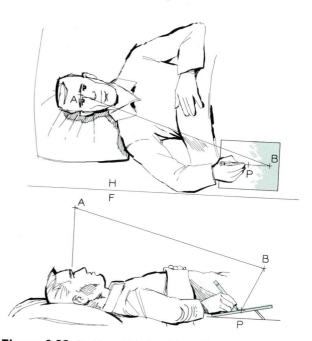

Figure 9.32 Problem 116. A writing tablet for the handicapped.

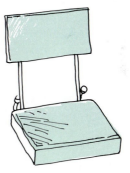

Figure 9.33 Problem 117. A portable seat.

118. (Fig. 9.34) **Display booth.** Design a portable display booth that can be used behind or with an 8-ft table. Your design should consist of panels that stand behind the table on which your company's name and product advertising can be integrated. It must be collapsible so that it can be easily carried by one person as airplane luggage.

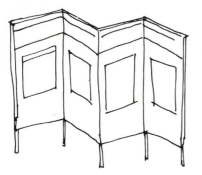

Figure 9.34 Problem 118. A portable display booth.

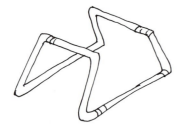

Figure 9.35 Problem 119. A push-up and sit-up exerciser.

119. (Fig. 9.35) **Exerciser.** Design a simple device that can be used specifically for aiding in doing pushups and sit-ups. It should sell for around $20 and store easily in the minimum of space.

Figure 9.36 Problem 120. A patio grill.

120. (Fig. 9.36) **Patio grill.** Design a portable charcoal grill for cooking on the patio. Consider how it would be cleaned, stored, and used. Study the competing products on the market to develop a marketable grill.

121. (Fig. 9.37) **Shop bench.** Design a shop bench that will be marketable, serve a need, be collapsible for ease of storage, and be adjustable. Could it also be designed to serve as a step stool? Design it to accommodate other accessories such as vises, anvils, and electrical appliances.

122. (Fig. 9.38) **Chimney cover.** Design a chimney cover that can be closed from inside the house for repelling rain and saving lost

Figure 9.37 Problem 121. A shop bench.

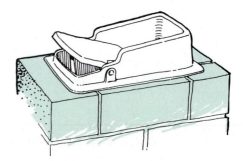

Figure 9.38 Problem 122. A chimney cover.

heat and cooling. Evaluate your design's market potential, and analyze its manufacturing cost.

123. (Fig. 9.39) **Punching bag platform.** Design a punching bag platform for supporting a speed bag. Your design should offer an easy means of adjusting the platform's height. It should be portable and easy to ship in a flat box.

Figure 9.40 Problem 124. A shopping caddy.

Figure 9.39 Problem 123. A punching bag platform.

124. (Fig. 9.40) **Shopping caddy.** Design a portable caddy that a shopper can use for carrying parcels. It should be as lightweight as possible.

125. **Computer workstation.** Figure 9.41 illustrates a unique design for a computer workstation to reduce discomfort of the user. Notice that the various components of the computer system have been positioned for best visibility and greatest comfort. Design a similar workstation, using this example as a guide.

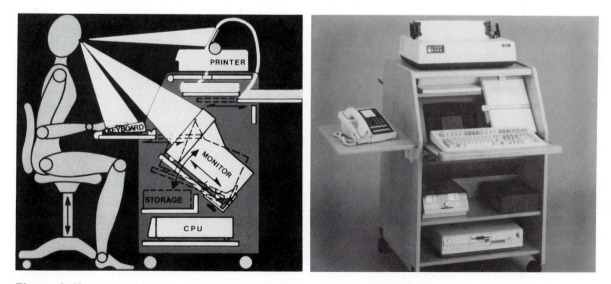

Figure 9.41 Problem 125. The Peanut Ultra View Workstation. (Courtesy of Continental Engineering Group, Inc.)

10
The Computer in Design and Drafting

10.1 Introduction

The use of computers in engineering and related fields is widespread, and great growth is still expected. It is increasingly important for students of engineering and technology to become familiar with the nature and prospects of computer technologies. We devote this chapter to computer usage in drafting and design. To familiarize you with how computer graphics can help apply the principles of this text, we emphasize techniques of computer-aided design drafting (CADD).

10.2 Computer-Aided Design

Computer-Aided Design (CAD) is the process of solving design problems with the aid of computers. This includes computer generation and modification of graphic images on a video display, printing these images as hard copy on a plotter or printer, analysis of design data, and electronic storage and retrieval of design information. Many CAD systems perform these func-

tions in an integrated fashion, which can increase the designer's productivity manyfold.

Computer-Aided Design Drafting (CADD), a subset of CAD, is the computer-assisted generation of working drawings and other engineering documents. The CADD user generates graphics by interactive communication with the computer. Graphics are displayed on a video display and can be converted into hard copy with a plotter or printer.

Most engineers agree that the computer does not change the nature of the design process and that it is simply a tool to improve efficiency and productivity. The designer and the CAD system should be viewed as a design team: The designer provides knowledge, creativity, and control, whereas the computer is able to generate accurate, easily modifiable graphics; to perform complex design analysis at great speeds; and to store and recall design information. Occasionally, the computer can augment or replace many of the engineer's other tools, but it is important to remember that it does not change the fundamental role of the designer.

Advantages of Using the Computer in Design and Drafting

Depending on the nature of the problem and the sophistication of the computer system, computer use can afford the designer or drafter several potential advantages.

1. **Easier creation and correction of working drawings.** In a CAD system, working drawings can be created using function commands and digitizing methods (assigning numerical coordinates). Complicated changes and corrections are made using a few keystrokes.
2. **Easier visualization of drawings.** In many systems, different views of the same part can be displayed quickly and easily. In systems with three-dimensional capabilities, a part can even be rotated on the CRT screen.
3. **Drawings can be stored and easily referenced for modification.** Modified designs can be made from one original in far less time than it would take using a manual approach. Design databases (libraries of designs) can be created in some systems. These databases can also store standard parts and symbols for easy recall. Many systems are configured so that information in the databases can be easily accessed by others in an organization, such as management or production personnel.
4. **Quick and convenient solution of computational design analysis problems.** Because the computer offers a tremendous advantage in ease of design analysis, the designer can rigorously analyze each design, thus speeding up the design refinement stage.
5. **Simulation and testing of designs.** Some systems enable the engineer to simulate the operation of a design and to perform tests and analyses in which the part is subjected to a variety of conditions or stresses. This capacity may improve or replace the process of building models and prototypes.
6. **Increased accuracy.** The accuracy of the computer lessens the chance for error. Many CAD systems are capable of detecting errors

and will inform the user when data or designs are incorrect.

10.3 Applications of Computer Graphics

Computer graphics is almost limitless in its applications to engineering and technical fields. Almost all graphical solutions that can be done with a pencil can be done with the computer, and often they can be done more productively. Applications can vary from three-dimensional modeling and finite element analysis to two-dimensional drawings and mathematical calculations.

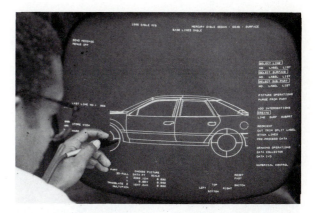

Figure 10.1 The Ford Taurus and Mercury Sable received more computer-aided design and engineering than any other car in the company's history. (Courtesy of Ford Motor Company.)

Automobile bodies are designed at the computer (Fig. 10.1). As part of this process, Fig. 10.2 shows a scanning device that records measurements of the life-size tape drawing of a new vehicle. Once recorded and stored in the computer, the data can be manipulated to form the three-dimensional shape of the body style on the computer monitor from which three-dimensional clay models are milled (Fig. 10.3).

Advanced applications, once the domain of large computer systems, can now be done on mi-

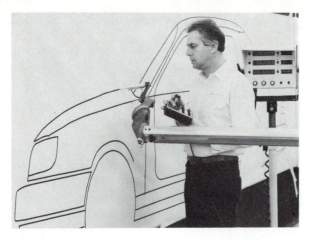

Figure 10.2 A scanning device is used to record measurements from a life-size tape drawing of a new vehicle. Once stored, this data can be used for other design functions. (Courtesy of Ford Motor Company.)

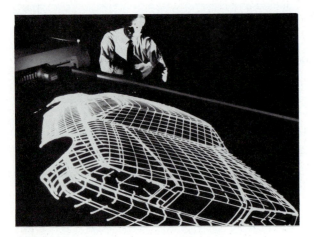

Figure 10.3 Once design data is stored in the computer's memory, it can be manipulated to obtain variations in body design. (Courtesy of Ford Motor Company.)

crocomputers. Figure 10.4 illustrates how a portion of the earth's surface was modeled and shown three dimensionally by using a microcomputer. Using simple keyboard commands, this site plan can be viewed from an infinite number of positions to obtain desired vantage points.

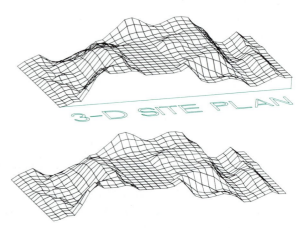

Figure 10.4 A three-dimensional representation of a site plan. The lower version is slightly different from the upper one since hidden lines on the surface have been hidden by the computer. (Courtesy of LANDCADD Inc.)

In Fig. 10.5, a perspective drawing of an urban area was plotted using microcomputer software. This program lets the viewer be positioned at any height and location in the site. The viewer can "walk through" the city by successively changing positions.

A microcomputer was used to produce the layer of the electronic circuit board shown in Fig.

Figure 10.5 An urban area has been plotted as a perspective drawing with hidden lines removed using MegaCADD's Design Board 3D. (Courtesy of MegaCADD, Inc.)

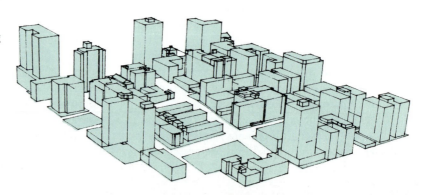

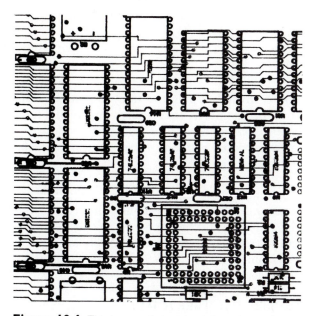

Figure 10.6 This is one of nine circuit board trace layers designed on the computer using Auto-Board System II.

10.6. Circuit board drawings are drawn as much as five times their actual size and then photographically reduced.

Figure 10.7 This automatic welding system for Chrysler LeBaron GTS and Dodge Lancer car bodies uses computer-controlled robots for consistent welds of all components in the unitized body structure. This assembly plant also features energy-efficient electric robot welders that require less maintenance and provide a high degree of accuracy. (Courtesy of Chrysler Corp.)

10.4 CAD/CAM

An important application of CAD lies in the field of manufacturing. **Computer-aided design/computer-aided manufacturing** (CAD/CAM) describes a system that can design a part or product, devise the essential production steps, and electronically communicate this information to manufacturing equipment like robots (Fig. 10.7). A CAD/CAM system offers many potential advantages over traditional manufacturing systems, including less design effort through the use of CAD and CAD databases, more efficient material use, reduced lead time, greater accuracy, and improved inventory functioning.

In **computer-integrated manufacturing** (CIM) a computer or system of computers coordinates all stages of manufacturing, which enables manufacturers to custom design products efficiently and economically.

10.5 Hardware Systems

Computer-aided design systems have three major components: the designer, hardware, and software. **Hardware** is the physical components of a computer system, and **software** is the programmer's instructions to the computer. The hardware of a computer graphics system includes the computer, terminal, input devices (digitizers, light pens), and output devices (plotters, printers) (Fig. 10.8).

Computer

Computers receive input from the user, execute the instructions contained in the input, and then produce some form of output. A sequence of instructions called a **program** controls the computer's activities. The part of the computer that follows the program's instructions is the **central processing unit** (CPU).

Figure 10.8 A Hewlett-Packard Vectra with a computer, keyboard, monitor, and mouse.

Computers are often classified by size. **Mainframes,** the largest computers, are big, fast, powerful, and expensive. **Minicomputers,** smaller and less costly than mainframes, are used by many small businesses. **Microcomputers,** the smallest computers, are widely used for both personal and business applications. Rapid advances in hardware technology and software capability, along with continually falling prices, have brought microcomputers into wide use in engi-

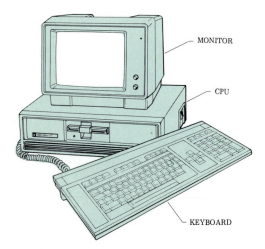

Figure 10.9 The basic components of a computer system: CPU, keyboard, and monitor (CRT).

neering graphics applications in industry, government, and education.

Terminal

The **terminal** allows the user to communicate with the computer. It typically consists of a keyboard, a cathode-ray tube (monitor), and the interconnections between these devices and the computer (Fig. 10.9).

Keyboard The keyboard allows the user to communicate with the computer through a set of alphanumeric and function keys. Keyboards generally resemble typewriters but include many other function keys, some of which may be user-defined.

Cathode-Ray Tube (CRT) A CRT is a video display device consisting of a tube with a phosphor-coated screen. An electron gun throws a beam that sweeps out rows, called **raster lines,** on the screen. Each raster line consists of a number of dots called **pixels.** Images are generated on the screen by turning pixels on and off. Raster-scanned CRTs refresh the picture display many times per second. One measure of the quality of the pictures that can be produced on a screen is **resolution,** the number of pixels per inch that can be drawn on the screen. Higher-quality graphics can be drawn on higher-resolution screens.

Raster-scan technology has largely replaced the **vector-refreshed tube** used in early video displays. In this type of display, each line in the picture is continuously redrawn by the computer. Because the display produced a collection of lines instead of a set of individual pixels, vector-display pictures are less realistic than raster-scanned displays.

Input Devices

Digitizer The digitizer is a graphics input device that can communicate information in a picture to the computer for display, storage, or modification (Fig. 10.10). The user lays a draw-

Figure 10.10 A digitizer board is made up of a set of coordinates that correspond to points on the CRT screen. (Courtesy of GTCO Corporation.)

ing or sketch on a digitizer board and scans it, thereby converting the picture to a digital or computerized form based on the xy coordinates of individual points. Since a digitizer can also input symbols, a fully labeled drawing can be represented on a CRT screen and stored for later use.

Light Pen   Digitized pictures on a screen can be created or modified point by point using a

Figure 10.11 A light pen allows the user to modify drawings by touching the CRT screen with it. (Courtesy of Ford Motor Company.)

light pen. The user places this device in contact with the CRT screen, thereby telling the computer the position of the pen on the screen (Fig. 10.11).

Mouse   A mouse is a device that can be rolled on a table top to move a cursor (a mark on the screen that indicates location) around a CRT screen. The mouse can be used to activate commands or change information on the screen (Fig. 10.12).

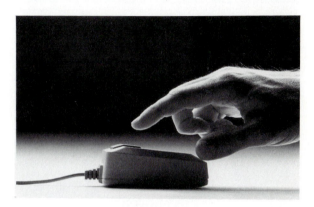

Figure 10.12 A mouse can move the cursor around a CRT screen. The user activates commands or makes changes by pressing the button(s). (Courtesy of Apollo Computer, Inc.)

Joy Stick   A joy stick allows the user to "steer" a cursor around the screen by tilting a lever in appropriate directions (Fig. 10.13).

Output Devices

Plotter   The plotter is a machine directed by the computer to make a drawing. Two types of plotters are the **flatbed** plotter and the **drum,** or **roll-feed,** plotter. In the flatbed plotter, paper is attached to the bed, and the pen is moved about the paper in raised or lowered position to complete the drawing (Fig. 10.14). In the drum plotter, a special type of paper is held between grit wheels and rolled over a rotating drum (Fig. 10.15). As the drum rotates, the pen suspended above the paper moves left or right along the drum.

Figure 10.13 A joy stick can be used to "steer" a cursor around a CRT screen. (Courtesy of Apple Computer, Inc.)

Figure 10.15 The Draft Master plotters used for plotting CAD applications. (Courtesy of Hewlett-Packard.)

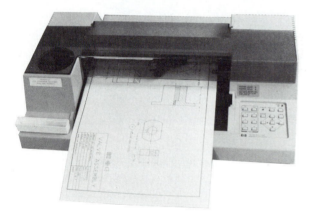

Figure 10.14 A Hewlett-Packard 7475 plotter for making Size A and Size B drawings.

However, the output of dot-matrix printers is of lower quality than some other printers, so they have limited usefulness in graphics applications.

Nonimpact printers form characters from a distance by using ink sprays, laser beams, photography, or heat. One example, the **thermal transfer plotter,** uses electrical fields to direct jets of ink to appropriate spots on the paper (Fig. 10.16). Like the dot-matrix printer, the ink-jet printer forms a pattern of dots, thus limiting its accuracy or resolution. One feature of this type of printer is that multicolor graphics can be generated by using multiple jet nozzles.

Printer   The printer is a device operated by the computer that makes images on paper. An **impact printer,** which works like a typewriter, forms characters by forcing typefaces to impact with an inked ribbon and paper. **Dot-matrix printers,** a popular, inexpensive type of impact printer, have printheads composed of a rectangle of pins, each of which can be raised or lowered to form a character. These patterns of pins are forced against a ribbon to make dotted characters on paper. Dot-matrix printers are useful for graphics because the patterns of dots can be made to correspond to lighted pixels on a CRT screen.

Figure 10.16 The PlotMaster thermal transfer plotter/printer produces fast, high-resolution color hard copy for computer-aided drawings. (Courtesy of CalComp.)

As printer technology makes rapid strides in cost reduction and output quality, it will be found increasingly useful for graphics applications. Still, the need for strict accuracy makes the use of plotters more attractive for many applications.

10.6 CAD Software for the Microcomputer

The availability of CAD software for the microcomputer continues to increase, and each upgrade is more powerful than the previous version. Before long, almost all computer graphics will be done on the microcomputer (Fig. 10.17).

The major advantages of CAD microcomputer software packages are cost and ease of upgrading when improvements in the software are made.

Among the more popular computer graphics software packages for the microcomputer are AutoCAD, VersaCAD, MegaCADD, Dyna-

Figure 10.17 A typical computer-graphics workstation comprised of a Hewlett-Packard Vectra with a color monitor, mouse, and AutoCAD software.

Perspective, and Cadkey. Most of these packages require a computer with at least a 20 Mb hard disk and 640 Kb of RAM. Figure 10.17 shows a typical IBM-compatible graphics system where a mouse is used to draw on the color monitor.

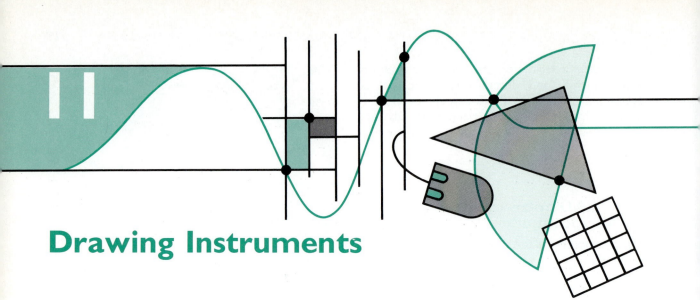

Drawing Instruments

11.1 Introduction

Preparing technical drawings is possible only by being knowledgeable of and skillful in the use of drafting instruments. Your skill and productivity will increase as you become more familiar with using the available tools. Persons with little artistic ability can produce professional technical drawings when they learn to use drawing instruments properly.

11.2 Pencil

Pencils may be the conventional wood pencil or the lead holder, which is a mechanical pencil (Fig. 11.1). Both types are identified by numbers and letters at their ends. Sharpen the end of the pencil

LEAD HOLDER

DRAFTING PENCIL

Figure 11.1 The mechanical pencil (lead holder) or the wood pencil can be used for mechanical drawing. The ends of these pencils are labeled to indicate the grade of the pencil lead.

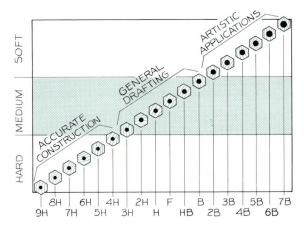

Figure 11.2 The hardest pencil lead is 9H, and the softest is 7B. Note the diameters of the hard leads are smaller than the soft leads.

opposite these markings so that you do not sharpen away the identity of the grade of lead.

Pencil grades range from the hardest, 9H, to the softest, 7B (Fig. 11.2). The pencils in the medium-grade range, 3H–3B, are most often used for drafting work.

It is important that your pencil be properly sharpened. This can be done with a small knife or a drafter's pencil sharpener, which removes the wood and leaves approximately $\frac{3}{8}$ inch of lead exposed (Fig. 11.3). The point can then be sharpened to a conical point with a sandpaper pad by stroking the sandpaper with the pencil point while revolving the pencil (Fig. 11.4). Excess graphite is wiped from the point with a cloth or tissue.

A pencil pointer used by professional drafters (Fig. 11.5) can be used to sharpen wood and mechanical pencils. Simply insert the pencil in the hole and revolve it to sharpen the lead. Other types of small hand-held point sharpeners are also available.

11.3 Papers and Drafting Media

Sizes The surface a drawing is made on must be carefully selected to yield the best results for a given application. Sheet sizes are specified by letters such as Size A, Size B, and so forth. These sizes, which are listed below, are multiples of either the standard $8\frac{1}{2}$-×-11-inch sheet or the 9-×-12-inch sheet.

	English		**Metric**	
A	$8\frac{1}{2}'' \times 11''$	$9'' \times 12''$	A4	210×297
B	$11'' \times 17''$	$12'' \times 18''$	A3	297×420
C	$17'' \times 22''$	$18'' \times 24''$	A2	420×594
D	$22'' \times 34''$	$24'' \times 36''$	A1	594×841
E	$34'' \times 44''$	$36'' \times 48''$	A0	841×1189

Detail Paper When drawings are **not** to be reproduced by the diazo process (which is a blueline print), an opaque paper, called **detail paper,** can be used as the drawing surface. The higher the rag content (cotton additive) of the paper, the better its quality and durability. Preliminary layouts can be drawn on detail paper and then traced onto the final tracing surface.

Tracing Paper A thin, translucent paper used for making detail drawings is **tracing paper** or **tracing vellum.** These papers permit the passage

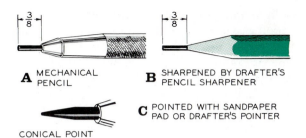

Figure 11.3 The drafting pencil should be sharpened to a tapered conical point (not a needle point) with a sandpaper pad or other type of sharpener.

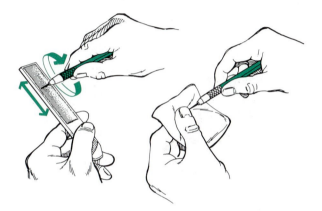

Figure 11.4 The drafting pencil is revolved about its axis as you stroke the sandpaper pad to form a conical point. The graphite is wiped from the sharpened point with a tissue or cloth.

Figure 11.5 The professional drafter often uses a pencil pointer of this type to sharpen pencils.

of light through them so that drawings can be reproduced by the diazo, or blue-line, process. The tracing papers that yield the best reproductions are most translucent. Vellum is tracing paper that has been chemically treated to improve its translucency, but it does not retain its original quality as long as do high-quality, untreated tracing papers.

Tracing Cloth   **Tracing cloth** is a permanent drafting medium for both ink and pencil drawings. It is made of cotton fabric that has been covered with a compound of starch to provide a tough, erasable drafting surface that yields excellent blue-line reproductions. More stable than paper, tracing cloth does not change its shape with variations in temperature and humidity as much as does tracing paper. Erasures can be repeatedly made on tracing cloth without damaging the surface, which is especially important when drawing with ink.

Polyester Film   An excellent drafting surface is polyester film, which is available under several trade names such as Mylar. It is more transparent, stable, and tough than paper or cloth. It is also waterproof and difficult to tear.

Mylar film is used for both pencil and ink drawings. Some films specify that a plastic-lead pencil be used, whereas others adapt well to standard lead pencils. Ink will not wash off with water and will not erase with a dry eraser; erasures can be made with a dampened hand-held eraser.

11.4   T-square and Board

The T-square and drafting board are the basic equipment used by the beginning drafter (Fig. 11.6). With its head in contact with the edge of the drawing board, the T-square can be moved for drawing parallel horizontal lines. Drawing paper should be attached to the board parallel to the blade of the T-square. The drafting board is made of basswood, which is lightweight but strong. Standard sizes of drafting boards are 12 × 14 inches, 15 × 20 inches, and 21 × 26 inches.

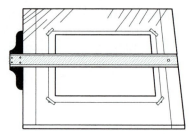

Figure 11.6 The T-square and drafting board are the basic tools used by the student drafter. The drawing paper is taped to the board with drafting tape.

Figure 11.7 The drafting machine is often used instead of the T-square and drafting board. (Courtesy of Keuffel & Esser Co., Morristown, N.J.)

Figure 11.8 The professional who uses a drafting station may work in an environment similar to the one shown here. (Courtesy of Martin Instrument Co.)

11.5 Drafting Machines

Although the T-square is used in industry and the classroom, most professional drafters prefer the **mechanical drafting machine** (Fig. 11.7). This machine is attached to the table top and has fingertip controls for drawing lines at any angle.

Figure 11.8 shows a modern, fully equipped drafting station. Today, many offices are equipped with computer graphics stations to supplement manual equipment and techniques (Fig. 11.9).

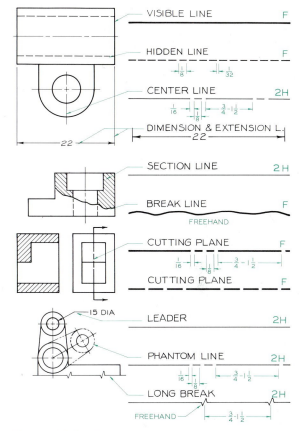

Figure 11.10 The alphabet of lines varies in width. The full-size lines are shown in the right column along with the recommended pencil grades for drawing them.

Figure 11.9 More and more professional work stations are being equipped with computer graphics equipment. (Courtesy of T&W Systems.)

11.6 Alphabet of Lines

The type of line produced by a pencil depends on the hardness of the lead, drawing surface, and technique of the drafter. Figure 11.10 shows examples of the standard lines, or **alphabet of lines**, and the recommended pencils for drawing the lines. These pencil grades may vary greatly with the drawing surface being used. Guidelines are very light lines (just dark enough to be seen) used to aid in lettering and laying out a drawing; a 4H pencil is recommended for drawing most guidelines.

11.7 Horizontal Lines

A horizontal line is drawn using the upper edge of your horizontal straightedge and, for the right-handed person, drawing the line from left to right (Fig. 11.11). Your pencil should be held in a vertical plane to make a 60° angle with the drawing surface. As horizontal lines are drawn, the pencil should be rotated about its axis to allow its point to wear evenly (Fig. 11.12). If necessary, lines can be darkened by drawing over them one or more times. For the best line, a small space should be left between the straightedge and the pencil point (Fig. 11.13).

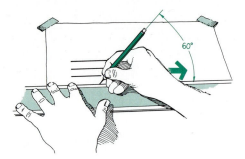

Figure 11.11 Horizontal lines are drawn with a pencil held in a plane perpendicular to the paper and at 60° to the surface. These lines are drawn left to right along the upper edge of the straightedge.

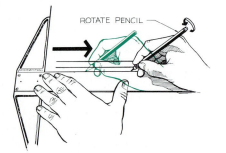

Figure 11.12 As the horizontal lines are drawn, the pencil should be rotated about its axis so that the point will wear down evenly.

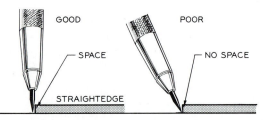

Figure 11.13 The pencil point should be held in a vertical plane and inclined 60° to leave a space between the point and the straightedge.

11.8 Vertical Lines

A triangle is used with a straightedge for drawing vertical lines. While the straightedge is held firmly with one hand, the triangle can be positioned where needed and the vertical lines drawn

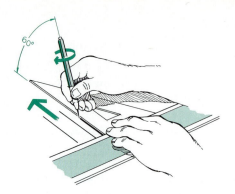

Figure 11.14 Vertical lines are drawn along the left side of a triangle in an upward direction with the pencil held in a vertical plane at 60° to the surface.

(Fig. 11.14). Vertical lines are drawn upward along the left side of the triangle while holding the pencil in a vertical plane at 60° to the drawing surface.

11.9 Drafting Triangles

The two most often used triangles are the 45° triangle and the 30°–60° triangle. The 30°–60° triangle is specified by the longer of the two sides adjacent to the 90° angle (Fig. 11.15). Standard sizes of 30°–60° triangles range, in 2-in. intervals, from 4 to 24 in.

The 45° triangle is specified by the length of the sides adjacent to the 90° angle. These range

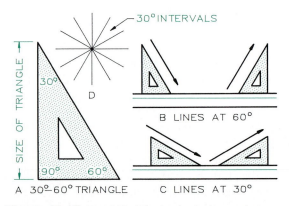

Figure 11.15 The 30°–60° triangle can be used to construct lines spaced at 30° intervals throughout 360°.

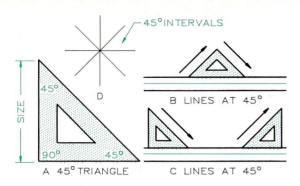

Figure 11.16 The 45° triangle can be used to draw lines at 45° intervals throughout 360°.

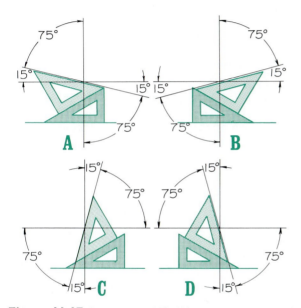

Figure 11.17 By using the 30°–60° triangle in combination with the 45° triangle, angles can be drawn at 15° intervals.

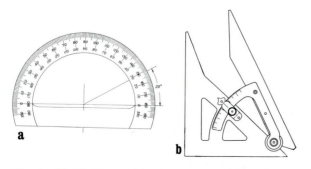

Figure 11.18 The semicircular protractor (a) can be used to measure angles. The adjustable triangle (b) can be used as a drawing edge and to measure angles.

from 4 to 24 in. at 2-in. intervals, but the 6- and 10-in. sizes are adequate for most classroom applications. Figure 11.16 shows the various angles that can be drawn with this triangle. By using the 45° and 30°–60° triangles in combination, angles can be drawn at 15° intervals throughout 360° (Fig. 11.17).

11.10 Protractor

When lines must be drawn or measured at multiples of other than 15°, a **protractor** is used (Fig. 11.18). Protractors are available as semicircles (180°) or circles (360°). An adjustable triangle serves as both a protractor and a drawing edge (Fig. 11.18b).

11.11 Parallel Lines

A series of lines can be drawn parallel to a given line by using a triangle and straightedge (Fig. 11.19). The 45° triangle is placed parallel to a given line and is held in contact with the straightedge (which may be another triangle). By holding the straightedge in one position, the triangle can be moved to various positions for drawing a series of parallel lines.

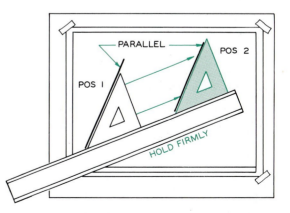

Figure 11.19 A straightedge and a 45° triangle can be used to draw a series of parallel lines. The straightedge is held firmly in position, and the triangle is moved from position 1 to position 2.

11.12   Perpendicular Lines

Perpendicular lines can be constructed by using either of the standard triangles. A 30°–60° triangle is used with a straightedge or another triangle to draw line 3–4 perpendicular to line 1–2 (Fig. 11.20). One edge of the triangle is placed parallel to line 1–2 in position 1 with the straightedge in contact with the triangle. By holding the straightedge in place, the triangle is rotated and moved to position 2 to draw the perpendicular line.

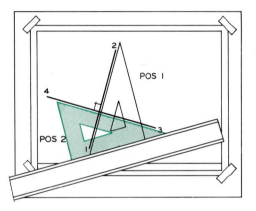

Figure 11.20 A 30°–60° triangle and a straightedge can be used to construct a line perpendicular to line 1–2. The triangle is aligned with line 1–2 and is then rotated to position 2 to construct line 3–4.

11.13   Irregular Curves

Curves that are not arcs must be drawn with an **irregular curve.** These plastic curves come in a variety of sizes and shapes, but the one shown in Fig. 11.21 is typical. In this figure, the irregular curve connects a series of points to form a smooth curve. The **flexible spline** is an instrument used for drawing long, irregular curves (Fig. 11.22). The spline is held in position by weights while the curve is drawn.

11.14   Erasing

Erasing should be done with the softest eraser that will serve the purpose. For example, ink erasers should not be used to erase pencil lines because ink erasers are coarse and may damage the surface of the paper. When working in small areas, an **erasing shield** can prevent your accidentally erasing adjacent lines (Fig. 11.23). All erasing should be followed by removing the "crumbs" with a dusting brush; do not use your hands for this, or you may smudge your drawing. A cordless model electric eraser (Fig. 11.24) uses several grades of erasers for erasing ink and pencil lines.

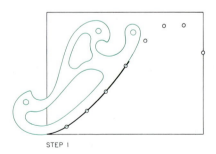

STEP 1

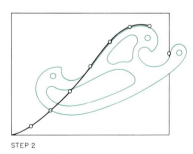

STEP 2

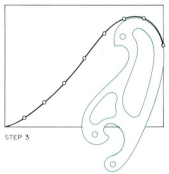

STEP 3

Figure 11.21 Use of the irregular curve.

Step 1 The irregular curve is positioned to pass through as many points as possible, and a portion of the curve is drawn.

Step 2 The irregular curve is positioned for drawing another portion of the connecting curve.

Step 3 The last portion is drawn to complete the smooth curve. Most irregular curves must be drawn in separate steps.

Figure 11.22 A flexible spline can be used for drawing large irregular curves. The spline is held in position by weights.

Figure 11.23 The erasing shield is used for erasing in tight spots. The dusting brush is used to remove the erased material. Brushing with the palm of your hand will smear the drawing.

Figure 11.24 This cordless electric eraser is typical of those used by professional drafters.

11.15 Scales

All engineering drawings require the use of scales to measure lengths, sizes, and so forth. Scales may be flat or triangular and are made of wood, plastic, or metal. Figure 11.25 shows triangular engineers' and architects' scales. Most scales are either 6 or 12 inches long. In this section, we cover the architects', engineers', and metric scale.

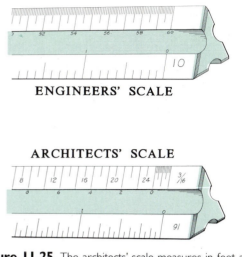

ENGINEERS' SCALE

ARCHITECTS' SCALE

Figure 11.25 The architects' scale measures in feet and inches, whereas the engineers' scale measures in decimal units.

Architects' Scale

The architects' scale is used to dimension and scale features encountered by the architect such as cabinets, plumbing, and electrical layouts. Most indoor measurements are made in feet and inches with an architects' scale. Figure 11.26 shows the basic form for indicating the scale being used. This form should be used in the title block or in

ARCHITECTS' SCALE $\qquad$ FROM END OF SCALE

BASIC FORM $SCALE: \dfrac{X}{X} = I'-0$

TYPICAL SCALES

SCALE: FULL SIZE (USE 16-SCALE)

SCALE: HALF SIZE (USE 16-SCALE)

SCALE: $3 = I'-0$	SCALE: $\frac{1}{4} = I'-0$
SCALE: $I\frac{1}{2} = I'-0$	SCALE: $\frac{3}{4} = I'-0$
SCALE: $\frac{1}{2} = I'-0$	SCALE: $\frac{3}{8} = I'-0$
SCALE: $\frac{3}{16} = I'-0$	SCALE: $\frac{1}{8} = I'-0$
SCALE: $\frac{3}{32} = I'-0$	SCALE: $I = I'-0$

Figure 11.26 The basic form for indicating the scale on the architects' scale and the variety of scales available.

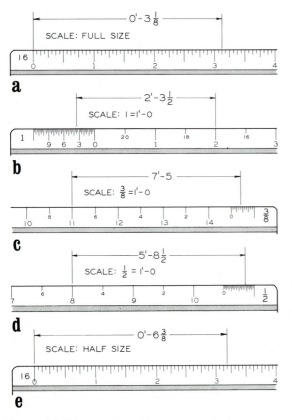

Figure 11.27 Examples of lines measured using an architects' scale.

some other prominent location on the drawing. Since the dimensions made with the architects' scale are in feet and inches, it is necessary to convert all dimensions to decimal equivalents (all feet or all inches) before the simplest arithmetic can be performed.

Scale: Full Size The 16 scale is used for measuring full-size lines (Fig. 11.27a). An inch on the 16 scale is divided into sixteenths to match the ruler used by the carpenter. This example is measured to be $3\frac{1}{8}''$. When the measurement is less than 1 ft, a zero may precede the inch measurements; note that the inch marks are omitted.

Scale: 1 = 1′-0 In Fig. 11.27b, a line is measured to its nearest whole foot (2 ft in this case), and the remainder is measured in inches at the

end of the scale ($3\frac{1}{2}$ in.) for a total of $2′-3\frac{1}{2}$. At the end of each architects' scale, a foot is divided into inches for measuring dimensions less than a foot. The scale $1'' = 1′-0$ is the same as saying 1 in. is equal to 12 in., or a $\frac{1}{12}$ size.

Scale: $\frac{3}{8} = 1′-0$ When this scale is used, $\frac{3}{8}$ in. represents 12 in. on a drawing. Figure 11.27c is measured to be $7′-5$.

Scale: $\frac{1}{2} = 1′-0$ A line is measured to be $5′-8\frac{1}{2}$ in Fig. 11.27d.

Scale: Half Size The 16 scale is used to measure or draw a line that is half size. This is sometimes specified as Scale: 6 = 12 (inch marks omitted). The line in Fig. 11.27e is measured to be $0′-6\frac{3}{8}$.

A couple of pointers: When marking measurements, hold your pencil vertically for the greatest accuracy (Fig. 11.28). And when indicating dimensions in feet and inches, they should be in the form shown in Fig. 11.29. (Notice the fractions are twice as tall as the whole numerals.)

Engineers' Scale

The engineers' scale is a decimal scale on which each division is a multiple of ten units. Because it is used for making drawings of outdoor engineering projects—streets, structures, land measurements, and other large topographical dimensions—it is sometimes called the civil engineers' scale.

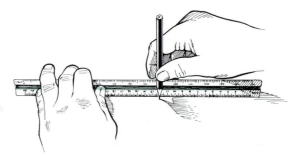

Figure 11.28 When marking off measurements along a scale, hold your pencil vertically for the most accurate measurement.

Figure 11.29 Inch marks are omitted according to current standards, but foot marks are shown. When the inch measurement is less than a whole inch, a leading zero is used. When representing feet, a zero is optional if the measurement is less than a foot.

ENGINEERS' SCALE   FROM END OF SCALE

BASIC FORM   SCALE: 1= XX

EXAMPLE SCALES

10	SCALE: 1=1–0'	SCALE: 1=1,000
20	SCALE: 1=200'	SCALE: 1=20 LB
30	SCALE: 1=3'	SCALE: 1=3,000'
40	SCALE: 1=4'	SCALE: 1=40'
50	SCALE: 1=50'	SCALE: 1=500'
60	SCALE: 1=6	SCALE: 1=0.6'

Figure 11.30 The basic form for indicating the scale on the engineers' scale and the variety of scales available.

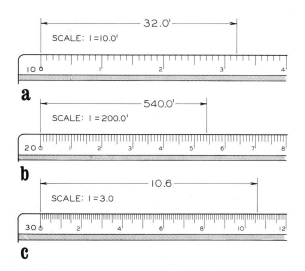

Figure 11.31 Examples of lines measured with the engineers' scale.

Since the measurements are in decimal form, it is easy to perform arithmetic operations; there is no need to convert from one unit to another, as there is when using the architects' scale. Figure 11.30 shows the form for specifying scales on the engineers' scale, for example, Scale: 1 = 10'. Each end of the scale is labeled 10, 20, 30 (and so on), which indicates the number of units per inch on the scale. Many combinations may be obtained by moving the decimal places of a given scale, as Fig. 11.30 shows.

10 Scale   In Fig. 11.31a, the 10 scale is used to measure a line drawn at the scale of 1 = 10'. The line is 32.0 ft long.

20 Scale   In Fig. 11.31b, the 20 scale is used to measure a line drawn at a scale of 1 = 200'. The line is 540.0 ft long.

30 Scale   In Fig. 11.31c, a line of 10.6 (inch marks omitted) is measured using the scale of 1 = 3.

Figure 11.32 shows the format for indicating measurements in feet and inches.

OMIT INCH MARKS   SHOW FT. MARKS
.13   0.13"   2.13   0.15'
GOOD   CROWDED   GOOD   GOOD

Figure 11.32 When using English units (inches), decimal fractions do not have leading zeros, and inch marks are omitted. Be sure to provide adequate space for decimal points between the numbers. Foot marks are shown.

English System

The English system (Imperial system) of measurement has been used in the United States, Britain, and Canada since these countries were established. The English system is based on arbitrary units of the inch, foot, cubit, yard, and mile (Fig. 11.33). Because there is no common relationship between these units, the system is cumbersome to use when simple arithmetic is performed; for example, finding the area of a rectangle that is 25 inches by $6\frac{3}{4}$ yards is a complex problem.

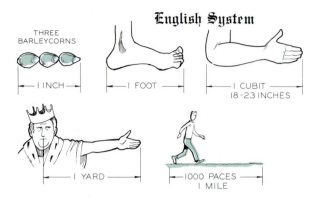

Figure 11.33 The units of the English system were based on arbitrary dimensions.

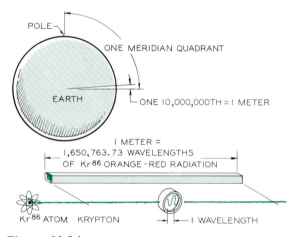

Figure 11.34 Originally based on the dimensions of the earth, the meter was later based on the wavelength of krypton-86. A meter is 39.37 inches.

Metric System—SI Units

The metric system was proposed by France in the fifteenth century. In 1793, the French National Assembly agreed that the meter (m) would be one ten-millionth of the meridian quadrant of the earth (Fig. 11.34). Fractions of the meter were expressed as decimal fractions. Debate continued until an international commission officially adopted the metric system in 1875. Since a slight error in the first measurement of the meter was found, the meter was later established as equal to 1,650,763.73 wavelengths of the orange-red light given off by krypton-86 (Fig. 11.34).

The international organization charged with the establishment and promotion of the metric system is the **International Standards Organization** (ISO). The system they have endorsed is called **Système International d'Unités** (International System of Units) and is abbreviated SI.

Several practical units of measurement have been derived from basic SI units (Table 11.1). Note that degrees Celsius (centigrade) is recommended over Kelvin, the official temperature measurement. When using Kelvin, the freezing and boiling temperatures are 273.15 K and 373.15 K, respectively.

Many SI units have prefixes to indicate placement of the decimal. Table 11.2 shows the more common of these. Figure 11.35 compares several English and SI units.

Table 11.1
These practical metric units are a few of those that are widely used because they are easier to deal with than the official SI units.

Parameter	Practical Units		SI = Equivalent
Temperature	Degrees Celsuis	°C	0°C = 273.15 K
Liquid Volume	Liter	l	$l = dm^3$
Pressure	Bar	BAR	BAR = 0.1 MPa
Mass Weight	Metric Ton	t	$t = 10^3 kg$
Land Measure	Hectare	ha	$ha = 10^4 m^2$
Plane Angle	Degree	°	$1° = \pi/180$ RAD

Table 11.2
The prefixes and abbreviations used to indicate the decimal placement for SI measurements.

Value	Prefix	Symbol
$1\ 000\ 000 = 10^{6}$	= Mega	M
$1\ 000 = 10^{3}$	= Kilo	k
$100 = 10^{2}$	= Hecto	h
$10 = 10^{1}$	= Deka	da
$1 = 10^{0}$		
$0.1 = 10^{-1}$	= Deci	d
$0.01 = 10^{-2}$	= Centi	c
$0.001 = 10^{-3}$	= Milli	m
$0.000\ 001 = 10^{-6}$	= Micro	μ

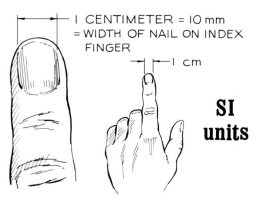

Figure 11.36 The width of the nail on your index finger is approximately equal to 1 centimeter, or 10 millimeters.

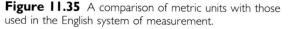

INCHES = 0.04 X CENTIMETERS

CENTIMETERS = 2.54 X INCHES

MILES = 1.6 X KILOMETERS

KILOMETERS = 0.63 X MILES

POUNDS = 0.45 X KILOGRAMS

KILOGRAMS = 2.21 X POUNDS

GALLONS = 0.26 X LITERS

LITERS = 3.8 X GALLONS

Figure 11.35 A comparison of metric units with those used in the English system of measurement.

METRIC SCALES

FROM END OF SCALE

BASIC FORM SCALE: 1= $\overline{XX}$

EXAMPLE SCALES

SCALE: 1:1 (1mm=1mm; 1cm=1cm)

SCALE: 1:2 (1mm=2mm; 1mm=20mm)

SCALE: 1:3 (1mm=30mm; 1mm=0.3mm)

SCALE: 1:4 (1mm=4mm; 1mm=40mm)

SCALE: 1:5 (1mm=5mm; 1mm=500mm)

SCALE: 1:6 (1mm=6mm; 1mm=60mm)

Figure 11.37 The basic form for indicating the scale on the metric scale and the variety of scales available.

11.16 Metric Scales

The basic unit of measurement on an engineering drawing is the **millimeter** (mm), which is one-thousandth of a meter, or one-tenth of a centimeter.

These units are understood unless otherwise specified on a drawing. The width of the fingernail of your index finger can serve as a conve-

nient gauge to approximate the dimension of one centimeter, or ten millimeters (Fig. 11.36). Figure 11.37 shows the form for indicating metric scales.

Decimal fractions are unnecessary on drawings dimensioned in millimeters; thus the dimensions are usually rounded off to whole numbers except for those measurements dimensioned with specified tolerances.

For metric units less than 1, a zero is placed in front of the decimal. In the English system, the zero is omitted from inch measurements (Fig. 11.38).

22.0   413.5   0.14
ZERO HERE

147   14.7   14.7
POOR-CROWDED   GOOD-SPACE

Figure 11.38 When decimal fractions are shown in metric units, a zero precedes the decimal. Be sure to allow adequate space for the decimal point when numbers with decimals are lettered.

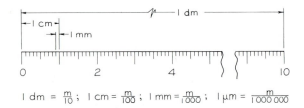

1 dm = $\frac{m}{10}$;   1 cm = $\frac{m}{100}$;   1 mm = $\frac{m}{1000}$;   1 μm = $\frac{m}{1000\,000}$

Figure 11.39 The dekameter is one tenth of a meter; the centimeter is one hundredth of a meter; a millimeter is one thousandth of a meter; and a micrometer is one millionth of a meter.

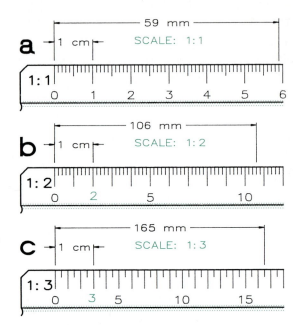

Figure 11.40 Examples of lines measured with metric scales.

Scale 1:1   The full-size metric scale (Fig. 11.39) shows the relationship between the metric units of the dekameter, centimeter, millimeter, and micrometer. There are 10 dekameters in a meter, 100 centimeters in a meter, 1000 millimeters in a meter, and 1,000,000 micrometers in a meter. A line of 59 mm is measured in Fig. 11.40a.

Scale 1:2   This scale is used when 1 mm represents 2 mm, 20 mm, 200 mm, and so forth. The line in Fig. 11.40b is 106 mm long.

Scale 1:3   A line of 165 mm is measured in Fig. 11.40c, where 1 mm represents 3 mm.

Other Scales

Many other metric (SI) scales are used: 1:250, 1:400, 1:500, and so on. The scale ratios mean one unit represents the number of units to the right of the colon. For example, 1:20 means 1 millimeter equals 20 mm, or 1 centimeter equals 20 cm, or 1 meter equals 20 m.

Metric Symbols

When drawings are made in metric units, this can be noted in the title block or elsewhere using the SI symbol (Fig. 11.41), which indicates Systéme International. The two views of the partial cone are used to denote whether the orthographic views were drawn in the U.S. system (third-angle projection) or the European system (first-angle projection).

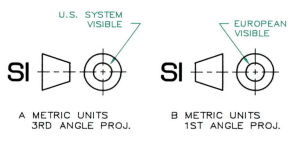

U.S. SYSTEM
VISIBLE

EUROPEAN
VISIBLE

SI

A METRIC UNITS
3RD ANGLE PROJ.

SI

B METRIC UNITS
1ST ANGLE PROJ.

Figure 11.41 The large SI indicates that the measurements are in metric units. The partial cones indicate that the views are arranged using (A) the third-angle projection (the U.S. system) or (B) the first-angle projection (the European system).

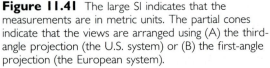

Scale Conversion

Appendix 2 gives tables for converting inches to millimeters, but this conversion can be performed by multiplying decimal inches by 25.4 to obtain millimeters.

An architect's scale must be multiplied by 12 to convert it to an approximate metric scale. For example, Scale: $\frac{1}{8} = 1'-0$ is the same as $\frac{1}{8}$ in. = 12 in. or 1 in. = 96 in. This scale closely approximates the metric scale of 1:100. Many of the scales used in the metric system cannot be converted to exact English scales, but the metric scale of 1:60 converts exactly to the scale of 1 = 5'.

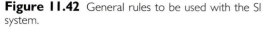

OMIT COMMAS AND GROUP INTO THREES
1 000 000 000 NOT   1,000,000,000

USE RAISED DOT FOR MULTIPLICATION
N•M   OR   NM

INDICATE DIVISION BY EITHER
kg/m   OR   kg m⁻¹

USE ZERO PRECEDING DECIMALS
0.72 mm NOT   .72 mm

INDICATE SI SCALES AS
SCALE: 1:2 SI

Figure 11.42 General rules to be used with the SI system.

Expression of Metric Units

Figure 11.42 gives the general rules for expressing SI units. Commas are not used between sets of zeros; instead, a space is left between them.

11.17 The Instrument Set

Figure 11.43 shows a basic set of drawing instruments. Although these can be bought separately, they are available as sets in cases similar to the one shown in Fig. 11.44.

Compass

The **compass** is used to draw circles and arcs in ink and in pencil (Fig. 11.45). To obtain good results with the compass, its pencil point must be

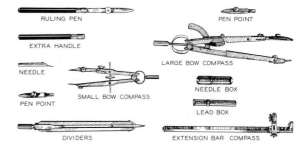

Figure 11.43 The parts of a set of drafting instruments.

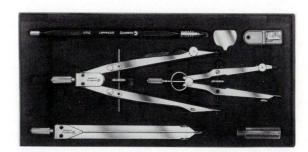

Figure 11.44 Instruments usually come as a cased set. (Courtesy of Gramercy Guild.)

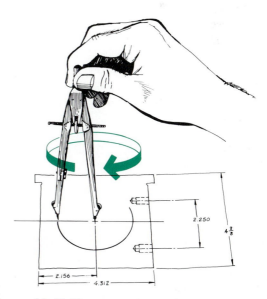

Figure 11.45 The compass is used for drawing circles.

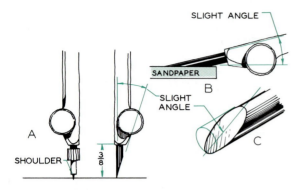

Figure 11.46 (A) The pencil point should be about the same length as the compass point. (B) The compass lead should be sharpened from the outside on a sandpaper pad. (C) The lead should be sharpened to a wedge point.

sharpened on its outside with a sandpaper board (Fig. 11.46). A bevel cut of this type gives the best all-around point for drawing a circle. When the compass point is set in the drawing surface, it should be inserted just enough for a firm set, not to the shoulder of the point. When the table top has a hard covering, several sheets of paper should be placed under the drawing to provide a seat for the compass point.

Bow compasses are provided in some sets (Fig. 11.47) for drawing small circles. For large

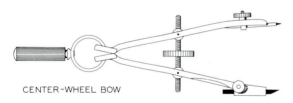

Figure 11.47 A small bow compass for drawing circles of about 1-in. radius.

Figure 11.48 Large circles can be drawn with the beam compass. Ink attachments are also available.

circles, bars are provided to extend the range of the large bow compass. A beam compass can be used (Fig. 11.48) for still larger circles. Small circles and ellipses can be effectively drawn with a circle template that is aligned with the perpendicular centerlines of the circle. The circle or ellipse is drawn with a pencil to match the other lines of the drawing (Fig. 11.49).

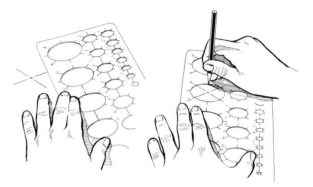

Figure 11.49 Circle templates can be used for drawing circles without the use of a compass. The circle or ellipse template is aligned with the centerlines.

Dividers

The **dividers** look much like a compass but are used for laying off and transferring dimensions onto a drawing. For example, equal divisions can be stepped off rapidly along a line (Fig. 11.50). As each measurement is made, a slight impression is made in the drawing surface with the dividers' points.

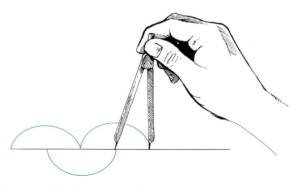

Figure 11.50 Dividers are used to step off measurements.

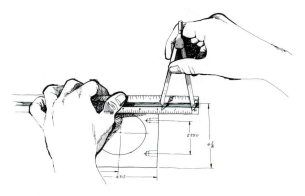

Figure 11.51 Dividers are also used to transfer dimensions from a scale to a drawing.

Dividers can be used to transfer dimensions from a scale to a drawing (Fig. 11.51) or to divide a line into a number of equal parts. Small bow dividers can be used for transferring smaller dimensions, such as the spacing between the guidelines for lettering (Fig. 11.52).

Proportional Dividers

Dimensions can be transferred from one scale to another by using a special type of dividers, the **proportional dividers.** The central pivot point can be moved to vary the ratio of the spacing at one end of the dividers to the ratio at the other end (Fig. 11.53).

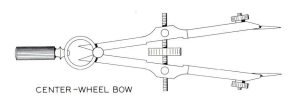

CENTER-WHEEL BOW

Figure 11.52 A type of a bow divider for transferring small dimensions, such as the spacing between guidelines for lettering.

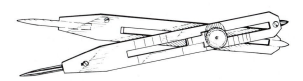

Figure 11.53 A proportional divider can be used for making measurements that are proportional to other dimensions.

11.18 Ink Drawing

Unlike pencil drawings, ink drawings remain dark and distinct. Ink lines also reproduce better than pencil.

Materials for Ink Drawing

A good grade of tracing paper can be used for ink drawings, but erasing errors may result in holes in the paper and the loss of your time. Therefore, tracing film or tracing cloth should be used, which will withstand many erasings and corrections.

When using drafting film, the drawing should be made on the matte surface according to the manufacturer's directions. A cleaning solution is available that can be used to prepare the surface for ink and to remove spots that might not properly take the ink. Tracing cloths need to be prepared for inking by applying a coating of powder or pounce to absorb oily spots that will otherwise repel an ink line. India ink—a dense, black carbon ink that is much thicker and faster drying than fountain pen ink—is used for engineering drawings.

Ruling Pen

The ruling pen should be inked with the spout on the cap of the ink bottle (Fig. 11.54). Experiment with your pen to learn the proper amount of ink

Figure 11.54 The pen is inked between the nibs with the spout on the ink bottle cap.

to apply to the nibs. When drawing horizontal lines, the ruling pen is held in the same position as a pencil (Fig. 11.55), maintaining a space between the nibs and straightedge.

An alternative pen that can be used is the technical ink fountain pen (Fig. 11.56). These pens come in sets with pen points of various sizes that are used for the alphabet of lines. Lines drawn with this type of pen dry faster than those drawn with ruling pens because the ink is applied in a thinner layer.

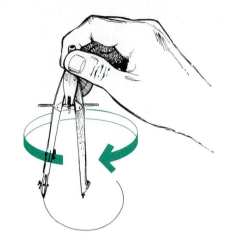

Figure 11.57 An attachment can be used with a large bow compass for drawing circles and arcs in ink.

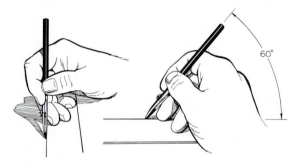

Figure 11.55 The ruling pen is held in a vertical plane at 60° to the drawing surface.

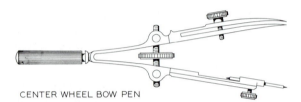

CENTER WHEEL BOW PEN

Figure 11.58 A small bow compass for drawing arcs in ink up to a radius of about one inch.

Figure 11.56 An India ink technical pen that can be used for making ink drawings.

11.58 shows a compass for drawing smaller circles. The spring bow compass can be used to draw pencil and ink circles about one-eighth inch in radius. Larger circles can be drawn using the extension bar with the large compass (Fig. 11.59). Special compasses are available for drawing circles with a technical fountain pen. These pens screw into an adapter that fits the compass.

Inking Compass

The inking compass is usually the same compass used for the circles drawn by pencil, with the inking attachment inserted in place of the pencil attachment. The circle can be drawn with one continuous line, as Fig. 11.57 shows. Figure

Templates

Various templates are available for drawing nuts and bolts, circles and ellipses, architectural symbols, and many other applications. Templates work best when used with technical fountain pens rather than the traditional ruling pen.

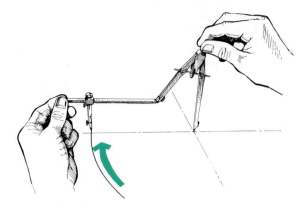

Figure 11.59 An extension bar can be used with a bow compass for drawing large arcs.

11.19 Solutions of Problems

The following formats are suggested for the layout of problem sheets. Most problems will be drawn on $8\frac{1}{2}$-×-11-inch sheets, as Fig. 11.60 shows. A title strip is suggested in this figure, with a border as shown. Guidelines should be drawn very lightly to be only faintly visible. The $8\frac{1}{2}$-×-11-inch sheet in the vertical format is called Size AV throughout the rest of this textbook. When this sheet is in the horizontal format, as Fig. 11.61 shows, it will be called Size AH.

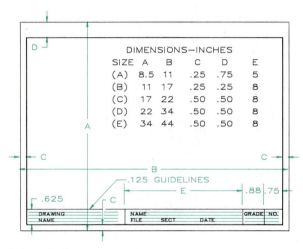

Figure 11.61 The format for Size AH (an 11" × 8½" sheet in a horizontal position) and the sizes of other sheets. The dimensions under columns A through E give the various layouts.

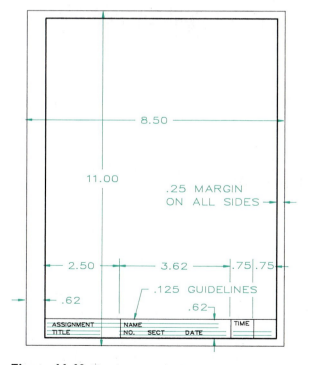

Figure 11.60 The format and title strip for a Size A sheet (8½" × 11") suggested for solving the problems at the end of each chapter. When the sheet is in the vertical format, it will be called a Size AV.

Figure 11.61 shows the standard sizes of sheets, from Size A through Size E. Figure 11.62 shows an alternative title strip for Sizes B, C, D, and E. Guidelines should always be used for lettering title strips.

A smaller title block and parts list is given above, which are placed in the lower right-hand corner of the sheet against the borders. When both are used on the same drawing, the parts list is placed directly above and in contact with the title block or title strip.

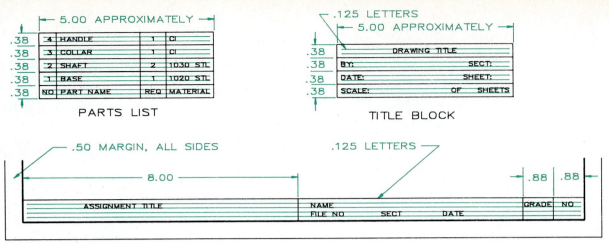

PARTS LIST

NO	PART NAME	REQ	MATERIAL
4	HANDLE	1	CI
3	COLLAR	1	CI
2	SHAFT	2	1030 STL
1	BASE	1	1020 STL

TITLE BLOCK

.125 LETTERS
5.00 APPROXIMATELY

DRAWING TITLE	
BY:	SECT:
DATE:	SHEET:
SCALE:	OF SHEETS

.50 MARGIN, ALL SIDES   .125 LETTERS

8.00   .88 .88

| ASSIGNMENT TITLE | NAME | GRADE | NO |
| | FILE NO   SECT   DATE | | |

Figure 11.62 A title strip that can be used on sheet Sizes B, C, D, and E instead of the one given in Fig. 11.61.

Problems

Construct the problems (Figs. 11.63–11.65) on Size AH (8½-×-11-inch) paper, plain or with a printed grid, using the format shown in Fig. 11.60. Use pencil or ink as assigned by your instructor. Two problems can be drawn per sheet using the scale of each square equals .125 in., or 3 mm. One double-size problem can be drawn per sheet.

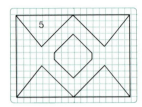

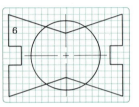

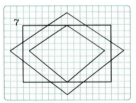

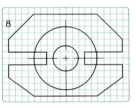

Figure 11.63 Problems 1–4.

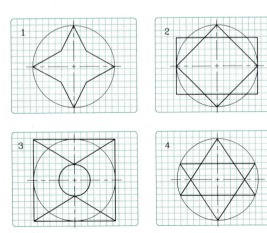

Figure 11.64 Problems 5–8.

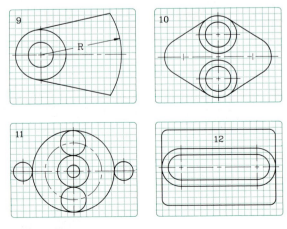

Figure 11.65 Problems 9–12.

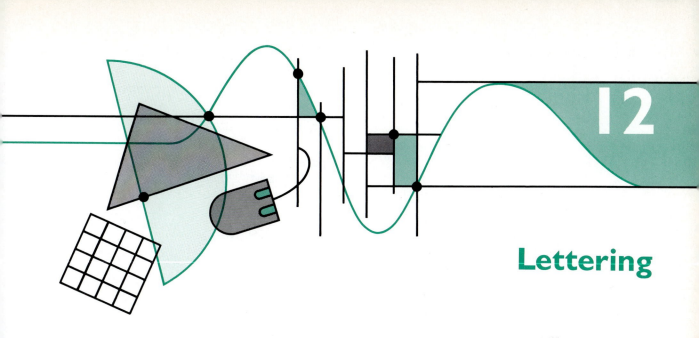

12

Lettering

12.1 Introduction

All drawings are supplemented with notes, dimensions, and specifications that must be lettered. The ability to construct legible freehand letters is an important skill to develop since it affects the usage and interpretation of a drawing.

12.2 Tools of Lettering

The best pencils for lettering on most surfaces are in the H–HB grade range, with an F pencil being the most commonly used grade. Some papers and films are coarser than others and may require a harder pencil lead. To give the desired line width, the point of the pencil should be slightly rounded (Fig. 12.1); a needle point will break off when pressure is applied.

When lettering, the pencil should be revolved slightly between your fingers as the strokes are being made so that the lead will wear down gradually and evenly. Bear down firmly to make letters black and bright for good reproduction. To prevent smudging your drawing while lettering,

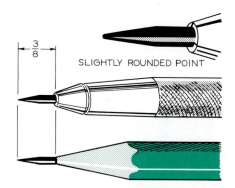

Figure 12.1 Good lettering begins with a properly sharpened pencil point. The point should be slightly rounded, not a needle point. The F pencil is usually the best grade for lettering.

place a sheet of paper under your hand to protect the drawing (Fig. 12.2).

12.3 Gothic Lettering

The type of lettering recommended for engineering drawings is **single-stroke Gothic lettering,** so called because the letters are made with a series

139

Figure 12.2 When lettering a drawing, use a protective sheet under your hand to prevent smudges. Your lettering will be best when you are working from a comfortable position; you may wish to turn your paper for the most natural strokes.

of single strokes, and the letter form is a variation of Gothic lettering.

Two general categories of Gothic lettering are **vertical** and **inclined** (Fig. 12.3). Although equally acceptable, they both should not be used on the same drawing.

Figure 12.3 Two types of Gothic lettering recommended by engineering standards are vertical and inclined lettering.

12.4 Guidelines

The most important rule of lettering is **use guidelines at all times.**

This applies whether you are lettering a paragraph or a single letter or numeral. Figure 12.4 shows the method of constructing and using guidelines. Use a sharp pencil in the 3H–5H grade range, and draw these lines lightly, just dark enough for them to be seen.

Most lettering is done with the capital letters $\frac{1}{8}$ in. (3 mm) high. The spacing between lines of lettering should be no closer than half the height of the capital letters, $\frac{1}{16}$ in. in this case.

Lettering Guides

The two instruments used most often for drawing guidelines are the **Braddock-Rowe lettering triangle** and the **Ames lettering instrument.**

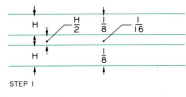

STEP 1

STEP 2

STEP 3

Figure 12.4 Lettering guidelines.

Step 1 Letter heights, *H,* are laid off, and light guidelines are drawn with a 4H pencil. The spacing between the lines should be no closer than *H*/2.

Step 2 Vertical guidelines are drawn as light, thin lines. These are randomly spaced to serve as visual guides for lettering.

Step 3 The letters are drawn with single strokes using a medium-grade pencil, H–HB. The guidelines need not be erased since they are drawn lightly.

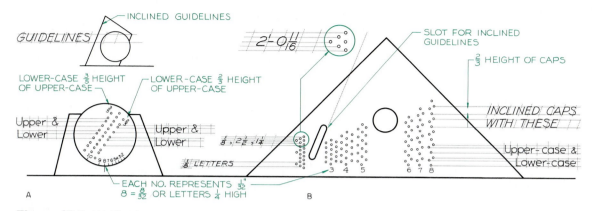

Figure 12.5 (A) The Ames lettering guide can be used to draw guidelines for vertical or inclined uppercase and lowercase letters. The dial is set to the desired number of thirty-seconds of an inch for the height of uppercase letters. (B) The Braddock-Rowe triangle can be used as a 45° triangle and as an instrument for constructing guidelines. The numbers designating the guidelines represent thirty-seconds of an inch.

The Braddock-Rowe triangle is pierced with sets of holes for spacing guidelines (Fig. 12.5b). The numbers under each set of holes represent thirty-seconds of an inch. For example, the numeral 4 represents $\frac{4}{32}$ in. or $\frac{1}{8}$ in. for making uppercase (capital) letters. Some triangles are marked for metric lettering in millimeters. In Fig. 12.5, intermediate holes are provided for guidelines for lowercase letters, which are not as tall as the capital letters.

With a horizontal straightedge held firmly in position, place the Braddock-Rowe triangle against its edge. A sharp 4H pencil is placed in one hole of the desired set of holes to contact the drawing surface, and the pencil point is guided across the paper to draw the guideline while the triangle slides against the straightedge. This is repeated as the pencil point is moved successively to each hole until the desired number of guidelines are drawn. An oblique slot for drawing guidelines for inclined lettering is cut in the triangle. Slanting guidelines are spaced randomly by eye.

The Ames lettering guide (Fig. 12.5a) is a similar device with a circular dial for selecting the proper spacing of guidelines. Again, the numbers around the dial represent thirty-seconds of an inch. The number 8 represents $\frac{8}{32}$ in. or guidelines for drawing capital letters that are $\frac{1}{4}$ in. tall.

12.5   Vertical Letters
Vertical Capital Letters

Figure 12.6 shows the capital letters for the single-stroke Gothic alphabet. Each letter is drawn inside a square box of guidelines to help you learn their correct proportions. Some letters require the full area of the box, some require less space, and a few require more space. Each straight-line stroke should be drawn as a single stroke; for example, the letter A is drawn with three single strokes. Letters composed of curves can best be drawn in segments; the letter O can be drawn by joining two semicircles to form the full circle.

Memorize the shape of each letter given in this alphabet. Small wiggles in your strokes will not detract from your lettering if the letter forms are correct. Figure 12.7 shows examples of poor lettering.

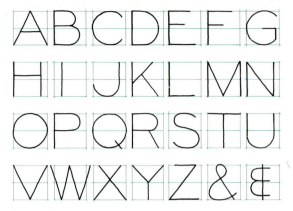

Figure 12.6 The uppercase letters used in single-stroke Gothic lettering. Each letter is drawn inside a square to help you learn their correct proportions.

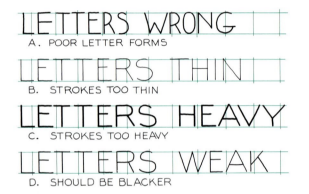

Figure 12.7 There are many ways to letter poorly. A few of them, and the reasons the lettering is inferior, are shown here.

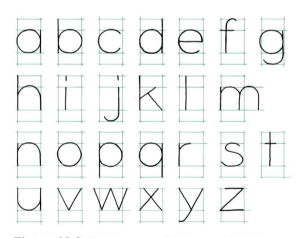

Figure 12.8 The lowercase alphabet used in single-stroke Gothic lettering. The body of each letter is drawn inside a square to help you learn the proportions.

Vertical Lowercase Letters

Figure 12.8 shows an alphabet of lowercase letters. Lowercase letters are either two thirds or three fifths as tall as the uppercase letters they are used with. Both of these ratios are labeled on the Ames guide, but only the two-thirds ratio is available on the Braddock-Rowe triangle.

Some lowercase letters have **ascenders** that extend above the body of the letter, such as the letter b; some have **descenders** that extend below the body, such as the letter y. The ascenders are the same length as the descenders.

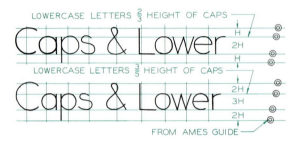

Figure 12.9 Uppercase and lowercase letters are sometimes used together. The ratio of the lowercase letters to the uppercase letters will be either two thirds or three fifths. The Ames guide has both, and the Braddock-Rowe triangle has only the three-fifths ratio.

The guidelines in Fig. 12.8 form perfect squares about the body of each letter to illustrate the proportions. Several of these letters have bodies that are perfect circles that touch all sides of the squares. In Fig. 12.9, capital and lowercase letters are used together.

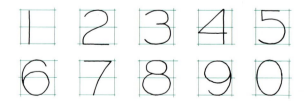

Figure 12.10 The numerals for single-stroke Gothic lettering. Each numeral is drawn inside a square to help you learn their proportions.

Vertical Numerals

Figure 12.10 shows vertical numerals for single-stroke Gothic lettering. Each number is enclosed in a square box of guidelines. Each number is also made the same height as the capital letters being used, usually $\frac{1}{8}$ in. high. The numeral, 0 (zero) is an oval, whereas the letter O is a perfect circle in vertical lettering.

12.6 Inclined Letters

Inclined Uppercase Letters

Inclined uppercase letters (capitals) have the same heights and proportions as vertical letters; the only difference is their 68° inclination (Fig. 12.11). Inclined guidelines should be drawn using the Braddock-Rowe triangle or the Ames guide.

Figure 12.11 The inclined uppercase alphabet for single-stroke Gothic lettering.

Inclined Lowercase Letters

Inclined lowercase letters are drawn in the same manner as vertical lowercase letters (Fig. 12.12). Ovals (ellipses) are used instead of the circles used in vertical lettering. The angle of inclination is 68°, the same as is used for uppercase letters.

Inclined Numerals

Figure 12.13 shows the inclined numerals that should be used with inclined lettering, and Fig.

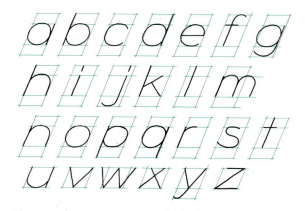

Figure 12.12 The inclined lowercase alphabet for single-stroke Gothic lettering. The body of each letter is drawn inside a rhombus to help you learn the proportions.

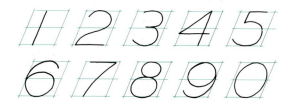

Figure 12.13 The inclined numerals for single-stroke Gothic lettering. Each number is drawn inside a rhombus to help you learn the proportions.

12.14 shows the use of inclined letters and numbers in combination. The guidelines in Fig. 12.14 were constructed using the Braddock-Rowe triangle.

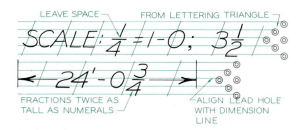

Figure 12.14 Inclined common fractions are twice as tall as single numerals. Inch marks are omitted when numerals are used to show dimensions.

12.7 Spacing Numerals and Letters

Common fractions are twice as tall as single numerals (Fig. 12.15). A separate set of holes for common fractions is given on the Braddock-Rowe triangle and on the Ames guide. These are equally spaced $\frac{1}{16}$ in. apart with the centerline being used for the fraction's crossbar.

When numbers are used with decimals, space should be provided for the decimal point (Fig. 12.16). Figure 12.16D shows the correct method of drawing common fractions, and Figs. 12.16E–G show several errors that are often encountered.

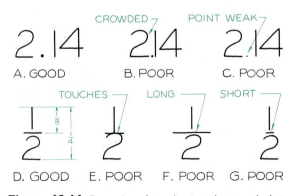

Figure 12.15 Common fractions are twice as tall as single numerals. Guidelines for these can be drawn by using the Ames guide or the Braddock-Rowe triangle.

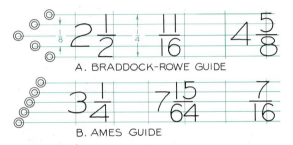

Figure 12.16 Examples of poorly spaced numerals that result in inferior lettering.

When letters are grouped together to spell words, the area between the letters should be approximately equal for the most pleasing result (Fig. 12.17). Figure 12.18 shows the incorrect use of guidelines and other violations of good lettering practice.

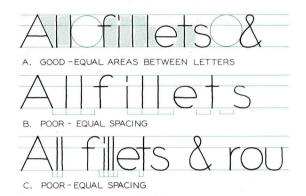

Figure 12.17 Proper spacing of letters is necessary for good lettering and appearance. The areas between letters should be approximately equal.

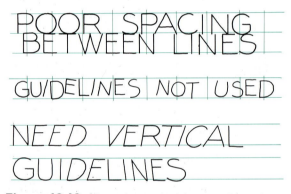

Figure 12.18 Always leave space between lines of lettering. After constructing guidelines, use them. Use vertical guidelines to improve the angle of your vertical strokes.

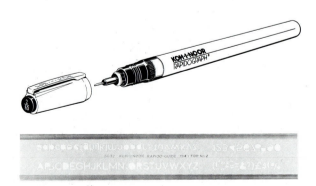

Figure 12.19 A typical India ink fountain pen and template that can be used for mechanical lettering. (Courtesy of Koh-I-Noor Rapidograph, Inc.)

12.8 Mechanical Lettering

Drawings and illustrations to be reproduced by a printing process are usually drawn in India ink. Several mechanical aids for ink lettering are available.

The Rapidograph lettering template (Fig. 12.19) can be placed against a fixed straightedge for aligning the letters. You move the template from position to position while drawing each letter (with a lettering pen) through the raised portion of the template where the holes form the letters.

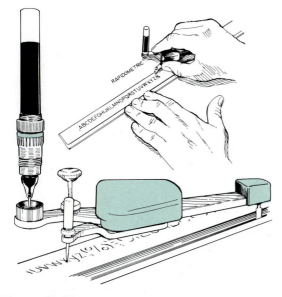

Figure 12.20 Templates and scribers of this type are available for mechanical lettering.

Another system of mechanical lettering uses a grooved template along with a scriber that follows the grooves and inks the letters on the drawing surface (Fig. 12.20). A standard India ink technical pen can be unscrewed from its barrel and attached to the scriber (Fig. 12.20). Many templates of varying styles of lettering and symbols are available for this system of mechanical lettering.

ABCDEFGHIJK
KLMNOPQRST
UVWXYZ&%01
23456789abc
defghijklmnop
rstuvwxyz

Figure 12.21 Characters drawn by the SIMPLEX font provided by AutoCAD.

12.9 Lettering by Computer

Lettering of all types can be done by using AutoCAD (see Chapter 17). Figure 12.21 shows letters and numbers made with the SIMPLEX font. The SIMPLEX font has smoother curves than the TXT font, which is the default font.

The COMPLEX font available from AutoCAD can be inclined at any angle, as Fig. 12.22 shows. This font is drawn using multiple strokes to form letters of varying widths to generate a Roman style.

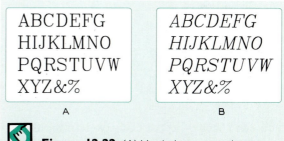

ABCDEFG HIJKLMNO PQRSTUVW XYZ&%	*ABCDEFG HIJKLMNO PQRSTUVW XYZ&%*
A	B

Figure 12.22 (A) Vertical uppercase letters drawn with the COMPLEX font. (B) Inclined uppercase letters drawn with the COMPLEX font. The angle of inclination is the angle with the vertical in a clockwise direction.

The COMPLEX font is used in Fig. 12.23 to illustrate some of the many styles of text that can be created using AutoCAD options. Text can be inclined, backward or upside-down, and it can be varied in width to any degree.

NARROW TEXT   0.8 WIDE COMPLEX

REGULAR WIDTH   REG. COMPLEX

WIDE TEXT   1.2 WIDE COMPLEX

VERY WIDE   1.8 WIDE

INCLINED TEXT   13° INCLINED

ƧᗡᴙAWᴋↃAᗺ   BACKWARDS COMPLEX

∩bᴚIᗡE ᗡOMᴎUPSIDE DOWN COMPLEX

Figure 12.23 Styles of lettering can be created to your specifications by changing the variables of the **STYLE** command.

When you begin a new drawing, use the following **STYLE** command unless you have previously SAVEd a different STYLE in a default drawing:

```
Command: STYLE (CR)
Text style name (or ?)
‹STANDARD›: STANDARD (CR)
Existing style.
Font file ‹TXT›:  TXT (CR)
Height ‹0.0000›:  0 (CR)
Width factor ‹1.00›:  1 (CR)
Obliquing angle ‹0›:  0 (CR)
Backwards? ‹N›  No (CR)
Upside-down? ‹N›  No (CR)
Vertical? ‹N›  No (CR)
STANDARD is now the current text
style.
```

You can change the Text style definitions with the STYLE command in the following manner:

```
Command: STYLE (CR) Text style name
(or ?): PRETTY (CR)
Font file ‹default›: SIMPLEX (CR)
Height ‹default›: 0 (CR)
Width factor (default): 1 or (CR)
Obliquing angle (default): 22
(CCW angle from vertical) (CR)
Backwards? ‹Y/N› N (CR)
Upside-down? ‹Y/N› N (CR)
```

The style, PRETTY, has been created, and it can be called by using the S (STYLE) option under the **TEXT** command. When prompted for the style name, enter PRETTY, and the font can be used with all the variables defined above. When asked for height, you must specify the lettering height when it is first used since it was specified to have a 0 height. The value you enter will remain the default height by pressing the space bar or (CR) when asked for the desired height. You can see that numerous styles can be created and named for ready use.

By using the SIMPLEX font and assigning a width factor of 1.30 to it, the lettering will closely approximate the form of hand lettering (Fig. 12.24).

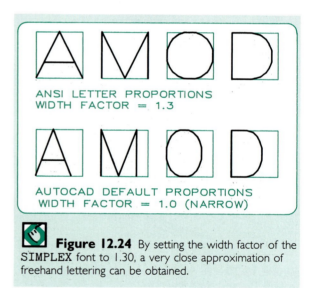

ANSI LETTER PROPORTIONS
WIDTH FACTOR = 1.3

AUTOCAD DEFAULT PROPORTIONS
WIDTH FACTOR = 1.0 (NARROW)

Figure 12.24 By setting the width factor of the SIMPLEX font to 1.30, a very close approximation of freehand lettering can be obtained.

12.10 Third-Party Software

Third-party vendors supply additional fonts and letter styles that can be used in conjunction with AutoCAD. One of these is CAD Lettering Systems, Inc., P.O. Box 832, Oldsmar, FL 34677, who markets **LETTEREASE** software that is shown in the following examples.

Figure 12.25 shows two lettering fonts, one in the outline mode and the other filled in with solid letters. These letters can be applied to a drawing in the same manner as other AutoCAD fonts.

CAD Lettering Systems
Third-party software
Oldsmar, FL

Figure 12.25 Third-party software developers provide additional fonts of text that can be used with AutoCAD, such as the **LETTEREASE** fonts shown here. (Courtesy of CAD Lettering Systems, Inc., PO Box 832, Oldsmar, FL 34677.)

Additional features include the positioning of text in an arc in clockwise or counterclockwise directions as shown in Fig. 12.26. Text can be made to conform to an irregular polyline as shown in Fig. 12.27. Many other options are available with this software, such as automatic spacing between letters, rotation, and changing options.

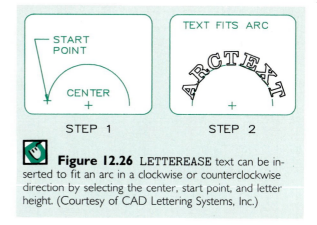

Figure 12.26 LETTEREASE text can be inserted to fit an arc in a clockwise or counterclockwise direction by selecting the center, start point, and letter height. (Courtesy of CAD Lettering Systems, Inc.)

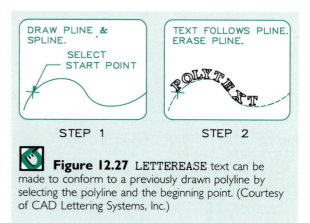

Figure 12.27 LETTEREASE text can be made to conform to a previously drawn polyline by selecting the polyline and the beginning point. (Courtesy of CAD Lettering Systems, Inc.)

Problems

Lettering problems are to be presented on Size AV ($8\frac{1}{2} \times 11$-inch) paper, plain or grid, using the format shown in Fig. 12.28.

1. Practice lettering the vertical uppercase alphabet shown in Fig. 12.28. Construct each letter three times: three A's, three B's, and so on. Use a medium-weight pencil—H, F, or HB.

2. Practice lettering vertical numerals and the lowercase alphabet as shown in Fig. 12.29. Construct each letter and numeral two times: two 1's, two 2's, and so on. Use a medium-weight pencil—H, F, or HB.

3. Practice lettering the inclined uppercase alphabet shown in Fig. 12.11. Construct each letter three times. Use a medium-weight pencil—H, F, or HB.

4. Practice lettering the vertical numerals and the lowercase alphabet shown in Figs. 12.8 and 12.10. Construct each letter two times. Use a medium-weight pencil—H, F, or HB.

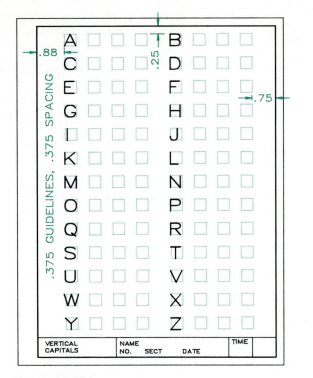

Figure 12.28 Problem 1. Construct each vertical uppercase letter three times.

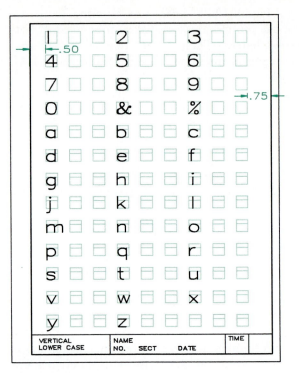

Figure 12.29 Problem 2. Construct each vertical numeral and lowercase letter two times.

5. Construct guidelines for $\frac{1}{8}$-in. capital letters starting $\frac{1}{4}$ in. from the top border. Each guideline should end $\frac{1}{2}$ in. from the left and right borders. Using these guidelines, letter the first paragraph of the text of this chapter. Use all vertical capitals. Spacing between the lines should be $\frac{1}{8}$ in.

6. Repeat Problem 5, but use all inclined capital letters. Use inclined guidelines to help you uniformly slant your letters.

7. Repeat Problem 5, but use vertical capitals and lowercase letters in combination. Capitalize only those words that are capitalized in the text.

8. Repeat Problem 5, but use inclined capitals and lowercase letters in combination. Capitalize only those words that are capitalized in the text.

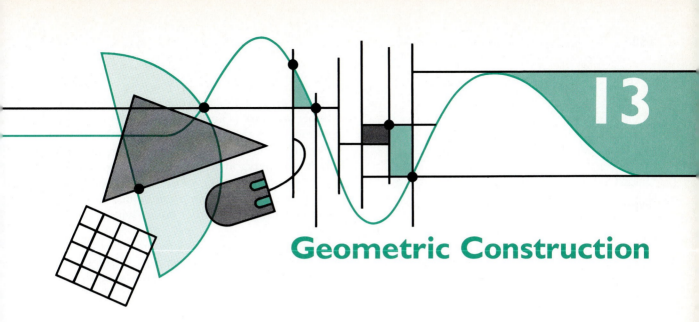

Geometric Construction

13.1 Introduction

Many graphical problems can be solved only by using geometry and geometric construction. Mathematics was an outgrowth of graphical construction, so the two areas are closely related. The proofs of many principles of plane geometry and trigonometry may be developed by using graphics. Moreover, graphical methods can be applied to algebra and arithmetic, and virtually all problems of analytical geometry can be solved graphically.

13.2 Angles

A fundamental requirement of geometric construction is the construction of lines that join at specified angles with each other. Figure 13.1 gives the definitions of various angles.

The unit of angular measurement is the degree, and a circle has 360 degrees. A degree (°) can be divided into 60 parts called minutes ('), and a minute can be divided into 60 parts called seconds ("). An angle of 15°32'14" is an angle of 15 degrees, 32 minutes, and 14 seconds.

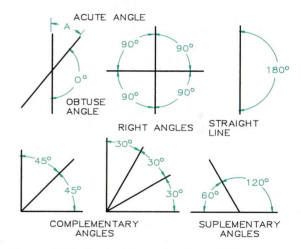

Figure 13.1 Standard types of angles and their definitions.

13.3 Triangles

The **triangle** is a three-sided polygon (or figure) that is named according to its shape. The four types of triangles are the **scalene, isosceles, equilateral,** and **right triangle** (Fig. 13.2). The sum of the angles inside a triangle is always 180°.

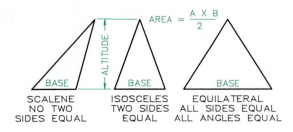

$$AREA = \frac{A \times B}{2}$$

ALTITUDE		
BASE	BASE	BASE
SCALENE NO TWO SIDES EQUAL	ISOSCELES TWO SIDES EQUAL	EQUILATERAL ALL SIDES EQUAL ALL ANGLES EQUAL

Figure 13.2 Types of triangles and their definitions.

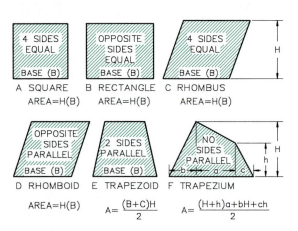

Figure 13.3 Types of quadrilaterals (four-sided plane figures).

13.4 Quadrilaterals

A **quadrilateral** is a four-sided figure of any shape. The sum of the angles inside a quadrilateral is 360°. Figure 13.3 shows the various types of quadrilaterals and the equations for the areas of these figures.

Figure 13.4 Regular polygons inscribed in circles.

13.5 Polygons

A **polygon** is a multisided plane figure of any number of sides. (The triangle is a three-sided polygon, and the quadrilateral is a four-sided polygon.) If the sides of the polygon are equal in length, the polygon is a **regular polygon.** Figure 13.4 shows four types of regular polygons. Note that a regular polygon can be inscribed in a circle and that all the corner points will lie on the circle.

Other regular polygons not pictured are the **heptagon** (7 sides), the **nonagon** (9 sides), the **decagon** (10 sides), and the **dodecagon** (12 sides). The sums of the angles inside any polygon can be found by the equation

$$S = (n - 2) \times 180°,$$

where *n* equals the number of sides of the polygon.

13.6 Elements of Circles

A circle can be divided into a number of parts, each of which has its own special name (Fig. 13.5). These terms, and others that deal with elements of circles, are used throughout this book.

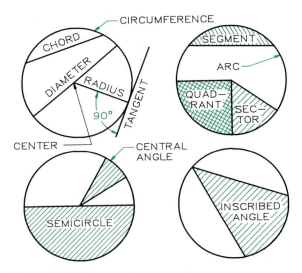

Figure 13.5 Definitions of the elements of a circle.

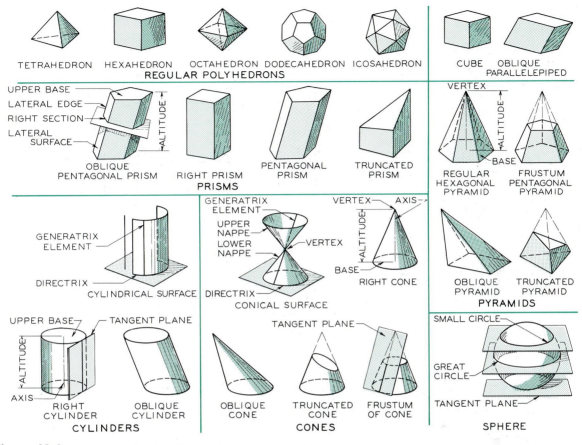

Figure 13.6 Types of geometric solids and their elements and definitions.

13.7 Geometric Solids

Figure 13.6 shows the various types of solid geometric shapes, along with their names and definitions.

Polyhedra   A multisided solid formed by intersecting planes is called a **polyhedron.** If the faces of a polyhedron are regular polygons, it is a **regular polyhedron.** The five regular polyhedra are the **tetrahedron** (4 sides), the **hexahedron** (6 sides), the **octahedron** (8 sides), the **dodecahedron** (12 sides), and the **icosahedron** (20 sides).

Prisms   A **prism** is a solid that has two parallel bases that are equal in shape. The bases are connected by sides that are parallelograms. The line from the center of one base to the center of the other is the **axis.** An axis that is perpendicular to the bases is an **altitude,** and the prism is a **right prism.** If the axis is not perpendicular to the base, the prism is an **oblique prism.** A prism that has been cut off to form a base that is not parallel to the other is called a **truncated prism.** A **parallelepiped** is a prism with a base that is either a rectangle or a parallelogram.

Pyramids   A **pyramid** is a solid with a polygon as a base and triangular faces that converge at a point called the **vertex.** The line from the vertex to the center of the base is the **axis.** If the axis is perpendicular to the base, it is the **altitude**

of the pyramid, and the pyramid is a **right pyramid.** If the axis is not perpendicular to the base, the pyramid is an **oblique pyramid.** A truncated pyramid is called a **frustum** of a pyramid.

Cylinders   A **cylinder** is formed by a line or element (called a **generatrix**) that moves about the circle while remaining parallel to its axis. The axis of a cylinder connects the centers of each end of a cylinder. If the axis is perpendicular to the bases, it is the **altitude** of a **right cylinder.** If the axis does not make a 90° angle with the base, the cylinder is an **oblique cylinder.**

Cones   A **cone** is also formed by a generatrix, one end of which moves about the circular base while the other end remains at a fixed point called the **vertex.** The line from the center of the base to the vertex is the **axis.** If the axis is perpendicular to the base, it is called the **altitude,** and the cone is a **right cone.** A truncated cone is called a **frustum** of a cone.

Spheres   A **sphere** is generated by the plane of a circle that is revolved about one of its diameters to form a solid. The ends of an axis through the center of the sphere are called **poles.**

13.8   Constructing Triangles

When three sides of a triangle are given, the triangle can be constructed by using a compass, as Fig. 13.7 shows. Only one triangle can be found when the sides are given by this method, called

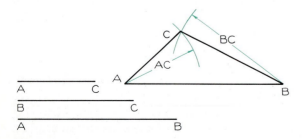

Figure 13.7 When three sides are given, a triangle can be drawn with a compass.

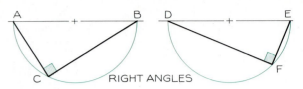

Figure 13.8 Any angle inscribed in a semicircle will be a right angle.

triangulation. A right triangle can be constructed by inscribing it in a semicircle, as Fig. 13.8 shows. Any triangle inscribed in a semicircle will always be a right triangle.

13.9   Constructing Polygons

A regular polygon (having equal sides) can be inscribed in a circle or circumscribed about a circle. When inscribed, all the corner points will lie along the circle (Fig. 13.9). For example, a 10-sided polygon is constructed by dividing the circle into 10 sectors and connecting the points to form the polygon.

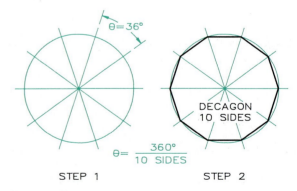

Figure 13.9 The regular polygon.

13.10   Hexagons

The **hexagon,** a 6-sided regular polygon, can be inscribed and circumscribed (Fig. 13.10). Hexagons are drawn with 30°–60° triangles either inside or outside the circles. Note that the circle represents the distance from corner to corner

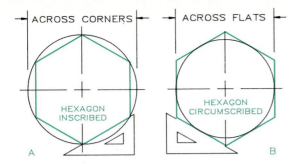

Figure 13.10 A circle can be (A) inscribed or (B) circumscribed to form a hexagon by using a 30°–60° triangle.

when inscribed, and from flat to flat when circumscribed.

13.11 Octagons

The **octagon,** an 8-sided regular polygon, can be inscribed in, or circumscribed about, a circle (Fig. 13.11) by using a 45° triangle. A second method inscribes the octagon inside a square (Fig. 13.12).

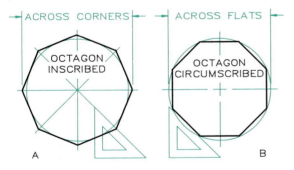

Figure 13.11 A circle can be (A) inscribed or (B) circumscribed to form an octagon with a 45° triangle.

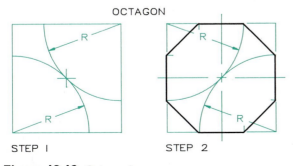

Figure 13.12 Octagon in a square.

13.12 Pentagons

The **pentagon,** a 5-sided regular polygon, can be inscribed in, or circumscribed about, a circle. Figure 13.13 shows another method of constructing a pentagon. This construction is performed with a compass and straightedge.

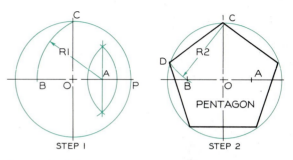

Figure 13.13 The pentagon.

Step 1 Bisect radius *OP* to locate point *A*. With *A* as the center and *AC* as the radius *R1*, locate point *B* on the diameter.

Step 2 With point *C* as the center and *BC* as the radius *R2*, locate point *D* on the arc. Line *CD* is the chord that can be used to locate the other corners of the pentagon.

Computer Method Polygons can be drawn using AutoCAD's POLYGON option under the DRAW command. One option allows you to give the number of sides, select the center, select the radius, and indicate whether the polygon is to be inscribed in, or circumscribed about, the circle (Fig. 13.14).

The second option allows you to select the endpoints of one edge of the polygon, and it will be drawn with the specified number of sides.

13.13 Bisecting Lines and Angles

Figure 13.15 shows two methods of finding the midpoint, or perpendicular bisector, of a line. The first method, which can be used to find the midpoint of an arc or a straight line, uses a compass to construct a perpendicular to a line. The second method uses a standard triangle.

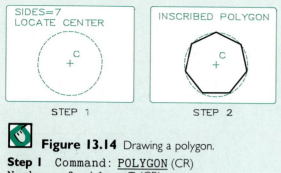

STEP 1   STEP 2

Figure 13.14 Drawing a polygon.

Step 1 Command: <u>POLYGON</u> (CR)
Number of sides: <u>7</u> (CR)
Edge‹Center of polygon› : (Select C.)
Inscribed in circle/Circumscribed
about circle (I/C): <u>I</u> (CR)

Step 2 Radius of circle: (Select radius with
cursor.) (Polygon is inscribed inside the imaginary circle.)

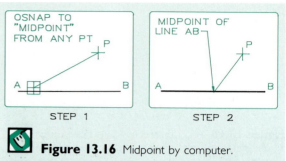

STEP 1   STEP 2

Figure 13.16 Midpoint by computer.

Step 1 Find the midpoint of AB in the following
manner:
Command: <u>LINE</u> (CR) From point: <u>P</u> (Locate P
anywhere.)
To point: <u>OSNAP</u> (Select **MIDPOINT** mode.)
(Select any point on line AB.)

Step 2 The line from point P will be drawn to the
midpoint of line AB.

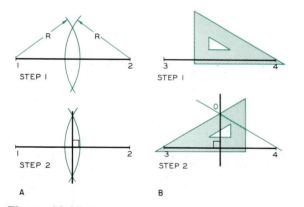

Figure 13.15 Bisecting a line. A line can be bisected by
using (A) a compass and any radius or (B) a standard
triangle and a straightedge.

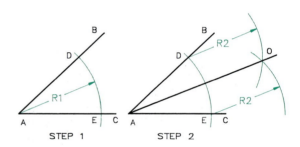

STEP 1   STEP 2

Figure 13.17 Bisecting an angle.

Step 1 Swing an arc of any radius to locate points D
and E.

Step 2 Draw two equal arcs from D and E to locate
point 0. Line A0 is the bisector of the angle.

Computer Method   The midpoint of a
line can be found by computer (Fig. 13.16) by using
the **MIDPOINT** mode of **OSNAP** and drawing a line
from any point, P, to the line. The line will automatically snap to the given line's midpoint. The angle in Fig. 13.17 can be bisected with a compass by
drawing three arcs.

Computer Method   A second computer
method is to draw an arc at any position between
the two given lines using their point of intersection
as the center (Fig. 13.18). Using the **DRAW** command and the **MIDPOINT** mode of **OSNAP**, a line is
drawn from any point to the arc, which will be the
arc's midpoint. The bisector can then be drawn
from vertex A to this midpoint.

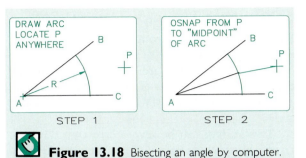

STEP 1   STEP 2

Figure 13.18 Bisecting an angle by computer.

Step 1 Using the `ARC` command, any radius, and center *A*, draw an arc that `SNAP`s to lines *AC* and *AB*.

Step 2 Command: `LINE` (CR) From point: (Select *P* anywhere.)
To point: `(OSNAP) Midpoint of` (Select arc.) (The line from *P* is drawn to the midpoint of the arc, which locates the bisector.)

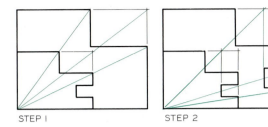

STEP 1   STEP 2

Figure 13.20 Enlargement of a figure.

Step 1 A proportional enlargement is made by using a series of diagonals drawn through a single point, the lower left-hand corner in this case.

Step 2 Additional diagonals are drawn to locate the other features of the object. This process can be reversed for reducing an object.

13.14 Revolution of Figures

Figure 13.19 demonstrates rotating a triangle about one of its points. Where the triangle is rotated about point 1 of line 1–4, point 4 is rotated to its desired position using a compass. Points 2 and 3 are found by triangulation by drawing arcs with radii 4–2 and 4–3 from 4′ to complete the rotated view.

13.15 Enlargement and Reduction of Figures

In Fig. 13.20, the small figure is enlarged by using a series of radial lines from the lower left-hand corner. The smaller figure is completed as a rectangle, and the larger rectangle is drawn proportional to the small one. The upper right-hand notch is located using the same technique. This method can also be used to reduce a larger drawing.

13.16 Division of Lines

It is often necessary to divide a line into several equal parts when a convenient scale is not available for this purpose. For example, suppose a 6-inch line must be divided into seven equal parts. The method shown in Fig. 13.21 is an efficient way to solve this problem.

An application of this principle is used for locating lines on a graph that are equally spaced (Fig. 13.22). A scale with the desired number of units (0 to 5) is laid across from left to right on the graph. Vertical lines are drawn through these points to divide the graph into five equal divisions.

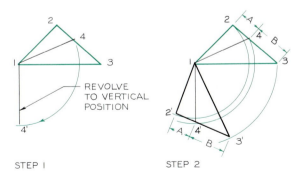

STEP 1   STEP 2

Figure 13.19 Rotation of a figure.

Step 1 A plane figure can be rotated about any point. Line 1–4 is rotated about point 1 to its desired position with a compass.

Step 2 Points 2′ and 3′ are located by measuring distances *A* and *B* from 4.

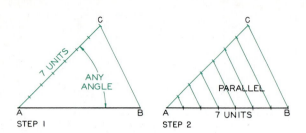

Figure 13.21 Division of line.

Step 1 Line *AB* is divided into seven equal divisions by constructing a line through *A* and dividing it into seven known units with your dividers. Point *C* is connected to point *B*.

Step 2 A series of lines are drawn parallel to *CB* to locate the divisions along line *AB*.

13.17   Arcs Through Three Points

An arc can be drawn through any three points by connecting the points with two lines (Fig. 13.23). Perpendicular bisectors are found for each line to locate the center at point *C*. The radius is drawn, and the lines *AB* and *BD* become chords of the circle.

This system can be reversed to find the center of a given circle or arc. Draw two chords that intersect at a point on the circumference and bisect them. The perpendicular bisectors will intersect at the center of the circle.

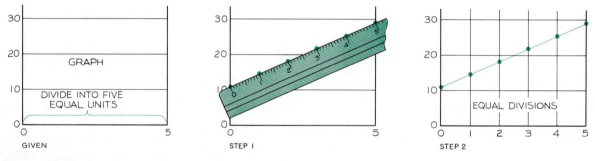

Figure 13.22 Division of a space.

Step 1 To divide a graph into five equal divisions along the *x* axis, a scale with five units of measurement that approximate the horizontal distance is laid across the graph. Align the 0 and 5 markings with the lines.

Step 2 Construct vertical lines through the points found in Step 1. This method can also be used to calibrate the divisions along the *y* axis.

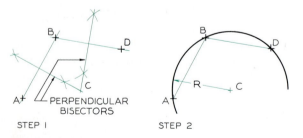

Figure 13.23 Arc through three points.

Step 1 Connect the points with two lines and find their perpendicular bisectors. The bisectors will intersect at the center, *C*.

Step 2 Using the center, *C*, and the distance to the points as the radius, construct the arc through the points.

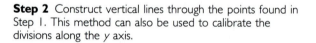   **Computer Method** Using the ARC command and the 3-point option, a counterclockwise arc can be drawn through any three points selected on the screen (Fig. 13.24).

13.18   Parallel Lines

A line can be drawn parallel to another by using either of the methods shown in Fig. 13.25. The first method uses a compass to draw two arcs to

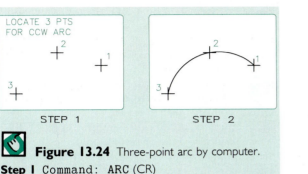

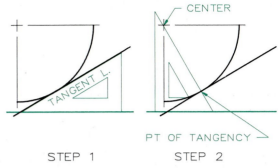

Figure 13.24 Three-point arc by computer.

Step 1 Command: ARC (CR)
Center/‹Start point›: (Locate pt. 1.)
Center/End ‹Second point›: (Locate pt. 2.)

Step 2 Endpoint: (Locate pt. 3.)
Arcs are drawn counterclockwise.

Figure 13.26 Locating a tangent point.

Step 1 Align your triangle with the tangent line while holding it firmly against a straightedge.

Step 2 Hold the straightedge in position, rotate the triangle, and construct a line through the center that is perpendicular to the line.

locate a parallel line that is the desired distance away. The second method requires constructing a perpendicular from a given line and measuring the distance, *R,* to locate the parallel, which is drawn with a straightedge.

13.19 Points of Tangency

A point of tangency is the theoretical point where a straight line joins an arc or where two arcs join. In Fig. 13.26, a line is tangent to an arc. The point of tangency is located by constructing a thin perpendicular to the line from the center of the arc. Figure 13.27 shows the conventional methods of marking points of tangency.

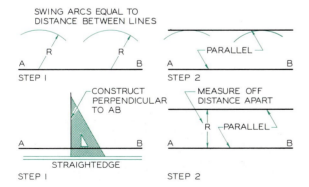

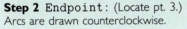

Figure 13.25 Drawing parallel lines.

Compass Method: Step 1 Draw 2 arcs from the line.

Step 2 Draw the parallel line tangent to the arcs.

Triangle method: Step 1 Draw a perpendicular to the line.

Step 2 Measure the distance (*R*) along the perpendicular and draw the parallel line with your drafting machine.

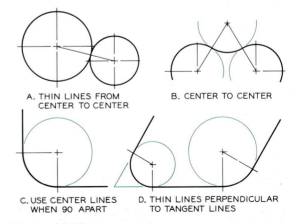

Figure 13.27 Thin lines that extend beyond the curves from the centers are used to mark points of tangency. These lines should always be shown in this manner.

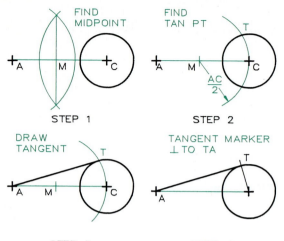

Figure 13.28 Line from a point tangent to an arc.

Step 1 Connect point A with center C. Locate point M by bisecting AC.

Step 2 Using point M as the center and MC as the radius, locate point T on the arc.

Step 3 Draw the line from A to T that is tangent to the arc of point T.

Step 4 Draw tangent marker from the center past the arc.

13.20 Line Tangent to an Arc

Figure 13.28 shows the method of finding the exact point of tangency between a line and an arc. Point A and the arc are given. Point A is connected to the center in Step 1, AC is bisected in Step 2, the tangent is drawn in Step 3, and T is located in Step 4. The point of tangency could also have been found by using a standard triangle, as Fig. 13.29 shows.

Computer Method A line can be drawn from a point tangent to an arc by using the TANGENT option of the OSNAP command (Fig. 13.30). When prompted for the second point on the line, the snap target is placed over the arc, and the tangent is drawn.

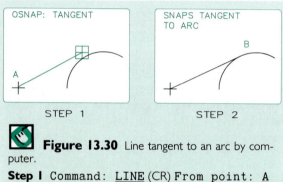

Figure 13.30 Line tangent to an arc by computer.

Step 1 Command: **LINE** (CR) From point: **A** To point: (OSNAP—Tangent mode) Tangent to:

Step 2 Select point on the arc, and line AB is drawn tangent to the arc.

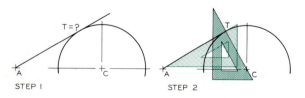

Figure 13.29 Line from a point tangent to an arc.

Step 1 A line can be drawn from point A tangent to the arc by eye.

Step 2 By rotating your triangle, the point of tangency can be located at the 90° angle with the line that passes through the center.

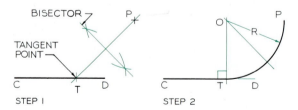

Figure 13.31 Arc through two points.

Step 1 If an arc must be tangent to a given line at T and pass through P, find the perpendicular bisector of the line TP.

Step 2 Construct a perpendicular to the line at T to intersect the bisector. The arc is drawn from center C with radius OT.

13.21 Arc Tangent to a Line From a Point

If an arc is to be constructed tangent to line *CD* at *T* (Fig. 13.31) and pass through point *P*, a perpendicular bisector of *TP* is drawn. A perpendicular to *CD* is drawn at *T* to locate the center at *C*. A similar problem in Fig. 13.32 requires you to draw an arc of a given radius that will be tangent to line *AB* and pass through point *P*. In this case, the point of tangency on the line is not known until the problem has been solved.

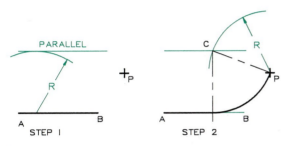

Figure 13.32 Arc tangent to a line and a point.

Step 1 When an arc of a given radius is to be drawn tangent to a line and through a point *P*, draw a line parallel to *AB* and *R* from it.

Step 2 Draw an arc from *P* with radius *R* to locate the center at *C*. The arc is drawn with radius *R* and center *C*.

Computer Method An arc can be drawn tangent to a line from the end of a line by selecting the ARC option and CONTINUE immediately after the line has been drawn (Fig. 13.33). The arc will begin at the end of the line, and the other end can be DRAGged to the location desired.

13.22 Arc Tangent to Two Lines

An arc of a given radius can be constructed tangent to two nonparallel lines. This method is shown in Fig. 13.34, where two lines form an acute angle. The same steps are used to find an arc tangent to two lines that form an obtuse angle

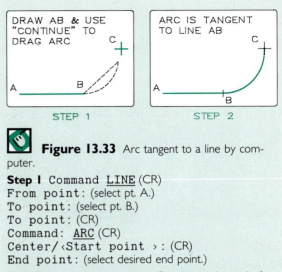

Figure 13.33 Arc tangent to a line by computer.

Step 1 Command <u>LINE</u> (CR)
From point: (select pt. A.)
To point: (select pt. B.)
To point: (CR)
Command: <u>ARC</u> (CR)
Center/‹Start point ›: (CR)
End point: (select desired end point.)

Step 2 Move cursor to point *C* to locate the end of the curve that is tangent to *AB* at *B*.

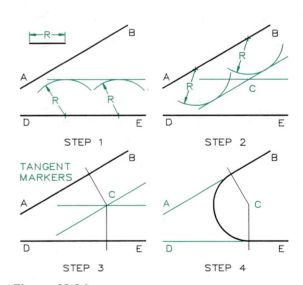

Figure 13.34 Arc tangent to two lines—acute angle.

Step 1 Construct a line parallel to *DE* with the radius of the specified arc *R*.

Step 2 Draw a second construction line parallel to *AB* to locate the center *C*.

Step 3 Thin lines are drawn from *C* perpendicular to *AB* and *DE* to locate the points of tangency. The tangent arc is drawn using the center *C*.

Step 4 Draw the tangent arc and darken your lines.

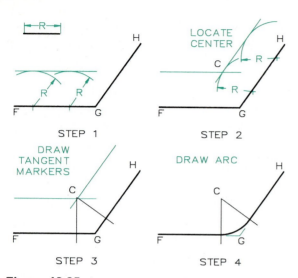

Figure 13.35 Arc tangent to two lines—obtuse angle.

Step 1 Using the specified radius R, construct a line parallel to FG.

Step 2 Construct a line parallel to GH that is distance R from it to locate center C.

Step 3 Construct thin lines from center C perpendicular to lines FG and GH to locate the points of tangency. Draw the arc using radius R and center C.

Step 4 Draw the tangent arc and darken your lines.

(Fig. 13.35). In both cases, the points of tangency are located with thin lines drawn from the centers through the points of tangency.

A different technique can be used to find an arc of a given radius that is tangent to perpendicular lines (Fig. 13.36). This method will work only for perpendicular lines.

Computer Method An arc can be drawn tangent to two nonparallel lines by using the FILLET command (Fig. 13.37). You may assign the radius length and select a point on each line; the arc will be drawn, and the lines will be automatically trimmed. If tangent points need to be shown, locate the center of the arc by using the CENTER option of the DIM: command. Lines can be 0–SNAPped perpendicular to AB and CD from the center, C.

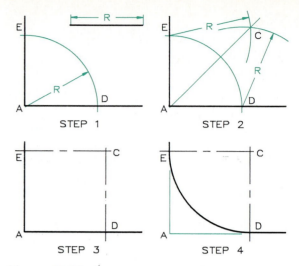

Figure 13.36 Arc tangent to perpendicular lines.

Step 1 Using the specified radius, R, locate points D and E by using center A.

Step 2 To locate point C, swing two arcs using the radius, R, that was used in Step 1.

Step 3 Locate the tangent points with lines from C.

Step 4 Draw the tangent arc and darken your lines.

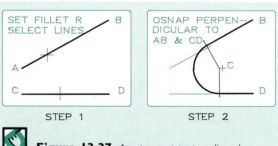

Figure 13.37 Arc tangent to two lines by computer.

Step 1 Command: FILLET (CR)
Polyline/Radius/‹Select two objects›: R (CR) (To define radius.)
Enter fillet radius ‹0.0000›: .75 (CR)
Command: (CR)
FILLET Polyline/Radius/‹Select two objects›:(Select Pts. on AB and CD.)

Step 2 Command: LINE (CR) From point: OSNAP—CENTER (CR) center of (Select point on the arc.) To point: PERPEND (CR) perpend to (Select point on AB, and a perpendicular is drawn to the point of tangency. Extend the line beyond AB. Locate the tangent point on CD in the same manner.)

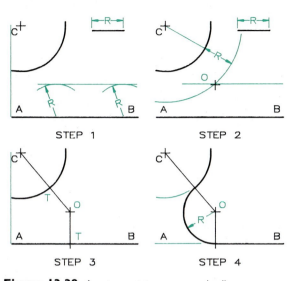

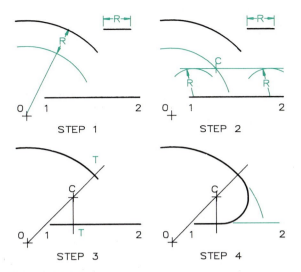

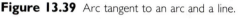

Figure 13.38 Arc tangent to an arc and a line.

Step 1 Draw a line parallel to *AB* that is *R* from it. Use thin construction lines.

Step 2 Add radius *R* to the extended radius from center *C*. Use this extended radius to locate center *O*.

Step 3 Lines *OC* and *OT* are drawn to locate the tangency points.

Step 4 Draw the tangent arc between the points of tangency with radius *R* and center *O*.

13.23 Arc Tangent to an Arc and a Line

Figure 13.38 shows the steps for constructing an arc tangent to an arc and a line. A variation of this principle of construction is shown in Fig. 13.39, where the arc is drawn parallel to an arc and line with the arc in a reverse position.

13.24 Arc Tangent to Two Arcs

A third arc is drawn tangent to two given arcs in Fig. 13.40. Thin lines are drawn from the centers to locate the points of tangency. This tangent arc is concave from the top. A convex arc can be drawn tangent to the given arcs if the radius of the arc is greater than the radius of either of the given arcs (Fig. 13.41).

Figure 13.39 Arc tangent to an arc and a line.

Step 1 Radius *R* is subtracted from the radius through center *O*. Draw a concentric arc with this shortened radius.

Step 2 A line parallel to 1–2 is drawn a distance of *R* from it to locate center *C*.

Step 3 The tangent points are located with lines from *O* through *C*, and through *C* perpendicular to 1–2.

Step 4 Draw the tangent arc between the tangent points with radius *R* and center *C*.

A variation of this problem is shown in Fig. 13.42, where an arc of a given radius is drawn tangent to the top of one arc and the bottom of the other. A similar problem is shown in Fig. 13.43, where an arc is drawn tangent to a circle and a larger arc.

Computer Method An arc is automatically drawn tangent to two arcs by using the FILLET command (Fig. 13.44). After entering the FILLET command, give the desired radius when prompted. You will be returned to the COMMAND mode. Press the carriage return (CR) to return to the FILLET command.

When prompted, select the two arcs with your cursor; the tangent arc will be drawn and the given arcs trimmed at the points of tangency. Locate the tangent points by drawing lines from the center to *C1* and *C2*.

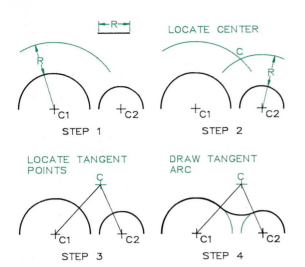

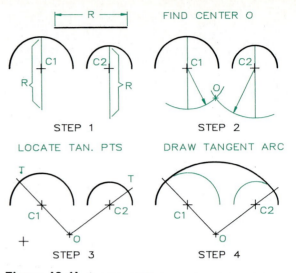

Figure 13.40 Arc tangent to two arcs—concave arc.

Step 1 The radius of one circle is extended, and the radius *R* is added to it. Use the extended radius to draw a concentric arc.

Step 2 Extend the radius of the other circle and add radius *R* to it. Use the extended radius to construct an arc to locate center *C*.

Step 3 Connect center *C* with centers *C1* and *C2* with thin black lines to locate points of tangency.

Step 4 Using the given radius, *R*, draw the tangent arc with center *C* between the two points of tangency.

Figure 13.41 Arc tangent to two arcs—convex arc.

Step 1 The radius of each arc is extended from the arc past its center, and the specified radius *R* is laid off from the arcs along these lines.

Step 2 The distance from each center to the ends of the extended radii are used for drawing two arcs to locate the center *O*.

Step 3 Thin lines from *O* through centers C_1 and C_2 locate the points of tangency.

Step 4 The tangent arc is drawn between the tangent points with radius *R* and center *O*.

13.25 Ogee Curves

The **ogee curve** is a double curve formed by tangent arcs. By constructing two arcs tangent to three intersecting lines, the ogee curve in Fig. 13.45 was found.

An ogee curve can be drawn between two parallel lines (Fig. 13.46) from points *B* to *C* by geometric construction. Figure 13.47 shows an alternative method of drawing an ogee curve that passes through points *B, E,* and *C*.

13.26 Rectifying Arcs

An arc is rectified when its true length is laid out along a straight line. Figure 13.48 shows a method of rectifying an arc. Another method of rectifying an arc uses the mathematical equation

for finding the circumference of the circle. Since a circle has 360°, the arc of a 30° sector is one twelfth of the full circumference (360 ÷ 30 = 12). Therefore, if the circumference is 12 inches, the 30° arc equals 1 inch.

13.27 Conic Sections

Conic sections are plane figures that can be described graphically as well as mathematically. They are formed by passing imaginary cutting planes through a right cone (Fig. 13.49).

13.28 Ellipses

The **ellipse** is a conic section formed by passing a plane through a right cone at an angle (Fig. 13.49). The ellipse is mathematically defined as

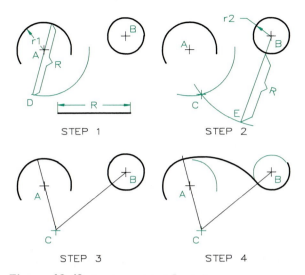

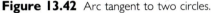

STEP 1   STEP 2

STEP 3   STEP 4

Figure 13.42 Arc tangent to two circles.

Step 1 Radius R is laid off from the arc along the extended radius to locate point D. Radius AD is used to draw a concentric arc.

Step 2 The radius through center B is extended, and radius R is added to it from point F. Radius BE is used to locate the center C.

Step 3 Thin lines from C through centers A and B locate points of tangency.

Step 4 The tangent arc is drawn between the tangent points with radius R and center C.

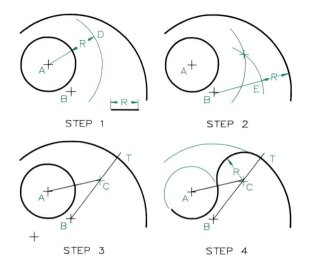

STEP 1   STEP 2

STEP 3   STEP 4

Figure 13.43 Arc tangent to two arcs.

Step 1 Radius R is added to the radius from center A. Radius AD is used to draw a concentric arc with center A.

Step 2 Radius R is subtracted from the radius through B. Radius BE is used to draw an arc to locate center C.

Step 3 Thin lines are drawn to connect the centers and locate the points of tangency.

Step 4 The tangent arc is drawn between the tangent points with radius R and center C.

the path of a point that moves in such a way that the sum of the distances from two focal points is a constant. The largest diameter of an ellipse is always the true length and is called the **major diameter.** The shortest diameter is perpendicular to the major diameter and is called the **minor diameter.**

The construction of an ellipse is found by revolving the edge view of a circle, as Fig. 13.50 shows. This ellipse could have been drawn using the ellipse template shown in Fig. 13.51. The angle between the line of sight and the edge view of the circle (or the one closest to this size) is the angle of the ellipse template that should be used. Ellipse templates are available in intervals of 5° and in variations in size of the major diameter of about $\frac{1}{8}$ inch (Fig. 13.52).

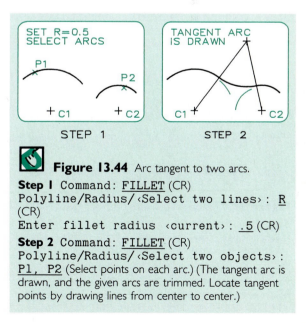

SET R=0.5
SELECT ARCS

P1

P2

+ C1   + C2

STEP 1

TANGENT ARC
IS DRAWN

C1 +   + C2

STEP 2

Figure 13.44 Arc tangent to two arcs.

Step 1 Command: FILLET (CR)
Polyline/Radius/‹Select two lines›: R (CR)
Enter fillet radius ‹current›: .5 (CR)

Step 2 Command: FILLET (CR)
Polyline/Radius/‹Select two objects›: P1, P2 (Select points on each arc.) (The tangent arc is drawn, and the given arcs are trimmed. Locate tangent points by drawing lines from center to center.)

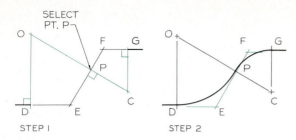

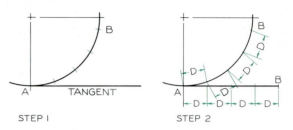

Figure 13.45 An ogee curve.

Step 1 To draw an ogee curve between two parallel lines, draw line *EF* at any angle. Locate a point of your choosing along *EF*, *P* in this case. Find the tangent points by making *FG* equal to *FP* and *DE* equal to *EP*. Draw perpendiculars at *G* and *D* to intersect the perpendicular at *O* and *C*.

Step 2 Using radii *CP* and *OP*, at centers *O* and *C*, draw the two tangent arcs to complete the ogee curve.

Figure 13.48 To rectify an arc.

Step 1 Construct a line tangent to the arc, and divide the arc into a series of equal divisions from *A* to *B*.

Step 2 The chordal distances, *D*, along the arc are laid out along the straight line until point *B* is located.

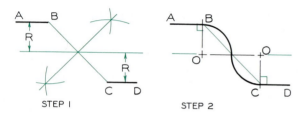

Figure 13.46 An ogee curve.

Step 1 To draw an ogee curve formed by two equal arcs passing through points *B* and *C*, draw a line between the points. Bisect the line *BC*, and draw a line parallel to *AB* and *CD* to find the radius, *R*.

Step 2 Construct perpendiculars at *B* and *C* to locate the centers at both points *O*. Draw the arcs to complete the ogee curve.

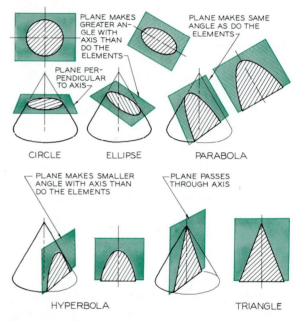

Figure 13.49 The conic sections are formed by passing cutting planes at various angles through right cones. The conic sections are the circle, ellipse, parabola, hyperbola, and triangle.

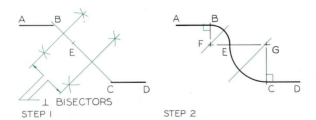
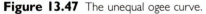

Figure 13.47 The unequal ogee curve.

Step 1 Two parallel lines are to be connected by an ogee curve that passes through *B* and *C*. Draw line *BC*, and select point *E* on the line. Bisect *BE* and *EC*.

Step 2 Construct perpendiculars at *B* and *C* to intersect the bisectors to locate centers *F* and *G*. Locate the points of tangency, and draw the ogee curve using radii *FB* and *GC*.

The ellipse can also be constructed inside a rectangle or parallelogram (Fig. 13.53), where a series of points is plotted to form an elliptical curve. Two circles can be used for constructing an ellipse by making the diameter of the large circle equal to the major diameter and the diameter of the small circle equal to the minor diameter (Fig. 13.54).

164

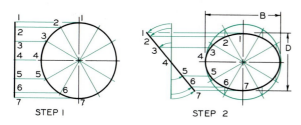

STEP 1   STEP 2

Figure 13.50 An ellipse by revolution.

Step 1 When the edge view of a circle is perpendicular to the projectors between its adjacent view, the view will be a true circle. Mark equally spaced points along the arc, and project them to the edge.

Step 2 Revolve the edge of the circle to the desired position, and project the points to the circular view, which will now appear as an ellipse. The points are projected vertically downward to new positions.

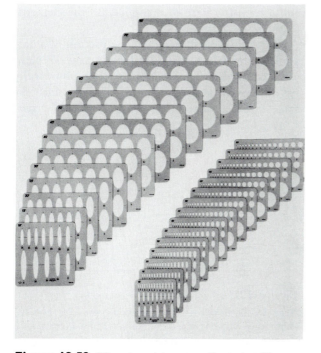

Figure 13.52 Ellipse templates are calibrated at 5° intervals from 15° to 60°. (Courtesy of Timely Products, Inc.)

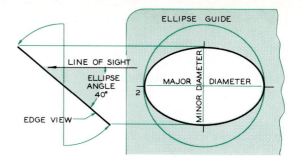

Figure 13.51 The ellipse template. When the edge view of a circle is revolved so that the line of sight between the two views is not perpendicular to the edge view, the circle will appear as an ellipse. The major diameter remains constant, but the minor diameter will vary. The angle between the line of sight and the edge view of the circle is the angle of the ellipse template.

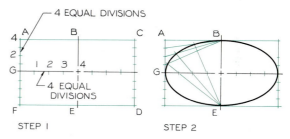

STEP 1   STEP 2

Figure 13.53 Ellipse—parallelogram method.

Step 1 An ellipse can be drawn inside a rectangle or parallelogram by dividing the horizontal centerline into the same number of equal divisions as the shorter sides, *AF* and *CD*.

Step 2 The construction of the curve in one quadrant is shown by using sets of rays from *E* and *B* to plot the points.

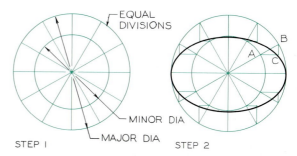

STEP 1   STEP 2

Figure 13.54 Ellipse—circle method.

Step 1 Two concentric circles are drawn with the large one equal to the major diameter and the small one equal to the minor diameter. Divide them into equal sectors.

Step 2 Plot points on the ellipse by projecting downward from the large curve to intersect horizontal construction lines drawn from the intersections on the small circle.

165

Computer Method Using the EL-LIPSE command, you can draw an ellipse by selecting the endpoints of the major diameter. The third point defines the length of minor radius; the ellipse is then drawn (Fig. 13.55).

Another option of the ELLIPSE command allows you to select the center, the end of the major radius, and the end of the minor radius; the ellipse is then drawn. Isometric ellipses can also be drawn in isometric pictorial drawings.

The mathematical equation of an ellipse is

$$\frac{x^2}{a^2} + \frac{y^2}{b^2} = 1, \quad \text{where } a, b \neq 0.$$

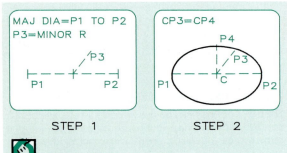

STEP 1   STEP 2

Figure 13.55 Ellipse—computer method.

Step 1 Command: ELLIPSE (CR)
‹Axis endpoint 1›/Center: (Select P1.)
Axis endpoint 2: (Select P2.)

Step 2 ‹Other axis distance›/Rotation:
(Select minor radius, P3, in any direction.)
(The ellipse is drawn. The option, Rotation, could have been used to specify the orientation of the major diameter.)

13.29 Parabolas

The **parabola** is mathematically defined as a plane curve, each point of which is equidistant from a straight line (called a **directrix**) and its focal point. The parabola is a conic section formed when the cutting plane makes the same

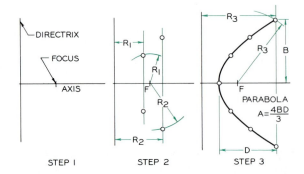

Figure 13.56 Parabola—mathematical method.

Step 1 Draw an axis perpendicular to a line (a directrix). Choose a point for the focus, F.

Step 2 Locate points by using a series of selected radii to plot points on the curve. For example, draw a line parallel to the directrix and R_2 from it. Swing R_2 from F to intersect the line and plot the point.

Step 3 Continue the process with a series of arcs of varying radii until an adequate number of points have been found to complete the curve.

angle with the base of a cone as do the elements of the cone.

Figure 13.56 shows the construction of a parabola using its mathematical definition. A parabolic curve can also be constructed geometrically by dividing the two perpendicular lines into the same number of divisions (Fig. 13.57). The par-

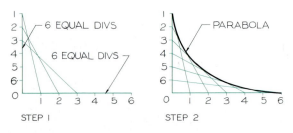

Figure 13.57 Parabola—tangent method.

Step 1 Construct two lines at a convenient angle, and divide each of them into the same number of divisions. Connect the points with a series of diagonals.

Step 2 When finished, construct the parabolic curve to be tangent to the diagonals.

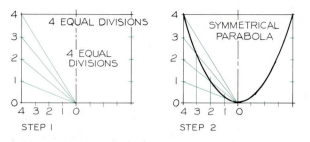

Figure 13.58 Parabola—parallelogram method.

Step 1 Construct a rectangle or parallelogram to contain the parabola, and locate its axis parallel to the sides through *O*. Divide the sides into equal divisions. Connect the divisions with point *O*.

Step 2 Construct lines parallel to the sides (vertical in this case) to locate the points along the rays from *O*. Draw the parabola.

abola is drawn through the plotted points with an irregular curve. Figure 13.58 shows a third method of construction using parallelograms.

The mathematical equation of the parabola is

$$y = ax^2 + bx + c, \quad \text{where } a \neq 0.$$

13.30 Hyperbolas

The **hyperbola** is a two-part conic section. Mathematically, it is defined as the path of a point that moves in such a way that the difference of its distances from two focal points is a constant. Figure 13.59 shows the construction of a hyperbola using this definition.

Figure 13.60 shows a second method of construction. Two perpendicular lines are drawn through point *B* as asymptotes. The hyperbolic curve becomes more nearly parallel and closer to the asymptotes as the hyperbola is extended, but the curve never merges with the asymptotes.

13.31 Spirals

The **spiral** is a coil that begins at a point and becomes larger as it travels around the origin. A spiral lies in a single plane. Figure 13.61 shows the steps for constructing a spiral.

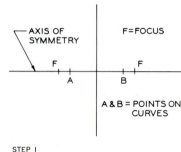

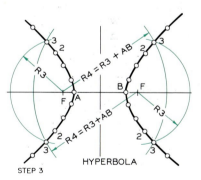

Figure 13.59 The hyperbola.

Step 1 A perpendicular is drawn through the axis of symmetry, and focal points *F* are located equidistant from it on both sides. Points on the curve, *A* and *B*, are located equidistant from the perpendicular at a location of your choice but between the focal points.

Step 2 Radius *R*1 is selected to draw arcs using focal points *F* as the centers. *R*1 is added to *AB* (the distance between the nearest points on the hyperbolas) to find *R*2. Radius *R*2 is used to draw arcs using the focal points as centers. The intersections of *R*1 and *R*2 establish points 2 on the hyperbola.

Step 3 Other radii are selected and added to distance *AB* to locate additional points in the same manner as described in Step 2. A smooth curve is drawn through the points to form the hyperbolic curves.

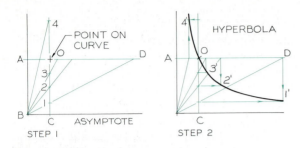

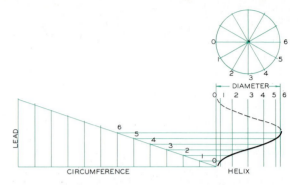

Figure 13.60 The equilateral hyperbola.

Step 1 Two perpendiculars are drawn through *B,* and any point *O* on the curve is located. Horizontal and vertical lines are drawn through *O.* Line *CO* is divided into equal divisions, and rays from *B* are drawn through them to the horizontal line.

Step 2 Horizontal construction lines are drawn from the divisions along line *OC,* and lines from *AD* are projected vertically to locate points 1' through 4' on the curve.

Figure 13.62 The helix. Divide the top view of the cylinder into equal divisions, and project them to the front view. Lay out the circumference and the height of the cylinder, which is the **lead.** Divide the circumference into the same number of equal parts by taking the measurements from the top view. Project the points along the inclined rise to their respective elements to find the helix.

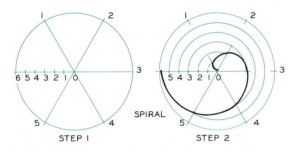

Figure 13.61 The spiral.

Step 1 Draw a circle and divide it into equal parts. The radius is divided into the same numer of equal parts (six in this example).

Step 2 By beginning on the inside, draw arc 0–1 to intersect radius 0–1. Then swing arc 0–2 to radius 0–2, and continue until the last point is reached at 6, which lies on the original circle.

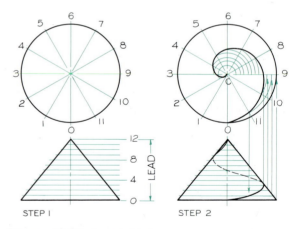

Figure 13.63 A conical helix.

Step 1 Divide the cone's base into equal parts. Pass a series of horizontal cutting planes through the front view of the cone. Use the same number as the divisions on the base (12 in this case).

Step 2 Project all the divisions along the front view of the cone to line *C* 9, and draw a series of arcs from center *C* to their respective radii in the top view to plot the points. Project the points to their respective cutting planes in the front view.

13.32 Helixes

The **helix** is a curve that coils around a cylinder or cone at a constant angle of inclination. Examples of helixes are corkscrews or threads on a screw. Figure 13.62 shows a helix constructed about a cylinder, and Fig. 13.63 shows a helix constructed about a cone.

Problems

These problems are to be solved on Size AV paper similar to the one shown in Fig. 13.64, where Problems 1A–1E are laid out. Each inch on the grid is equal to 0.20 inches; therefore, use your engineers' 10 scale to lay out the problems. By equating each grid to 5 mm, you can use your full-size metric scale to lay out and solve the problems.

Show your construction and mark all points of tangency, as discussed in the chapter.

1. (Fig. 13.64) a) Draw triangle ABC using the given sides.

b–c) Inscribe an angle in the semicircles with the vertexes at point P.

d) Inscribe a three-sided regular polygon inside the circle.

e) Circumscribe a four-sided regular polygon about the circle.

2. (Fig. 13.65) a) Circumscribe a hexagon about the circle.

b) Inscribe a hexagon in the circle.

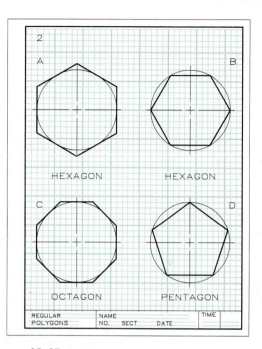

Figure 13.65 Problems 2A–2D. Construction of regular polygons.

c) Circumscribe an octagon about the circle.

d) Construct a pentagon inside the circle using the compass method.

3. (Fig. 13.66) a) Bisect the lines.

b) Bisect the angles.

c) Rotate the triangle 60° in a clockwise direction about point A.

d) Enlarge the given shape to the size indicated by the diagonal.

4. (Fig. 13.67) a) Divide AB into seven equal parts. Draw the construction line through A for your construction.

b) Divide the space between the two vertical lines into four equal divisions. Draw three vertical lines at these divisions that are equal in length to the given lines.

c) Construct an arc with radius R that is tangent to the line at J and passes through point P.

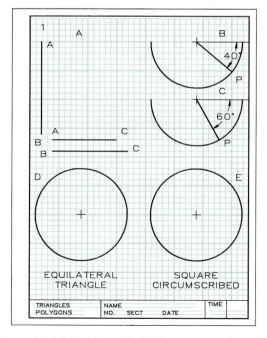

Figure 13.64 Problems 1A–1E. Basic constructions.

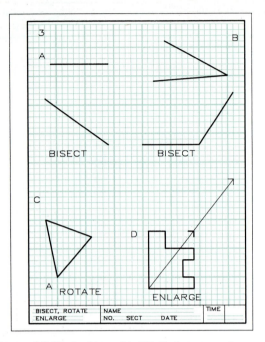

Figure 13.66 Problems 3A–3D. Basic constructions.

d) Construct an arc with radius R that is tangent to the line and passes through P.

5. (Fig. 13.68) a) Construct a line from P that is tangent to the semicircle. Locate the points of tangency. Use the compass method.

b–d) Construct arcs with the given radii tangent to the lines.

6. (Fig. 13.69) a–d) Construct arcs that are tangent to the arcs or lines. The radii are given for each problem.

7. (Fig. 13.70) a–d) Construct ogee curves that connect the ends of the given lines and pass through points P where given.

8. (Fig. 13.71) a–b) Using the given radii, connect the given arcs with a tangent arc as indicated in the sketches.

9. (Fig. 13.72) a–b) Rectify the arc along the given line by dividing the circumference into equal divisions and laying them off with your dividers.

c) Construct an ellipse inside the rectangular layout.

d) Construct an ellipse inside the large circle. The small circle represents the minor diameter.

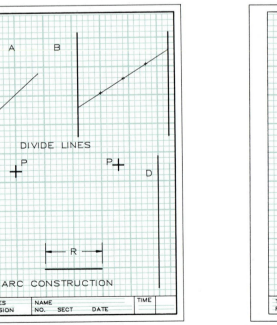

Figure 13.67 Problems 4A–4D. Tangency construction.

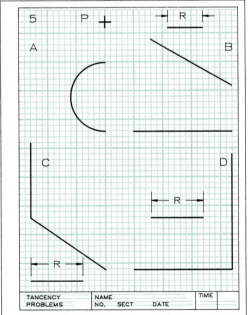

Figure 13.68 Problems 5A–5D. Tangency construction.

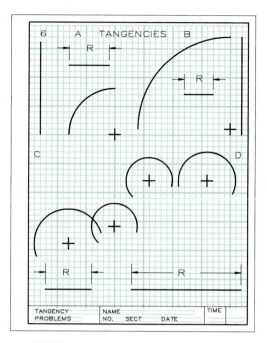

Figure 13.69 Problems 6A–6D. Tangency construction.

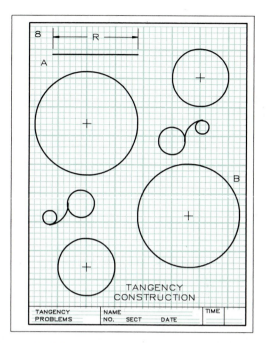

Figure 13.71 Problems 8A–8B. Tangency construction.

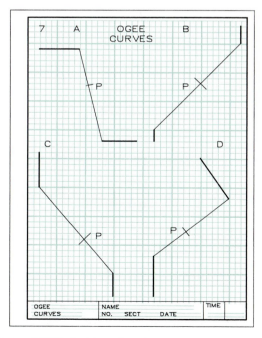

Figure 13.70 Problems 7A–7D. Ogee curve construction.

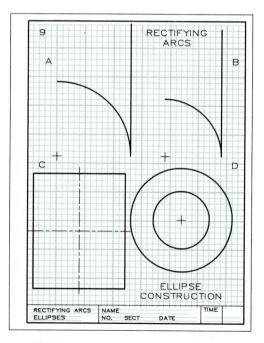

Figure 13.72 Problems 9A–9D. Rectifying an arc, ellipse construction.

10. (Fig. 13.73) a) Construct an ellipse inside the circle when the edge view has been rotated 45° as shown.

b) Using the focal point *F* and the directrix, plot and draw the parabola formed by these elements.

11. (Fig. 13.74) a) Using the focal point *F,* points *A* and *B* on the curve, and the axis of symmetry, construct the hyperbolic curve.

b) Construct a hyperbola that passes through 0. The perpendicular lines are asymptotes.

c) Construct a spiral by using the four divisions marked along the radius.

12. (Fig. 13.75) a–b) Construct helixes that have a rise equal to the heights of the cylinder and cone. Show construction and the curve in all views.

13–21. (Figs. 13.76–13.84) Construct these problems on Size A sheets, one problem per sheet. Select the proper scale that will best fit the problem to the sheet. Mark all points of tangency and strive for good line quality.

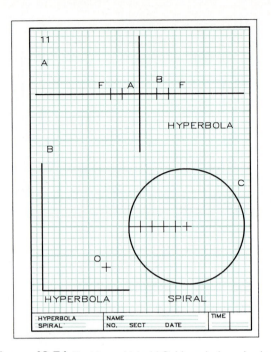

Figure 13.74 Problems 11A–11C. Hyperbola and spiral construction.

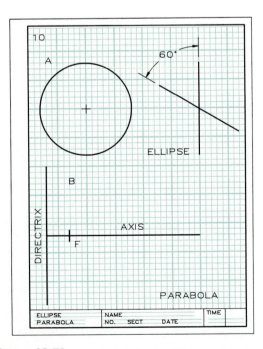

Figure 13.73 Problems 10A–10B. Ellipse and parabola construction.

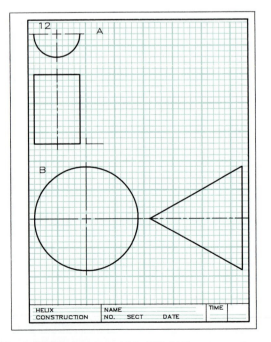

Figure 13.75 Problems 12A–12B. Helix construction.

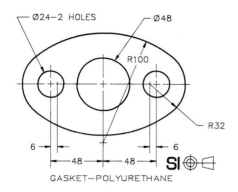

Figure 13.76 Problem 13. Gasket.

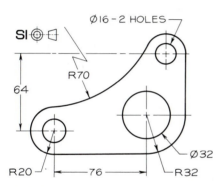

Figure 13.77 Problem 14. Lever crank.

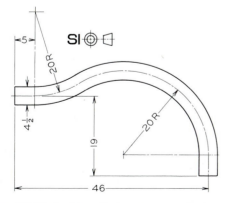

Figure 13.78 Problem 15. Road tangency.

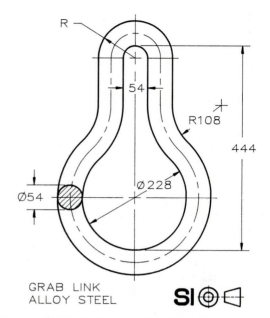

Figure 13.79 Problem 16. Grab link.

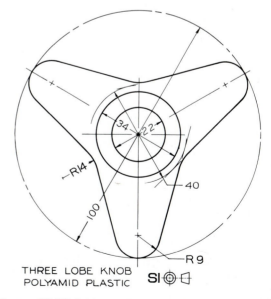

Figure 13.80 Problem 17. Three-lobe knob.

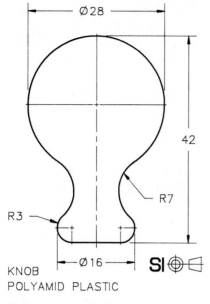

Ø28

42

R7

R3

Ø16

KNOB
POLYAMID PLASTIC

SI

Figure 13.81 Problem 18. Knob.

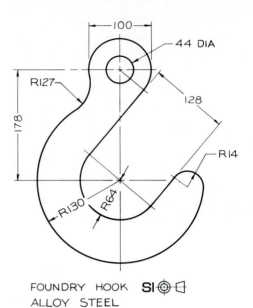

100

44 DIA

R127

178

128

R14

R130 R64

FOUNDRY HOOK SI
ALLOY STEEL

Figure 13.82 Problem 19. Foundry hook.

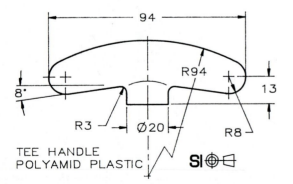

94

R94

8°

13

R3 Ø20 R8

TEE HANDLE
POLYAMID PLASTIC SI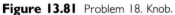

Figure 13.83 Problem 20. Tee handle.

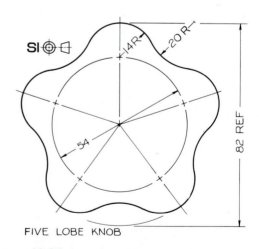

SI

14R 20 R

54

82 REF

FIVE LOBE KNOB

Figure 13.84 Problem 21. Five-lobe knob.

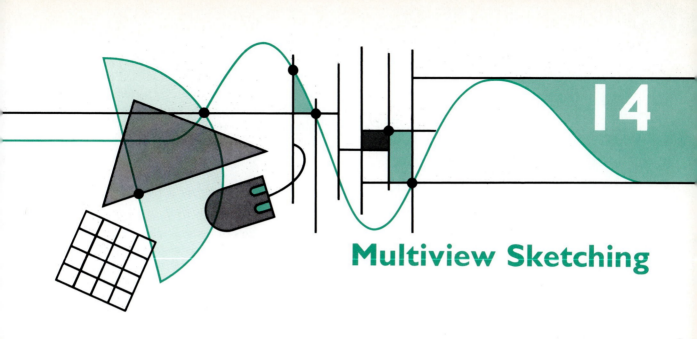

Multiview Sketching

14.1 The Purpose of Sketching

Sketching is as much a thinking process as it is a communication technique. Designers develop their ideas by making many sketches and revising them before finally arriving at the desired solution.

Sketching should be a rapid method of drawing. Designers who develop their sketching skills can assign drafting work to assistants who can then prepare the finished drawings from their sketches. If sketches are not sufficiently clear to communicate the ideas to someone else, the designer likely has not thought out the solution well enough. The ability to communicate by any means is a great asset, and sketching is one of the more powerful techniques.

14.2 Shape Description

Figure 14.1 shows a pictorial of an object with the top, front, and right-side views of the object. Each view is two-dimensional.

This system is called **orthographic,** or **multiview projection.** In multiview projection, it is important that the views be located as shown in

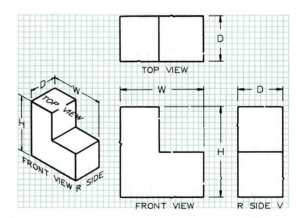

Figure 14.1 The three views—top, front, and right side—describe the object by multiview projection.

Fig. 14.1. The top view is placed over the front view since both views share the dimension of width. The side view is placed to the right of the front view where these views share the dimension of height. The distance between the views can vary, but the views must be positioned so that they project from each other as shown here.

Figure 14.2 shows several examples of poorly arranged views. Although these individual views are correct, they are hard to interpret because the

175

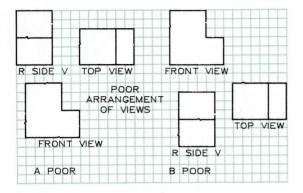

Figure 14.2 Poor arrangement of views. (A) These views are sketched incorrectly; the views are scrambled. (B) These views are incorrectly positioned also.

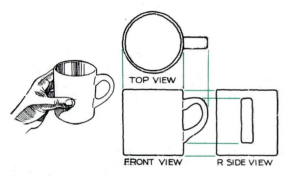

Figure 14.3 Three orthographic views of a cup.

views are not placed in their standard positions.

Figure 14.3 shows three views of a coffee cup using the principles of multiview projection. To simplify the views, hidden lines have been omitted.

14.3 Six-View Drawings

Six principal views may be found for any object by using the rules of orthographic, or multiview, projection. The directions of sight for the six orthographic views are shown in Fig. 14.4, where the views are drawn in their standard positions. The width dimension is common to the top, front, and bottom views. Height is common to the right-side, front, left-side, and rear views.

Seldom will an object be so complex as to require six orthographic views, but if six views are needed, they should be arranged as shown in this figure.

14.4 Sketching Techniques

Sketching means **freehand drawing** without the use of instruments or straightedges. Medium-weight pencils, such as H, F, or HB grades, are

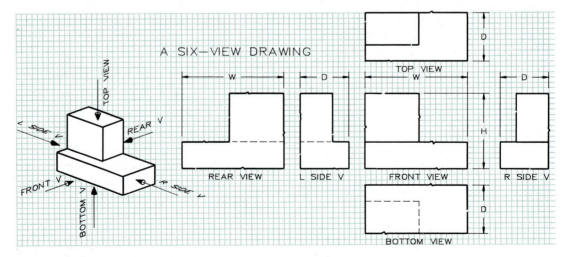

Figure 14.4 Six principal views can be sketched by looking at the object in the directions indicated by the lines of sight. Note how the dimensions are placed on the views. Height (*H*) is shared by all four of the horizontally positioned views.

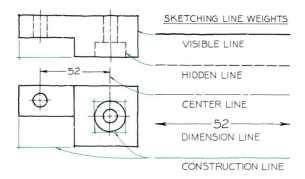

SKETCHING LINE WEIGHTS
VISIBLE LINE
HIDDEN LINE
CENTER LINE
DIMENSION LINE
CONSTRUCTION LINE

Figure 14.5 These lines are examples of those that you should sketch with an F or an HB pencil when drawing views of an object. Some lines are thinner than others, and all except construction lines are black.

the best pencils for sketching. Figure 14.5 shows the standard lines used in multiview drawing and their respective line weights.

By sharpening the pencil point to match the desired line width, you will be able to use the same grade of pencil for all lines (Fig. 14.6). A line drawn freehand should have a freehand appearance; no attempt should be made to give the line the appearance of one drawn by instruments. However, sketching technique can be improved

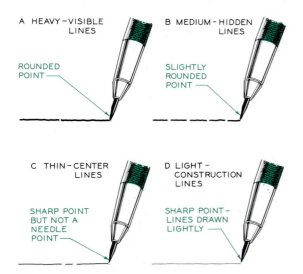

A HEAVY-VISIBLE LINES — ROUNDED POINT
B MEDIUM-HIDDEN LINES — SLIGHTLY ROUNDED POINT
C THIN-CENTER LINES — SHARP POINT BUT NOT A NEEDLE POINT
D LIGHT-CONSTRUCTION LINES — SHARP POINT-LINES DRAWN LIGHTLY

Figure 14.6 The alphabet of lines that are sketched freehand are all made with the same pencil grade (F or HB). The variation in the lines is achieved by varying the sharpness of the pencil point.

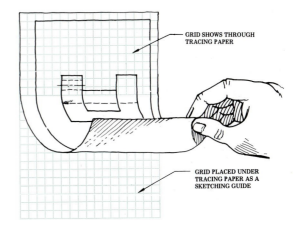

GRID SHOWS THROUGH TRACING PAPER
GRID PLACED UNDER TRACING PAPER AS A SKETCHING GUIDE

Figure 14.7 To aid you in freehand sketching, a grid can be placed under a sheet of tracing paper.

by using a printed grid on sketching paper or by overlaying a printed grid with translucent tracing paper (Fig. 14.7) so that the grid can be seen through the paper.

When a freehand sketch is made, some lines will be vertical and others horizontal or angular. If you do not tape your drawing to the table top, you will be able to position the sheet for the most comfortable strokes, which are (for the right-handed drafter) from left to right (Fig. 14.8). Finally, for the best effect, the lines sketched to form the various views should intersect as indicated in Fig. 14.9.

14.5 Three-View Sketching

Figure 14.10 shows three views of an object with height, width, and depth dimensions given. Each view is labeled. Figure 14.11 shows the steps of drawing three orthographic views of a similar part on a printed grid. The most commonly used combination of views are the front, top, and right-side views, as shown in this figure. First, the overall dimensions of the object are sketched. Then the slanted surface is drawn in the top view and projected to the other views. Lastly, the lines are darkened; the views labeled; and the overall dimensions of height, width, and depth applied.

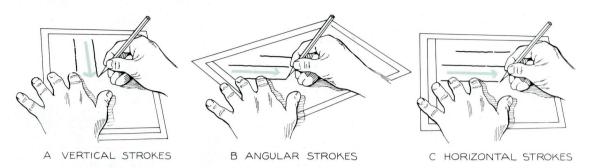

Figure 14.8 Freehand sketching techniques. (A) Vertical lines should be sketched in a downward direction. (B) Angular strokes can be sketched left to right if you rotate your sheet slightly. (C) Horizontal strokes are made best in a left-to-right direction. Always sketch from a comfortable position, and turn you paper if necessary.

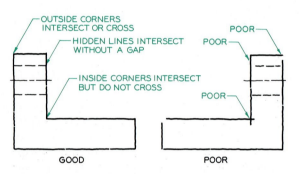

Figure 14.9 For good sketches, follow these examples of good technique. Compare the good drawing with the poor one.

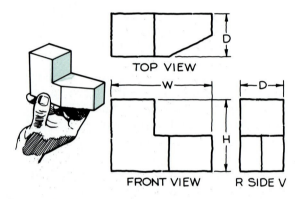

Figure 14.10 The standard arrangement of drawing three views of an object with its dimensions and labels.

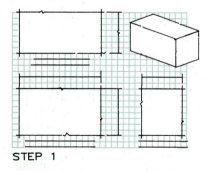

STEP 1

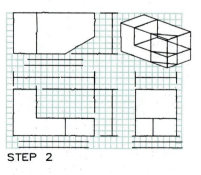

STEP 2

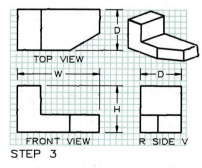

STEP 3

Figure 14.11 Three-view sketching.

Step 1 Block in the views by using the overall dimensions. Allow proper spacing for labeling and dimensioning the views.

Step 2 Remove the notches, and project from view to view.

Step 3 Check your layout for correctness; then darken the lines, and complete the labels and dimensions.

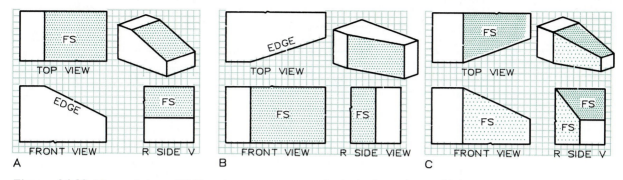

Figure 14.12 Views of planes. (A) The plane appears as an edge in the front view, and it is foreshortened in the top and side views. (B) The plane is an edge in the top view and foreshortened in the front and side views. (C) These two planes appear foreshortened in the right-side view. Each appears as an edge in the top and front views.

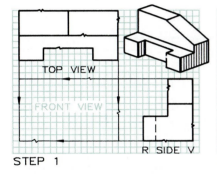

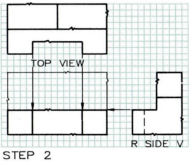

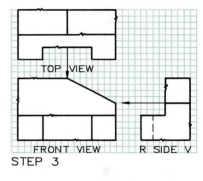

Figure 14.13 Missing front views.

Step 1 When two views are given and the third is required, begin by projecting the overall dimensions from the top and right-side views.

Step 2 The various features of the object are sketched using construction lines.

Step 3 The features are completed, the views checked, and the lines darkened to the proper line quality.

When surfaces are slanted, they will not appear true shape in the principal views of orthographic projection (Fig. 14.12). Surfaces that do not appear true size either are **foreshortened** or appear as **edges.** In Fig. 14.12C, two planes of the object are slanted; thus both appear foreshortened in the right-side view.

A good exercise for analyzing the given views is to find the missing view when two views are given (Fig. 14.13). The right-side view is found in Fig. 14.14, where the top and front views are given. The right-side view has the depth in common with the top view and the height in common with the front view. Knowing this enables us to block in the side view in Step 1. The side view is developed in Step 2 and completed in Step 3.

Another exercise is to complete the views when some or all of them have missing lines (Fig. 14.15).

Figure 14.16 shows three views of an aircraft bracket, which contains several curved corners.

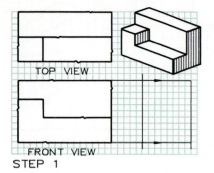

STEP 1

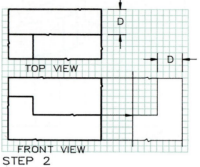

STEP 2

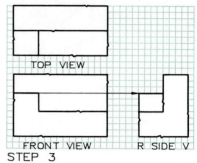

STEP 3

Figure 14.14 Missing side view.

Step 1 To find the right-side view when the top and front views are given, block in the view with the overall dimensions.

Step 2 Develop the features of the view by analyzing the views together. Use light construction lines.

Step 3 Check the views for correctness, and darken the lines to their proper line weight.

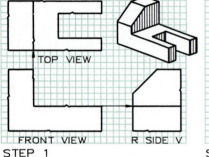

STEP 1

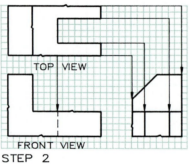

STEP 2

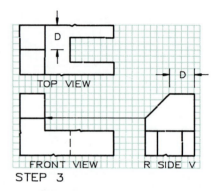

STEP 3

Figure 14.15 Missing lines.

Step 1 Lines may be missing in all views in this type of problem. The first missing line is found by projecting the edges of the planes from the front view.

Step 2 The notch in the top view is projected to the front and side views. The line in the front view is hidden.

Step 3 The line formed by the beveled surface is found in the front view by projecting from the side view.

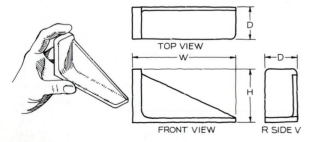

Figure 14.16 Three orthographic views of an aircraft part.

14.6 Circular Features

Centerlines indicate the features are true circles or cylinders (Figure 14.17). In circular views, centerlines cross to indicate the center of the circle. Centerlines consist of short dashes, which should cross in the circular views, spaced at 1-inch intervals along the line. If a centerline coincides with an object line—visible or hidden—the centerline should be omitted since the object lines are more important.

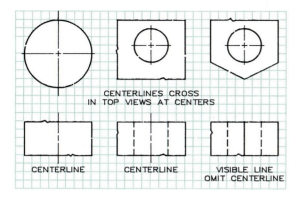

Figure 14.17 Centerlines are used to indicate the centers of circles and the axes of cylinders. They are drawn as very thin lines. When they coincide with visible or hidden lines, centerlines are omitted.

The application of centerlines is shown in Fig. 14.18, where they indicate whether or not circles and arcs are concentric (share the same centers). The centerline should extend beyond the arc by about $\frac{1}{8}$ inch. Figure 14.19 shows the correct manner of applying centerlines. The circular view clearly indicates that the cylinders are concentric since each shares the same centerlines.

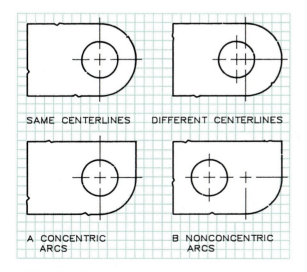

Figure 14.18 The centerline should extend beyond the last arc that has the same center. When the arcs are not concentric, separate centerlines should be drawn.

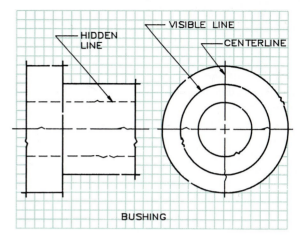

Figure 14.19 Here you can see the application of centerlines of concentric cylinders and the relative weight of hidden lines, visible lines, and centerlines.

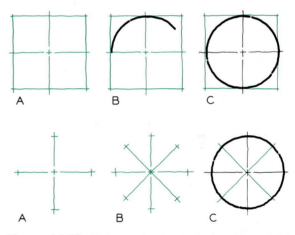

Figure 14.20 Circles can be sketched using either of the construction methods shown here. The use of guidelines is essential to freehand sketching of circles and arcs.

Sketching Circles

Circles can be sketched by using light guidelines along with centerlines (Fig. 14.20). Since it is difficult to draw a freehand circle in one continuous line, short arcs are drawn using the guidelines. If you fail to become reasonably skilled at sketching circles, use a circle template or compass to lightly draw the circle or arc; then darken the line freehand to match the other lines of your sketch.

Figure 14.21 shows the construction of three orthographic views with circular features. The circular features are located with centerlines and guidelines in Step 2; then they are sketched in and darkened in Step 3.

A similar example is given in Fig. 14.22, where the object consists of circular features and arcs. The circles should be drawn first so that their corresponding rectangular views (such as the hidden hole in the top view) can be found by projecting from the circular view.

14.7 Isometric Sketching

Another type of three-dimensional pictorial is the **isometric drawing,** which may be drawn on a specially printed grid composed of a series of lines making 60° angles with one another (Fig. 14.23). The squares in the orthographic views can be laid off along the isometric grid, as Step 1 shows. The notch is located in the same manner in Steps 2 and 3 to complete the isometric pictorial.

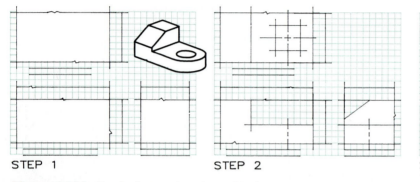

STEP 1

STEP 2

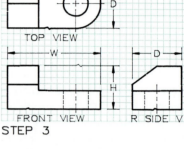

STEP 3

Figure 14.21 Circular features in orthographic views.

Step 1 To draw orthographic views of the object shown, begin by blocking in the overall dimensions. Leave room for the labels and dimensions.

Step 2 Construct the centerlines and the squares about the centerlines in which the circles will be drawn. Show the slanted surface in the side view.

Step 3 Sketch the arcs, and darken the final lines of the views. Label the views, and show the dimensions of W, D, and H.

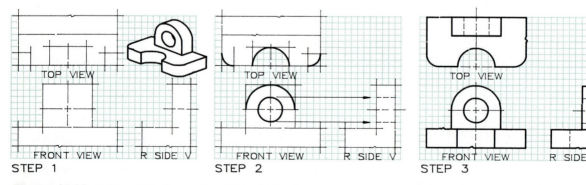

STEP 1

STEP 2

STEP 3

Figure 14.22 Orthographic views with multiple circular features.

Step 1 When sketching orthographic views with circular features, you should begin by sketching the centerlines and guidelines.

Step 2 Using the guidelines, sketch the circular features. These can be darkened as they are drawn if they will be final lines.

Step 3 The outlines of the circular features are found by projecting from the views found in Step 2. All final lines are darkened.

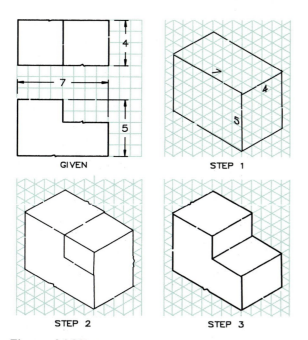

Figure 14.23 Isometric pictorial sketching.

Step 1 When orthographic views of a part are given, an isometric pictorial can be sketched on a printed isometric grid. Construct a box using the overall given dimensions.

Step 2 The notch can be located by measuring over five squares as shown in the orthographic views. The notch is measured four squares downward.

Step 3 The pictorial is completed, and the lines are darkened. This is a three-dimensional pictorial, whereas the orthographic views are two dimensional views.

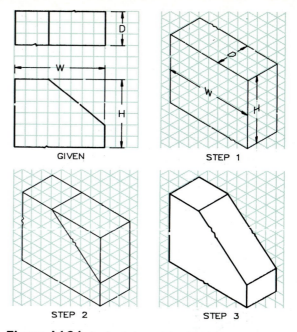

Figure 14.24 Angles in isometric pictorials.

Step 1 Begin by drawing a box using the overall dimensions given in the orthographic views. Count the squares, and transfer them to the isometric grid.

Step 2 Angles cannot be measured with a protractor. Angles will be either larger or smaller than their true measurements. Find each end of the angle by measuring along the axes.

Step 3 The ends of the angles are connected. Dimensions can be measured only in directions parallel to the three axes.

Angles cannot be measured with a protractor in isometric pictorials; they must be drawn by measuring coordinates along the three axes of the printed grid. In Fig. 14.24, the ends of the angular slope are located by measuring the direction of width and height. When an object has two sloping planes that intersect (Fig. 14.25), it is necessary to draw the sloping planes one at a time to find point *B*. The line from *A* to *B* is the line of intersection between the two planes.

Circles in Isometric Pictorials

Circles, which will appear as ellipses in isometric pictorials, can be sketched by locating their centerlines as shown in Step 1 of Fig. 14.26. The center must be equidistant from the top, bottom, and end of the front view.

Figure 14.27 shows three methods of constructing elliptical views of circles.

Figure 14.28 shows the technique of sketching ellipses in isometric to represent a cylinder. This same technique is used to draw an object with a semicircular end in Fig. 14.29.

Figure 14.30 shows how a part composed of several cylindrical forms is sketched isometrically. When an isometric grid is not used, the axes of the isometric sketch are positioned 120° apart. In other words, the height dimension is vertical, and the width and depth dimensions make 30° angles with the horizontal direction on your paper.

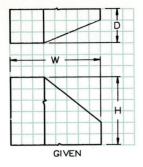 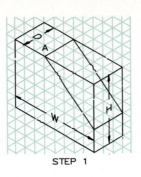 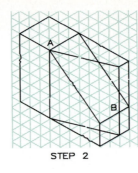 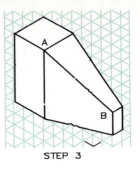

| GIVEN | STEP 1 | STEP 2 | STEP 3 |

Figure 14.25 Double angles in isometric pictorials.

Step 1 When part of an object has a double angle, begin by constructing the overall box and finding one of the angles.

Step 2 Find the second angle that locates point B. Point A will connect to point B to give the intersection line.

Step 3 The final lines are darkened. Line AB is the line of intersection between the two sloping planes.

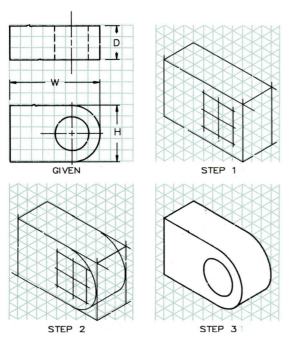

| GIVEN | STEP 1 |
| STEP 2 | STEP 3 |

Figure 14.26 Circles in isometric pictorials.

Step 1 Construct a box using the overall orthographic dimensions. Draw the centerlines and a square (rhombus) of guidelines around the circular hole.

Step 2 Draw the pictorial views of the arcs tangent to the boxes formed by the guidelines. These arcs will appear elliptical rather than circular.

Step 3 Construct the hole, and darken the lines. Hidden lines are normally omitted in pictorial sketches.

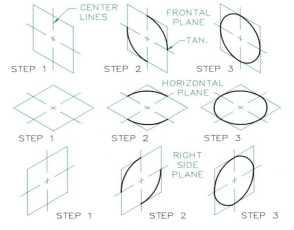

Figure 14.27 Sketching ellipses. The methods of sketching ellipses on each of the three isometric surfaces are shown. A rhombus is drawn on each plane with guidelines that are parallel to the sides of the surface and the ellipses are sketched tangent to the guidelines.

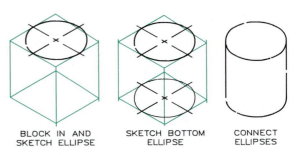

| BLOCK IN AND SKETCH ELLIPSE | SKETCH BOTTOM ELLIPSE | CONNECT ELLIPSES |

Figure 14.28 To make an isometric sketch of a cylinder, the ends are blocked in with rhombuses that are connected by the axis of the cylinder. Ellipses are sketched tangent to the rhombuses, and the sides of the cylinders are drawn.

184

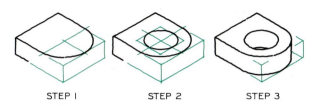

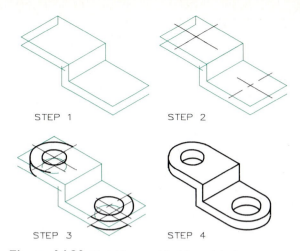

STEP 1   STEP 2

STEP 3   STEP 4

Figure 14.29 Sketching circular features.

Step 1 Block in the semicircular end of the part and sketch the arc.

Step 2 Block in the cylindrical hole and sketch the ellipse.

Step 3 Draw the partial ellipse to represent the bottom of the hole.

Figure 14.30 Sketching an object isometrically.

Step 1 Block in isometric of the object.

Step 2 Locate centerlines of the holes.

Step 3 Sketch the circular features of the ends of the part.

Step 4 Show the bottoms of the holes and darken your lines.

Problems

These sketching problems should be drawn on Size A ($8\frac{1}{2}$-×-11-inch) paper, with or without a printed grid. A typical format for this size sheet is shown in Fig. 14.31, where a 0.20″ grid is given. (This grid can be converted to an approximate metric grid by equating each square to 5 mm.) All sketches and lettering should be neatly executed by applying the principles covered in this chapter. Figures 14.32 and 14.33 contain the problems and instructions.

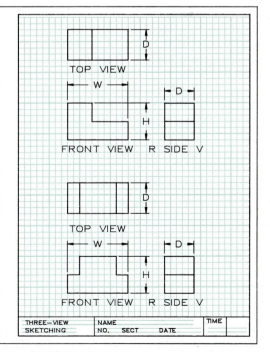

Figure 14.31 The layout of a Size A sheet for sketching problems.

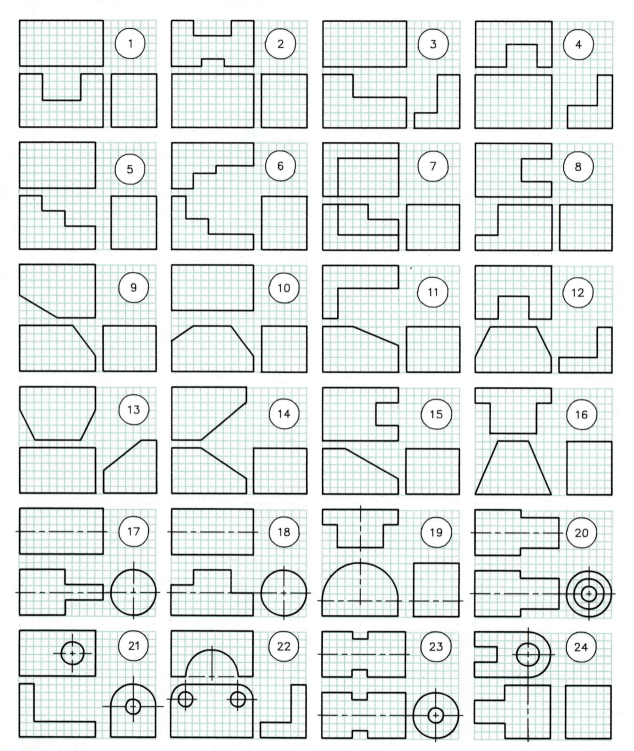

Figure 14.32 On Size A paper, sketch top, front, and right-side views of the problems assigned. Two problems can be drawn on each sheet. Give the overall dimensions of *W, D,* and *H,* and label each view.

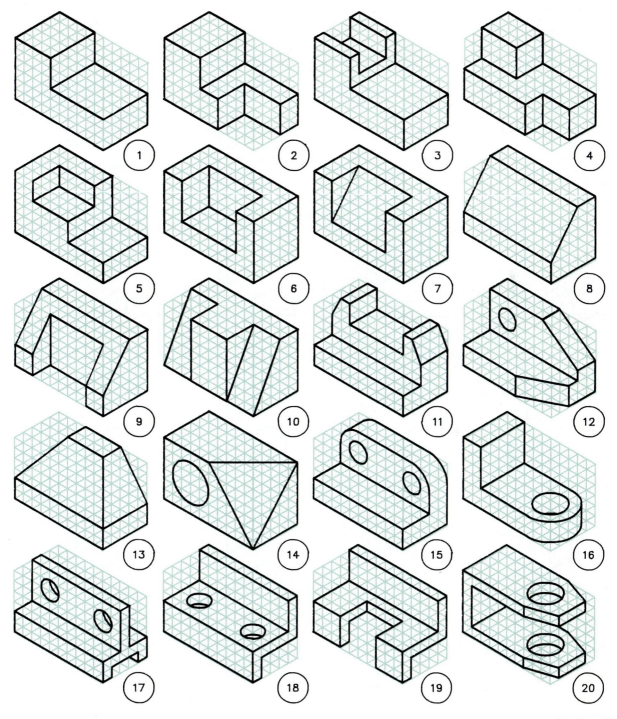

Figure 14.33 Multiview problems and isometric sketching. On Size A paper, sketch the top, front, and right-side views of the problems assigned. Supply the lines that may be missing from all views. Then sketch isometric pictorials of the object assigned, two per sheet.

(continued)

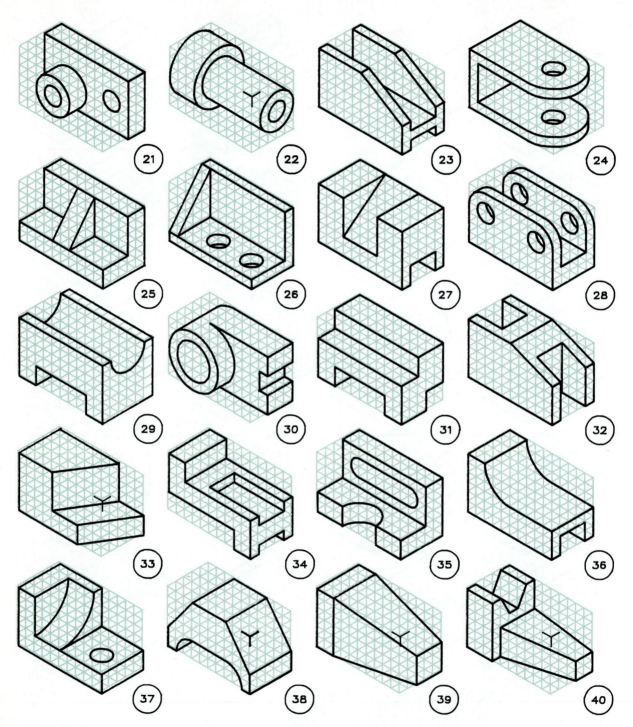

Multiview Drawing with Instruments

15.1 Introduction

Multiview drawing, the system of representing three-dimensional objects by separate views arranged in a standard manner, is readily understood by the technical community. Because multiview drawings are usually executed with instruments and drafting aids, they are often called **mechanical drawings.** They are called **working,** or **detail drawings,** when dimensions and notes are added to complete the specifications of the parts that have been drawn.

15.2 Orthographic Projection

The artist is likely to represent objects impressionistically, but the drafter must represent them precisely. The method of preparing a precise, detailed, clearly understood drawing is **orthographic projection,** or **multiview drawing.**

> By **orthographic projection,** the views of an object are projected perpendicularly onto projection planes with parallel projectors (Fig. 15.1).

The front view is projected onto a vertical frontal plane with parallel projectors. The resulting front view is two dimensional since it has no depth and lies in a single plane described by two dimensions, width and height.

The top view of the same object is projected onto a horizontal projection plane perpendicular to the frontal projection plane in Fig. 15.2A. The right-side view is projected onto a vertical profile plane perpendicular to both the horizontal and frontal planes (Fig. 15.2B).

Imagine that the same object has been enclosed in a glass box showing the frontal, horizontal, and profile projection planes (Fig. 15.3). While in the glass box, the object's views are projected onto the projection planes. Then the box is opened into the plane of the drawing surface to give the standard positions for the three orthographic views.

A similar example of this principle is shown by the object in the projection box in Fig. 15.4. The three views are positioned in the same manner: the top view over the front view and the right-side view to the right of the front view.

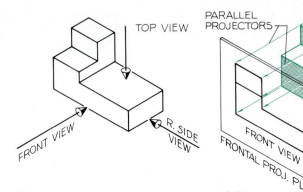

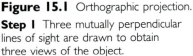

STEP 1

STEP 2

STEP 3

FRONT VIEW
ORTHOGRAPHIC VIEW

Figure 15.1 Orthographic projection.

Step 1 Three mutually perpendicular lines of sight are drawn to obtain three views of the object.

Step 2 The frontal plane is a vertical plane on which the front view is projected with parallel projectors perpendicular to the frontal plane.

Step 3 The resulting view is the front view of the object. This is a two-dimensional orthographic view.

> The three principal projection planes of orthographic projection are the **horizontal** (H), **frontal** (F), and **profile** (P) planes.

Any view projected onto one of these principal planes is a **principal view.** The dimensions of an object used to show its three-dimensional form are height (H), width (W), and depth (D).

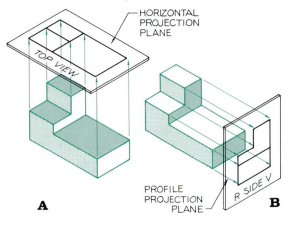

Figure 15.2 (A) The top view is projected onto a horizontal projection plane. (B) The right-side view is projected onto a vertical profile plane perpendicular to the horizontal and frontal planes.

15.3 Alphabet of Lines

Using proper line weights will greatly improve a drawing's readability and appearance. All lines should be drawn dark and dense, as if drawn with ink. Only by their width should the lines vary—except, that is, for guidelines and construction lines, which are drawn very lightly for lettering and laying out drawings.

The lines of an orthographic view are labeled, along with the suggested pencil grades for drawing them, in Fig. 15.5. The lengths of dashes in hidden lines and centerlines are drawn longer as the size of a drawing increases. Figure 15.6 gives additional specifications for these lines.

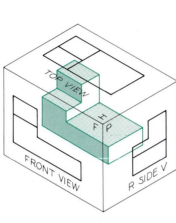
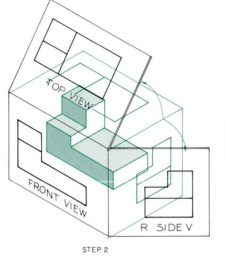
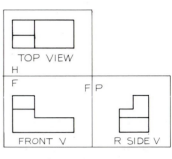

STEP 1

STEP 2

STEP 3

Figure 15.3 Glass-box theory.

Step 1 Imagine that the object has been placed inside a box formed by the horizontal, frontal, and profile planes onto which the top, front, and right-side views have been projected.

Step 2 The three projection planes are then opened into the plane of the drawing surface.

Step 3 The three views are positioned with the top view over the front view and the right-side view to the right. The planes are labeled H, F, and P at the fold lines.

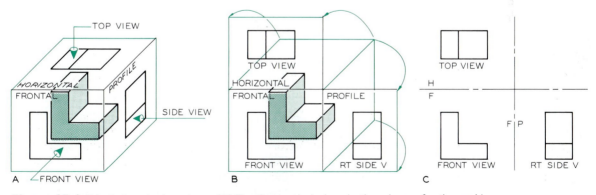

A   FRONT VIEW

B

C

Figure 15.4 Principal projection planes. (A) The three principal projection planes of orthographic projection can be thought of as planes of a glass box. (B) The views of an object are projected onto the projection planes, which are opened into the plane of the drawing surface. (C) The outlines of the planes are omitted. The fold lines are drawn and labeled.

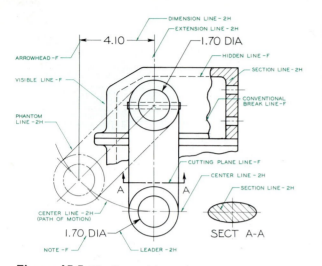

Figure 15.5 The line weights and suggested pencil grades recommended for orthographic views.

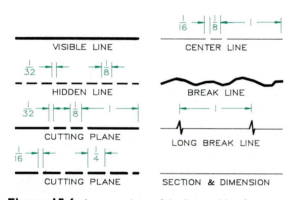

Figure 15.6 A comparison of the line weights for orthographic views. These dimensions will vary for different sizes of drawings and should be approximated by eye.

 Computer Lines Smaller computer graphics plotters vary the widths of lines on a drawing by multiple-pen accessories that hold pens of different point widths, usually 0.7 mm and 0.3 mm wide. AutoCAD provides a command, LTSCALE, that varies dash and space lengths in dashed lines (Fig. 15.7).

A comparison of lines drawn with LTSCALEs of 0.4 and 0.2 and pen widths of 0.7 mm and 0.3 mm is given. Once the drawing has been com-

pleted, the line scales are changed with the LTSCALE command by changing one number (from 0.4 to 0.2 in this example), and the views are automatically redrawn to show the revised lines (Fig. 15.8).

 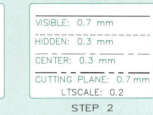

 Figure 15.7 Line weights by computer.

Step 1 These lines were drawn using AutoCAD's standard LINETYPEs and two pens. The LTSCALE factor was 0.4, which affects the spacing between the dashes and the lengths of the dashes.

Step 2 By changing the LTSCALE factor from 0.4 to 0.2, the dashes and the spaces between them are reduced in size.

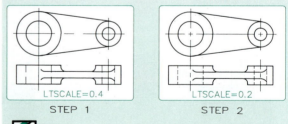

Figure 15.8 Line weights by computer.

Step 1 This two-view drawing was made using an LTSCALE of 0.4, which resulted in the omission of dashes in centerlines and hidden lines with dashes that were too long.

Step 2 By activating LTSCALE and giving a factor of 0.2, the views are redrawn, and centerlines and hidden lines appear with the appropriate dashes.

15.4 Six-View Drawings

If you visualize an object placed inside a glass box, you will see there are two horizontal planes, two frontal planes, and two profile planes (Fig. 15.9).

> Therefore, the maximum number of principal views that can be used to represent an object is six.

The top and bottom views are projected onto horizontal planes, the front and rear views onto frontal planes, and the right- and left-side views onto profile planes.

To draw the views on a sheet of paper, imagine the glass box is opened up into the plane of the drawing paper; the views will then appear as shown in Fig. 15.10. The top view is placed over, and the bottom view under the front view; the right-side view is to the right of, and the left-side

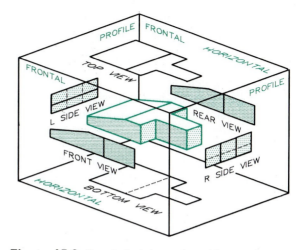

Figure 15.9 Six principal views of an object can be drawn in orthographic projection. You can imagine that the object is in a glass box with the views projected onto the six planes.

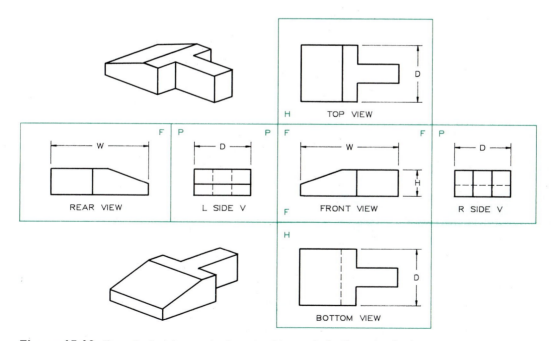

Figure 15.10 Once the box is completely opened into a single plane, the six views are positioned to describe the object. The outlines of the planes are usually omitted. They are shown here to help you relate this figure to Fig. 15.9.

view to the left of the front view; the rear view is to the left of the left-side view.

Height, width, and depth are three dimensions of an object necessary to give its size. The standard arrangement of the six views allows some of the views to share dimensions by projection. For example, the height dimension, which is shown only once between the front and right-side views, applies to the four horizontally arranged views. Furthermore, the width dimension is placed between the top and front views, but it also applies to the bottom view.

Projectors align the views both horizontally and vertically about the front view in Fig. 15.10. Each side of the fold lines of the glass box is labeled H, F, or P (horizontal, frontal, or profile) to identify the projection planes on a given side of the fold lines.

15.5 Three-View Drawings

Because three views are usually adequate to describe an object, the most commonly used orthographic arrangement is the **three-view draw-**

ing, consisting of front, top, and right-side views. The object used in the previous example is shown placed in a glass box in Fig. 15.11, which is opened onto the plane of the drawing surface. The resulting three-view arrangement is shown in Fig. 15.12, where the views are labeled and dimensioned.

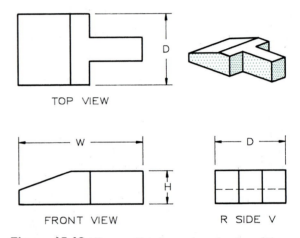

Figure 15.12 The resulting three-view drawing of the object from Fig. 15.11.

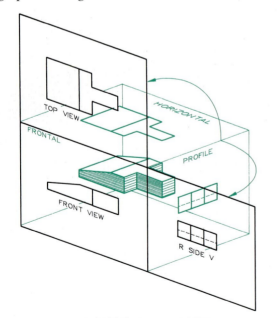

Figure 15.11 Three-view drawings are commonly used for describing small machine parts. The glass box is used to illustrate how the views are projected onto their projection planes.

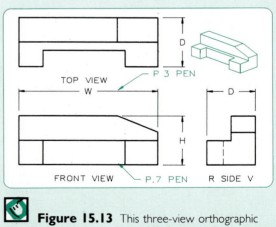

Figure 15.13 This three-view orthographic drawing was made by AutoCAD and an A-B size plotter using two pen sizes (0.7 mm and 0.3 mm). The same principles of orthographic projection are used whether drawings are made manually or by computer.

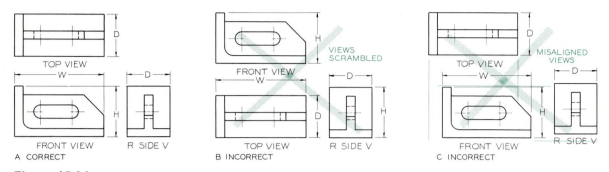

Figure 15.14 Positioning orthographic views. (A) This is a correct arrangement of views, labels, and dimensions. The views project from each other in proper alignment. (B) These views are scrambled into unconventional positions, making it hard to interpret them. Dimensions are needlessly repeated. (C) These views are misaligned, so they do not project from one to the other. This is an incorrect arrangement.

Computer Views The three-view drawing of a part (Fig. 15.13) was drawn by using two pen widths, 0.7 mm and 0.3 mm, and the DRAW and DIMension commands of AutoCAD.

15.6 Arrangement of Views

Figure 15.14A shows the standard positions for a three-view drawing: The top and side views are projected directly from the front view. The views are properly labeled and dimensioned. Views that are arranged in a nonstandard sequence (Fig 15.14B) and that do not project from view to view (Fig. 15.14C) are incorrect. Figure 15.15 emphasizes these rules of arrangement and dimensioning.

15.7 Selection of Views

When drawing an object by orthographic projection, you should select the views with the fewest hidden lines. In Fig. 15.16, Example 1, the right-side view is preferred to the left-side view because it has fewer hidden lines.

Although the three-view arrangement of top, front, and right-side views is the most commonly used, the top, front, and left-side views is equally acceptable (Fig. 15.16, Example 2) if the left-side view has fewer hidden lines than the right-side view.

The most descriptive view is usually selected as the front view. Some objects have standard views that are regarded as the front view, top view, and so forth. A chair, for example, has front and top views that are recognized as such by everyone; therefore, a chair's accepted front view should be used as the orthographic front view.

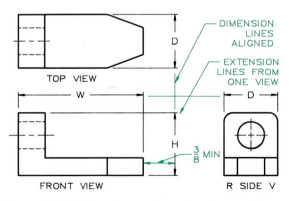

Figure 15.15 Dimension and extension lines used in three-view orthographic projection should be aligned. Notice that extension lines are drawn from only one view when dimensions are placed between two views.

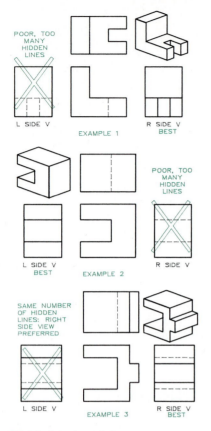

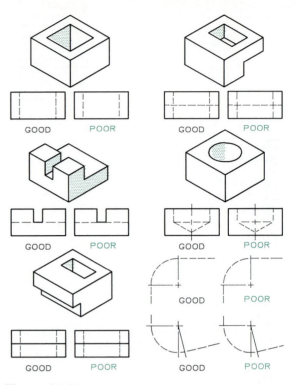

Figure 15.16 Selection of views.

Example 1 In orthographic projection, you should select the sequence of views with the fewest hidden lines.

Example 2 The left-side view has fewer hidden lines; therefore, this view is selected over the right-side view.

Example 3 When both views have an equal number of hidden lines, the right-side view is traditionally selected.

Figure 15.17 When lines intersect in orthographic projection, they should intersect as shown in these examples.

15.8 Line Techniques

As drawings become more complex, you will encounter more instances of lines overlapping and intersecting in ways similar to those shown in Fig. 15.17. This illustration shows the techniques of handling most types of intersecting lines. Methods of drawing hidden lines composed of lines and arcs are also shown.

You should become familiar with the order of precedence (priority) of lines (Fig. 15.18): The most important line is the visible object line; it is

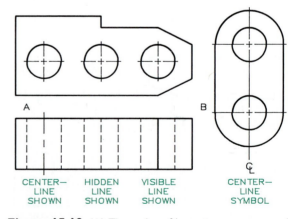

Figure 15.18 (A) The order of importance (precedence) of lines is visible lines, hidden lines, and centerlines. (B) The symbol made of the letters *C* and *L* is used to label a centerline when needed on symmetrical parts.

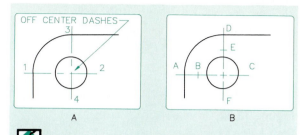

Figure 15.19 Centerlines by computer. (A) When centerlines for a rounded corner are drawn from 1 to 2 and from 3 to 4, the center dashes do not cross at the center of the hole. (B) Centerline dashes can be made to cross at the center by drawing a line from A to B to C, where the BC segment extends equally beyond the center on both sides. Line DF is drawn from D to E to F in the same manner.

shown regardless of any other line lying behind it. Of next importance is the hidden object line, which takes precedence over the centerline.

Centerlines by Computer   Figure 15.19 shows a method of ensuring that centerline dashes cross properly when drawn by computer. Centerlines must be drawn in connected segments (Fig. 15.19B).

15.9 Point Numbering

The method of numbering points and lines of an object will be helpful to you in constructing orthographic views. For example, Fig. 15.20 shows an object that has been numbered to aid in the construction of the missing front view when the top and side views are given. By projecting selected points from the top and side views, the front view of the object can be found.

15.10 Line and Planes

An orthographic view of a line can appear **true length, foreshortened,** or as a **point** (Fig. 15.21). When a line appears true length, it must be parallel to the reference line in the previous view. A plane in orthographic projection can appear **true size, foreshortened,** or as an **edge** (Fig. 15.21).

15.11 Alternate Arrangement of Views

Although the right-side view is usually placed to the right of the front view (Fig. 15.22), the side view can be projected from the top view (Fig. 15.23); this is advisable when the object has a much larger depth than height.

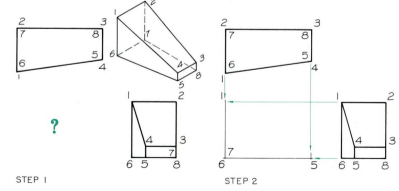
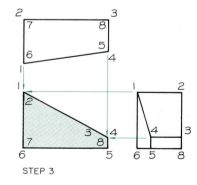
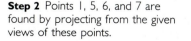

Figure 15.20 Point numbering.
Step 1 When a missing orthographic view is to be drawn, it is helpful to number the points in the given views.

Step 2 Points 1, 5, 6, and 7 are found by projecting from the given views of these points.

Step 3 The plotted points are connected to form the missing front view.

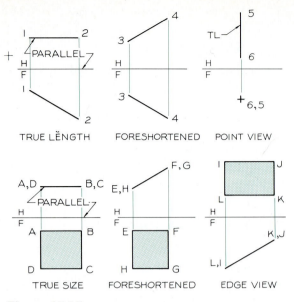

TRUE LENGTH	FORESHORTENED	POINT VIEW

TRUE SIZE	FORESHORTENED	EDGE VIEW

Figure 15.21 Lines and planes. A line will appear in orthographic projection as true length, foreshortened, or a point. A plane in orthographic projection will appear as true size, foreshortened, or an edge.

15.12 Laying Out Three-View Drawings

The depth dimension applies to both the top and side views, but these views are usually positioned where this dimension does not project between them (Fig. 15.24). The depth dimension can be

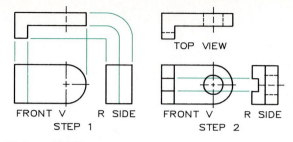

Figure 15.22 Three views.

Step 1 The three views are blocked in. Notice that the depth dimension in the top view is the same as the depth in the right-side view.

Step 2 The circular hole and its centerlines are drawn in all views. The notch is drawn in the side view and projected to the top and front views.

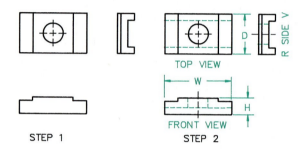
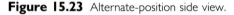

Figure 15.23 Alternate-position side view.

Step 1 The right-side view of this part is projected from the top view to save space since the depth dimension is large.

Step 2 The views are completed, and dimensions and labels are added.

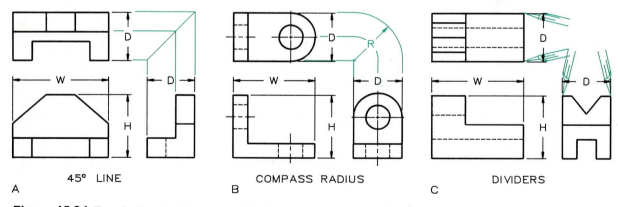

A 45° LINE	B COMPASS RADIUS	C DIVIDERS

Figure 15.24 Transferring depth dimensions. (A) The depth dimension can be projected from the top view to the right-side view by constructing a 45° line positioned as shown. (B) The depth dimension can be projected from the top view to the side view by using a compass and center point. (C) The depth dimension can be transferred from the top view to the side view by using dividers—the most desirable method.

198

graphically transferred by using a 45° line, an arc, or a pair of dividers. Figures 15.25–15.29 show examples of three-view orthographic drawings of objects.

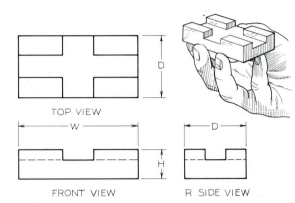

Figure 15.25 A three-view drawing of an object.

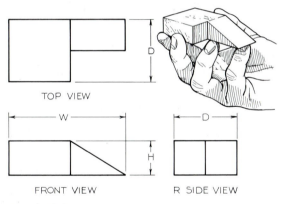

Figure 15.26 A three-view drawing of an object.

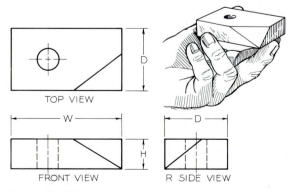

Figure 15.27 A three-view drawing of an object.

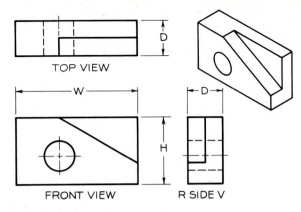

Figure 15.28 A three-view drawing of an object.

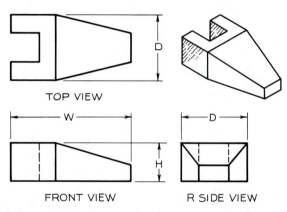

Figure 15.29 A three-view drawing of an object.

Three-View Drawings by Computer Figure 15.30 illustrates the steps for constructing three views of an object by computer methods. The overall outlines of the views are drawn with the LINE command, and the views are labeled with the TEXT command in Step 1. Other lines are added in Step 2 by projecting from view to view. In Step 3, the remaining visible and hidden lines are drawn.

15.13 Two-View Drawings

It is good economy of time and space to use only the views necessary to depict an object. Figure 15.31 shows objects that require only two views.

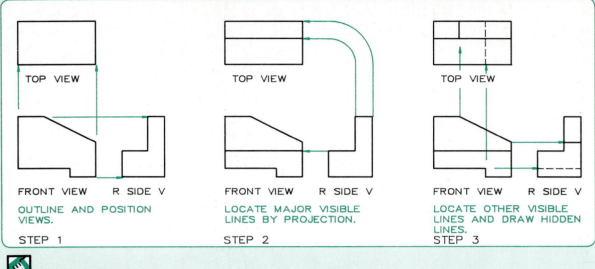

FRONT VIEW   R SIDE V

OUTLINE AND POSITION VIEWS.

STEP 1

FRONT VIEW   R SIDE V

LOCATE MAJOR VISIBLE LINES BY PROJECTION.

STEP 2

FRONT VIEW   R SIDE V

LOCATE OTHER VISIBLE LINES AND DRAW HIDDEN LINES.

STEP 3

Figure 15.30 Three views by computer.

Step 1 Using the LINE command, draw the outlines of the views and label them.

Step 2 By orthographic projection, draw other visible lines.

Step 3 Draw the remaining visible and hidden lines. Change LAYERs before drawing hidden lines.

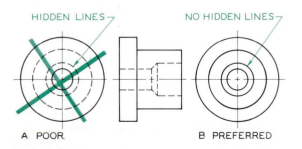

TWO—VIEW DRAWINGS

A   B

Figure 15.31 These objects can be adequately described with two orthographic views.

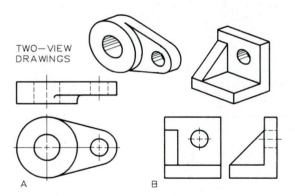

HIDDEN LINES   NO HIDDEN LINES

A POOR   B PREFERRED

Figure 15.32 Cylindrical objects can be depicted with two views. Always select views with the fewest hidden lines.

Cylindrical objects need only two views, as Fig. 15.32 shows. Because it is preferable to select the views with the fewest hidden lines, the right-side view is the better view in this example.

Two-View Drawings by Computer Figure 15.33 shows the steps in making a two-view drawing of an object. The ARC and FIL-LET commands are used to draw the semicircular feature and the rounded corners in Step 2.

The MIRROR command can be used to reduce construction by drawing only half (or one quarter) of a view and then mirroring the drawing to give the other symmetrical half (Fig. 15.34).

The object in Fig. 15.35, which is composed of arcs and tangent lines, is constructed as a half top view, which is mirrored along line *AB*. An arc is drawn with the FILLET command in Step 2.

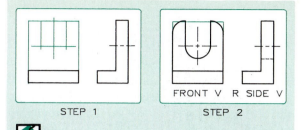

STEP 1   STEP 2

 Figure 15.33 A two-view drawing by computer.

Step 1 The front and side views are drawn using the overall dimensions.

Step 2 The FILLET command is used to draw the corners in the front view, and the ARC command is used to draw the semicircular arc.

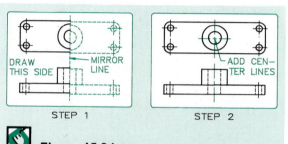

STEP 1   STEP 2

 Figure 15.34 Mirroring by computer.

Step 1 To save drawing time, the top and front views are drawn as half views, and MIRRORed about the centerline.

Step 2 Centerlines that coincide with the MIRROR line should be drawn after mirroring to prevent the centerline from being drawn twice.

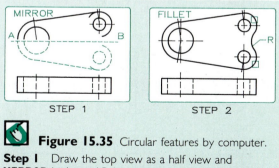

STEP 1   STEP 2

 Figure 15.35 Circular features by computer.

Step 1 Draw the top view as a half view and MIRROR it about *AB*. Locate tangent points by using the PERPendicular option of the OSNAP command.

Step 2 The arc (radius R) is drawn with the FILLET command. You must select the radius and points on the arcs that the fillet is to be tangent to.

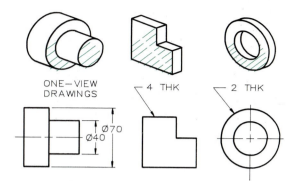

Figure 15.36 Objects that are cylindrical or of uniform thickness can be described with only one orthographic view and supplementary notes.

15.14 One-View Drawings

Cylindrical parts and those with a uniform thickness can be described in one view. In these cases, notes explain the missing feature or dimension (Fig. 15.36).

15.15 Incomplete and Removed Views

The right- and left-side views of the part in Fig. 15.37 would be hard to interpret if all hidden lines were shown as specified by the rules of orthographic projection. Therefore, it is best to

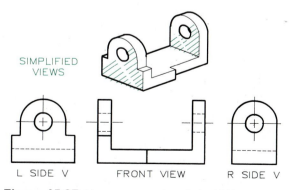

Figure 15.37 Unnecessary and confusing hidden lines are omitted in the side views to improve their clarity.

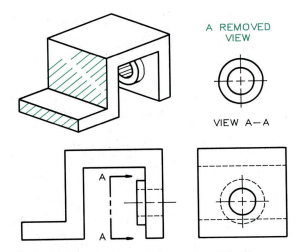

A REMOVED VIEW

VIEW A—A

Figure 15.38 A removed view, indicated by the directional arrows, can be used to draw views in new hard-to-see locations.

omit lines that confuse a clear understanding of the views.

Often it is difficult to show a feature because of its location. Standard views can be confusing when lines overlap other features. The view indicated by the directional arrows in Fig. 15.38 is more clearly shown when removed to an isolated position.

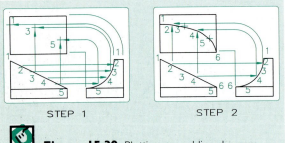

STEP 1   STEP 2

Figure 15.39 Plotting curved lines by computer.

Step 1 A series of points is found in the front and side views by orthographic projection. Points 1, 3, and 5 are projected to the top view.

Step 2 Points 2 and 4 are projected to the top view, and a `PLINE` curve is drawn to `FIT` the points to complete the top view.

15.16 Curve Plotting

An irregular curve can be drawn by following the rules of orthographic projection, as Fig. 15.39 shows. Plotting begins by locating points along the curve in two given views. These points are projected to the top view where each point is located, and the points are then connected by a smooth curve. In Fig. 15.40, an ellipse is plotted in the top view by projecting from the front and side views. You will find it helpful to number points on curves that are being located by projection.

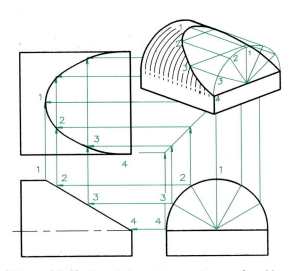

Figure 15.40 The ellipse in the top view was found by numbering points in the front and side views and then projecting them to the top view.

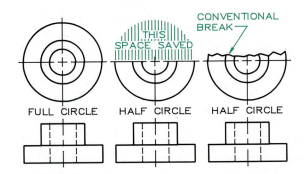

FULL CIRCLE   HALF CIRCLE   HALF CIRCLE

Figure 15.41 To save space and drawing time, the top view of a cylindrical part can be drawn as a partial view using either of these methods.

15.17 Partial Views

A partial view can be used to save time and space when the parts are symmetrical or cylindrical. By omitting the rear of the circular top view in Fig. 15.41, space can be saved without sacrificing clarity. Or, to make it more apparent that a portion of the view has been omitted, a break may be used.

15.18 Conventional Revolutions

The readability of an orthographic view may be improved if the rules of projection are violated.

Established violations of rules that are customarily made for the sake of clarity are called **conventional practices.**

When holes are symmetrically spaced in a circular plate, as shown in Fig. 15.42, it is conventional practice to show them at their true radial distance from the center of the plate. This requires an imagined revolution of the holes in the

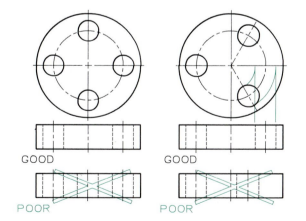

Figure 15.42 Revolving holes.

Left: A true projection of equally spaced holes gives a misleading impression that the center hole passes through the center of the plate.

Right: A conventional view is used to show the true radial distances of the holes from the center by revolution. The third hole is omitted.

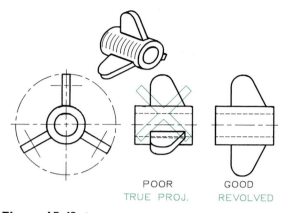

Figure 15.43 Symmetrically positioned external features, such as these lugs, are revolved to their true-size positions for the best views.

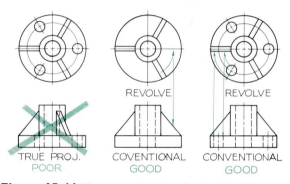

Figure 15.44 The conventional methods of revolving holes and ribs in combination for improved clarity.

top view. This principle of revolution also applies to symmetrically positioned features, such as the three lugs on the outside of the part in Fig. 15.43. Figure 15.44 shows the conventional and desired method of drawing holes and ribs in combination.

Another conventional practice is illustrated in Fig. 15.45, where an inclined arm is revolved to a horizontal position in the front, so that it can be drawn as true size in the top view. The revolution of the part is not drawn since it is an imagined revolution.

Figure 15.46 shows other parts whose views are improved by revolution. It is desirable to show the top-view features at 45°, so they will not coincide with the centerlines. The front views

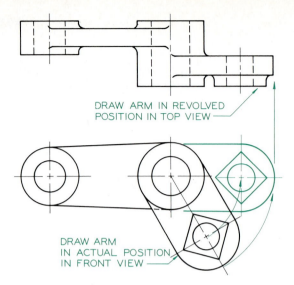

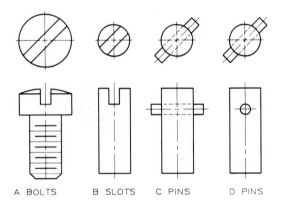

DRAW ARM IN REVOLVED
POSITION IN TOP VIEW

DRAW ARM
IN ACTUAL POSITION
IN FRONT VIEW

Figure 15.45 The arm in the front view is imagined to be revolved so that its true length can be drawn in the top view. This is an accepted conventional practice.

A BOLTS   B SLOTS   C PINS   D PINS

Figure 15.46 Parts like these are drawn at 45° angles in the top views, but the front views are drawn to show the details as revolved views.

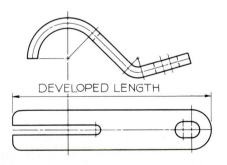

DEVELOPED LENGTH

Figure 15.47 Objects that have been shaped by bending thin stock can be shown as true-size developed views.

are drawn by imagining the features have been revolved. A closely related type of conventional view is the true-size developed view where a bent piece of material is drawn as if it were flattened out (Fig. 15.47).

15.19 Intersections

In orthographic projection, the intersection between planes results in a line that describes the object. In Fig. 15.48, examples of views are shown where lines may or may not be required.

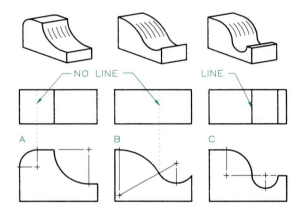

NO LINE   LINE

A   B   C

Figure 15.48 Object lines are drawn only where there are sharp intersections or where arcs are tangent at their centerlines, as in part C.

Figure 15.49 shows the standard types of intersections between cylinders. Figures 15.49A and B are conventional intersections, which means they are approximations drawn for ease of construction. Figure 15.49C shows a true intersection between cylinders of equal diameters. Figures 15.50 and 15.51 show similar intersections, and Fig. 15.52 shows the types of conventional intersections formed by holes in cylinders.

15.20 Fillets and Rounds

Fillets and **rounds** are rounded corners used on castings, such as the body of the Collet Index Fixture shown in Fig. 15.53. A fillet is an inside

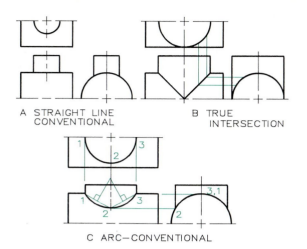

A STRAIGHT LINE CONVENTIONAL   B TRUE INTERSECTION

C ARC–CONVENTIONAL

Figure 15.49 The conventional methods of showing intersections between cylinders. Except for part C, these are approximations.

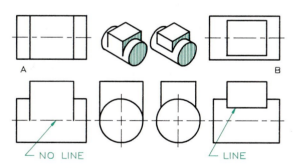

NO LINE   LINE

Figure 15.50 Conventional intersections between prisms and cylindrical shapes.

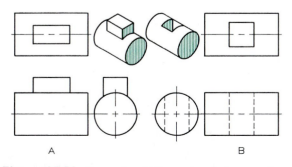

A   B

Figure 15.51 Conventional intersections between prisms and cylindrical shapes.

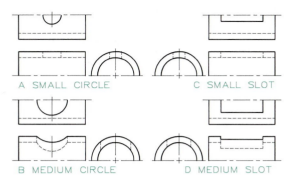

A SMALL CIRCLE   C SMALL SLOT

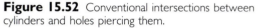

B MEDIUM CIRCLE   D MEDIUM SLOT

Figure 15.52 Conventional intersections between cylinders and holes piercing them.

Figure 15.53 The edges of this Collet Index Fixture are rounded to form fillets and rounds. The surface of the casting is rough except where it has been machined. (Courtesy of Hardinge Brothers Inc.)

rounding, and a round is an external rounding. The radii of fillets and rounds may be many sizes, but they are usually about $\frac{1}{4}$ inch. Fillets and rounds are used on castings for added strength and improved appearance.

A casting will have square corners only when its surface has been finished, which is the process of machining away part of the surface to a smooth finish (Fig. 15.54B).

Finished surfaces are indicated by placing a **finish mark** (V) on the edge views of the finished surfaces whether the edges are visible or hidden. Figure 15.55 shows alternative finish marks.

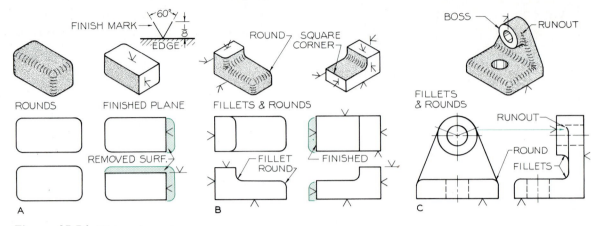

Figure 15.54 Fillets and rounds.

(A) When a surface has been finished by machining, rounds are removed, and the corners are squared. A finish mark is indicated by a V placed on the edge view of the finished surfaces. (B) A fillet is a rounded inside corner. The rounds are removed when the outside surfaces are finished. The fillets can be seen only in the front view. (C) The views of an object with fillets and rounds must be drawn in a way that calls attention to them.

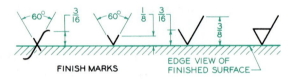

Figure 15.55 Alternative finish marks are applied to the edge views of the finished surface, whether hidden or visible.

Note (in Fig. 15.54C) that a **boss** is a raised cylindrical feature that is thickened to receive a shaft or to be threaded, and that the curve formed by a fillet at a point of tangency is a **runout.** Figure 15.56 shows the techniques of showing fillets and rounds on orthographic views.

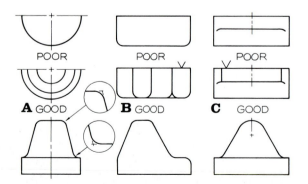

Figure 15.56 Examples of conventionally drawn fillets and rounds.

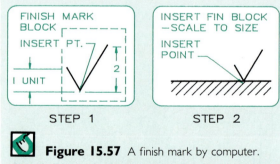

STEP 1   STEP 2

Figure 15.57 A finish mark by computer.

Step 1 Draw a finish mark that is 1 inch high. Make a BLOCK of it named FIN. Select the insertion point to be the point of the vee.

Step 2 INSERT the block, FIN, by selecting the point on the edge view of the finished surface. Scale the block to 0.125, and its height will be $\frac{1}{8}$ inch high.

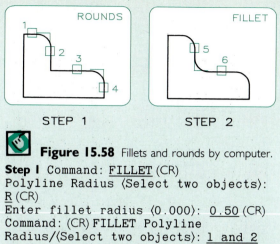

Figure 15.58 Fillets and rounds by computer.

Step 1 Command: <u>FILLET</u> (CR)
Polyline Radius ⟨Select two objects⟩:
<u>R</u> (CR)
Enter fillet radius ⟨0.000⟩: <u>0.50</u> (CR)
Command: (CR) FILLET Polyline
Radius/⟨Select two objects⟩: <u>1 and 2</u>
(Repeat and select 3 and 4 to draw round.)

Step 2 Press Return, and select points 5 and 6 on inside lines to construct a fillet. The straight lines that have the fillets and rounds applied to them are trimmed to their proper lengths.

A computer-drawn finish mark can be saved as a BLOCK that can be used repeatedly and rapidly (Fig. 15.57). If the finish mark is drawn 1 in. high, it will be easy to scale it when INSERTed. For example, a scale factor of 0.125 will reduce it to $\frac{1}{8}$ in. By saving the BLOCK as a WBLOCK, it can be used on different drawing files, not just the one it was created on.

Fillets and rounds can be drawn by computer as shown in Fig. 15.58. The object is first drawn with square intersections, and the FILLET command is used to round the corners for both fillets and rounds.

Figure 15.59 shows a comparison of intersections and runouts of parts with and without fillets and rounds. Large runouts are constructed as an eighth of a circle with a compass, as Fig. 15.60 shows. Small runouts can be drawn with a circle

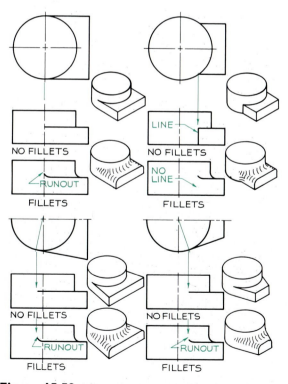

Figure 15.59 Intersections between features of objects. Intersections with fillets have runouts.

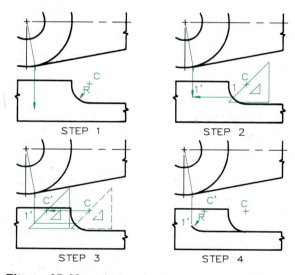

Figure 15.60 Plotting runouts.

Step 1 Find the point of tangency in the top view and project it to the front view.

Step 2 A 45° triangle is used to find point 1, which is projected to point 1'.

Step 3 The 45° triangle is moved to locate point C', which is on the horizontal projector from the center of the C.

Step 4 The radius of the fillet is used to draw the runout with C' as its center. The runout arc is equal to one eighth of a circle.

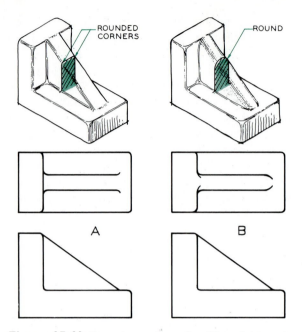

Figure 15.61 Runouts are shown for differently shaped ribs. Part A has fillets, and part B has rounded edges.

template. Runouts on orthographic views will reveal much about the details of an object. For example, the runout in the top view of Fig. 15.61A tells us the rib has rounded corners, whereas the top view of Fig. 15.61B tells us the rib is completely round. Figures 15.62 and 15.63 illustrate methods of showing other types of intersections.

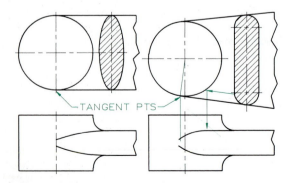

Figure 15.62 Conventional runouts of different cross-sectional shapes.

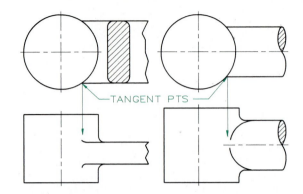

Figure 15.63 Conventional runouts of different cross-sectional shapes.

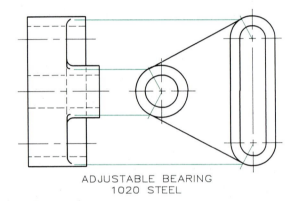

ADJUSTABLE BEARING
1020 STEEL

Figure 15.64 A typical application of runouts on a part with fillets and rounds.

Computer Method   Figure 15.64 shows a drawing of a part with runouts at its tangent points. The runouts are plotted with the ARC command after the tangent points have been projected from the right-side view.

15.21 Left-Hand and Right-Hand Views

Two parts are often required that are "mirror images" of each other (Fig. 15.65). The drafter can reduce drawing time by drawing views of only

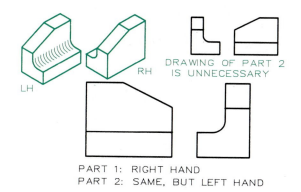

DRAWING OF PART 2
IS UNNECESSARY

RH

LH

PART 1: RIGHT HAND
PART 2: SAME, BUT LEFT HAND

Figure 15.65 When left- and right-hand mirror parts are needed, only one view is drawn and labeled. The other view need not be drawn but should be indicated by a note.

one of the parts and labeling these views. A note can be added to indicate that the other matching part has the same dimensions.

15.22 First-Angle Projection

The examples in this chapter are presented as third-angle projections, where the top view is placed over the front view, and the right-side view is placed to the right of the front view. This method is used extensively in the United States, Britain, and Canada. Most of the rest of the industrial world uses **first-angle** projection.

The first-angle system is illustrated in Fig. 15.66, where an object is placed above the horizontal plane and in front of the frontal plane. When these projection planes are opened onto the surface of the drawing paper, the front view is

projected over the top view, and the left-side view is placed to the right of the front view.

It is important that the angle of projection be indicated on a drawing to aid in the interpretation of the views. This is done by placing a truncated cone in or near the title block (Fig. 15.67). When metric units of measurement are used, the cone and SI symbol are placed together on the drawing.

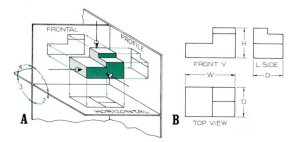

Figure 15.66 First-angle projection. (A) The first angle of projection is used by many of the countries that use the metric system. You imagine that the object is placed above the horizontal plane and in front of the frontal plane. (B) The views are drawn in this location, which is different from the third-angle of projection used in the United States.

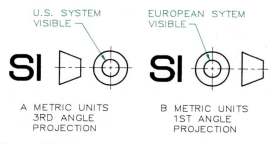

Figure 15.67 The angle of projection used to prepare a set of drawings is indicated by a truncated cone, which is placed in or near the title block of a drawing.

Problems

The following problems are to be drawn as orthographic views on Size A or Size B paper, as assigned by your instructor.

1–7. (Figs. 15.68–15.74) Draw the given views using the dimensions provided, and then construct the missing top, front, or right-side views. Use Size A sheets,

and draw one or two problems per sheet.

8–17. (Figs. 15.75–15.84) Complete the three views of the objects. Each square grid is equal to 0.20 inches or 6 mm. Two problems can be placed on an A-size sheet (AV format). Label the views and show the overall dimensions as W, D, and H.

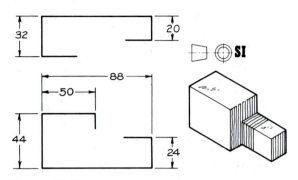

Figure 15.68 Problem 1. Guide block.

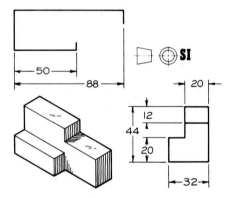

Figure 15.69 Problem 2. Double step.

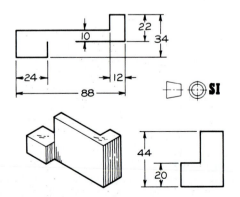

Figure 15.70 Problem 3. Adjustable stop.

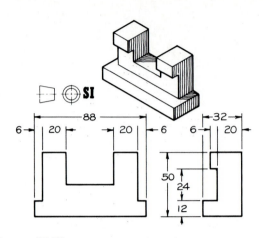

Figure 15.71 Problem 4. Lock catch.

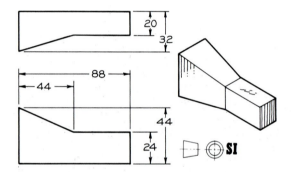

Figure 15.72 Problem 5. Two-way adjuster.

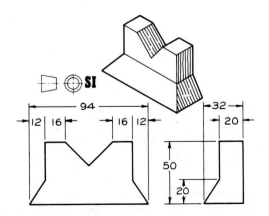

Figure 15.73 Problem 6. Vee block.

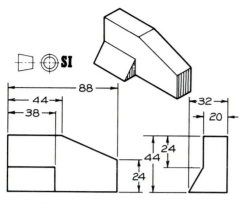

Figure 15.74 Problem 7. Filler.

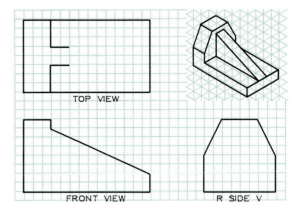

Figure 15.77 Problem 10. Angle block.

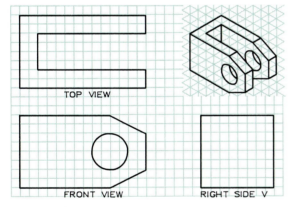

Figure 15.75 Problem 8. Clevis.

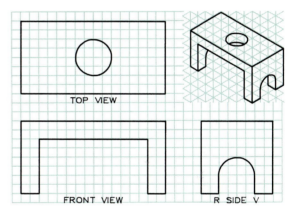

Figure 15.78 Problem 11. Fixture guide.

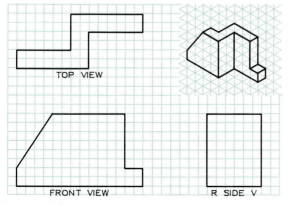

Figure 15.76 Problem 9. Corner box.

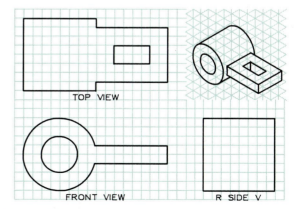

Figure 15.79 Problem 12. Lifting ring.

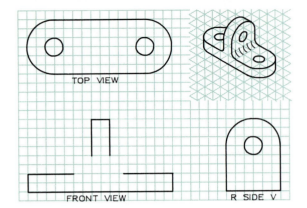

Figure 15.80 Problem 13. Pivot piece.

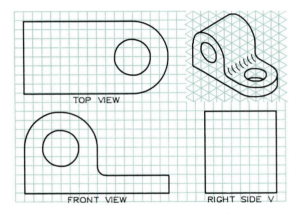

Figure 15.81 Problem 14. Bushing.

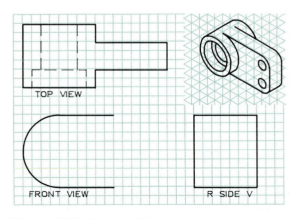

Figure 15.82 Problem 15. Journal.

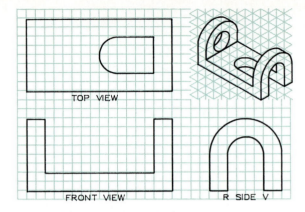

Figure 15.83 Problem 16. End guide.

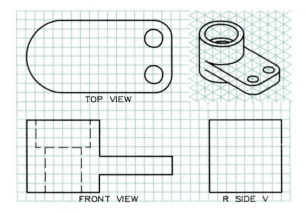

Figure 15.84 Problem 17. Left journal.

18–48. (Figs. 15.85–15.115) Construct the necessary orthographic views to describe the objects in these figures. Draw the views on Size A or Size B sheets. Label the views and show the overall dimensions of *W, D,* and *H.*

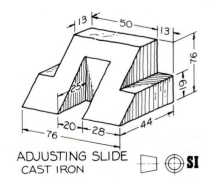

ADJUSTING SLIDE
CAST IRON

Figure 15.85 Problem 18. Adjusting slide.

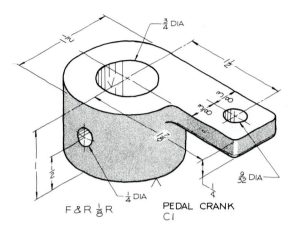

Figure 15.86 Problem 19. Pedal crank.

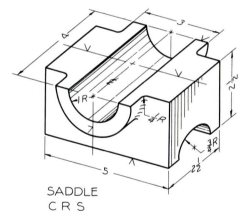

SADDLE
C R S

Figure 15.87 Problem 20. Saddle.

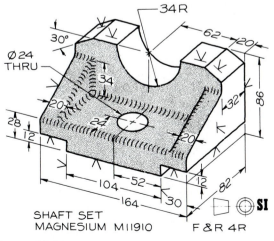

SHAFT SET
MAGNESIUM M11910

Figure 15.88 Problem 21. Shaft set.

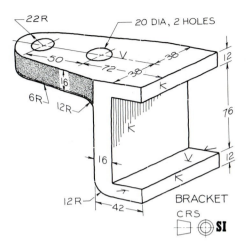

BRACKET
C R S

Figure 15.89 Problem 22. Bracket.

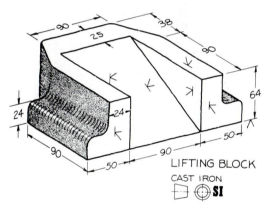

LIFTING BLOCK
CAST IRON

Figure 15.90 Problem 23. Lifting block.

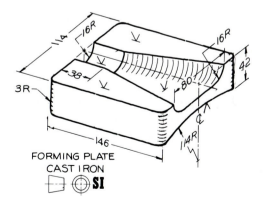

FORMING PLATE
CAST IRON

Figure 15.91 Problem 24. Forming plate.

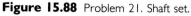

214

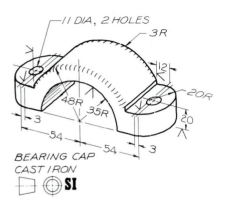

Figure 15.92 Problem 25. Bearing cap.

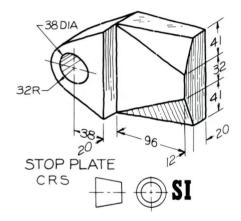

STOP PLATE
C R S
SI

Figure 15.93 Problem 26. Stop plate.

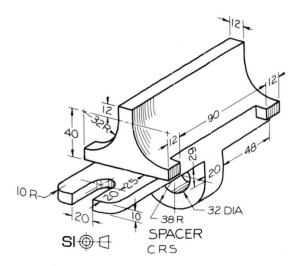

SPACER
C R S

Figure 15.94 Problem 27. Spacer.

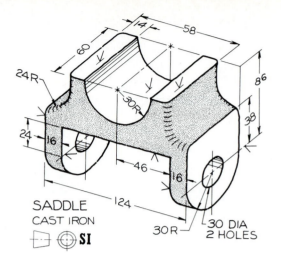

SADDLE
CAST IRON
SI

Figure 15.95 Problem 28. Saddle.

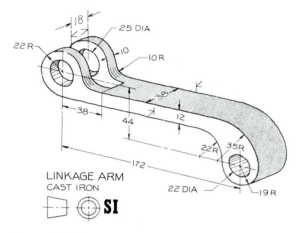

LINKAGE ARM
CAST IRON
SI

Figure 15.96 Problem 29. Linkage arm.

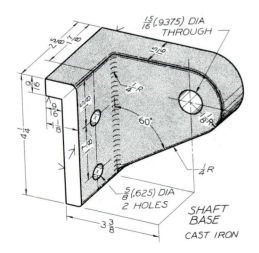

SHAFT
BASE
CAST IRON

Figure 15.97 Problem 30. Shaft base.

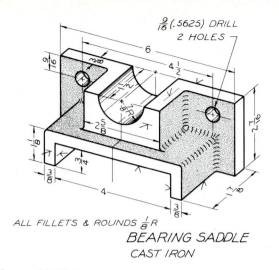

$\frac{9}{16}$ (.5625) DRILL
2 HOLES

ALL FILLETS & ROUNDS $\frac{1}{8}$ R

BEARING SADDLE
CAST IRON

Figure 15.98 Problem 31. Bearing saddle.

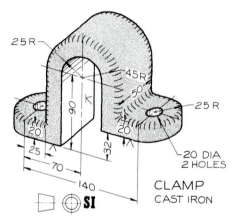

25 R

45 R

25 R

20 DIA
2 HOLES

CLAMP
CAST IRON

 **SI**

Figure 15.99 Problem 32. Clamp.

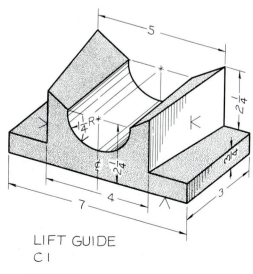

LIFT GUIDE
C I

Figure 15.100 Problem 33. Lift guide.

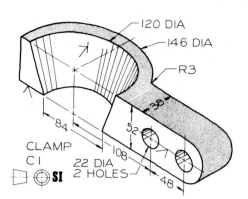

120 DIA
146 DIA
R3

CLAMP
C I

22 DIA
2 HOLES

 **SI**

Figure 15.101 Problem 34. Clamp.

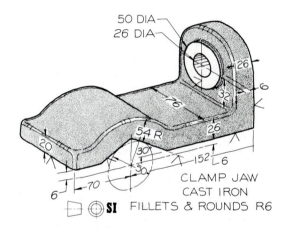

50 DIA
26 DIA

54 R

CLAMP JAW
CAST IRON
SI FILLETS & ROUNDS R6

Figure 15.102 Problem 35. Clamp jaw.

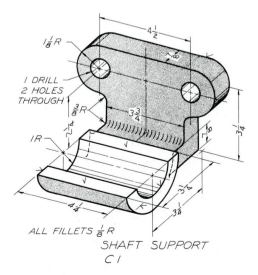

$1\frac{1}{8}$ R

1 DRILL
2 HOLES
THROUGH

$\frac{3}{8}$ R

1 R

ALL FILLETS $\frac{1}{8}$ R

SHAFT SUPPORT
C I

Figure 15.103 Problem 36. Shaft support.

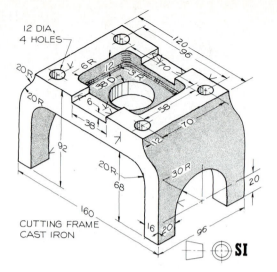

Figure 15.104 Problem 37. Cutting frame.

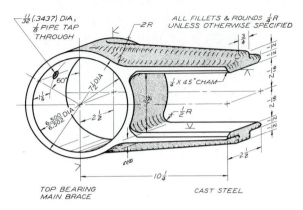

Figure 15.107 Problem 40. Top bearing.

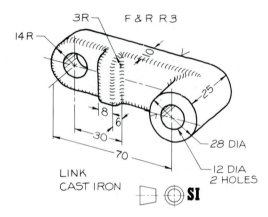

Figure 15.105 Problem 38. Link.

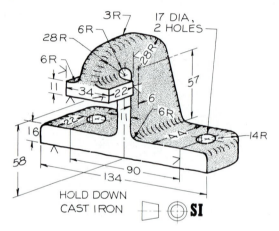

Figure 15.108 Problem 41. Hold down.

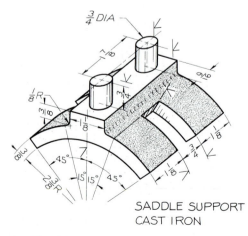

Figure 15.106 Problem 39. Saddle support.

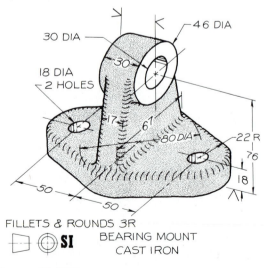

Figure 15.109 Problem 42. Bearing mount.

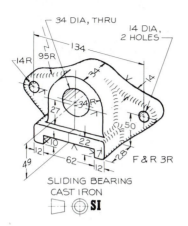

Figure 15.110 Problem 43. Sliding bearing.

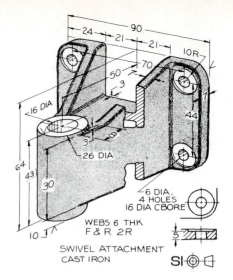

Figure 15.113 Problem 46. Swivel attachment.

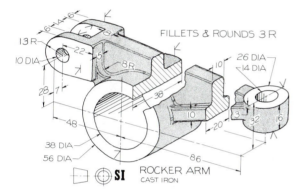

Figure 15.111 Problem 44. Rocker arm.

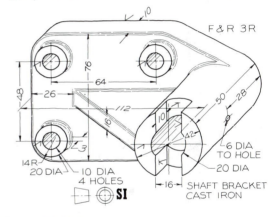

Figure 15.114 Problem 47. Shaft bracket.

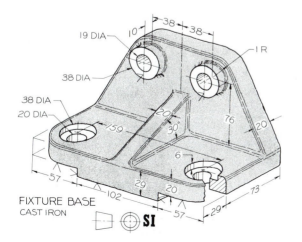

Figure 15.112 Problem 45. Fixture base.

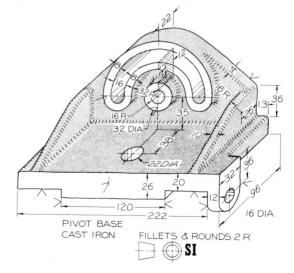

Figure 15.115 Problem 48. Pivot base.

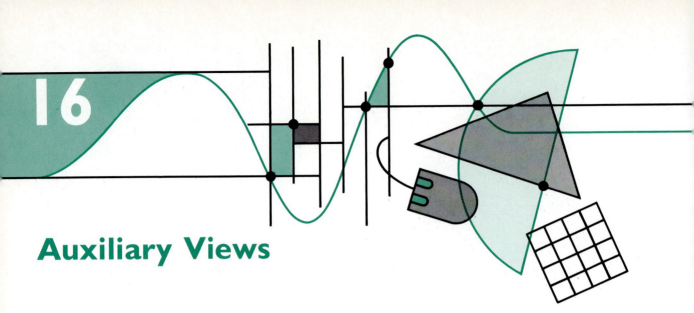

Auxiliary Views

16.1 Introduction

A plane not parallel to one of the principal projection planes is a nonprincipal plane that will **not** appear true size in a principal view; it can be found true size only on an **auxiliary plane** parallel to the nonprincipal plane. This nonprincipal view is called an **auxiliary view.**

An inclined surface of an object (Fig. 16.1) does not appear true size in the top view because it is not parallel to the horizontal projection plane. However, the inclined surface will appear true size if an auxiliary view is projected perpendicularly from the edge view of the plane in the front view.

16.2 Folding-Line Approach

The three principal orthographic planes are the **frontal** (F), **horizontal** (H), and **profile** (P) planes.

> A primary auxiliary plane is perpendicular to one of the principal planes but oblique to the other two, and a **primary auxiliary view** is projected from a **primary orthographic view.**

Auxiliary planes can be thought of as planes that fold from principal planes, as Fig. 16.2 shows. The plane in Fig. 16.2A folds from the frontal plane to make a 90° angle with it. The fold line between the two planes is labeled F-1, where F is an abbreviation for frontal, and 1 rep-

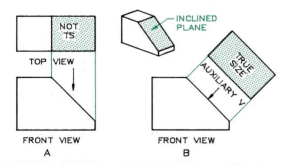

Figure 16.1 When a surface appears as an inclined edge in a principal view, it can be found true size by an auxiliary view. In part A, the top view is foreshortened, but this plane is true size in part B, the auxiliary view.

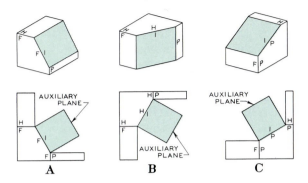

Figure 16.2 A primary auxiliary plane can be folded from the frontal (A), horizontal (B), or profile (C) planes. The fold lines are labeled F-1, H-1, and P-1.

resents **first,** or **primary,** auxiliary plane. Figures 16.2B and 16.2C illustrate the positions for auxiliary planes that fold from the horizontal and profile planes.

16.3 Auxiliaries Projected from the Top View

The inclined plane in Fig. 16.3 is an edge in the top view, and it is perpendicular to the horizontal plane. An auxiliary plane can be drawn parallel to the inclined surface, and the view projected onto it will be a true-size view of it.

> A surface must appear as an **edge** in a principal view before it can be found as true size in a primary auxiliary view.

A similar example is the object shown in the glass box in Fig. 16.4. Since the inclined surface in the top view appears as an edge, it can be found true size in a primary auxiliary view. The height *(H)* is transferred from the front view to the auxiliary view since both views are measured

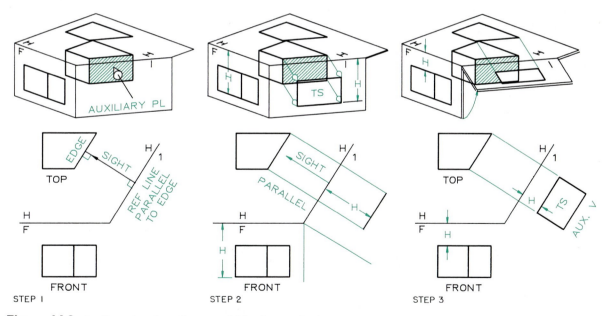

Figure 16.3 Auxiliary view from the top—folding-line method.

Required Find the inclined surface true size by an auxiliary view.

Step 1 Draw the line of sight perpendicular to the edge view of the surface. Draw the H-1 line parallel to the edge. Draw the HF reference line between the top and front views.

Step 2 Project from the edge view of the inclined surface parallel to the line of sight. Use the H dimensions from the front view to locate a line in the auxiliary view.

Step 3 Locate the other corners of the inclined surface by projecting to the auxiliary view. Locate the points by transferring the height (H) dimensions from the front view to the auxiliary view.

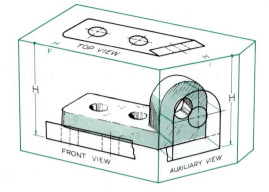

Figure 16.4 A pictorial showing the relationship of the auxiliary projection plane is used to find the true-size view of the inclined surface.

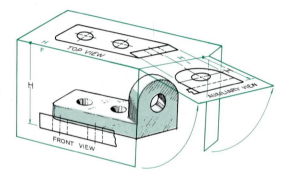

Figure 16.5 The auxiliary plane is opened into the plane of the top view by revolving it about the H-1 fold line.

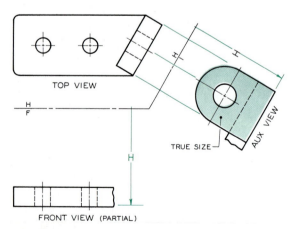

Figure 16.6 When the object is drawn on a sheet of paper, it is laid out in this manner. The front view is drawn as a partial view since the omitted part is shown true size in the auxiliary view.

from the same horizontal plane. The auxiliary plane is rotated about the H-1 fold line into the plane of the top view in Fig. 16.5.

When drawn on a sheet of paper, the drawing of this object would appear as shown in Fig. 16.6. The front view is shown as a partial view because the omitted portion would have been hard to draw, and it would not have been true size.

16.4 Auxiliaries from the Top View—Folding-Line Method

Figure 16.7 shows the steps of constructing an auxiliary view projected from the top view to find the true-size view of the inclined surface. Since the inclined surface appears as an edge in the top view, it can be found true size in a primary auxiliary view. The line of sight is drawn perpendicular to the edge view, and the fold line is drawn parallel to the edge. Height *(H)* is transferred from the front view.

The fold line, drawn thin but black, is labeled H-1. It is also helpful to number or letter points on the views. The projectors are construction lines, and they should be drawn with a hard pencil (3H–4H) just dark enough to be seen.

Computer Method Using computer graphics, the true-size view of a plane is found (Fig. 16.8) by projecting from the top view, where the inclined plane appears as an edge. With AutoCAD's ROTATE option of the SNAP command, the grid is rotated so that it is parallel to the edge view of the plane in the top view.

16.5 Auxiliaries from the Top View—Reference-Line Method

A similar method of locating an auxiliary view uses a reference plane instead of the fold line (Fig. 16.9). Instead of placing a fold line between the top and front views, a reference plane is passed

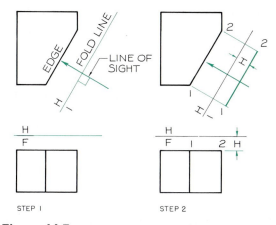

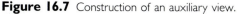

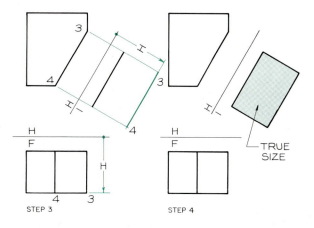

STEP 1

STEP 2

STEP 3

STEP 4

Figure 16.7 Construction of an auxiliary view.

Step 1 The line of sight is drawn perpendicular to the edge view of the inclined surface. The H-1 fold line is drawn parallel to the edge view of the inclined surface. An H-F fold line is drawn between the given views.

Step 2 Points 1 and 2 are found by transferring the height (*H*) dimensions from the front view to the auxiliary view.

Step 3 Points 3 and 4 are found in the same manner using the dimensions of height (*H*).

Step 4 The corner points are connected to complete the true-size view of the inclined plane.

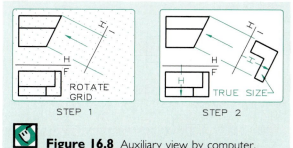

STEP 1

STEP 2

Figure 16.8 Auxiliary view by computer.

Step 1 To perpendicularly project an auxiliary from the edge of the inclined surface in the top view, the grid is rotated, using the **SNAP** command and **ROTATE** option. The H-1 reference line is drawn parallel to the edge, and projectors are drawn perpendicular to the H-1 line.

Step 2 The auxiliary view is found by transferring height dimensions measured in the front view and transferring them perpendicularly from the H-1 line. The auxiliary view is the true-size view of the inclined surface.

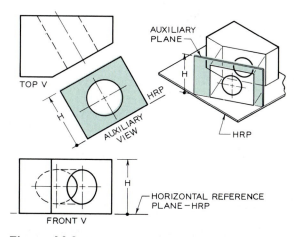

Figure 16.9 A horizontal reference plane can be used instead of the folding-line technique to construct an auxiliary view. Instead of placing the reference plane between the top and auxiliary views, it is placed outside the auxiliary view.

through the bottom of the front view. The height *(H)* dimensions are measured upward from the reference plane instead of downward from a fold line.

The reference plane shown as a horizontal edge in the front view is a **horizontal reference plane** (HRP). The HRP will appear as an edge view of the inclined surface from which the auxiliary is projected. The auxiliary view will lie between the HRP and the top view.

16.6   Auxiliaries from the Front View—Folding-Line Method

A plane that appears as an edge in the front view (Fig. 16.10) can be found true size in a primary auxiliary view projected from the front view.

Fold line F-1 is drawn parallel to the edge view of the inclined plane in the front view at a convenient location.

The line of sight is drawn perpendicular to the edge view of the inclined plane in the front view. Observed from this direction, the frontal plane appears as an edge; therefore, the measurements perpendicular to the frontal plane—the depth *(D)* dimensions—will be seen true length. Depth dimensions are transferred from the top view to the auxiliary view by using a pair of dividers.

The object in Fig. 16.11 is enclosed in a glass box, and an auxiliary plane is constructed parallel to the inclined plane. When drawn on a sheet of paper, the object appears as shown in Fig. 16.12. The top and side views are drawn as partial views since the auxiliary view eliminates the need for

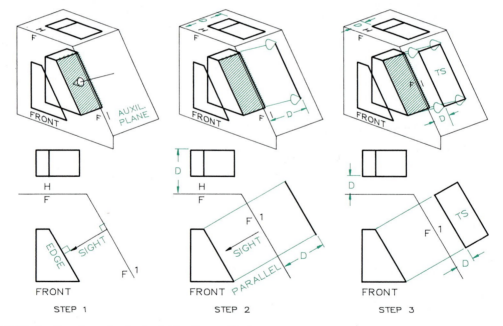

Figure 16.10 Auxiliary from the front—folding-line method.

Required Find the inclined surface true size by an auxiliary view.

Step 1 Draw the line of sight perpendicular to the edge of the plane and draw the F1 reference parallel to the edge. Draw the HF fold line between the top and front views.

Step 2 Project from the edge view of the inclined surface parallel to the line of sight. Use the D dimensions from the top view to locate a line in the auxiliary view.

Step 3 Locate the other corners of the inclined surface by projecting to the auxiliary view. Locate the points by transferring the depth (D) dimensions from the top to the auxiliary view.

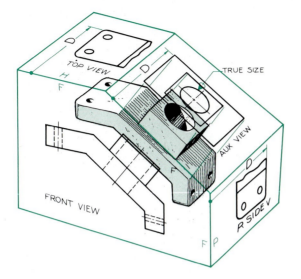

Figure 16.11 A pictorial showing the relationship of the projection planes used to find the true-size view of the inclined surface of the object.

seeing their complete views. The auxiliary view, which is located by using the depth dimension measured from the edge view of the frontal projection plane, shows the surface's true size.

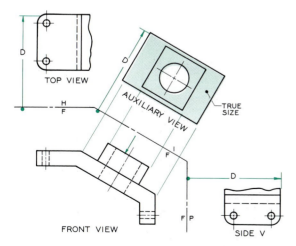

Figure 16.12 The layout and construction of an auxiliary view of the object shown in Fig. 16.11.

Computer Method   The inclined surface that appears as an edge in the front view (Fig. 16.13) is found true size using the LISP commands PAR–ALLEL and TRANSFER (see Section 28.2). While in AutoCAD's drafting mode, type <u>PARALLEL</u> to draw the reference line parallel to edge *AB*.

In Step 2, type <u>TRANSFER</u> to obtain prompts for transferring measurements from the top view to the auxiliary view the same manner as you would use a pair of dividers. In Step 3, connect the circular points to complete the auxiliary view.

16.7 Auxiliaries from the Front View—Reference-Plane Method

The object in Fig. 16.14 has an inclined surface that appears as an edge in the front view; therefore, this plane can be found true size in a primary auxiliary view.

It is advantageous to use a reference plane that passes through the center of the symmetrical top view. Because the reference plane is a frontal plane, it is called a **frontal reference plane** (FRP) in the auxiliary view. The FRP is located parallel to the edge view of the inclined plane and through the center of the auxiliary view.

16.8 Auxiliaries from the Profile View—Folding-Line Method

Since the inclined surface in Fig. 16.15 appears as an edge in the profile plane, it can be found true size in a primary auxiliary view projected from the profile view. The auxiliary fold line, P-1, is drawn parallel to the edge view of the inclined surface.

A line of sight perpendicular to the auxiliary plane will see the profile plane as an edge. Therefore, width *(W)* dimensions, which are transferred from the front view to the auxiliary view, will appear true length in the auxiliary view.

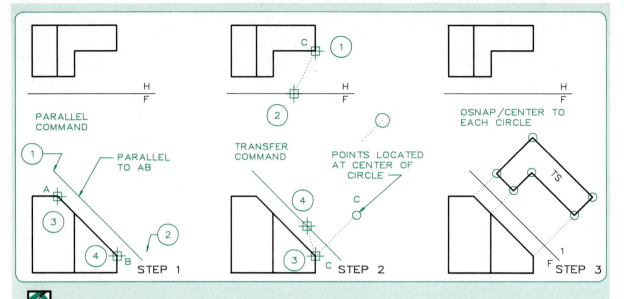

Figure 16.13 Auxiliary view by computer.

Step 1 While in AutoCAD, type **PARALLEL** and you are prompted for the first end of the reference line (1) and its approximate second end (2). You are then prompted for the beginning points and endpoints of the line (*AB*) that the reference line is parallel to.

Step 2 Type **TRANSFER**, and you are prompted to select a point in the top view (1) and its distance from the reference line (2) to be transferred. You are then asked for the front view of the point to be projected from (3) and the reference plane (4). The point is projected to the auxiliary view and located with a circle.

Step 3 Continue using **TRANSFER** to locate the other corner points of the inclined plane. Connect the points using the **CENTER** option of the **OSNAP** command to snap to the centers of the circles. Circle points can be **ERASE**d when their centers have been connected to complete the auxiliary view.

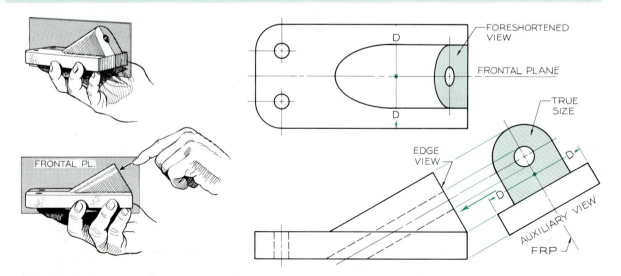

Figure 16.14 Since the inclined surface of this part is symmetrical, it is advantageous to use a frontal reference plane (FRP) that passes through the object. The auxiliary view is projected perpendicularly from the edge view of the plane in the front view. The FRP appears as an edge in the auxiliary view, and depth (*D*) dimensions are made on each side of it to locate points on the true-size view of the inclined surface.

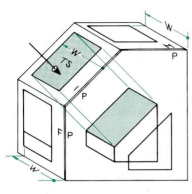

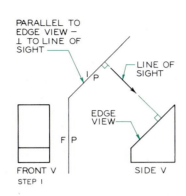

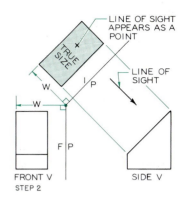

Figure 16.15 Auxiliary from the side—folding-line method.

Given An object with an inclined surface.

Required Find the inclined surface true size by an auxiliary view.

Step 1 Draw a line of sight perpendicular to the edge view of the inclined surface. Draw the P-1 fold line parallel to the edge view, and draw the F-P fold line between the given views.

Step 2 Project the corners of the edge parallel to the line of sight. Locate the corners by measuring perpendicularly from the P-1 fold line with width (*W*) dimensions.

16.9   Auxiliaries from the Profile View—Reference-Plane Method

The object in Fig. 16.16 has two inclined surfaces that appear as edges in the right-side view, the profile view. These inclined surfaces are found true size by using a **profile reference plane** (PRP) that is a vertical edge in the front view.

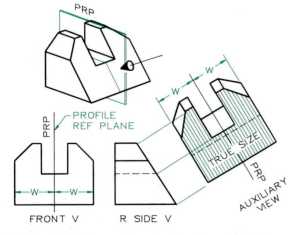

Figure 16.16 An auxiliary view is projected from the right-side view by using a profile reference plane (PRP). The auxiliary view shows the true-size view of the inclined surface.

To find the inclined surface's true size, an auxiliary view is drawn. The profile reference plane is positioned at the far outside of the auxiliary view instead of between the profile and auxiliary views, as in the folding-line method. By transferring the width *(W)* dimensions from the edge view of the PRP in the front view to the auxiliary view, the view is found.

16.10   Auxiliaries of Curved Shapes

When an auxiliary view is drawn to show a curve that is not a true arc, a series of points must be plotted. The cylinder in Fig. 16.17 has a beveled surface that appears as an edge in the front view.

Since the cylinder is symmetrical, it is beneficial to use an FRP through the object so that dimensions can be measured on both sides of it. Points are located about the circular right-side view, and these are projected to the edge view of the surface in the front view. The FRP is located parallel to the edge view of the plane, and the points are projected perpendicularly from the edge view of the plane. Dimensions *A* and *B* are shown as examples for plotting points in the aux-

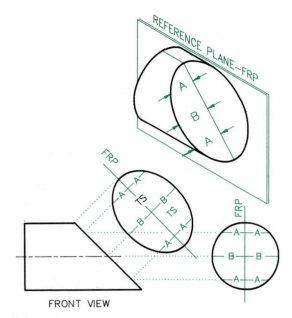

Figure 16.17 This auxiliary view requires that a series of points be plotted. Since the object is symmetrical, the reference plane (FRP) is positioned through its center.

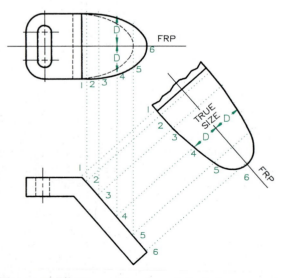

Figure 16.18 The auxiliary view of this surface required that a series of points be located in the given view and then projected to the auxiliary view.

iliary view. To construct a smooth elliptical curve, more points than shown are needed.

An irregular curve is found as an outline of a true-size surface in Fig. 16.18. Points are located on the curve in the top view and are projected to the front view. These points are found in the auxiliary view by plotting each of them using the depth *(D)* dimensions transferred from the top view.

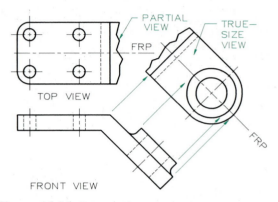

Figure 16.19 This combination of views is used to represent an object although the top and side views are partial views. The foreshortened portions of the object have been omitted. The FRP reference line passes through the center of the object in the top view since the object is symmetrical about this line.

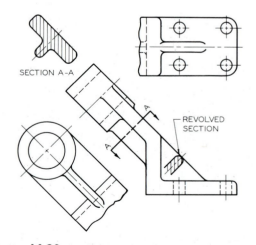

Figure 16.20 Auxiliary section A-A is projected from the cutting plane labeled A-A to show the cross section of the object.

16.11 Partial Views

An auxiliary view is a supplementary view, so some views of an orthographic arrangement can be drawn as partial views. The object in Fig. 16.19 shows a complete front view and partial auxiliary and side views. The partial views are easier to draw and are more functional without sacrificing clarity.

16.12 Auxiliary Sections

Figure 16.20 shows a section through a part that is projected as an auxiliary view. The section is labeled A-A, and a cutting plane passing through the object is labeled A-A. The cutting plane shows where the sectional view was projected from and where the object was cut to show the section. The auxiliary section provides a good description of the part that cannot be readily understood from the given principal views.

16.13 Secondary Auxiliary Views

> A **secondary auxiliary view** is projected from a primary auxiliary view

In Fig. 16.21, the inclined plane is found as an edge view in the primary auxiliary view, and then a line of sight perpendicular to the edge is drawn. The second auxiliary view shows the inclined surface true size. Remember that it is necessary to project from an edge view of a plane before it can be found true size.

The problem in Fig. 16.22 is a secondary auxiliary projection in which the point view of a diagonal of a cube is found. When this is found, the three surfaces of the cube are equally foreshortened. The secondary auxiliary view is an **isometric projection,** which is the basis for isometric pictorial drawing.

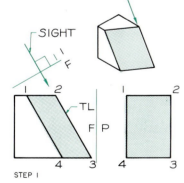
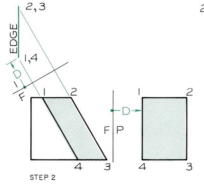
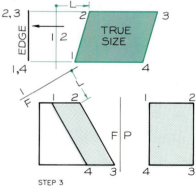

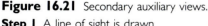

Figure 16.21 Secondary auxiliary views.

Step 1 A line of sight is drawn parallel to the true-length view of a line on the oblique surface. The folding line F-1 is drawn perpendicular to the line of sight.

Step 2 The primary auxiliary view of the oblique surface is an edge view. Depth (D) is used to locate a point in the primary auxiliary view.

Step 3 A line of sight is drawn perpendicular to the edge view, and a secondary auxiliary view is projected in this direction. Dimension L is used to locate one of the points in the true-size view.

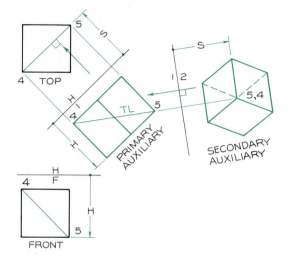

Figure 16.22 A point view of a diagonal of a cube is found in the secondary auxiliary view. Line 4–5 is found true length in the primary auxiliary view and then as a point in the secondary auxiliary view, which is an isometric projection of the cube.

Since auxiliary views are supplementary views, they can be partial views if all features are sufficiently shown. The object in Fig. 16.23 is shown as a series of orthographic views, all of which are partial views.

16.14   Elliptical Features

Occasionally, circular shapes will project as ellipses, which can be drawn using any of the techniques introduced in Chapter 13 once the necessary points have been plotted. The most convenient method of drawing ellipses is with an ellipse guide (template). The angle of the ellipse

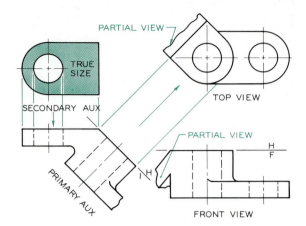

Figure 16.23 An example of a secondary auxiliary view projected from a partial auxiliary view.

guide is the angle the line of sight makes with the edge view of the circular feature. In Fig. 16.24, the angle is found to be 45°, so the right-side view is drawn as a 45° ellipse.

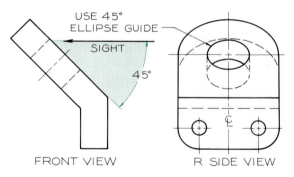

Figure 16.24 The ellipse guide angle is the angle that the line of sight makes with the edge view of the circular feature. The ellipse guide for the right-side view is 45°.

 **Problems**

The following problems are to be solved on Size A or Size B sheets, as assigned by your instructor.

1–10. (Fig. 16.25) Using the example layout, change the top and front views by substituting the top views given at the right in place of the one given in the example. The angle of inclination in the front view is 45° for all problems, and the height is 1½ inches in the front view. Construct auxiliary views that show the inclined surface true size. Draw two problems per Size A sheet.

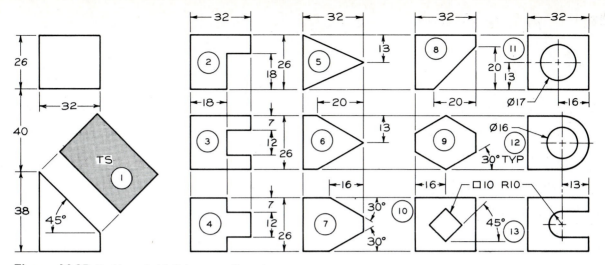

Figure 16.25 Problems 1–10. Primary auxiliary views.

11–32. (Figs. 16.26–16.47) Draw the necessary primary and auxiliary views to describe the parts shown. Draw one per Size A or Size B sheet, as assigned. Adjust the scale of each to fit the space on the sheet.

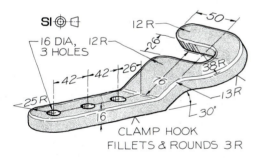

Figure 16.26 Problem 11. Clamp hook.

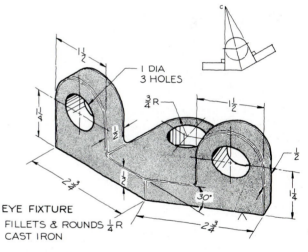

EYE FIXTURE

FILLETS & ROUNDS $\frac{1}{4}$ R
CAST IRON

Figure 16.28 Problem 13. Eye fixture.

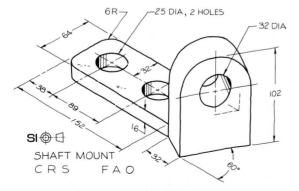

SHAFT MOUNT
C R S   F A O

Figure 16.27 Problem 12. Shaft mount.

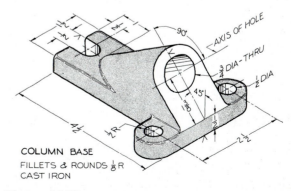

COLUMN BASE

FILLETS & ROUNDS $\frac{1}{8}$ R
CAST IRON

Figure 16.29 Problem 14. Column base.

230

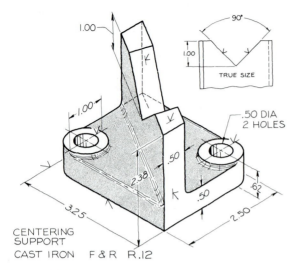

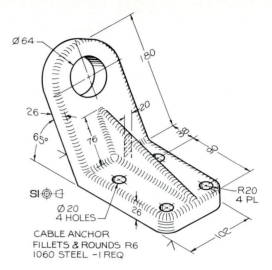

CENTERING
SUPPORT
CAST IRON F & R R.12

Figure 16.30 Problem 15. Centering support.

CABLE ANCHOR
FILLETS & ROUNDS R6
1060 STEEL – 1 REQ

Figure 16.33 Problem 18. Cable anchor.

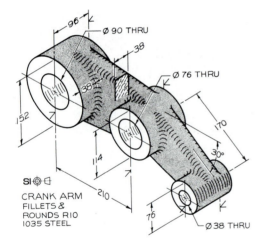

CRANK ARM
FILLETS &
ROUNDS R10
1035 STEEL

Figure 16.31 Problem 16. Crank arm.

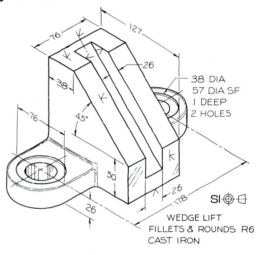

WEDGE LIFT
FILLETS & ROUNDS R6
CAST IRON

Figure 16.34 Problem 19. Wedge lift.

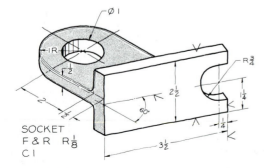

SOCKET
F & R R⅛
C I

Figure 16.32 Problem 17. Socket.

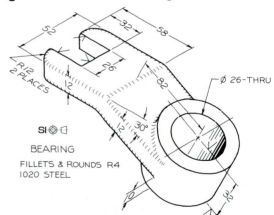

BEARING
FILLETS & ROUNDS R4
1020 STEEL

Figure 16.35 Problem 20. Bearing.

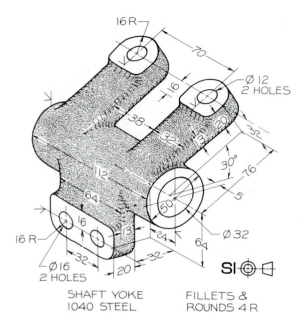

SHAFT YOKE
1040 STEEL

FILLETS &
ROUNDS 4 R

Figure 16.36 Problem 21. Shaft yoke.

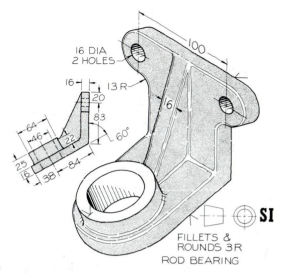

FILLETS &
ROUNDS 3 R
ROD BEARING

Figure 16.38 Problem 23. Rod bearing.

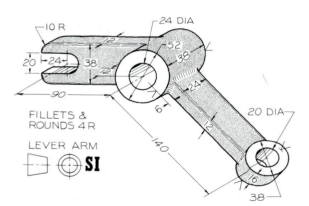

FILLETS &
ROUNDS 4 R

LEVER ARM

Figure 16.37 Problem 22. Lever arm.

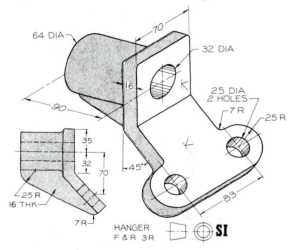

HANGER
F & R 3R

Figure 16.39 Problem 24. Hanger.

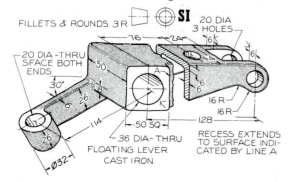

FILLETS & ROUNDS 3R

FLOATING LEVER
CAST IRON

RECESS EXTENDS
TO SURFACE INDI-
CATED BY LINE A

Figure 16.40 Problem 25. Floating lever.

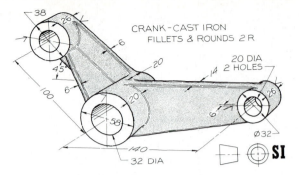

Figure 16.41 Problem 26. Crank.

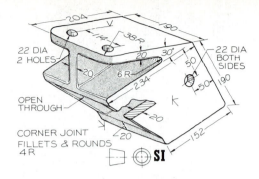

Figure 16.45 Problem 30. Corner joint.

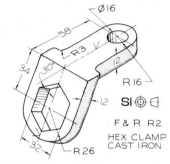

Figure 16.42 Problem 27. Hexagon angle.

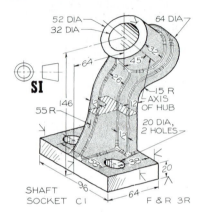

Figure 16.46 Problem 31. Shaft socket.

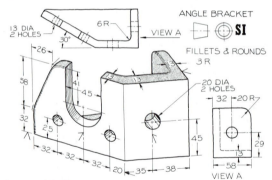

Figure 16.43 Problem 28. Angle bracket.

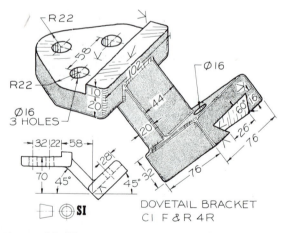

Figure 16.47 Problem 32. Dovetail bracket.

33–34. (Figs. 16.48–16.49) Lay out these orthographic views on Size B sheets, and complete the auxiliary and primary views.

35. (Fig. 16.50) Construct orthographic views of the given object, and using secondary auxiliary views, draw auxiliary views that give the true-size views of the inclined surfaces. Draw one per Size B sheet.

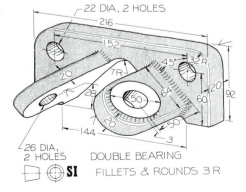

Figure 16.44 Problem 29. Double bearing.

232

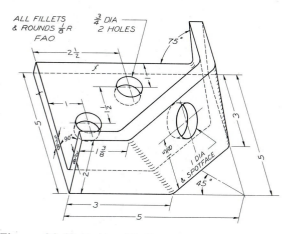

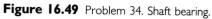

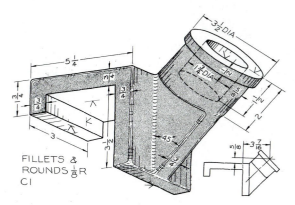

Figure 16.48 Problem 33. Corner connector.

Figure 16.49 Problem 34. Shaft bearing.

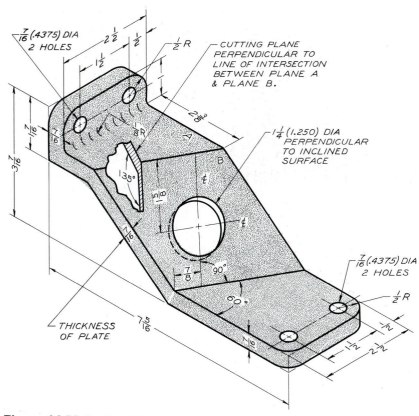

Figure 16.50 Problem 35. Oblique support.

17

Sections

17.1 Introduction

Standard orthographic views that show all hidden lines may confuse the true details of an object. This shortcoming can often be improved by cutting away part of the object and looking at the cross-sectional view. Such a cutaway view is called a **section.**

A section is shown pictorially in Fig. 17.1, where an imaginary cutting plane is passed through the object to show its internal features. Figure 17.1A shows the standard top and front views, and Fig. 17.1B shows the method of drawing a section. The front view has been converted to a **full section,** and the cut portion is cross-hatched. Hidden lines have been omitted since they are not needed. The cutting plane is drawn as a heavy line with short dashes at intervals; this can be thought of as a knife-edge cutting through the object.

Figure 17.2 shows two types of cutting planes. Either is acceptable although the example in Fig. 17.2A is more often used. The spacing of the dashes depends on the size of the drawing. The weight of the cutting plane is the same as that of a visible object line. Letters can be placed

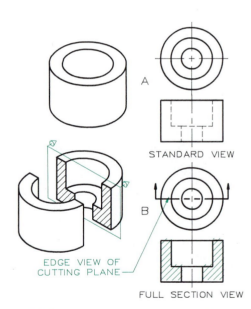

Figure 17.1 A comparison of a regular orthographic view with a full-section view showing the internal and external features of the same object.

at each end of the cutting plane to label the sectional view, such as section B-B in Fig. 17.2B.

Figure 17.3 shows the three basic views that appear as sections, with their respective cutting planes. Each cutting plane has perpendicular arrows pointing in the direction of the line of sight for the section. For example, the cutting plane in Fig. 17.3A passes through the top view, the front of the top view is removed, and the line of sight is toward the remaining portion of the top view. The top view will appear as a section when the cutting plane passes through the front view and the line of sight is downward (Fig. 17.3B). When the cutting plane passes through the front view (Fig. 17.3C), the right-side view will be a section.

17.2 Sectioning Symbols

Figure 17.4 shows the symbols used to distinguish between different materials in sections. Although the symbols can be used to indicate the

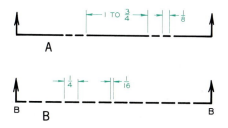

Figure 17.2 Typical cutting-plane lines used to represent sections. The cutting plane marked B-B will produce a section labeled B-B.

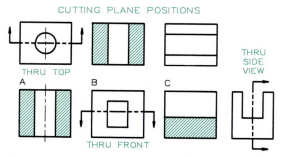

Figure 17.3 The three standard positions of cutting planes that pass through views to result in sectional views in the (A) top, (B) front, and (C) side views. The arrows point in the direction of the line of sight for each section.

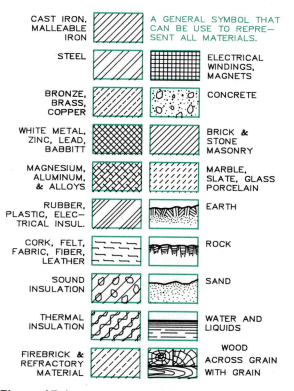

Figure 17.4 The symbols used for lining parts in section. The cast-iron symbol can be used for any material.

materials within a section, it is advisable to provide supplementary notes specifying the materials to avoid misinterpretation.

Cast-iron symbols are usually drawn with a 2H pencil with lines slanted at 30°, 45°, or 60° angles and spaced about $\frac{1}{16}$ inch apart.

> The **cast-iron symbol** of evenly spaced section lines can be used to represent **any material** and is the most often used sectioning symbol.

Computer Method   Figure 17.5 shows a few of the many cross-sectional symbols available from AutoCAD. The spacing between the lines and dash lengths may be varied by changing the pattern scale factor.

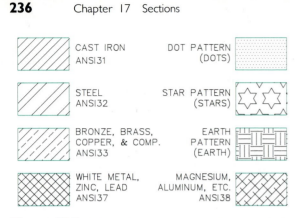

Figure 17.5 A few of the sectioning symbols provided by AutoCAD. The pattern scale can be used to vary the spacing of the lines and dashes.

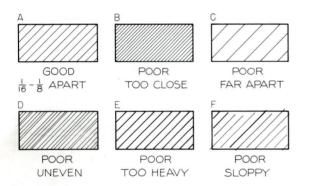

Figure 17.6 Good section lines are thin and $\frac{1}{16}$ to $\frac{1}{8}$ inch apart. Some typical lining errors are shown.

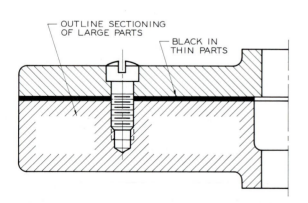

Figure 17.7 Thin parts are blacked in, and large areas are section-lined around their outlines to save time and effort.

The proper spacing of section lines is shown in Fig. 17.6A, where the lines are evenly spaced. The other parts of the figure show common errors of section lining.

Extremely thin parts such as sheet metal, washers, and gaskets (Fig. 17.7) are sectioned by completely blacking in the areas rather than by using section lines. Large parts are sectioned with an **outline section** to save time and effort. The section lines are drawn closer together in small parts rather than in larger parts.

Sectioned areas should be lined with symbols that are neither parallel nor perpendicular to the outlines of the parts lest they be confused for serrations or other machining treatments of the surface. (Fig. 17.8).

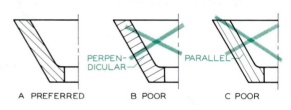

Figure 17.8 Section lines should be drawn so that they are neither parallel nor perpendicular to the outline of a part.

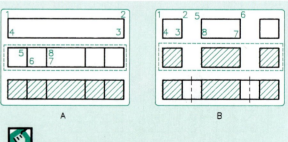

Figure 17.9 Sections by computer. (A) If lines were drawn from 1 to 2 to 3 to 4 and division lines were drawn from 5 to 6 and 7 to 8, the resulting section would ignore the intermediate lines (5–6, for example) when the area is windowed. The section lines will extend to the perimeter, 1–2–3–4. (B) If lines are drawn in segments to stop at the corners of the sectioned areas, and then windowed, the section lines will fill the areas as intended. The areas can now be connected and centerlines added to complete the sectional view.

Computer Method   Figure 17.9 shows the method of drawing areas to be section-lined using AutoCAD's HATCH command. The areas to be sectioned must be drawn with lines terminating at each corner point, as Fig. 17.9B shows.

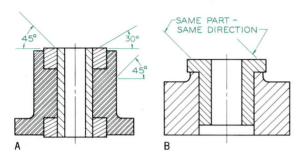

Figure 17.10 (A) Section lines of the same part should be drawn in the same direction. (B) Section lines of different parts should be drawn at varying angles to distinguish the parts.

17.3   Sectioning Assemblies

When an assembly of several parts is sectioned, it is important that the section lines be drawn at varying angles to distinguish the parts (Fig. 17.10). Using different material symbols in an assembly also helps distinguish the materials of the

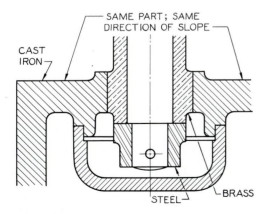

Figure 17.11 A typical assembly in section with well-defined parts and correctly drawn section lines.

parts. The same part is cross-hatched at the same angle and with the same symbol even though the part may be separated into different areas (Fig. 17.10B). Section lines are effectively used in Fig. 17.11 to identify the parts of the assembly.

17.4   Full Sections

> A **full section** is a view formed by passing a cutting plane fully through an object and removing half of it.

In Fig. 17.12, an object is drawn as two orthographic views in which hidden lines are shown. The front view can be drawn as a full section by passing a cutting plane fully through the top view and removing the front portion. The arrows on the cutting plane indicate the direction of sight, and the front view is then section-lined to give the full section.

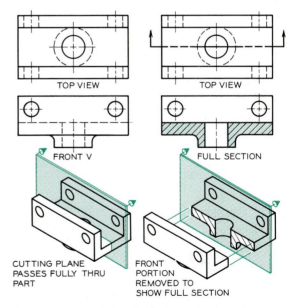

Figure 17.12 A full section is formed by a cutting plane that passes completely through the part. The cutting plane is shown passing through the top view, and the direction of sight is indicated by the arrows at each end. The front view is converted to a sectional view to give a clear understanding of the internal features.

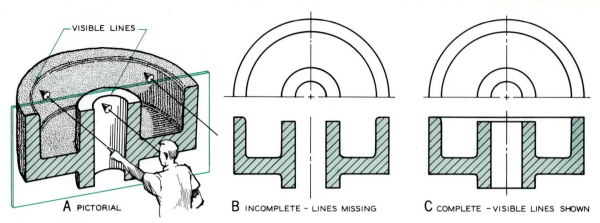

A PICTORIAL   B INCOMPLETE – LINES MISSING   C COMPLETE – VISIBLE LINES SHOWN

Figure 17.13 Full section–cylindrical part. (A) When a full section is passed through an object, you will see lines behind the sectioned area. (B) If only the sectioned area were shown, the view would be incomplete. (C) Visible lines behind the sectioned area must be shown also.

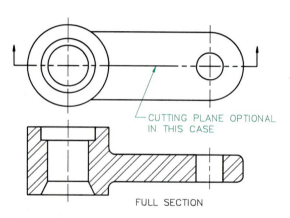

CUTTING PLANE OPTIONAL IN THIS CASE

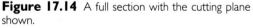

FULL SECTION

Figure 17.14 A full section with the cutting plane shown.

A full section through a cylindrical part is shown in Fig. 17.13A, where half the object is removed. A common mistake in constructing sectional views is omitting the visible lines behind the cutting plane, as in Fig. 17.13B. Figure 17.13C shows the correctly drawn sectional view.

> Hidden lines are omitted in all sectional views unless they are considered necessary to provide a clear understanding of the view.

Figure 17.14 is an example of a part whose front view is shown as a full section. Likewise, the part in Fig. 17.15 illustrates a front view that appears as a full section. Lines behind the cutting plane are shown as visible lines.

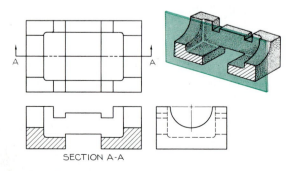

SECTION A-A

Figure 17.15 A full section, section A-A, is used to supplement the given views of the object.

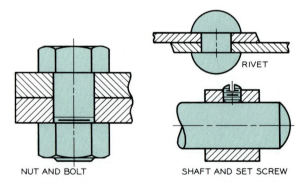

RIVET

NUT AND BOLT   SHAFT AND SET SCREW

Figure 17.16 These parts are not section-lined even though the cutting plane passes through them.

238

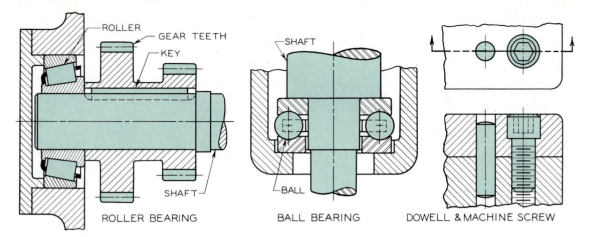

ROLLER BEARING   BALL BEARING   DOWELL & MACHINE SCREW

Figure 17.17 These parts are not section-lined even though cutting planes pass through them.

17.5 Parts Not Section-Lined

Many standard parts like nuts and bolts, rivets, shafts, and set screws, are not section-lined even though the cutting plane passes through them (Fig. 17.16). Since these parts have no internal features, sections through them would be of no value. Other parts not section-lined are roller bearings, ball bearings, gear teeth, dowels, pins, and washers (Fig. 17.17).

17.6 Ribs in Section

Ribs are not section-lined when the cutting plane passes flatwise through them, as in Fig. 17.18A, since this would give a misleading impression of

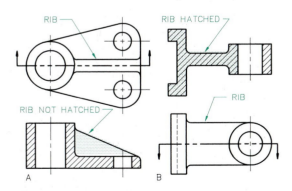

Figure 17.18 A rib cut in a flatwise direction by a cutting plane is not section-lined. Ribs are section-lined when cutting planes pass perpendicularly through them, as shown in part B.

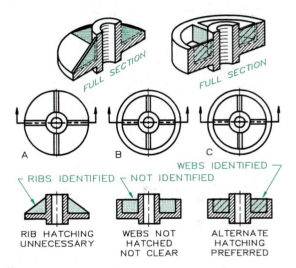

Figure 17.19 Outside ribs in section are not section-lined. Poorly identified webs, as in part B, should be identified by alternating section lines with cross-hatching, as in part C.

the rib. But a rib is section-lined when the cutting plane passes through it and shows its true thickness (Fig. 17.18B).

Figure 17.19 shows an alternative method of section-lining webs and ribs. The ribs are not section-lined since the cutting plane passes flatwise through them (Fig. 17.19A). The webs are symmetrically spaced about the hub (Fig. 17.19B). As a rule, webs are not cross-hatched, but this would leave them unidentified; therefore, it is better to use the **alternate sectioning** technique,

239

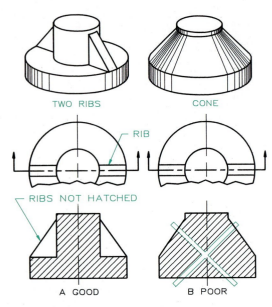

Figure 17.20 When ribs are not section-lined, the view is more descriptive of the part. Partial views are used in the top views to save space. The front part of the top view is removed when the front view is a section.

which extends every other section line through the webs (Fig. 17.19C).

The ribs in Fig. 17.20A are not section-lined and thus afford a more descriptive view of the part. If the ribs had been section-lined, the section would have given the impression that the part was solid and conical, as Fig. 17.20B shows.

17.7 Half Sections

A **half section** is a view that results from passing a cutting plane halfway through an object and removing a quarter of it to show external and internal features.

A half section is most often used with symmetrical parts, and with cylinders in particular. A cylindrical part in Fig. 17.21A is shown as a pictorial half section in Fig. 17.21B. The method of drawing the orthographic half section is shown in Fig. 17.21C, where both the internal and external

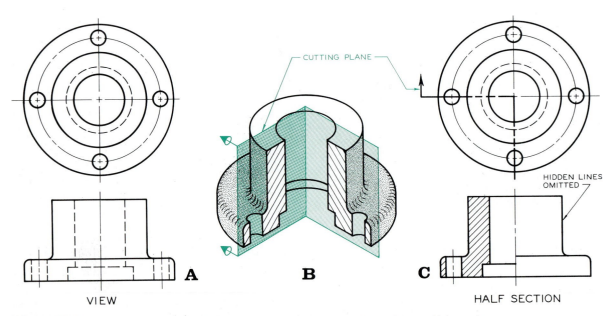

Figure 17.21 The cutting plane of a half section passes halfway through the object, which results in a sectional view that shows half the outside and half the inside of the object. Hidden lines are omitted unless they are needed to clarify the view.

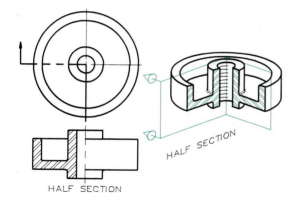

Figure 17.22 A half section of pulley shown orthographically and pictorially.

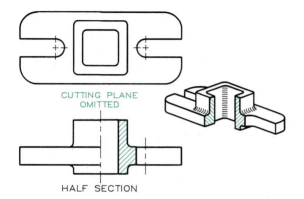

Figure 17.23 When it is obvious where the cutting plane is located, it is unnecessary to show it, as in this example.

features can be seen. Figure 17.22 shows a half section of a pulley.

The half section in Fig. 17.23 has been drawn without showing the cutting plane, which is permissible if it is obvious where the cutting plane was passed through the object. Instead of using an object line, centerlines separate the sectional half from the half that appears as an external view.

17.8 Partial Views

Figure 17.24 shows a conventional method of representing symmetrical views. A half view is sufficient when it is drawn adjacent to the sec-

tional view (Fig. 17.24A). In full sections, the removed half is the portion nearest the section (Fig. 17.24B). When drawing half views (nonsectional views), the removed half of the partial view is the half away from the adjacent view (Fig. 17.24A).

When partial views are drawn with half sections, either the near or the far halves of the partial views can be omitted.

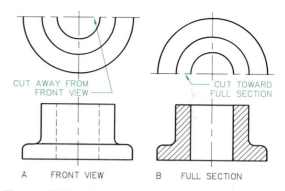

Figure 17.24 Half views can be used for symmetrical objects to conserve space and drawing time. In part A, the omitted portion of the view is away from the view. In part B, the omitted portion of the view is toward the full section. The omitted half can be toward or away from the section in the case of a half section.

17.9 Offset Sections

An **offset section** is a full section in which the cutting plane is offset to pass through important features.

Figure 17.25 shows an offset section where the plane is offset to pass through the large hole and one of the small holes. The second part of the figure shows the method of drawing the offset section orthographically. The cut formed by the offset is not shown in the section since this is an imaginary cut. The object in Fig. 17.26 also lends itself to representation by an offset section.

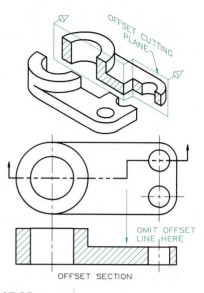

Figure 17.25 Offset section. The cutting plane may need to be offset to pass through features of a part. The offset cutting plane is shown in the top view, and the front view is shown as if it were a full section.

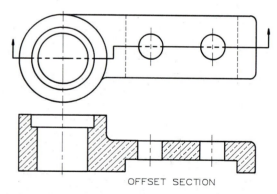

Figure 17.26 An offset section.

17.10   Revolved Sections

A **revolved section** is used to describe a cross section of a part by revolving it about an axis of revolution and placing it on the view centered on the axis of revolution.

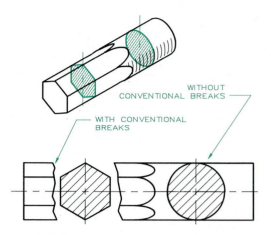

Figure 17.27 Revolved sections can be drawn with or without conventional breaks; either method is acceptable.

For example, revolved sections are used to indicate cross sections of the parts in Fig. 17.27. Revolved sections are shown with and without conventional breaks.

A more advanced type of revolved section is illustrated in Fig. 17.28, where a cutting plane is passed through the object (Step 1). The plane is imagined to be revolved in the top view to give a true-size revolved section in the front view (Step 2). It would also have been permissible to use conventional breaks on each side of the revolved section.

Figure 17.29 shows typical revolved sections. These sections provide a method of giving a part's cross section without drawing another complete orthographic view.

17.11   Removed Sections

A **removed section** is a revolved section that has been removed from the view where it was revolved (Fig. 17.30).

Centerlines are used as axes of rotation to show where the sections were taken from. Removed sections may be necessary where room does not

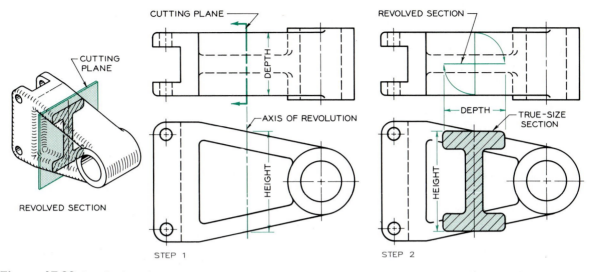

Figure 17.28 Revolved section.

Step 1 An axis of revolution is shown in the front view. The cutting plane would appear as an edge in the top view if it were shown.

Step 2 The vertical section in the top view is revolved so that the section can be seen true size in the front view. Object lines are not drawn through the revolved section.

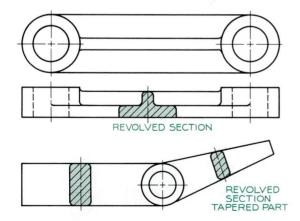

Figure 17.29 The revolved sections given here are helpful in describing the cross-sections of the two parts without using additional orthographic views.

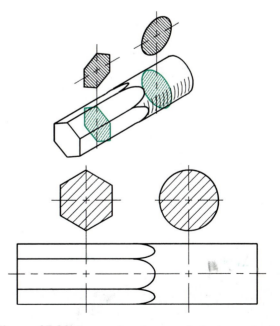

Figure 17.30 Removed sections are similar to revolved sections, but they have been removed outside the object along an axis of revolution.

permit revolution on the given view (Fig. 17.31A); instead, the cross section must be removed from the view (Fig. 17.31B).

Removed sections do not have to be positioned directly along an axis of revolution adjacent to the view from where the sections were taken. Instead, cutting planes can be labeled at each end, as Fig. 17.32 shows, to specify the sections. For example, the plane labeled with an A at each end is used to label section A-A; section B-B is similarly found.

When a set of drawings consists of many pages, removed sections may be put on different

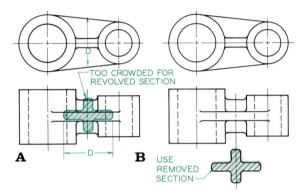

Figure 17.31 Removed sections can be used where space does not permit the use of revolved sections.

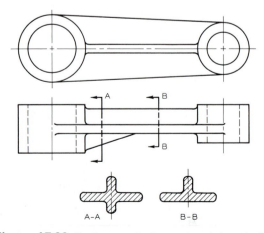

Figure 17.32 Sections can be lettered at each end of a cutting plane, such as A-A. This removed section can then be shown elsewhere on the drawing and is designated section A-A.

Figure 17.33 If it is necessary to remove a section to another page in a set of drawings, each end of the cutting plane can be labeled with a letter and a number. The letters refer to section A-A, and the numbers mean this section is on page 7.

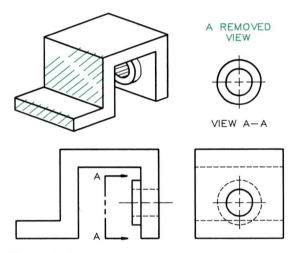

Figure 17.34 A removed view (not a section) can also be used to view a part from an unconventional direction.

sheets. In this case, a cutting plane may be labeled as shown in Fig. 17.33.

As Fig. 17.34 shows, removed views can also be used to provide inaccessible orthographic views (nonsectional views).

17.12 Broken-Out Sections

A **broken-out section** is used to show interior features by breaking away a portion of a view.

A portion of the object in Fig. 17.35 is broken out to reveal details of the wall thickness that better explain the drawing. The irregular lines representing breaks are **conventional breaks.**

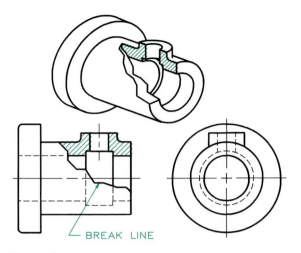

Figure 17.35 A broken-out section drawn by computer shows internal features by using a conventional break.

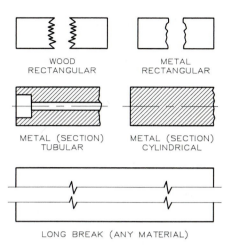

Figure 17.37 These conventional breaks indicate that a portion of an object has been broken away.

17.13 Phantom (Ghost) Sections

A **phantom** or **ghost section** is used to depict parts as if they were viewed by an x ray.

In Fig. 17.36, the cutting plane is drawn in the usual manner, but the section lines are drawn as dashed lines. If the object had been shown as a regular full section, the circular hole through the

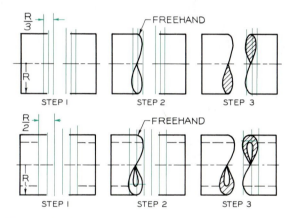

Figure 17.38 Conventional breaks in cylindrical and tubular sections can be drawn freehand with the aid of the guidelines shown. The radius, *R,* is used to establish the width of both "figure 8's."

front surface could not have been shown in the same view.

17.14 Conventional Breaks

Figure 17.37 shows examples of **conventional breaks.** The "figure 8" breaks, which are used for cylindrical and tubular parts, can be drawn freehand (Fig. 17.38). They can be drawn with a compass when drawn to a large scale (Fig. 17.39).

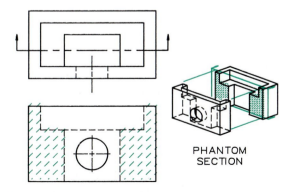

Figure 17.36 Phantom sections give an "x-ray" view of an object. The section lines are shown as dashed lines, which makes it possible to show the section without removing the hole in the front of the part.

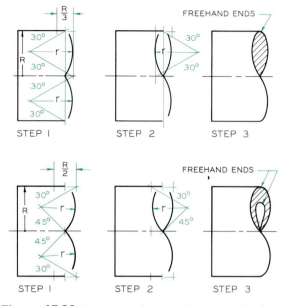

STEP 1   STEP 2   STEP 3

Figure 17.39 The steps of constructing conventional breaks of solid and tubular shapes with instruments.

One use of conventional breaks is to shorten a long piece that has a uniform cross section. The long part in Fig. 17.40 has been shortened and drawn at a larger scale for more clarity by using conventional breaks (Fig. 17.40). The dimension specifies the true length of the part, and the

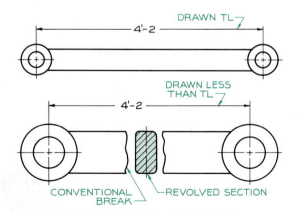

Figure 17.40 By using conventional breaks and a revolved section, this part can be drawn at a larger scale that is easier to read.

breaks indicate that a portion of the length has been removed.

17.15 Conventional Revolutions

Figure 17.41 shows three conventional sections. The center hole is omitted in Fig. 17.41A since it does not pass through the center of the circular plate. However, the hole in Fig. 17.41B does pass through the plate's center and is sectioned accordingly. In Fig. 17.41C, although the cutting plane does not pass through one of the symmetrically spaced holes in the top view, the hole is revolved to the cutting plane to show the recommended full section.

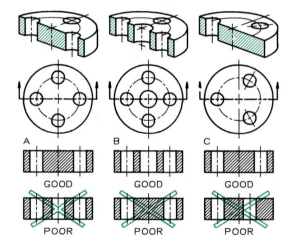

Figure 17.41 Symmetrically spaced holes in a circular plate should be revolved in their sectional views to show them at their true radial distance from the center. In part A, no hole is shown at the center, but a hole is shown at the center in part B since the hole is through the center of the plate. In part C, one of the holes is rotated to the cutting plane so that the sectional view will be symmetrical and more descriptive.

When ribs are symmetrically spaced about a hub (Fig. 17.42), it is conventional practice to revolve them to where they will appear true size in either a view or a section. Figure 17.43 shows a full section that shows both ribs and holes revolved to their true-size locations.

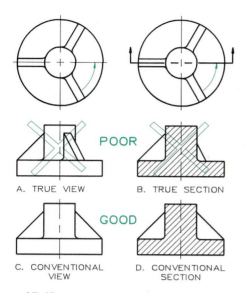

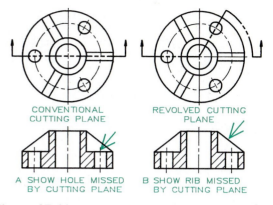

Figure 17.44 Symmetrically located ribs are shown in section in revolved positions to show the ribs true size. Foreshortened ribs are omitted.

Figure 17.42 Symmetrically located ribs are shown revolved in both orthographic and sectional views as a conventional practice.

The cutting plane can be positioned as Fig. 17.44 shows. Even though the cutting plane does not pass through the ribs and holes in Fig. 17.44A, the sectional view should be drawn as shown in Fig. 17.44B, where the cutting plane is revolved. The cutting plane can be drawn in either position.

In the same manner as ribs in section, symmetrically spaced spokes are rotated and not section-lined (Fig. 17.45).

Only the revolved, true-size spokes are drawn; the intermediate spokes are omitted.

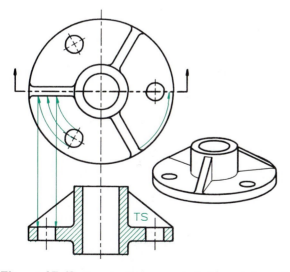

Figure 17.43 A part with symmetrically located ribs and holes is shown in section with both ribs and holes rotated to the cutting plane.

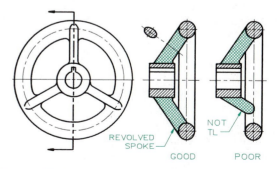

Figure 17.45 Symmetrically positioned spokes are revolved to show the spokes true size in section. Spokes are not section-lined in section.

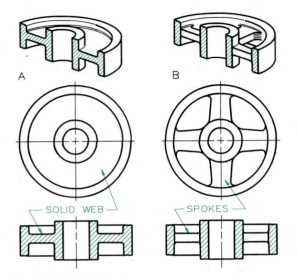

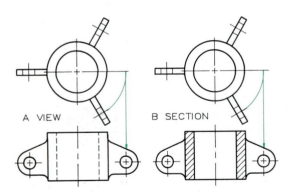

Figure 17.46 Solid webs in sections of the type in part A are section-lined. Spokes are not section-lined when the cutting plane passes through them, as in part B.

Figure 17.47 Symmetrically spaced lugs (flanges) are revolved to show the front view (A) and the sectional view (B) as symmetrical.

If the spokes in Fig. 17.46B had been section-lined, the cross section of the part would be confused with the part in Fig. 17.46A, where there are no spokes but a continuous web.

The lugs symmetrically positioned about the central hub of the object in Fig. 17.47 are revolved to show they are true size in both views and sections. A more complex object involving the same principle of rotation can be seen in Fig. 17.48, where the oblique arm is drawn in the sec-

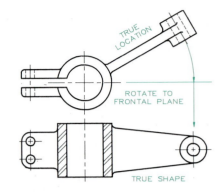

Figure 17.48 A part with an oblique feature attached to the circular hub is revolved so that it will appear true shape in the front view, which is a sectional view.

tion as if it had been revolved to the centerline in the top view and then projected to the sectional view.

17.16 Auxiliary Sections

Auxiliary sections can be used to supplement the principal views used in orthographic projections, as Fig. 17.49 shows. Auxiliary cutting plane A-A is passed through the front view, and the auxiliary view is projected from the cutting plane as indicated by the sight arrows. Section A-A gives the cross-sectional description of the part.

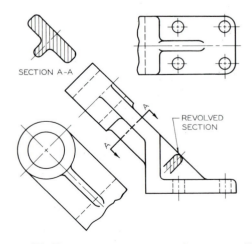

Figure 17.49 Sectional views can be shown as auxiliary views for added clarity.

Problems

These problems can be solved on Size A or Size B sheets.

problems per Size A sheet. Each grid equals 0.20 inch or 5 mm. Complete the front views as full sections.

1–24. (Fig. 17.50) Full sections: Draw two of these

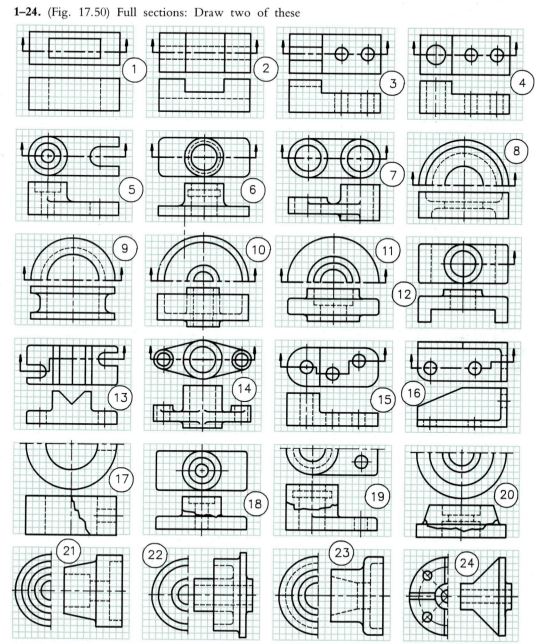

Figure 17.50 Problems 1–24. Introductory sections.

25–29. (Fig. 17.51) Construct views of the fixtures assigned on A-size sheets and use the appropriate sections to give them the best clarity. The grid is spaced at 0.20 in. or 5 mm. Select the best scale and sheet size for each problem. (Courtesy of Jergens Inc., Cleveland, Ohio.)

30–34. (Figs. 17.52–17.56) Full sections: Complete the drawings as full sections. Draw one problem per Size A sheet. Each grid equals 0.20 inch or 5 mm. Show the cutting planes when they are not given.

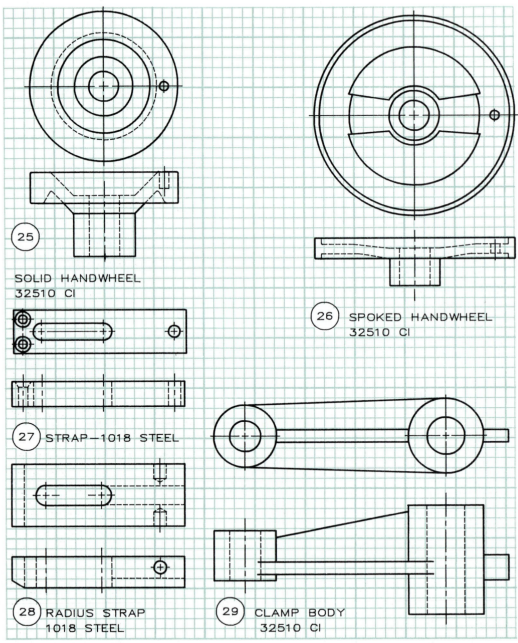

25 SOLID HANDWHEEL
32510 CI

26 SPOKED HANDWHEEL
32510 CI

27 STRAP—1018 STEEL

28 RADIUS STRAP
1018 STEEL

29 CLAMP BODY
32510 CI

Figure 17.51 Problems 25–29.

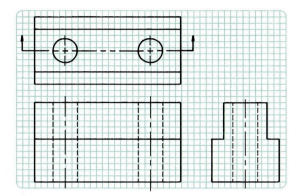

Figure 17.52 Problem 30. Full section.

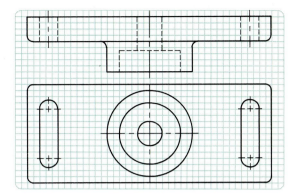

Figure 17.55 Problem 33. Full section.

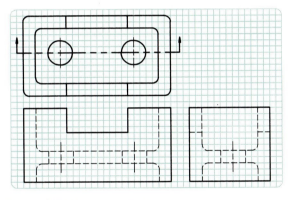

Figure 17.53 Problem 31. Full section.

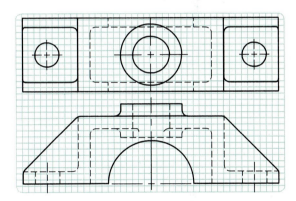

Figure 17.56 Problem 34. Full section.

35–36. (Figs. 17.57–17.58) Half sections: Complete the drawings as half sections. Draw one problem per Size A sheet. Each grid equals 0.20 inch or 5 mm. Show the cutting planes when they are not given.

37–38. (Figs. 17.59–17.60) Offset sections: Complete the drawings as offset sections. Draw one problem per Size A sheet. Each grid equals 0.20 inch or 5 mm. Show the cutting planes.

39. (Fig. 17.61) Full section: Complete the partial view as a full section. Draw the views on a Size A sheet. Each grid equals 0.20 inch or 5 mm. Show the cutting plane.

40. (Fig. 17.62) Assembly: Complete the front view as a full section of the assembled parts. Draw the views on a Size A sheet. Each grid equals 0.20 inch or 5 mm. Show the cutting plane.

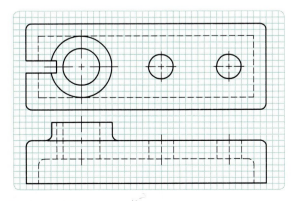

Figure 17.54 Problem 32. Full section.

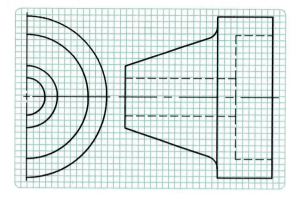

Figure 17.57 Problem 35. Half section.

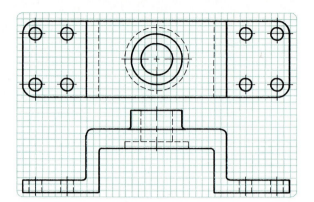

Figure 17.60 Problem 38. Offset section.

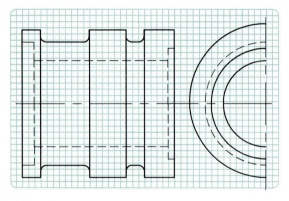

Figure 17.58 Problem 36. Half section.

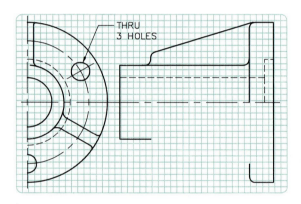

THRU
3 HOLES

Figure 17.61 Problem 39. Full section.

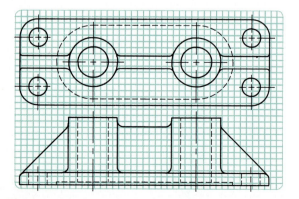

Figure 17.59 Problem 37. Offset section.

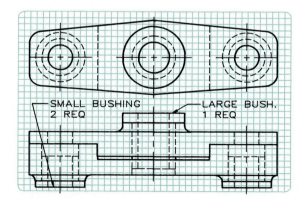

SMALL BUSHING
2 REQ

LARGE BUSH.
1 REQ

Figure 17.62 Problem 40. Full section in assembly.

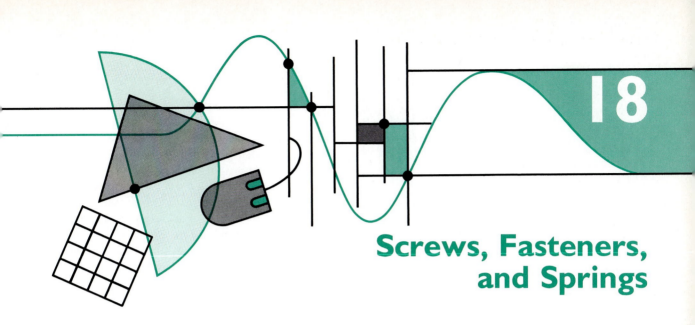

Screws, Fasteners, and Springs

18.1 Threaded Fasteners

Screw threads provide a fast and easy method of fastening two parts together and of exerting a force that can be used for adjustment of movable parts. For a screw thread to function, there must be an **internal thread** and an **external thread.**

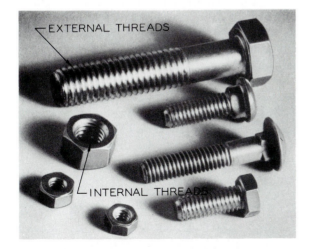

Figure 18.1 Examples of (A) external threads (bolts) and (B) internal threads (nuts). (Courtesy of Russell, Burdsall & Ward Bolt and Nut Co.)

Internal threads may be tapped inside a part such as a motor block or, more commonly, a nut. Whenever possible, the nuts and bolts used in industrial projects should be stock parts that can be obtained from many sources. This reduces manufacturing expenses and improves the interchangeability of parts.

Progress has been made toward establishing standards that will unify threads in this country and abroad by the introduction of metric standards. Other efforts have led to the adoption of the Unified Screw thread by the United States, Britain, and Canada (ABC Standards), which is a modification of both the American Standard thread and the Whitworth thread.

18.2 Definitions of Thread Terminology

External Thread is a thread on the outside of a cylinder, such as a bolt (Fig. 18.1).

Internal Thread is a thread cut on the inside of a part, such as a nut (Fig. 18.1).

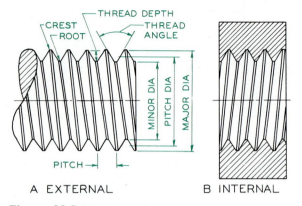

Figure 18.2 Thread terminology.

Major Diameter is the largest diameter on an internal or external thread (Fig. 18.2).

Minor Diameter is the smallest diameter that can be measured on a screw thread (Fig. 18.2).

Pitch Diameter is the diameter of an imaginary cylinder passing through the threads at the points where the thread width is equal to the space between the threads (Fig. 18.2).

Lead is the distance a screw will advance when turned 360°.

Pitch is the distance between crests of threads. Pitch is found mathematically by dividing 1 in. by the number of threads per inch of a particular thread (Fig. 18.2).

Crest is the peak edge of a screw thread (Fig. 18.2).

Thread Angle is the angle between threads cut by the cutting tool, usually 60° (Fig. 18.2).

Root is the bottom of the thread cut into a cylinder (Fig. 18.2).

Thread Form is the shape of the thread cut into a threaded part.

Thread Series is the number of threads per inch for a particular diameter, grouped into three series: coarse, fine, extra fine, and there are eight

constant-pitch thread series. Coarse series provides rapid assembly, and extra-fine series provides fine adjustment.

Thread Class is a closeness of fit between two mating threaded parts. Class 1 represents a loose fit and Class 3 a tight fit.

Right-Hand Thread is a thread that will assemble when turned clockwise. A right-hand thread slopes downward to the right on an external thread when the axis is horizontal, and in the opposite direction on an internal thread.

Left-Hand Thread is a thread that will assemble when turned counterclockwise. A left-hand thread slopes downward to the left on an external thread when the axis is horizontal, and in the opposite direction on an internal thread.

18.3 Thread Specifications (English System)

Form

Thread form is the shape of the thread cut into a part, as illustrated in Fig. 18.3. The Unified form, a combination of the American National and British Whitworth, is the most widely used because it is a standard in several countries. The Unified form is signified by UN in abbreviations and thread notes, and the American National form is signified by N.

Another thread form, the Unified National Rolled, abbreviated UNR, was introduced into the 1974 ANSI standards. This designation is specified only for external threads—there is no

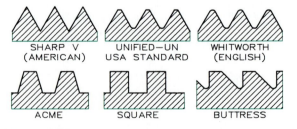

Figure 18.3 Standard thread forms.

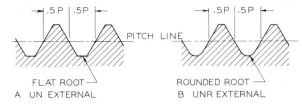

Figure 18.4 (A) The UN external thread has a flat root (rounded root is optional). (B) The UNR has a rounded root formed by rolling. The UNR form does not apply to internal threads.

UNR designation for internal threads. The UN form has a flat root (rounded root is optional) in Fig. 18.4A, whereas the UNR thread *must* have a rounded root formed by rolling, as shown in Fig. 18.4B.

The transmission of power is achieved by using the **Acme, square,** and **buttress** threads, which are commonly used in gearing and other pieces of machinery (Fig. 18.3). The **sharp V** thread is used for set screws and in applications where friction in assembly is desired.

Series

Thread series, which is closely related to thread form, designates the type of thread specified for a given application. Eleven standard series of threads are listed under the American National form and the Unified National (UN/UNR) form. There are three series, with abbreviations coarse (C), fine (F), and extra fine (EF), and eight series with constant pitches (4, 6, 8, 12, 16, 20, 28, and 32 threads per inch).

A Unified National form for a **coarse-series** thread is specified as UNC or UNRC, which is a combination of form and series in a single note. Similarly, an American National form for a coarse thread is written NC. The **coarse-thread** series (UNC/UNRC or NC) is suitable for bolts, screws, nuts, and general use with cast iron, soft metals, or plastics when rapid assembly is desired. The **fine-thread** series (NF or UNF/UNRF) is suitable for bolts, nuts, or screws when a high degree of tightening is required. The **extra-fine** series (UNEF/UNREF or NEF) is suitable for sheet metal, thin nuts, fer-

rules, or couplings when length of engagement is limited and is used for applications that will have to withstand high stresses.

The 8-thread series (8 UN), 12-thread series, (12 N or 12 UN/UNR), and 16-thread series (16 N or 16 UN/UNR) are threads with a uniform pitch for large diameters. The 8 UN is used as a substitute for the coarse-thread series on diameters larger than 1 in. when a medium-pitch thread is required. The 12 UN is used on diameters larger than $1\frac{1}{2}$ in., with a thread of a medium-fine pitch as a continuation of the fine-thread series. The 20 UN is used on diameters larger than $1\frac{11}{16}$ in., with threads of an extra-fine pitch as a continuation of the extra-fine series.

Class of Fit

Thread classes are used to indicate the tightness of fit between a nut and bolt or any two mating threaded parts. Classes of fit are indicated by the numbers 1, 2, or 3 followed by the letters *A* or *B*. For UN forms, the letter *A* represents an external thread, whereas the letter *B* represents an internal thread. These letters are omitted when the American National form (N) is used.

Class 1A and 1B threads are used on parts that require assembly with a minimum of binding.

Class 2A and 2B threads are general-purpose threads for bolts, nuts, screws, and nominal applications in the mechanical field and are widely used in the mass-production industries.

Class 3A and 3B threads are used in precision assemblies where a close fit is desired to withstand stresses and vibration.

Single and Multiple Threads

A **single thread** (Fig. 18.5A) is a thread that will advance the distance of its pitch in one full revolution of 360°; in other words, its pitch is equal to its lead. In the drawing of a single thread, the crest line of the thread will slope $\frac{1}{2}P$ since only 180° of the revolution is visible in a single view.

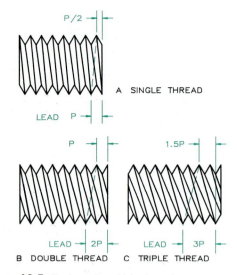

LEAD   P

P   1.5P

LEAD   2P   LEAD   3P
B DOUBLE THREAD   C TRIPLE THREAD

Figure 18.5 Single and multiple threads.

A double thread is composed of two threads, resulting in a lead equal to 2P, meaning that the threaded part will advance a distance of 2P in a single revolution of 360° (Fig. 18.5B). The crest line of a double thread will slope a distance equal to P in the view in which 180° can be seen.

Similarly, a triple thread will advance 3P in 360° with a crest line slope of $1\frac{1}{2}P$ in the view in which 180° of the cylinder is visible (Fig. 18.5C). The lead of a double thread is 2P, and the lead of a triple thread is 3P. Multiple threads are used wherever quick assembly is required.

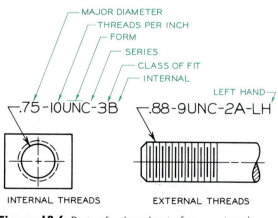

INTERNAL THREADS   EXTERNAL THREADS

Figure 18.6 Parts of a thread note for an external thread.

Thread Notes

Drawings of threads are only symbolic representations and are inadequate unless accompanying notes give the thread specifications (Fig. 18.6). For a double or triple thread, the word *DOUBLE* or *TRIPLE* is included in the note; for a left-hand thread, the letters *LH* are included.

Figure 18.7 shows the UNR thread note for the external thread. (UNR does not apply to internal threads.) When inches are used as the unit of measurement, thread notes can be written as common fractions, but decimal fractions are preferred.

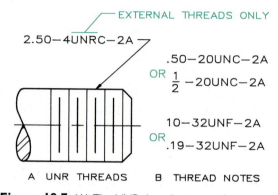

A UNR THREADS   B THREAD NOTES

Figure 18.7 (A) The UNR thread notes apply to external threads only. (B) Notes can be given as decimal fractions or common fractions.

18.4 Using Thread Tables

Appendix 14 gives the UN/UNR thread table, and Table 18.1 shows part of this table. If an external thread (bolt) with a $1\frac{1}{2}$-inch diameter is to have a "fine" thread, it will have 12 threads per inch. Therefore, the thread note can be written

$1\frac{1}{2}$–12 UNF–2A   or   1.500–12 UNF–2A.

If the thread were an internal one (nut), the thread note would be the same, but the letter B would be used instead of the letter A.

A constant-pitch thread series can be selected for the larger diameters. The constant-pitch thread notes are written with the abbreviations C, F, and EF omitted. For example, a $1\frac{3}{4}$-inch diameter bolt with a fine thread could be noted in

Table 18.1														
American National Standard unified inch screw threads (UN and UNR thread form)★														
Sizes		Basic Major Diameter	Series with Graded Pitches			Threads per Inch — Series with Constant Pitches								Sizes
Primary	Secondary		Coarser UNC	Fine UNF	Extra fine UNEF	UN	6 UN	8 UN	12 UN	16 UN	20 UN	28 UN	32 UN	
1		1.0000	8	12	20	—	—	UNC	UNF	16	UNEF	28	32	1
	1 1/16	1.0625	—	—	18	—	—	8	12	16	20	28	—	1 1/16
1 1/8		1.1250	7	12	18	—	—	8	UNF	16	20	28	—	1 1/8
	1 3/16	1.1875	—	—	18	—	—	8	12	16	20	28	—	1 3/16
1 1/4		1.2500	7	12	18	—	—	8	UNF	16	20	28	—	1 1/4
	1 5/16	1.3125	—	—	18	—	—	8	12	16	20	28	—	1 5/16
1 3/8		1.3750	6	12	18	—	UNC	8	UNF	16	20	28	—	1 3/8
	1 7/16	1.4375	—	—	18	—	6	8	12	16	20	28	—	1 7/16
1 1/2		1.5000	6	12	18	—	UNC	8	UNF	16	20	28	—	1 1/2
	1 9/16	1.5625	—	—	18	—	6	8	12	16	20	—	—	1 9/16

★By using this table, a diameter of 1½ inches that is to be threaded with a fine thread would have the following thread note: 1½-12 UNF-2A.

Source: Courtesy of ANSI; B1.1.

constant-pitch series as

$$1\tfrac{3}{4}\text{–12 UN–2A}$$

or

$$1.750\text{–12 UN–2A.}$$

Table 18.1 can also be used for the UNR thread form (for external threads only) by substituting UNR for UN; for example, UNREF for extra fine.

18.5 Metric Thread Specifications (ISO)

Metric thread specifications are recommended by the ISO (International Organization for Standardization). Thread specifications can be given with a **basic designation,** which is suitable for general applications, or with the **complete designation,** which is used where detailed specifications are needed.

Basic Designation

Figure 18.8 shows examples of metric screw thread notes. Each note begins with the letter *M,* which designates the note as a metric note, followed by the diameter in millimeters, and the

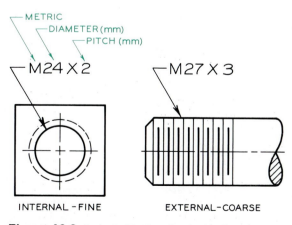

Figure 18.8 Basic designations for metric threads.

pitch in millimeters separated by ×, the multiplication sign. The pitch can be omitted in notes for coarse threads, but U.S. standards prefer that it be shown. Table 18.2 shows the commercially available ISO threads recommended for general use. Appendix 17 gives additional ISO specifications.

Complete Designation

For some applications it is necessary to show a complete thread designation (Fig. 18.9). The first part of this note is the same as the basic designation; however, the note also has a tolerance class designation separated by a dash. The 5g represents the pitch diameter tolerance, and 6g represents the crest diameter tolerance.

The numbers 5 and 6 are **tolerance grades** (variations from the basic diameter). Grade 6 is commonly used for a medium general-purpose thread that is nearly equal to class 2A and 2B of

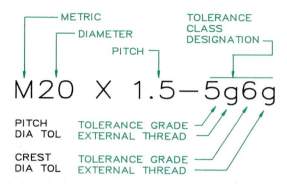

Figure 18.9 A complete designation note for metric threads.

the Unified system. Grades less than 6 are used for fine-quality fits and short lengths of engagement. Grades greater than 6 are recommended for coarse-quality fits and long lengths of engagement. Table 18.3 gives the tolerance grades for internal and external threads for the pitch diameter and the major and minor diameters.

Table 18.2								
Basic thread designations for commercial series of ISO metric threads								
Nominal Size (mm)	Pitch P (mm)	Basic Thread Designation*	Nominal Size (mm)	Pitch P (mm)	Basic Thread Designation*	Nominal Size (mm)	Pitch P (mm)	Basic Thread Designation*
1.6	0.35	M1.6		1.25	M8		2.5	M22
1.8	0.35	M1.8	8	1	M8 × 1	22	1.5	M22 × 1.5
2	0.4	M2		1.5	M10		3	M24
2.2	0.45	M2.2	10	1.25	M10 × 1.25	24	2	M24 × 2
2.5	0.45	M2.5		1.75	M12		3	M27
3	0.5	M3	12	1.25	M12 × 1.25	27	2	M27 × 2
3.5	0.6	M3.5		2	M14		3.5	M30
4	0.7	M4	14	1.5	M14 × 1.5	30	2	M30 × 2
4.5	0.75	M4.5		2	M16		3.5	M33
5	0.8	M5	16	1.5	M16 × 1.5	33	2	M33 × 2
6	1	M6		2.5	M18		4	M36
7	1	M7	18	1.5	M18 × 1.5	36	3	M36 × 3
				2.5	M20		4	M39
			20	1.5	M20 × 1.5	39	3	M39 × 3

*U.S. practice is to include the pitch symbol even for the coarse pitch series. Basic descriptions shown are as specified in ISO Recommendations.

Source: Courtesy of Greenfield Tap and Die Corp.

Table 18.3
Tolerance grades, ISO threads

External Thread		Internal Thread	
Major Diameter (d_1)	Pitch Diameter (d_2)	Minor Diameter (D_1)	Pitch Diameter (D_2)
—	3	—	—
4	4	4	4
—	5	5	5
6	6	6	6
—	7	7	7
8	8	8	8
—	9	—	—

Grade 6 is medium; smaller numbers are finer, and larger numbers are coarser.

Source: Courtesy of ANSI; B1.

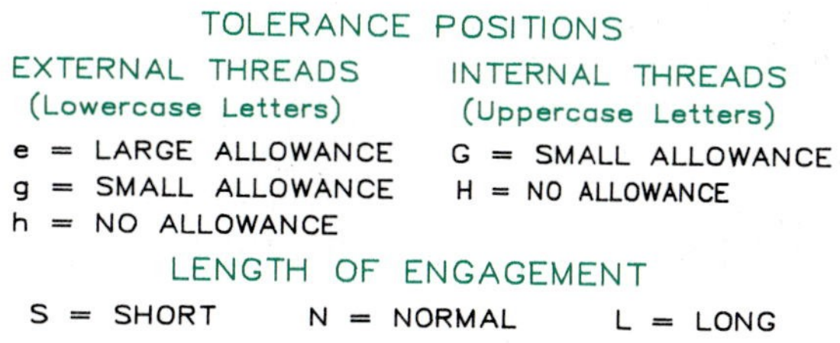

TOLERANCE POSITIONS

EXTERNAL THREADS (Lowercase Letters)	INTERNAL THREADS (Uppercase Letters)
e = LARGE ALLOWANCE	G = SMALL ALLOWANCE
g = SMALL ALLOWANCE	H = NO ALLOWANCE
h = NO ALLOWANCE	

LENGTH OF ENGAGEMENT

S = SHORT   N = NORMAL   L = LONG

Figure 18.10 Symbols used to represent tolerance grade, position, and class.

e,g, and *n* represent large allowance, small allowance, and no allowance, respectively. (**Allowance** is the variation from the basic diameter.) Uppercase letters designate internal threads (nuts). On internal threads, *G* designates small allowance, and *H* designates no allowance. The letters are placed after the tolerance grade number. For example, 5g designates a medium tolerance with small allowance for the pitch diameter of an external thread, and 6H designates a medium tolerance with no allowance for the minor diameter of an internal thread.

Tolerance classes are fine, medium, and coarse, as listed in Table 18.4. These classes of fit

The letters following the grade numbers designate **tolerance positions** (external or internal). Lowercase letters represent external threads (bolts), as Fig. 18.10 shows. The lowercase letters

Table 18.4
Preferred tolerance classes, ISO threads★

Quality	External Threads (bolts)									Internal Threads (nuts)					
	Tolerance position e (large allowance)			Tolerance position g (small allowance)			Tolerance position h (no allowance)			Tolerance position G (small allowance)			Tolerance position H (no allowance)		
	Length of engagement			Length of engagement			Length of engagement			Length of engagement			Length of engagement		
	Group S	Group N	Group L	Group S	Group N	Group L	Group S	Group N	Group L	Group S	Group N	Group L	Group S	Group N	Group L
Fine							3h4h	4h	5h4h				4H	5H	6H
Medium		6e	7e6e	5g6g	6g	7g6g	5h6h	6h	7h6h	5G	6G	7G	5H	6H	7H
Coarse					8g	9g8g					7G	8G		7H	8H

★In selecting tolerance class, select first from the large bold print, second from the medium-size print, and third from the small-size print. Classes shown in boxes are for commercial threads.

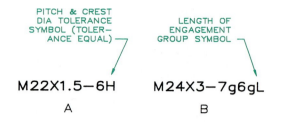

Figure 18.11 (A) When both pitch and crest diameter tolerance grades are the same, the tolerance class symbol is shown only once. (B) Letters *S, N,* and *L* are used to indicate the length of the thread engagement.

are combinations of tolerance grades, tolerance positions, and lengths of engagement—short (S), normal (N), and long (L). The length of engagement can be determined by referring to Appendix 16. Once it has been decided to use a fine, medium, or coarse class of fit for a particular application, the specific designation should be selected, first from the classes shown in large print in Table 18.4, second from the classes shown in medium-size print, and third from the classes shown in small print. Classes shown in boxes are for commercial threads.

Figure 18.11 shows variations in the complete designation thread notes. The tolerance class symbol is written 6H if the crest and pitch diameters have identical grades (Fig. 18.11A). Since an uppercase *H* is used, this is an internal thread. Where considered necessary, the length-of-engagement symbol may be added to the tolerance class designation (Fig. 18.11B).

Designations for the desired fit between mating threads can be specified as shown in Fig. 18.12. A slash is used to separate the tolerance

Figure 18.12 A slash mark is used to separate the tolerance class designations of mating internal and external threads.

class designations of the internal and external threads.

Additional information about ISO threads may be obtained from *ISO Metric Screw Threads,* a booklet of standards published by ANSI. These standards were used as the basis for most of this section.

18.6 Thread Representation

Three major types of thread representations are the **simplified, schematic,** and **detailed** (Fig. 18.13). The detailed representation is the most realistic approximation of a thread's appearance, and the simplified representation is the least realistic.

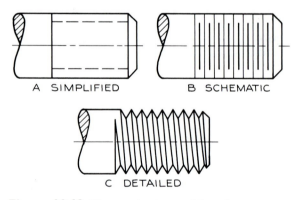

Figure 18.13 Three major types of thread representations.

18.7 Detailed UN/UNR Threads

Figure 18.14 shows examples of detailed representations of internal and external threads. Instead of helical curves, straight lines indicate crest and root lines.

Figure 18.15 shows the construction of a detailed thread representation.

Metric threads should be drawn in the same manner, using the pitch given in millimeters in the metric thread table, or an expanded pitch, if needed.

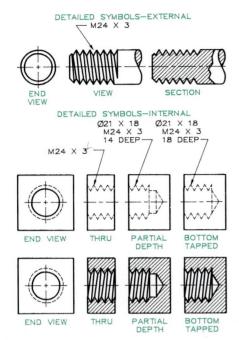

Figure **18.14** Detailed thread representations of external and internal threads. Use Appendix 15 to determine tap drill sizes and convert to metric units.

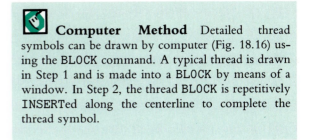

Computer Method Detailed thread symbols can be drawn by computer (Fig. 18.16) using the BLOCK command. A typical thread is drawn in Step 1 and is made into a BLOCK by means of a window. In Step 2, the thread BLOCK is repetitively INSERTed along the centerline to complete the thread symbol.

18.8 Detailed Square Threads

Figure 18.17 shows the method of drawing a detailed representation of a square thread to give an approximation of a square thread.

In Step 1, the major diameter is laid out. The number of threads per inch is taken from Appendix 18. The pitch *(P)* is found by dividing 1 in. by the number of threads per inch, but this pitch is usually enlarged. Distances of *P/2* are marked off with dividers. Steps 2, 3, and 4 are then completed, and a thread note is added.

Square internal threads are drawn in a similar manner (Fig. 18.18). The threads in the section view are drawn in a slightly different way. The

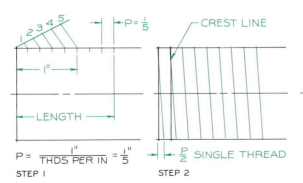
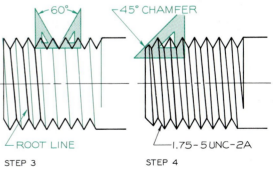

Figure **18.15** Detailed thread representation.

Step 1 To draw a detailed representation of a 1.75–5 UNC–2A thread, the pitch is determined by dividing 1″ by the number of threads per inch, 5 in this case. The pitch is laid off the length of the thread. Pitch is usually enlarged.

Step 2 Since this is a right-hand thread, the crest lines slope downward to the right equal to ½P. The crest lines will be final lines drawn with an H or F pencil.

Step 3 The root lines are found by constructing 60° vees between the crest lines. The root lines are drawn from the bottom of the vees. Root lines are parallel to each other, but not to crest lines.

Step 4 A 45° chamfer is constructed at the end of the thread from the minor diameter. Strengthen all lines, and add a thread note.

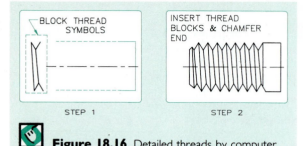

STEP 1   STEP 2

Figure 18.16 Detailed threads by computer.

Step 1 A typical detailed thread symbol is drawn and BLOCKed by using a window.

Step 2 The BLOCK is repetitively INSERTed along the centerline. A chamfer is drawn at the left end, and the right end is completed.

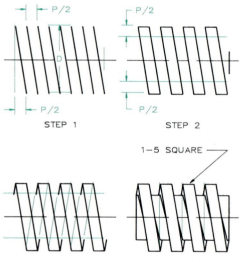

STEP 1   STEP 2

1–5 SQUARE

STEP 3   STEP 4

Figure 18.17 Drawing the square thread.

Step 1 Lay out the major diameter. Space the crest lines $\frac{1}{2}P$ apart. Slope them downward to the right for right-hand threads.

Step 2 Connect every other pair of crest lines. Find the minor diameter by measuring $\frac{1}{2}P$ inward from the major diameter.

Step 3 Connect the opposite crest lines with light construction lines. This will establish the profile of the thread form.

Step 4 Connect the inside crest lines with light construction lines to locate the points on the minor diameter where the thread wraps around the minor diameter. Darken the final lines.

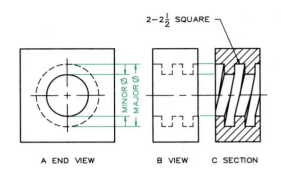

2–2½ SQUARE

A END VIEW   B VIEW   C SECTION

Figure 18.18 Internal square threads.

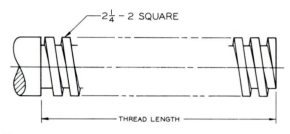

2¼ – 2 SQUARE

THREAD LENGTH

Figure 18.19 Conventional method of showing square threads without drawing each thread.

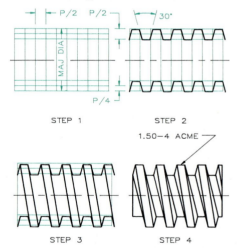

STEP 1   STEP 2

1.50–4 ACME

STEP 3   STEP 4

Figure 18.20 Drawing the Acme thread.

Step 1 Lay out the major diameter and thread length, and divide the shaft into equal divisions $\frac{1}{2}P$ apart. Locate the minor and pitch diameters using distances $\frac{1}{2}P$ and $\frac{1}{4}P$.

Step 2 Draw construction lines at 15° angles with the vertical along the pitch diameter as shown to make a total angle of 30°.

Step 3 Draw the crest lines across the screw.

Step 4 Darken the lines, draw the root lines, and add the thread note to complete the drawing.

thread note for an internal thread is placed in the circular view whenever possible, with the leader pointing toward the center. When a square thread is long, it can be represented using the symbol shown in Fig. 18.19.

18.9   Detailed Acme Threads

Figure 18.20 shows the method of preparing detailed drawings of Acme threads. In Step 1, the length and the major diameter are laid out with light construction lines, and the pitch is found by dividing 1 in. by the number of threads per inch. Steps 2, 3, and 4 complete the thread representation, and the thread note is added.

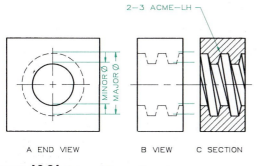

Figure 18.21 Internal Acme threads.

Figure 18.22 Cutting an Acme thread on a lathe. (Courtesy of Clausing Corp.)

Figure 18.21 shows internal Acme threads. In the section view, left-hand internal threads are sloped so that they look the same as right-hand external threads.

Figure 18.22 shows a shaft that is being threaded with Acme threads on a lathe as the tool travels the length of the shaft.

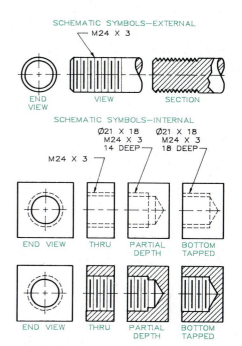

Figure 18.23 Schematic representations of external and internal threads as views and sections. Use Appendix 15 to determine tap drill sizes and convert to metric units.

18.10   Schematic Threads

Figure 18.23 shows schematic representations of internal and external threads. Since the schematic representation is easy to draw and gives a good symbolic representation of threads, it is the most often used thread symbol. Left-hand threads must be specified in the thread note.

Figure 18.24 shows the method of constructing schematic threads, and Fig. 18.25 shows the method of drawing metric threads using schematic representations. The pitch (in millimeters)

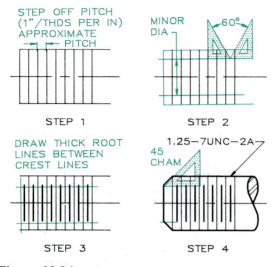

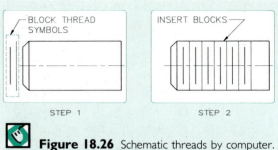

STEP 1   STEP 2

Figure 18.26 Schematic threads by computer.

Step 1 Draw the outline of the threaded shaft with a chamfered end. Draw typical minor and major diameters, and make them into a **BLOCK** by using a window.

Step 2 Repetitively **INSERT** the **BLOCK**s of the threads along the centerline to complete the thread representation.

Figure 18.24 Drawing schematic threads.

Step 1 Lay out the major diameter, and divide the shaft into divisions of a distance of approximately *P* apart. Draw these crest lines as thin lines.

Step 2 Find the minor diameter by drawing a 60° angle between two crest lines on each side.

Step 3 Draw heavy root lines between the crest lines.

Step 4 Chamfer the end of the thread from the minor diameter, and give a thread note.

given in the metric thread table is used as the approximate distance to separate the crest lines.

Computer Method Schematic threads can be drawn by computer as shown in Fig. 18.26. In Step 1, a typical major and minor diameter is drawn, and a **BLOCK** is made of them by a window. In Step 2, the **BLOCK** of the threads are repetitively inserted to complete the thread symbols. The **COPY** command can be used instead of a **BLOCK**.

18.11 Simplified Threads

Figure 18.27 illustrates the use of simplified representations with notes to specify thread details. Of the three types of thread representations, this is the easiest to draw. Hidden lines can be positioned by eye to approximate the minor diameter. Figure 18.28 shows the steps in constructing a simplified thread drawing.

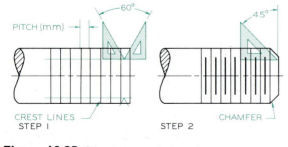

Figure 18.25 Schematic metric threads.

Step 1 The pitch of metric threads can be taken directly from the metric tables, which can be used to find the minor diameter.

Step 2 The root lines are drawn in heavy between the crest lines. The end of the thread is chamfered.

18.12 Drawing Small Threads

Instead of drawing small threads to exact measurements, minor diameters can be drawn smaller to separate the root and crest lines (Fig. 18.29).

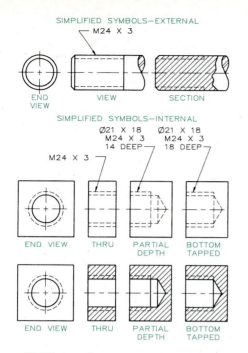

SIMPLIFIED SYMBOLS—EXTERNAL
M24 X 3

END VIEW   VIEW   SECTION

SIMPLIFIED SYMBOLS—INTERNAL

Ø21 X 18   Ø21 X 18
M24 X 3   M24 X 3
14 DEEP   18 DEEP

M24 X 3

END VIEW   THRU   PARTIAL DEPTH   BOTTOM TAPPED

END VIEW   THRU   PARTIAL DEPTH   BOTTOM TAPPED

Figure 18.27 Simplified thread representations of external and internal threads. Use Appendix 15 to determine tap drill sizes and convert to metric units.

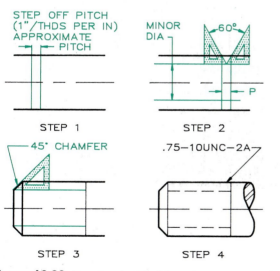

STEP OFF PITCH
(1"/THDS PER IN)
APPROXIMATE
PITCH

STEP 1

MINOR DIA   60°
P

STEP 2

45° CHAMFER

STEP 3

.75—10UNC—2A

STEP 4

Figure 18.28 Drawing simplified threads.

Step 1 Lay out the major diameter. Find the approximate pitch *(P),* and lay out two lines a distance *P* apart.

Step 2 Find the minor diameter by constructing a 60° angle between the two lines on both sides.

Step 3 Draw a 45° chamfer from the minor diameter to the major diameter.

Step 4 Show the minor diameter as dashed lines. Add a thread note.

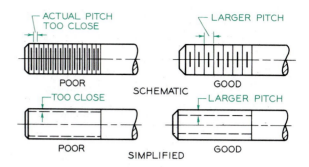

ACTUAL PITCH TOO CLOSE   LARGER PITCH

POOR   SCHEMATIC   GOOD

TOO CLOSE   LARGER PITCH

POOR   SIMPLIFIED   GOOD

Figure 18.29 Simplified and schematic threads should be drawn using approximate dimensions if the actual dimensions would result in lines drawn too close together.

This procedure makes the thread easier to draw and read. Schematic threads are drawn with crest lines farther apart to prevent crowding of lines. For both internal and external threads, a thread note completes the symbolic drawing by giving the necessary specifications.

18.13 Nuts and Bolts

Nuts and bolts come in many forms and sizes for different applications (Fig. 18.30). Figure 18.31 shows the more common types of threaded fasteners. A **bolt** is a threaded cylinder with a head

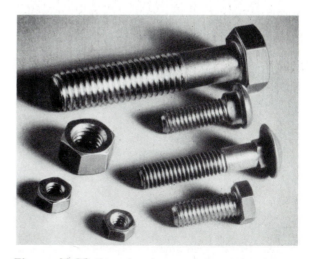

Figure 18.30 Examples of nuts and bolts. (Courtesy of Russell, Burdsall & Ward Bolt and Nut Co.)

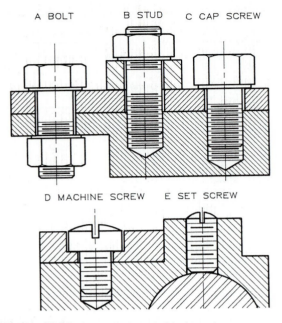

A BOLT   B STUD   C CAP SCREW

D MACHINE SCREW   E SET SCREW

Figure 18.31 Types of threaded bolts and screws.

that is used with a nut for holding two parts together (Fig. 18.31A). A **stud** does not have a head but is screwed into one part with a nut attached to the other threaded end (Fig. 18.31B). A **cap screw** is similar to a bolt, but it does not have a nut; instead, it is screwed into a member with internal threads (Fig. 18.31C). A **machine screw** is similar to, but smaller than, a cap screw

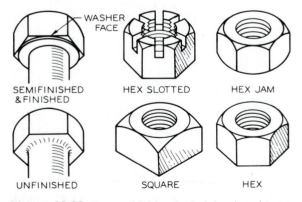

SEMIFINISHED & FINISHED   HEX SLOTTED   HEX JAM

WASHER FACE

UNFINISHED   SQUARE   HEX

Figure 18.32 Types of finishes for bolt heads and types of nuts.

(Fig. 18.31D). A **set screw** is used to secure one member with respect to another, usually to prevent a rotational movement (Fig. 18.31E).

Figure 18.32 shows the types of heads used on standard bolts and nuts. These heads are used on both **regular** and **heavy** bolts; the thickness of the head is the primary difference between the two types. Heavy-series bolts have the thicker heads and are used at points where bearing loads are heaviest. Bolts and nuts are either **finished** or **unfinished.** Figure 18.32 shows an unfinished head; that is, none of the surfaces of the head are machined. The finished head has a washer face that is $\frac{1}{64}$ inch thick to provide a circular boss on the bearing surface of the bolt head or the nut.

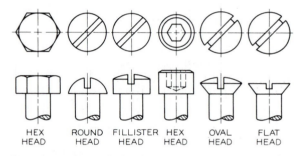

HEX HEAD   ROUND HEAD   FILLISTER HEAD   HEX HEAD   OVAL HEAD   FLAT HEAD

Figure 18.33 Common types of bolt and screw heads.

Other standard forms of bolt and screw heads (Fig. 18.33) are used primarily on cap screws and machine screws. Finished nuts have washer faces for more accurate assembly. A hexagon **jam nut** does not have a washer face, but it is chamfered on both sides.

A properly dimensioned bolt is shown in Fig. 18.34. Although ANSI tables in Appendixes 19–24 indicate the standard bolt lengths and their corresponding thread lengths, the following can be used as a general guide for square- and hexagon-head bolts:

• Hexagon bolt lengths are available in $\frac{1}{4}$-in. increments up to 8 in. long, in $\frac{1}{2}$-in. increments from 8 to 20-in. long, and in 1-in. increments from 20 to 30 in. long.

.50–13UNC–2A X 2.25 LONG
REG HEX HEAD CAP SCREW

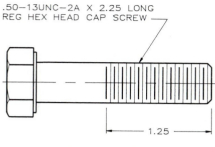

|← 1.25 →|

A DIMENSIONED BOLT

Figure 18.34 A properly dimensioned and noted bolt.

- Square-head bolt lengths are available in $\frac{1}{8}$-in. increments from $\frac{1}{2}$ to $\frac{3}{4}$ in. long, in $\frac{1}{4}$-in. increments from $\frac{3}{4}$ to 5 in. long, in $\frac{1}{2}$-in. increments from 5 to 12 in. long, and in 1-in. increments from 12 to 30 in. long.
- The lengths of the threads on both hexagon-head and square-head bolts up to 6 in. long can be found by the formula: Thread length $= 2D + \frac{1}{4}$ in., where D is the diameter of the bolt. The threaded length for bolts more than 6 in. long can be found by the formula: Thread length $= 2D + \frac{1}{2}$ in.
- The threads for bolts can be coarse, fine, or 8-pitch threads. It is understood that the class of fit for bolts and nuts will be 2A and 2B if no class is specified.
- Standard square-head and hexagon-head bolts are designated by notes in one of the following forms:

  $\frac{3}{8}$–16 × 1$\frac{1}{2}$ SQUARE BOLT—STEEL;
  $\frac{1}{2}$–13 × 3 HEX CAP SCREW—SAE GRADE 8—STEEL;
  0.75 × 5.00 HEX LAG SCREW—STEEL.

The numbers represent bolt diameter, threads per inch (omit for lag screws), length, name of screw, and material. It is understood that these will have a class 2 fit.

- Nuts are designated by notes in one of the following forms:

  $\frac{1}{2}$–13 SQUARE NUT—STEEL;
  $\frac{3}{4}$–16 HEAVY HEX NUT, SAE GRADE 5—STEEL;

1.00–8 HEX THICK SLOTTED NUT—CORROSION RESISTANT STEEL.

When nuts are *not* noted as HEAVY, they are assumed to be REGULAR. The class of fit is assumed to be 2B for nuts when not noted.

18.14 Drawing Square Bolt Heads

Detailed tables are available in Appendix 19 for square bolt heads and nuts. Usually, it is sufficient to draw nuts and bolts with only general proportions.

The first step in drawing a bolt head or nut is to determine whether it is to be **across corners** or **across flats**—that is, are the outlines at either side of the view going to represent corners or edge views of flat surfaces of the part? Nuts and bolts should be drawn across corners for the best representation (Fig. 18.35).

18.15 Drawing Hexagon Bolt Heads

Figure 18.36 shows an example of constructing the head of a hexagon-head bolt across corners.

The major diameter of the bolt is D. The thickness of the head is drawn equal to $\frac{2}{3}D$. The top view of the head is drawn as a circle with a radius of $\frac{3}{4}D$. This proportionality based on D (the major diameter) is sufficient for drawing bolt heads in most applications.

Computer Method A bolt head can be drawn, as shown in Step 1 of Fig. 18.37, by using a bolt diameter of 1 inch to make a UNIT BLOCK after the drawing has been completed. Step 2 shows how the BLOCK can be reduced and rotated to any position. To draw a thread diameter of 0.50 inch, a size factor of 0.50 is assigned when prompted by the BLOCK and INSERT commands.

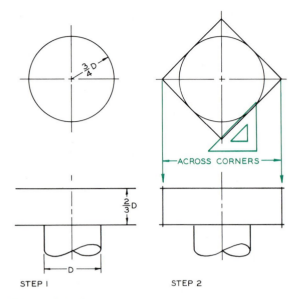
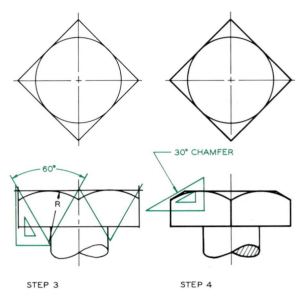

STEP 1 STEP 2 STEP 3 STEP 4

Figure 18.35 Drawing the square head.

Step 1 Draw the diameter of the bolt. Use major diameter (*D*) to establish the head diameter and thickness.

Step 2 Draw the top view of the square head with a 45° triangle to give an across-corners view.

Step 3 Show the chamfer in the front view by using a 30°–60° triangle to find the centers for the radii.

Step 4 Show a 30° chamfer tangent to the arcs in the front view. Strengthen the lines.

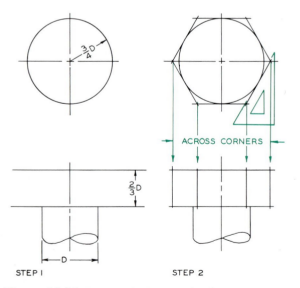
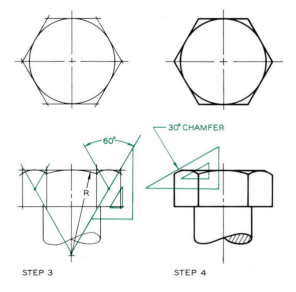

STEP 1 STEP 2 STEP 3 STEP 4

Figure 18.36 Drawing the hexagon head.

Step 1 Draw the major (*D*) diameter of the bolt and use it to establish the head diameter and thickness.

Step 2 Construct a hexagon with a 30°–60° triangle to give an across-corners view.

Step 3 Find arcs in the front view to show the chamfer of the head.

Step 4 Draw a 30° chamfer tangent to the arcs in the front view. Strengthen the lines.

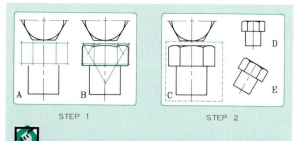

STEP 1   STEP 2

Figure 18.37 Bolt head by computer.

Step 1 A hexagon head is drawn using the steps of geometric construction covered in Fig. 18.36. The size of the head is based on a bolt diameter of 1 inch to form a UNIT BLOCK.

Step 2 The drawing is made into a BLOCK by using a window. The block can be INSERTed at any size and positioned as shown at D and E.

18.16 Drawing Nuts

The drawing of a square and a hexagon nut across corners is the same as the drawing of a bolt head across corners. The only variation is the thickness of the nut. The REGULAR nut thickness is $\frac{7}{8}D$, and the HEAVY nut thickness is D (D = major diameter).

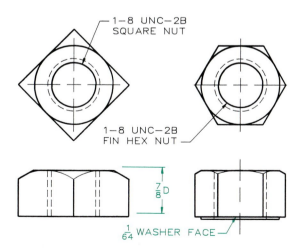

Figure 18.38 Drawings of hexagon and square nuts are constructed in the same manner as drawings of bolt heads. Notes are added to give nut specifications.

Figure 18.38 shows square and hexagon nuts drawn across corners. In the front view, hidden lines indicate threads. Since it is understood that nuts are threaded, these hidden lines may be omitted in general applications.

A $\frac{1}{64}$-inch washer face is shown on the hexagon nut. It is usually drawn thicker than $\frac{1}{64}$ inch to make the face more noticeable in the drawing. Thread notes are placed in the circular views where possible. Since the note for square nut is not labeled HEAVY, it is a REGULAR square nut. The hexagon nut is similar except that it is a finished hexagon nut.

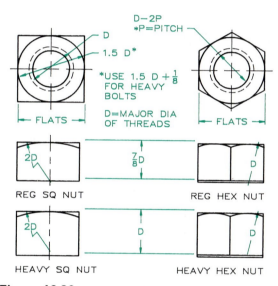

Figure 18.39 Examples of hexagon and square nuts drawn across flats. Notes are added to give nut specifications. Square nuts are always unfinished.

Nuts can be drawn across flats if doing so improves the drawing (Fig. 18.39). For regular nuts, the distance across flats is $1\frac{1}{2} \times D$ (D = major diameter of the thread); for heavy nuts, this distance is $1\frac{5}{8} \times D$. The top views are drawn in the same manner as in across-corners drawings except that they are positioned to give different front views.

Examples of dimensioned and noted nuts are shown in Fig. 18.40.

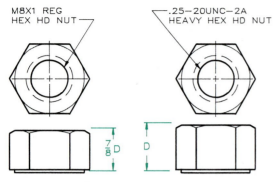

Figure 18.40 Examples of dimensioned and noted nuts.

18.17 Drawing Nuts and Bolts in Combination

The rules followed when drawing nuts and bolts separately also apply when drawing nuts and bolts in assembly (Fig. 18.41). The major diameter, D, of the bolt is used as the basis for other dimensions. The note is added to give the specifications of the nut and bolt. The bolt heads are drawn across corners, and the nuts are drawn across flats. The half-end views show how the front views were found by projection.

18.18 Cap Screws

Cap screws are used to hold two parts together without a nut since one of these parts has a threaded cylindrical hole. The other part is drilled with an oversized hole so that the cap screw will pass through it freely. When the cap screw is tightened, the two parts are held securely together.

Figure 18.42 shows the standard types of cap screws. Appendixes 20–22 give the dimensions of several types of cap screws. In Fig. 18.42, the cap screws are drawn on a grid to show their proportions, which can be used for drawing cap screws of all sizes, ranging in diameter from No. 0 (0.060) to No. 12.

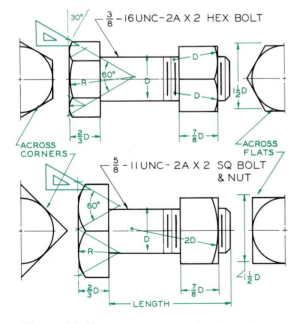

Figure 18.41 Construction of nuts and bolts in assembly.

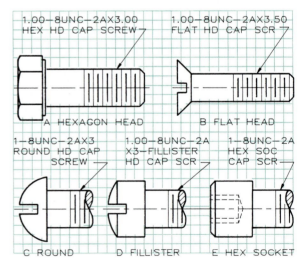

Figure 18.42 These proportions of the standard types of cap screws can be used for drawing cap screws of all sizes. Notes provide typical specifications.

18.19 Machine Screws

Machine screws are smaller than most cap screws, usually less than 1 inch in diameter. The machine screw is screwed into either another part or a nut. Machine screws are fully threaded when their length is 2 inches or shorter.

Figure 18.43 shows a few of the common machine screws and their notes. Appendix 23 gives the dimensions of round-head machine screws. The four types of machine screws in Fig. 18.43 are drawn on a grid to give proportions of their heads in relation to the major diameter of the screw. Machine screws range in diameter from No. 0 (0.060 inch) to $\frac{3}{4}$ inch.

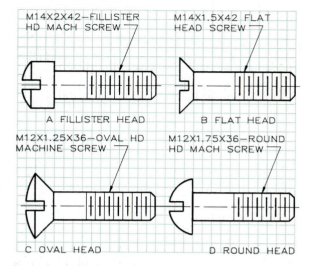

Figure 18.43 Standard types of machine screws. These proportions can be used for drawing machine screws of all sizes.

18.20 Set Screws

Set screws or keys are used to secure parts like pulleys and handles on a shaft. Figure 18.44 shows various types of set screws, and Table 18.5 shows their dimensions.

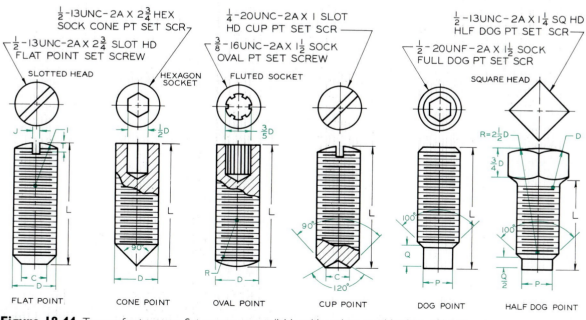

Figure 18.44 Types of set screws. Set screws are available with various combinations of heads and points. Notes give their specifications, and Table 18.5 gives dimensions.

Table 18.5

Dimensions for the set screws shows in Fig. 18.44 (dimensions in inches)

D	I	J	T	R	C		P		Q	q
Nominal Size	Radius of Headless Crown	Width of Slot	Depth of Slot	Oval Point Radius	Diameter of Cup and Flat Points		Diameter of Dog Point		Length of Dog Point	
					Max	Min	Max	Min	Full	Half
5 0.125	0.125	0.023	0.031	0.094	0.067	0.057	0.083	0.078	0.060	0.030
6 0.138	0.138	0.025	0.035	0.109	0.047	0.064	0.092	0.087	0.070	0.035
8 0.164	0.164	0.029	0.041	0.125	0.087	0.076	0.109	0.103	0.080	0.040
10 0.190	0.190	0.032	0.048	0.141	0.102	0.088	0.127	0.120	0.090	0.045
12 0.216	0.216	0.036	0.054	0.156	0.115	0.101	0.144	0.137	0.110	0.055
$\frac{1}{4}$ 0.250	0.250	0.045	0.063	0.188	0.132	0.118	0.156	0.149	0.125	0.063
$\frac{5}{16}$ 0.3125	0.313	0.051	0.076	0.234	0.172	0.156	0.203	0.195	0.156	0.078
$\frac{3}{8}$ 0.375	0.375	0.064	0.094	0.281	0.212	0.194	0.250	0.241	0.188	0.094
$\frac{7}{16}$ 0.4375	0.438	0.072	0.109	0.328	0.252	0.232	0.297	0.287	0.219	0.109
$\frac{1}{2}$ 0.500	0.500	0.081	0.125	0.375	0.291	0.270	0.344	0.344	0.250	0.125
$\frac{9}{16}$ 0.5625	0.563	0.091	0.141	0.422	0.332	0.309	0.391	0.379	0.281	0.140
$\frac{5}{8}$ 0.625	0.625	0.102	0.156	0.469	0.371	0.347	0.469	0.456	0.313	0.156
$\frac{3}{4}$ 0.750	0.750	0.129	0.188	0.563	0.450	0.425	0.563	0.549	0.375	0.188

Source: Courtesy of ANSI; B18.6.2.

Set screws are available in combinations of points and heads. The shaft against which the set screw is tightened may have a flat surface to give a good bearing surface for the set screw point; if so, a dog or flat point would be most effective to press against the flat surface. The cup point gives good friction when applied to a round shaft. Appendixes 26–28 give specifications for set screws.

18.21   Miscellaneous Screws

Figure 18.45 shows a few of the many types of specialty bolts, each of which has its own special application. Figure 18.45 also shows further examples of specialty bolts and screws.

Figure 18.46 shows three types of wing screws. They are available in incremental lengths of $\frac{1}{8}$ inch and are used to join parts assembled and disassembled by hand. Figure 18.47 shows two types of thumb screws, which serve the same purpose as wing screws, and Fig. 18.48 shows three types of wing nuts that can be used with various types of screws.

18.22   Wood Screws

A wood screw is pointed and has a sharp thread of coarse pitch for insertion into wood. Figure 18.49 shows the three most common types of wood screws and their proportions in relation to their major diameters.

Sizes of wood screws are specified by single numbers, such as 0, 6, or 16. From 0 to 10, each digit represents a different size (1, 2, 3, and so

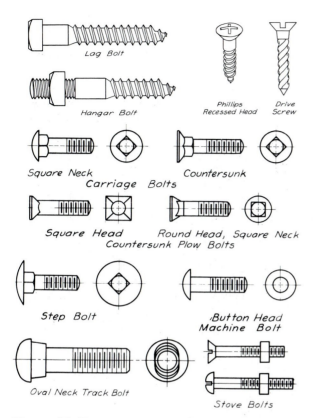

Figure 18.45 Miscellaneous types of bolts and screws.

on). Beginning at 10, only even-numbered sizes are standard, that is, 10, 12, 14, 16, 18, 20, 22, and 24. The following formula relates these numbered sizes to the actual diameter of the screws:

$$\text{Actual DIA} = 0.060 + \text{screw number} \times 0.013.$$

For example, the diameter of a No. 5 wood screw is

$$0.060 + 5(0.013) = 0.125.$$

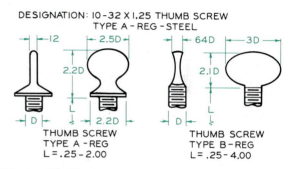

Figure 18.47 Thumb screw proportions for screw diameters of about $\frac{1}{4}$″. These proportions can be used to draw thumb screws of any screw diameter. Type A is available in diameters of 6, 8, 10, 12, 0.25″, 0.313″, and 0.375″. Type B thumb screws are available in diameters of 6 to 0.50″.

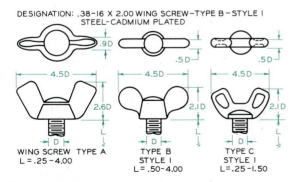

Figure 18.46 Wing screw proportions for screw diameters of about $\frac{5}{16}$″. These proportions can be used to draw wing screws of any size diameter. Type A is available in screw diameters of 4, 6, 8, 10, 12, 0.25″, 0.313″, 0.375″, 0.438″, 0.50″, and 0.625″; Type B in diameters of 10 to 0.625″; and Type C in diameters of 6 to 0.375″.

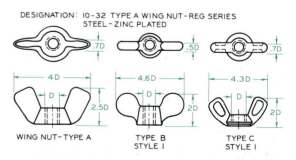

Figure 18.48 Wing nut proportions for screw diameters of $\frac{3}{8}$″. These proportions can be used to draw thumb screws of any size. Type A wing nuts are available in screw diameters (in inches) of 3, 4, 5, 6, 8, 10, 12, 0.25″, 0.313″, 0.375″, 0.438″, 0.50″, 0.583″, 0.625″, and 0.75″; Type B nuts are available in sizes from 5 to 0.75″; and Type C nuts in sizes from 4 to 0.50″.

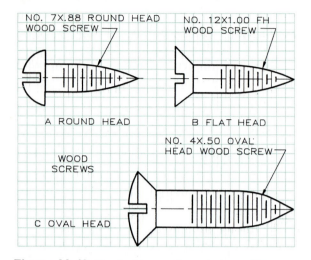

Figure 18.49 Standard types of wood screws. The proportions shown here can be used for drawing wood screws of all sizes.

18.23 Tapping a Hole

A threaded hole is called a **tapped hole,** and the tool used to cut the threads is called a **tap.** Figure 18.50 shows the types of taps available for threading small holes by hand.

The **taper, plug,** and **bottoming** hand taps are identical in size, length, and measurement;

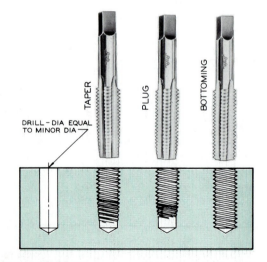

Figure 18.50 Three types of taps for threading internal holes.

only the chamfered portion of their ends is different. The taper tap has a long chamfer (8 to 10 threads), the plug tap has a shorter chamfer (3 to 5 threads), and the bottoming tap has the shortest chamfer (1 to $1\frac{1}{2}$ threads).

When tapping by hand in open or "through" holes, the taper should be used for coarse threads since it ensures straighter starting. The taper tap is also recommended for the harder metals. The plug tap can be used in soft metals for fine-pitch threads. When a hole is tapped to the very bottom, all three taps—taper, plug, and bottoming—should be used in this order.

Notes can be added to specify the depth of the drilled hole and the depth of the threads. For example, a note reading 7/8 DIA–3 DEEP × 1–8 UNC–2A × 2 DEEP means the hole will be drilled deeper than it is threaded, and the last usable thread will be 2 in. deep in the hole. Note that the drill point has a 120° angle.

18.24 Washers, Lock Washers, and Pins

Washers, called **plain washers,** are used with nuts and bolts to improve the assembly and strength of the fastening. Plain washers are noted on a working drawing in the following manner:

0.938 × 1.750 × 0.134 TYPE A PLAIN WASHER

These numbers represent the washer's inside diameter, outside diameter, and thickness, respectively (see Appendix 36).

A **lock washer** prevents a nut or cap screw from loosening as a result of vibration or movement. Common types of lock washers are the **external-tooth lock washer** and the **helical-spring lock washer.** Figure 18.51 shows other types of locking washers and devices.

Appendix 36 gives tables for regular and extra–heavy-duty helical-spring lock washers, which are designated with a note in the following form:

HELICAL-SPRING LOCK WASHER-
$\frac{1}{4}$ REGULAR—PHOSPHOR BRONZE.

(The $\frac{1}{4}$ is the washer's inside diameter.) Tooth

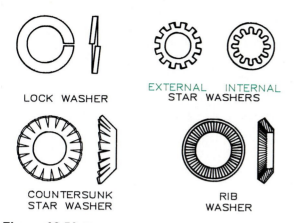

Figure 18.51 Types of lock washers and locking devices.

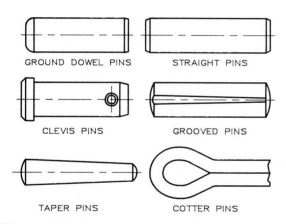

Figure 18.52 Types of pins used to fix parts together.

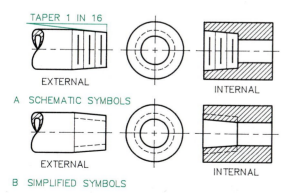

Figure 18.53 (A) Schematic techniques of representing pipe threads. (B) Simplified techniques of representing pipe threads.

lock washers are designated with notes in one of the following forms:

INTERNAL-TOOTH LOCK WASHER-
¼–TYPE A—STEEL;
EXTERNAL-TOOTH LOCK WASHER-
562–TYPE B—STEEL.

Straight pins and taper pins (Fig. 18.52) are used to fix parts together; Appendix 34 gives dimensions for them. Another locking device is the cotter pin; Appendix 37 gives tables of specifications for them.

18.25 Pipe Threads

Pipe threads are used in connecting pipes, tubing, and lubrication fittings. The most commonly used pipe thread is tapered at a ratio of 1 to 16, but straight pipe threads are available. Tapered pipe threads will only engage for an effective length determined by the formula:

$$L = (0.80D + 6.8)P,$$

where D is the outside diameter of the pipe, and P is the pitch of the thread. Figure 18.53 shows methods of representing tapered threads.

The following abbreviations are associated with pipe threads:

N = National	G = Grease
P = Pipe	I = Internal
T = Taper	M = Mechanical
C = Coupling	L = Locknut
S = Straight	H = Hose coupling
F = Fuel and oil	R = Railing fittings

Combining these abbreviations gives the following ANSI symbols:

NPT	= National pipe taper
NPTF	= National pipe thread (dryseal— for pressure-tight joints)
NPS	= Straight pipe thread
NPSC	= Straight pipe thread in couplings
NPSI	= National pipe straight internal thread
NPSF	= Straight pipe thread (dryseal)
NPSM	= Straight pipe thread for mechanical joints
NPSL	= Straight pipe thread for locknuts and locknut pipe threads

NPSH = Straight pipe thread for hose couplings and nipples

NPTR = Taper pipe thread for railing fittings

To specify a pipe thread in note form, the nominal pipe diameter (the internal diameter), the number of threads per inch, and the symbol that denotes the type of thread are given; for example,

$$1\tfrac{1}{4} – 11\tfrac{1}{2}\ \text{NPT} \quad \text{or} \quad 3 – 8\ \text{NPTR}$$

Appendix 9 gives these specifications. Figure 18.54 shows examples of external and internal thread notes. Dryseal threads, which may be straight or tapered, are used in applications where a pressure-tight joint is required without the use of a lubricant or sealer.

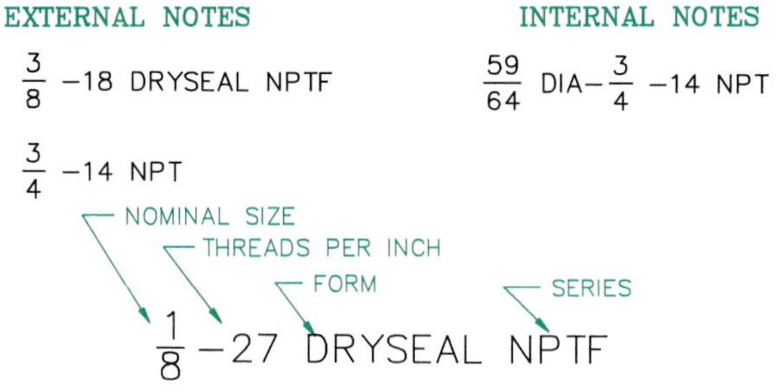

EXTERNAL NOTES

$\dfrac{3}{8}$ –18 DRYSEAL NPTF

$\dfrac{3}{4}$ –14 NPT

INTERNAL NOTES

$\dfrac{59}{64}$ DIA– $\dfrac{3}{4}$ –14 NPT

NOMINAL SIZE
THREADS PER INCH
FORM
SERIES

$\dfrac{1}{8}$ –27 DRYSEAL NPTF

Figure 18.54 Typical pipe-thread notes.

18.26 Keys

Keys are used to attach parts to shafts in order to transmit power to pulleys, gears, or cranks. The four types of keys shown pictorially and orthographically in Fig. 18.55 are the most commonly used. Notes must be given for the keyway, key, and keyseat, as shown in Fig. 18.55 (parts A, C, E, and G). These notes are typical of those used to give key specifications. Appendixes 31–33 give dimensions for various types of keys.

18.27 Rivets

Rivets are fasteners used to join thin overlapping materials in a permanent joint. The rivet is inserted in a hole slightly larger than the diameter of the rivet, and the headless end is formed into the specified shape by applying pressure to the projecting end. This forming operation is done when the rivets are either hot or cold, depending on the application.

Figure 18.56 shows typical shapes and proportions of small rivets. These rivets vary in diameter from $\tfrac{1}{16}$ to $1\tfrac{3}{4}$ inches. Rivets are used extensively in pressure-vessel fabrication, heavy structures such as bridges and buildings, and sheet–metal construction.

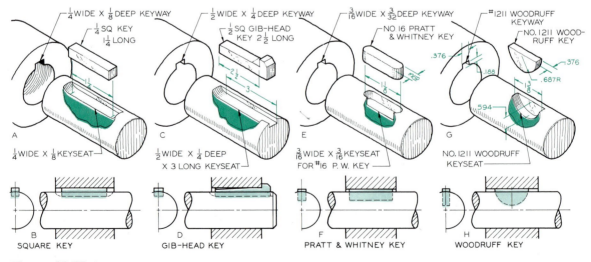

Figure 18.55 Standard keys used to hold parts on a shaft.

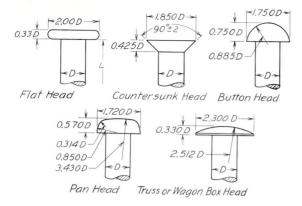

Figure 18.56 Types and proportions of small rivets. Small rivets have shank diameters up to $\frac{1}{2}''$.

Figure 18.57 shows the standard symbols recommended by ANSI for representing rivets. Rivets that are driven in the shop are **shop rivets,** and those assembled on the job at the site are **field rivets.**

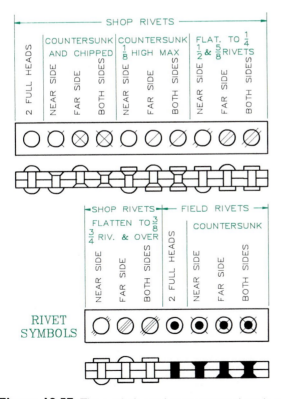

Figure 18.57 The symbols used to represent rivets in a drawing.

18.28 Springs

Some of the more commonly used types of springs are **compression, torsion, extension, flat,** and **constant force.** Figure 18.58 shows the **single-line** conventional representation of the first three types. Also shown are the types of ends that can be used on compression springs and the simplified single-line representation of coil springs.

Figure 18.59 shows a typical working drawing of a compression spring. The ends of the spring are drawn by using the **double-line** representation, and conventional lines are used to indicate the undrawn portion of the spring. The diameter and free length of the spring are given on the drawing, and the remaining specifications are given a table near the drawing.

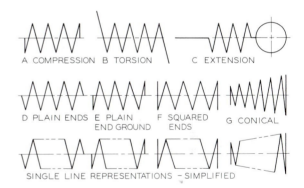

Figure 18.58 Single-line representations of various types of springs.

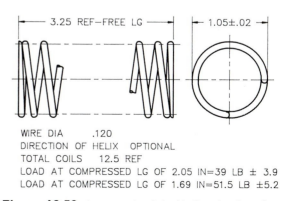

WIRE DIA   .120
DIRECTION OF HELIX   OPTIONAL
TOTAL COILS   12.5 REF
LOAD AT COMPRESSED LG OF 2.05 IN=39 LB ± 3.9
LOAD AT COMPRESSED LG OF 1.69 IN=51.5 LB ±5.2

Figure 18.59 A conventional double-line drawing of a compression spring and its specifications.

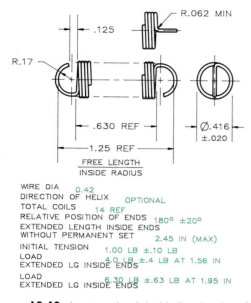

WIRE DIA   0.42
DIRECTION OF HELIX   OPTIONAL
TOTAL COILS   14 REF
RELATIVE POSITION OF ENDS   180° ±20°
EXTENDED LENGTH INSIDE ENDS
WITHOUT PERMANENT SET   2.45 IN (MAX)
INITIAL TENSION   1.00 LB ±.10 LB
LOAD   4.0 LB ±.4 LB AT 1.56 IN
EXTENDED LG INSIDE ENDS
LOAD   6.30 LB ±.63 LB AT 1.95 IN
EXTENDED LG INSIDE ENDS

Figure 18.60 A conventional double-line drawing of an extension spring and its specifications.

A working drawing of an extension spring (Fig. 18.60) is similar to that of a compression spring. In a drawing of a helical torsion spring (Fig. 18.61), angular dimensions must be shown to specify the initial and final positions of the spring as torsion is applied to it. All springs require a table of specifications to describe their details.

18.29 Drawing Springs

Springs may be drawn as schematic representations using single lines (Fig. 18.62). Each example is drawn by laying out the diameter of the coils and lengths of the springs, and then the number of active coils are drawn by using the diagonal-line method. In Fig. 18.62B, the two end coils are "dead" coils, and only five are active. Figure 18.62C shows a drawing of an extension spring. Where more realism is desired, a double-line drawing of a thread can be made, as Fig. 18.63 shows.

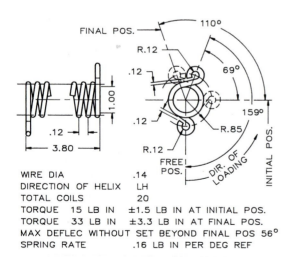

WIRE DIA   .14
DIRECTION OF HELIX   LH
TOTAL COILS   20
TORQUE 15 LB IN ±1.5 LB IN AT INITIAL POS.
TORQUE 33 LB IN ±3.3 LB IN AT FINAL POS.
MAX DEFLEC WITHOUT SET BEYOND FINAL POS 56°
SPRING RATE   .16 LB IN PER DEG REF

Figure 18.61 A conventional double-line drawing of a helical torsion spring and its specifications.

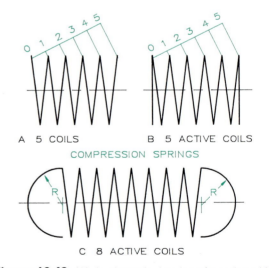

Figure 18.62 (A) A schematic drawing of a spring with four active coils. The diagonal-line method is used to divide it into four equally spaced coils. (B) A spring with six coils, but only four of them are active. (C) An extension spring with eight active coils.

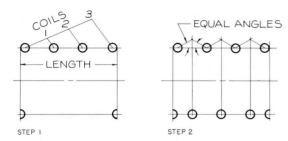
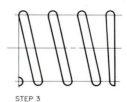

STEP 1   STEP 2   STEP 3   STEP 4

Figure 18.63 Detailed drawing of a spring.

Step 1 Lay out the diameter and length of the spring, and locate the coils by the diagonal-line technique.

Step 2 Locate the coils on the lower side along the bisectors of the spaces between the coils on the upper side.

Step 3 Connect the coils on each side. This is a right-hand coil; a left-hand spring would slope in the opposite direction.

Step 4 Construct the back side of the spring and the end coils to complete the detailed drawing of a compression spring.

Problems

These problems are to be drawn and solved on Size A sheets. Each grid equals 0.20 inch or 5 mm.

1. (Fig. 18.64) Draw detailed representations of Acme threads with major diameters of 2 inches. Show both external and internal threads as views and sections. Give a thread note using Appendix 18.

2. Repeat Problem 1, but draw internal and external detailed representations of square threads.

3. Repeat Problem 1, but draw internal and external detailed thread representations of Unified National threads. Give a thread note for a coarse thread with a class 2 fit.

4. Using the notes in Fig. 18.65, draw detailed representations of the internal threads and holes in section. Provide thread notes on each as specified.

5. Repeat Problem 4, but use schematic thread symbols.

6. Repeat Problem 4, but use simplified thread symbols.

7. Using the partial views in Fig. 18.66, draw external, internal, and end views of the full-size threaded parts using detailed thread symbols. Apply thread notes for UNC threads with a class 2 fit.

8. Repeat Problem 7, but use schematic thread symbols.

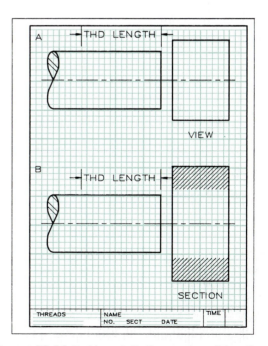

Figure 18.64 Problems 1–3. Construction of thread symbols.

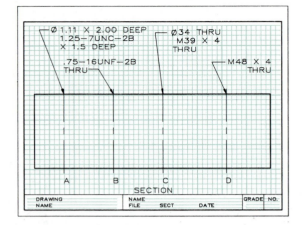

Figure 18.65 Problems 4–6. Internal threads.

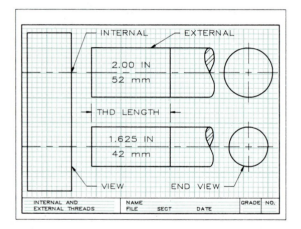

Figure 18.66 Problems 7–9. Internal and external threads.

9. Repeat Problem 7, but use simplified symbols.

10. (Fig. 18.67) Complete the drawing of the finished hexagon-head bolt and a heavy hexagon nut. The bolt head and nut are to be drawn across corners. Use detailed thread symbols. Provide thread notes in either English or metric forms as assigned.

11. Repeat Problem 10, but draw the nut and bolt as having unfinished square heads. Use schematic thread symbols.

12. Repeat Problem 10, but draw the bolt with a regular finished hexagon head across flats, using simpli-

fied thread symbols. Draw the nut across flats also, and provide thread notes for both.

13. Use the notes in Fig. 18.68 to draw the screws in section and complete the sectional view showing all cross-hatching. Use detailed thread symbols, and apply thread notes to the parts.

14. Repeat Problem 13, but use schematic thread symbols.

15. Repeat Problem 13, but use simplified thread symbols.

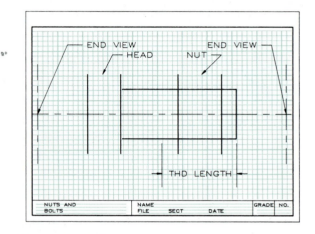

Figure 18.67 Problems 10–12. Nuts and bolts in assembly.

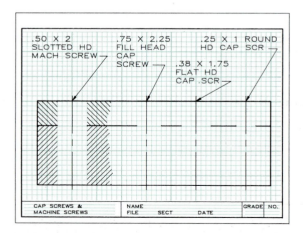

Figure 18.68 Problems 13–15. Cap screws and machine screws.

Figure 18.69 Problem 16. Design involving threaded parts.

16. (Fig. 18.69) The pencil pointer has a $\frac{1}{4}$ in. shaft that fits into a bracket designed to clamp onto a desk top. A set screw holds the shaft in position. Make a drawing of the bracket, estimating its dimensions. Show the details and the method of using the set screw to hold the shaft, and provide a thread note.

17. (Fig. 18.70) On axes *A* and *B*, construct hexagon-head cap screws (across flats) with UNC threads, and a class 2 fit. The cap screws should not reach the bottoms of the threaded holes. Convert the view to a half section.

18. (Fig. 18.71) On axes *A* and *B*, draw studs with a hexagon-head nut (across flats) that hold the two parts together. The studs are to be fine series with a class 2 fit, and they should not reach the bottom of the threaded hole. Provide a thread note. Show the view as a half section.

19. (Fig. 18.72) Draw a 2.00-inch (50-mm) diameter hexagon-head bolt with its head across flats using schematic symbols. Draw a plain washer and regular nut (across corners) at the right end. Design the size of the opening in the part at the left end to hold the bolt head so that it will not turn. Use a UNC thread with a series 2 fit. Give a thread note.

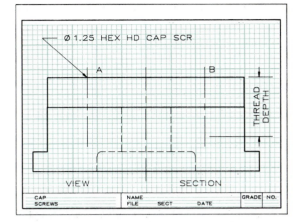

Figure 18.70 Problem 17.

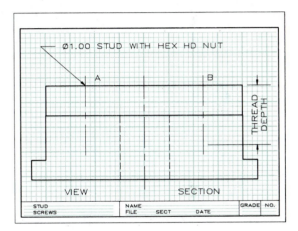

Figure 18.71 Problem 18.

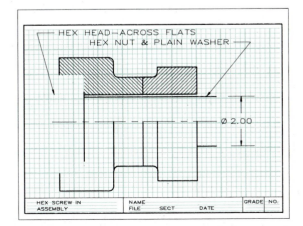

Figure 18.72 Problem 19.

20. (Fig. 18.73) Draw a 2.00-inch (50-mm) diameter hexagon-head cap screw that holds the two parts together. Determine the length of the bolt, and show the threads using schematic thread symbols. Give a thread note.

21. (Fig. 18.74) The part at A is held on the shaft by a square key, and the part at B is held on the shaft by gib-head key. Using Appendix 31, complete the drawings and give the necessary notes.

22. Repeat Problem 21, but use Woodruff keys, one with a flat bottom and the other with a round bottom. Using Appendix 32, complete the drawings and give the necessary notes.

23–26. (Fig. 18.75) Using Table 18.6, make a double-line drawing of the spring assigned.

27–30. Repeat Problem 23, but draw the springs using single-line representations.

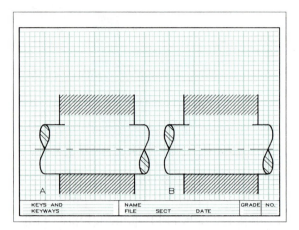

Figure 18.74 Problems 21–22.

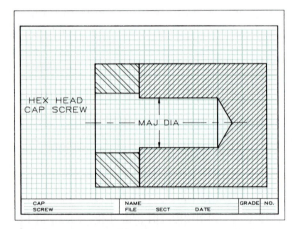

Figure 18.73 Problem 20.

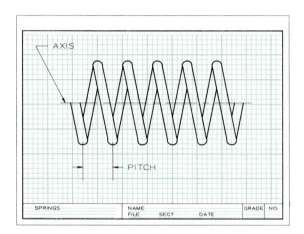

Figure 18.75 Problems 23–30.

Table 18.6 Problems 23–26						
Problem	**No. of Turns**	**Pitch**	**Size Wire**	**Inside Diameter**	**Outside Diameter**	
23	4	1	No. 4 = 0.2253	3		RH
24	5	$\frac{3}{4}$	No. 6 = 0.1920		2	LH
25	6	$\frac{5}{8}$	No. 10 = 0.1350	2		RH
26	7	$\frac{3}{4}$	No. 7 = 0.1770		$1\frac{3}{4}$	LH

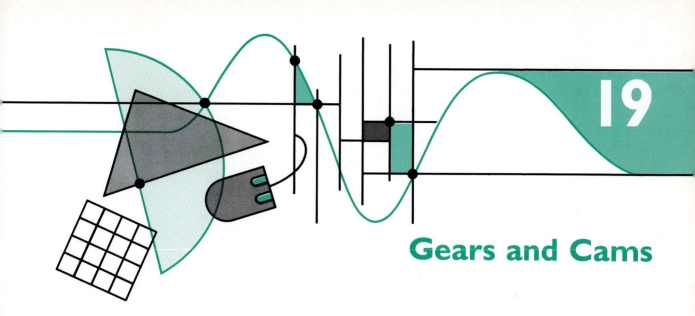

19

Gears and Cams

19.1 Introduction to Gears

Gears are toothed wheels whose circumferences mesh together to transmit force and motion from one gear to the next. In this chapter, we discuss the three most common types—**spur gears, bevel gears,** and **worm gears** (Fig. 19.1).

19.2 Spur Gear Terminology

The spur gear is a circular gear with teeth cut around its circumference. Two mating spur gears can transmit power from a shaft to another, parallel shaft.

Figure 19.1 The three basic types of gears are (A) spur gears, (B) bevel gears, and (C) worm gears. (Courtesy of the Process Gear Co.)

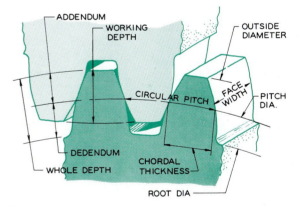

Figure 19.2 Gear terminology for spur gears.

When the two meshing gears are unequal in diameter, the smaller gear is called the **pinion,** and the larger one the **gear.**

The following terms describe the parts of a spur gear. Figure 19.2 shows many of these features. The corresponding formulas for each feature are also given.

Pitch Circle (PC) is the imaginary circle of a gear if it were a friction wheel without teeth that contacted the pitch circle of another friction wheel.

Pitch Diameter (PD) is the diameter of the pitch circle: $PD = N/DP$, where N is the number of teeth, and DP is the diametral pitch.

Diametral Pitch (DP) is the ratio between the number of teeth on a gear and its pitch diameter. For example, a gear with 20 teeth and a 4-in. pitch diameter will have a diametral pitch of 5, which means there are 5 teeth per inch of a diameter: $DP = N/PD$, where N is the number of teeth.

Circular Pitch (CP) is the circular measurement from one point on a tooth to the corresponding point on the next tooth measured along the pitch circle: $CP = 3.14/DP$.

Center Distance (CD) is the distance from the center of a gear to its mating gear's center: $CD = (N_P + N_S)/(2DP)$, where N_P and N_S are the number of teeth in the pinion and spur, respectively.

Addendum (A) is the height of a gear above its pitch circle: $A = 1/DP$.

Dedendum (D) is the depth of a gear below the pitch circle: $D = 1.157/DP$.

Whole Depth (WD) is the total depth of a gear tooth: $WD = A + D$.

Working Depth (WKD) is the depth to which a tooth fits into a meshing gear: $WKD = 2/DP$, or $WKD = 2A$.

Circular Thickness (CRT) is the circular distance across a tooth measured along the pitch circle: $CRT = 1.57/DP$.

Chordal Thickness (CT) is the straight-line distance across a tooth at the pitch circle: $CT = PD (\sin 90°/N)$, where N is the number of teeth.

Face Width (FW) is the width across a gear tooth parallel to its axis. This is a variable dimension, but it is usually three to four times the circular pitch: $FW = 3$ to $4(CP)$.

Outside Diameter (OD) is the maximum diameter of a gear across its teeth: $OD = PD + 2A$.

Root Diameter (RD) is the diameter of a gear measured from the bottom of its gear teeth: $RD = PD - (2D)$.

Pressure Angle (PA) is the angle between the line of action and a line perpendicular to the centerline of two meshing gears. Angles of 14.5° and 20° are standard for involute gears.

Base Circle (BC) is the circle from which an involute tooth curve is generated or developed: $BC = PD (\cos PA)$.

19.3 Tooth Forms

The most common gear tooth is an **involute tooth** with a 14.5° pressure angle. The 14.5° angle is the angle of contact between two gears when the tangents of both gears pass through the point of contact. Gears with pressure angles of 20° and 25° are also used. Gear teeth with larger pressure angles are wider at the base and thus stronger than the standard 14.5° teeth.

19.4 Gear Ratios

The diameters of two meshing spur gears establish ratios that are important to the function of the gears (Fig. 19.3).

If the radius of a gear is twice that of its pinion (the small gear), the diameter is twice that of the pinion, and the gear has twice as many teeth as the pinion. The pinion must make twice as many turns as the larger gear; that is, the revolutions per minute (RPM) of the pinion is twice that of the larger gear.

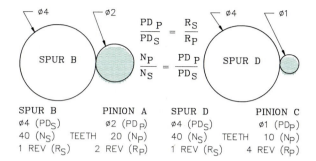

SPUR B	PINION A	SPUR D	PINION C
ø4 (PD$_S$)	ø2 (PD$_P$)	ø4 (PD$_S$)	ø1 (PD$_P$)
40 (N$_S$) TEETH	20 (N$_P$)	40 (N$_S$) TEETH	10 (N$_P$)
1 REV (R$_S$)	2 REV (R$_P$)	1 REV (R$_S$)	4 REV (R$_P$)

Figure 19.3 Ratios between meshing spur gears.

The relationship between two meshing spur gears can be developed in a formula by finding the velocity of a point on the small gear that is equal to $\pi PD \times$ RPM of the pinion. The velocity of a point on the large gear equals $\pi PD \times$ RPM of the spur. Since the velocity of points on each gear must be equal, the equation may be written as

$$\pi PD_P(\text{RPM}) = \pi PD_S(\text{RPM});$$

therefore,

$$\frac{PD_P}{PD_S} = \frac{\text{RPM}_S}{\text{RPM}_P}.$$

If the radius of the pinion is 1 inch, the radius of the spur 4 in., and the RPM of the pinion 20, then the RPM of the spur can be found as follows:

$$\frac{2(1)}{2(4)} = \frac{\text{RPM}_S}{20\ \text{RPM}_P}$$

$$\text{RPM}_S = \frac{2(20)}{2(4)} = 5\ \text{RPM}.$$

The RPM of the spur is 5, or one fourth the RPM of the pinion.

The number of teeth on each gear is proportional to the diameters of a pair of meshing gears. This relationship can be written

$$\frac{N_P}{N_S} = \frac{PD_P}{PD_S},$$

where N_P and N_S are the number of teeth on the pinion and spur, respectively, and PD_P and PD_S are their pitch diameters.

19.5 Spur Gear Calculations

Before a working drawing of a gear can be started, the drafter must perform calculations to determine the gear's dimensions.

Problem 1 Calculate the dimensions for a spur gear that has a pitch diameter of 5 inches, a diametral pitch of 4, and a pressure angle of 14.5°. (The diametral pitch is the same for meshing gears.)

Solution

Number of teeth: $PD \times DP = 5 \times 4 = 20$.

Addendum: $\frac{1}{4} = 0.25''$.

Dedendum: $= 1.157/4 = 0.2893''$.

Circular thickness: $1.5708/4 = 0.3927''$.

Outside diameter: $(20 + 2)/4 = 5.50''$.

Root diameter: $5 - 2(0.2893) = 4.421''$.

Chordal thickness:
$5(\sin 90°/20) = 5(0.079) = 0.392''$.

Chordal addendum:
$0.25 + [0.3927^2/(4 \times 5)] = 0.2577''$.

Face width: $3.5(0.79) = 2.75$.

Circular pitch: $3.14/4 = 0.785''$.

Working depth:
$0.6366 \times (3.14/4) = 0.4997''$.

Whole depth: $0.250 + 0.289 = 0.539''$.

These dimensions can be used to draw the spur gear and to provide specifications necessary for its manufacture.

Problem 2 shows the method of determining the design information for two meshing gears when their working ratios are known.

Problem 2 Find the number of teeth and other specifications for a pair of meshing gears with a driving gear that turns at 100 RPM and a driven gear that turns at 60 RPM. The diametral pitch for each is 10, and the center-to-center distance between the gears is 6 inches.

Solution

Step 1 Find the sum of the teeth on both gears:

Total teeth = 2 × (c-to-c distance) × DP
= 2 × 6 × 10 = 120 teeth.

Step 2 Find the number of teeth for the driving gear:

$$\frac{\text{Driver RPM}}{\text{Driven RPM}} + 1 = \frac{100}{60} + 1 = 2.667;$$

$$\frac{\text{Total teeth}}{\dfrac{100}{60} + 1} = \frac{120}{2.667} = 45 \text{ teeth.}^\star$$

Step 3 Find the number of teeth for the driven gear:

Total teeth −
teeth on driver = teeth on driven gear
120 − 45 = 75 teeth.

Step 4 Other specifications for the gears can be calculated as shown in Problem 1 by using the formulas in Section 19.2.

19.6 Drawing Spur Gears

Figure 19.4 shows a conventional drawing of a spur gear. Since it is so time consuming, the teeth need not be drawn. It is possible to omit the circular view and to show only a sectional view of the gear with a table of dimensions called **cutting data.** Circular centerlines are drawn to represent the root circle, pitch circle, and outside circle of the gear in the circular view.

A table of dimensions is a necessary part of a gear drawing, as Fig. 19.5 shows. These data can be calculated by formula or taken from tables of standards in gear handbooks such as *Machinery's Handbook.*

★The number of teeth must be a whole number since there cannot be fractional teeth on a gear. It may be necessary to adjust the center distance to yield a whole number of teeth.

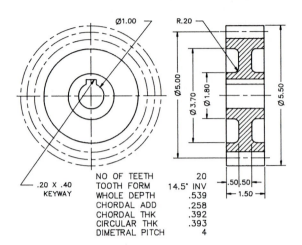

NO OF TEETH	20
TOOTH FORM	14.5° INV
WHOLE DEPTH	.539
CHORDAL ADD	.258
CHORDAL THK	.392
CIRCULAR THK	.393
DIMETRAL PITCH	4

Figure 19.4 A detail drawing of a spur gear with a table of values to supplement the dimensions shown on the drawing.

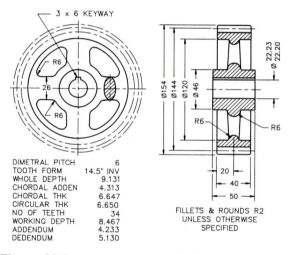

DIMETRAL PITCH	6
TOOTH FORM	14.5° INV
WHOLE DEPTH	9.131
CHORDAL ADDEN	4.313
CHORDAL THK	6.647
CIRCULAR THK	6.650
NO OF TEETH	34
WORKING DEPTH	8.467
ADDENDUM	4.233
DEDENDUM	5.130

Figure 19.5 A computer-drawn detail drawing of a spur gear.

19.7 Bevel Gear Terminology

Bevel gears are gears whose axes intersect at angles. Although the angle of intersection is usually 90°, other angles are also used. The smaller of the two bevel gears is the **pinion,** as in spur gearing.

Figure 19.6 illustrates the terminology of bevel gearing. The corresponding formulas for

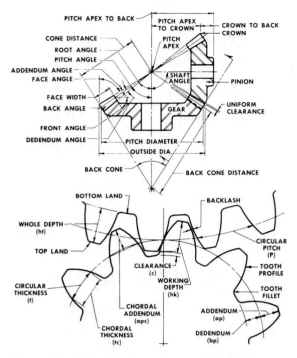

Figure 19.6 The terminology and definitions of bevel gears. (Courtesy of Philadelphia Gear Corp.)

each feature are given below. Gear handbooks can also be used for finding these dimensions.

Pitch Angle of Pinion (Small Gear) (PA_p):

$$\tan PA_p = \frac{N_p}{N_g},$$

where N_g and N_p are the number of teeth on the gear and pinion, respectively.

Pitch Angle of Gear (PA_g):

$$\tan PA_g = \frac{N_g}{N_p}.$$

Pitch Diameter (PD) is the number of teeth (N) divided by the diametral pitch *(DP)*: *PD = N/P*.

Addendum (A) is measured at the large end of the tooth: *A = 1/DP*.

Dedendum (D) is measured at the large end of the tooth: *D = 1.157/DP*.

Whole Tooth Depth (WD): *WD = 2.157/DP*.

Thickness of Tooth (TT) is measured at the pitch circle: *TT = 1.571/DP*.

Diametral Pitch: *DP = N/PD*, where *N* is the number of teeth.

Addendum Angle (AA) is the angle formed by the addendum and pitch cone distance:

$$\tan AA = \frac{A}{PCD}.$$

Angular Addendum: *AK = COS PA × A*.

Pitch Cone Distance (PCD): *PCD = PD/(2 × sin PA)*.

Dedendum Angle (DA) is the angle formed by the dedendum and the pitch cone distance:

$$\tan DA = \frac{D}{PCD}.$$

Face Angle (FA) is the angle between the gear's centerline and the top of its teeth: *FA = 90° − (PCD + AA)*.

Cutting Angle (or Root Angle) (CA) is the angle between the gear's axis and the roots of the teeth: *CA = PCD − D*.

Outside Diameter (OD) is the greatest diameter of a gear across its teeth: *OD = PD + 2A*.

Apex to Crown Distance (AC) is the distance from the crown of the gear to the apex of the cone measured parallel to the axis of the gear: *AC = OD/(2 tan FA)*.

Chordal Addendum (CA):

$$CA = A + \frac{TT^2 \cos PA}{4(PD)}.$$

Chordal Thickness (CT) is measured at the large end of the tooth:

$$CT = PD \times \sin\frac{90°}{N}.$$

Face Width (FW) can vary, but it is recommended that it be approximately equal to the pitch cone distance divided by 3: $FW = PCD/3$.

19.8   Bevel Gear Calculations

Problem 3 demonstrates how the formulas in Section 19.7 are used. Some of the formulas result in the same specifications that apply to both the gear and pinion.

Problem 3   Two bevel gears intersect at right angles. They have a diametral pitch of 3, 60 teeth on the gear, 45 teeth on the pinion, and a face width of 4 inches. Find the dimensions of the gear.

Solution

Pitch cone angle of gear:

$$\tan PCA = 60/45 = 1.33;$$
$$PCA = 53°7'.$$

Pitch cone angle of pinion:

$$\tan PCA = 45/60;$$
$$PCA = 36°52'.$$

Pitch diameter of gear: $60/3 = 20.00''$.

Pitch diameter of pinion: $45/3 = 15.00''$.

The following formulas are the same for both the gear and the pinion:

Addendum: $\frac{1}{3} = 0.333''$.

Dedendum: $1.157/3 = 0.3857''$.

Whole depth: $2.157/3 = 0.719''$.

Tooth thickness on pitch circle:

$$1.571/3 = 0.5237''.$$

Pitch cone distance:

$$20/(2 \sin 53°7') = 12.5015''.$$

Addendum angle:

$$\tan AA = 0.333/12.5015 = 1°32'.$$

Dedendum angle:

$$DA = 0.3857/12.5015 = 0.0308 = 1°46'.$$

Face width: $PCD/3 = 4.00''$.

The following formulas must be applied separately to the gear and pinion:

Chordal addendum of gear:

$$0.333'' + \frac{0.5237^2 \times \cos 53°7'}{4 \times 20} = 0.336''.$$

Chordal addendum of pinion:

$$0.333'' + \frac{0.5237^2 \times \cos 36°52'}{4 \times 15} = 0.338''.$$

Chordal thickness of gear:

$$\sin\frac{90°}{60} \times 20'' = 0.524''.$$

Chordal thickness of pinion:

$$\sin\frac{90°}{45} \times 15'' = 0.523''.$$

Face angle of gear:

$$90° - (53°7' + 1°32') = 35°21'.$$

Face angle of pinion:

$$90° - (36°52' + 1°32') = 51°36'.$$

Cutting angle of gear:

$$53°7' - 1°46' = 51°21'.$$

Cutting angle of pinion:

$$36°52' - 1°46' = 35°6'.$$

Angular addendum of gear:

$$0.333'' \times \cos 53°7' = 0.1999''.$$

Angular addendum of pinion:

$$0.333'' \times \cos 36°52' = 0.2667''.$$

Outside diameter of gear:

$$20'' + 2(0.1999'') = 20.4000''.$$

Outside diameter of pinion:

$$15'' + 2(0.2667'') = 15.533''.$$

Apex-to-crown distance of gear:

$$\frac{20.400''}{2} \times \tan 35°7' = 7.173''.$$

Apex-to-crown distance of pinion:

$$\frac{15.533''}{2} \times \tan 51°36' = 9.800''.$$

19.9 Drawing Bevel Gears

The dimensions calculated in Section 19.8 are used to lay out the bevel gears in a detail drawing. Many of the calculated dimensions would be difficult to measure on a drawing within a high degree of accuracy; therefore, it is important to provide a table of cutting data for each gear.

Figure 19.7 shows the steps of drawing the bevel gears. The finished drawings are shown

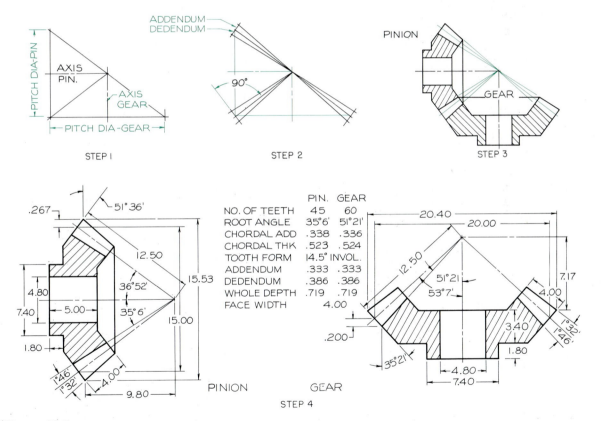

Figure 19.7 Construction of bevel gears.

Step 1 Lay out the pitch diameters and axes of the two bevel gears.

Step 2 Draw construction lines to establish the limits of the teeth by using the addendum and dedendum dimensions.

Step 3 Draw the pinion and gear using the specified dimensions or those calculated by formula.

Step 4 Complete the detail drawings of both gears, and provide a table of cutting data.

with a combination of dimensions and a table of dimensions.

19.10 Worm Gears

A worm gear is composed of a threaded shaft called a **worm** and a circular gear called a **spider** (Fig. 19.8). The worm is revolved, which causes the spider to revolve about its axis. Figures 19.8 and 19.9 illustrate the following terminology.

Worm Specifications and Formulas

Linear Pitch (P) is the distance from one thread to the next measured parallel to the worm's axis: $P = L/N$, where N is the number of threads: 1 if a single thread, 2 if a double thread, and so on.

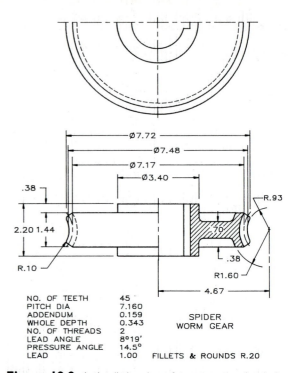

NO. OF TEETH	45
PITCH DIA	7.160
ADDENDUM	0.159
WHOLE DEPTH	0.343
NO. OF THREADS	2
LEAD ANGLE	8°19′
PRESSURE ANGLE	14.5°
LEAD	1.00

SPIDER
WORM GEAR

FILLETS & ROUNDS R.20

Figure 19.9 A detail drawing of a worm gear (spider) and the table of cutting data.

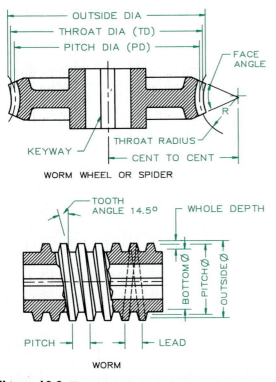

Figure 19.8 The terminology and definitions of worm gears.

Lead (L) is the distance a thread advances in a turn of 360°.

Addendum of Tooth (AW): $AW = 0.3183P$.

Pitch Diameter (PDW):

$$PDW = OD - 2AW,$$

where OD is the outside diameter.

Whole Depth of Tooth (WDT):

$$WDT = 0.6866 \times P.$$

Bottom Diameter of Worm (BD):

$$BD = OD - 2WDT.$$

Width of Thread at Root (WT): $WT = 0.31P$.

Minimum Length of Worm (MLW):

$$MLW = \sqrt{8PDS \times AW},$$

where *PDS* is the pitch diameter of the spider.

Helix Angle of Worm (HA):

$$\cot \beta = \frac{3.14(PDW)}{L}.$$

Outside Diameter (OD): $OD = PD + 2A$.

Spider Specifications and Formulas

Pitch Diameter of Spider (PDS):

$$PDS = \frac{N(P)}{3.14},$$

where *N* is the number of teeth on the spider.

Throat Diameter of Spider (TD):

$$TD = PDS + 2A.$$

Radius of Spider Throat (RST):

$$RST = \frac{OD \text{ of worm}}{2} - 2A.$$

Face Angle (FA) may be selected to be between 60° and 80° for the average application.

Center-to-Center Distance (CD) is measured between the worm and spider:

$$CD = \frac{PDW + PDS}{2}.$$

Outside Diameter of Spider (ODS):

$$ODS = TD + 0.4775P.$$

Face Width of Gear (FW): $FW = 2.38(P) + 0.25$.

19.11 Worm Gear Calculations

Problem 4 has been solved for a worm gear by using the formulas in Section 19.10.

Problem 4 Calculate the specifications for a worm and worm gear (spider). The gear has 45 teeth, and the worm has an outside diameter of 2.50 inches. The worm has a double thread and a pitch of 0.5 inch.

Solution

Lead: $L = 0.5'' \times 2 = 1''$.

Worm addendum:

$$AW = 0.3183P = 0.1592''.$$

Pitch diameter of worm:

$$PDW = 2.50'' - 2(0.1592'') = 2.1818''.$$

Pitch diameter of gear:

$$PDS = (45'' \times 0.5)/3.14 = 7.166''.$$

Center distance between worm and gear:

$$CD = \frac{(2.182'' + 7.166'')}{2} = 4.674''.$$

Whole depth of worm tooth:

$$WDT = 0.687 \times 0.5'' = 0.3433''.$$

Bottom diameter of worm:

$$BD = 2.50'' - 2(0.3433'') = 1.813''.$$

Helix angle of worm:

$$\cot \beta = \frac{3.14(2.1816)}{1} = 8°19'.$$

Width of thread at root:

$$WT = 0.31(1) = 0.155''.$$

Minimum length of worm:

$$MLW = \sqrt{8(0.1592)(7.1656)} = 3.02''.$$

Throat diameter of gear:

$$TD = 7.1656'' + 2(0.1592'') = 7.484''.$$

Radius of gear throat:

$$RST = (2.5/2) - (2 \times 0.1592) = 0.9318''.$$

Face width:

$$FW = 2.38(0.5) + 0.25 = 1.44''.$$

Outside diameter of gear:

$$ODS = 7.484 + 0.4775(0.5) = 7.723''.$$

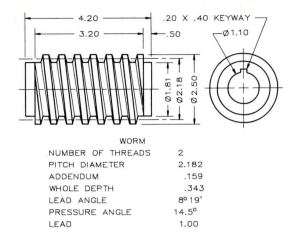

WORM	
NUMBER OF THREADS	2
PITCH DIAMETER	2.182
ADDENDUM	.159
WHOLE DEPTH	.343
LEAD ANGLE	8° 19'
PRESSURE ANGLE	14.5°
LEAD	1.00

Figure 19.10 A detail drawing of a worm using the dimensions calculated.

19.12 Drawing Worm Gears

The worm and worm wheel (spider) are drawn and dimensioned as shown in Figs. 19.9 and 19.10. The specifications derived by the formulas in Section 19.11 must be used for scaling, laying out the drawings, and providing cutting data.

19.13 Introduction to Cams

Plate cams are irregularly shaped machine elements that produce motion in a single plane, usually up and down (Fig. 19.11). As the cam revolves about its center, the variation in the cam's shape produces a rise or fall in the follower that is in contact with it. The shape of the cam is determined graphically before the preparation of manufacturing specifications.

Figure 19.11 Examples of machined cams. (Courtesy of Ferguson Machine Co.)

Cams utilize the principle of the inclined wedge, with the surface of the cam causing a change in the slope of the plane, thereby producing the desired motion.

19.14 Cam Motion

Cams are designed primarily to produce (1) **uniform** or linear motion (2) **harmonic motion,** (3) **gravity motion,** (uniform acceleration), or (4) **combinations** of these.

Uniform Motion

Uniform motion is shown in the displacement diagram in Fig. 19.12A to represent the motion of the cam follower as the cam rotates through 360°. The uniform-motion curve has sharp corners, which indicates abrupt changes of velocity and causes the follower to bounce. Therefore,

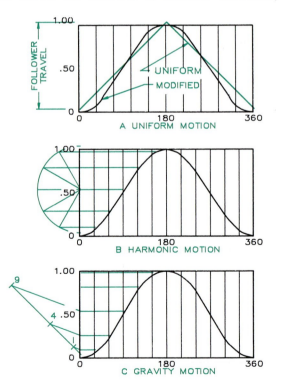

Figure 19.12 Displacement diagrams of three standard motions: Uniform, harmonic, and gravity.

uniform motion is usually modified with arcs that smooth this change of velocity. The radius of the modifying arc is varied up to a radius of one half the total displacement, depending on the speed of operation.

Harmonic Motion

Harmonic motion, plotted in Fig. 19.12B, is a smooth, continuous motion based on the change of position of points on the circumference of a circle. At moderate speeds, this displacement gives a smooth operation.

Gravity Motion

Gravity motion (uniform acceleration) is used for high-speed operation (Fig. 19.12C). The variation of displacement is analogous to the force of gravity, with the difference in displacement being 1, 3, 5, 5, 3, 1 based on the square of the number; for instance, $1^2 = 1$; $2^2 = 4$; $3^2 = 9$ to give a uniform acceleration. This motion is repeated in reverse order for the remaining half of the motion of the follower. Intermediate points can be found by squaring fractional increments such as $(2.5)^2$.

Cam Followers

Three basic types of cam followers are the **flat surface, roller,** and **knife edge** (Fig. 19.13). The flat-surface and knife-edge followers are limited to use with slow-moving cams where minor force will be exerted during rotation. The

roller follower is used to withstand higher speeds.

19.15 Construction of a Plate Cam

Plate Cam—Harmonic Motion

Figure 19.14 shows the steps of constructing a plate cam with harmonic motion. Before designing a cam, the drafter must know the motion of the follower, rise of the follower, diameter of the base circle, and direction of rotation. The specifications for the cam pictured in Fig. 19.14 are given graphically in the displacement diagram.

Plate Cam—Gravity Motion

The steps of constructing a cam with gravity motion are the same as in the previous example except for a different displacement diagram and a knife-edge follower. Figure 19.15 shows the graphic layout of the problem.

19.16 Construction of a Cam with an Offset Follower

The cam in Fig. 19.16 is designed to produce harmonic motion through 360°. This motion is plotted directly from the follower rather than from a displacement diagram.

A semicircle is drawn with its diameter equal to the total motion of the follower. The base circle is drawn to pass through the center or roller of the follower. The centerline of the follower is extended downward, and a circle is drawn tangent to the extension with its center at the center of the base circle. The small circle is divided into 30° intervals to establish points through which construction lines will be drawn tangent to the circle.

The distances from tangent points to the position points along the path of the follower are laid out along the tangent lines drawn at 30° intervals. These points can be located by measuring from the base circle, as shown in the figure, where point 3 was located distance X from the base circle. The circular roller is drawn in all views, and the profile of the cam is drawn to be tangent to the rollers at all positions.

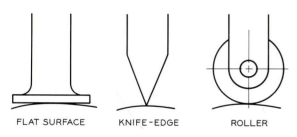

FLAT SURFACE   KNIFE-EDGE   ROLLER

Figure 19.13 Three basic types of cam followers are the flat surface, roller, and knife edge.

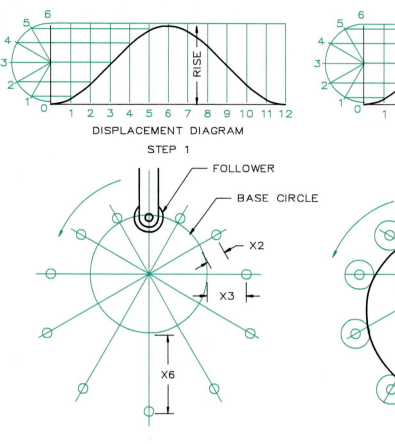

DISPLACEMENT DIAGRAM
STEP 1

DISPLACEMENT DIAGRAM
STEP 2

FOLLOWER

BASE CIRCLE

HARMONIC—MOTION
PLATE CAM

STEP 3

STEP 4

Figure 19.14 Construction of a plate cam with harmonic motion.

Step 1 Construct a semicircle whose diameter equals the rise of the follower. Divide the semicircle into the same number of divisions as there are between 0° and 180° on the horizontal axis of the displacement diagram. Plot the displacement curve.

Step 2 Distances of rise and fall ($X1$, $X2$, $X3$, $X6$) at each interval will be measured from the base circle.

Step 3 Construct the base circle, and draw the follower. Divide the circle into the same number of sectors as there are divisions on the displacement diagram. Transfer distances from the displacement diagram to the respective radial lines of the circle, measuring outward from it.

Step 4 Draw circles to represent the positions of the roller as the cam revolves in a counterclockwise direction. Draw the cam profile tangent to all the rollers to complete the drawing.

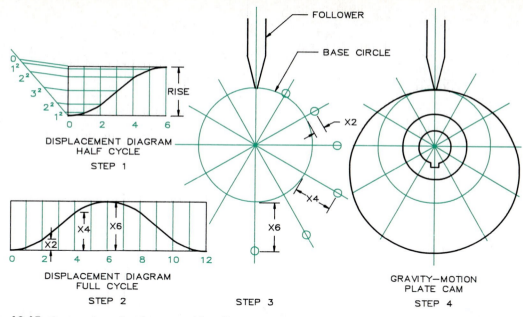

DISPLACEMENT DIAGRAM
HALF CYCLE

STEP 1

RISE

FOLLOWER

BASE CIRCLE

X2

X4

X6

DISPLACEMENT DIAGRAM
FULL CYCLE

STEP 2

STEP 3

GRAVITY—MOTION
PLATE CAM

STEP 4

X4 X6 X2

Figure 19.15 Construction of a plate cam with uniform acceleration.

Step 1 Construct a displacement diagram to represent the rise of the follower. Divide the horizontal axis into angular increments of 30°. Draw a construction line through point 0; locate the 1^2, 2^2, and 3^2 divisions, and project them to the vertical axis to represent half the rise.

Step 2 Use the same construction to find the right half of the symmetrical curve.

Step 3 Construct the base circle, and draw the knife-edge follower. Divide the circle into the same number of sectors as there are divisions in the displacement diagram. Transfer distances from the displacement diagram to their respective radial lines of the base circle, measuring outward from the base circle.

Step 4 Connect the points found in Step 3 with a smooth curve to complete the cam profile. Show also the cam hub and keyway.

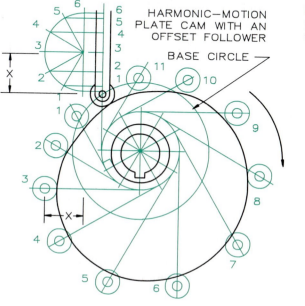

HARMONIC—MOTION
PLATE CAM WITH AN
OFFSET FOLLOWER

BASE CIRCLE

Figure 19.16 Construction of a plate cam with an offset roller follower.

Problems

Gears

Use Size A sheets ($8\frac{1}{2} \times 11$ inches) for the following gear problems. Select the most appropriate scale so that the drawings will use the available space.

1–5. Calculate the dimensions for the following spur gears, and make a detail drawing of each. Give the dimensions and cutting data for each gear. Provide any other dimensions needed.

Problem	Gear Teeth	Diametral Pitch	14.5° Involute
1	20	5	"
2	30	3	"
3	40	4	"
4	60	6	"
5	80	4	"

6–10. Calculate the gear sizes and number of teeth using the ratios and data below.

Problem	RPM Pinion	RPM Gear	Center to Center	Diametral Pitch
6	100 (driver)	60	6.0"	10
7	100 (driver)	50	8.0"	9
8	100 (driver)	40	10.0"	8
9	100 (driver)	35	12.0"	7
10	100 (driver)	25	14.0"	6

11–20. Make detail drawings of each of the gears for which calculations were made in Problems 6–10. Provide a table of cutting data and other dimensions needed to complete the specifications.

21–25. Calculate the specifications for the bevel gears that intersect at 90°, and make detail drawings of each with the necessary dimensions and cutting data.

Problem	Diametral Pitch	No. of Teeth on Pinion	No. of Teeth on Gear
21	3	60	15
22	4	100	40
23	5	100	60
24	6	100	50
25	7	100	30

26–30. Calculate the specifications for worm gears, and make detail drawings of each, with the necessary dimensions and cutting data.

Problem	No. of Teeth in Spider Gear	Outside DIA of Worm	Pitch of Worm	Thread of Worm
26	45	2.50	0.50	double
27	30	2.00	0.80	single
28	60	3.00	0.80	double
29	30	2.00	0.25	double
30	80	4.00	1.00	single

Cams

Use Size B sheets (11×17 inches) for the following cam problems. The standard dimensions are base circle, 3.50 in.; roller follower, 0.60-in. diameter; shaft, 0.75-in. diameter; hub, 1.25-in. diameter; direction of rotation, clockwise. The follower is positioned vertically over the center of the base circle. Lay out the problems and displacement diagrams as shown in Fig. 19.17.

31. Draw a plate cam with a knife-edge follower for uniform motion and a rise of 1.00 in.

32. Draw a displacement diagram and a cam that will give a modified uniform motion to a knife-edge follower with a rise of 1.7 in. Modify the uniform motion with an arc of one quarter the rise in the displacement diagram.

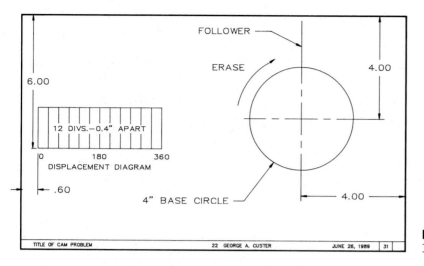

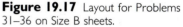

Figure 19.17 Layout for Problems 31–36 on Size B sheets.

33. Draw a displacement diagram and a cam that will give a harmonic motion to a roller follower with a rise of 1.60 in.

34. Draw a displacement diagram and a cam that will give a harmonic motion to a knife-edge follower with a rise of 1.00 in.

35. Draw a displacement diagram and a cam that will give uniform acceleration to a knife-edge follower with a rise of 1.70 in.

36. Draw a displacement diagram and a cam that will give a uniform acceleration to a roller follower with a rise of 1.40 in.

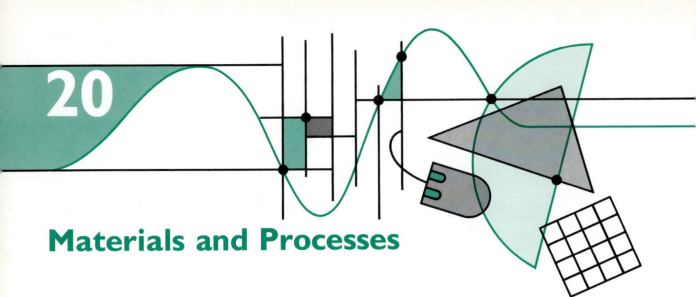

20

Materials and Processes

20.1 Introduction

Metallurgy, the study of metals, is a complex area that is constantly changing as new processes and alloys are developed (Fig. 20.1). The guidelines for designating various types of metals have been standardized by three associations: the American Iron and Steel Institute (AISI), the Society of Automotive Engineers (SAE), and the American Society for Testing Materials (ASTM).

20.2 Iron★

Metals that contain iron, even in small quantities, are called **ferrous metals.** The three types of iron are **gray iron, white iron,** and **ductile iron.**

Cast Iron is iron that is melted and poured into a mold to form it; it is used in the production of machine parts. Though cheaper and easier to machine than steel, iron does not have steel's ability to withstand shock and force.

Figure 20.1 This furnace operator is pouring an aluminum alloy of manganese into ingots (shown at the right) that will be remelted and cast. (Courtesy of the Aluminum Co. of America.)

Gray Iron contains flakes of graphite, which results in low strength and ductility but makes the material easy to machine. Gray iron resists vibrations better than other types of iron. Table 20.1 gives types of gray iron with two designations and their typical applications.

★This section on iron was developed by Dr. Tom Pollock, a metallurgist at Texas A&M University.

Table 20.1
Numbering and applications of gray iron

ATSM Grade (1000 psi)	SAE Grade	Typical Uses
ASTM 25 CI	G 2500 CI	Small engine blocks, pump bodies, transmission cases, clutch plates
ASTM 30 CI	G 3000 CI	Auto engine blocks, flywheels, heavy casting
ASTM 35 CI	G 3500 CI	Diesel engine blocks, tractor transmission cases, heavy and high-strength parts
ASTM 40 CI	G 4000 CI	Diesel cylinders, pistons, camshafts

White Iron contains carbide particles that are very hard and brittle, which enables it to withstand wear and abrasion. There are no designated grades of white iron, but there are differences in composition from one supplier to another. White iron is used for parts on grinding and crushing machines, digging teeth on earthmovers and mining equipment, and wear plates on reciprocating machinery used in textile mills.

Ductile Iron (also called **nodular** or **spheroidized** iron) contains tiny spheres of graphite,

Table 20.2
Numbering and applications of ductile iron

Grade	Typical Uses
60–40–18 CI	Valves, steam fittings, chemical plant equipment, pump bodies
65–45–12 CI	Machine components that are shock loaded, disc brake calipers
80–55–6 CI	Auto crankshafts, gears, rollers
100–70–3 CI	High-strength gears and machine parts
120–90–2 CI	Very high-strength gears, rollers, and slides

making it stronger and tougher than most types of gray iron but also more expensive to produce. The numbering system for ductile iron is given by three sets of numbers, as shown below.

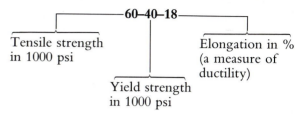

Table 20.2 gives commonly used alloys of ductile iron and their applications.

Malleable Iron is made from white iron by a heat-treatment process that converts carbides into carbon nodules (similar to ductile iron). The numbering system for designating the grades of malleable iron is shown below.

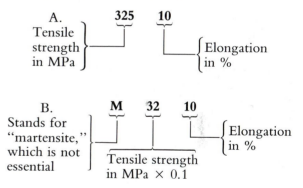

Table 20.3 gives some of the commonly used grades of malleable iron and their applications.

20.3 Steel

Steel is an alloy of iron and carbon and often contains other constituents such as manganese, chromium, or nickel. Carbon (usually between 0.20% and 1.50%) is the ingredient that has the greatest effect on the grade of the steel. Broadly, the three types of steel are **plain carbon steels, free-cutting carbon steels,** and **alloy steels.** Table 20.4 gives the types of steels and their designations by four-digit numbers. The first digit

Table 20.3
Numbering and applications of
malleable iron

ASTM Grade	Typical Uses
35018 CI	Marine and railroad valves and fittings, "black-iron" pipe fittings (similar to 60–40–18 ductile CI)
45006 CI	Machine parts (similar to 80–55–6 ductile CI)
M3210 CI	Low-stress components, brackets
M4504 CI	Crankshafts, hubs
M7002 CI	High-strength parts, connecting rods, universal joints
M8501 CI	Wear-resistant gears and sliding parts

indicates the type of steel: 1 is carbon steel, 2 is nickel steel, and so on. The second digit gives the percentage content of the material represented by the first digit. The last two or three digits give the percentage of carbon in the alloy, where 100 equals 1%, and 50 equals 0.50%.

Some frequently used SAE steels are 1010, 1015, 1020, 1030, 1040, 1070, 1080, 1111, 1118, 1145, 1320, 2330, 2345, 2515, 3130, 3135, 3240, 3310, 4023, 4042, 4063, 4140, and 4320.

20.4 Copper

Copper, one of the first metals discovered, can be easily formed and bent without breakage. Because it is highly resistant to corrosion and highly conductive, it is used in the manufacture of pipes, tubing, and electrical wiring. It is also an excellent roofing and screening material since it withstands the weather well.

Table 20.4
Numbering and applications of steel

Type of Steel	Number	Applications
Carbon steels		
Plain carbon	10XX	Tubing, wire, nails
Resulphurized	11XX	Nuts, bolts, screws
Manganese steel	13XX	Gears, shafts
Nickel steel	23XX	Keys, levers, bolts
	25XX	Carburized parts
Nickel-chromium	31XX	Axles, gears, pins
	32XX	Forgings
	33XX	Axles, gears
Molybdenum steel	40XX	Gears, springs
Chromium-molybdenum	41XX	Shafts, tubing
Nickel-chromium	43XX	Gears, pinions
Nickel-molybdenum	46XX	Cams, shafts
	48XX	Roller bearings, pins
Chromium steel	51XX	Springs, gears
	52XX	Ball bearings
Chromium vanadium	61XX	Springs, forgings
Silicon manganese	92XX	Leaf springs

Source: Courtesy of the Society of Automotive Engineers.

Table 20.5 Numbering designations for wrought aluminum and aluminum alloys		
Composition	**Alloy Number**	**Applications**
Aluminum (99% pure)	1XXX	Tubing, tank cars
Aluminum alloys Copper	2XXX	Aircraft parts, screws, rivets
Manganese	3XXX	Tanks, siding, gutters
Silicon	4XXX	Forging, wire
Magnesium	5XXX	Tubes, welded vessels
Magnesium and silicon	6XXX	Auto body, pipe
Zinc	7XXX	Aircraft structures
Other elements	8XXX	

Copper has several alloys, including brasses, tin bronzes, nickel silvers, and copper nickels. **Brass** is an alloy of copper and zinc, and **bronze** is an alloy of copper and tin. Copper and copper alloys can be easily finished by buffing or plating. These alloys can be joined by soldering, brazing, or welding and can be easily machined and used for casting.

Wrought copper has properties that permit it to be formed by hammering. A few of the numbered designations of wrought copper are C11000, C11100, C11300, C11400, C11500, C11600, C10200, C12000, and C12200.

20.5 Aluminum

Aluminum is a corrosion-resistant, lightweight metal that has applications for many industrial products. Most materials called aluminum are actually aluminum alloys, which are stronger than pure aluminum.

The types of wrought aluminum alloys are designated by four digits, as Table 20.5 shows. The first digit, from 2 through 9, indicates the alloying element that is combined with aluminum. The second digit indicates modifications of the original alloy or impurity limits. The last two digits identify the other alloying materials or indicate the aluminum purity.

Table 20.6 shows a four-digit numbering system used to designate types of cast aluminum and aluminum alloys. The first digit indicates the alloy group. The next two digits identify the aluminum alloy or aluminum purity. The number 1 to the right of the decimal point represents the aluminum form: XX.0 indicates castings, XX.1 indicates ingots with a specified chemical composition, and XX.2 indicates ingots with a specified chemical composition other than the XX.1 ingot, whereas 0 represents aluminum for casting. **Ingots** are blocks of cast metal to be remelted, and **billets** are castings of aluminum to be formed by forging.

Table 20.6 Aluminum casting and ingot designations	
Composition	**Alloy Number**
Aluminum (99 % pure)	1XX.X
Aluminum alloys Copper	2XX.X
Silicon with copper and/or magnesium	3XX.X
Silicon	4XX.X
Magnesium	5XX.X
Magnesium and silicon	6XX.X
Zinc	7XX.X
Tin	8XX.X
Other elements	9XX.X

Source: Courtesy of the Society of Automotive Engineers.

20.6 Magnesium

Magnesium is a light metal available in an inexhaustible supply since it is extracted from seawater and natural brines. Approximately half the weight of aluminum, magnesium is an excellent material for aircraft parts, clutch housing, crankcases for air-cooled engines, and applications where lightness is desirable.

Magnesium is used for die and sand castings, extruded tubing, sheet metal, and forging. Magnesium and its alloys can be joined by bolting, riveting, or welding. Some numbered designations of magnesium alloys are M10100, M11630, M11810, M11910, M11912, M12390, M13320, M16410, and M16620.

20.7   Properties of Materials

All materials have properties that designers must use to their best advantage. The following terms describe these properties:

Ductility   is a softness present in some materials, such as copper and aluminum, that permits them to be formed by stretching (drawing) or hammering without breaking. Wire is made of ductile materials that can be drawn through a die.

Brittleness   is a characteristic of metals that will not stretch without breaking, such as cast irons and hardened steels.

Malleability   is the ability of a metal to be rolled or hammered without breaking.

Hardness   is the ability of a metal to resist being dented when it receives a blow.

Toughness   is the property of being resistant to cracking and breaking while remaining malleable.

Elasticity   is the ability of a metal to return to its original shape after being bent or stretched.

20.8   Heat Treatment of Metals

The properties of metals can be changed by various forms of heat treating. Steels are affected to a greater extent by heat treating than are other materials.

Hardening   is performed by heating steel to a prescribed temperature and then quenching it in oil or water.

Quenching   is the process of rapidly cooling heated metal by immersing it in liquids, gases, or solids (such as sand, limestone, or asbestos).

Tempering   is the process of reheating previously hardened steel and then cooling it, usually by air. This increases the steel's toughness.

Annealing   is the process of heating and cooling metals to soften them, release their internal stresses, and make them easier to machine.

Normalizing   is achieved by heating metals and letting them cool in air to relieve their internal stresses.

Case Hardening   is the process of hardening a thin outside layer of a metal. The outer layer is placed in contact with carbon or nitrogen compounds that are absorbed by the metal as it is heated; afterward, the metal is quenched.

Flame Hardening   is the method of hardening by heating a metal to within a prescribed temperature range with a flame and then quenching the metal.

20.9   Castings

Two major methods of forming shapes are **casting** and **pressure forming.** Casting involves the preparation of a mold into which is poured molten metal that cools and forms the part. The types of casting, which differ in the way the molds are made, are **sand casting, permanent-mold casting, die casting,** and **investment casting.**

Sand Casting

In the first step of sand casting, a wood or metal form or pattern is made that is representative of the final part to be cast. The pattern is placed in a metal box called a **flask,** and molding sand is packed around the pattern. When the pattern is withdrawn from the sand, it leaves a void form-

Figure 20.2 A large casting of an aircraft's landing-gear mechanism is being removed from its mold. (Courtesy of Cameron Iron Works.)

Since the patterns are placed in sand and then withdrawn before the metal is poured, the sides of the patterns must be tapered for ease of withdrawal from the sand (Fig. 20.3). This taper is called **draft.** The amount of draft depends on the depth of the pattern in the sand; for most applications, it varies from 2° to 8°. To compensate for shrinkage that will occur when the metal cools, patterns are made oversize.

Because the sand casting has a rough surface, it is common practice to machine portions of a casting by drilling, grinding, or shaping (Fig. 20.4). The pattern should be made larger in these areas to compensate for removal of metal.

Fillets and rounds are used at intersections to increase the strength of a casting and because it is difficult to form square corners by the sand-casting process.

ing the mold. Molten metal is poured into the mold through sprues, or gates. After cooling, the casting is removed and cleaned (Fig. 20.2).

Cores, parts formed in sand, are placed within a mold to leave holes or hollow portions within the finished casting. Once the casting has been formed, the cores can be broken apart and removed, leaving behind the desired void within the casting.

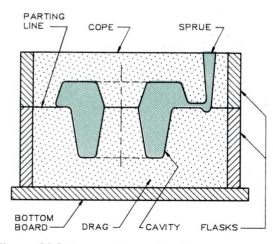

Figure 20.3 A two-section sand mold.

Figure 20.4 This casting of the outer cylinder of an aircraft's landing gear is being bored on a horizontal boring mill. (Courtesy of Cameron Iron Works.)

Permanent-Mold Casting

Permanent molds are made for the mass production of parts. They are generally made of cast iron and coated to prevent fusing with the molten metal poured into them (Fig. 20.5).

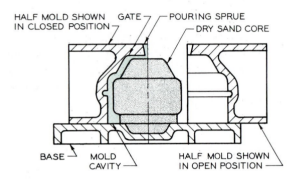

Figure 20.5 Permanent molds are made of metal for repetitive usage. Here, a sand core made from another mold is placed in the permanent mold to give a void within the casting.

Die Casting

Die castings are used for the mass production of parts made of aluminum, magnesium, zinc alloys, copper, or other materials. Made by forcing molten metal into dies (or molds) under pressure, die castings can be produced at a low cost, at close tolerances, and with good surface qualities. The same general principles recommended for sand castings—using fillets and rounds, allowing for shrinkage, and specifying draft angles—apply to die castings (Fig. 20.6).

Investment Casting

Investment casting is a process used to produce complicated parts that would be difficult to form by any other method. This technique is used to form the intricate shapes of artistic sculptures.

Since a new pattern must be used for each investment casting, a mold or die is made to cast a wax master pattern. The wax pattern, which will be identical to the finished casting, is placed inside a container, and plaster or sand is poured (or invested) around the pattern. Once the invest-

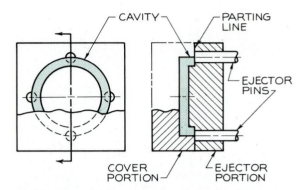

Figure 20.6 A die for casting a simple part. Unlike the sand casting, the metal is forced into the die to form the casting.

Figure 20.7 Three stages of manufacturing a turbine fan. *(Top left)* The blank is formed by forging. *(Bottom left)* It is machined. *(Right)* The fan blades are attached to their machined slots. (Courtesy of Avco Lycoming.)

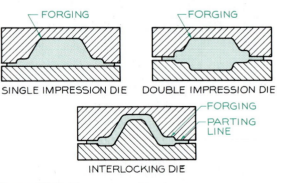

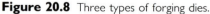

Figure 20.8 Three types of forging dies.

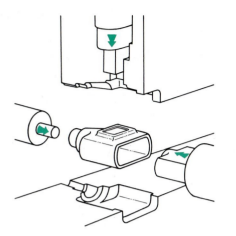

Figure 20.9 Auxiliary rams can be used to form internal features on a part. (Courtesy of General Motors Corp.)

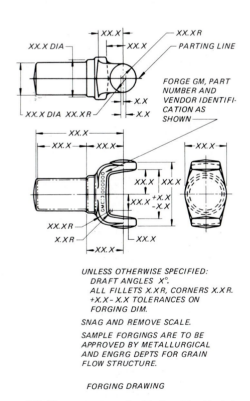

UNLESS OTHERWISE SPECIFIED:
DRAFT ANGLES X°.
ALL FILLETS X.XR, CORNERS X.XR.
+X.X - X.X TOLERANCES ON
FORGING DIM.

SNAG AND REMOVE SCALE.

SAMPLE FORGINGS ARE TO BE
APPROVED BY METALLURGICAL
AND ENGRG DEPTS FOR GRAIN
FLOW STRUCTURE.

FORGING DRAWING

Figure 20.10 A drawing of a forging. The blank is forged oversize to allow for machining operations that will remove metal from it. (Courtesy of General Motors Corp.)

ment has cured, the wax pattern is melted, leaving a hollow cavity that will serve as the mold for the molten metal. When filled and set, the plaster or sand is broken away from the finished investment casting.

20.10 Forgings

Forging is the process of shaping or forming heated metal by hammering or squeezing it into a die. Drop forges and press forges are used to hammer the metal (called billets) into the forging dies by multiple blows. The resulting forging possesses high strength and a resistance to loads and impacts (Fig. 20.7).

When preparing forging drawings, the following must be considered: (1) draft angles and parting lines, (2) fillets and rounds, (3) forging tolerances, (4) extra material for machining, and (5) heat treatment of the finished forging. Some of the standard steels used for forging are designated by the SAE numbers 1015, 1020, 1025, 1045, 1137, 1151, 1335, 1340, 4620, 5120, and 5140. Iron, copper, and aluminum can also be forged.

Figure 20.8 shows examples of dies. A single-impression die gives an impression on one side of the parting line between the mating dies; a double-impression die gives an impression on both sides of the parting line; and the interlocking dies give an impression that may cross the parting line on either side. Figure 20.9 shows an object that is forged with auxiliary rams to hollow the forging, and Fig. 20.10 shows a drawing of a forged part.

Rolling

Rolling is a type of forging in which the stock is rolled between two rollers to give it a desired shape. Rolling can be done at right angles to the axis of the part or parallel to its axis (Fig. 20.11). If a high degree of shaping is required, the stock is usually heated before rolling. If the forming requires only a slight change in configuration, the rolling can be performed when the metal is cold; this is called **cold rolling.**

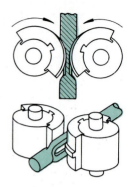

Figure 20.11 Features on parts may be formed by rolling. In this example, parts are being rolled parallel and perpendicular to their axes. (Courtesy of General Motors Corp.)

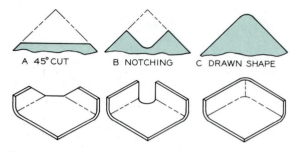

A 45° CUT   B NOTCHING   C DRAWN SHAPE

Figure 20.12 Box-shaped parts formed by stamping. (A) A corner cut of 45° permits folding flanges and may require no further trim. (B) Notching has the same effect as the 45° cut, but it is often more attractive. (C) A continuous corner flange requires that the blank be developed so that it can be drawn into shape.

20.11   Stamping

Stamping is a method of forming flat metal stock into three-dimensional shapes. The first step of stamping is to cut out the shapes, called **blanks,** to be bent. Blanks are formed into shape by bending and pressing them against forms. Figure 20.12 shows examples of box-shaped parts, and Fig. 20.13 shows a flange stamping. Holes in stampings are made by punching, extruding, or piercing (Fig. 20.14).

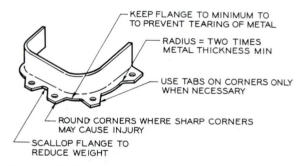

KEEP FLANGE TO MINIMUM TO TO PREVENT TEARING OF METAL

RADIUS = TWO TIMES METAL THICKNESS MIN

USE TABS ON CORNERS ONLY WHEN NECESSARY

ROUND CORNERS WHERE SHARP CORNERS MAY CAUSE INJURY

SCALLOP FLANGE TO REDUCE WEIGHT

Figure 20.13 A sheet metal flange design with notes calling attention to design details.

20.12   Plastics and Miscellaneous Materials

Figure 20.15 shows commonly used plastics and materials. The weights and yields of the materials are given along with examples of their applications.

20.13   Machining Operations

After the metal has been formed, machining operations must be performed to complete the part. The following machines are often used: **lathe, drill press, milling machine, shaper,** and **planer.**

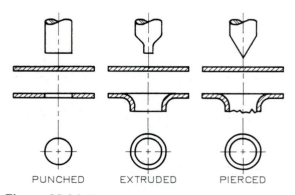
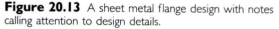

PUNCHED   EXTRUDED   PIERCED

Figure 20.14 Three methods of forming holes in sheet metal. (Courtesy of General Motors Corp.)

E=EXCELLENT
G=GOOD
F=FAIR
P=POOR
A=ADHESIVES

	MACHINABILITY	FORMABILITY	CASTABILITY	WELDABILITY	CORROSION RES.	ABRASION RES.	LB./CU. FT	YIELD: 1000 PSI	TYPICAL APPLICATIONS
NYLON	E	G	G		G	E	73	15	HELMETS, GEARS, DRAWER SLIDES, HINGES, BEARINGS
ABS PLASTIC	G	G	G		E	G	66	66	LUGGAGE, BOAT HULLS, TOOL HANDLES, PIPE FITTINGS
POLYETHLENE	G	F	G	A	F	F	58	2	CHEMICAL TUBING, CONTAINERS, ICE TRAYS, BOTTLES
POLYPROPYLENE	G	G	G	A	E	G	56	5.3	CARD FILES, COSMETIC CASES, AUTO PEDALS, LUGGAGE
POLYSTYRENE	G	E	G	A	P	G	67	7	JUGS, CONTAINERS, FURNITURE, LIGHTED SIGNS
POLYURETHANE	G	G	G	A	G	E	74	6	SOLID TIRES, GASKETS, BUMPERS, SYNTHETIC LEATHERS
POLYVINYLE CHLORIDE	E	E	G	A	G	G	78	4.8	RIGID PIPE AND TUBING, HOUSE SIDING, PACKAGING
ACRYLIC	G	G	E	A	E	F	74	9	AIRCRAFT WINDOWS, TV PARTS, LENSES, SKYLIGHTS
EPOXY	F	G	G		E	G	69	17	CIRCUIT BOARDS, BOAT BODIES, COATINGS FOR TANKS
SILICONE	F	G	G		G	G	109	28	FLEXIBLE FUEL HOSES, HEART VALVES, GASKETS
GLASS	F	G			F	F	160	10+	BOTTLES, WINDOWS, TUMBLERS, CONTAINERS
FIBERGLASS	G		E	A	G		109	20+	BOATS, SHOWER STALLS, AUTO BODIES, CHAIRS, SIGNS

Figure 20.15 Specifications for commonly used plastics and materials.

The Lathe

The lathe shapes cylindrical parts while rotating the work piece between the centers of the lathe (Fig. 20.16). The more fundamental operations performed on the lathe are **turning, facing, drilling, boring, reaming, threading,** and **undercutting** (Fig. 20.17).

Turning forms a cylinder by a tool that advances against and moves parallel to the cylinder being turned (Fig. 20.18).

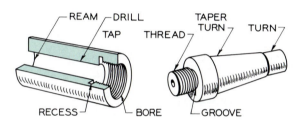

Figure 20.17 The fundamental operations performed on a lathe are illustrated on the two parts above.

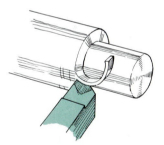

Figure 20.18 Turning is the most basic operation performed on the lathe. A continuous chip is removed by a cutting tool as the part is rotated.

Facing forms flat surfaces perpendicular to the axis of rotation of the part being rotated by the lathe.

Drilling is performed by mounting a drill in the tail stock of the lathe and rotating the work while the bit is advanced into the part (Fig. 20.19).

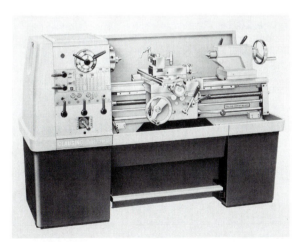

Figure 20.16 A typical metal lathe that holds and rotates the work piece between the centers of the lathe. (Courtesy of the Clausing Corp.)

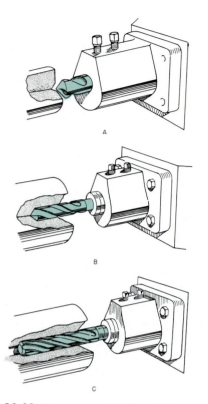

Figure 20.19 Three steps of drilling a hole in the end of a cylinder: (A) start drilling, (B) twist drilling, and (C) core drilling, which enlarges the previously drilled hole to the required size.

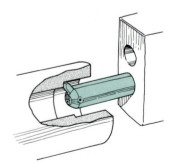

Figure 20.20 Boring is the method of enlarging holes that are larger than available drill bits. The cutting tool is attached to the boring bar on the lathe.

Boring   makes large holes that are too big to be drilled. Large holes are bored by enlarging smaller drilled holes (Fig. 20.20).

Reaming   removes only a few thousandths of an inch of material inside a drilled hole to bring it to its required level of tolerance. Conical and cylindrical reaming can be performed on the lathe (Fig. 20.21).

Threading   of external and internal holes can be done on the lathe. The die used for cutting internal holes is called a **tap** (Fig. 20.22).

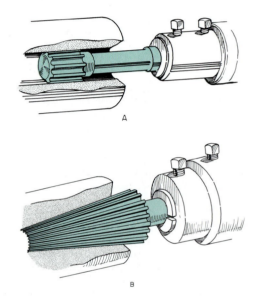

Figure 20.21 Fluted reamers for cylindrical holes (A), and for conical holes (B), can be used to finish inside cylindrical and conical holes within a few thousandths of an inch.

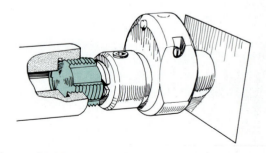

Figure 20.22 External and internal threads (shown here) can be cut on a lathe. The die used to cut the internal threads is called a tap. A recess has been formed at the end of the threaded hole prior to threading.

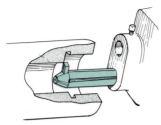

Figure 20.23 A recess (undercut) can be formed by using the boring bar with the cutting tools attached as shown. As the boring bar is moved off center of the axis, the tool will form the recess.

Undercutting cuts a recess inside a cylindrical hole with a tool mounted on a boring bar. The groove is cut as the tool advances from the center of the axis of revolution into the part (Fig. 20.23).

The turret lathe is a programmable lathe that can perform sequential operations, such as drilling a series of holes, boring them, and then reaming them. The turret is mounted to rotate each tool into position for its particular operation (Fig. 20.24).

Figure 20.24 A turret lathe that performs a sequence of operations.

The Drill Press

The drill press is used to drill small- and medium-sized holes into stock that is held on the bed of the press by a fixture or clamp (Fig. 20.25). The drill press can be used for **counterdrilling, countersinking, counterboring, spotfacing,** and **threading** (Fig. 20.26).

Broaching Cylindrical holes can be converted into square or hexagonal holes by using a **broach.** The broach has a series of teeth along its axis, beginning with teeth that are nearly the size

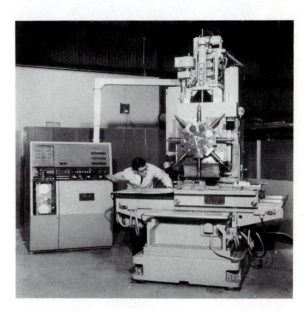

Figure 20.25 A multiple-head drill press that can be programmed to perform a series of operations in a desired sequence.

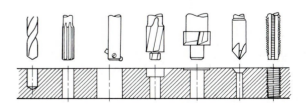

Figure 20.26 The basic operations that can be performed on the drill press are (*left to right*): drilling, reaming, boring, counterboring, spotfacing, countersinking, and tapping (threading).

Figure 20.27 This milling machine is being used to cut a groove in the work piece. (Courtesy of the Brown & Sharpe Mfg. Co.)

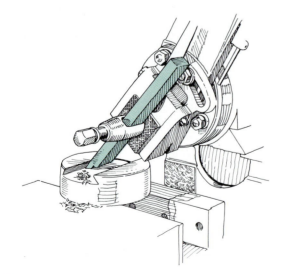

Figure 20.28 The shaper moves back and forth across the part, removing metal as it advances. It can be used to finish surfaces, cut slots, and for many other operations.

of the hole to be broached and tapering to the size of the finished hole to be broached. The broach is forced through the hole, with each tooth cutting more from the hole as it passes through.

The Milling Machine

The milling machine uses a variety of cutting tools, rotated about a shaft (Fig. 20.27), to form different grooved slots, threads, and gear teeth. The milling machine can cut irregular grooves in cams and finish surfaces on a part within a high degree of tolerance.

The Shaper

The shaper is a machine that holds the work piece stationary while the cutter passes back and forth across the work to finish the surface or to cut a groove one stroke at a time (Fig. 20.28). With each stroke of the cutting tool, the material is shifted slightly so as to align the part for the next overlapping stroke.

The Planer

Unlike the shaper, which holds the work piece stationary, the planer passes the work under the cutters (Fig. 20.29). Like the shaper, the planer

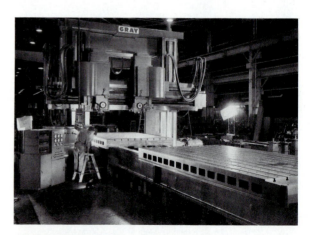

Figure 20.29 The planer has stationary cutters, and the work is fed past them to finish larger surfaces. This planer has a 30-foot bed. (Courtesy of Gray Corp.)

can cut grooves or slots and finish surfaces that must meet tolerance specifications.

20.14 Surface Finishing

Surface finishing is the process of finishing a surface to the desired uniformity. It may be accomplished by several methods, including **grinding, polishing, lapping, buffing,** and **honing.**

Grinding   finishes of a flat surface by holding it against a rotating abrasive wheel (Fig. 20.30). Grinding is used to smooth surfaces and to sharpen edges used for cutting, such as drill bits.

Polishing   is performed in the same manner as grinding except the polishing wheel is flexible since it is made of felt, leather, canvas, or fabric.

Lapping   produces very smooth surfaces. The surface to be finished is held against a **lap,** a large, flat surface coated with a fine abrasive powder. As the lap rotates, the surface is finished. Lapping is done only after the surface has been previously finished by a less accurate technique like grinding or polishing. Cylindrical parts can be lapped by using a lathe with the lap.

Buffing   removes scratches from a surface with a rotating buffer wheel made of wool, cotton, or

Figure 20.30 The upper surface of this part is being ground to a smooth finish by a grinding wheel. (Courtesy of the Clausing Corp.)

other fabric. Sometimes the buffer is a cloth or felt belt that is applied to the surface being buffed. To enhance the buffing, an abrasive mixture is applied to the buffed surface from time to time.

Honing   finishes the outside or inside of holes within a high degree of tolerance. As it is passed through the holes, the honing tool is rotated to produce the sort of finishes found in gun barrels, engine cylinders, and other products requiring a high degree of smoothness.

21

Dimensioning

21.1 Introduction

Working drawings are dimensioned drawings used to describe the details of a part or project so that construction can be performed according to specifications. When properly applied, dimensions and notes will supplement the drawings so that they can be used as legal contracts for construction.

The techniques of dimensioning presented in this chapter are based primarily on the standards of the American National Standards Institute (ANSI), especially Y14.5M, *Dimensioning and Tolerancing for Engineering Drawings*. Various industrial standards from companies such as General Motors Corporation have also been used.

21.2 Dimensioning Terminology

The guide slide in Fig. 21.1 illustrates some of the terms of dimensioning.

Dimension Lines are thin lines (2H–4H pencil) with arrows at each end. Numbers placed near their midpoints specify a part's size.

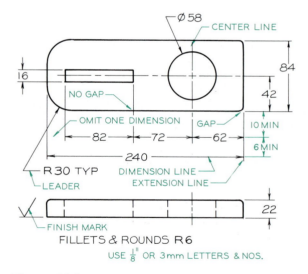

Figure 21.1 This typical working drawing is dimensioned in millimeters.

Extension Lines are thin lines (2H–4H pencil) that extend from a view of an object for dimensioning the part. The arrowheads of dimension lines end at these lines.

Centerlines are thin lines (2H–4H pencil) used to locate the centers of cylindrical parts such as cylindrical holes.

Leaders are thin lines (2H–4H pencil) drawn from a note to a feature the note applies to.

Arrowheads are placed at the ends of dimension lines and leaders to indicate their endpoints. Arrowheads are drawn the same length as the height of the letters or numerals, usually $\frac{1}{8}$ inch. Figure 21.2 shows the form of the arrowhead.

Dimension Numbers are placed near the middle of the dimension line and are usually $\frac{1}{8}$-in. high; units of measurement (″, IN, or mm) are omitted.

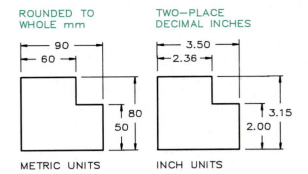

Figure 21.3 In the metric system, millimeters are rounded to the nearest whole number. In the English system, inches are carried to two decimal places even for whole numbers like 1.00.

Examples are shown in Fig. 21.4, where the units are given in millimeters, decimal inches, and fractional inches.

> Dimensions in millimeters are usually rounded off to **whole numbers** without decimal fractions.

When a metric dimension is less than a millimeter, a zero precedes the decimal point, but no zero precedes a decimal point when inches are used. When using decimal inches, show all dimensions with two-place decimal fractions even if the last numbers are zeros.

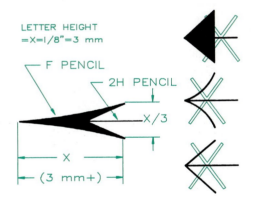

Figure 21.2 Arrowheads are drawn as long as the height of the letters used on a drawing. They are one third as wide as they are long.

21.3 Units of Measurement

The two most commonly used units of measurement are the decimal inch, in the **English** (imperial) system, and the millimeter, in the **metric** (SI) system.

The inch in its common fraction form can be used, but it is preferable to give fractions in decimal form. Common fractions make arithmetic hard to perform. Figure 21.3 compares dimensions in millimeters with those in inches.

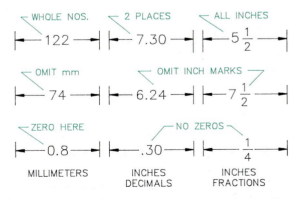

Figure 21.4 A comparison of the application of SI units with English units with decimal and common fractions.

Units of measurement are omitted from the dimension numbers since they are normally understood to be in millimeters or inches. For example, 112, not 112 mm, and 67, not 67″ or 5′–7″.

Architects use a combination of feet and inches, but the inch units are omitted (for example, 7′–2). Engineers use feet and decimal fractions of feet to dimension large-scale projects such as road designs (for example, 252.7′ where feet units *are* shown).

21.4   English/Metric Conversions

Dimensions in inches can be converted to millimeters by multiplying by 25.4. Similarly, dimensions in millimeters can be converted to inches by dividing by 25.4.

For most applications, the millimeter does not need more than a one-place decimal. When millimeters are found by conversion from inches

- The last digit retained in a conversion of either mm or inches is unchanged if it is followed by a number less than 5; for example, 34.43 is rounded off to 34.4.
- The last digit to be retained is increased by 1 if it is followed by a number greater than 5; 34.46 is rounded off to 34.5.
- The last digit to be retained is unchanged if it is even and followed by exactly 5; 34.45 is rounded off to 34.4.
- The last digit to be retained is increased by 1 if it is odd and followed by exactly 5; 34.75 is rounded off to 34.8.

21.5   Dual Dimensioning

Some drawings require that both metric and English units be given on each dimension. This **dual dimensioning** can be shown by placing the

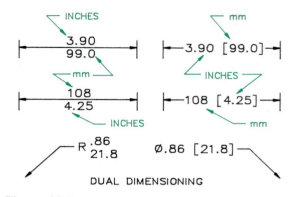

Figure 21.5 If the drawing was originally made in inches, the equivalent measurement in millimeters is placed under or to the right of the inches in brackets in dual-dimensioning. If the drawing was originally made in millimeters, the equivalent measurement in inches is placed under or to the right. When inches are converted to millimeters, the millimeters may need to be written as decimal fractions.

inch equivalent of millimeters either under or over the other units (Fig. 21.5).

A second method of dual dimensioning uses brackets placed on either side of the converted dimensions (Fig. 21.5). Do not mix these two methods on the same drawing.

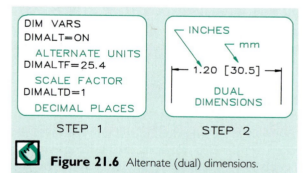

Figure 21.6 Alternate (dual) dimensions.

Step 1   Set the Dim Vars to DIMALT to set dual dimensions to ON, assign the scale factor (DIMALTF), and specify the number of decimal places (DIMALTD).

Step 2   Linear dimensions are found using the same steps as shown in Fig. 21.29. The dimension in brackets is the metric equivalent to the inch dimensions.

Computer Method As Fig. 21.6 shows, the dimensioning variable `DIMALT` must be set to `ON` to obtain alternate (dual) dimensions in brackets following the primary units used. The variable `DI-MALTF` (scale factor) is set to a value representing the multiplier by which the first dimension is changed. `DIMALTD` is used to assign the desired number of decimal places for the second dimension.

Dimensions are then selected using the `DIM` command in the same way single-value dimensions are found (see Fig. 21.29).

21.6 Metric Designation

The metric system is the Système Internationale d'Unités and is denoted by the abbreviation SI (Fig. 21.7). This system uses the first-angle of projection, which locates the front view over the top view and the right-side view to the left of the front view.

When drawings are made for international circulation, it is customary to use one of the sym-

bols shown in Figs. 21.7C and 21.7D to designate the angle of projection. Either the letters *SI* or the word *METRIC,* written prominently on the drawing or in the title block, indicates the measurements are metric.

21.7 Aligned and Unidirectional Numbers

The two methods of positioning dimension numbers on a dimension line are the **aligned** and **unidirectional** methods. The unidirectional system is more widely accepted since it is easier to apply numerals positioned horizontally (Fig. 21.8A). The aligned system aligns the numerals with the dimension lines (Fig. 21.8B).

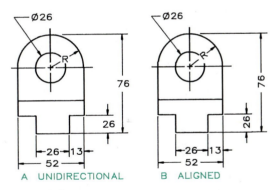

Figure 21.8 (A) When dimensions are positioned so that all of them read from the bottom of the page, they are unidirectional. (B) Aligned dimensions are positioned to read from the bottom and right side of the sheet.

The numbers must be readable from the bottom or right side of the page.

Figure 21.9 shows examples of aligned dimensions on angular dimension lines. Avoid placing aligned dimensions in the "trouble zone" since these numerals would read from the left instead of the right side of the sheet.

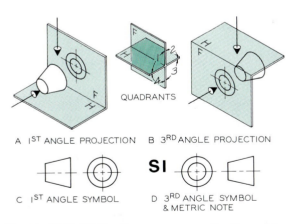

Figure 21.7 (A) The European system of orthographic projection places the top view under the front view. (B) The American system places the front view under the top view. (C) These symbols indicate first-angle projection. (D) These symbols indicate third-angle projection. SI indicates that metric units are used.

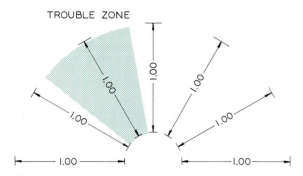

TROUBLE ZONE

Figure 21.9 Numbers on angular dimension lines should not be placed in the trouble zone. To do so would cause the numbers to be read from the left side of the sheet rather than from the right side and bottom.

Computer Method   A mode of DIMTIH (dimensioning text inside dimension lines is horizontal) under Dim Vars of the DIM command of AutoCAD must be set to OFF for aligned dimensions and to ON for unidirectional dimensions (Fig. 21.8). The DIMTOH mode controls the positioning of text that lies outside the dimension line in cases where the numerals do not fit within a short dimension line. When DIMTOH is ON, the dimensions will be horizontal; when OFF, the dimensions will be aligned with the direction of the dimension line.

21.8   Placement of Dimensions

It is good practice to dimension the most descriptive views.

The front view in Fig. 21.10 is more descriptive than the top view, so the front view should be dimensioned. Dimensions should be applied to views in an organized manner (Fig. 21.11). Locate the dimension lines by beginning with the smaller ones to avoid crossing dimension and extension lines.

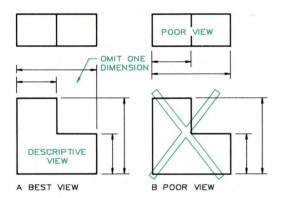

Figure 21.10 Place the dimensions on the most descriptive views where the true outline of the object can be seen.

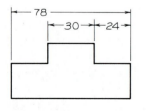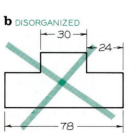

Figure 21.11 Dimensions should be placed on the views in a well-organized manner to make them as readable as possible.

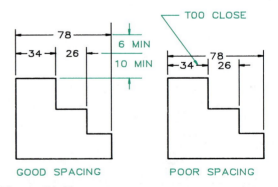

Figure 21.12 The first row of dimensions should be placed at least 0.40 in. (10 mm) from the view, and successive rows should be at least 0.25 in. (6 mm) from the first row. If greater spaces are used, these proportions should still be maintained.

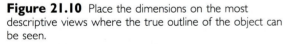

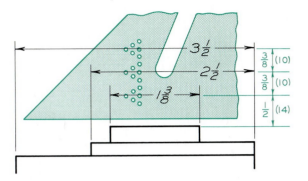

Figure 21.13 When common fractions are used, the center holes of the triangular arrangements on the Braddock-Rowe triangle are aligned with the dimension lines to automatically space the lines.

Always leave at least 0.40 in. (10 mm) between the object and first row of dimensions (Fig. 21.12). Successive rows of dimensions should be equal and at least 0.25 in. (6 mm) apart.

If greater spaces are used, apply these same general proportions. The Braddock-Rowe lettering guide triangle can be used to space the dimension lines (Fig. 21.13).

The positioning of dimensions is based on the letter height, as Fig. 21.14 shows. These proportions are the **minimum** values.

Computer Method Figure 21.15 shows dimensioning variables and their minimum settings. Computer variables can be set at these proportions, and all of them can be changed at once using DIM-SCALE. For example, DIMSCALE=25.4 would convert inch proportions to millimeter proportions.

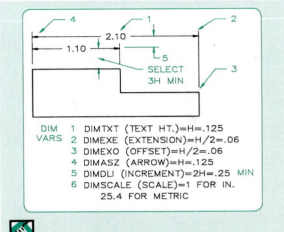

Figure 21.15 The assignment of dimensioning variables (Dim Vars) to be applied by AutoCAD is shown here. Once set, these variables remain set with the file in use. DIMSCALE can be used to enlarge or reduce these proportions.

Figure 21.16 shows other dimensioning variables. Chapter 38 covers variables not covered in this chapter.

When a row of dimensions is placed on a drawing, one of the dimensions is omitted since the overall dimension supplements the omitted dimension (Fig. 21.17A). A dimension that needs to be given as a reference dimension is either placed in parentheses or followed by the abbreviation REF to indicate it is a reference dimension.

Figure 21.18 shows the recommended techniques of dimensioning features, and Fig. 21.19 shows examples of the placement of extension

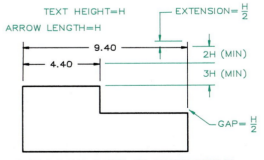

DIMENSIONING BASED ON LETTER HEIGHT

Figure 21.14 The proportions for placing dimensions on a drawing are based on the letter height, usually 0.125" or 3 mm.

```
DIM VARS  DEFAULT
DIMALT    OFF     Alternate units selected
DIMALTD   2       Alternate unit decimal places
DIMALTF   25.4    Alternate unit scale facator
DIMAPOST          Default suffix for alternate text
DIMASO    OFF     Create associative dimensions
DIMASZ    .12     Arrow size
DIMBLK            Arrow block name
DIMBLK1           First arrow block name
DIMBLK2           Second arrow block name
DIMCEN    -.05    Center mark size
DIMDLE    .00     Dimension line extension
DIMDLI    .38     Dim. line increment for continuation
DIMEXE    .12     Extension beyond dimension line
DIMEXO    .06     Extension line offset
DIMLFAC   1.00    Length factor
DIMLIM    OFF     Generates dimen tolerance limits
DIMPOST           Character suffix after dimensions
DIMRND    .00     Rounding value for distances
DIMSAH    OFF     Separate arrow heads at each end
DIMSCALE  1.00    Scale factor for all dim. variables
DIMSE1    OFF     Suppress first extension line
DIMSE2    OFF     Suppress second extension line
DIMSHO    OFF     Shows new dimens. while dragging
DIMSOXD   OFF     Suppresses outside dimen. lines
DIMTAD    OFF     Text placed above dimension lines
DIMTIH    ON      Text inside ext. lines is horizontal
DIMTIX    OFF     Text forced inside extension lines
DIMTM     ,00     Minus tolerance value
DIMTOFL   OFF     Forces dim. line inside/text outside
DIMTOH    ON      Text outside ext lines is horizontal
DIMTOL    OFF     Applies tolerances to dimensions
DIMTP     .00     Plus tolerance value
DIMTSZ    .00     Assigns sizes to ticks
DIMTVP    .00     Locates text over or under dim. line
DIMTXT    .12     Text height
DIMZIN    0       Suppress 0 inches in feet & inches
```

Figure 21.16 The dimensioning variables (Dim Vars) shown here are covered partly in this chapter and partly in Chapter 38.

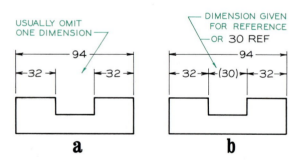

Figure 21.17 (A) One intermediate dimension is customarily omitted since the overall dimension provides this measurement. (B) If all the intermediate dimensions are given, one should be placed in parentheses to indicate that it has been given as a reference dimension.

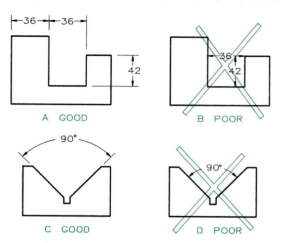

Figure 21.18 Dimensioning rules. (A) Placing the dimensions outside a part is the preferred practice. (B) Dimension lines should not be used as extension lines. (C) Angles are dimensioned with arcs and extension lines. (D) The angle should not be placed inside the angular cut.

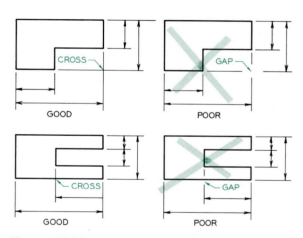

Figure 21.19 Extension lines extend from the edge of an object, leaving a small gap. They do not have gaps where they cross object lines or other extension lines.

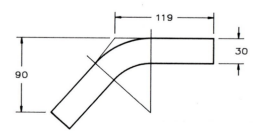

Figure 21.20 A curved surface is dimensioned by locating the theoretical point of intersection with extension lines.

lines. Extension lines may cross other extension lines or object lines; they are also used to locate theoretical points outside curved surfaces (Fig. 21.20).

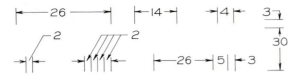

Figure 21.21 Where room permits, numerals and arrows should be placed inside the extension lines. Other placements are shown as the spacing becomes smaller.

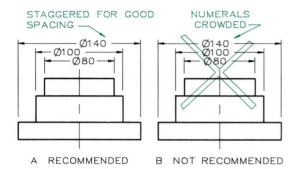

Figure 21.22 Dimensioning numerals should be staggered when close spacing tends to crowd them.

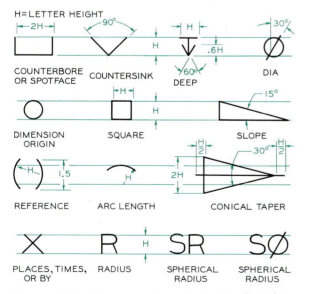

Figure 21.23 These symbols can be used to dimension parts. The proportions of the symbols are based on the letter height, *H*, which is usually $\frac{1}{8}$ inch.

21.9 Dimensioning in Limited Spaces

Figure 21.21 shows examples of dimensioning in limited spaces. Regardless of space limitations, the numerals should not be drawn smaller than they appear elsewhere on the drawing. Where dimension lines are closely grouped, the numerals should be staggered to make them more readable (Fig. 21.22).

21.10 Dimensioning Symbology

Figure 21.23 shows several symbols used in dimensioning. The sizes of the symbols are based on the height of the lettering used in dimensioning an object, usually $\frac{1}{8}$ inch. Symbols reduce the time in preparing notes and add to the clarity of a dimension.

21.11 Computer Dimensioning

The rules of dimensioning must be followed whether you are using computer graphics or manual techniques. The AutoCAD program offers many features to use in applying the rules. For example, Fig. 21.24 shows several combina-

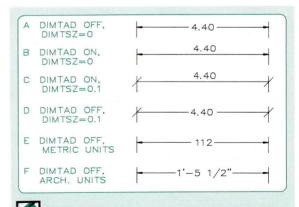

Figure 21.24 The dimension lines illustrate the effects of using the different `Dim Vars` (dimensioning variables) in AutoCAD.

tions of Dim Vars, which are modes of the DIM command.

Text can be placed inside the dimension line (DIMTAD: OFF) or above the dimension line (DIM-TAD: ON). Arrowheads can be placed at the ends of dimension lines DIMASZ>0, or tick marks (slashes) can be used when DIMTSZ is set to a value greater than 0, usually about half the letter height of text. UNITS can be selected to be architectural (feet and inches), metric (no decimal fractions), decimal inches (two or more decimal fractions), or engineering units (feet and decimal inches).

An important Dim Vars in dimensioning is DIMSCALE, which can be used to change all variables of dimensioning to allow for changes in scales (Fig. 21.25). DIMSCALE is set at a default of 1.00, which contains the lengths of arrows, text size, and extension line offsets. When invoked, DIMSCALE will not change previously drawn dimensions, only those applied afterward.

The text STYLE that was last used will be used as the dimensioning text. If you had set the text to a specified height, this height would be used in dimensioning whether or not DIMSCALE was used. For DIMSCALE to change text height, you should set the text height for the STYLE

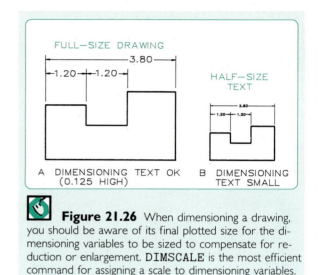

Figure 21.26 When dimensioning a drawing, you should be aware of its final plotted size for the dimensioning variables to be sized to compensate for reduction or enlargement. DIMSCALE is the most efficient command for assigning a scale to dimensioning variables.

being used at 0 (zero). You will be prompted for the letter height each time, which allows you either to accept the default last used or to assign another height. In this case, DIMSCALE will enlarge or reduce text height along with the other variables.

You must apply DIMSCALE to a dimensioned drawing when it will be plotted at a different scale (Fig. 21.26). For example, if a drawing is to be plotted half size, you would want to use a DIMSCALE of 2.00 so that the variables would appear full size when reduced.

AutoCAD automatically performs many dimensioning operations, but occasionally you will want to modify the placement of arrows and text. Figure 21.27 shows an edited dimension where MOVE and ERASE commands are used.

21.12   Dimensioning Prisms

Figure 21.28 illustrates the following rules for dimensioning prisms:

1. Dimensions should extend from the most **descriptive views** (Fig. 21.28A).
2. Dimensions that apply to two views should be placed **between** them (Fig. 21.28A).

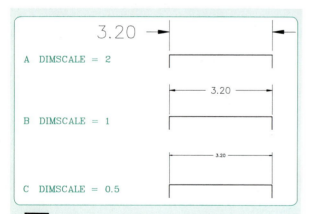

Figure 21.25 DIMSCALE, a subcommand under DIM:, can be used to change dimensioning variables by inputting a single factor. The text sizes, arrows, offsets, and extensions are all changed at one time.

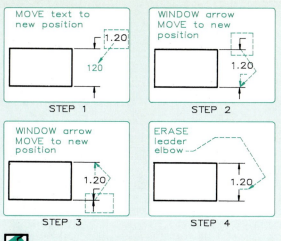

 Figure 21.27 Editing dimensions.

Step 1 You may wish to MOVE a dimension text within the extension lines.

Step 2 Arrows can be MOVEd inside extension lines also.

Step 3 A second arrow is windowed and MOVEd inside the extension lines.

Step 4 The unneeded leader extension is ERASEd to complete the modified dimension.

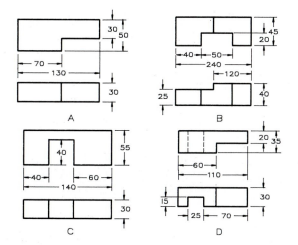

Figure 21.28 Dimensioning prisms (metric). (A) Dimensions should extend from the most descriptive view and be placed between the views they apply to. (B) One intermediate dimension is not given. Extension lines may cross object lines. (C) It is permissible to dimension a notch inside the object if this improves clarity. (D) Whenever possible, dimensions should be placed on visible lines, not hidden lines.

3. The first row of dimension lines should be placed at **least 0.40 in.** (10 mm) from the view. Successive rows should be placed at **least 0.25** in. (6 mm) apart.
4. Extension lines may **cross** each other and other lines, but dimension lines should **not cross** unless absolutely necessary.
5. To dimension a part in its most descriptive view, dimensions may have to be placed in **more** than one view (Fig. 21.28B).
6. Dimension lines may be placed **inside** a notch to improve clarity (Fig. 21.28C).
7. Whenever possible, dimensions should be applied to **visible** lines rather than to hidden lines (Fig. 21.28D).
8. Dimensions should **not be repeated,** and unnecessary information should not be given.

Computer Method The part in Fig. 21.29 is dimensioned by entering the DIM command and selecting the LINEAR and HORIZONTAL options.

The extension lines can be obtained automatically by responding to the First extension line origin prompt with a (CR). You will get the prompt Select line, arc, or circle. Point to the line to be measured, and locate the dimension line when prompted.

If the dimensioning variable DIMASO is set to ON before dimensions are applied to a part, the dimensions will be associative, that is, they will change in value as the size of the part is STRETCHed (Fig. 21.30). Associative dimensions can also be erased as a single unit (extension lines, dimension lines, arrows, and numerals).

When DIMSHO is ON, the dimensioning numerals will be recomputed dynamically as the part is dragged to a new size.

21.13 Dimensioning Angles

Angles can be dimensioned either by using coordinates to locate the ends of angular lines or planes or by using angular measurements in degrees (Fig. 21.31).

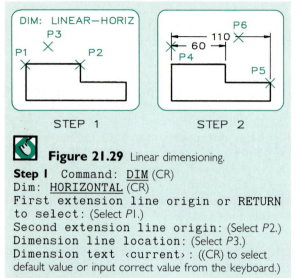

STEP 1   STEP 2

**Figure 21.29** Linear dimensioning.

Step 1 Command: <u>DIM</u> (CR)
Dim: <u>HORIZONTAL</u> (CR)
First extension line origin or RETURN
to select: (Select P1.)
Second extension line origin: (Select P2.)
Dimension line location: (Select P3.)
Dimension text <current>: ((CR) to select
default value or input correct value from the keyboard.)

Step 2 Dim: (CR)
HORIZONTAL
First extension line origin or RETURN
to select: (Select P4.)
Second extension line origin: (Select P5.)
Dimension line location: (Select P6.)
Dimension text <current>: ((CR) to select
default or input correct value from the keyboard.)

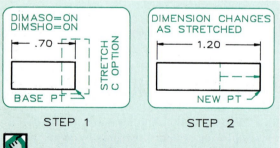

STEP 1   STEP 2

Figure 21.30 Associative dimensions.

Step 1 Set Dim Vars, DIMASO, and DIMSHO, to
ON, and the dimensions will be associated with the size
of the part they are applied to. Use the C option of the
STRETCH command to window the part and its dimension.

Step 2 A base point is selected, and a new point is
selected. As the size of the part is dragged, the dimensions will be recalculated dynamically during the dragging.

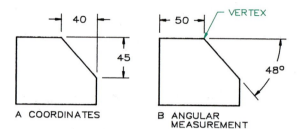

A COORDINATES   B ANGULAR
                        MEASUREMENT

Figure 21.31 (A) Angular planes can be dimensioned
using coordinates. (B) Angles can be measured by locating
the vertex and measuring the angle in degrees. Angles can
be specified in degrees, minutes, and seconds when
accuracy is essential.

Units for angular measurements are degrees,
minutes, and seconds (Fig. 21.31B). There are 60
minutes in a degree and 60 seconds in a minute.
Seldom will angular measurements need to be
measured to the nearest second.

Computer Method By selecting the
ANGULAR option under the DIM command, an angle
can be dimensioned as shown in Fig. 21.32. If room
is not available for the arrows between the extension lines, they will be drawn outside the extension
lines.

21.14 Dimensioning Cylinders

Cylinders that are dimensioned may be either
solid cylinders or cylindrical holes.

> Solid cylinders are dimensioned in their **rectangu-
> lar views** by using **diameters,** not radii.

All diametral dimensions should be preceded
by the symbol ∅ (Fig. 21.33), which indicates the
dimension is a diameter. The English system often uses the abbreviation DIA following the diametral dimension.

Parts having several cylinders, which are
concentric, are dimensioned with diameters, beginning with the smallest cylinder (Fig. 21.33C).

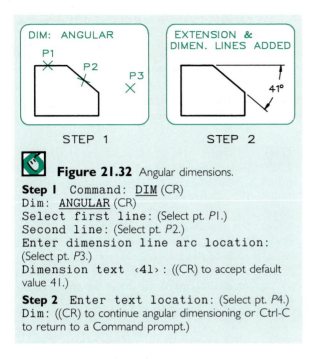

Figure 21.32 Angular dimensions.

Step I Command: <u>DIM</u> (CR)
Dim: <u>ANGULAR</u> (CR)
Select first line: (Select pt. *P1*.)
Second line: (Select pt. *P2*.)
Enter dimension line arc location:
(Select pt. *P3*.)
Dimension text ‹41› : ((CR) to accept default
value 41.)

Step 2 Enter text location: (Select pt. *P4*.)
Dim: ((CR) to continue angular dimensioning or Ctrl-C
to return to a Command prompt.)

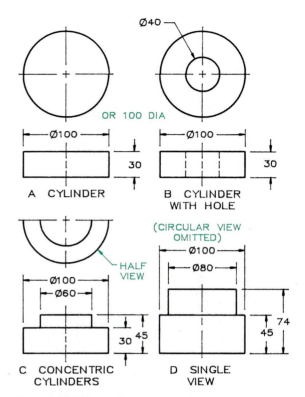

Figure 21.33 (A) It is preferred that cylinders be dimensioned in their rectangular views using a diameter rather than a radius. (B) Dimensions should be placed between the views when possible, and holes should be dimensioned with leaders. (C) Dimension concentric cylinders beginning with the smallest one. (D) The circular view can be omitted if DIA or ∅ is used with the diametral dimensions.

A cylindrical part may be sufficiently dimensioned with only one view if ∅ or DIA is used with the diametral dimension (Fig. 21.33D).

21.15 Measuring Cylindrical Parts

Cylindrical parts are dimensioned with diameters rather than radii because diameters are easier to measure. An internal cylindrical hole is measured with an internal micrometer caliper (Fig. 21.34). Likewise, an external micrometer caliper can be used for measuring the outside diameters of a part (Fig. 21.35). Measuring a **diameter** rather than a **radius** makes it possible to measure during machining when the part is held between centers on a lathe.

21.16 Cylindrical Holes

Cylindrical holes may be dimensioned by one of the methods shown in Fig. 21.36. The preferred method of dimensioning cylindrical holes is to

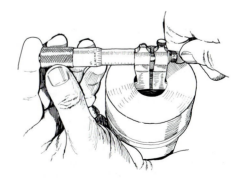

Figure 21.34 An internal micrometer caliper for measuring internal cylindrical diameters.

Figure 21.35 An external micrometer caliper for measuring the diameter of a cylinder.

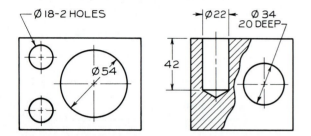

Figure 21.36 Several acceptable methods of dimensioning cylindrical holes and shapes. The symbol ∅ is placed in front of the diametral dimension.

draw a **leader** from the **circular view,** and then add the dimension preceded by ∅ (Fig. 21.37) or followed by DIA. Sometimes the note DRILL or BORE is added to specify the shop operation, but current standards prefer the use of DIA instead.

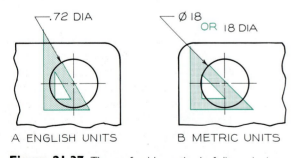

Figure 21.37 The preferable method of dimensioning cylindrical holes is with a leader, the symbol ∅, and a dimension to indicate that the dimension is a diameter. Previous standards recommended the abbreviation DIA after the dimension.

To illustrate various methods of dimensioning, a part containing cylindrical features is dimensioned in Fig. 21.38 in both the circular and rectangular views. Figure 21.39 shows three methods of dimensioning holes. The depth of a hole is its **usable** depth, not the depth to the point left by the drill bit.

Computer Method Circles can be dimensioned in their circular views as shown in Fig. 21.40. The dimension lines will begin with the point selected on the circle and pass through the center. The diameter symbol will be placed in front of the dimension numerals. Small circles will be dimensioned with the arrows inside the circle and the dimension numerals outside connected by a leader.

Even smaller circles will have the arrows and dimension outside the circle.

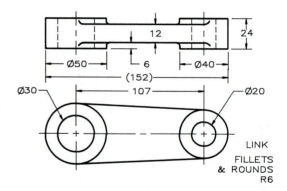

Figure 21.38 This part, composed of cylindrical features, has been dimensioned using several approved methods. (F&R R6 means that fillets and rounds have a 6 radius.)

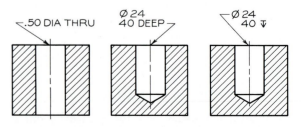

Figure 21.39 Three methods of dimensioning holes with combinations of notes and symbols.

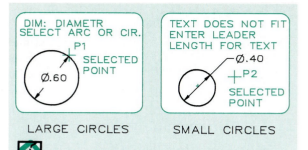

DIM: DIAMETR
SELECT ARC OR CIR.

P1

SELECTED
POINT

Ø.60

TEXT DOES NOT FIT
ENTER LEADER
LENGTH FOR TEXT

Ø.40

P2

SELECTED
POINT

LARGE CIRCLES

SMALL CIRCLES

Figure 21.40 Dimensioning circles.

Step 1 Select the `Diametr` option under the `DIM` command to dimension a circle. Select `P1` on the arc (an endpoint of the dimension), and the dimension is drawn. You have the option to replace the dimension measured by the computer with a different value.

Step 2 When text does not fit, you receive the prompt "`Text does not fit, Enter leader length for text.`" Select a point, `P2`, and the leader and dimension are drawn.

21.17 Pyramids, Cones, and Spheres

Figures 21.41A–C show three methods of dimensioning pyramids. The **pyramids** in Figs. 21.41B and 21.41C are truncated (that is, the apex has been cut off and replaced by a plane).

Figures 21.41D and 21.41E show two acceptable methods of dimensioning **cones.**

A **sphere,** if it is complete, is dimensioned by using its diameter (Fig. 21.41F); a radius is used if it is not a complete sphere (Fig. 21.41G). Only one view is necessary to describe a sphere.

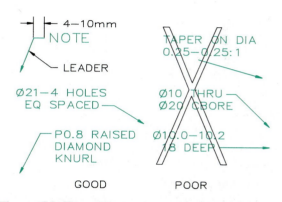

Figure 21.42 Leaders from notes should begin with a horizontal bar from the first or last word of the note, not from the middle of the note.

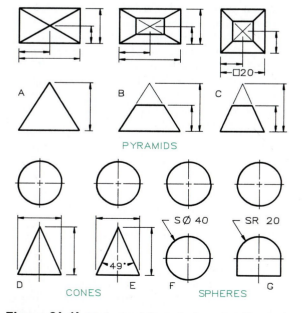

Figure 21.41 Methods of dimensioning pyramids, cones, and spheres.

21.18 Leaders

Leaders are used to apply notes and dimensions to a feature they describe. As illustrated in Fig. 21.37, leaders are drawn at a standard angle of a triangle. Figure 21.42 shows examples of notes using leaders. The leader should be drawn from either the first word of the note or the last word of the note and begin with a short horizontal line from the note.

Figure 21.43 shows applications of leaders on a part. A dot is used instead of an arrowhead when the note applies to a surface that does not appear as an edge.

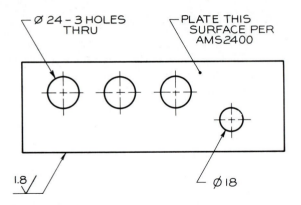

Figure 21.43 Examples of notes with leaders applied to a part.

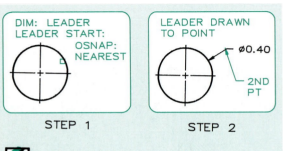

 Figure 21.44 Leaders.

Step 1 Command: <u>DIM</u> (CR)
Dim: <u>LEADER</u> (CR)
Leader start: (OSNAP to PI using NEAREST option.)

Step 2 To point: (Select next point.) (CR)
Dimension text ‹0.40› : ((CR) to accept default value.) (The leader and dimension are drawn.)
Dim: ((CR) to continue leader dimensioning or Ctrl-C to return to a command prompt.)

 Computer Method By selecting the LEADER option under the DIM command, a leader can be drawn that begins with the arrow end (Fig. 21.44). To ensure that the arrow touches the circle, use OSNAP and NEAREST to snap to the circumference. You will be prompted To point until (CR) is pressed. Then you will be shown the diameter of the circle, which you can accept by pressing (CR) or override by typing in a different value.

If the previous option was DIAMETR, the dimension will be preceded with the diameter symbol. If the previous option was RADIUS, the dimension will be preceded with an R, the radius symbol.

21.19 Dimensioning Arcs

Cylindrical parts **less than a full circle** are dimensioned with **radii,** as Fig. 21.45 shows. Current standards recommend that radii be dimensioned with an **R preceding** the dimension, such as R10. Previous standards recommended that the R follow the dimension, such as 10R. Thus both methods are seen in practice.

When the arc being dimensioned is long, it may be dimensioned with a false radius, as Fig. 21.46A shows. A zigzag is placed in the radius to indicate that it is not a true radius.

 Computer Method By using the RADIUS option of the DIM command, arcs can be dimensioned by selecting a point on the arc, which will be the starting point of the arrow (Fig. 21.47). If room permits, the dimension will be placed between the center and the arrow.

For smaller arcs, the arrows will be placed inside, and dimensions will be placed outside. When room is not available for arrows inside, both the arrow and dimension will be placed outside. Decimal values will be preceded with a zero unless you override them by typing in the value without a preceding zero. If this is done, the R must be typed in also.

21.20 Fillets and Rounds and TYP

Fillets and **rounds** are rounded corners conventionally used on castings. A fillet is an internal rounding, and a round is an external rounding.

A note may be placed on the drawing to eliminate the need for repetitive dimensioning of fillets and rounds. The note may read **ALL FILLETS AND ROUNDS R6.** If most, but not

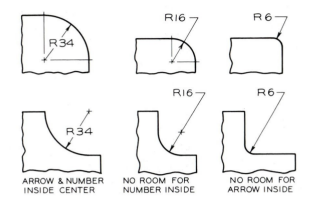

ARROW & NUMBER INSIDE CENTER

NO ROOM FOR NUMBER INSIDE

NO ROOM FOR ARROW INSIDE

Figure 21.45 When space permits, the dimension and arrow should be placed between the center and the arc. If room is not available for the number, the arrow is placed between the center and the arc with the number on the outside. If there is no room for the arrow, both the dimension and arrow are placed outside the arc with a leader.

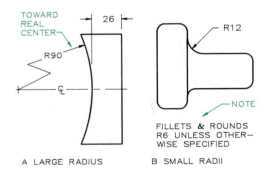

A LARGE RADIUS

B SMALL RADII

Figure 21.46 (A) When a radius is very long, it may be shown with a false radius and a zigzag to indicate that it is not true length. It should end on the centerline of the true center. (B) Fillets and rounds may be noted to reduce repetitive dimensions of small arcs.

all, of the fillets and rounds have equal radii, the note may read **ALL FILLETS AND ROUNDS R6 UNLESS OTHERWISE SPECIFIED.** In this case, only the fillets and rounds of different radii are dimensioned (Fig. 21.46B). The notes may be abbreviated, for example, **F&R R10.**

Fillets and rounds should be dimensioned with short, simple leaders (Fig. 21.48A) rather than with long, confusing leaders (Fig. 21.48B).

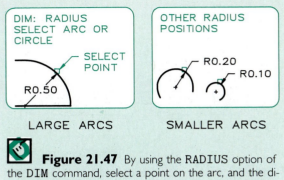

LARGE ARCS

SMALLER ARCS

Figure 21.47 By using the RADIUS option of the DIM command, select a point on the arc, and the dimension is drawn with its arrow located at the point selected. The dimension is preceded by an R. Smaller arcs are dimensioned with the dimensions outside the arc, or because of limited space, with the arcs and dimensions outside the arc.

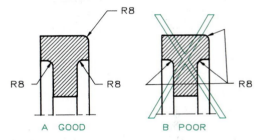

A GOOD

B POOR

Figure 21.48 When several arcs are dimensioned, using separate leaders is preferable to extending the leaders.

Repetitive features on a drawing may be noted as shown in Fig. 21.49. The note TYPICAL or TYP means although only one of these features is dimensioned, the dimensions are typical of those undimensioned. The note PLACES is sometimes used to specify the number of places that a similar feature appears.

21.21 Curved Surfaces

An irregular shape composed of several tangent arcs of varying sizes (Fig. 21.50) can be dimensioned by using a series of radii.

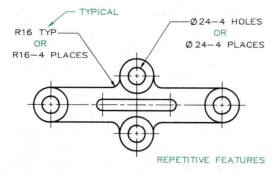

Figure 21.49 Notes can be used to indicate that similar features and dimensions are repeated on drawings without having to dimension them individually.

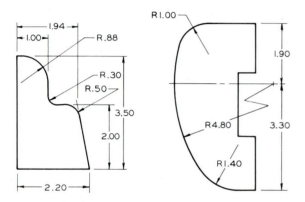

Figure 21.50 Examples of dimensioned parts composed of a series of tangent parts.

When the curve is irregular rather than composed of arcs (Fig. 21.51), the coordinate method can be used to locate a series of points along the curve from two datum lines. The drafter must use judgment to determine the proper spacing for the points. Extension lines may be placed at an angle to provide additional space for showing dimensions.

An irregular curve that is symmetrical about an axis is shown dimensioned in Fig. 21.52. Dimension lines are used as extension lines, which is an acceptable violation of rules.

21.22 Symmetrical Objects

Symmetrical objects may be dimensioned as shown in Fig. 21.53A, where it is assumed that the dimensions are each centered about the centerline, abbreviated CL. The better method is shown in Fig. 21.53B, where the assumption is eliminated.

21.23 Finished Surfaces

Many parts are formed as castings in a mold that gives the parts' exterior surfaces a rough finish. If the part is designed to come in contact with another surface, the rough finish must be ma-

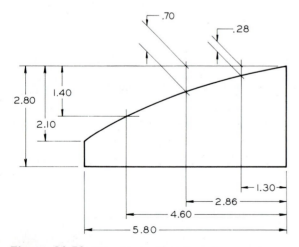

Figure 21.51 This object with an irregular curve is dimensioned by using coordinates.

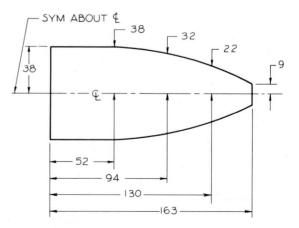

Figure 21.52 This symmetrical part is dimensioned about its centerline.

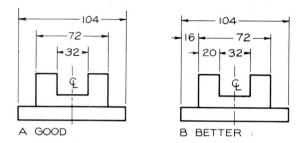

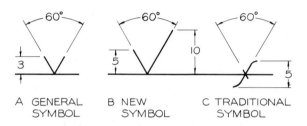

Figure 21.53 (A) Symmetrical parts may be dimensioned about their centerlines as shown here. (B) The better way of dimensioning symmetrical parts is shown here.

Figure 21.54 Finish marks indicate that a surface has been machined to a smooth surface. (A) The traditional V can be used for general applications. (B) The new finish mark is related to surface texture. (C) The *f* is also used to indicate finished surfaces.

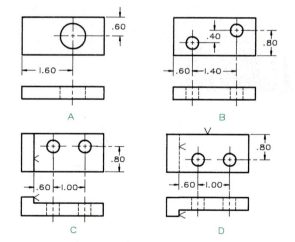

Figure 21.55 Location dimensions. (A) Cylindrical holes should be located in their circular views from two surfaces of the object. (B) When more than one hole is to be located, they should be located from center to center. (C) Holes should be located from finished surfaces. (D) Holes should be located in the circular view and from finished surfaces even if the finished surfaces are hidden.

chined by grinding, shaping, lapping, or similar process.

To indicate that a surface is to be finished, **finish marks** are drawn on the surface where it appears as an **edge** (Fig. 21.54).

> Finish marks should be repeated in **every view** where the finished surface appears as an edge even if it is a **hidden line.**

Figure 21.54 shows three methods of drawing finish marks. The simple V mark is used in general cases. The uneven V (Fig. 21.54), a newly recommended symbol, is related to surface texture; we discuss it in Chapter 22. When an object is finished on all surfaces, the note FINISHED ALL OVER (abbreviated **FAO**) is placed on the drawing.

21.24 Location Dimensions

Location dimensions are used to locate the **positions, not the sizes,** of geometric elements, such as cylindrical holes (Fig. 21.55). The sizes of the holes are omitted for clarity. The centers of the holes are located with coordinates in the circular view when possible.

> Holes should be **located** from **finished surfaces** since holes can be located more accurately from a smooth, machined surface than from a rough, unfinished one.

This rule is followed even if the finished surface is a hidden line, as in Fig. 21.56.

Location dimensions should be placed on views where both dimensions can be shown (Fig. 21.57). Cylinders are located in their circular views (Figs. 21.57B and 21.58).

21.25 Location of Holes

When holes must be located accurately, the dimensions should originate from a common datum plane on the part to reduce errors in mea-

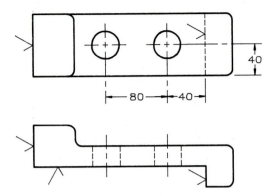

Figure 21.56 Cylindrical holes are located from center to center in their circular view and from a finished surface if one is available.

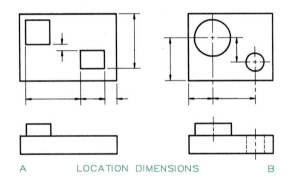

Figure 21.57 Prisms (A) and cylinders (B) are located with coordinates in the view where both coordinates can be seen.

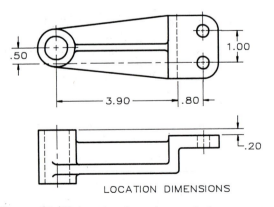

Figure 21.58 Location dimensions applied to a part to locate its geometric features.

surement (Figs. 21.59A and 21.59B). When several holes in a series are to be equally spaced, as in Fig. 21.59C, a note specifying this can be used to locate the holes. The first and last holes of the series are determined by the usual location dimensions.

Holes through circular plates may be located by coordinates or a note (Fig. 21.60). When a note is used, the diameter of the imaginary circle passing through the centers of the holes must be given. This circle is called the **bolt circle** or **circle of centers.**

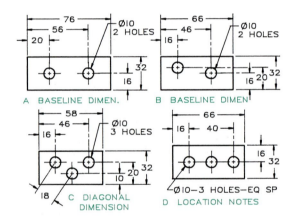

Figure 21.59 Location of holes. (A and B) Holes can be more accurately located if measured from common datum. (C) A diagonal dimension can be used to locate a hole of this type from another hole's center. (D) A note can be used to specify the spacing between the centers of equally spaced holes.

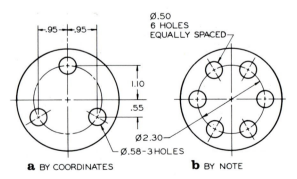

Figure 21.60 Holes may be located in circular plates by (a) coordinates or (b) notes.

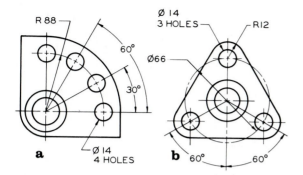

Figure 21.61 Methods of locating cylindrical holes on a concentric arc (a) and on a full circle (b).

A similar method of locating holes is the **polar system** illustrated in Fig. 21.61. The radial distances from the point of concurrency and their angular measurements (in degrees) between the holes are used to locate the centers.

21.26 Objects with Rounded Ends

Objects with rounded ends should be dimensioned from end to end (Fig. 21.62A). The radius, shown as R without a dimension, specifies the end is an arc. Since the height is given, the radius is understood to be half the height.

If the object is dimensioned from center to center (Fig. 21.62B), the overall dimension should be given as a reference dimension (3.20) to eliminate calculating the overall dimension. In this case, the radii must be given.

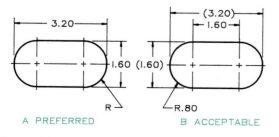

Figure 21.62 (A) The preferred method of dimensioning objects with rounded ends. (B) The less desirable method.

A part with partially rounded ends is dimensioned in Fig. 21.63A. The overall dimension and radii are given so that their centers may be located. When an object has a rounded end that is less than a semicircle (Fig. 21.63B), location dimensions must be used to locate the center of the arc.

Slots with rounded ends are dimensioned in Fig. 21.64. Only one slot is dimensioned in Fig. 21.64A, with a note indicating there are two

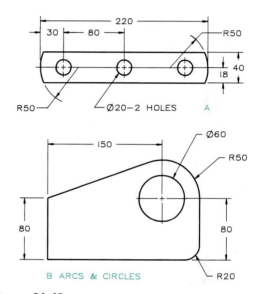

Figure 21.63 Examples of dimensioned parts with rounded ends and cylindrical features.

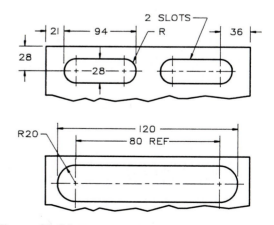

Figure 21.64 Methods of dimensioning slots.

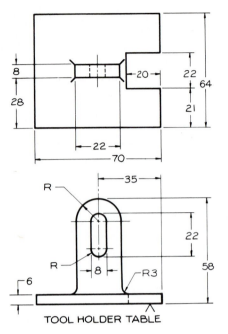

Figure 21.65 Examples of arcs and slots.

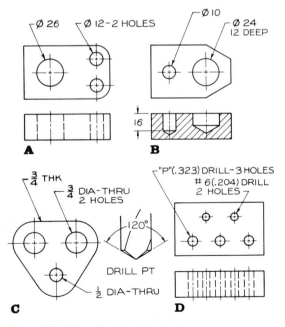

Figure 21.66 (A and B) Cylindrical holes may be dimensioned by either of these methods. (C and D) When only one view is given, it is necessary to note the THRU holes.

slots. The slot in Fig. 21.64B is dimensioned giving the overall dimension and the two arcs, which are understood to apply to both ends. The distance between the centers is given as a reference dimension.

The dimensioned views of the tool holder table in Fig. 21.65 show examples of arcs and slots. To prevent dimension lines from crossing, it is often necessary to place dimensions in a view less descriptive than might be desired.

21.27 Machined Holes

Machined holes are made or refined by a machine operation, such as drilling or boring (Fig. 21.66). It is preferable to give the diameter of the hole with the symbol ∅ in front of the dimension (∅ 32); however, the note 32 DRILL may be used in some cases. You will also see diameters dimensioned as XX DIA since this was the standard previously recommended.

Drilling is the most common method of machining holes. The depth of a drilled hole can be specified in the note, or it may be dimensioned

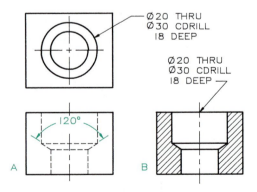

Figure 21.67 Counterdrilling notes gives the specifications for a larger hole drilled inside a smaller hole. The 120° angle is not required as a dimension since this is the standard angle of a drill point. It is preferable to note the counterdrill with a leader from the circular view (A). If only the rectangular view is drawn, the counterdrill can be noted as shown at (B).

in the rectangular view. The depth of a drilled hole is **dimensioned** as the **usable part** of the hole; the conical point is disregarded (Fig. 21.66B).

Counterdrilling   is drilling a large hole inside a smaller drilled hole to enlarge it (Fig. 21.67). The 120° dimension indicates the angle of the drill point; it need not be dimensioned on the drawing.

Countersinking   is the process of forming a conical hole and is often used with flat-head screws (Fig. 21.68). The diameter of the countersunk hole (the maximum diameter on the surface) and the angle of the countersink are given in the notes. Countersinking is also used to provide center holes in shafts, spindles, and other cylindrical parts held between the centers of a lathe (Fig. 21.69).

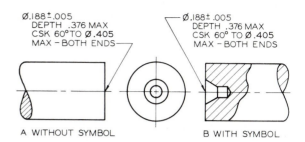

Figure 21.69 Methods of specifying countersinks in the ends of cylinders for mounting them on a lathe between centers.

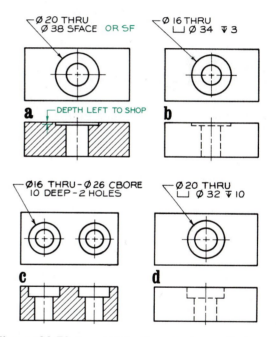

Figure 21.70 (a and b) Spotfaces can be specified as shown here. The depth of the spotface can be specified if needed. (c and d) Counterbores are dimensioned by giving the diameters of both holes and the depth of the larger hole.

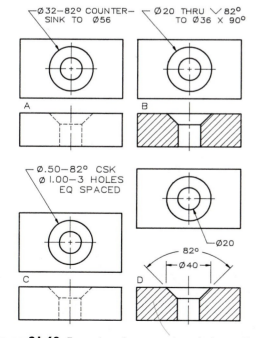

Figure 21.68 Examples of notes and symbols specifying countersunk holes.

Spotfacing   is a machining process used to finish the surface around the top of a hole to provide a level seat for a washer or fastener head (Figs. 21.70a and 21.70b). The spotfacing tool in Fig. 21.71 has spotfaced a boss (a raised cylindrical element).

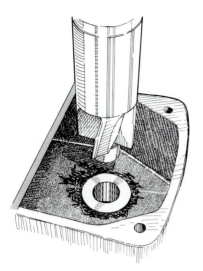

Figure 21.71 This tool is spotfacing the cylindrical boss to provide a smooth seat for a bolt head.

Boring is a machine operation for making large holes. It is usually performed on a lathe with a boring bar (Fig. 21.72).

Counterboring is the process of enlarging the diameter of a drilled hole (Fig. 21.70C and D). The bottoms of counterbored holes are flat with no taper as in counterdrilled and countersunk holes.

Reaming is the operation of finishing or slightly enlarging holes that have been drilled or bored. This operation uses a ream similar to a drill bit.

21.28   Chamfers

Chamfers are beveled edges that are made on cylindrical parts, such as shafts and threaded fasteners. They eliminate sharp edges and facilitate the assembly of parts.

When a chamfer angle is 45°, a note can be used in either of the forms shown in Fig. 21.73A. When the chamfer angle is other than 45°, the angle and length are given, as Fig. 21.73B shows. Chamfers can also be specified at the openings of holes (Fig. 21.74).

Figure 21.72 Boring a large hole on a lathe with a boring bar. (Courtesy of Clausing Corp.)

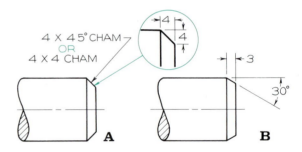

Figure 21.73 (A) Chamfers can be dimensioned by using either type of note when the angle is 45°.
(B) If the angle is other than 45°, it should be dimensioned as shown here.

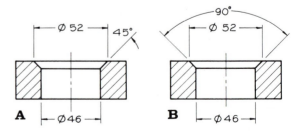

Figure 21.74 Inside chamfers are dimensioned by using one of these methods.

21.29 Keyseats

A **keyseat** is a slot cut into a shaft for aligning the shaft with a pulley or collar mounted on it. Figure 21.75 shows the method of dimensioning a keyway and keyseat. The double dimensions are tolerances, which we discuss in Chapter 22.

21.30 Knurling

Knurling is the operation of cutting diamond-shaped or parallel patterns on cylindrical surfaces for gripping, decoration, or a press fit between two parts that will be permanently assembled.

A **diamond knurl** and **straight knurl** are drawn and dimensioned in Fig. 21.76. Knurls should be dimensioned with specifications that give type, pitch, and diameter.

The abbreviation DP in Fig. 21.76 means diametral pitch, the ratio of the number of grooves on the circumference (N) to the diameter (D), which is found by the equation $DP = N/D$. The preferred diametral pitches for knurling are 64 DP, 96 DP, 128 DP, and 160 DP. For diameters of 1 in., knurling of 64 DP, 96 DP, 128 DP, and 160 DP will have 64, 96, 128, and 160 teeth, respectively, on the circumference. The note $P\ 0.8$ means the knurling grooves are 0.8 mm apart. Calculations for knurling must be made using inches, with conversion to millimeters made afterward.

Knurls for press fits are specified with diameters before knurling and with the minimum diameter after knurling. A simplified method of representing knurls is shown in Fig. 21.77, where notes are used and the knurls are not drawn.

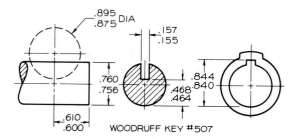

Figure 21.75 Methods of dimensioning slots in a shaft and a slot for a Woodruff key that will hold the part on the shaft. Appendix 33 gives the dimensions for these features.

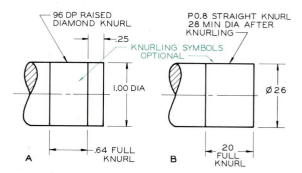

Figure 21.77 Knurls need not be drawn if they are dimensioned as shown here.

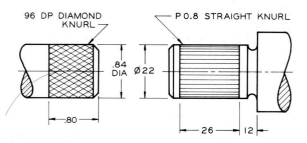

Figure 21.76 A diamond knurl with a diametral pitch of 96 and a straight knurl where the linear pitch *(P)* is 0.8 mm. Pitch is the distance between the grooves on the circumference.

21.31 Necks and Undercuts

A **neck** is a recess cut into a cylindrical part. Where cylinders of different diameters join (Fig. 21.78), a neck ensures that the part assembled on the smaller shaft will fit flush against the shoulder of the larger cylinder.

Undercuts, which are similar to necks (Fig. 21.79A), ensure that a part fitting in the corner of the part will fit flush against both surfaces; they also permit space for trash to drop out of the way when entrapped in the corner. An undercut could also be a recessed neck inside a cylindrical

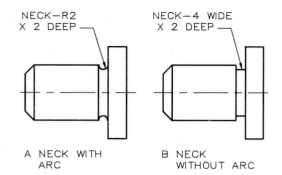

A NECK WITH ARC

B NECK WITHOUT ARC

Figure 21.78 Necks are recesses in cylinders that are used where cylinders of different sizes join together.

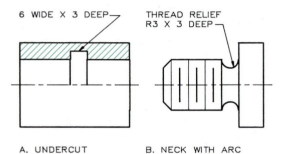

A. UNDERCUT

B. NECK WITH ARC

Figure 21.79 (A) An undercut can be dimensioned as shown here. (B) A thread relief, which is a type of neck, is dimensioned here.

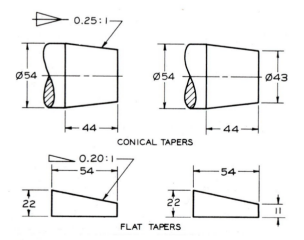

CONICAL TAPERS

FLAT TAPERS

Figure 21.80 Examples of conical and flat tapers. Taper is the ratio of the diameters (or heights) at each end of a sloping surface to the length of the taper.

hole. A thread relief (Fig. 21.79B) is used to square the threads where they intersect a larger cylinder.

21.32 Tapers

Tapers can be either conical surfaces or flat planes (Fig. 21.80). A taper can be dimensioned by the diameter or width at each end of the taper, the length of the tapered feature, or the rate of taper.

Taper is the ratio of the difference in the diameters of a cone to the distance between two diameters (Fig. 21.80). Taper can be expressed as inches per inch (.25 per inch), inches per foot (3.00 per foot), or millimeters per millimeter (0.25:1).

Flat taper is the ratio of the difference in the heights at each end of a feature to the distance between the heights. Flat taper can be expressed as inches per inch (.20 per inch), inches per foot (2.40 per foot), or millimeters per millimeter (0.20:1), as Fig. 21.80 shows.

21.33 Dimensioning Sections

Sections are dimensioned in the same manner as regular views (Fig. 21.81). Most of the principles of dimensioning covered in this chapter have been applied to the part shown in this figure.

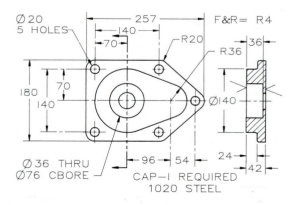

Figure 21.81 An example of a computer-drawn part dimensioned with a variety of dimensioning principles.

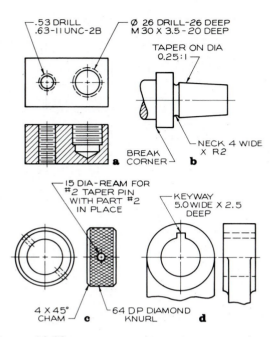

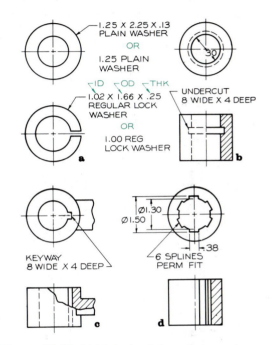

Figure 21.82 (a and b) Threaded holes are sometimes dimensioned by giving the tap drill size in addition to the thread specifications. The tap drill size is not required, but it is permissible. (b) This part is dimensioned to indicate a neck, taper, and break corner, which is a slight round to remove the sharpness from a corner. (c) This collar has a knurl note, chamfer note, and note indicating the insertion of a #2 taper pin. (d) This part has a note for dimensioning a keyway.

21.34 Miscellaneous Notes

A variety of notes are used on detail drawings to provide information and specifications that would otherwise be difficult to represent (Figs. 21.82 and 21.83).

Notes are placed horizontally on the sheet; if the notes lie on the same line, short dashes should be used between them. The abbreviations shown in these notes can be used to save space. Appendix 1 lists the standard abbreviations.

Figure 21.83 (a) Methods of dimensioning washers and lock washers. These dimensions can be found in Appendixes 35 and 36. (b) An undercut is dimensioned. (c) A keyway is dimensioned. (d) A spline inside a hole is dimensioned.

Problems

1–24. (Figs. 21.84–21.85) These problems are to be solved on Size A paper, one per sheet, if they are drawn full size. If drawn double size, use Size B paper. The views are drawn on a 0.20″ (5 mm) grid.

You will need to vary the spacing between the views to provide adequate room for the dimensions. It would be a good idea to sketch the views and dimensions to determine the required spacing before laying out the problems with instruments.

Supply lines that may be missing in all views.

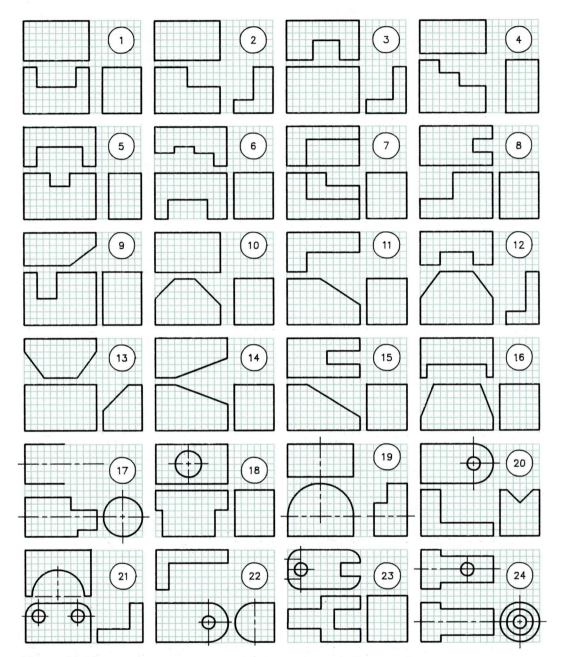

Figure 21.84 Problems 1–24(a). Lay out the views, supply missing lines, and dimension them.

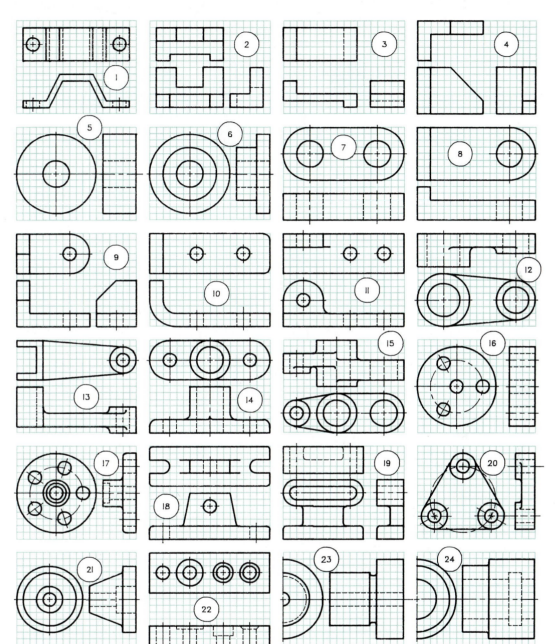

Figure 21.85 Problems 1–24(b). Lay out the views, supply missing lines, and dimension them.

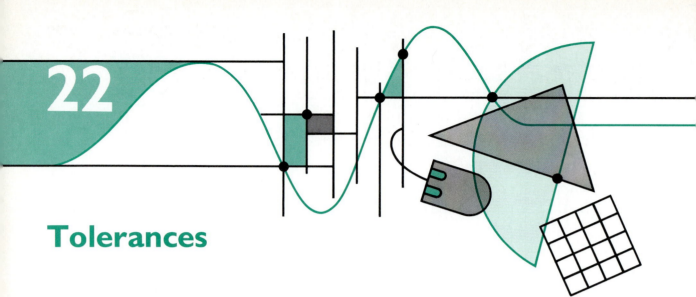

22

Tolerances

22.1 Introduction

Today's technologies require increasingly exact dimensions. What is more, many of today's parts are made by different companies in different locations; therefore, these parts must be specified so that they will be interchangeable.

The techniques of dimensioning parts to ensure interchangeability is called **tolerancing.** Each dimension is allowed a certain degree of variation within a specified zone, or a **tolerance.** For example, a part's dimension might be expressed as 100 ± 0.50, which yields a tolerance of 1.00 mm.

> Dimensions should be given as **large a tolerance** as possible without interfering with the function of the part to reduce production costs. Manufacturing to close tolerances is expensive.

22.2 Tolerance Dimensions

Figure 22.1 shows several acceptable methods of specifying tolerances. When plus-and-minus tolerancing is used, tolerances are applied to a **basic**

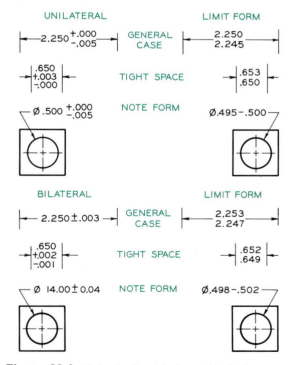

Figure 22.1 Methods of positioning and indicating tolerances in unilateral, bilateral, and limit forms.

dimension. When dimensions allow variation in only one direction, the tolerancing is **unilateral.** Tolerancing that permits variation in either direction from the basic dimension is **bilateral.**

Tolerances may also be given in the form of **limits;** that is, two dimensions are given that represent the largest and smallest sizes permitted for a feature of the part.

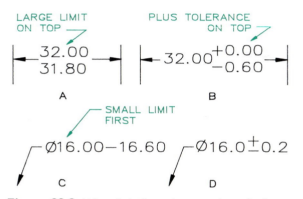

Figure 22.2 When limit dimensions are given, the large limits are placed either above or to the right of the small limits. In plus-and-minus tolerancing, the plus limits are placed above the minus limits.

Figure 22.2 shows the customary methods of indicating toleranced dimensions, and Fig. 22.3 shows the positioning and spacing of numerals of toleranced dimensions.

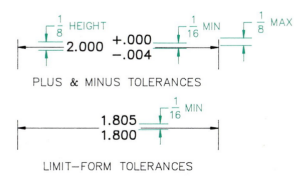

Figure 22.3 Positioning and spacing of numerals used to specify tolerances.

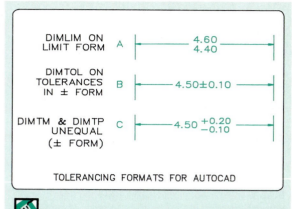

Figure 22.4 AutoCAD will give toleranced dimensions in the forms shown here. In all cases, the DIM-TOL mode must be ON. If DIMLIM is ON, tolerances will be applied in limit form; when OFF, tolerances will be in plus-and-minus form.

Tolerances by Computer Using Auto-CAD, toleranced dimensions can be shown in limit form or plus-and-minus form (Fig. 22.4). Both DIMLIM and DIMTOL must be turned ON to obtain dimensions in limit form. The tolerances are assigned to DIMTM and DIMTP modes under the DIM: command.

In addition to linear dimensions, diametral and radial dimensions are automatically given with either a circle symbol or an R preceding the dimensions as Figs. 22.5 and 22.6 show.

Angular measurements can be toleranced using the plus-and-minus form or the limit form as shown in Fig. 22.7. The computer measures the angle and automatically computes the upper and lower limits.

You will want to edit a toleranced dimension when you change the limits given by the automatic process. By a series of erasures and moves (Fig. 22.8), you will be able to make the desired changes.

22.3 Mating Parts

Mating parts are parts that fit together within a prescribed degree of accuracy (Fig. 22.9). The upper piece is dimensioned with two measure-

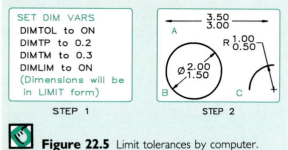

STEP 1   STEP 2

Figure 22.5 Limit tolerances by computer.

Step 1 Set DIM: variables (DIM VARS) as shown for tolerances to be given in LIMIT form.

Step 2 Linear dimensions will be given in limit form, diametral dimensions will be given as limits preceded by Ø, and radial dimensions will be preceded by R.

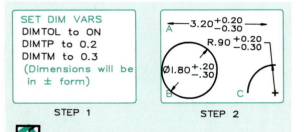

STEP 1   STEP 2

Figure 22.6 Plus-and-minus tolerances by computer.

Step 1 Set DIM: variables (DIM VARS) as shown with DIMLIM set to OFF.

Step 2 Linear dimensions will be given as a basic diameter followed by plus-and-minus tolerances, diametral dimensions will be given as linear dimensions preceded by Ø, and radial dimensions will be preceded by R.

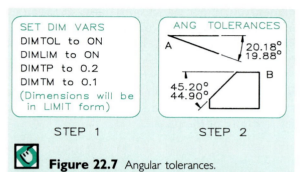

STEP 1   STEP 2

Figure 22.7 Angular tolerances.

Step 1 Set DIM VARS as shown above. Be sure the UNITS command is used to assign the desired number of decimal places for angular units beforehand.

Step 2 Select the lines forming the angles, and the toleranced measurements will be given in limit form. Angular measurements can be specified by the UNITS command as decimal degrees, minutes and seconds, grads, radians, or surveyor's units.

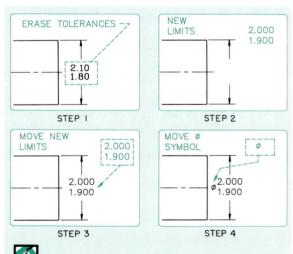

Figure 22.8 Editing tolerances by computer.

Step 1 When you wish to change a toleranced dimension, begin by erasing the given tolerances.

Step 2 Using TEXT, draw the new limits of tolerance in a convenient location.

Step 3 Using a window, MOVE the new limits to the dimension line.

Step 4 Locate a diameter symbol, Ø, by typing % % C, and move it in front of the two new limits.

ments that indicate the upper and lower limits of the size. The notch is slightly larger, allowing the parts to be assembled with a clearance fit.

Figure 22.10A shows an example of mating cylindrical parts, and Fig. 22.10B illustrates the meaning of the tolerance dimensions. The size of the shaft can vary in diameter from 1.500 in. (maximum size) to 1.498 in. (minimum size). The difference between these limits on a single

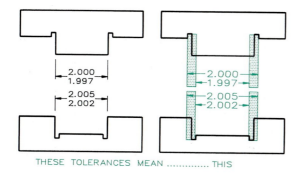

THESE TOLERANCES MEAN .... THIS

Figure 22.9 Each of these mating parts has a tolerance of 0.003″ (variation in size). The allowance between the assembled parts (tightest fit) is 0.002″.

Tolerance is the difference between the limits prescribed for a single part. The tolerance of the shaft in Fig. 22.10 is 0.002 in.

Limits of Tolerance are the extreme measurements permitted by the maximum and minimum sizes of a part. The limits of tolerance of the shaft in Fig. 22.10 are 1.500 and 1.498.

Allowance is the tightest fit between two mating parts. The allowance between the largest shaft and the smallest hole in Fig. 22.10 is 0.003 (negative for an interference fit).

Nominal Size is an approximate size that is usually expressed with common fractions. The nominal sizes of the shaft and hole in Fig. 22.10 are 1.50 in. or $1\frac{1}{2}$ in.

Basic Size is the exact theoretical size from which limits are derived by the application of plus-and-minus tolerances. There is no basic diameter if this is expressed with limits.

Actual Size is the measured size of the finished part.

Fit signifies the type of fit between two mating parts when assembled. There are four types of fit: clearance, interference, transition, and line.

Clearance Fit is a fit that gives a clearance between two assembled mating parts. The fit between the shaft and the hole in Fig. 22.10 is a clearance fit that permits a minimum clearance of 0.003 in. and a maximum clearance of 0.007 in.

Interference Fit is a fit that results in an interference between the two assembled parts. The shaft in Fig. 22.11A is larger than the hole, so it requires a force or press fit, which has an effect similar to welding the two parts.

Transition Fit can result in either an interference or a clearance. The shaft in Fig. 22.11B can be either smaller or larger than the hole and still be within the prescribed tolerances.

part is a tolerance of 0.002 in. The dimensions of the hole in Fig. 22.10A are given with limits of 1.503 and 1.505, for a tolerance of 0.002 (the difference between the limits as shown in Fig. 22.10B).

22.4 Terminology of Tolerancing

The meaning of most of the terms used in tolerancing can be seen by referring to Fig. 22.10.

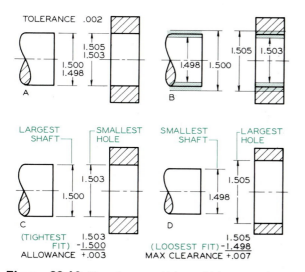

Figure 22.10 The allowance (tightest fit) between these assembled parts is +0.003″. The maximum clearance is 0.007″.

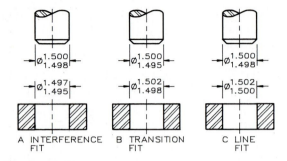

A INTERFERENCE FIT   B TRANSITION FIT   C LINE FIT

Figure 22.11 Types of fits between mating parts. The clearance fit is not shown.

Line Fit   can result in a contact of surfaces or a clearance between them. The shaft in Fig. 22.11C can have contact or clearance when the limits are approached.

Selective Assembly   is a method of selecting and assembling parts by trial and error. Using this method, parts can be assembled that have greater tolerances and produced at a reduced cost. This hand-assembly process is a compromise between a high degree of manufacturing accuracy and an ease of assembly of interchangeable parts.

Single Limits   are dimensions designated by either MIN (minimum) or MAX (maximum), not by both (Fig. 22.12). Depths of holes, lengths, threads, corner radii, chamfers, and so on are sometimes dimensioned in this manner.

22.5   Basic Hole System

Widely used by industry, the basic hole system of dimensioning holes and shafts gives the required allowance between two assembled parts. The

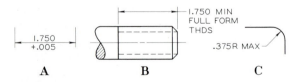

Figure 22.12 Single tolerances can be given in some applications in MAX or MIN form.

smallest hole is taken as the basic diameter from which the limits of tolerance and allowance are applied. It is advantageous to use the hole diameter as the basic dimension because many of the standard drills, reamers, and machine tools are designed to give standard hole sizes.

If the smallest diameter of a hole is 1.500 in., the allowance (0.003 in this example) can be subtracted from this diameter to find the diameter of the largest shaft (1.497 in.). The smallest limit for the shaft can then be found by subtracting the tolerance from 1.497 in.

22.6   Basic Shaft System

Some industries use the basic shaft system of applying tolerances to dimensions of shafts since many shafts come in standard sizes. In this system, the largest diameter of the shaft is used as the basic diameter from which the tolerances and allowances are applied.

If the largest permissible shaft is 1.500 in., the allowance can be added to this dimension to yield the smallest possible diameter of the hole into which the shaft must fit. Therefore, if the parts are to have an allowance of 0.004 in., the smallest hole would have a diameter of 1.504 in.

22.7   Metric Limits and Fits

In this section, we cover the metric system as recommended by the International Standards Organization (ISO), which has been presented in ANSI B4.2. These fits usually apply to cylinders—holes and shafts. These tables can also be used to determine the fits between any parallel surfaces, such as a key in a slot.

Metric Definitions of Limits and Fits

Figure 22.13 illustrates some of the definitions given below.

Basic Size   is the size from which the limits or deviations are assigned. Basic sizes, usually di-

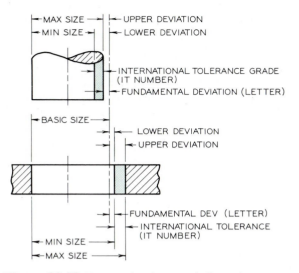

Figure 22.13 Terms related to metric fits and limits.

Basic Size (mm)		Basic Size (mm)		Basic Size (mm)	
First Choice	Second Choice	First Choice	Second Choice	First Choice	Second Choice
1		10		100	
	1.1		11		110
1.2		12		120	
	1.4		14		140
1.6		16		160	
	1.8		18		180
2		20		200	
	2.2		22		220
2.5		25		250	
	2.8		28		280
3		30		300	
	3.5		35		350
4		40		400	
	4.5		45		450
5		50		500	
	5.5		55		550
6		60		600	
	7		70		700
8		80		800	
	9		90		900
				1000	

Table 22.1 Preferred sizes

ameters, should be selected from Table 22.1 under the First Choice column.

Deviation   is the difference between the hole or shaft size and the basic size.

Upper Deviation   is the difference between the maximum permissible size of a part and its basic size.

Lower Deviation   is the difference between the minimum permissible size of a part and its basic size.

Fundamental Deviation   is the deviation closest to the basic size. In the note 40H7, the H (an uppercase letter) represents the fundamental deviation for a hole. In the note 40g6, the g (a lowercase letter) represents the fundamental deviation for a shaft.

Tolerance   is the difference between the maximum and minimum allowable sizes of a single part.

International Tolerance (IT) Grade   is a group of tolerances that vary in accordance with

the basic size and provide a uniform level of accuracy within a given grade. In the note 40H7, the 7 represents the IT grade. There are 18 IT grades: IT01, IT0, IT1, . . . , IT16.

Tolerance Zone   is the zone that represents the tolerance grade and its position in relation to the basic size. This is a combination of the fundamental deviation (represented by a letter) and the international tolerance grade (IT number). In note 40H8, the H8 indicates the tolerance zone.

Hole Basis   is a system of fits based on the minimum hole size as the basic diameter. The

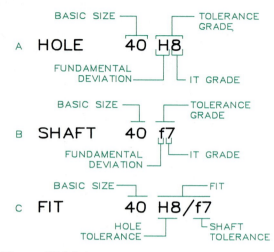

Figure 22.14 Symbols and their definitions as applied to holes and shafts.

fundamental deviation for a hole basis system is an uppercase letter, *H,* for example (Fig. 22.14).

Shaft Basis is a system of fits based on the maximum shaft size as the basic diameter. The fundamental deviation for a shaft basis system is a lowercase letter, *f,* for example (Fig. 22.14).

Clearance Fit is a fit that results in a clearance between two assembled parts under all tolerance conditions.

Interference Fit is a fit between two parts that requires they be forced together when assembled.

Transition Fit is a fit that results in either a clearance or an interference fit between two assembled parts.

		Table 22.2 Description of preferred fits	
	ISO Symbol		**Description**	
	Hole Basis	Shaft Basis		
Clearance fits ↕	H11/c11	C11/h11	**Loose running fit** for wide commercial tolerances or allowances on external members	More clearance ↑
	H9/d9	D9/h9	**Free running fit** not for use where accuracy is essential, but good for large temperature variations, high running speeds, or heavy journal pressures	
	H8/f7	F8/h7	**Close running fit** for running on accurate machines and for accurate location at moderate speeds and journal pressures	
	H7/g6	G7/h6	**Sliding fit** not intended to run freely but to move and turn freely and locate accurately	
Transition fits ↕	H7/h6	H7/h6	**Locational clearance fit** provides snug fit for locating stationary parts but can be freely assembled and disassembled	
	H7/k6	K7/h6	**Locational transisition fit** for accurate location; a compromise between clearance and interference	
	H7/n6	N7/h6	**Locational transition fit** for more accurate location where greater interference is permissible	
Interference fits ↕	H7/p6★	P7/h6	**Locational interference fit** for parts requiring rigidity and alignment with prime accuracy of location but without special bore pressure requirements	More interference ↓
	H7/s6	S7/h6	**Medium drive fit** for ordinary steel parts or shrink fits on light sections; the tightest fit usable with cast iron.	
	H7/u6	U7/h6	**Force fit** suitable for parts that can be highly stressed or for shrink fits where the heavy pressing forces required are impractical	
		★Transition fit for basic sizes in range from 0 through 3 mm.	

Tolerance Symbols are notes used to communicate the specifications of tolerance and fit (Fig. 22.14). The **basic size** is the primary dimension the tolerances are determined from; therefore, it is the first part of the symbol. It is followed by the fundamental deviation letter and the IT number to give the tolerance zone. **Uppercase letters** are used to indicate the fundamental deviation for **holes,** and **lowercase letters** are used for **shafts.**

Figure 22.15 shows three methods of specifying tolerance information. Parenthetical information is for reference only. Appendixes 44–47 give the upper and lower limits.

22.8 Preferred Sizes and Fits

Table 22.1 shows the preferred basic sizes for computing tolerances. Under the First Choice heading, each number increases by about 25% of the preceding number. Each number in the Second Choice column increases by about 12%. To reduce expenses, you should, where possible, select basic diameters from the first column since these correspond to standard stock sizes for round, square, and hexagonal metal products.

Preferred fits for clearance, transition, and interference fits are shown in Table 22.2 for hole basis and shaft basis fits. The tables in Appendixes 44–47 correspond to these fits.

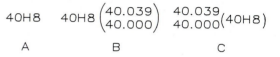

Figure 22.15 Three methods of giving tolerance symbols. The numbers in parentheses are for reference.

Preferred Fits—Hole Basis System Figure 22.16 illustrates the symbols used to show the possible combinations of fits when using the hole basis system. There is a **clearance fit** between the two parts, a **transition fit,** and an **interference fit.** This technique of representing fits is used in Fig. 22.17 to show a series of fits for a hole basis system. Note that the lower deviation of the hole is zero; in other words, the smallest size of the hole is the basic size. The different sizes of the shafts give a variety of fits from c11 to u6, where there is a maximum of interference. These fits correspond to those given in Table 22.2.

Preferred Fits—Shaft Basis System Figure 22.18 shows the preferred fits based on the shaft basis system, where the largest shaft size is the basic diameter. The variation in the fit between the parts is caused by varying the size of the holes, which results in a range from a clearance fit of C11/h11 to an interference fit of U7/h6.

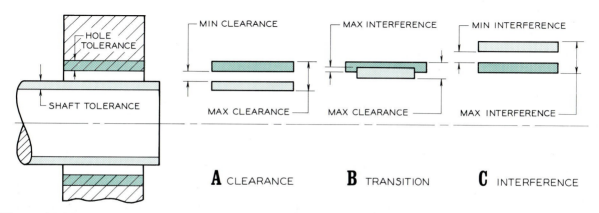

Figure 22.16 Types of fits. (A) A clearance fit. (B) A transition fit where there can be an interference or a clearance. (C) An interference fit where the parts must be forced together.

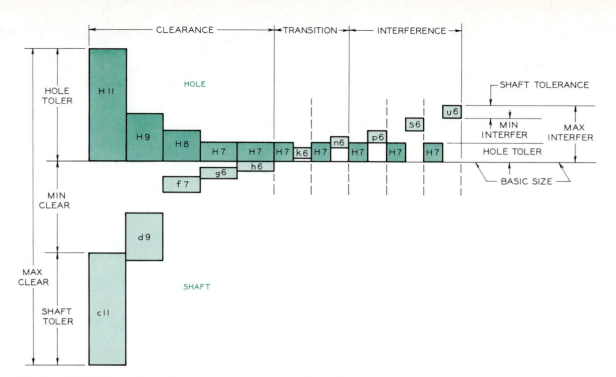

Figure 22.17 The preferred fits for a shaft basis system. These fits correspond to those given in Table 22.2

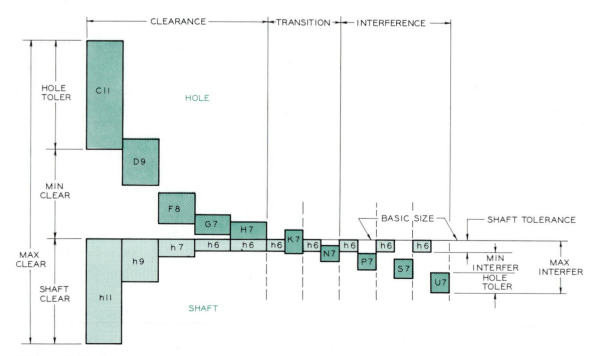

Figure 22.18 The preferred fits for a shaft basis system. These fits correspond to those given in Table 22.2.

22.9 Example Problems— Metric System

The following problems are given and solved as examples of determining the sizes and limits and the applications of the proper symbols to mating parts. The solution of these problems requires using the tables in Appendix 44, the table of preferred sizes (Table 22.1), and the table of preferred fits (Table 22.2).

Example 1 (Fig. 22.19)

Given: Hole basis system, close running fit, basic diameter = 49 mm.

Solution: Use a basic diameter of 50 mm (Table 22.1) and fit of H8/f7 (Table 22.2).

Hole: Find the upper and lower limits of the hole in Appendix 44 under H8 and across from 50 mm. These limits are 50.000 and 50.039 mm.

Shaft: The upper and lower limits of the shaft are found under f7 and across from 50 mm in Appendix 44. These limits are 49.950 and 49.975 mm.

Symbols: Figure 22.19 shows the methods of applying toleranced dimensions to the hole and shaft.

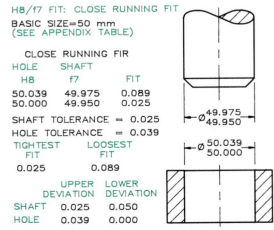

Figure 22.19 The method of calculating limits and fits between a shaft and a hole in metric units.

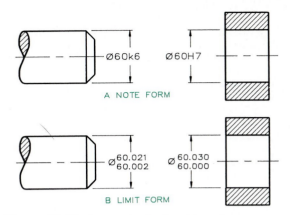

Figure 22.20 Two methods of applying tolerance symbols to a transition fit.

Example 2 (Fig. 22.20)

Given: Hole basis system, locational transition fit, basic diameter = 57 mm.

Solution: Use a basic diameter of 60 mm (Table 22.1) and a fit of H7/k6 (Table 22.2).

Hole: Find the upper and lower limits of the hole in Appendix 45 under H7 and across from 60 mm. These limits are 60.000 and 60.030 mm.

Shaft: The upper and lower limits of the shaft are found under k6 and across from 60 mm in Appendix 45. These limits are 60.021 and 60.002 mm.

Symbols: Figure 22.20 shows two methods of applying the tolerance symbols to a drawing.

Example 3 (Fig. 22.21)

Given: Hole basis system, medium drive fit, basic diameter = 96 mm.

Solution: Use a basic diameter of 100 mm (Table 22.1) and a fit of H7/s6 (Table 22.2).

Hole: Find the upper and lower limits of the hole in Appendix 45 under H7 and across from 100 mm. These limits are 100.035 and 100.000 mm.

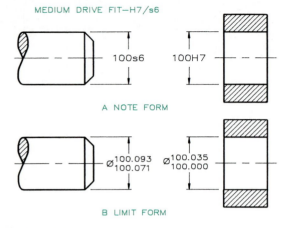

Figure 22.21 Two methods of applying tolerance symbols to an interference fit.

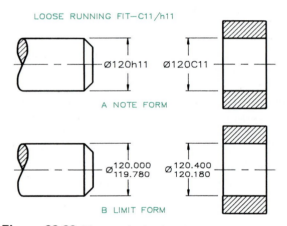

Figure 22.22 Two methods of applying tolerance symbols to a clearance fit.

```
FIT: H8/f7    Ø 45 BASIC

FROM APPENDIX
                    HOLE LIMITS      45.039
HOLE    SHAFT                        45.000
H8       f7

0.039   -0.025    SHAFT LIMITS      44.975
0.000   -0.050                      44.950
```

Figure 22.23 The limits of a nonstandard diameter, 45 mm, and an H8/f7 fit are calculated by using values from Appendixes 48 and 49.

Shaft: The upper and lower limits of the shaft are found under s6 and across from 100 mm in Appendix 45. These limits are found to be 100.093 and 100.071 mm. From the appendix, the tightest fit is an interference of 0.093 mm, and the loosest fit is an interference of 0.036 mm. An interference is indicated by a minus sign in front of the numbers.

Example 4   (Fig. 22.22)

Given: Shaft basis system, loose running fit, basic diameter = 116 mm.

Solution: Use a basic diameter 120 mm (Table 22.1) and a fit of C11/h11 (Table 22.2).

Hole: Find the upper and lower limits of the hole in Appendix 46 under C11 and across from 120 mm. These limits are 120.400 and 120.180 mm.

Shaft: The upper and lower limits of the shaft are found under h11 and across from 120 mm in Appendix 46. These limits are 119.780 and 120.000 mm.

22.10   Preferred Metric Fits—Nonpreferred Sizes

Limits of tolerances for preferred fits, shown in Table 22.2, can be calculated for nonstandard sizes. Limits of tolerances appear in Appendix 48 for nonstandard hole sizes and in Appendix 49 for nonstandard shaft sizes.

The hole and shaft limits for an H8/f 7 fit and a 45-mm DIA are calculated in Fig. 22.23. The tolerance limits of 0.000 and 0.039 mm for an H8 hole are taken from Appendix 48 across from the size range of 40–50 mm. The tolerance limits of −0.025 and −0.050 mm for the shaft are taken from Appendix 49. The limits of sizes for the hole and shaft are calculated by applying these limits of tolerance to the 45-mm basic diameter.

22.11 Standard Fits—English Units

The ANSI B4.1 standard specifies a series of fits between cylindrical parts that are based on the basic hole system in inches. The types of fit covered in this standard are

RC—Running and sliding fits

LC—Clearance locational fits

LT—Transition locational fits

LN—Interference locational fits

FN—Force and shrink fits

Appendixes 38–42 list these five types of fit, each of which has several classes.

Running and Sliding Fits (RC) are fits which provide a similar running performance, with suitable lubrication allowance throughout the range of sizes. The clearance for the first two classes (RC 1 and RC 2), used chiefly as slide fits, increases more slowly with diameter size than that of other classes so that accurate location is maintained even at the expense of free relative motion.

Locational Fits (LC, LT, LN) are intended to determine only the location of the mating parts; they may provide rigid or accurate location (interference fits) or some freedom of location (clearance fits). Locational fits are divided into three groups: **clearance fits** (LC), **transition fits** (LT), and **interference fits** (LN).

Force Fits (FN) are special types of interference fits, typically characterized by maintenance of constant bore pressures throughout the range of sizes. The interference therefore varies almost directly with diameter, and the difference between its minimum and maximum values is small enough to maintain the resulting pressures within reasonable limits.

Figure 22.24 illustrates how to use the values from the tables in Appendix 38 for an RC 9 fit. The basic diameter for the hole and shaft is 2.5000 in., which is between 1.97 and 3.15 in.

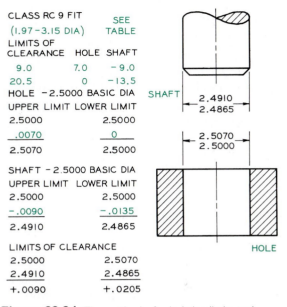

Figure 22.24 The method of calculating limits and allowances for an RC 9 fit between a shaft and a hole. The basic diameter is 2.5000 inches.

given in the last column of the table. Since all limits are in thousandths, the values can be converted by moving the decimal point three places to the left; for example, +0.7 is +0.0007 in.

The upper and lower limits of the shaft (2.4910 and 2.4865 in.) are found by subtracting the two limits (−0.0090 and −0.0135 in.) from the basic diameter. The upper and lower limits of the hole (2.5007 and 2.5000 in.) are found by adding the two limits (+0.007 and 0.000 in.) to the basic diameter.

When the two parts are assembled, the tightest fit (+0.0090 in.) and the loosest fit (+0.0205 in.) are found by subtracting the maximum and minimum sizes of the holes and shafts. These values are provided in the first column of the table as a check on the limits.

The same method (but different tables) is used for calculating the limits for all types of fit. Plus values indicate clearance, and minus values indicate interference between the assembled parts.

To convert these values to millimeters multiply inches by 25.4, or use metric tables, instead.

22.12 Chain Dimensions

When parts are dimensioned to locate surfaces or geometric features by a chain of dimensions (Fig. 22.25A), variations may occur that exceed the tolerances specified. As successive measurements are made, with each based on the preceding one, the tolerances may accumulate, as Fig. 22.25A shows. For example, the tolerance between surfaces *A* and *B* is 0.002; between *A* and *C,* 0.004; and between *A* and *D,* 0.006.

This accumulation of tolerances can be eliminated by measuring from a single plane called a **datum plane.** A datum plane is usually on the object, but it could be on the machine used to make the part. Since each of the planes in Fig. 22.25B was located with respect to a single datum, the tolerances between the intermediate planes are a uniform 0.002, which represents the maximum tolerance.

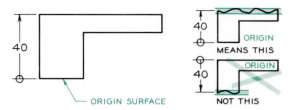

Figure 22.26 This method is used to indicate the origin surface for locating one feature of a part with respect to another.

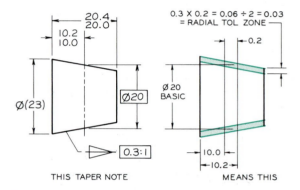

Figure 22.27 Taper is indicated with a combination of tolerances and taper symbols. The variation in diameter at any point is 0.06 mm, or 0.03 mm in radius.

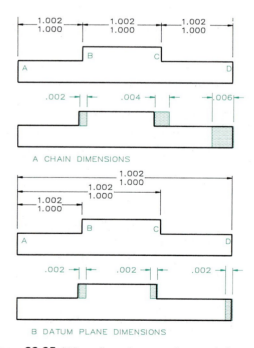

A CHAIN DIMENSIONS

B DATUM PLANE DIMENSIONS

Figure 22.25 When dimensions are given as chain dimensions, the tolerances can accumulate to give a variation of 0.006″ at *D* instead of 0.002″. When dimensioned from a single datum, the variations of *X* and *Y* cannot deviate more than the specified 0.002″ from the datum.

22.13 Origin Selection

Sometimes there is a need to specify a surface as the origin for locating another surface. An example is shown in Fig. 22.26, where the origin surface is the shorter one at the base of an object. The result is that the angular variation permitted is less for the longer surface than it would be if the origin plane had been the longer surface.

22.14 Conical Tapers

Taper is a ratio of the difference in the diameters of two circular sections of a cone to the distance between the sections. Figure 22.27 shows a method of specifying a conical taper by giving a basic diameter and basic taper. The basic diameter of 20 mm is located midway in the length of the cone with a toleranced dimension. Figure 22.27 also shows how to find the radial tolerance zone.

22.15 Tolerance Notes

All dimensions on a drawing are toleranced either by the rules previously discussed or by a note placed in or near the title block. For example, the note TOLERANCE $\pm\frac{1}{64}$ (or its decimal equivalent, 0.40 mm) might be given on a drawing for less critical dimensions.

Some industries may give dimensions in inches where decimals are carried out to two, three, and four places. A note for dimensions with two and three decimal places might be given on the drawing as TOLERANCES XX.XX ±0.10; XX.XXX ±0.005. Tolerances of four places would be given directly on the dimension lines.

The most common method of noting tolerances is to give as large a tolerance as feasible in a note, such as TOLERANCES ±0.05 (±1 mm when using metrics), and to give the tolerances for the mating dimensions that require smaller tolerances on the dimension lines.

Angular tolerances should be given in a general note in or near the title block, such as ANGULAR TOLERANCES $\pm0.5°$ or $\pm30'$. Angular tolerances less than this should be given on the drawing where these angles are dimensioned. Figure 22.28 shows techniques of tolerancing angles.

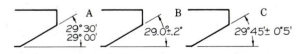

Figure 22.28 Angles can be toleranced by any of these techniques using limits or the plus-and-minus method.

22.16 General Tolerances—Metric

All dimensions on a drawing are understood to have tolerances, and the amount of tolerance must be noted. In this section, which is based on

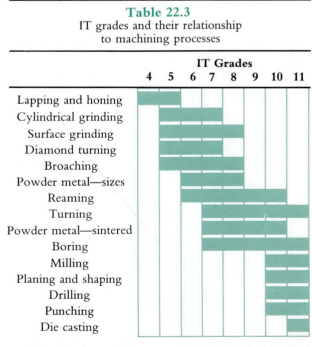

Figure 22.29 International tolerance (IT) grades and their applications.

the metric system as outlined in ANSI B4.3, we use the **millimeter** as the unit of measurement.

Tolerances may be specified by applying them directly to the dimensions, giving them in specification documents, or giving them in a general note on the drawing.

Linear Dimensions may be toleranced by indicating $\pm$ one half of an international tolerance (IT) grade as given in Appendix 43. The appropriate IT grade can be selected from the graph in Fig. 22.29.

IT grades for mass-produced items range from IT12 through IT16. When the machining process is known, the IT grades can be selected from Table 22.3.

Table 22.3
IT grades and their relationship to machining processes

	IT Grades							
	4	5	6	7	8	9	10	11
Lapping and honing								
Cylindrical grinding								
Surface grinding								
Diamond turning								
Broaching								
Powder metal—sizes								
Reaming								
Turning								
Powder metal—sintered								
Boring								
Milling								
Planing and shaping								
Drilling								
Punching								
Die casting								

Table 22.4
Fine, medium, and coarse series: general tolerance—linear dimensions

		Variations (mm)						
Basic Dimensions (mm)		0.5 to 3	Over 3 to 6	Over 6 to 30	Over 30 to 120	Over 120 to 315	Over 315 to 1000	Over 1000 to 2000
Permissible variations	Fine series	±0.05	±0.05	±0.1	±0.15	±0.2	±0.3	±0.5
	Medium series	±0.1	±0.1	±0.2	±0.3	±0.5	±0.8	±1.2
	Coarse series		±0.2	±0.5	±0.8	±1.2	±2	±3

General tolerances using IT grades may be expressed in a note as follows:

UNLESS OTHERWISE SPECIFIED ALL UNTOLERANCED DIMENSIONS ARE $\pm\dfrac{IT14}{2}$.

This means a tolerance of ±0.700 mm is allowed for a dimension between 315 and 400 mm. The value of the tolerance is listed in Appendix 43 as 1.400.

Table 22.4 shows recommended tolerances for **fine, medium,** and **coarse series** for dimensions of graduated sizes. A medium tolerance, for example, can be specified by the following note:

GENERAL TOLERANCES SPECIFIED IN ANSI B4.3 MEDIUM SERIES APPLY.

DIMENSIONS IN mm

GENERAL TOLERANCE
UNLESS OTHERWISE SPECIFIED THE FOLLOWING TOLERANCES ARE APPLICABLE

LINEAR	OVER TO	0.5 6	6 30	30 120	120 315	315 1000	1000 2000
TOL	±	0.1	0.2	0.3	0.5	0.8	1.2

Figure 22.30 This table is a *medium* series of values taken from Table 22.4. It is placed on the working drawing to provide the tolerances for various ranges of sizes.

DIMENSIONS IN mm

GENERAL TOLERANCE
UNLESS OTHERWISE SPECIFIED THE FOLLOWING TOLERANCES ARE APPLICABLE

LINEAR	OVER TO	– 120	120 315	315 1000	1000 –
TOL	ONE DECIMAL ±	0.3	0.5	0.8	1.2
	NO DECIMALS ±	0.8	1.2	2	3

Figure 22.31 This table of tolerances can be placed on a drawing to indicate the tolerances for dimensions with one or no decimal places.

This same information can be given on the drawing in a table by selecting the grade—medium in this example—from Table 22.4 and giving it as shown in Fig. 22.30.

General tolerances can be expressed in a table that gives the tolerances for dimensions with one or no decimal places (Fig. 22.31).

Another method of giving general tolerances is a note in the following form:

UNLESS OTHERWISE SPECIFIED ALL UNTOLERANCED DIMENSIONS ARE ±0.8 mm.

This method should be used only where the dimensions on a drawing have slight differences in size.

Angular Tolerances are expressed (1) as an **angle** in decimal degrees or in degrees and minutes, (2) as a taper expressed in **percentage** (number of millimeters per 100 mm), or (3) as **milliradians.** A milliradian is found by multiplying the degrees of an angle by 17.45. The suggested tolerances for each of these units are shown in Table 22.5 and are based on the length of the shorter leg of the angle.

General angular tolerances may be given on the drawing with a note in the following form:

UNLESS OTHERWISE SPECIFIED THE
GENERAL TOLERANCES IN
ANSI B4.3 APPLY.

A second method shows a portion of Table 22.5 on the drawing using the units desired (Fig. 22.32). A third method is a note with a single tolerance such as:

UNLESS OTHERWISE SPECIFIED THE
GENERAL ANGULAR TOLERANCES
ARE ±0°30′ (or ±0.5°).

22.17 Geometric Tolerances

Geometric tolerancing is a term used to describe tolerances that specify and control form, profile, orientation, location, and runout on a dimensioned part. The basic principles of this area of tolerancing are standardized by the ANSI Y14.5M–1982 Standards and the Military Standards (Mil-Std) of the U.S. Department of Defense.

ANGULAR TOLERANCE				
LENGTH OF SHORTER LEG - mm	UP TO 10	OVER 10 TO 50	OVER 50 TO 120	OVER 120 TO 400
TOL	±1°	±0° 30′	±0° 20′	±0° 10′

Figure 22.32 This table can be placed on a drawing to indicate the general tolerances for angles that were extracted from Table 22.5.

These standards are based on the metric system with the millimeter as the unit of measurement. Inch units with decimal fractions can be used instead of millimeters if needed.

22.18 Symbology of Geometric Tolerances

Figure 22.33 shows the various symbols used to specify geometric characteristics of dimensioned drawings. Additional features and their proportions are shown in Fig. 22.34, where the letter height *(H)* is used as the basis of the proportions. On most drawings, a $\frac{1}{8}$-in. or 3-mm letter height is recommended. Figure 22.35 shows other examples of feature control symbols.

22.19 Limits of Size

Three terms used to specify the limits of size of a part when applying geometric tolerances are **maximum material condition (MMC), least**

Table 22.5 General tolerance—angles and tapers					
Length of the Shorter Leg (mm)		Up to 10	Over 10 to 50	Over 50 to 120	Over 120 to 400
Permissible variations	In degrees and minutes	±1°	±0°30′	±0°20′	±0°10′
	In millimeters per 100 mm	±1.8	±0.9	±0.6	±0.3
	In milliradians	±18	±9	±6	±3

	TOLERANCE	CHARACTERISTIC	SYMBOL
INDIVIDUAL FEATURES	FORM	STRAIGHTNESS	—
		FLATNESS	▱
		CIRCULARITY	○
		CYLINDRICITY	⌭
INDIVIDUAL OR RELATED FEATURES	PROFILE	PROFILE OF A LINE	⌒
		PROFILE OF A SURFACE	⌓
RELATED FEATURES	ORIENTATION	ANGULARITY	∠
		PERPENDICULARITY	⊥
		PARALLELISM	//
	LOCATION	POSITION	⊕
		CONCENTRICITY	◎
	RUNOUT	CIRCULAR RUNOUT	↗
		TOTAL RUNOUT	⌰

Figure 22.33 These symbols are used to specify the geometric characteristics of a dimensioned part.

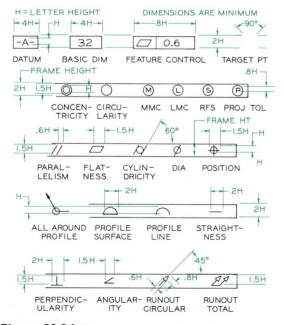

Figure 22.34 The general proportions of notes and symbols used in feature control symbols.

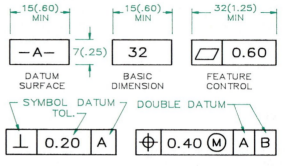

FEATURE CONTROL SYMBOLS

Figure 22.35 Examples of symbols used to indicate datum planes, basic dimensions, and feature control symbols.

material condition (LMC), and **regardless of feature size (RFS).**

MMC indicates a part is made with the maximum amount of material. For example, the shaft in Fig. 22.36 is at MMC when it has the largest permitted diameter of 24.6 mm. The hole is at MMC when it has the most material or the smallest diameter of 25.0 mm.

LMC indicates a part has the least amount of material. The shaft in Fig. 22.36 is at LMC when it has the smallest diameter of 24.0 mm. The hole is at LMC when it has the largest diameter of 25.6 mm.

RFS indicates tolerances apply to a geometric feature regardless of the size it may be, from MMC to LMC.

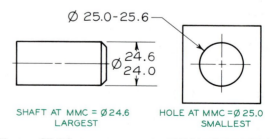

SHAFT AT MMC = Ø24.6 LARGEST   HOLE AT MMC = Ø 25.0 SMALLEST

Figure 22.36 A shaft is at MMC when it is at the largest size permitted by its tolerance. A hole is at MMC when it is at the smallest size.

22.20   Three Rules of Tolerances

Three general rules of tolerancing geometric features should be followed in this type of dimensioning.

Rule 1 (Individual Feature Size) When only a tolerance of size is specified on a part, the limits of size prescribe the amount of variation permitted in its geometric form. In Fig. 22.37, the forms of the shaft and hole are permitted to vary within the tolerance of size indicated by the dimensions.

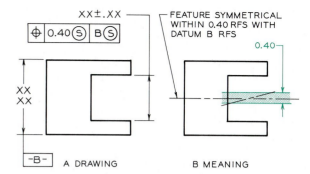

Figure 22.38 Tolerances of position should include a note of M, S, or L to indicate maximum material condition, regardless of feature size or least material condition.

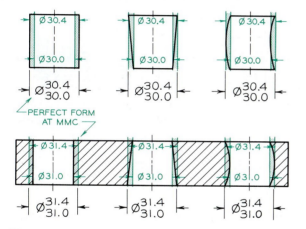

Figure 22.37 When only a tolerance of size is specified on a part, the limits prescribe the form of the part, as shown in these shafts and holes with the same limits of tolerance.

Rule 2 (Tolerances of Position) When a tolerance of position is specified on a drawing, RFS, MMC, or LMC must be specified with respect to the tolerance, datum, or both. The specification of symmetry of the part in Fig. 22.38 is based on a tolerance at RFS from a datum at RFS.

Rule 3 (All Other Geometric Tolerances) RFS applies for all other geometric tolerances for individual tolerances and datum references if no modifying symbol is given in the feature control symbol. If a feature is to be at MMC, it must be specified.

22.21   Three-Datum Plane Concept

A datum plane is used as the origin of a part's dimensioned features that have been toleranced. Datum planes are usually associated with manufacturing equipment, such as machine tables, or with locating pins.

Three mutually perpendicular datum planes are used to dimension a part accurately. For example, the part in Fig. 22.39 is placed in contact with the primary datum plane at its base where three points must make contact with the datum. The part is further related to the secondary plane with two contacting points. The third (tertiary) datum is contacted by a single point on the object.

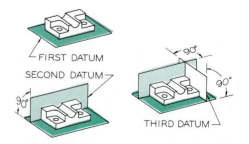

Figure 22.39 When an object is referenced to a primary datum plane, it contacts the plane with its three highest points. The vertical surface contacts the secondary vertical datum plane with two points. The third datum plane is contacted by one point on the object. The datum planes are listed in this order in the feature control symbol.

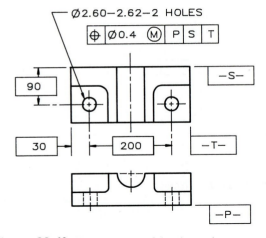

Figure 22.40 The sequence of the three-plane reference system (shown in Fig. 22.39) is labeled where the planes appear as edges. Note that the primary datum plane, *P*, is listed first in the feature control symbol; the secondary plane, *S*, is next; and the tertiary plane, *T*, is last.

The priority of these datum planes is noted on the drawing of the part by feature control symbols, as Fig. 22.40 shows. The primary datum is surface *P*, the secondary is surface *S*, and the tertiary is surface *T*. Examples of feature control symbols are given in Fig. 22.41, where the primary, secondary, and tertiary datum planes are listed in order of priority.

22.22 Cylindrical Datum Features

Figure 22.42 illustrates a part with a cylindrical datum feature that is the axis of a true cylinder. Primary datum *K* establishes the first datum. Datum *M* is associated with two theoretical planes—the second and third in a three-plane relationship.

The two theoretical planes are represented on the drawing by perpendicular centerlines. The intersection of the centerlines coincides with the datum axis. All dimensions originate from the datum axis, which is perpendicular to datum *K*; the two intersecting datum planes indicate the direction of measurements in an *x* and *y* direction.

The sequence of the datum reference in the feature control symbol is significant to the manufacturing and inspection processes. The part in

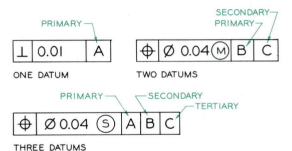

Figure 22.41 Feature control symbols may indicate from one to three datum planes listed in order of priority.

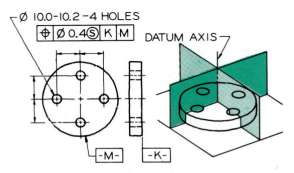

Figure 22.42 These true-position holes are located with respect to primary datum *K* and datum *M*. Since datum *M* is a circle, this implies that the holes are located about two intersecting datum planes formed by the crossing centerlines in the circular view satisfying the three-plane concept.

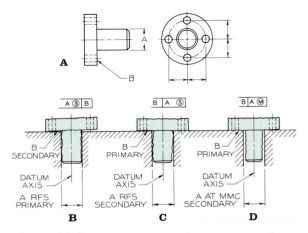

Figure 22.43 Three examples that illustrate the effects of selection of the datum planes in order of priority and the effect of RFS and MMC.

Fig. 22.43 is dimensioned with an incomplete feature control symbol; it does not specify the primary and secondary datum planes. The schematic drawing in Fig. 22.43B illustrates the effect of specifying the diameter A as the primary datum plane and surface B as the secondary datum plane. This means the part is centered about cylinder A by mounting the part in a chuck, mandrel, or centering device on the processing equipment, which centers the part at RFS. Surface B is assembled to contact at least one point of the third datum plane.

If surface B were specified as the primary datum feature, it would be assembled to contact datum plane B in at least three points. The axis of datum feature A will be gauged by the smallest true cylinder that is perpendicular to the first datum that will contact surface A at RFS.

In Fig. 22.43D, plane B is specified as the primary datum feature, and cylinder A is specified as the secondary datum feature at MMC. The part is mounted on the processing equipment where at least three points on feature B are in contact with datum B. The second and third planes intersect at the datum axis to complete the three–plane relationship. Using the modifier to specify MMC gives a more liberal tolerance zone than otherwise would be acceptable when RFS was specified.

22.23   Datum Features at RFS

When dimensions of size are applied to a part at RFS, the datum is established by contacting surfaces on the processing equipment with surfaces of the part. Variable machine elements, such as chucks or center devices, are adjusted to fit the external or internal features of a part and thereby establish datums.

Primary Diameter Datums   For an external cylinder (shaft), the datum axis is the axis of the smallest circumscribed cylinder that contacts the cylindrical feature of the part. That is, the largest diameter of the part will make contact with the

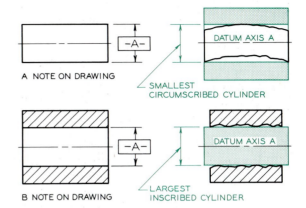

Figure 22.44   The datum axis of a shaft is the smallest circumscribed cylinder that contacts the shaft. The datum axis of a hole is the centerline of the largest inscribed cylinder that contacts the hole.

smallest contacting cylinder of the machine element that holds the part (Fig. 22.44).

For an internal cylinder (hole), the datum axis is the axis of the largest inscribed cylinder that contacts the inside of the hole. That is, the smallest diameter of the hole will make contact with the largest cylinder of the machine element inserted in the hole (Fig. 22.44).

Primary External Parallel Datums   The datum for external features is the center plane between two parallel planes, at their minimum separation, that contact the planes of the object (Fig. 22.45A). These are planes of a viselike device that holds the part; therefore, the planes of the part are at maximum separation, whereas the planes of the device are at minimum separation.

Primary Internal Parallel Datums   The datum for internal features is the center plane between two parallel planes, at their maximum separation, that contact the planes of the object (Fig. 22.45B). This is the condition in which the slot is at its smallest opening size.

Secondary Datums   The secondary datum (axis or center plane) for both external and internal diameters or distances between parallel planes is found as covered in the previous two para-

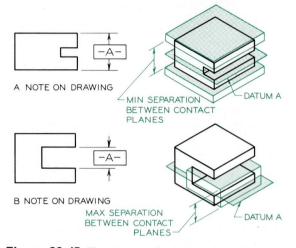

A NOTE ON DRAWING

MIN SEPARATION
BETWEEN CONTACT
PLANES

DATUM A

B NOTE ON DRAWING

MAX SEPARATION
BETWEEN CONTACT
PLANES

DATUM A

Figure 22.45 The datum plane for external parallel surfaces is the center plane between two contacting parallel planes at their minimum separation. The datum plane for internal parallel surfaces is the center plane between two contacting parallel surfaces at their maximum separation.

graphs but with an additional requirement: The contacting cylinder of the contacting parallel planes must be perpendicularly oriented to the primary datum. Figure 22.46 illustrates how datum *B* is the axis of a cylinder. This principle also can be applied to parallel planes.

Tertiary Datums   The third datum (axis or center plane) for both external and internal features is found as covered in the previous three paragraphs but with an additional requirement: The contacting cylinder or parallel planes must be angularly oriented to the secondary datum. Datum *C* in Fig. 22.46 is the tertiary datum plane.

22.24 Datum Targets

Instead of using a plane surface as a datum, specified datum targets are indicated on the surface of a part where the part is supported by spherical or pointed locating pins. The symbol X indicates target points that are supported by locating pins at specified points (Fig. 22.47). Datum target

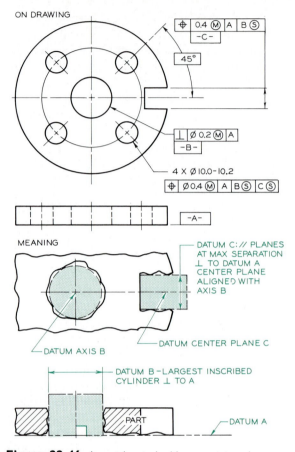

ON DRAWING

MEANING

DATUM AXIS B

DATUM CENTER PLANE C

DATUM C: // PLANES
AT MAX SEPARATION
⊥ TO DATUM A
CENTER PLANE
ALIGNED WITH
AXIS B

DATUM B – LARGEST INSCRIBED
CYLINDER ⊥ TO A

PART

DATUM A

Figure 22.46 A part located with respect to primary, secondary, and tertiary datum planes.

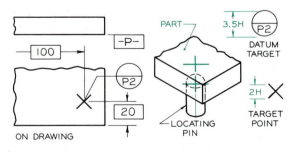

ON DRAWING

PART

LOCATING
PIN

DATUM
TARGET

TARGET
POINT

Figure 22.47 Target points from which a datum point is established are located with an X and a target symbol.

symbols are placed outside the outline of the part with a leader directed toward the target. When the target is on the near (visible) surface, the leaders are solid lines; when the target is on the far (invisible) surface (see Fig. 22.47), the leaders are hidden.

Three target points are required to establish the primary datum plane, two for the secondary, and one for the tertiary. The target symbol in Fig. 22.47 is labeled *P2* to match the designation of the primary datum, *P*. Were the other two points shown, they would be labeled *P1* and *P2* to establish the primary datum.

A datum target line is specified in Fig. 22.48 for a part supported on a datum line instead of on a datum point. An X and a phantom line are used to locate the line of support.

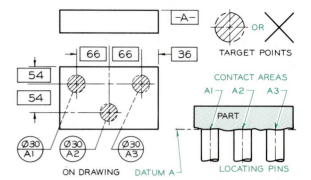

Figure 22.49 Target points with areas are located with basic dimensions and target symbols that give the diameters of the targets. Hidden leaders indicate that the targets are on the hidden side of the plane.

22.25 Tolerances of Location

Tolerances of location deal with **position, concentricity,** and **symmetry.**

Position Tolerancing

Whereas toleranced location dimensions give a square tolerance zone for locating the center of a hole, **true-position dimensions** locate the exact position of a hole's center, about which a circular tolerance zone is given (Fig. 22.50). **Basic dimensions** are exact untoleranced dimensions

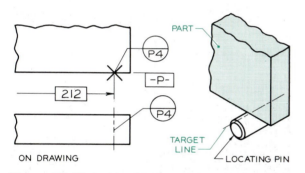

Figure 22.48 An X and a phantom line are used to locate target lines on a drawing.

Target areas are specified for cases where spherical or pointed locating pins are inadequate to support a part. The diameters of the targets are specified with cross-hatched circles surrounded by phantom lines, as Fig. 22.49 shows. The target symbols give both the diameter of the targets and their number designations. The X could also be used to indicate target areas.

The part is located on its datum plane by placing it on the three locating pins with 30-mm diameters (Fig. 22.49). Leaders from the target areas to the target symbols are hidden, indicating the targets are on the hidden side of the object.

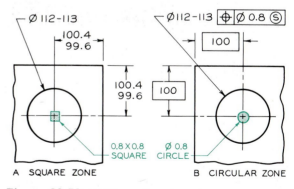

Figure 22.50 (A) These dimensions give a square tolerance zone for the axis of the hole. (B) Basic dimensions locate the true center of the circle about which a circular tolerance zone of 0.8 mm is specified.

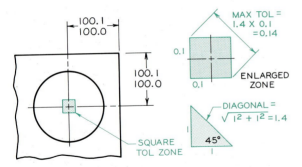

Figure 22.51 The toleranced-coordinate method of dimensioning gives a square tolerance zone. The diagonal of the square exceeds the specified tolerance by a factor of 1.4.

used to locate true positions indicated by boxes drawn around them (Fig. 22.50B).

In both methods, the diameters of the holes are toleranced by notes. The true-position method (Fig. 22.50B) uses a feature control symbol to specify the diameter of the circular tolerance zone inside which the center of the hole must lie. A circular zone gives a more uniform tolerance of the hole's true position than a square.

In Fig. 22.51, you can see an enlargement of the square tolerance zone that results from using toleranced coordinates to locate a hole's center. The diagonal across the square zone is greater than the specified tolerance by a factor of 1.4. The true-position method shown in Fig. 22.52 can have a larger circular tolerance zone by a fac-

tor of 1.4 and still have the same degree of accuracy of position as the specified 0.1 square zone.

If the toleranced coordinate method could accept a variation of 0.014 across the diagonal of the square tolerance zone, the true-position tolerance should be acceptable with a circular zone of 0.014, which is a greater tolerance than the square zone permitted (Fig. 22.51). True-position tolerances can be applied by symbol, as Fig. 22.52 shows.

The circular tolerance zone specified in the circular view of a hole is assumed to extend the full depth of the hole. Therefore, the tolerance zone for the centerline of the hole is a cylindrical zone inside which the axis must lie. Since the size of the hole and its position are both toleranced, these two tolerances are used to establish the diameter of a gauge cylinder used to check for the conformance of the holes to the specifications (Fig. 22.53).

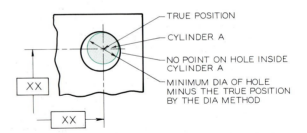

Figure 22.53 When a hole is located at true position at MMC, no element of the hole will be inside the imaginary cylinder A.

By subtracting the true-position tolerance from the hole at MMC (the smallest permissible hole), the circle is found. This zone represents the least favorable condition when the part is gauged or assembled with a mating part. When the hole is not at MMC, it is larger and permits greater tolerance and easier assembly.

Gauging a Two-Hole Pattern

Gauging is a technique of checking dimensions to determine whether they have met the specifications of tolerance (Fig. 22.54). The two holes, which are positioned 26.00 mm apart with a basic

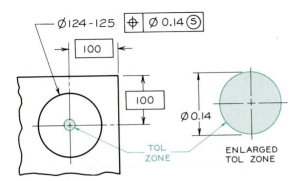

Figure 22.52 The true-position method of tolerancing gives a circular tolerance zone with equal variations in all directions from the true axis of the hole.

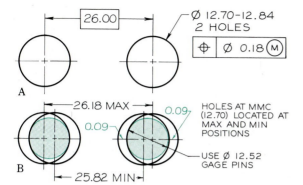

Figure 22.54 (A) When two holes are located at true position at MMC, they may be gauged with pins 12.52 mm in diameter that are located 26.00 mm apart. (B) Two holes, located at MMC, as shown at A can be gauged with pins 12.52 mm in diameter that are located 26.00 mm apart.

dimension, have limits of 12.70 and 12.84 for a tolerance of 0.14 and are located at true position within a diameter of 0.18. The gauge pin diameter is calculated to be 12.52 mm (the smallest hole's size minus the true-position tolerance), as Fig. 22.54B shows. This means two pins with diameters of 12.52 mm spaced exactly 26.00 mm apart could be used to check the diameters and positions of the holes at MMC, the most critical

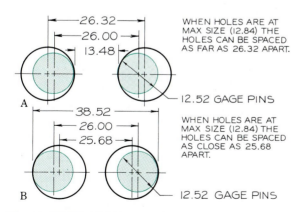

Figure 22.55 (A) When two holes are at their maximum size, the centers of the holes can be spaced as far as 26.32 mm apart and still be acceptable. (B) The holes can be placed as close as 25.68 mm apart when they are at maximum size.

size. If the pins can be inserted in the holes, the holes are properly sized and located.

When the holes are not at MMC—that is, when they are larger than their minimum size—these gauge pins will permit a greater range of variation (Fig. 22.55). When the holes are at their maximum size of 12.84 mm, they can be located as close as 25.68 mm from center to center or as far apart as 26.32 mm from center to center.

Concentricity

Concentricity is a feature of location because it specifies the relationship of one cylinder with another since both share the same axis. In Fig. 22.56, the large cylinder is flagged as datum A, which means the large diameter is used as the datum for measuring the variation of the smaller cylinder's axis.

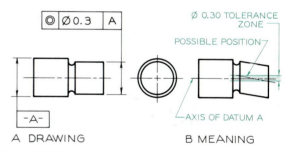

Figure 22.56 Concentricity is a tolerance of location. The feature control symbol specifies that the smaller cylinder should be concentric to cylinder A, within 0.3 mm about the axis of A.

Feature control symbols will be used to specify concentricity and other geometric characteristics throughout the remainder of this chapter (Fig. 22.57).

Symmetry

Symmetry is also a feature of location. A part or feature is symmetrical when it has the same contour and size on opposite sides of a central plane. A symmetry tolerance locates features with respect to a datum plane (Fig. 22.58). The feature

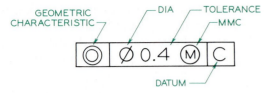

Figure 22.57 A typical feature control symbol. This one indicates that a surface is concentric to datum *C* within a diameter of 0.4 mm at MMC.

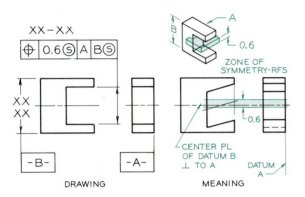

Figure 22.58 Symmetry is a tolerance of location that specifies a part's feature be symmetrical about the center plane between parallel surfaces of the part.

control symbol notes that the notch is symmetrical about datum *B* within a zone of 0.6 mm.

22.26   Tolerances of Form

Flatness

A surface is flat when all its elements are in one plane. A feature control symbol is used to specify flatness within a 0.4 mm zone in Fig. 22.59. No

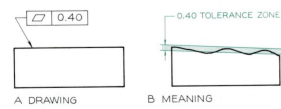

Figure 22.59 Flatness is a tolerance of form that specifies two parallel planes inside of which the object's surface must lie.

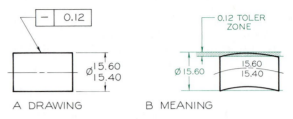

Figure 22.60 Straightness is a tolerance of form that indicates the elements of a surface are straight lines. The symbol must be applied where the elements appear as straight lines.

point on the surface may vary more than 0.40 from the highest to the lowest point on the surface.

Straightness

A surface is straight if all its elements are straight lines. A feature control symbol is used to specify straightness of a cylinder in Fig. 22.60. A total of 0.12 mm is permitted as the elements are gauged in a vertical plane parallel to the axis of the cylinder.

Roundness

A surface of revolution (a cylinder, cone, or sphere) is round when all points on the surface intersected by a plane are equidistant from the

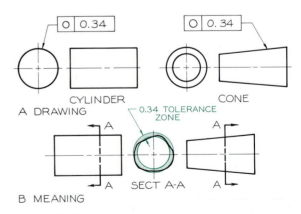

Figure 22.61 Roundness is a tolerance of form that indicates a cross section through a surface of revolution is round and lies within two concentric circles.

axis. A feature control symbol is used to specify roundness of a cone and cylinder in Fig. 22.61. This symbol permits a tolerance of 0.34 mm on the radius of each part. Figure 22.62 specifies the roundness of a sphere.

Cylindricity

A surface of revolution is cylindrical when all its elements form a cylinder. A cylindricity tolerance zone is specified in Fig. 22.63, where a tolerance of 0.54 mm is permitted on the radius of the cylinder. Cylindricity is a combination of tolerances of roundness and straightness.

22.27   Tolerances of Profile

Profile tolerancing is used to specify tolerances about a contoured shape formed by arcs or irregular curves. Profile can apply to a surface or a single line.

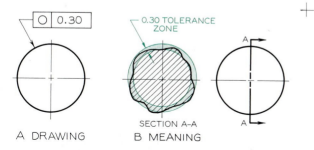

Figure 22.62 Roundness of a sphere is indicated in this manner, which means any cross section through it is round within the specified tolerance in the symbol.

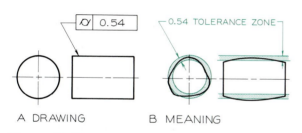

Figure 22.63 Cylindricity is a tolerance of form that indicates the surface of a cylinder lies within an envelope formed by two concentric cylinders.

The surface in Fig. 22.64 is given a unilateral profile tolerance because it can only be smaller than the points located. Figure 22.64B shows examples of specifying bilateral and unilateral tolerance zones.

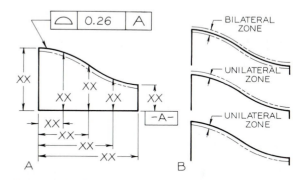

Figure 22.64 Profile is a tolerance of form used to tolerance irregular curves of planes. (A) The curving plane is located by coordinates. (B) The tolerance is located by any of these methods.

A profile tolerance for a single line can be specified as shown in Fig. 22.65, where the curve is formed by tangent arcs whose radii are given as basic dimensions. The radii are permitted to vary by plus or minus 0.10 mm about the basic radii.

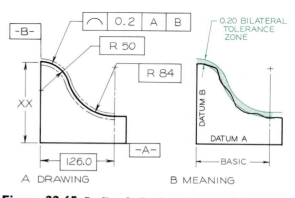

Figure 22.65 Profile of a line is a tolerance of form that specifies the variation allowed from the path of a line. Here, the line is formed by tangent arcs. The tolerance zone may be either bilateral or unilateral, as Fig. 22.64B shows.

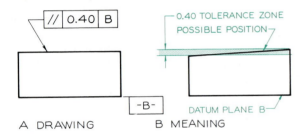

Figure 22.66 Parallelism is a tolerance of form that specifies a plane is parallel to another within specified limits. Plane *B* is the datum plane in this figure.

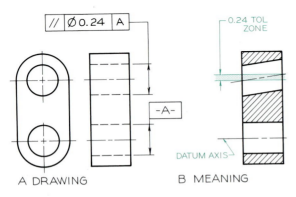

Figure 22.67 Parallelism of one centerline to another can be specified by using the diameter of one of the holes as the datum.

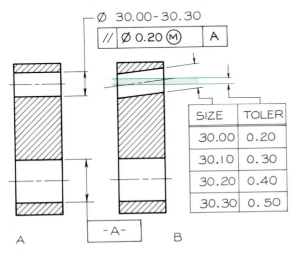

SIZE	TOLER
30.00	0.20
30.10	0.30
30.20	0.40
30.30	0.50

Figure 22.68 The most critical tolerance will exist when features are at MMC. At (A), the upper hole must be parallel to the hole used as datum *A* within 0.20 DIA. As the hole approaches its maximum size of 30.30 mm, the tolerance zone approaches 0.50 mm as shown at (B).

22.28 Tolerances of Orientation

Tolerances of orientation include **parallelism, perpendicularity,** and **angularity.**

Parallelism

A surface or line is parallel when all its points are equidistant from a datum plane or axis. Two types of parallelism are

1. A tolerance zone between planes parallel to a datum plane within which the axis or surface of the feature must lie (Fig. 22.66). This tolerance also controls flatness.
2. A cylindrical tolerance zone parallel to a datum feature within which the axis of a feature must lie (Fig. 22.67).

The effect of specifying parallelism at MMC can be seen in Fig. 22.68, where the modifier *M* is given in the feature control symbol. Tolerances of form apply at RFS when not specified. Specifying parallelism at MMC means the axis of the cylindrical hole must vary no more than 0.20 mm when the holes are the smallest permissible size.

As the hole approaches its upper limit of 30.30, the tolerance zone increases until it reaches 0.50 DIA. Therefore, a greater variation is given at MMC than at RFS.

Perpendicularity

Figure 22.69 specifies the perpendicularity of two planes. Note that datum plane *C* is flagged, and the feature control symbol is applied to the per-

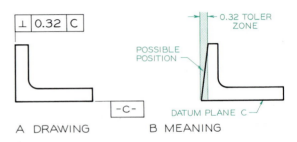

Figure 22.69 Perpendicularity is a tolerance form that gives a tolerance zone for a plane perpendicular to a specified datum plane.

pendicular surface. A hole is specified as perpendicular to a surface in Fig. 22.70, where surface A is indicated as the datum plane.

Angularity

A surface or line is angular when it is at a specified angle (other than 90°) from a datum or an axis. The angularity of a surface is specified in Fig. 22.71, where the angle is given a basic dimension of 30°. The angle is permitted to vary within a tolerance zone of 0.25 mm about the angle.

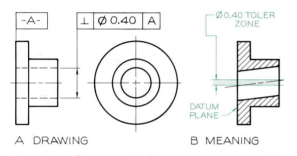

Figure 22.70 Perpendicularity can apply to the axis of a feature, such as the centerline of a cylinder.

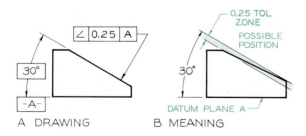

Figure 22.71 Angularity is a tolerance of form that specifies the tolerance for an angular surface with respect to a datum plane. The 30° angle is a true angle, a basic angle. The tolerance of 0.25 mm is applied to this basic angle.

22.29 Tolerances of Runout

Runout tolerance is a means of controlling the functional relationship between one or more parts to a common datum axis. The features controlled

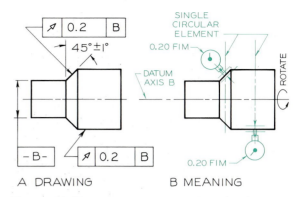

Figure 22.72 Runout tolerance of a surface, a composite of several tolerance-of-form characteristics, is used to specify concentric cylindrical parts. The part is mounted on the datum axis and is gauged as it is rotated about it.

by runout are surfaces of revolution about an axis and surfaces perpendicular to the axis.

The datum axis is established by using a functional cylindrical feature that rotates about the axis, such as diameter B in Fig. 22.72. When the part is rotated about this axis, the features of rotation must fall within the prescribed tolerance at **full indicator movement (FIM).**

The two types of runout are circular runout and total runout. One arrow in the feature control symbol indicates circular runout, and two arrows indicate total runout.

Circular Runout   (one arrow) is measured by rotating an object about its axis for 360° to determine whether a circular cross section at any point exceeds the permissible runout tolerance. This same technique is used to measure the amount of wobble existing in surfaces perpendicular to the axis of rotation.

Total Runout   (two arrows) is used to specify cumulative variations of circularity, straightness, coaxiality, angularity, taper, and profile of a surface (Fig. 22.73). Total runout is applied to all circular and profile positions as the part is rotated 360°. When applied to surfaces perpendicular to the axis, total runout controls variations in perpendicularity and flatness.

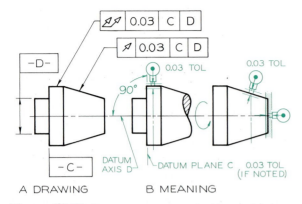

A DRAWING   B MEANING

Figure 22.73 The runout tolerance in this example is measured by mounting the object on the primary datum plane, surface *C*, and the secondary datum plane, cylinder *D*. The cylinder and conical surface is gauged to determine if it conforms to a tolerance zone of 0.03 mm. The end of the cone could have been noted to specify its runout (perpendicularity to the axis).

The part shown in Fig. 22.74 is dimensioned by using a composite of several techniques of geometric tolerancing.

22.30   Surface Texture

Because the surface texture of a part will affect its function, it must be precisely specified. Figure 22.75 illustrates most of the terms of surface texture defined below.

Surface Texture   is the variation in the surface, including roughness, waviness, lay, and flaws.

Roughness   describes the finest of the irregularities in the surface. These are usually caused by

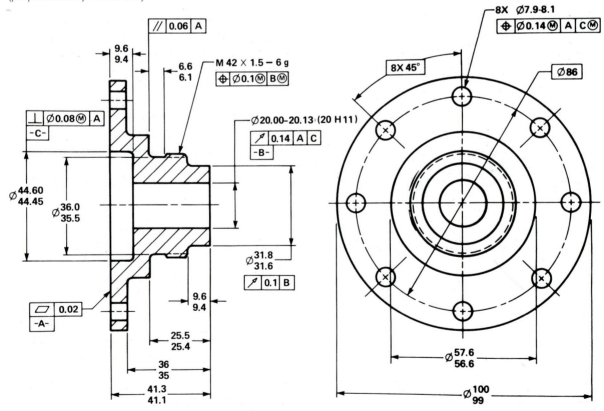

Figure 22.74 A part dimensioned with a combination of notes and symbols to describe its geometric features. (Courtesy of ANSI Y14.5M–1982.)

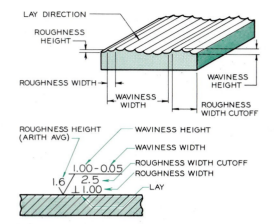

Figure 22.75 Characteristics of surface texture.

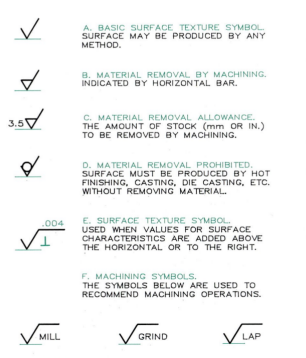

Figure 22.76 Surface control symbols for specifying surface finish.

the manufacturing process used to smooth the surface.

Roughness Height is the average deviation from the mean plane of the surface. It is measured in microinches (μin) or micrometers (μm), respectively millionths of an inch and of a meter.

Roughness Width is the width between successive peaks and valleys that forms the roughness measured in microinches or micrometers.

Roughness Width Cutoff is the largest spacing of repetitive irregularities that includes average roughness height (measured in inches or millimeters). When not specified, a value of 0.8 mm (0.030 in.) is assumed.

Waviness is a widely spaced variation that exceeds the roughness width cutoff. Roughness may be regarded as superimposed on a wavy surface. Waviness is measured in inches or millimeters.

Waviness Height is the peak-to-valley distance between waves. It is measured in inches or millimeters.

Waviness Width is the spacing between peaks or wave valleys measured in inches or millimeters.

Lay is the direction of the surface pattern and is determined by the production method used.

Flaws are irregularities or defects that occur infrequently or at widely varying intervals on a surface. These include cracks, blow holes, checks, ridges, scratches, and the like. Unless otherwise specified, the effect of flaws is not included in roughness height measurements.

Contact Area is the surface that will make contact with its mating surface.

Figure 22.76 shows the symbols used to specify surface texture. The point of the V must

1.6 — ROUGHNESS AVERAGE RATING (MAXIMUM) IN MICROINCHES OR MICROMETERS.

 1.6 / 0.8 — ROUGHNESS AVERAGE RATING (MAXIIMUM AND MINIMUM) IN MICROINCHES OR MICROMETERS

 0.005−5 — MAXIMUM WAVINESS HEIGHT (1ST NUMBER) IN MILLIMETERS OR INCHES.
MAXIMUM WAVINESS HEIGHT RATING (2ND NUMBER) IN MILLIMETERS OR INCHES.

3.5 — AMOUNT OF STOCK PROVIDED FOR MATERIAL REMOVAL IN MILLIMETERS OR INCHES.

1.6 — REMOVAL OF MATERIAL IS PROHIBITED.

0.8 ⊥ — LAY DIRECTION IS PERPENDICULAR TO THIS EDGE OF THE SURFACE.

0.8 / 2.5 — ROUGHNESS LENGTH OR CUTTOFF RATING IN mm OR INCHES BELOW THE HORIZON— TAL. WHEN NO VALUE IS SHOWN, USE 0.8mm (0.03 IN).

0.8 ⊥ 0.5 — ROUGHNESS SPACING (MAXIMUM) IN mm OR INCHES IS PLACED TO THE RIGHT OF THE LAY SYMBOL.

Figure 22.77 Values can be added to surface control symbols for more precise specifications. These may be in combinations other than those shown here.

touch the edge view of the surface, an extension line from the surface, or a leader pointing to the surface being specified.

In Fig. 22.77, values of surface texture that can be applied to surface texture symbols, individually or in combination, are given. The roughness height values are related to manufacturing processes used to finish the surface (Fig. 22.78).

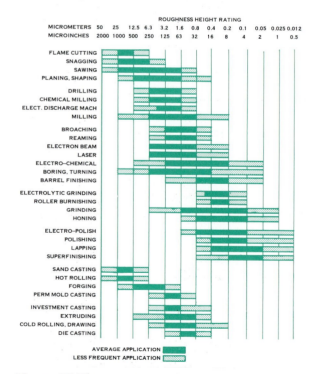

Figure 22.78 The surface roughness heights produced by various types of production methods are shown here in micrometers (microinches). (Courtesy of the General Motors Corp.)

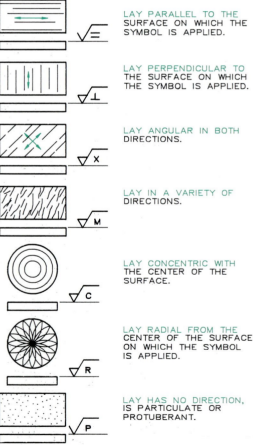

LAY PARALLEL TO THE SURFACE ON WHICH THE SYMBOL IS APPLIED.

LAY PERPENDICULAR TO THE SURFACE ON WHICH THE SYMBOL IS APPLIED.

LAY ANGULAR IN BOTH DIRECTIONS.

LAY IN A VARIETY OF DIRECTIONS.

LAY CONCENTRIC WITH THE CENTER OF THE SURFACE.

LAY RADIAL FROM THE CENTER OF THE SURFACE ON WHICH THE SYMBOL IS APPLIED.

LAY HAS NO DIRECTION, IS PARTICULATE OR PROTUBERANT.

Figure 22.79 These symbols are used to indicate the direction of lay with respect to the surface where the control symbol is placed.

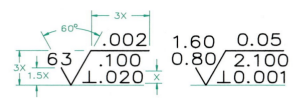

Figure 22.80 Examples of fully specified surface control symbols.

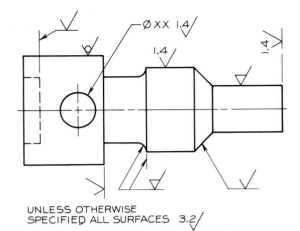

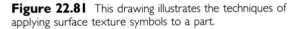

UNLESS OTHERWISE
SPECIFIED ALL SURFACES 3.2/

Lay symbols that indicate the direction of texture of a surface (Fig. 22.79) can be incorporated into surface texture symbols (Fig. 22.80). Figure 22.81 shows a part with a variety of surface texture symbols.

Figure 22.81 This drawing illustrates the techniques of applying surface texture symbols to a part.

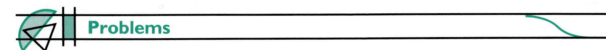

Problems

These problems can be solved on Size A sheets. The problems are laid out on a grid of 0.20 inches (5 mm).

Cylindrical Fits

1. (Fig. 22.82) Construct the drawing of a shaft and hole as shown (it need not be drawn to scale), give the limits for each diameter, and complete the table of values. Use a basic diameter of 1.00 in. (25 mm) and a class RC 1 fit or a metric fit of H8/f 7.

2. Same as Problem 1, but use a basic diameter of 1.75 in. (45 mm) and a class RC 9 fit or a metric fit of H11/c11.

3. Same as Problem 1, but use a basic diameter of 2.00 in. (51 mm) and a class RC 5 fit or a metric fit of H9/d9.

4. Same as Problem 1, but use a basic diameter of 12.00 in. (305 mm) and a class LC 11 fit or a metric fit of H7/h6.

5. Same as Problem 1, but use a basic diameter of 3.00 in. (76 mm) and a class LC 1 fit or a metric fit of H7/h6.

6. Same as Problem 1, but use a basic diameter of 8.00 in. (203 mm) and a class LC 1 fit or a metric fit of H7/k6.

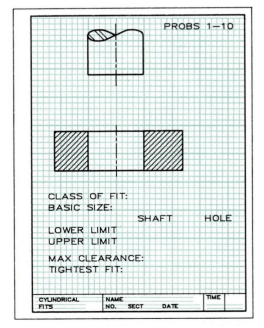

Figure 22.82 Problems 1–10.

7. Same as Problem 1, but use a basic diameter of 102 in. (2591 mm) and a class LN 3 fit or a metric fit of H7/n6.

8. Same as Problem 1, but use a basic diameter of 11.00 in. (279 mm) and a class LN 2 fit or a metric fit of H7/p6.

9. Same as Problem 1, but use a basic diameter of 6.00 in. (152 mm) and a class FN 5 fit or a metric fit of H7/s6.

10. Same as Problem 1, but use a basic diameter of 2.60 in. (66 mm) and a class FN 1 fit or a metric fit of H7/u6.

Tolerances of Position

11. (Fig. 22.83) On Size A paper, make an instrument drawing of the part shown. Locate the two holes with a size tolerance of 1.00 mm and a position tolerance of 0.50 DIA. Show the proper symbols and dimensions for this arrangement.

12. Same as Problem 11, but locate three holes using the same tolerances for size and position.

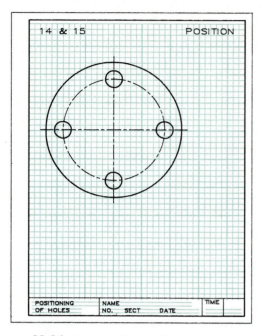

Figure 22.84 Problems 14 and 15.

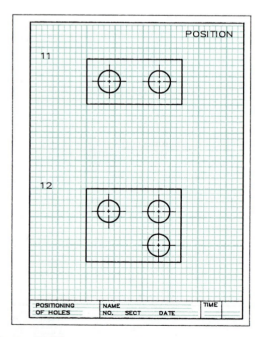

Figure 22.83 Problems 11–13.

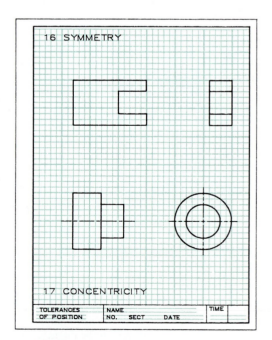

Figure 22.85 Problems 16 and 17.

13. Give the specifications for a two-pin gauge that can be used to gauge the correctness of the two holes specified in Problem 11. Make a sketch of the gauge and show the proper dimensions on it.

14. (Fig. 22.84) Using positioning tolerances, locate the holes and properly note them to provide a size tolerance of 1.50 mm and a locational tolerance of 0.60 DIA.

15. Same as Problem 14, but locate six equally spaced, equally sized holes using the same tolerances of position.

16. (Fig. 22.85) Using a feature control symbol and the necessary dimensions, indicate that the notch is symmetrical to the left-hand end of the part within 0.60 mm.

17. (Fig. 22.85) Using a feature control symbol and the necessary dimensions, indicate that the small cylinder is concentric with the large one (the datum cylinder) within a tolerance of 0.80.

18. (Fig. 22.86) Using a feature control symbol and the necessary dimensions, indicate that the elements of the cylinder are straight within a tolerance of 0.20 mm.

19. (Fig. 22.86) Using a feature control symbol and the necessary dimensions, indicate that surface *A* of the object is flat within a tolerance of 0.08 mm.

20–22. (Fig. 22.87) Using feature control symbols and the necessary dimensions, indicate that the cross sections of the cylinder, cone, and sphere are round within a tolerance of 0.40 mm.

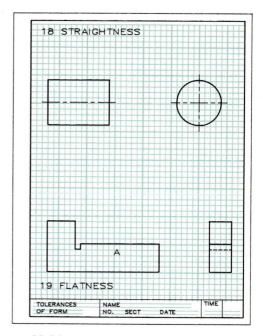

Figure 22.86 Problems 18 and 19.

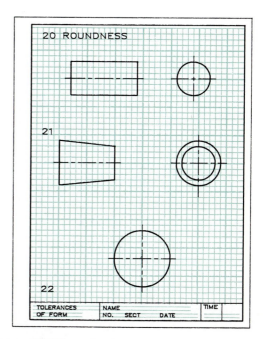

Figure 22.87 Problems 20–22.

23. (Fig. 22.88) Using a feature control symbol and the necessary dimensions, indicate that the profile of the irregular surface of the object lies within a bilateral or unilateral tolerance zone of 0.40 mm.

24. (Fig. 22.88) Using a feature control symbol and the necessary dimensions, indicate that the profile of the line formed by tangent arcs lies within a bilateral or unilateral tolerance zone of 0.40 mm.

25. (Fig. 22.89) Using a feature control symbol and the necessary dimensions, indicate that the cylindricity of the cylinder is 0.90 mm.

26. (Fig. 22.89) Using a feature control symbol and the necessary dimensions, indicate that the angularity tolerance of the inclined plane is 0.7 mm from the bottom of the object, the datum plane.

27. (Fig. 22.90) Using a feature control symbol and the necessary dimensions, indicate that surface *A* of the object is parallel to datum *B* within 0.30 mm.

28. (Fig. 22.90) Using a feature control symbol and the necessary dimensions, indicate that the small hole is parallel to the large hole, the datum, within a tolerance of 0.80 mm.

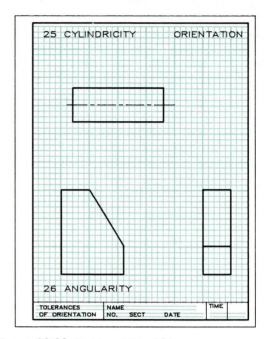

Figure 22.89 Problems 25 and 26.

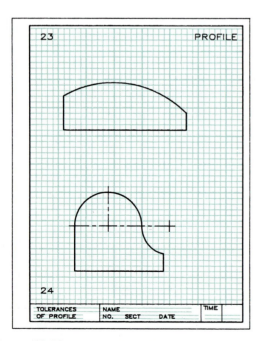

Figure 22.88 Problems 23 and 24.

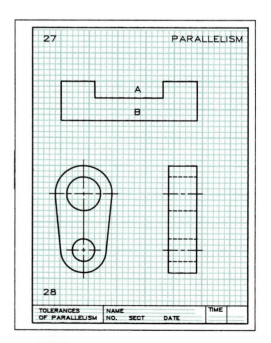

Figure 22.90 Problems 27 and 28.

29. (Fig. 22.91) Using a feature control symbol and the necessary dimensions, indicate that the vertical surface *B* is perpendicular to the bottom of the object, the datum *C*, within a tolerance of 0.20 mm.

30. (Fig. 22.91) Using a feature control symbol and the necessary dimensions, indicate that the hole is perpendicular to datum *A* within a tolerance of 0.08 mm.

31. (Fig. 22.92) Using a feature control symbol and cylinder *A* as the datum, indicate that the conical feature has a runout of 0.80 mm.

32. (Fig. 22.92) Using a feature control symbol with cylinder *B* as the primary datum and surface *C* as the secondary datum, indicate that surfaces *D, E,* and *F* have a runout of 0.60 mm.

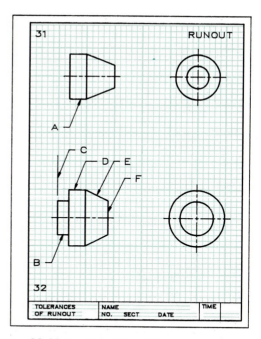

Figure 22.92 Problems 31 and 32.

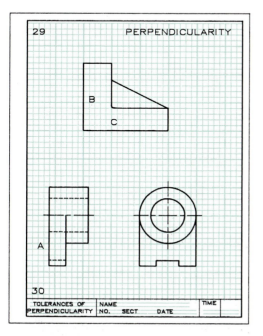

Figure 22.91 Problems 29 and 30.

23

Welding

23.1 Introduction

Welding is the process of permanently joining metal by heating a joint to a suitable temperature with or without applying pressure and with or without using filler material.

The welding practices in this chapter comply with the standards developed by the American Welding Society and the American National Standards Institute (ANSI). We also refer to the drafting standards used by General Motors Corporation.

> Advantages of welding over other methods of fastening include (1) simplified fabrication, (2) economy, (3) increased strength and rigidity, (4) ease of repair, (5) creation of gas- and liquid-tight joints, and (6) reduction in weight and size.

Figure 23.1 shows various types of welding processes, but the three main types are **gas welding, arc welding,** and **resistance welding.**

Gas Welding is a process in which gas flames are used to melt and fuse metal joints. Gases like

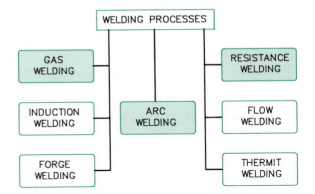

Figure 23.1 The three main types of welding processes are gas welding, arc welding, and resistance welding.

acetylene or hydrogen are mixed in a welding torch and burned with air or oxygen (Fig. 23.2). The oxyacetylene method is widely used for repair work and field construction.

Most oxyacetylene welding is done manually with a minimum of equipment. Filler material in the form of welding rods is used to deposit metal at the joint as it is heated. Most metals, except for low- and medium-carbon steels, require

GAS WELDING PRINCIPLE

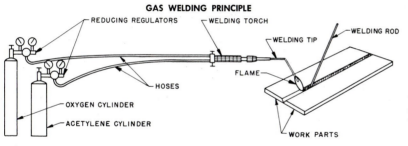

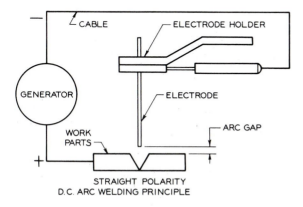

Figure 23.2 The gas welding process burns gases like oxygen and acetylene in a torch to apply heat to a joint. The welding rod supplies the filler material. (Courtesy of General Motors Corp.)

fluxes to aid the process of melting and fusing the metals.

Arc Welding is a process that uses an electric arc to heat and fuse the joints (Fig. 23.3). Pressure is sometimes required in addition to heat. The filler material is supplied by a consumable or nonconsumable electrode through which the electric arc is transmitted. Metals well-suited to arc welding are wrought iron, low- and medium-carbon steels, stainless steel, copper, brass, bronze, aluminum, and some nickel alloys.

Flash Welding is a form of arc welding, but it is similar to resistance welding in that both pressure and electric current are used to join two pieces (Fig. 23.4). The two pieces are brought together, and an electric current causes heat to build up between them. As the metal burns, the current is turned off, and the pressure between the pieces is increased to fuse them together.

Figure 23.3 In arc welding, either AC or DC current is passed through an electrode to heat the joint.

Resistance Welding is a group of processes where metals are fused both by the heat produced from the resistance of the parts to the flow of an electric current and by the pressure applied. Fluxes and filler materials are normally not used. All resistance welds are either lap- or butt-type welds.

Fig. 23.5 illustrates how resistance spot welding is performed on a lap joint. The two parts are lapped and pressed together, and an electric current fuses the parts where they join. A series of spots spaced at intervals, called **spot welds,** are used to secure the parts. Table 23.1 suggests the

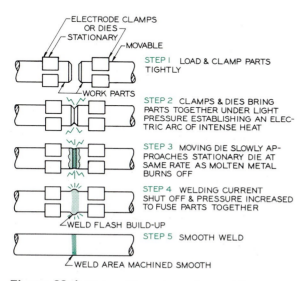

Figure 23.4 Flash welding, a type of arc welding, uses a combination of electric current and pressure to fuse two parts.

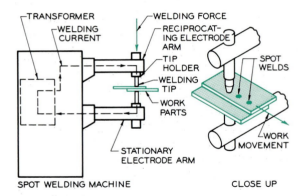

Figure 23.5 Resistance spot welding can be used to join lap and butt joints.

Table 23.1		
Recommended resistance welding processes		
Material	**Spot Welding**	**Flash Welding**
Low-carbon mild steel		
SAE 1010	R*	R
SAE 1020	R	R
Medium-carbon steel		
SAE 1030	R	R
SAE 1050	R	R
Wrought alloy steel		
SAE 4130	R	R
SAE 4340	R	R
High-alloy austenitic stainless steel		
SAE 30301–30302	R	R
SAE 30309–30316	R	R
Ferritic and martensitic stainless steel		
SAE 51410–51430	S	S
Wrought heat-resisting alloys†		
19–9–DL	S	S
16–25–6	S	S
Cast iron	NA	NR
Gray iron	NA	NR
Aluminum and aluminum alloys	R	S
Nickel and nickel alloys	R	S

*R—Recommended   NR—Not recommended
S—Satisfactory   NA—Not applicable
†For composition, see *American Society of Metals Handbook*.
Source: Courtesy of General Motors Corp.

welding processes that can be used for different materials.

23.2 Weld Joints

Figure 23.6 shows five standard weld joints. The **butt joint** can be joined with the square groove, V-groove, bevel groove, U-groove, and J-groove welds. The **corner joint** can be joined with these welds and with the fillet weld. The **lap joint** can be joined with the bevel groove, J-groove, fillet, slot, plug, spot, projection, and seam welds. The **edge joint** uses the same welds as the lap joint along with the square groove, V-groove, U-groove, and seam welds. The **tee joint** can be joined by the bevel groove, J-groove, and fillet welds.

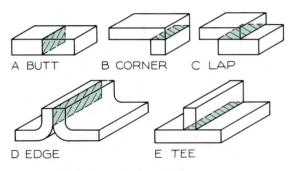

Figure 23.6 The standard weld joints.

23.3 Welding Symbols

The specification of welds on a working drawing is done by symbols. If a drawing has a general note such as ALL JOINTS WELDED or WELDED THROUGHOUT, the designer has transferred the design responsibility to the welder. Welding is too important to be left to chance; it must be thoroughly specified.

A welding symbol is used to provide specifications on a drawing (Fig. 23.7). This example gives the symbol in its complete form, which is seldom needed.

The scale of the welding symbol is shown in Fig. 23.8, where it is drawn on a $\frac{1}{8}$ in. (3 mm)

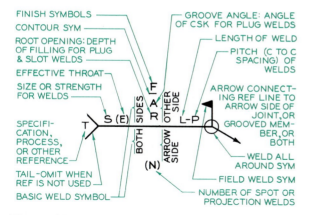

Figure 23.7 The welding symbol. Usually it is modified to a simpler form.

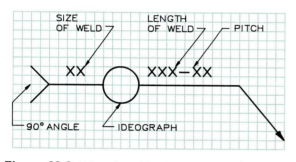

Figure 23.8 When the grid is drawn full size ($\frac{1}{8}$″ or 3 mm), the size of the welding symbol can be determined.

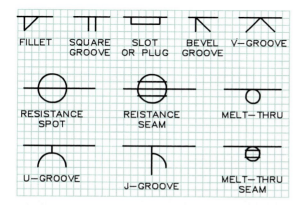

Figure 23.9 The sizes of the ideographs are shown on the $\frac{1}{8}$″ (3 mm) grid. These sizes are proportionately equal to the size of the welding symbol shown in Fig. 23.8.

grid. Its size can be scaled using these same proportions. The lettering used is the standard height of $\frac{1}{8}$ in. or 3 mm.

> The **ideograph** is the symbol that denotes the type of weld desired. Generally, the ideograph depicts the cross section of the weld.

Figure 23.9 shows the most often used ideographs. They are drawn to scale on the $\frac{1}{8}$ in. (3 mm) grid to represent their full size when added to the welding symbol.

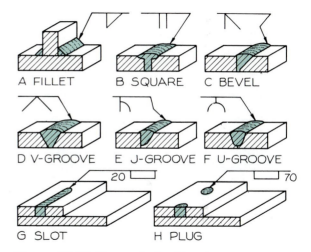

Figure 23.10 The standard welds and their corresponding ideographs.

23.4 Types of Welds

Figure 23.10 shows commonly used welds, along with their corresponding ideographs. The fillet weld is a built-up weld at the angular intersection between two surfaces. The square, bevel, V-groove, J-groove, and U-groove welds all have grooves, and the weld is placed inside these grooves. Slot and plug welds have intermittent holes or openings where the parts are welded. Holes are unnecessary when resistance welding is used.

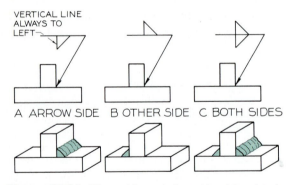

Figure 23.11 Fillet welds are indicated by abbreviated symbols. When the ideograph is below the horizontal line, it refers to the arrow side; when it is above the line, it refers to the other side.

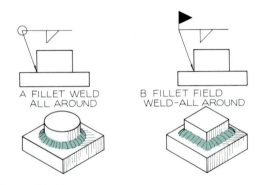

Figure 23.12 Symbols for indicating fillet welds all around two types of parts.

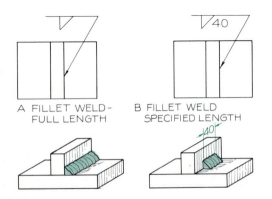

Figure 23.13 (A) Symbol for indicating full-length fillet welds. (B) Symbol for indicating fillet welds less than full length.

23.5 Application of Symbols

In Fig. 23.11A, the fillet ideograph is placed below the horizontal line of the symbol, indicating the weld is to be at the joint on the **arrow side,** the side of the arrow.

> The vertical leg of the ideograph is **always** on the left side.

Placing the ideograph above the horizontal line indicates that the weld is to be on the **other side**—that is, the joint on the other side of the part away from the arrow. When the part is to be welded on both sides, the ideograph shown in Fig. 23.11C is used. It is permissible to omit the tail and other specifications from the symbol when detailed specifications are given elsewhere.

A single arrow is often used to specify a weld that is to be all around two joining parts (Fig. 23.12); a circle, 6 mm in diameter, placed at the bend in the leader of the symbol denotes this. If the welding is to be done in the field (on the site rather than in the shop), a black circle, 3 mm in diameter, can be used to denote this joint.

A fillet weld that is to be the full length of the two parts may be specified as Fig. 23.13 shows. Since the ideograph is on the lower side of the horizontal line, the weld will be on the arrow side. A fillet weld that is to be less than full length may be specified as shown in Fig. 23.13B, where 40 represents the weld's length in millimeters.

Fillet welds of different lengths and positioned on both sides may be specified as Fig. 23.14A shows. The dimension on the lower side of the horizontal gives the length of the weld on the arrow side, and the dimension on the upper side gives the length on the other side.

Intermittent welds are welds of a given length that are spaced uniformly apart from center to center by a distance called the **pitch.** In Fig. 23.14B, the welds are on both sides, are 60 mm long, and have pitches of 124 mm; this can be indicated by the symbol shown. Intermittent welds that are to be staggered to alternate posi-

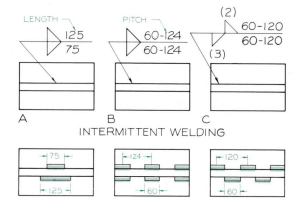

Figure 23.14 Symbols for specifying varying and intermittent welds.

23.6 Groove Welds

Figure 23.15 shows the more standard groove welds. When the depth of the grooves, angle of the chamfer, and root openings are not given on a symbol, they must be specified elsewhere on the drawings or in supporting documents. In Figs. 23.15B and 23.15E, the depth of the chamfer of the prepared joint is given in parentheses under the size dimension of the weld, which takes into account the penetration of the weld beyond this chamfer. If the size of the joint equals the depth of the prepared joint, only one number is given.

When the chamfer is different on each side of the joint, it can be noted with the symbol shown in Fig. 23.15C. If the spacing between the two parts, the **root opening,** is to be specified, this is done by placing its dimensions between the groove angle number and the weld ideograph, as Fig. 23.15E shows.

A bevel weld is a groove beveled from one of the parts being joined; therefore, the symbol must indicate which part is to be beveled (Fig. 23.15E). To call attention to this operation, the leader from the symbol is bent and aimed toward the part to be beveled. This practice also applies to J-welds where one side is grooved and the other is not (Fig. 23.16).

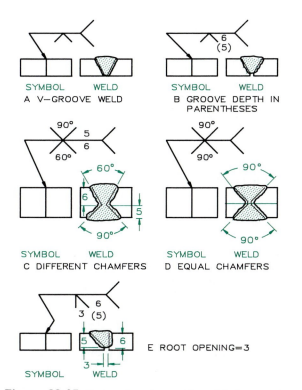

Figure 23.15 Types of groove welds and their general specifications.

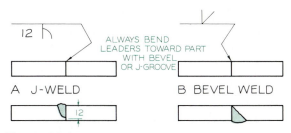

Figure 23.16 J-welds and bevel welds are specified by bent arrows pointing to the side of the joint to be grooved.

23.7   Surface Contoured Welds

Contour symbols are used to indicate which of the three types of contours, **flush, concave,** or **convex,** is desired on the surface of the weld. Flush contours are smooth with the surface or flat across the hypotenuse of a fillet weld. Concave contours bulge inward with a curve, and convex contours bulge outward with a curve (Fig. 23.17).

It is often necessary to finish the weld by a supplementary process to bring it to the desired contour. These processes, which may be indicated by their abbreviations, are chipping (C), grinding (G), hammering (H), machining (M), rolling (R), and peening (P). Figure 23.18 shows examples of these.

23.8   Seam Welds

A seam weld joins two lapping parts with either a continuous weld or a series of closely spaced spot welds. The process used for seam welds must be given by abbreviations placed in the tail of the weld symbol. The ideograph for a resistance weld is about 12 mm in diameter and is placed with the horizontal line of the symbol through its center. Figure 23.19 shows the weld's width, length, and pitch.

When the seam weld is made by arc welding, the diameter of the ideograph is about 6 mm and is placed on the upper or lower side of the symbol's horizontal line to indicate whether the seam will be applied to the arrow side or other side (Fig. 23.19B). When the length of the weld is omitted from the symbol, it is understood that the seam weld extends between abrupt changes in the seam or as it is dimensioned.

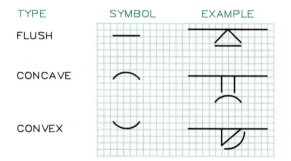

Figure 23.17 The contour symbols used to specify the surface finish of a weld.

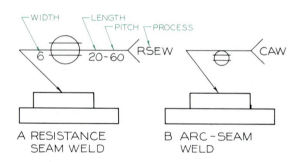

Figure 23.19 The process used for (A) resistance seam welds and (B) arc-seam welds is indicated in the tail of the symbol. The arc weld must specify arrow side or other side in the symbol.

Spot welds are similarly specified with ideographs, and specifications are given by diameter, number of welds, and pitch between the welds. The process, resistance spot welding (RSW), is noted in the tail of the symbol (Fig. 23.20A). For arc welding, the arrow side or other side must be indicated by a symbol (Fig. 23.20B). Also note the abbreviation of the welding process. Table 23.2 gives the abbreviations of various welding processes.

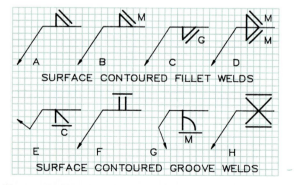

Figure 23.18 Examples of contoured symbols and letters of finishing applied to them.

		Table 23.2			
		Welding process symbols			

CAW	Carbon arc welding	FRW	Friction welding	PGW	Pressure gas welding
CW	Cold welding	FW	Flash welding	RB	Resistance brazing
DB	Dip brazing	GMAW	Gas metal arc welding	RPW	Projection welding
DFW	Diffusion welding	GTAW	Gas tungsten welding	RSEW	Resistance seam welding
EBW	Electron beam welding	IB	Induction brazing	RSW	Resistance spot welding
ESW	Electroslag welding	IRB	Infrared brazing	RW	Resistance welding
EXW	Explosion welding	OAW	Oxyacetylene welding	TB	Torch brazing
FB	Furnace brazing	OHW	Oxyhydrogen welding	UW	Upset welding
FOW	Forge welding				

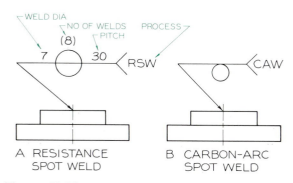

Figure 23.20 The process used for (A) resistance spot welds and (B) arc spot welds is indicated in the tail of the symbol. The arc weld must specify arrow side or other side in the symbol.

23.9 Built-Up Welds

When the surface of a part is to be enlarged, or **built-up,** by welding, this can be indicated by a symbol (Fig. 23.21). The width of the built-up

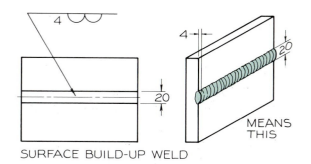

SURFACE BUILD-UP WELD

Figure 23.21 The method of applying a symbol to a built-up weld on a surface.

weld is dimensioned in the view, and the height of the weld above the surface is specified in the symbol to the left of the ideograph. The radius of the circular segment is 6 mm.

23.10 Brazing

Like welding, brazing is a method of joining pieces of metal. The process entails heating the joints above 800°F, and distributing by capillary action a nonferrous filler material, with a melting point below the base materials, between the closely fit parts.

Before brazing, the parts must be cleansed and the joints fluxed. The brazing filler is added before or just as the joints are heated beyond the filler's melting point. After the filler material has melted, it is allowed to flow between the parts to form the joint. As Fig. 23.22 shows, there are two basic brazing joints: lap joints and butt joints.

Brazing is used to hold parts together, to provide gas- and liquid-tight joints, to ensure electrical conductivity, and to aid in repair and salvage. Brazed joints will withstand more stress, higher temperature, and more vibration than soft-soldered joints.

23.11 Soft Soldering

Soldering is the process of joining two metal parts with a third metal that melts below the temperature of the metals being joined. Solders

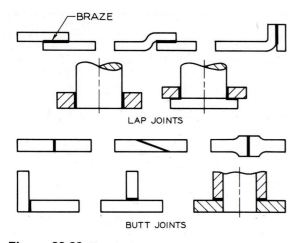

Figure 23.22 Two basic types of brazing joints: lap joints and butt joints.

are alloys of nonferrous metals that melt below 800°F. Widely used in the automotive and electrical industries, soldering is one of the basic techniques of welding and is often done by hand with a soldering iron like the one shown in Fig. 23.23. The iron is placed on the joint to heat it and to melt the solder. Figure 23.23 shows the method of indicating a soldered joint.

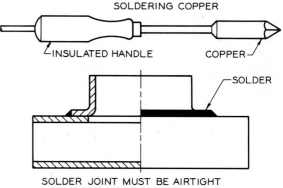

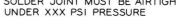

Figure 23.23 A typical hand-held soldering iron used to soft-solder two parts together and the method of indicating a soldered joint on a drawing.

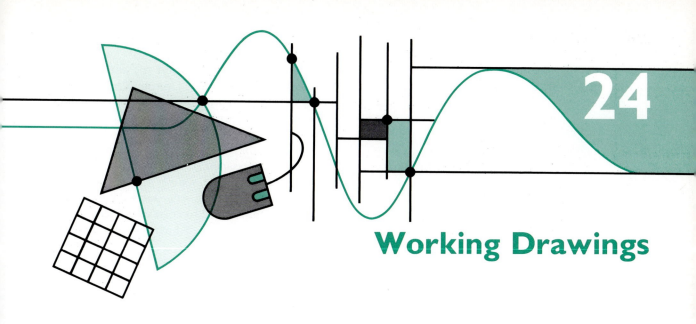

24

Working Drawings

24.1 Introduction

Working drawings are the drawings a design is constructed from. Depending on the complexity of the project, a set of working drawings may contain any number of sheets, from only one to more than one hundred. It is important to give the number of sheets in the set on each sheet, for example, sheet 2 of 6, sheet 3 of 6, and so on.

The written instructions that accompany working drawings are called **specifications.** When the project can be represented on several sheets, the specifications are often written on the drawings to consolidate the information into a single format. Much of the work in preparing working drawings is done by the drafter, but the designer, who is usually an engineer, is responsible for their correctness.

A working drawing is often called a **detail drawing** because it describes and gives the dimensions of the details of the parts being presented.

All the principles of orthographic projection and all the techniques of graphical presentation are used to communicate the details in a working drawing.

24.2 Working Drawings— Inch System

The inch is the basic unit of the English system, which is being superseded by the metric system.

The pulley (Fig. 24.1) is represented by the working drawings shown in Fig. 24.2. Dimensions and notes give the information needed to construct the pieces. This drawing is dimensioned in millimeters.

The clamp in Fig. 24.3 is represented by computer-drawn working drawings in Figs. 24.4–24.6. The clamp is dimensioned with decimal inches, which are preferable to common fractions although both systems are still widely used. Using decimal inches makes it possible to handle arithmetic with greater ease than is possible with fractions.

Inch marks are omitted from dimensions on a working drawing since it is understood that the units are in inches. Several dimensioned parts are shown on each sheet as orthographic views. The

arrangement of these parts on the sheet has no relationship to how they fit together; they are simply positioned to take advantage of the available space. Each part is given a number and name for identification. The material that each part is made of is indicated along with other notes to explain any necessary manufacturing procedures.

The orthographic assembly given on sheet 3 (Fig. 24.6) explains how the parts fit together. The parts are numbered to correspond to the part numbers in the parts list, which serves as a bill of materials.

Figure 24.1 This pulley and setscrew are detailed in the working drawing in Fig. 24.2.

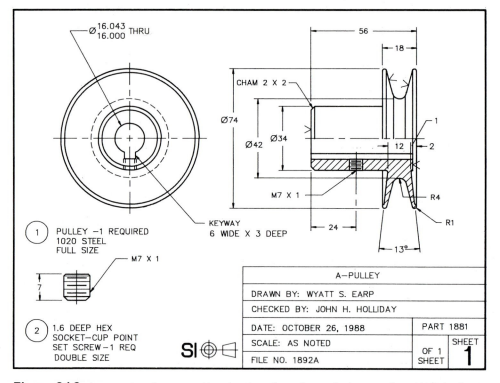

Figure 24.2 A computer-drawn working drawing of a pulley and setscrew dimensioned using the metric system.

Figure 24.3 A revolving clamp assembly manufactured to hold parts while they are being machined.

24.3 Working Drawings—Metric System

The left-end handcrank (Fig. 24.7) is detailed on two Size B sheets in Figs. 24.8 and 24.9. Since the millimeter is the basic unit of the metric system, all dimensions are measured to the nearest whole millimeter, with no decimal fractions except where tolerances are shown. Metric abbreviations after the numerals are omitted from dimensions on a working drawing since it is understood from the SI symbol near the title block that the units are metric.

If you have trouble relating to the length of a millimeter, remember that the fingernail of your index finger is about 10 mm wide.

Figure 24.9, an orthographic assembly of the left-end handcrank, shows how the parts are put

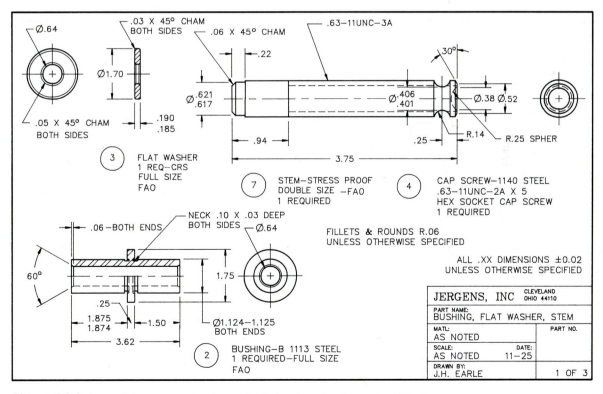

Figure 24.4 A set of three computer-drawn detail drawings showing parts of the clamp assembly. (Courtesy of Jergens, Inc.)

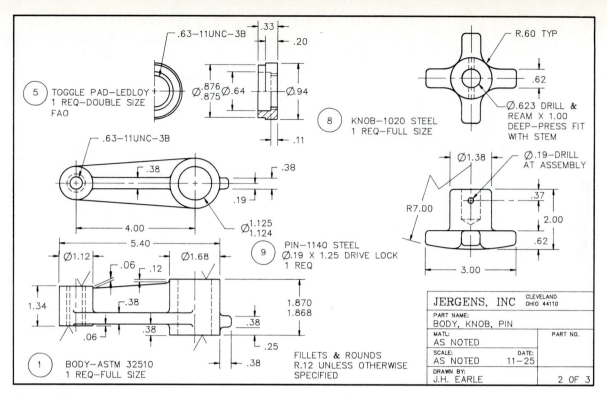

Figure 24.5 A continuation of Fig. 24.4. (Courtesy of Jergens, Inc.)

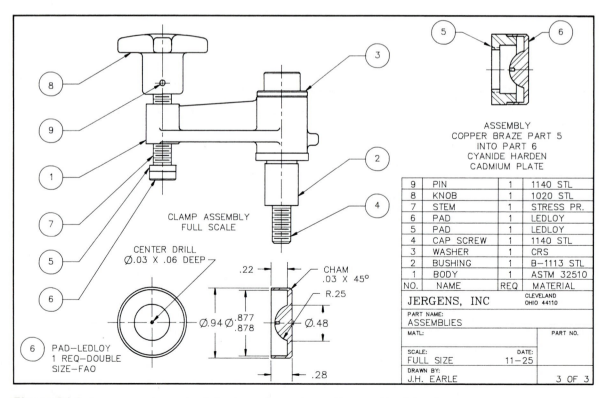

Figure 24.6 A computer-drawn detail drawing and an orthographic assembly of the clamp assembly.

Figure 24.7 The left-end handcrank detailed in the working drawings in Figs. 24.8 and 24.9.

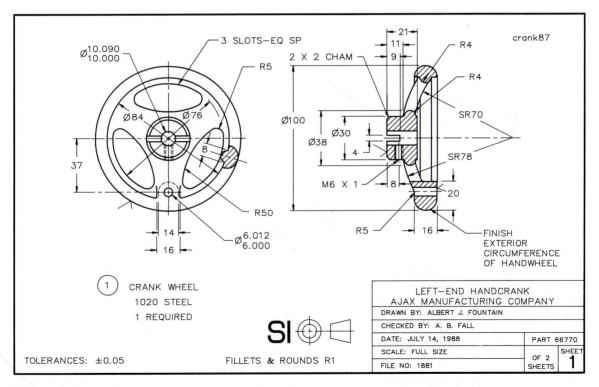

Figure 24.8 A computer-drawn working drawing of the crank wheel of the left-end handcrank. Dimensions are in millimeters.

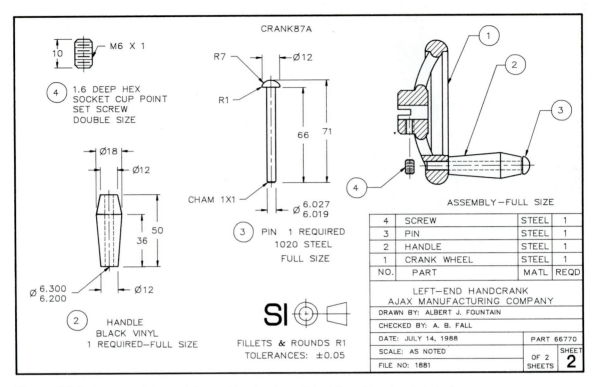

Figure 24.9 The second sheet of the working drawings of the left-end handcrank, including an assembly drawing.

together. The numbers in the balloons refer to the numbers in the parts list, which is above the title block.

Figure 24.10 shows a lifting device used to level heavy equipment such as lathes and milling machines, and Figs. 24.11 and 24.12 show working drawings of the parts of this product. The SI symbol indicates that the dimensions are in millimeters, and the truncated cone indicates that the views are drawn using third-angle projection. Figure 24.12 shows the parts in assembly.

24.4 Working Drawings— Dual Dimensions

Some working drawings are dimensioned in both inches and millimeters. Figure 24.13 shows an example, where the dimensions in parentheses or brackets are millimeters. The units may also be given in inches first and then converted to millimeters. Converting from one unit to the other will result in fractional units that must be rounded off. An explanation of the system used should be noted in the title block.

Figure 24.10 A Lev-L-Line lifting device used to level heavy machinery. This product is the basis of the working drawings in Figs. 24.11 and 24.12. (Courtesy of Unisorb Machinery Installation Systems.)

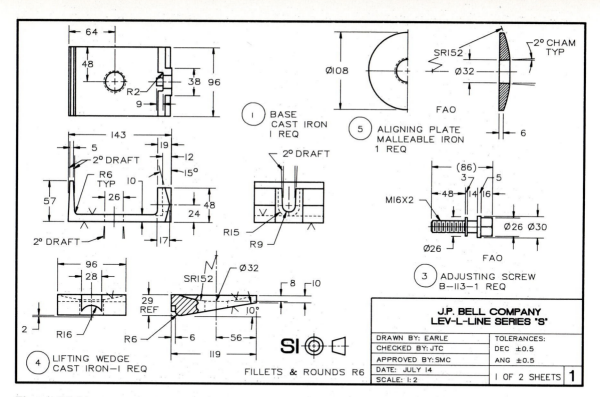

Figure 24.11 A working drawing of the lifting device dimensioned in SI units. (Courtesy of Unisorb Machinery Installation Systems.)

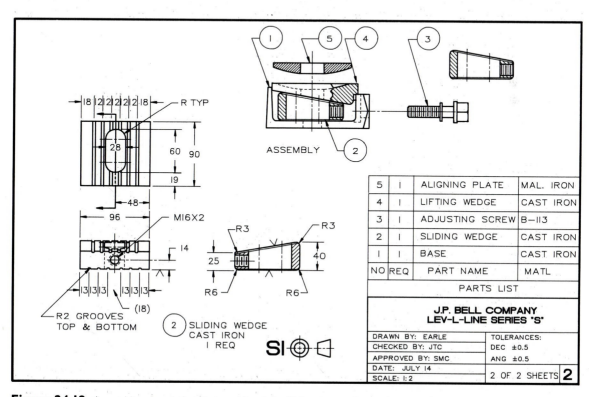

Figure 24.12 A working drawing and assembly of the lifting device dimensioned in SI units. (Courtesy of Unisorb Machinery Installation Systems.)

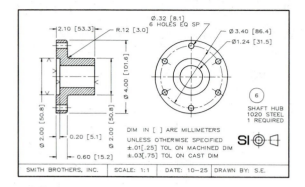

Figure 24.13 A dual-dimensioned drawing. The dimensions are given in millimeters, and their equivalents in inches are given in brackets.

24.5 Laying Out a Working Drawing

A typical working drawing is laid out by computer by beginning with the border (if a printed border is not available), as shown in Fig. 24.14. A margin of at least 0.25 in. (7 mm) should be allowed between the edge of the sheet and the borderlines. Space for the title block in the lower right-hand corner of the sheet is outlined to ensure that the drawing does not occupy this area.

A freehand sketch is made in Step 1 to determine the necessary views, the number of dimensions, and their placement so that the proper scale can be selected. In Step 2, the views are drawn close together since this makes projection from view to view easier.

In Step 3, the views are MOVEd to make room for notes and dimensions. In Step 4, the dimen-

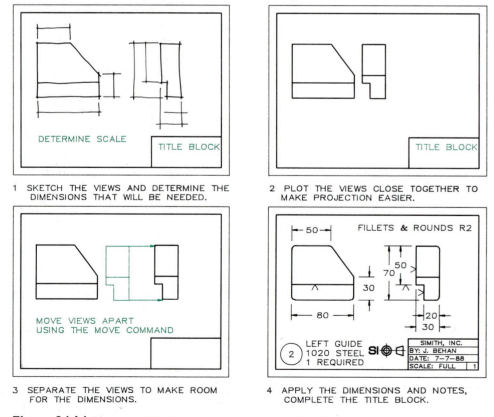

Figure 24.14 The steps of laying out a working drawing begin with a freehand sketch to determine the views and dimensions needed.

DRAWING SHEET SIZES

ENGLISH SIZES				METRIC SIZES	
A	11	X	8.5	A4	297 X 210
B	17	X	11	A3	420 X 297
C	22	X	17	A2	594 X 420
D	34	X	22	A1	841 X 594
E	44	X	34	A0	1189 X 841

Figure 24.15 The standard sheet sizes for working drawings dimensioned in inches.

sions, notes, SI symbol, and title block are added to complete the drawing.

When making a drawing by pencil on paper or film, it is more efficient to first lay out the views and dimensions on a different sheet of paper, and then overlay the drawing with vellum or film for tracing the final drawing. Guidelines for lettering must be drawn for each dimension and note.

Figure 24.15 shows the standard sheet sizes for working drawings. Papers, films, cloths, and reproduction materials are available in these modular sizes.

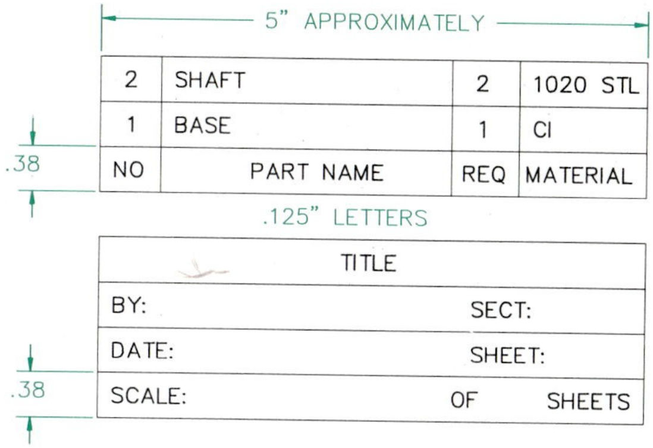

24.6   Title Blocks and Parts Lists

Figure 24.16 shows a parts list and title block suitable for student assignments. Title blocks are usually in the lower right-hand corner of the drawing sheet against the borders and usually contain the **title** or **part name, drafter, date, scale, company,** and **sheet** number. Other information such as **tolerances, checkers,** and **materials** may also be shown. The parts list should be placed directly over the title block.

Figure 24.17 shows a title block used by General Motors. A note to the left of the title block lists John F. Brown as the inventor. Two associates were asked to date and sign the drawings as witnesses of his work. This procedure establishes the ownership of the ideas and dates their development in case this becomes an issue in obtaining a patent.

Another example of a title block (Fig. 24.18) is typical of those used by various industries. Re-

2	SHAFT	2	1020 STL
1	BASE	1	CI
NO	PART NAME	REQ	MATERIAL

TITLE	
BY:	SECT:
DATE:	SHEET:
SCALE:	OF SHEETS

5" APPROXIMATELY · .38 · .125" LETTERS · .38

Figure 24.16 A typical parts list and title block suitable for most student assignments.

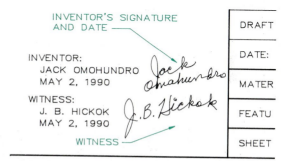

INVENTOR'S SIGNATURE AND DATE

INVENTOR:
JACK OMOHUNDRO
MAY 2, 1990

WITNESS:
J. B. HICKOK
MAY 2, 1990

WITNESS

DRAFT
DATE:
MATER
FEATU
SHEET

Figure 24.17 A General Motors title block witnessed by an associate. (Courtesy of General Motors Corp.)

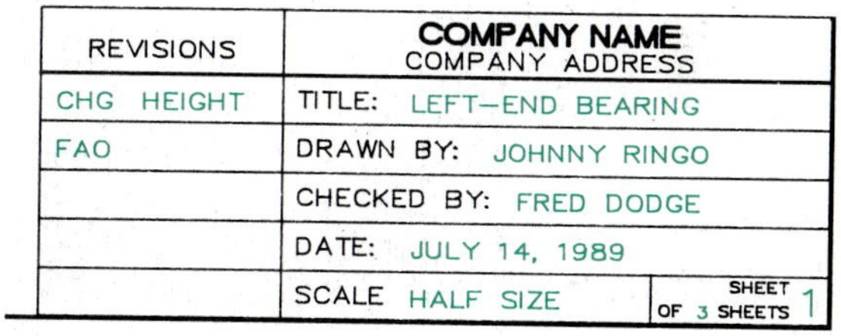

REVISIONS	COMPANY NAME COMPANY ADDRESS	
CHG HEIGHT	TITLE: LEFT—END BEARING	
FAO	DRAWN BY: JOHNNY RINGO	
	CHECKED BY: FRED DODGE	
	DATE: JULY 14, 1989	
	SCALE HALF SIZE	SHEET OF 3 SHEETS 1

Figure 24.18 An example of a title block with a revision block.

vision blocks list any modifications that will improve the design.

Figure 24.19 shows a computer shortcut for filling in a title block that will be used on several sheets. The title block can be drawn only once and filled in with dummy values to establish the positions of the text. The CHANGE command can be used to replace the dummy entries with updated values when the block is inserted or copied into a new drawing.

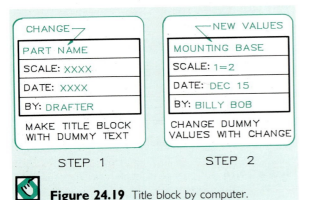

STEP 1   STEP 2

Figure 24.19 Title block by computer.

Step 1   Draw the title block and add dummy text values using the desired text style and size. Make a **BLOCK** of the title block.

Step 2   Insert the title block against the lower bottom and right borders. **EXPLODE** the **BLOCK** and, using the **CHANGE** command, change the dummy values to the desired ones.

A similar shortcut is the design of a title block in which ATTRIBUTES are used (see Section 38.74). Here, you will be prompted for the entries into the title block at the time of its INSERTion.

24.7   Scale Specification

If all drawings are the same scale, the scale of a working drawing should be indicated in or near the title block. If several drawings are made at different scales, the scales should be indicated on the drawings.

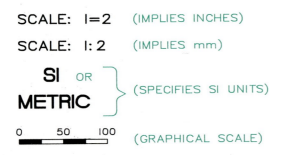

SCALE: 1=2   (IMPLIES INCHES)

SCALE: 1:2   (IMPLIES mm)

SI OR METRIC   (SPECIFIES SI UNITS)

0   50   100   (GRAPHICAL SCALE)

Figure 24.20 Methods of specifying scales and metric units on working drawings.

Figure 24.20 shows several methods of indicating scales. When the colon is used (for example, 1:2), the metric system is implied; when the equal sign is used (for example, 1 = 2), the English system is implied. The SI symbol or METRIC designation on a drawing specifies that the units of measurement are millimeters.

In some cases, a graphical scale with calibrations is given that permits the interpretation of linear units by transferring dimensions with your dividers from the drawing to the scale.

24.8   Tolerances

General notes can be given on working drawings to specify the tolerances of dimensions. Figure 24.21 shows a table of values where a check can be made to indicate whether the units will be in inches or millimeters. As the figure shows, plus-or-minus tolerances are given in the blanks, under the number of digits, and under each decimal fraction. For example, this table specifies that each dimension with two-place decimals will have a tolerance of ± 0.01 in.

Angular tolerances can also be given in general notes. (See Chapter 22 for more detailed examples of using tolerance notes.)

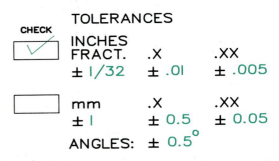

CHECK	TOLERANCES		
✓	INCHES FRACT. $\pm 1/32$	.X $\pm .01$	.XX $\pm .005$
	mm ± 1	.X ± 0.5	.XX ± 0.05
	ANGLES: $\pm 0.5°$		

Figure 24.21 General tolerance notes given on working drawings to specify the tolerances permitted on dimensions.

24.9   Part Names and Numbers

Each part should be given a name and number (Fig. 24.22). The letters and numbers should be $\frac{1}{8}$ in. (3 mm) high. The part numbers are placed

Figure 24.22 Each part of a working drawing should be named and numbered for listing in the parts list.

24.10 Checking a Drawing

The people who check drawings have special qualifications enabling them to suggest revisions and modifications that will result in a better product at less cost. The checker may be a chief drafter experienced in this type of work or the engineer or designer who originated the project. In larger companies, the drawings are reviewed by the various shops involved to determine whether the most efficient production methods are specified for each part.

Checkers never check the original drawing; instead, they note corrections with a colored pencil on a **diazo print** (a blue-line print). The print is returned to the drafter for revision of the original, and another print is made for approval.

In Fig. 24.23, the various modifications made by checkers are labeled with letters that are cir-

inside circles, called **balloons,** which are drawn approximately four times the height of the numbers.

On the working drawings, the part numbers should be placed near the parts to clarify which part they are associated with. On assembly drawings, balloons are especially important since the numbers of the parts refer to the same numbers in the parts list.

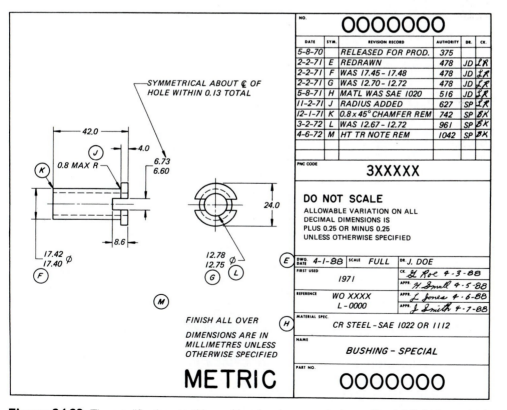

Figure 24.23 The modifications to this working drawing are noted near the details to be revised. The letters in balloons are cross-referenced in the revision table. (Courtesy of General Motors Corp.)

EVALUATION BY INSTRUCTOR   100%

Points
Earned

TITLE BLOCK (5 Points)		
Student's name	1	_____
Checker	1	_____
Date	1	_____
Scale	1	_____
Sheet number	1	_____
ORTHOGRAPHIC DETAILS (19 points)		
Proper views	10	_____
Spacing of views	5	_____
Part names and numbers	2	_____
Correct sections & conventions	2	_____
DESIGN INFORMATION (12 Points)		
Tolerances	5	_____
Finish marks	2	_____
Thread notes	5	_____
DIMENSIONING (20 Points)		
Proper arrowheads	3	_____
Spacing of dimension lines	3	_____
Adequate dimensions	4	_____
Fillets and rounds noted	2	_____
Hole notes	2	_____
Inch marks omitted	2	_____
Dimensions from best views	4	_____
DRAFTSMANSHIP (19 Points)		
Line weights	6	_____
Lettering	6	_____
Neatness	3	_____
Reproduction quality	4	_____
GENERAL NOTES (5 Points)		
SI symbol	2	_____
Third angle symbol	2	_____
General tolerance note	1	_____
ASSEMBLY (15 Points)		
Descriptive views	6	_____
Clarity	3	_____
Parts list	4	_____
Part numbers	2	_____
PRESENTATION (5 Points)		
Stapled	2	_____
Trimmed	1	_____
Folded	1	_____
Grade sheet attached	1	_____
Total	100	

Figure 24.24 A checklist for evaluating a working-drawing assignment.

cled and placed near the revisions. Changes are listed and dated in the revision record by the drafter.

Checkers check for the soundness of the design and its functional characteristics. They are also responsible for the drawing's completeness and quality, its readability and clarity, and its lettering and drafting techniques. Quality of lettering is especially important since the shop person must rely on lettered notes and dimensions.

The best way for students to check their drawings is to make a rapid scale drawing of the part from the working drawings. Figure 24.24 shows a grading scale for checking the working drawings prepared by students. This list can be used as an outline for reviewing working drawings to ensure that the major requirements have been met.

24.11 Drafter's Log

Drafters should keep a **log,** a record that shows all changes made during the project. As the project progresses, the changes, dates, and people involved should be recorded for reference and later review. Calculations are often made during a drawing's preparation. If they are lost or poorly done, it may be necessary to do them again; therefore, they should be made a permanent part of the log.

24.12 Assembly Drawings

After the parts have been made according to the specifications of the working drawings, they will be assembled (Fig. 24.25) by following the directions of an **assembly drawing.**

Figure 24.25 An assembly drawing is used to explain how the parts of a product such as this Ford tractor are assembled. (Courtesy of Ford Motor Co.)

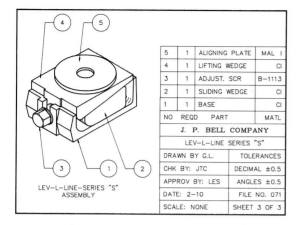

5	1	ALIGNING PLATE	MAL I
4	1	LIFTING WEDGE	CI
3	1	ADJUST. SCR	B–1113
2	1	SLIDING WEDGE	CI
1	1	BASE	CI
NO	REQD	PART	MATL

J. P. BELL COMPANY

LEV–L–LINE SERIES "S"

DRAWN BY G.L.	TOLERANCES
CHK BY: JTC	DECIMAL ±0.5
APPROV BY: LES	ANGLES ±0.5
DATE: 2–10	FILE NO. 071
SCALE: NONE	SHEET 3 OF 3

LEV–L–LINE–SERIES "S"
ASSEMBLY

Figure 24.26 An isometric assembly drawing that shows the parts of the lifting device fully assembled. Dimensions are usually omitted from an assembly drawing, and a parts list is given.

Two general types of assembly drawings are **orthographic assemblies** and **pictorial assemblies**. Dimensions are usually omitted from assembly drawings.

The lifting device in Fig. 24.10 is shown as an isometric assembly in Fig. 24.26. Each part is

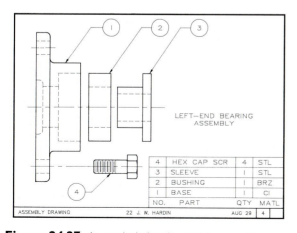

4	HEX CAP SCR	4	STL
3	SLEEVE	I	STL
2	BUSHING	I	BRZ
I	BASE	I	CI
NO.	PART	QTY	MATL

ASSEMBLY DRAWING   22 J. W. HARDIN   AUG 29   4

LEFT–END BEARING
ASSEMBLY

Figure 24.27 An exploded orthographic assembly illustrating how parts are put together.

numbered with a balloon and leader to cross-reference them to the parts list, where more information about each part is given.

Figure 24.27 shows an **orthographic exploded assembly.** In many applications, the assembly of parts can be more easily understood when the parts are shown "exploded" along their centerlines. In this example, the views are shown as regular views with some lines shown as hidden lines.

The same assembly of the part is shown in Fig. 24.28 as an orthographic assembly where the parts are in their assembled positions. The views have been sectioned to make them more easily understandable.

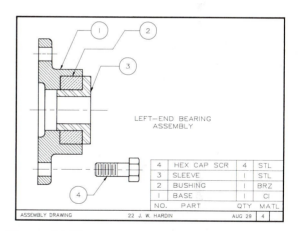

LEFT–END BEARING
ASSEMBLY

4	HEX CAP SCR	4	STL
3	SLEEVE	I	STL
2	BUSHING	I	BRZ
I	BASE	I	CI
NO.	PART	QTY	MATL

ASSEMBLY DRAWING   22 J. W. HARDIN   AUG 29   4

Figure 24.28 A sectioned orthographic assembly that shows the parts in their assembled positions except for the exploded bolt.

The **outline assembly** in Fig. 24.29 shows how various components are connected. Each part is composed of subassemblies not shown in detail. Only general dimensions are given that might be of value in connecting the pump into its overall system.

The brake-pedal assembly in Fig. 24.30 is an **exploded pictorial assembly** that shows how the parts fit together. Special balloons are used to give a variety of information in a standard manner.

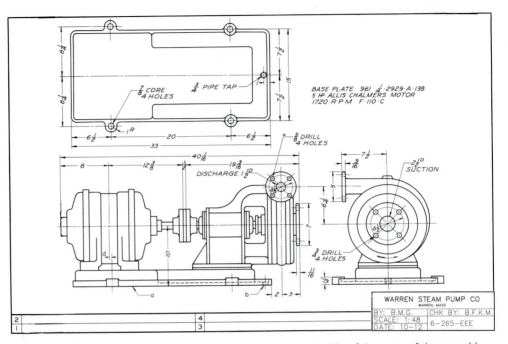

Figure 24.29 An outline assembly showing the general relationship of the parts of the assembly and their overall dimensions.

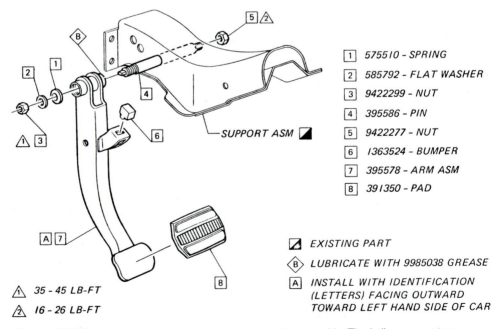

1	575510 – SPRING
2	585792 – FLAT WASHER
3	9422299 – NUT
4	395586 – PIN
5	9422277 – NUT
6	1363524 – BUMPER
7	395578 – ARM ASM
8	391350 – PAD

▨ EXISTING PART

Ⓑ LUBRICATE WITH 9985038 GREASE

Ⓐ INSTALL WITH IDENTIFICATION (LETTERS) FACING OUTWARD TOWARD LEFT HAND SIDE OF CAR

△1 35 - 45 LB-FT

△2 16 - 26 LB-FT

Figure 24.30 An exploded pictorial assembly of a pedal assembly. The balloons are cross-referenced to a legend that gives a variety of information. (Courtesy of General Motors Corp.)

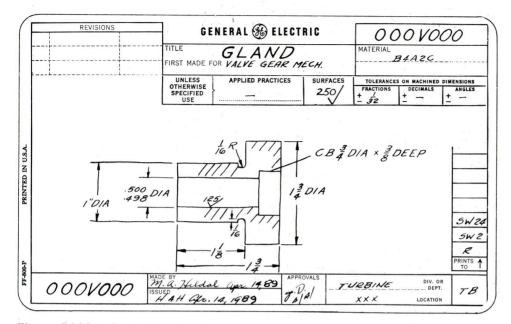

REVISIONS		GENERAL ⊕ ELECTRIC	O O O V O O O
		TITLE GLAND	MATERIAL B4A2C
		FIRST MADE FOR VALVE GEAR MECH.	

UNLESS OTHERWISE SPECIFIED USE	APPLIED PRACTICES —	SURFACES 250√	TOLERANCES ON MACHINED DIMENSIONS		
			FRACTIONS + $\frac{1}{32}$ −	DECIMALS + −	ANGLES + −

PRINTED IN U.S.A.

$\frac{1}{16}R$

$CB \frac{3}{4} DIA \times \frac{3}{8} DEEP$

$\frac{.500}{.498} DIA$

1" DIA

.125

$1\frac{3}{4} DIA$

$\frac{1}{16}$

$1\frac{1}{8}$

$1\frac{3}{4}$

SW24

SW2

R

PRINTS TO ↑

FF-800-P

O O O V O O O

MADE BY M. A. Hildal apr. 14, 89
ISSUED H A H Apr. 14, 1989

APPROVALS

TURBINE

XXX

DIV. OR DEPT. TB

LOCATION

Figure 24.31 A freehand working drawing with the essential dimensions can be as adequate as an instrument-drawn detail drawing. (Courtesy of General Electric Co.)

24.13 Freehand Working Drawings

A freehand sketch can serve the same purpose as an instrument drawing, provided the part is simple and the essential dimensions are given (Fig. 24.31). The same principles of working-drawing construction should be followed when instruments are used.

24.14 Castings and Forged Parts

The two parts shown in Fig. 24.32 illustrate the difference between a forged part and a machined part. A **forging** is a rough, oversize form made by hammering the metal into shape or pressing it between two forms (called **dies**). The forging is then machined to its finished dimensions and tolerances.

A **casting,** like a forging, is a general shape that must be machined so that it will fit with other parts in an assembly. The casting is formed by pouring molten metal into a mold formed by

Figure 24.32 The top part is a "blank" that has been forged. When the forging has been machined, it will look like the bottom part.

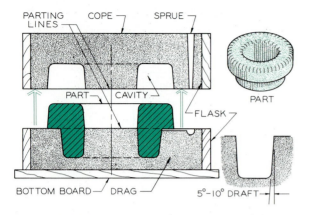

Figure 24.33 A two-part sand mold is used to produce a casting. A draft of from 5° to 10° is needed to permit withdrawal of the pattern from the sand. The casting must be machined to size it within specified tolerances.

a pattern made slightly larger than the finished part (Fig. 24.33). For the pattern to be removable from the sand that forms the mold, its sides must be tapered. This taper of from 5° to 10° is called the **draft** (Fig. 24.33).

In some industries, casting and forging drawings are made separate from machine drawings (Fig. 24.34). But more often, these drawings are combined into one drawing with the understanding that the forgings and castings must be made oversize to allow for the removal of material by the machining operations.

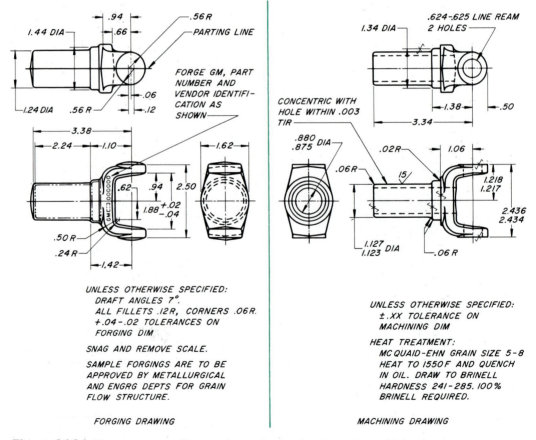

Figure 24.34 Two separate working drawings, a forging drawing and a machining drawing, are used to give the details of the same part. Often, this information is combined into a single drawing. (Courtesy of General Motors Corp.)

24.15 Sheet Metal Drawings

Parts made of sheet metal are formed by bending or stamping. The flat metal patterns for these parts must be developed graphically. Figure 24.35 shows an example of a sheet metal part, where the angles and radii of the bends are given. The note B.D. means to bend downward, and B.U. means to bend upward.

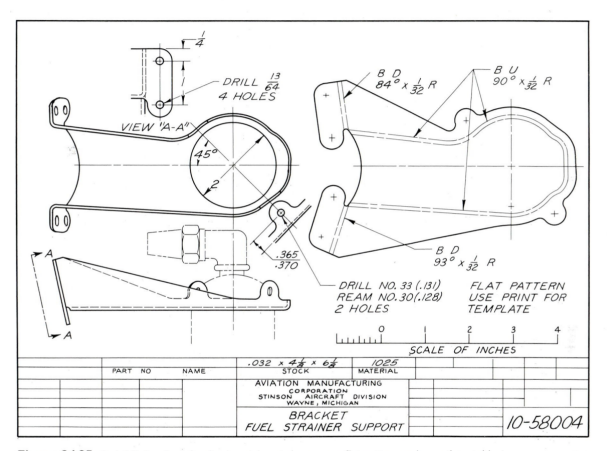

Figure 24.35 A detail drawing of a sheet metal part shown as a flat pattern and as orthographic views when bent into shape.

 **Problems**

The following problems (Figs. 24.36–24.84) are to be drawn on the sheet sizes assigned or those suggested. Each problem should be drawn with the appropriate dimensions and notes to fully describe the parts and assemblies being drawn.

Working drawings may be made on film or tracing vellum in ink or pencil. Select a suitable title block, and complete it using good lettering practices. Some problems will require more than one sheet to show all the parts properly.

Assemblies should be prepared with a parts list where there are several parts. These may be orthographic or pictorial assemblies, either exploded, partially exploded, or assembled.

The dimensions given in the problems do not always represent good dimensioning practices because of space limitations; however, the dimensions given are usually adequate for you to complete the detail drawings. In some cases, there may be omitted dimensions that you must approximate using your own judgment. When making the detail drawings, strive to provide all the necessary information, notes, and dimensions to describe the views completely. Use any of the previously covered principles, conventions, and techniques to present the views with the maximum clarity and simplicity.

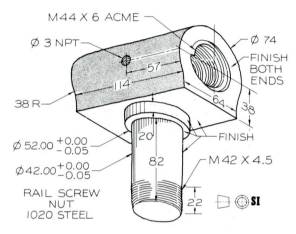

Figure 24.36 Make a detail drawing on a Size B sheet.

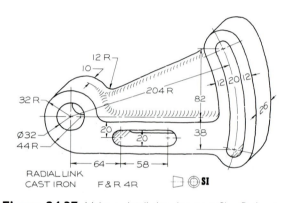

Figure 24.37 Make a detail drawing on a Size B sheet.

Figure 24.38 Make a detail drawing on a Size C sheet.

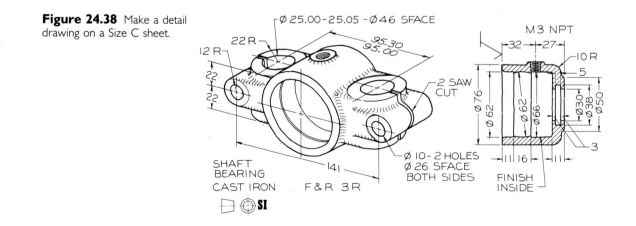

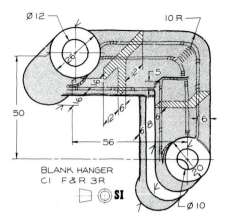

Figure 24.39 Make a detail drawing on a Size B sheet.

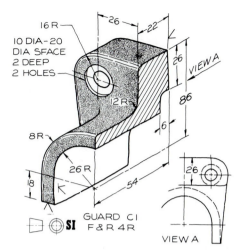

Figure 24.41 Make a detail drawing on a Size B sheet.

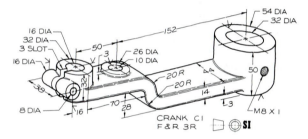

Figure 24.42 Make a detail drawing on a Size B sheet.

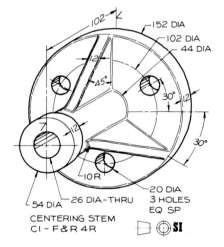

Figure 24.40 Make a detail drawing on a Size B sheet.

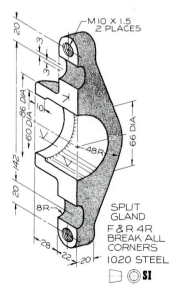

Figure 24.43 Make a detail drawing on a Size B sheet.

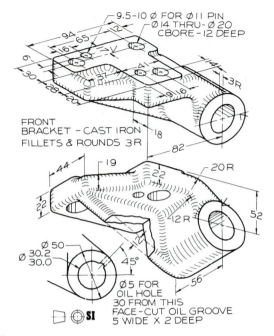

9.5–10 Ø FOR Ø11 PIN
Ø 14 THRU– Ø 20
CBORE – 12 DEEP

94
65
16
6
30
28
20
37
41
14
3R

FRONT
BRACKET – CAST IRON
FILLETS & ROUNDS 3R
18
82

44
19
22
20 R
1
22
12 R
52

Ø 50
Ø 30.2
30.0
45°
56
Ø 5 FOR
OIL HOLE
30 FROM THIS
FACE–CUT OIL GROOVE
5 WIDE X 2 DEEP

Figure 24.44 Make a detail drawing on a Size B sheet.

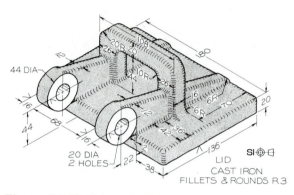

44 DIA
108
20R 20
26
44
10R
12
26
180
16
6R
70
6R
20
6
44
68
16
42
36
2
20 DIA
2 HOLES
22
136
38

LID
CAST IRON
FILLETS & ROUNDS R.3

Figure 24.46 Make a detail drawing on a Size B sheet.

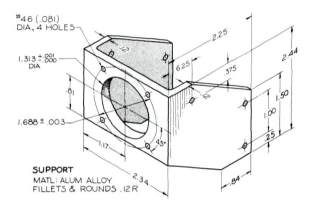

#46 (.081)
DIA, 4 HOLES
2.25
12
1.313 +.001 −.000
DIA
6.25
.375
2.44
.81
1.50
1.00
.25
1.688 ± .003
1.17
45°
.84

SUPPORT
MATL: ALUM ALLOY
FILLETS & ROUNDS .12R
2.34

Figure 24.47 Make a detail drawing on a Size B sheet using (A) decimal inches or (B) inches converted to millimeters.

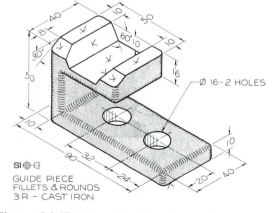

40
8
10
50
30°
60°10
10
16
50
Ø 16 – 2 HOLES
10
10
90
32
24
20
40

GUIDE PIECE
FILLETS & ROUNDS
3R – CAST IRON

Figure 24.45 Make a detail drawing on a Size B sheet.

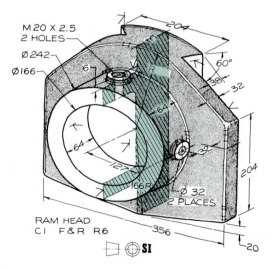

M 20 X 2.5
2 HOLES
Ø 242
Ø 166
6
204
60°
38
32
64
64
32
127
166R
Ø 32
2 PLACES
204

RAM HEAD
CI F & R R6
356
20

Figure 24.48 Make a detail drawing on a Size B sheet.

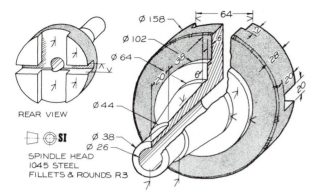

REAR VIEW

⬦ ⊚ **SI**

SPINDLE HEAD
1045 STEEL
FILLETS & ROUNDS R3

Figure 24.49 Make a detail drawing on a Size B sheet.

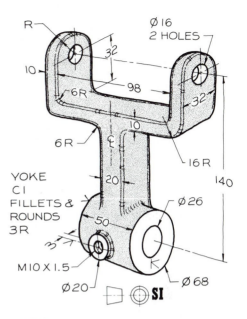

Figure 24.51 Make a detail drawing on a Size B sheet.

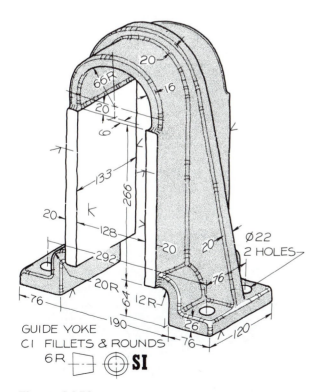

GUIDE YOKE
CI FILLETS & ROUNDS
6R ⬦ ⊚ **SI**

Figure 24.50 Make a detail drawing on a Size B sheet.

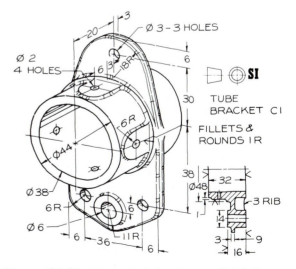

Figure 24.52 Make a detail drawing on a Size B sheet.

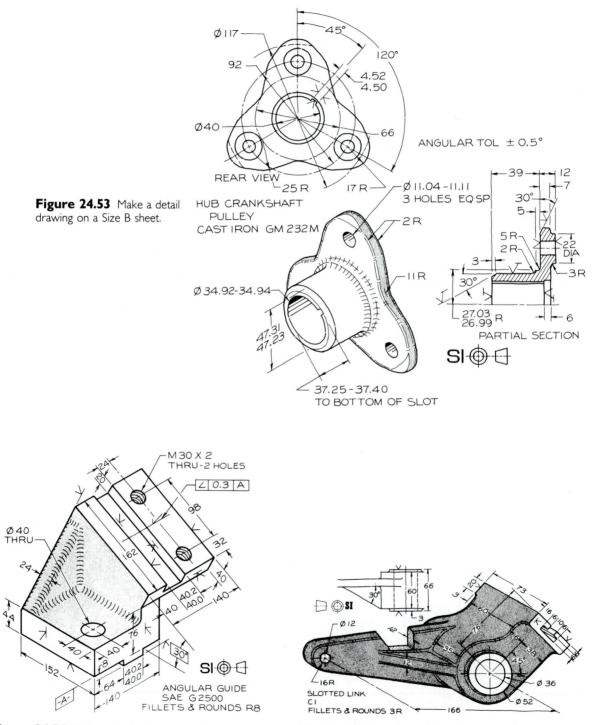

Figure 24.53 Make a detail drawing on a Size B sheet.

REAR VIEW

HUB CRANKSHAFT
PULLEY
CAST IRON GM 232M

ANGULAR TOL ± 0.5°

PARTIAL SECTION

TO BOTTOM OF SLOT

M 30 X 2
THRU-2 HOLES

ANGULAR GUIDE
SAE G 2500
FILLETS & ROUNDS R8

Figure 24.54 Make a detail drawing on a Size B sheet.

SLOTTED LINK
CI
FILLETS & ROUNDS 3R

Figure 24.55 Make a detail drawing on a Size B sheet.

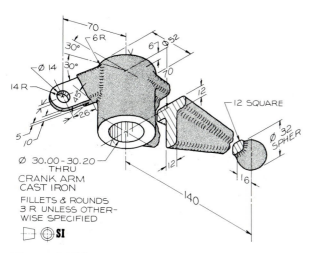

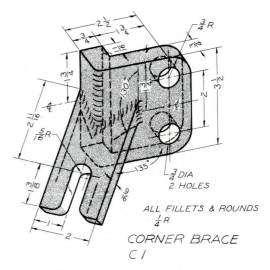

Figure 24.56 Make a detail drawing on a Size B sheet.

Figure 24.58 Make a detail drawing on a Size B sheet. Convert the fractional inches to (A) decimal inches or (B) millimeters.

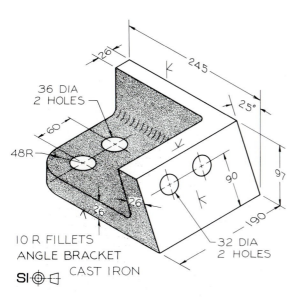

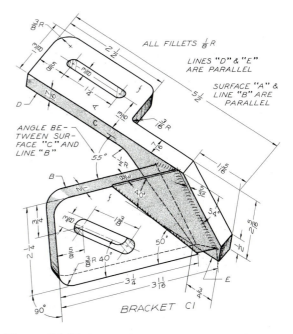

Figure 24.57 Make a detail drawing on a Size A sheet.

Figure 24.59 Make a detail drawing on a Size C sheet. Convert the fractional inches to (A) decimal inches or (B) millimeters.

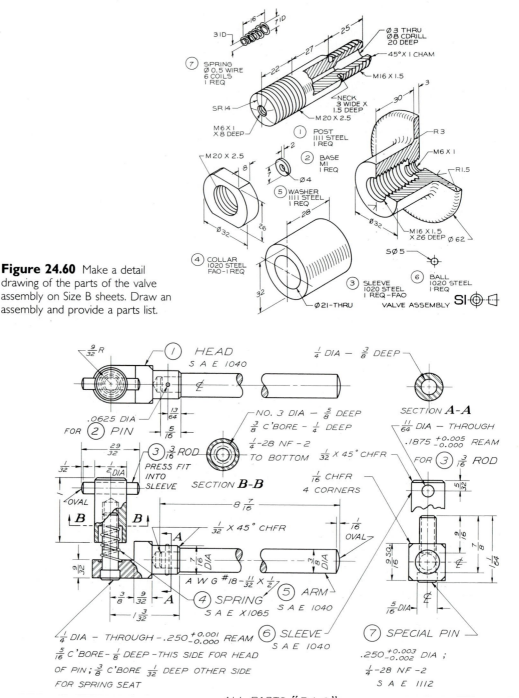

Figure 24.60 Make a detail drawing of the parts of the valve assembly on Size B sheets. Draw an assembly and provide a parts list.

Figure 24.61 Make a detail drawing of the parts of the cut-off crank on Size B sheets. Convert the fractional inches to (A) decimal inches or (B) millimeters. Draw an assembly and provide a parts list.

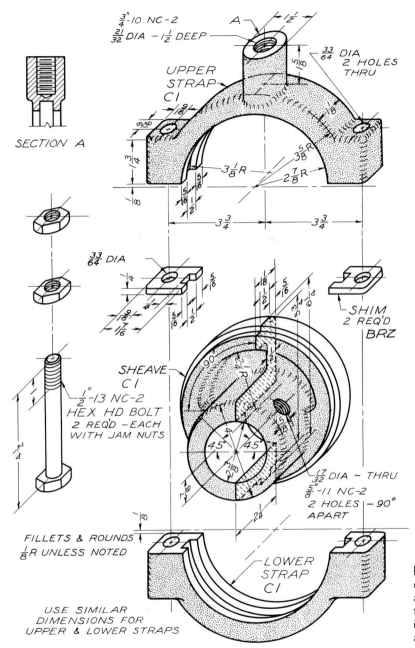

$\frac{3}{4}"$-10 NC-2
$\frac{21}{32}$ DIA - $1\frac{1}{2}$ DEEP

A

$1\frac{1}{2}$

$\frac{33}{64}$ DIA
2 HOLES
THRU

$\frac{5}{8}$

UPPER
STRAP-
CI

$\frac{9}{16}$

$\frac{9}{16}$

$\frac{1}{8}$

$1\frac{3}{4}$

$3\frac{1}{8}$R

$3\frac{3}{8}$R

$2\frac{7}{8}$R

$\frac{1}{8}$

$\frac{5}{16}$

$1\frac{5}{16}$

$1\frac{1}{2}$

$3\frac{3}{4}$

$3\frac{3}{4}$

SECTION A

$\frac{33}{64}$ DIA

$\frac{1}{4}$

$\frac{5}{16}$

$\frac{9}{16}$

$\frac{7}{16}$

$\frac{5}{16}$

$\frac{1}{2}$

$\frac{1}{8}$

$\frac{5}{16}$

$1\frac{1}{2}$

$5\frac{3}{4}$

6

SHIM
2 REQ'D
BRZ

SHEAVE
CI

$\frac{1}{2}"$-13 NC-2
HEX HD BOLT
2 REQ'D - EACH
WITH JAM NUTS

90°

$\frac{1}{4}$R

$1\frac{5}{16}$

$45°$ $45°$

$3\frac{1}{8}$

$7\frac{1}{8}$

$2\frac{1}{4}$

$\frac{17}{32}$ DIA - THRU
$\frac{5}{8}"$-11 NC-2
2 HOLES - 90°
APART

$\frac{1}{8}$

FILLETS & ROUNDS
$\frac{1}{8}$R UNLESS NOTED

LOWER
STRAP
CI

USE SIMILAR
DIMENSIONS FOR
UPPER & LOWER STRAPS

Figure 24.62 Make a detail drawing of the parts of the journal assembly on Size B sheets. Convert the fractional inches to (A) decimal inches or (B) millimeters. Draw an assembly and provide a parts list.

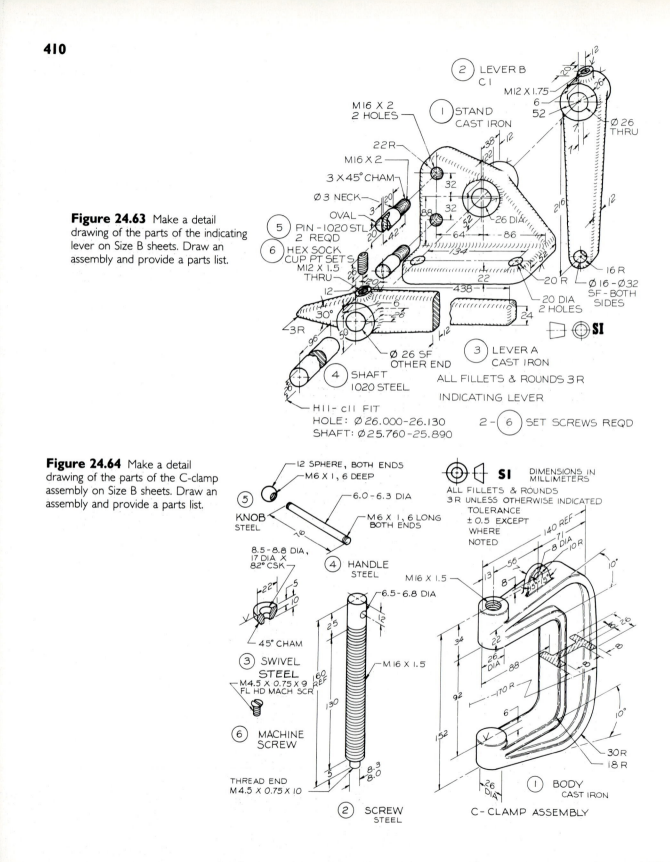

Figure 24.63 Make a detail drawing of the parts of the indicating lever on Size B sheets. Draw an assembly and provide a parts list.

Figure 24.64 Make a detail drawing of the parts of the C-clamp assembly on Size B sheets. Draw an assembly and provide a parts list.

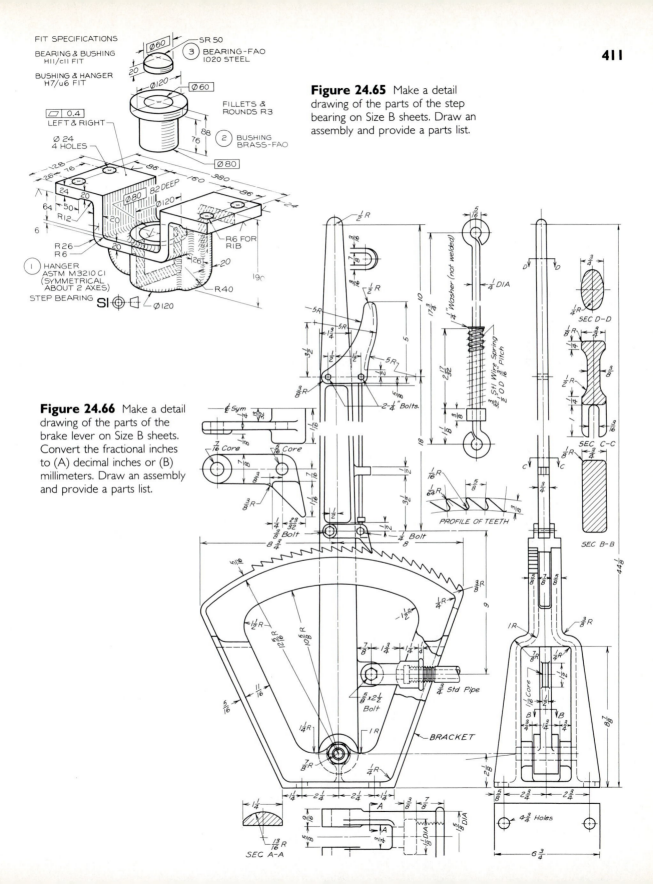

Figure 24.65 Make a detail drawing of the parts of the step bearing on Size B sheets. Draw an assembly and provide a parts list.

Figure 24.66 Make a detail drawing of the parts of the brake lever on Size B sheets. Convert the fractional inches to (A) decimal inches or (B) millimeters. Draw an assembly and provide a parts list.

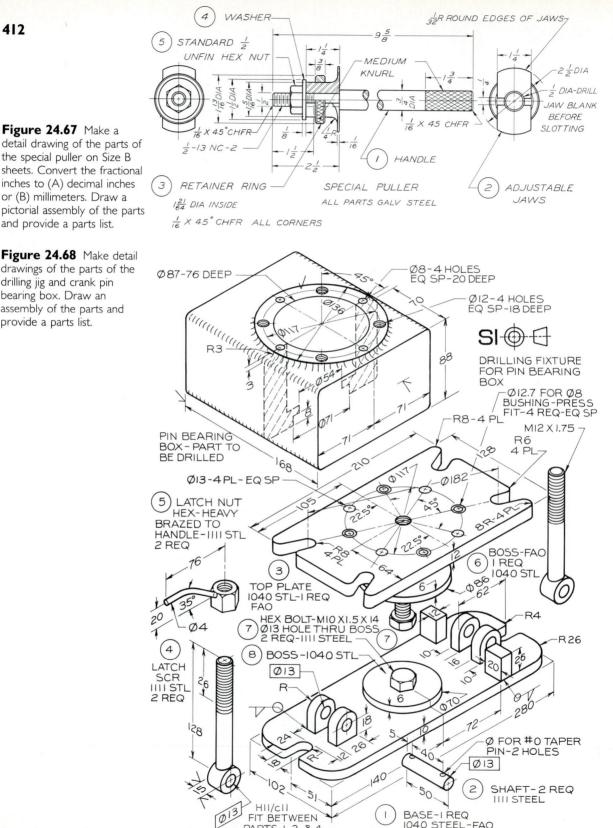

Figure 24.67 Make a detail drawing of the parts of the special puller on Size B sheets. Convert the fractional inches to (A) decimal inches or (B) millimeters. Draw a pictorial assembly of the parts and provide a parts list.

Figure 24.68 Make detail drawings of the parts of the drilling jig and crank pin bearing box. Draw an assembly of the parts and provide a parts list.

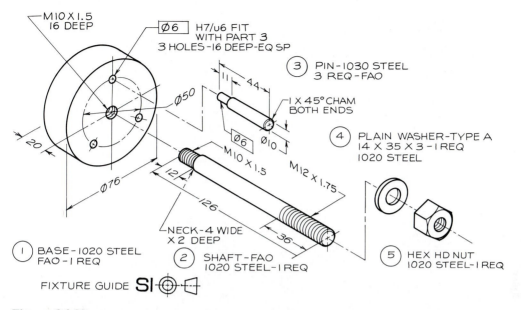

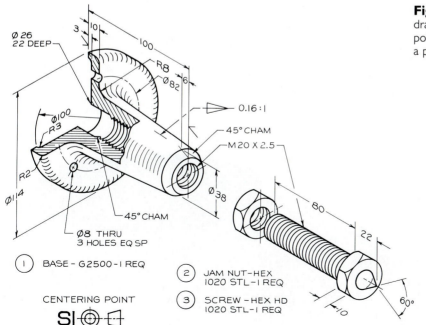

Figure 24.69 Make detail drawings of the parts of the fixture guide. Draw an assembly and provide a parts list.

Figure 24.70 Make a detail drawing of the parts of the centering point. Draw an assembly and provide a parts list.

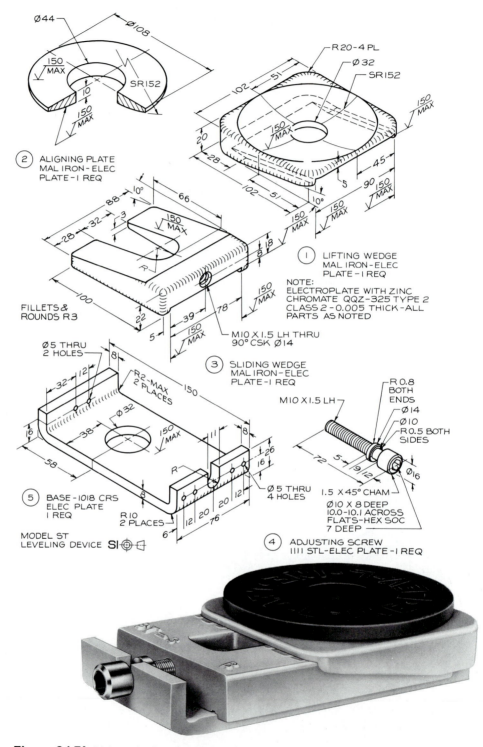

Ø44
Ø108
150 MAX
10
150 MAX
SR152

② ALIGNING PLATE
MAL IRON – ELEC
PLATE – I REQ

R 20 – 4 PL
Ø 32
SR152
102
51
20
28
102
51
10°
45
5
90
150 MAX
150 MAX
150 MAX
150 MAX
150 MAX

10°
88
66
150 MAX
32
28
3
8
18
8
R
100
22
5
39
78
150 MAX
150 MAX

FILLETS &
ROUNDS R3

① LIFTING WEDGE
MAL IRON – ELEC
PLATE – I REQ

NOTE:
ELECTROPLATE WITH ZINC
CHROMATE QQZ–325 TYPE 2
CLASS 2 – 0.005 THICK – ALL
PARTS AS NOTED

M10 X 1.5 LH THRU
90° CSK Ø14

③ SLIDING WEDGE
MAL IRON – ELEC
PLATE – I REQ

Ø 5 THRU
2 HOLES
8
32
12
R2 – MAX
2 PLACES
150
150 MAX
6
Ø 32
38
11
8
58
26
16
R
8
R 10
2 PLACES
12
20
20
12
6
76
Ø 5 THRU
4 HOLES

⑤ BASE – 1018 CRS
ELEC PLATE
I REQ

MODEL ST
LEVELING DEVICE SI ⊕ ⊟

M10 X1.5 LH
R 0.8
BOTH
ENDS
Ø14
Ø10
R 0.5 BOTH
SIDES
72
5
9
2
Ø16
1.5 X 45° CHAM
Ø10 X 8 DEEP
10.0 – 10.1 ACROSS
FLATS – HEX SOC
7 DEEP

④ ADJUSTING SCREW
IIII STL – ELEC PLATE – I REQ

Figure 24.71   Make a detail drawing of the parts of the leveling device. Draw an assembly and provide a parts list.

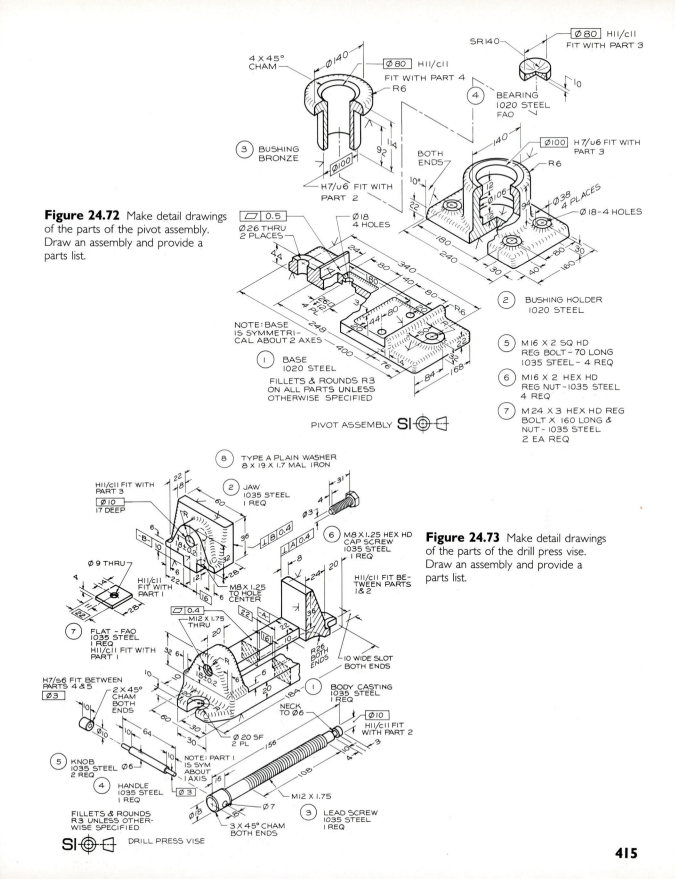

Figure 24.72 Make detail drawings of the parts of the pivot assembly. Draw an assembly and provide a parts list.

Figure 24.73 Make detail drawings of the parts of the drill press vise. Draw an assembly and provide a parts list.

415

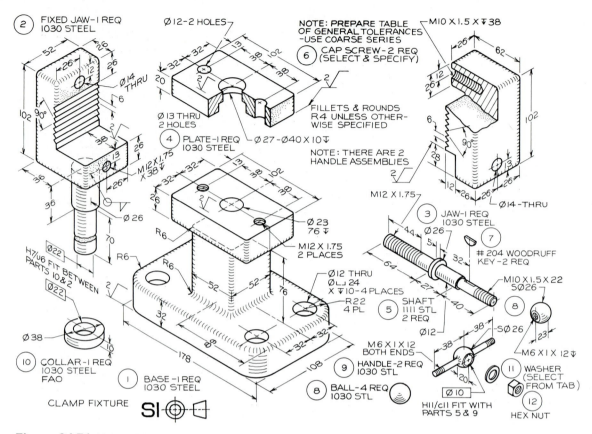

Figure 24.74 Make detail drawings of the parts of the clamp fixture. Draw an assembly and provide a parts list.

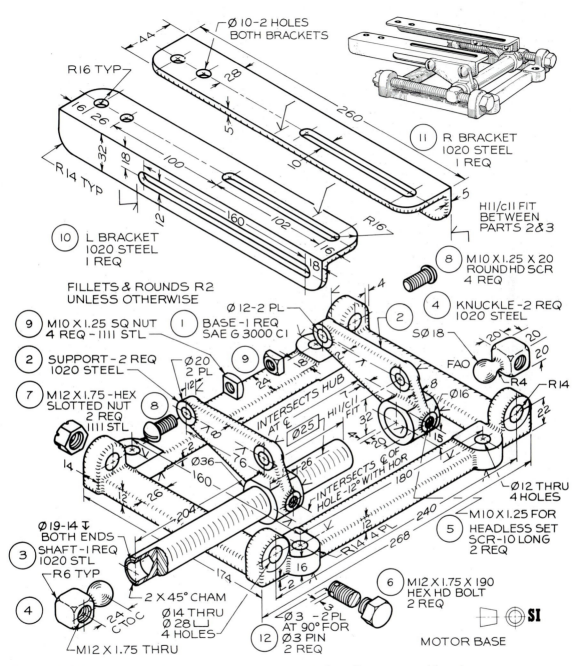

Ø 10–2 HOLES
BOTH BRACKETS

44

R16 TYP

16

26

28

5

260

260

R14 TYP

32

18

100

160

102

10

R16

16

18

⑪ R BRACKET
1020 STEEL
1 REQ

5

H11/C11 FIT
BETWEEN
PARTS 2 & 3

⑩ L BRACKET
1020 STEEL
1 REQ

FILLETS & ROUNDS R2
UNLESS OTHERWISE

⑧ M10 X 1.25 X 20
ROUND HD SCR
4 REQ

④ KNUCKLE – 2 REQ
1020 STEEL

S Ø 18
FAO

20   20
20
R4   R14

Ø 12–2 PL

① BASE –1 REQ
SAE G 3000 C1

②

⑨ M10 X 1.25 SQ NUT
4 REQ –1111 STL

② SUPPORT – 2 REQ
1020 STEEL

Ø 20
2 PL

⑨

12

24

INTERSECTS HUB
AT ₵

Ø 25   H11/C11
FIT

8

32

Ø 16

22

⑦ M12 X 1.75 –HEX
SLOTTED NUT
2 REQ
1111 STL

⑧

80

2

76

INTERSECTS ₵ OF
HOLE–12° WITH HOR

26

15

180

14

Ø 36

160

12   26

204

12   4
20

Ø 12 THRU
4 HOLES

M10 X 1.25 FOR
HEADLESS SET
SCR–10 LONG
2 REQ

⑤

240

R14 4 PL   268

12

16

Ø 19–14 ⤓
BOTH ENDS
SHAFT–1 REQ
1020 STL

R6 TYP

④

2 X 45° CHAM

174

24
CTOC

Ø 14 THRU
Ø 28 ⌴
4 HOLES

M12 X 1.75 THRU

16

2

3

Ø 3 –2 PL
AT 90° FOR
⑫ Ø 3 PIN
2 REQ

⑥ M12 X 1.75 X 190
HEX HD BOLT
2 REQ

MOTOR BASE

SI

Figure 24.75 Make a detail drawing of the parts of the motor base. Draw an assembly and provide a parts list.

418

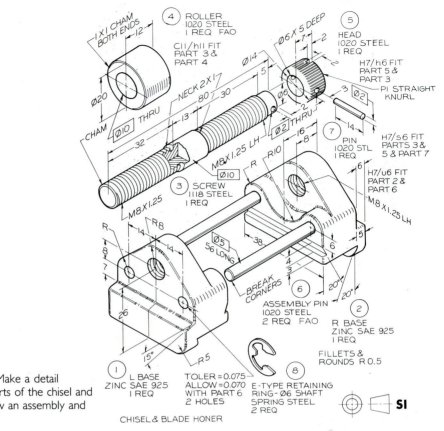

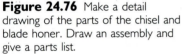

Figure 24.76 Make a detail drawing of the parts of the chisel and blade honer. Draw an assembly and give a parts list.

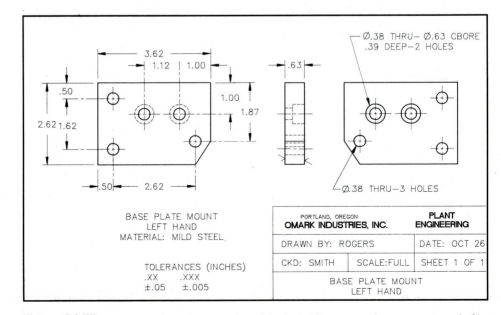

Figure 24.77 Duplicate the working drawing of the base plate mount, by computer or drafting instruments, on a Size B sheet.

Figure 24.78 The ball crank assembly detailed in the working drawings in Fig. 24.79.

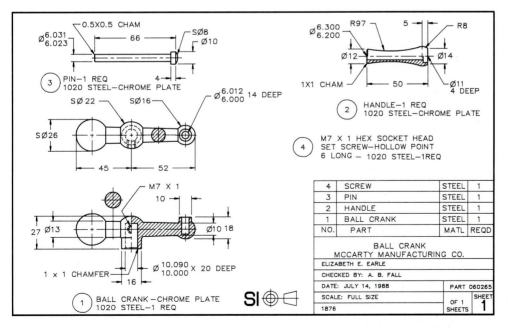

Figure 24.79 Make a working drawing of the ball crank, by computer or drafting instruments, on a Size B sheet. On a second Size B sheet, make an assembly drawing of the parts of the ball crank.

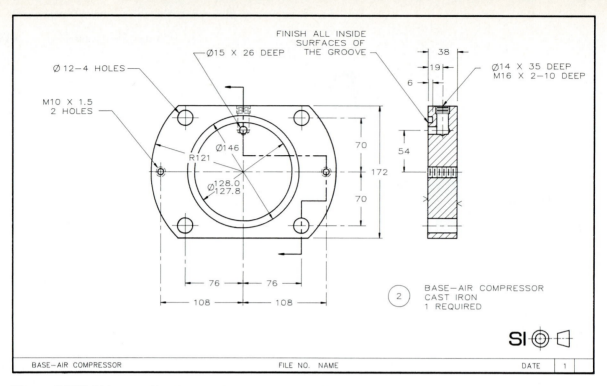

Figure 24.80 Make a working drawing of the air compressor base, by computer or drafting instruments, on a Size B sheet.

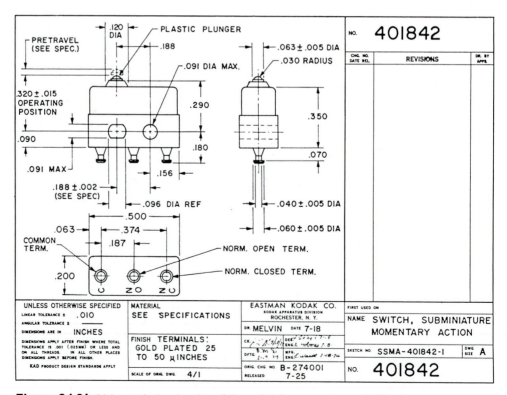

Figure 24.81 Make an A-size drawing of the switch by computer or drafting instruments. Notice that the views are enlarged by a factor of 4.

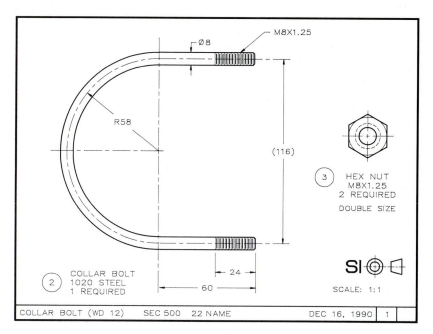

COLLAR BOLT (WD 12)   SEC 500   22 NAME   DEC 16, 1990   1

Figure 24.82 Make a working drawing of pipe hanger that is shown in three A-size drawings: Fig. 24.82, 24.83 and 24.84. Draw full size, using metric units. An assembly of the parts is shown in Fig. 24.84.

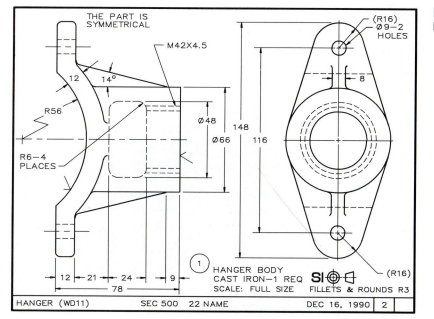

HANGER (WD11)   SEC 500   22 NAME   DEC 16, 1990   2

Figure 24.83 Sheet 2 of the pipe hanger working drawing.

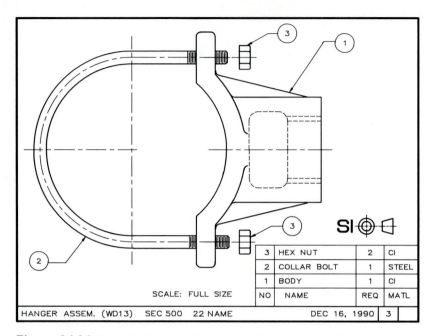

Figure 24.84 Sheet 3 of the working drawing shows the assembly of the pipe hanger.

The following problems are examples of working drawings of products requiring several sheets.

Angle Vise

The angle vise shown in Fig. 24.85 is detailed in four sheets (Fig. 24.86 through Fig. 24.89). Figure 24.90 (Sheet 5) gives an assembly drawing and a parts list.

As an exercise in draftsmanship, reproduce these drawings on B-size sheets by computer or by manual methods using your drawing instruments.

Figure 24.85 Angle vise.

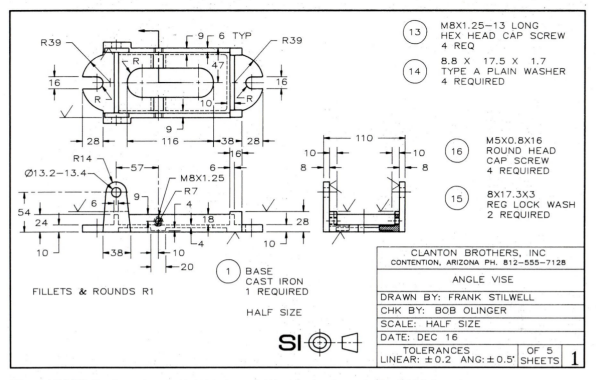

Figure 24.86 Duplicate the working drawings of the angle vise shown in Figs. 24.86 through 24.90 on B-Size sheets. Figure 24.90 is an assembly drawing.

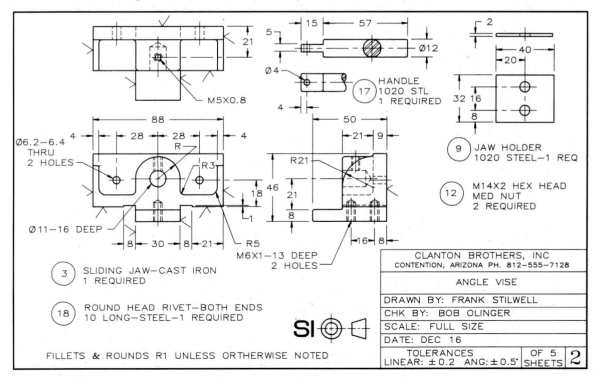

Figure 24.87 Angle vise working drawings. See Fig. 24.86 caption for instructions.

423

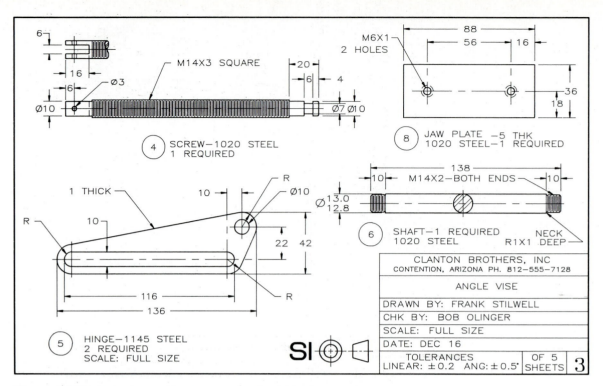

Figure 24.88 Angle vise working drawings. See Fig. 24.86 caption for instructions.

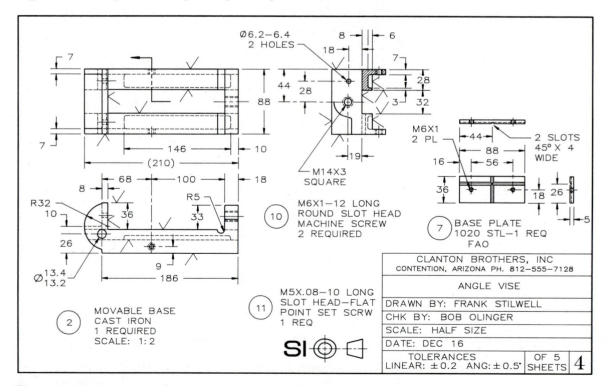

Figure 24.89 Angle vise working drawings. See Fig. 24.86 caption for instructions.

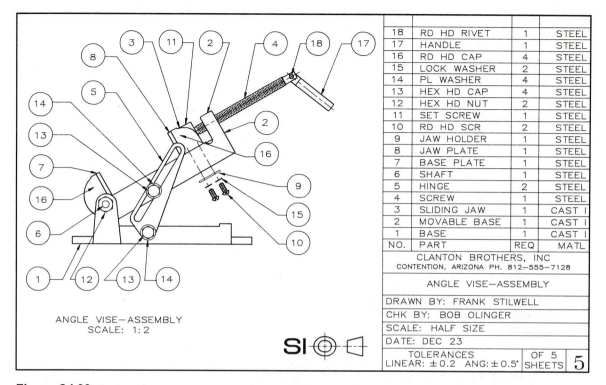

18	RD HD RIVET	1	STEEL
17	HANDLE	1	STEEL
16	RD HD CAP	4	STEEL
15	LOCK WASHER	2	STEEL
14	PL WASHER	4	STEEL
13	HEX HD CAP	4	STEEL
12	HEX HD NUT	2	STEEL
11	SET SCREW	1	STEEL
10	RD HD SCR	2	STEEL
9	JAW HOLDER	1	STEEL
8	JAW PLATE	1	STEEL
7	BASE PLATE	1	STEEL
6	SHAFT	1	STEEL
5	HINGE	2	STEEL
4	SCREW	1	STEEL
3	SLIDING JAW	1	CAST I
2	MOVABLE BASE	1	CAST I
1	BASE	1	CAST I
NO.	PART	REQ	MATL

CLANTON BROTHERS, INC
CONTENTION, ARIZONA PH. 812–555–7128

ANGLE VISE–ASSEMBLY

DRAWN BY: FRANK STILWELL
CHK BY: BOB OLINGER
SCALE: HALF SIZE
DATE: DEC 23

TOLERANCES
LINEAR: ±0.2 ANG: ±0.5°   OF 5 SHEETS   5

ANGLE VISE–ASSEMBLY
SCALE: 1:2

SI

Figure 24.90 Angle Vise assembly. See Fig. 24.86 caption for instructions.

Design/Drafting Problems

The problems that follow (Figs. 24.91–24.95) give orthographic, scale drawings of fixture assemblies produced by Jergens, Incorporated. Additional information is given in tabular form that will help you better understand the details and specifications of these assemblies.

By using the grid and graphics scale to determine dimensions, draw the necessary views of the individual parts, with dimensions, tolerances, and notes to make a working drawing from which the parts can be made. Place tolerances on the parts in limit form using the fits given in the tables. Use your judgement to determine other tolerances that are not given. Design the parts to function properly should you not fully understand them as they are shown.

Make an assembly drawing with a parts list to describe how the parts fit together.

Select the proper scale in order for the drawings to fit on B-Size sheets. Larger sheets may be used if assigned.

Trailer Hitch

Make a working drawing of the parts of the trailer hitch on B-size sheets, by hand or by computer (Fig. 24.96). You must refer to the tables in your book to find the sizes of the washers. Make an assembly drawing with a parts list.

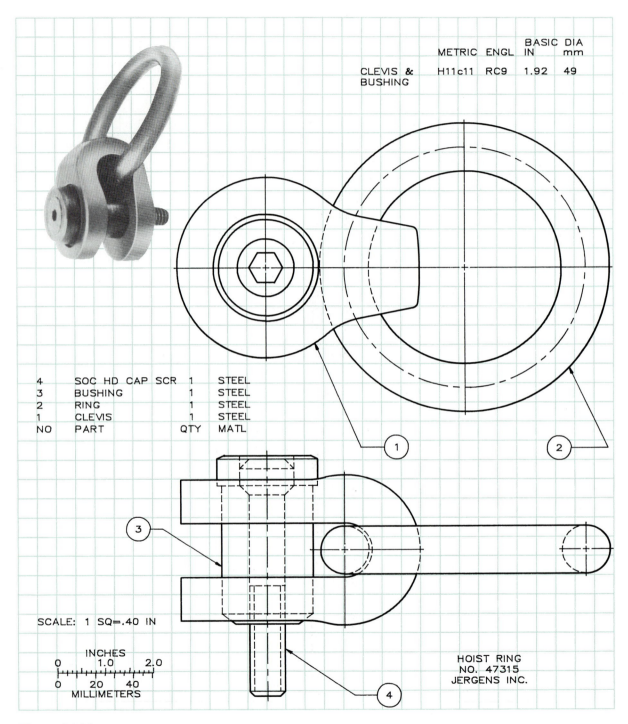

		METRIC	ENGL	BASIC DIA	
				IN	mm
CLEVIS & BUSHING		H11c11	RC9	1.92	49

NO	PART	QTY	MATL
4	SOC HD CAP SCR	1	STEEL
3	BUSHING	1	STEEL
2	RING	1	STEEL
1	CLEVIS	1	STEEL

SCALE: 1 SQ=.40 IN

INCHES
0   1.0   2.0

0   20   40
MILLIMETERS

HOIST RING
NO. 47315
JERGENS INC.

Figure 24.91 Clevis and bushing. (Courtesy of Jergens Inc.)

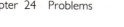

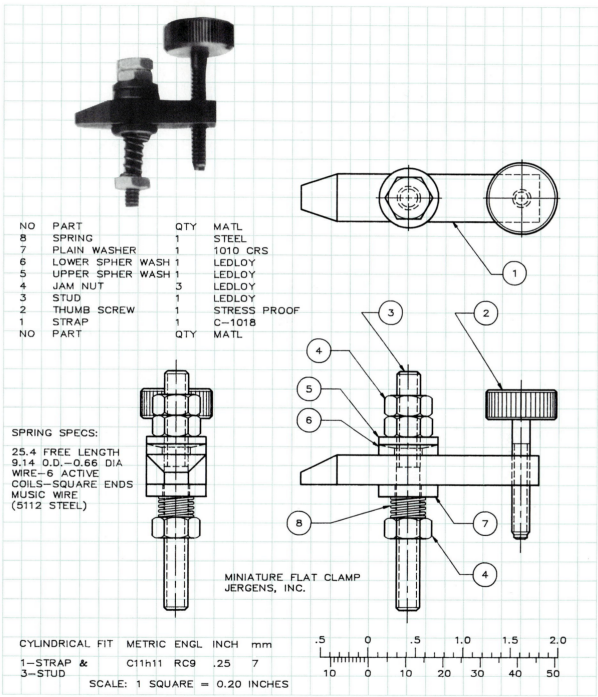

NO	PART	QTY	MATL
8	SPRING	1	STEEL
7	PLAIN WASHER	1	1010 CRS
6	LOWER SPHER WASH	1	LEDLOY
5	UPPER SPHER WASH	1	LEDLOY
4	JAM NUT	3	LEDLOY
3	STUD	1	LEDLOY
2	THUMB SCREW	1	STRESS PROOF
1	STRAP	1	C—1018
NO	PART	QTY	MATL

SPRING SPECS:

25.4 FREE LENGTH
9.14 O.D.—0.66 DIA
WIRE—6 ACTIVE
COILS—SQUARE ENDS
MUSIC WIRE
(5112 STEEL)

MINIATURE FLAT CLAMP
JERGENS, INC.

CYLINDRICAL FIT	METRIC ENGL	INCH	mm
1—STRAP & 3—STUD	C11h11 RC9	.25	7

SCALE: 1 SQUARE = 0.20 INCHES

Figure 24.92 Miniature flat clamp. (Courtesy of Jergens Inc.)

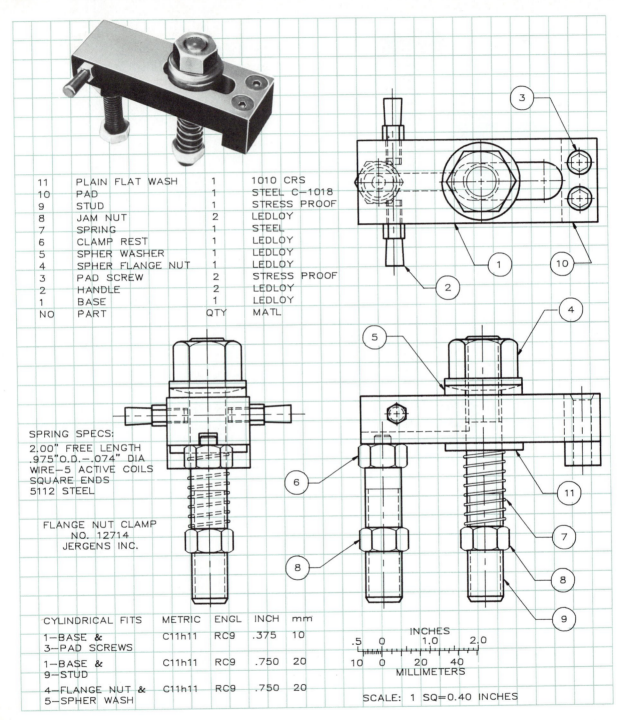

11	PLAIN FLAT WASH	1	1010 CRS
10	PAD	1	STEEL C–1018
9	STUD	1	STRESS PROOF
8	JAM NUT	2	LEDLOY
7	SPRING	1	STEEL
6	CLAMP REST	1	LEDLOY
5	SPHER WASHER	1	LEDLOY
4	SPHER FLANGE NUT	1	LEDLOY
3	PAD SCREW	2	STRESS PROOF
2	HANDLE	2	LEDLOY
1	BASE	1	LEDLOY
NO	PART	QTY	MATL

SPRING SPECS:
2.00" FREE LENGTH
.975"O.D.–.074" DIA
WIRE–5 ACTIVE COILS
SQUARE ENDS
5112 STEEL

FLANGE NUT CLAMP
NO. 12714
JERGENS INC.

CYLINDRICAL FITS	METRIC	ENGL	INCH	mm
1–BASE & 3–PAD SCREWS	C11h11	RC9	.375	10
1–BASE & 9–STUD	C11h11	RC9	.750	20
4–FLANGE NUT & 5–SPHER WASH	C11h11	RC9	.750	20

INCHES
.5 0   1.0   2.0
10 0   20   40
MILLIMETERS

SCALE: 1 SQ=0.40 INCHES

Figure 24.93 Flange nut clamp. (Courtesy of Jergens Inc.)

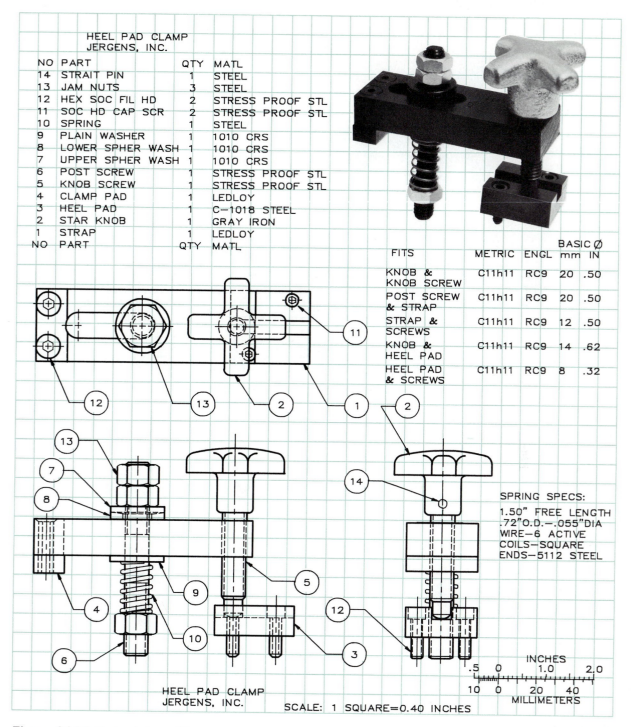

HEEL PAD CLAMP
JERGENS, INC.

NO	PART	QTY	MATL
14	STRAIT PIN	1	STEEL
13	JAM NUTS	3	STEEL
12	HEX SOC FIL HD	2	STRESS PROOF STL
11	SOC HD CAP SCR	2	STRESS PROOF STL
10	SPRING	1	STEEL
9	PLAIN WASHER	1	1010 CRS
8	LOWER SPHER WASH	1	1010 CRS
7	UPPER SPHER WASH	1	1010 CRS
6	POST SCREW	1	STRESS PROOF STL
5	KNOB SCREW	1	STRESS PROOF STL
4	CLAMP PAD	1	LEDLOY
3	HEEL PAD	1	C—1018 STEEL
2	STAR KNOB	1	GRAY IRON
1	STRAP	1	LEDLOY
NO	PART	QTY	MATL

FITS	METRIC	ENGL	BASIC Ø mm	IN
KNOB & KNOB SCREW	C11h11	RC9	20	.50
POST SCREW & STRAP	C11h11	RC9	20	.50
STRAP & SCREWS	C11h11	RC9	12	.50
KNOB & HEEL PAD	C11h11	RC9	14	.62
HEEL PAD & SCREWS	C11h11	RC9	8	.32

SPRING SPECS:
1.50" FREE LENGTH
.72"O.D.—.055"DIA
WIRE—6 ACTIVE
COILS—SQUARE
ENDS—5112 STEEL

INCHES
.5 0   1.0   2.0
10 0   20   40
MILLIMETERS

HEEL PAD CLAMP
JERGENS, INC.
SCALE: 1 SQUARE=0.40 INCHES

Figure 24.94 Heel pad clamp. (Courtesy of Jergens Inc.)

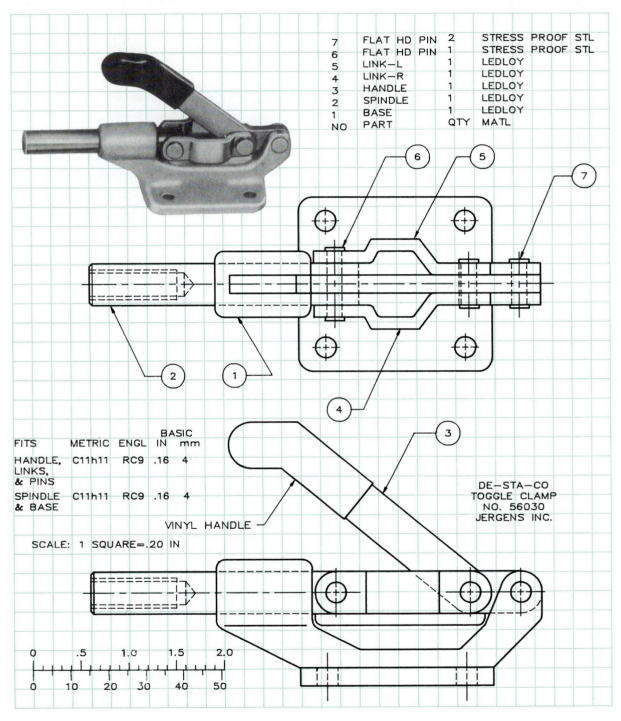

7	FLAT HD PIN	2	STRESS PROOF STL
6	FLAT HD PIN	1	STRESS PROOF STL
5	LINK—L	1	LEDLOY
4	LINK—R	1	LEDLOY
3	HANDLE	1	LEDLOY
2	SPINDLE	1	LEDLOY
1	BASE	1	LEDLOY
NO	PART	QTY	MATL

FITS	METRIC	ENGL	BASIC IN	mm
HANDLE, LINKS, & PINS	C11h11	RC9	.16	4
SPINDLE & BASE	C11h11	RC9	.16	4

VINYL HANDLE

DE—STA—CO
TOGGLE CLAMP
NO. 56030
JERGENS INC.

SCALE: 1 SQUARE=.20 IN

0   .5   1.0   1.5   2.0
0   10   20   30   40   50

Figure 24.95 Toggle clamp. (Courtesy of Jergens Inc.)

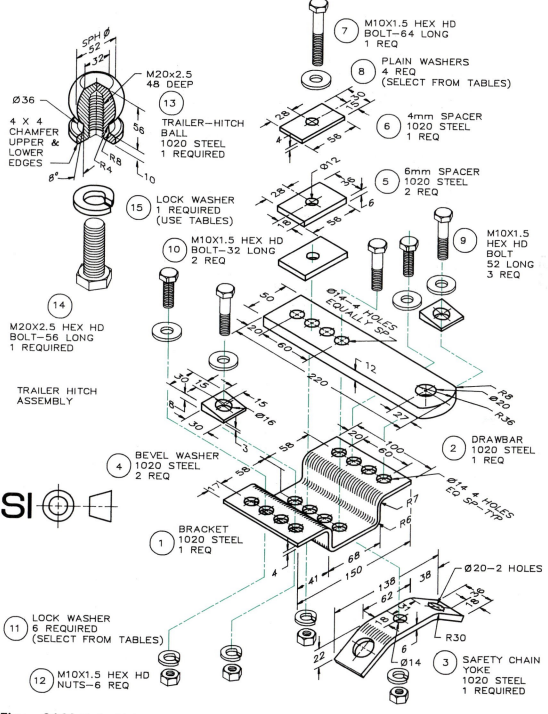

SPH Ø
52
32

Ø36

4 X 4
CHAMFER
UPPER &
LOWER
EDGES

8°

M20x2.5
48 DEEP

(13)

TRAILER—HITCH
BALL
1020 STEEL
1 REQUIRED

56

R8
R4

10

(15) LOCK WASHER
1 REQUIRED
(USE TABLES)

(14)

M20X2.5 HEX HD
BOLT—56 LONG
1 REQUIRED

TRAILER HITCH
ASSEMBLY

(10) M10X1.5 HEX HD
BOLT—32 LONG
2 REQ

(4) BEVEL WASHER
1020 STEEL
2 REQ

30 15
8
30 3

15
Ø16

17
58
58

(1) BRACKET
1020 STEEL
1 REQ

(11) LOCK WASHER
6 REQUIRED
(SELECT FROM TABLES)

(12) M10X1.5 HEX HD
NUTS—6 REQ

SI

(7) M10X1.5 HEX HD
BOLT—64 LONG
1 REQ

(8) PLAIN WASHERS
4 REQ
(SELECT FROM TABLES)

30
15
28
4
56

(6) 4mm SPACER
1020 STEEL
1 REQ

Ø12

28
36
6
18
56

(5) 6mm SPACER
1020 STEEL
2 REQ

(9) M10X1.5
HEX HD
BOLT
52 LONG
3 REQ

50

Ø14—4 HOLES
EQUALLY SP

20
60
12
220

R8
Ø20
R36
27

(2) DRAWBAR
1020 STEEL
1 REQ

20
100
60

Ø14 4 HOLES
EQ SP—TYP

R7
R6

58
4
41
150
68

138
62
Ø20—2 HOLES
36
18
18
R30
6
22
Ø14

(3) SAFETY CHAIN
YOKE
1020 STEEL
1 REQUIRED

Figure 24.96 Trailer hitch.

25

Reproduction Methods and Drawing Shortcuts

25.1 Introduction

So far, we have discussed the processes of preparing drawings and specifications through the working-drawing stage where a detailed drawing is completed on tracing film or paper. Now the drawing must be reproduced, folded, and prepared for filing or transmittal to the drawing's users.

25.2 Reproduction of Working Drawings

A drawing made by a drafter is of little use in its original form. It would be impractical for the original to be handled by checkers and, even more so, by workers in the field or shop. The drawing would quickly be damaged or soiled, and no copy would be available as a permanent record of the job. Therefore, reproduction of drawings is necessary so that copies can be available for use by the people concerned.

The most often used processes of reproducing engineering drawings are (1) **diazo printing,** (2) **blueprinting,** (3) **microfilming,** (4) **xerography,** and (5) **photostatting.**

Diazo Printing

The diazo print is more correctly called a whiteprint or blue-line print than a blueprint, since it has a white background and blue lines. Other colors of lines are available depending on the type of paper used.

> Both blueprinting and diazo printing require that the original drawing be made on semitransparent tracing paper, cloth, or film that will allow light to pass through the drawing.

The paper the copy is made on, the diazo paper, is chemically treated so that it has a yellow tint on one side. To prevent spoilage, this paper must be stored away from heat and light.

The tracing paper or film drawing is placed face up on the yellow side of the diazo paper and is run through the diazo-process machine, which exposes the drawing to a built-in light. The light passes through the tracing paper and burns out the yellow chemical on the diazo paper except where the drawing lines have shielded the paper from the light. After exposure, the diazo paper is a duplicate of the original drawing except that the lines are light yellow and not permanent.

The diazo paper is then passed through the developing unit of the diazo machine where the yellow lines are developed into permanent blue lines by exposure to ammonia fumes. Figure 25.1 shows a typical diazo printer-developer, sometimes called a white printer.

The speed at which the drawing passes under the light determines the darkness of the copy. A slow speed burns out more of the yellow and produces a clear white background; however, some of the lighter lines of the drawing may be lost. Most diazo copies are made at a somewhat faster speed to give a light tint of blue in the background and stronger lines in the copy. Ink drawings give the best reproductions.

Figure 25.1 A typical whiteprinter that operates on the diazo process. (Courtesy of Blu-Ray, Inc., Essex, CT)

Blueprinting

Blueprints are made with paper that is chemically treated on one side. As in the diazo process, the tracing-paper drawing is placed in contact with the chemically treated side of the paper and exposed to light. The exposed blueprint paper is washed in clear water and coated with a solution of potassium dichromate. Then the print is washed again and dried. The wet sheets can be hung on a line to dry, or they can be dried by equipment made for this purpose.

Microfilming

Microfilming is a photographic process that converts large drawings into film copies—either aperture cards or roll film. Drawings must be photographed on either 16-mm or 35-mm film. Figure 25.2 shows a camera and copy table.

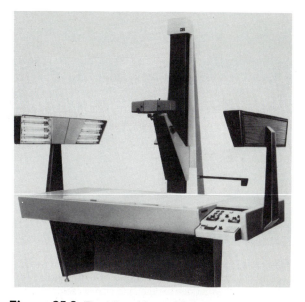

Figure 25.2 The Micro-Master 35-mm camera and copy table are used for microfilming engineering drawings. (Courtesy of Keuffel & Esser Co., Morristown, NJ)

The roll film or aperture cards can be placed in a microfilm enlarger-printer (Fig. 25.3), where the individual drawings can be viewed on a built-in screen. The selected drawings can then be

Figure 25.3 The Bruning 1200 microfilm enlarger-printer makes drawings up to 18″ × 24″ from aperture cards and roll film. (Courtesy of Bruning Co.)

Figure 25.4 This Xerox 7080 Engineering Print system accepts original drawings sized A to E; makes prints sized A to C as fast as 58 per minute; and stamps, folds, and sorts prints automatically.

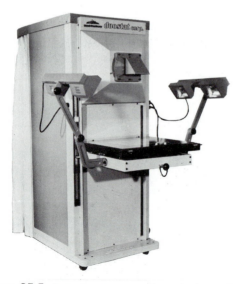

Figure 25.5 A camera-processor for enlarging and reducing drawings to be reproduced as photostats. (Courtesy of the Duostat Corp.)

printed from the film to give standard-size drawings. Microfilm copies are usually smaller than the original drawings to save paper and make the drawings easier to use.

Microfilming makes it possible to eliminate large, bulky files of drawings since hundreds of drawings can be stored in miniature size on a small amount of film. The aperture cards shown in Fig. 25.3 are data processing cards that can be cataloged and recalled by a computer to make them accessible with a minimum of effort.

Xerography

Xerography is an electrostatic process of duplicating drawings on ordinary, unsensitized paper. Originally developed for business and clerical uses, xerography has more recently been used for the reproduction of engineering drawings.

One advantage of the xerographic process is its ability to make reduced copies of drawings (Fig. 25.4). The Xerox 2080 can reduce a 24-$\times$-36-inch drawing to 8 $\times$ 10 inches.

Photostatting

Photostatting is a method of enlarging or reducing drawings using a camera. Figure 25.5 shows a combination camera and processor used for photographing drawings and producing high-contrast copies.

The drawing or artwork is placed under the glass of the exposure table, which is lit by built-in lamps. The image can be seen on the glass inside the darkroom where it is exposed on photographically sensitive paper. The negative paper that has been exposed to the image is placed in contact with receiver paper, and the two are fed through the developing solution to obtain a photostatic copy.

These high-contrast reproductions are often used to prepare artwork that is to be printed by offset printing presses. Photostatting can also be used to make reproductions on transparent films and for reproducing halftones (photographs with tones of gray).

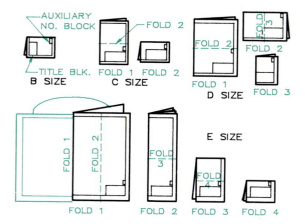

Figure 25.6 Standard folds for engineering drawing sheets. The final size in each case is $8\frac{1}{2}$" $\times$ 11".

25.3 Assembling the Drawings

Once the prints have been finished, the original drawings should be stored in a flat file for future use and updating. The original drawings should not be folded, and their handling should be kept to a minimum.

The printed drawings, on the other hand, are usually folded for transmittal from office to office. Figure 25.6 shows the methods of folding Sizes B, C, D, and E sheets; in each case, the final size after folding is $8\frac{1}{2} \times 11$ inches (or 9 $\times$ 12 inches). The drawings are folded so that the title blocks are positioned at the top and lower right of the drawing to allow them to be easily retrieved from a file.

An alternate method of folding B-size sheets, often used for student assignments, is shown in Fig. 25.7. By folding and stapling in this manner, the drawings can be kept in a three-ring note-book.

The basic rules of assembling drawings are given in Fig. 25.8.

25.4 Overlay Drafting Techniques

Valuable drafting time can be saved by taking advantage of current processes and materials that use a series of overlays to separate parts of a sin-

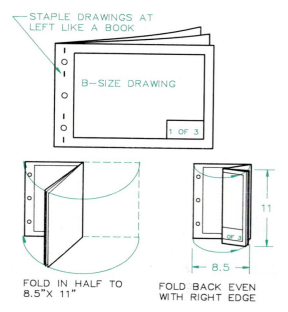

Figure 25.7 A set of B-size drawings can be assembled by stapling, punching, and folding as shown here. The title block is visible at the right and the drawings can be kept in a three-ring notebook.

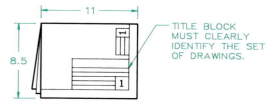

WORKING DRAWING CHECKLIST

1. STAPLE ALONG LEFT EDGE, LIKE A BOOK. USE SEVERAL STAPLES, NEVER JUST ONE.

2. FOLD WITH DRAWING ON OUTSIDE.

3. FOLD DRAWINGS AS A SET, NOT ONE AT A TIME SEPARATELY.

4. FOLD TO AN 8.5" $\times$ 11" MODULAR SIZE.

5. THE TITLE BLOCK MUST BE VISIBLE AFTER FOLDING.

6. SHEETS OF A SET SHOULD BE UNIFORM IN SIZE.

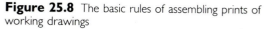

Figure 25.8 The basic rules of assembling prints of working drawings

gle drawing. Engineers and architects often work from a single site plan or floor plan. For example, the floor plan of a building will be used for the electrical plan, furniture arrangement plan, air-conditioning plan, floor materials plan, and so on. It would be expensive to retrace the plan for each application.

A series of overlays can be used in a system referred to as **pin drafting,** where accurately spaced holes are punched in the polyester drafting film at the top edge of the sheets. These holes are aligned on pins attached to a metal strip that match the holes punched in the film (Fig. 25.9). The pins ensure accurate alignment of registration of a series of sheets, and the polyester ensures stability of the material since it does not stretch or sag with changes in humidity.

The set of overlays could be attached by the alignment pins or taped together and run through a diazo machine for full-size prints. Another reproduction operation is the use of a flat-bed process camera (Fig. 25.10), which photographs and reduces the drawings to $8\frac{1}{2} \times 11$ inches.

Figure 25.9 In the pin system, separate overlays are aligned by seven pins mounted on metal strips. (Courtesy of Keuffel & Esser Co., Morristown, NJ)

Figure 25.10 The process camera used to reduce and enlarge engineering drawings is the heart of the pin system. (Courtesy of Keuffel & Esser Co., Morristown, NJ)

25.5   Paste-on Photos

When several repetitive drawings are needed, it is often more economical to use photographic reproductions of the drawings on transparent film. These features can be pasted into position on the master drawing. (Although the term, pasted, is commonly used to describe this attachment, tape is actually used when the drawings are reproduced on transparent film. If the reproductions are made as opaque photostats, rubber cement may be used.)

The office arrangements shown in Fig. 25.11 are transparencies that have been photographically duplicated from a single drawing. These photographically reproduced drawings can be applied to drawing and save drafting time.

25.6   Stick-on Materials

Several companies market stick-on symbols, screens, and lettering that can be applied to drawings to save time and improve the appearance of drawings. The three standard types of materials are stick-ons, burnish-ons, and tape-ons.

Stick-on symbols or letters are printed on thin plastic sheets backed with a sticky adhesive. The symbols are cut out with a razor-sharp blade and affixed to the drawing (Fig. 25.12). This material is available in glossy and matte finishes.

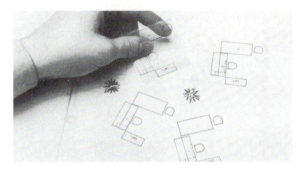

Figure 25.11 When drawings are to be used repetitively, it may be more economical to reproduce them than to redraw them. (Courtesy of Eastman Kodak Co.)

Burnish-on symbols are applied by placing the entire sheet over the drawing and burnishing the desired symbol into place with a rounded-end object such as the end of a pen cap.

Figure 25.12 Stick-on lettering can be applied to a drawing by cutting the letters from the plastic sheet, applying them to the drawing in alignment with a guideline, and burnishing them to the drawing. (Courtesy of Graphic Products Corp.)

Tape-ons are colored adhesive tapes in varying widths used in the preparation of graphs and charts. Tapes are also used to represent wide lines on large drawings.

Sheets of symbols can be custom-printed for users who have repetitive needs for trademarks and other often-used symbols (Fig. 25.13).

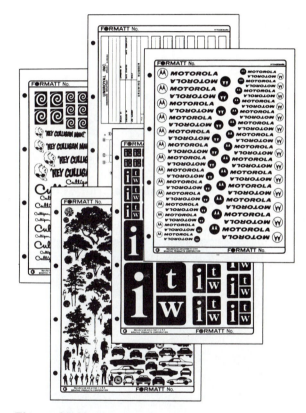

Figure 25.13 Stick-on symbols are available in a wide range of styles, and they can be custom-designed to suit the client's needs. (Courtesy of Graphic Products Corp.)

25.7 Photo Drafting

To clarify assembly details, it is sometimes worthwhile to build a model for photographing. This is especially true in the piping industry, where complex refineries are built as models, then photographed, noted, and reproduced as photodrawings.

Figure 25.14 shows the steps of preparing a photodrawing, where it is desired to specify certain parts of a sprocket-and-chain assembly. The negative is used to produce a positive on polyester drafting film. The notes can be lettered on this drawing film to complete the master drawing (Step 4). The master drawing can then be used to make diazo prints, or it can be microfilmed or reproduced photographically to the desired size.

Step 1: Make a halftone print of the photograph.

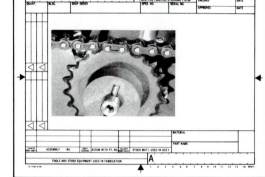

Step 3: Make a positive reproduction on matte film.

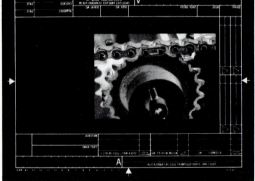

Step 2: Tape the halftone print to a drawing form, and photograph it to produce a negative.

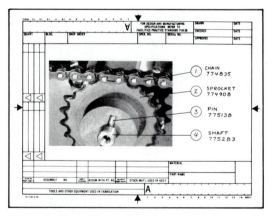

Step 4: Now draw in your callouts—and the job is done.

Figure 25.14 The steps in making a photodrawing. (Courtesy of Eastman Kodak Co.)

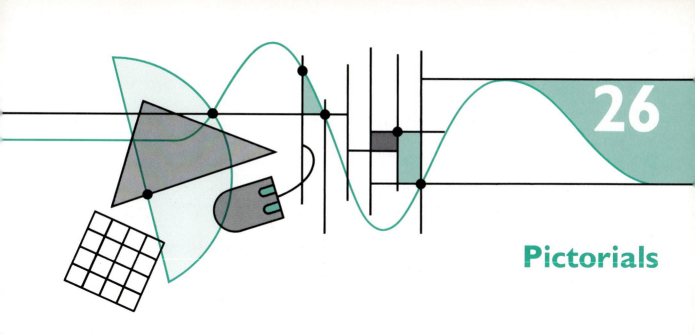

Pictorials

26.1 Introduction

A **pictorial** is an effective means of communicating an idea. Pictorials are especially helpful if a design is unique or the person to whom it is being explained has difficulty interpreting multiview drawings.

Pictorials, sometimes called **technical illustrations,** are widely used to describe various products in catalogs, parts manuals, and maintenance publications.

26.2 Types of Pictorials

The four commonly used types of pictorials are (1) **obliques,** (2) **isometrics,** (3) **axonometrics,** and (4) **perspectives** (Fig. 26.1).

Oblique Pictorials are three-dimensional pictorials on a plane of paper drawn by projecting from the object with parallel projectors that are oblique to the picture plane (Fig. 26.1B).

Isometric and Axonometric Pictorials are three-dimensional pictorials on a plane of paper drawn by projecting from the object to the pic-

ture plane (Fig. 26.1A). The parallel projectors are perpendicular to the picture plane.

Perspective Pictorials are drawn with projectors that converge at the viewer's eye and make varying angles with the picture plane (Fig. 26.1C).

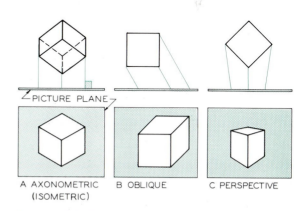

Figure 26.1 Types of projection systems for pictorials. (A) Axonometric pictorials are formed by parallel projectors perpendicular to the picture plane. (B) Obliques are formed by parallel projectors oblique to the picture plane. (C) Perspectives are formed by converging projectors that make varying angles with the picture plane.

439

26.3   Oblique Projections

Oblique projections are the basis of oblique drawings. In Fig. 26.2, Step 1, a number of lines of sight are drawn through point 2 of line 1–2. Each line of sight makes a 45° angle with the picture plane, which creates a cone with its apex at 2, and each element on the cone makes a 45° angle with the plane. A variety of projections of line 1–2′ on the picture plane can be seen in the front view (Fig. 26.2, Step 2). Each of these projections of 1–2′ is equal in length to the true length of 1–2. This is a **cavalier oblique projection** because all projectors make a 45° angle with the picture plane, and measurements along the receding axis can be made true length and in any direction.

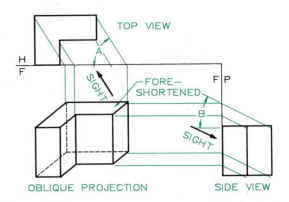

Figure 26.3 An oblique projection can be drawn at any angle of sight to obtain an oblique pictorial. However, the line of sight should not make an angle less than 45° with the picture plane. This would result in a receding axis longer than true length, thereby distorting the pictorial.

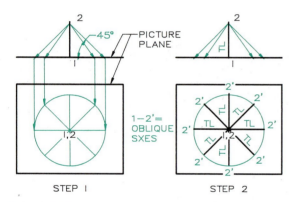

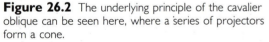

Figure 26.2 The underlying principle of the cavalier oblique can be seen here, where a series of projectors form a cone.

Step 1 Each element makes a 45° angle with the picture plane.

Step 2 Thus the projected lengths of 1–2′ are equal in length to line 1–2, which is perpendicular to the picture plane.

The top and side views are given as orthographic views, and the front view is drawn as an oblique projection by using the two views of a selected line of sight (Fig. 26.3). Projectors are drawn from the object parallel to the lines of sight to locate their respective points in the oblique view.

26.4   Oblique Drawings

Oblique projections are seldom used in the manner just illustrated.

> Instead, based on these principles there are three basic types of **oblique drawings:** (1) **cavalier,** (2) **cabinet,** and (3) **general** (Fig. 26.4).

In each case, the angle of the receding axis can be at any angle between 0° and 90° (Fig. 26.4). Measurements along the receding axes of the **cavalier oblique** are true length (full scale). The **cabinet oblique** has measurements along the receding axes reduced to half length. The **general oblique** has measurements along the receding axes reduced to between half and full length.

Figure 26.5 shows three examples of cavalier obliques of a cube. Each has a different angle for the receding axes, but the measurements along the receding axes are true length. Figure 26.6 compares cavalier with cabinet obliques.

26.5   Constructing Obliques

An oblique should be drawn by constructing a box using the overall dimensions of height, width, and depth with light construction lines. In

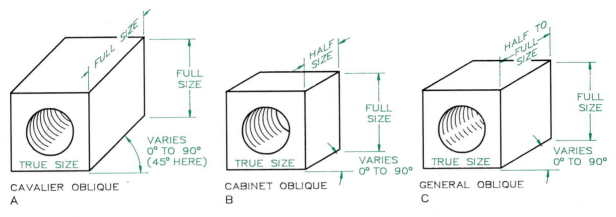

Figure 26.4 Types of obliques. A. The cavalier oblique can be drawn with a receding axis at any angle, but the measurements along this axis are true length. B. The cabinet oblique can be drawn with a receding axis at any angle, but the measurements along this axis are half size. C. The general oblique can be drawn with a receding axis at any angle, but the measurements along this axis can vary from half to full size.

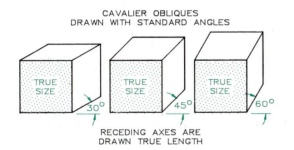

Figure 26.5 A cavalier oblique is usually drawn with the receding axis at the standard angles of the drafting triangles. Each gives a different view of a cube.

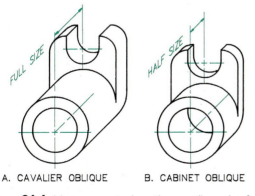

Figure 26.6 Measurements along the receding axis of a cavlier oblique are full size; those in a cabinet oblique are half size.

Fig. 26.7, the front view is drawn true size in Step 1. This will be a cavalier oblique. True measurements can be made parallel to the three axes. To complete the oblique, the notches are removed from the blocked-in construction box. These measurements are transferred from the given orthographic views with your dividers.

26.6 Angles in Oblique

Angular measurements can be made on the true-size plane of an oblique. However, angular measurements will not be true size on the other two planes of the oblique.

To construct an angle in an oblique, coordinates must be used (Fig. 26.8). The sloping surface of 30° must be found by locating the vertex of the angle H distance from the bottom. The inclination is found by measuring the distance of D along the receding axis to establish the slope. This angle is not equal to the 30° angle given in the orthographic view.

You can see in Fig. 26.9 that a true angle can be measured on a true-size surface. In Fig. 26.9B, angles along the receding planes are either smaller or larger than their true angles.

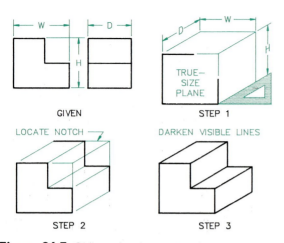

GIVEN
STEP 1

TRUE-SIZE PLANE

LOCATE NOTCH

DARKEN VISIBLE LINES

STEP 2
STEP 3

Figure 26.7 Oblique drawing construction.

Step 1 The front surface of the oblique is drawn as a true-size plane. The receding axis is drawn at a convenient angle, and depth is found by using the true distance of *D* taken from the given views.

Step 2 The notch in the front plane is drawn and projected to the rear plane.

Step 3 The lines are darkened to complete the cavalier oblique.

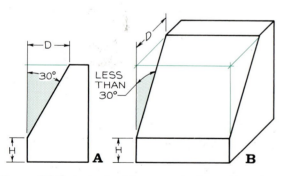

LESS THAN 30°

Figure 26.8 Angles in oblique must be located by using coordinates; they cannot be measured true size except on a true-size plane.

It requires less effort and gives a better appearance when obliques are drawn where angles will appear true size, as shown in Fig. 26.9A rather than in Fig. 26.9B.

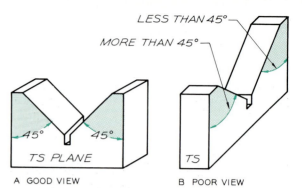

LESS THAN 45°

MORE THAN 45°

45° 45°
TS PLANE TS
A GOOD VIEW B POOR VIEW

Figure 26.9 The best view takes advantage of the ease of construction offered by oblique drawings. The view in part B is less descriptive and more difficult to construct than the view in part A.

26.7 Cylinders in Oblique

The major advantage of obliques is that circular features can be drawn as true circles when parallel to the picture plane (Fig. 26.10).

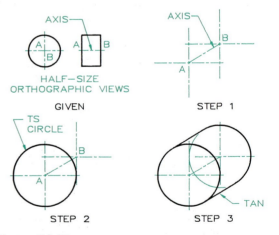

AXIS

HALF-SIZE ORTHOGRAPHIC VIEWS
GIVEN

AXIS
STEP 1

TS CIRCLE

STEP 2

TAN
STEP 3

Figure 26.10 An oblique drawing of a cylinder.

Step 1 Draw axis *AB*, and locate the centers of the circular ends of the cylinder at *A* and *B*. Since the axis is drawn true length, this will be a cavalier oblique.

Step 2 Draw a true-size circle with its center at *A* using a compass or computer-graphics techniques.

Step 3 Draw the other circular end with its center at *B*, and connect the circles with tangent lines parallel to the axis, *AB*.

The centerlines of the circular end at *A* are drawn, and the receding axis is drawn at any desired angle. The end at *B* is located by measuring along the axis. Circles drawn at each end using centers *A* and *B* are connected with tangent lines parallel to the axis.

These same principles are used to construct an object with semicircular features (Fig. 26.11). The oblique is positioned to take advantage of the option of drawing circular features as true circles. Centers *B* and *C* are located for drawing two semicircles (Step 2).

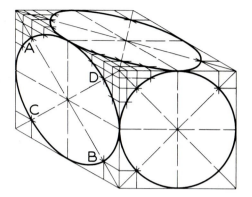

Figure 26.12 Circular features on the faces of a cavalier oblique of a cube appear as two ellipses and one true circle.

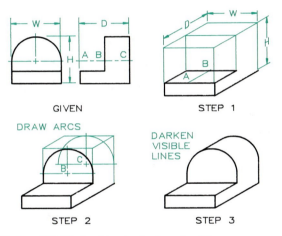

Figure 26.11 Semicircular features in oblique.

Step 1 Block in the overall dimensions of the cavalier oblique with light construction lines, and ignore the semicircular features.

Step 2 Locate centers *B* and *C*, and draw arcs with a compass or computer tangent to the sides of the construction boxes.

Step 3 Connect the arcs with lines tangent to each arc and parallel to axis *BC*; then darken the lines.

26.8 Circles in Oblique

Although circular features will be true size on a true-size plane of an oblique pictorial, circular features on the other two planes will appear as ellipses (Fig. 26.12).

A more frequently used technique of drawing elliptical views in oblique is the four-center ellipse method (Fig. 26.13). A rhombus is drawn

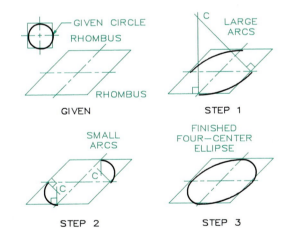

Figure 26.13 Four-center ellipse in oblique.

Given The circle to be drawn in oblique is blocked in with a square that is tangent to the circle at four points. This square will appear as a rhombus on the oblique plane.

Step 1 Construction lines are drawn perpendicularly from the points of tangency to locate the centers for drawing two segments of the ellipse.

Step 2 The centers for the two remaining arcs are located with perpendiculars drawn from adjacent tangent points.

Step 3 When the four arcs have been drawn, the final result is an approximate ellipse.

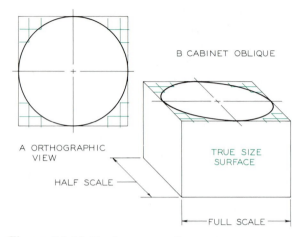

Figure 26.14 The four-center ellipse technique cannot be used to locate circular shapes on the foreshortened surface of a cabinet oblique. These ellipses must be plotted with coordinates.

that would be tangent to the circle at four points. Perpendicular construction lines are drawn from where the centerlines cross the sides of the rhombus. As Steps 2 and 3 show, this construction is made to locate four centers that are used to draw four arcs that form the ellipse in oblique.

The four-center ellipse method will not work for the cabinet or general oblique. In these cases, coordinates must be used to locate a series of points on the curve. Figure 26.14 shows coordinates in orthographic view and cabinet oblique. The coordinates parallel to the receding axis are reduced to half size; horizontal coordinates are drawn full size. The ellipse can be drawn with an irregular curve or ellipse template that approximates the plotted points.

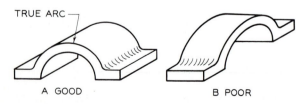

Figure 26.15 An oblique should be positioned to enable circular features to be drawn most easily.

Whenever possible, oblique drawings of objects with circular features should be positioned with circles on true-size planes so that they can be drawn as true circles. The view in Fig. 26.15A is better than the one in Fig. 26.15B because it gives a more descriptive view of the part and is easier to draw.

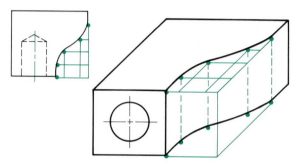

Figure 26.16 Coordinates are used to establish irregular curves in oblique. The lower curve is found by projecting the points downward a distance equal to the height of the oblique.

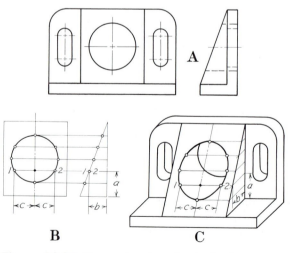

Figure 26.17 The construction of a circular feature on an inclined surface (A) must be found by (B) plotting points using three coordinates, a, b, and c, to locate points 1 and 2 in this example. (C) The plotted points are connected to complete the elliptical feature.

26.9 Curves in Oblique

Irregular curves in oblique must be plotted point by point using coordinates (Fig. 26.16). The coordinates are transferred from the orthographic view to the oblique view, and the curve is drawn through these points with an irregular curve.

If the object has a uniform thickness, the lower curve can be found by projecting vertically downward from the upper points a distance equal to the height of the object.

In Fig. 26.17, the elliptical feature on the inclined surface was found by using a series of coordinates to locate the points along its curve. These points are then connected by using an irregular curve or ellipse template of approximately the same size.

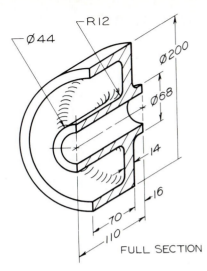

Figure 26.19 Oblique pictorials can be drawn as sections and dimensioned to serve as working drawings.

26.10 Oblique Sketching

Understanding the mechanical principles of oblique construction is essential for sketching obliques. As Fig. 26.18 shows, guidelines are helpful in developing a sketch. These guidelines should be drawn lightly so that they will not need to be erased when the finished lines of the sketch are darkened. When sketching on tracing vellum, it is helpful if a printed grid is placed under the vellum to provide guidelines.

26.11 Dimensioned Obliques

A dimensioned full-section oblique is given in Fig. 26.19, where the interior features and dimensions are shown.

In oblique pictorials, numerals and lettering should be applied using either the aligned method, in which the numerals are aligned with the dimensioned lines, or the unidirectional method, in which the numerals are all positioned in a single direction regardless of the direction of the dimension lines (Fig. 26.20). Notes connected with leaders are positioned horizontally in both methods.

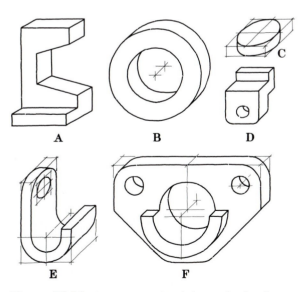

Figure 26.18 Obliques may be drawn as freehand sketches by using the same principles used for instrument pictorials. Use light construction lines to locate the more complex features.

26.12 Isometric Projections

An **isometric projection** is a type of axonometric projection in which parallel projectors are perpendicular to the picture plane, and the diag-

445

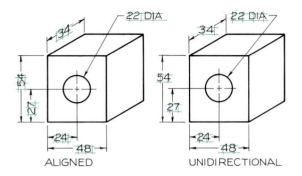

Figure 26.20 Oblique pictorials can be dimensioned by using either of these methods of applying numerals to the dimension lines.

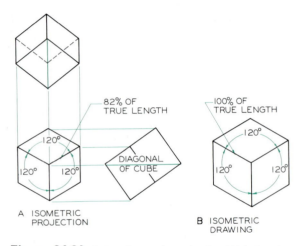

Figure 26.21 A true isometric projection (A) is found by constructing a view that shows the diagonal of a cube as a point. An isometric drawing (B) is not a true projection since the dimensions are drawn true size rather than reduced as in a projection.

onal of a cube is seen as a point (Fig. 26.21). The three axes are spaced 120° apart, and the sides are foreshortened to 82% of their true length.

Isometric, which means equal measurement, is used to describe this type of projection since the planes are equally foreshortened.

An **isometric drawing** is similar to an **isometric projection** except that it is not a true ax-

onometric projection but an approximate method of drawing a pictorial. Instead of reducing the measurements along the axes 82%, they are drawn true length (Fig. 26.21B).

Figure 26.22 compares an isometric projection with an isometric drawing. By using the isometric drawing instead of the isometric projection, pictorials can be measured using standard scales, the only difference being the 18% increase in size. Isometric drawings are used more often than isometric projections.

The axes of isometric drawings are separated by 120° (Fig. 26.23). Although one of the axes is usually drawn vertically, this is not necessary.

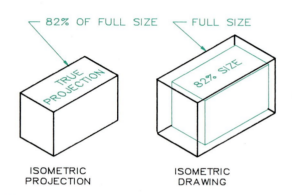

Figure 26.22 The isometric projection is foreshortened to 82% of full size. The isometric drawing is drawn full size for convenience.

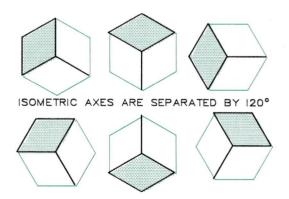

Figure 26.23 Isometric axes are spaced 120° apart, but they can be revolved into any position.

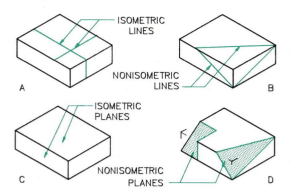

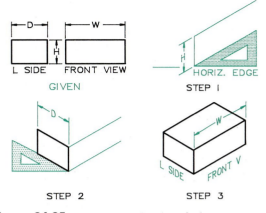

Figure 26.24 (A) True measurements can be made along isometric lines (parallel to the three axes). (B) Nonisometric lines cannot be measured true length. (C) Three isometric planes are indicated; no nonisometric planes are in this drawing. (D) Nonisometric planes are planes inclined to any of the three isometric planes of a cube.

Figure 26.25 An isometric drawing of a box.

Step 1 Use a 30°–60° triangle and a horizontal straight edge to construct a vertical line equal to the height (H), and draw two isometric lines through each end.

Step 2 Draw two 30° lines, and locate the depth (D) by transferring this dimension from the given views.

Step 3 Locate the width (W) of the object, and complete the surfaces of the isometric box.

Constructing Isometric Drawings

An isometric drawing is begun by drawing three axes 120° apart. Lines parallel to these axes are called **isometric lines** (Fig. 26.24A). True measurements can be made along isometric lines but not along nonisometric lines (Fig. 26.24B).

The three surfaces of a cube in isometric are called **isometric planes** (Fig. 26.24C). Planes parallel to these planes are isometric planes, and planes not parallel to them are nonisometric planes.

To draw an isometric, you will need a scale and a 30°–60° triangle (Fig. 26.25). Begin by constructing a plane of the isometric using the dimensions of height (H) and depth (D). In Step 3, the third dimension, width (W), is used to complete the isometric drawing.

All isometric drawings should be blocked in using light guidelines (Fig. 26.26) and overall dimensions of W, D, and H. Other dimensions can be taken from the given views and measured along the isometric axes to locate notches and portions removed from the blocked–in drawing.

Figure 26.27 shows a more complex isometric. Again, the object is blocked in using H, W, and D, and portions of the block are removed to complete the isometric drawing.

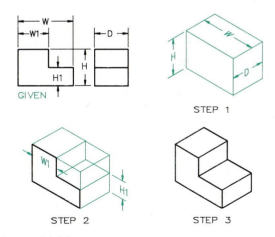

Figure 26.26 Layout of a simple isometric drawing.

Step 1 Construct an isometric drawing of a box by using the overall dimensions W, D, and H taken from the given view.

Step 2 Locate the notch in the box by using dimensions W1 and H1 taken from the given orthographic views.

Step 3 Darken the lines to complete the isometric drawing.

$\frac{6}{24} \times 12$

$\frac{6}{2} = 3$

$\frac{18}{24} \times 12$

$\frac{18}{12}$

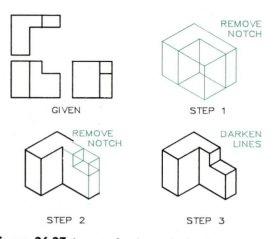

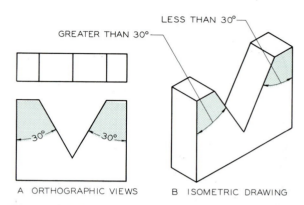

Figure 26.27 Layout of an isometric drawing.

Step 1 The overall dimensions of the object are used to block in the object with light lines. The notch is removed.

Step 2 The second notch is removed.

Step 3 The lines are darkened to complete the isometric drawing.

26.13   Angles in Isometric

Angles cannot be measured true size in an isometric drawing since the surfaces of an isometric are not true size. Angles must be located by using coordinates measured along isometric lines (Fig. 26.28). Lines *AD* and *BC* are equal in length in the orthographic view, but they are shorter and longer than true length in the isometric drawing.

A similar example can be seen in Fig. 26.29, where two angles are drawn in isometric. The

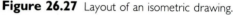

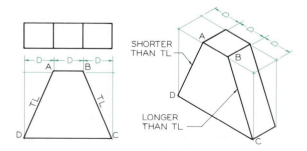

Figure 26.28 Inclined surfaces must be found by using coordinates measured along the isometric axes. The lengths of angular lines will not be true length in isometric.

Figure 26.29 Angles in isometric must be found by using coordinates. Angles will not appear true size in isometric.

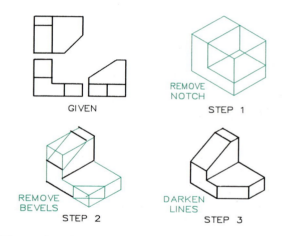

Figure 26.30 Inclined planes in isometric.

Step 1 The object is lightly blocked in using the overall dimensions, and the notch is removed.

Step 2 The ends of the inclined plane are located using measurements parallel to the isometric axes.

Step 3 The lines are darkened to complete the isometric drawing.

equal-size angles in orthographic are less than and greater than true size in the isometric drawing.

An isometric drawing of an object with an inclined surface is drawn in three steps in Fig. 26.30. The object is blocked in pictorially, and portions are removed.

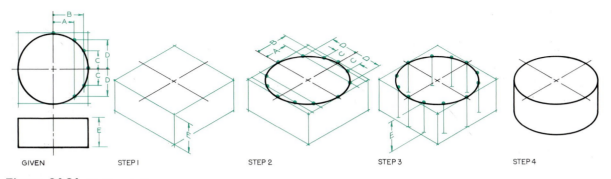

GIVEN STEP 1 STEP 2 STEP 3 STEP 4

Figure 26.31 Plotting circles.

Step 1 The cylinder is blocked in using the overall dimensions. The centerlines locate the points of tangency of the ellipse.

Step 2 Coordinates are used to locate points on the circumference of the circle.

Step 3 The lower ellipse is found by dropping each point a distance equal to the height of the cylinder, *E.*

Step 4 The two ellipses can be drawn with an irregular curve and connected with tangent lines to complete the cylinder.

26.14 Circles in Isometric

Three methods of constructing circles in isometric drawings are (1) **point plotting,** (2) **four-center ellipse construction,** and (3) **ellipse templates.**

Point Plotting

Point plotting is a method where a series of points on a circle are located by coordinates in the *X* and *Y* directions. The coordinates are transferred to the isometric drawing to locate the points on the ellipse one at a time. A series of points located on a circle can be located in an isometric drawing by using **coordinates** parallel to the isometric axes (Fig. 26.31).

The cylinder is blocked in and drawn pictorially, with the centerlines added in Step 1. Coordinates *A, B, C,* and *D* are used in Step 2 to locate points on the ellipse and then are connected using an irregular curve.

The lower ellipse located on the bottom plane of the cylinder can be found by using a second set of coordinates. The most efficient method is by measuring the distance, *E,* vertically beneath each point that was located on the upper ellipse (Step 3). A plotted ellipse is a true ellipse; it is equivalent to a 35° ellipse on an iso-

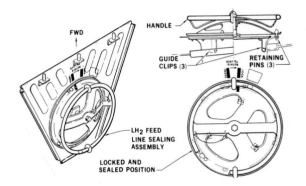

Figure 26.32 An example of parts drawn by using ellipses in isometric to represent circles. This is a handwheel proposed for use in an orbital workshop to be launched into space.

metric plane. An example of a design composed of circular features drawn in isometric is the handwheel shown in Fig. 26.32.

Four-Center Ellipse Construction

The **four-center ellipse** method can be used to construct an approximate ellipse in isometric by using four arcs drawn with a compass (Fig. 26.33). The four-center ellipse is drawn by blocking in the orthographic view of the circle with a square tangent to the circle at four points. The

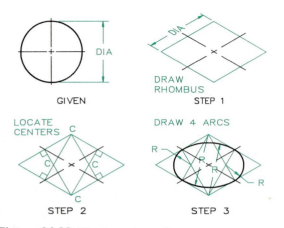

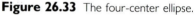

Figure 26.33 The four-center ellipse.

Step 1 The diameter of the given circle is used to draw a rhombus and centerlines.

Step 2 Light construction lines are drawn perpendicularly from the midpoints of each side to locate four centers.

Step 3 Four arcs are drawn from each center to represent an ellipse tangent to the rhombus.

four centers are found by constructing perpendiculars to the sides of the rhombus at the midpoints of the sides (Step 2). The four arcs are drawn to give the completed four-center ellipse (Step 3). This method can be used to draw ellipses on any of the three isometric planes, since each is equally foreshortened (Fig. 26.34).

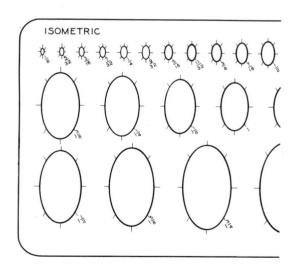

Figure 26.34 Four-center ellipses can be drawn on all three surfaces of an isometric drawing.

Figure 26.35 shows that the four-center ellipse is only an approximate ellipse when compared with a true ellipse.

Ellipse Templates

A specially designed **isometric ellipse template** can be used for drawing ellipses in isometric (Fig. 26.36).

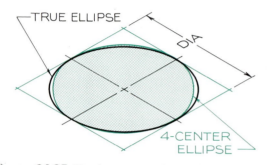

Figure 26.35 The four-center ellipse is not a true ellipse but an approximate ellipse.

Figure 26.36 The isometric template is designed to reduce drafting time. The isometric diameters of the ellipses are not the major diameters of the ellipses but are diameters parallel to the isometric axes.

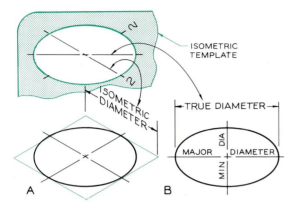

Figure 26.37 (A) The diameter of a circle in isometric is measured along the direction of the isometric axes. Therefore, the major diameter of an isometric ellipse is greater than the measured diameter. (B) The minor diameter is perpendicular to the major diameter.

The diameters of the ellipses on the template are measured along the direction of the isometric lines since this is how diameters are measured in an isometric drawing (Fig. 26.37). The maximum diameter across the ellipse is the **major diameter,** which is a true diameter. Thus the size of the diameter marked on the template is less than the ellipse's major diameter, the true diameter, since

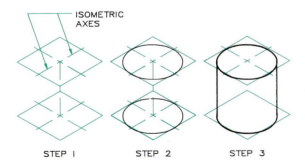

STEP 1   STEP 2   STEP 3

Figure 26.38 Cylinder: four-center method.

Step 1 A rhombus is drawn in isometric at each end of the cylinder's axis.

Step 2 A four-center ellipse is drawn within each rhombus.

Step 3 Lines are drawn tangent to each rhombus to complete the isometric drawing.

isometric drawings are drawn 18% larger than true projections.

The isometric ellipse template can be used to draw an ellipse by constructing the centerlines of the ellipse in isometric and aligning the ellipse template with these isometric lines (Fig. 26.37A).

26.15   Cylinders in Isometric

A cylinder can be drawn in isometric by using the four-center ellipse method (Fig. 26.38). A rhombus is drawn at each end of the cylinder's axis with the centerlines drawn as isometric lines (Step 1). The ellipses are drawn using the four-center ellipse method at each end (Step 2). The ellipses are then connected with tangent lines, and the lines are darkened (Step 3).

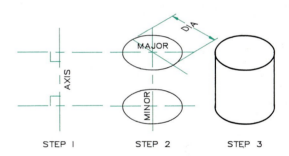

STEP 1   STEP 2   STEP 3

Figure 26.39 Cylinder: ellipse template.

Step 1 The axis of the cylinder is drawn to its proper length, and perpendiculars are drawn at each end.

Step 2 The elliptical ends are drawn by aligning the major diameter with the perpendiculars at the ends of the axis. The isometric diameter of the isometric ellipse template is given along the isometric axis.

Step 3 The ellipses are connected with tangent lines to complete the isometric drawing. Hidden lines are omitted.

A cylinder can also be drawn using the ellipse template (Fig. 26.39). The axis of the cylinder is drawn, and perpendiculars are constructed at each end (Step 1). Since the axis of a right cylinder is perpendicular to the major diameter of its elliptical end, the ellipse template is positioned as shown (Step 2). The ellipses are drawn at each end and are connected with tangent lines (Step 3).

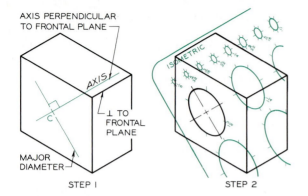

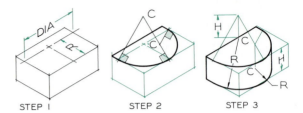

STEP I   STEP 2   STEP 3

Figure 26.41 Semicircular features.

Step 1 Objects with semicircular features can be drawn by blocking in the objects as if they had square ends. The centerlines are drawn to locate the centers and tangent points.

Step 2 Perpendiculars are drawn from each point of tangency to locate two centers. These are used to draw half a four-center ellipse.

Step 3 The lower surface can be drawn by lowering the centers by the distance of *H,* the thickness of the part. The same radii are used with these new centers to complete the isometric.

Figure 26.40 Cylinders in isometric.

Step 1 The center of the hole with a given diameter is located on a face of the isometric drawing. The axis of the cylinder is drawn from the center parallel to the isometric axis that is perpendicular to the plane of the circle. The major diameter is drawn perpendicular to the axis.

Step 2 The $1\frac{3}{8}''$ ellipse template is used to draw the ellipse by aligning the major and minor diameters with the guidelines on the template.

To construct a cylindrical hole in the block (Fig. 26.40), begin by locating the center of the hole on the isometric plane. The axis of the cylinder is drawn parallel to the isometric axis that is perpendicular to this plane through its center (Step 1). The ellipse template is aligned with the major and minor diameters to complete the elliptical view of the cylindrical hole (Step 2).

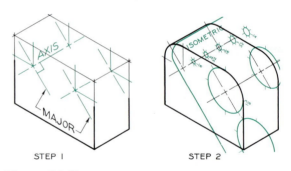

STEP I   STEP 2

Figure 26.42 Rounded corners.

Step 1 To construct rounded corners of an object, the centerlines of the ellipse are drawn.

Step 2 The elliptical corners are drawn with an ellipse template.

26.16 Partial Circular Features

When an object has a semicircular end, as in Fig. 26.41, the four-center ellipse method can be used with only two centers to draw half the circle (Step 2). To draw the lower ellipse at the bottom of the object, the centers are projected downward a distance of *H,* the height of the object. These centers are used with the same radii that were used on the upper surface to draw the arcs.

In Fig. 26.42, an object with rounded corners is blocked in, and the centerlines are located at each rounded corner (Step 1). An ellipse template is used to construct the rounded corners (Step 2).

The rounded corners could have been constructed by the four-center ellipse method or by plotting points on the arcs.

A similar drawing involving the construction of ellipses is the conical shape in Fig. 26.43. The ellipses are blocked in at the top and bottom surfaces (Step 1), and by using a template or the four-center method, the ellipses are drawn (Step 3), and tangent lines are drawn (Step 4).

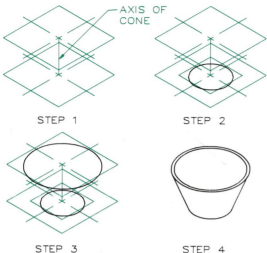

Figure 26.43 Construction of a cone in isometric.

Step 1 The axis of the cone is drawn and the larger end is blocked in at the top and bottom.

Step 2 The smaller end of the cone is blocked in.

Step 3 Ellipse are drawn in at both ends using an ellipse template or the four-center method

Step 4 The wall thickness of the cone is drawn and the lines are darkened to complete the drawing.

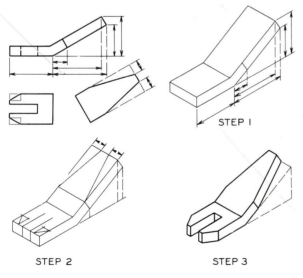

Figure 26.44 Inclined surfaces in isometric must be located by using coordinates laid off parallel to the isometric axes. True angles cannot be measured in isometric drawings (Step 1). Coordinates of the tapered sides are located with coordinates in Step 2 and the lines are strengthened in Step 3.

26.17 Measuring Angles

Angles in isometric may be located by coordinates, but they cannot be measured with a protractor since they will not appear true size.

The coordinate method is illustrated in Fig. 26.44 where horizontal and vertical coordinates (in the X- and Y-directions) are used to locate key points on the orthographic drawing. These coordinates are then transferred to the isometric drawing to develop the features of the inclined surface.

26.18 Curves in Isometric

Irregular curves must be plotted point by point, using coordinates to locate each point. Points A through F are located in the orthographic view with coordinates of width and depth (Fig. 26.45). These coordinates are transferred to the isometric view of the blocked-in part (Step 1) and are connected with an irregular curve.

Each point on the upper curve is projected downward for a distance of H, the height of the part, to locate points on the lower curve. The points are connected with an irregular curve to complete the isometric.

26.19 Ellipses on Nonisometric Planes

When an ellipse lies on a nonisometric plane, like the one shown in Fig. 26.46, points on the ellipse can be plotted to locate it. Coordinates are located in the orthographic views and then transferred to the isometric as shown in Steps 1 and 2. The plotted points can be connected with an irregular curve, or an ellipse template can be selected that will approximate the plotted points.

26.20 Machine Parts in Isometric

Figure 26.47 shows orthographic and isometric views of a spotface, countersink, and boss. The isometric drawings of these features can be drawn

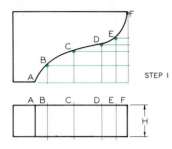

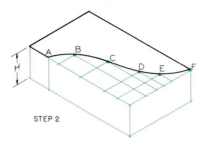

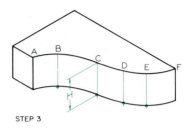

Figure 26.45 Plotting irregular curves.

Step 1 Draw two coordinates to locate a series of points on the irregular curve. These coordinates must be parallel to the standard W, D, and H dimensions.

Step 2 Block in the shape using overall dimensions. Locate points on the irregular curve using the coordinates from the orthographic views.

Step 3 Since the object has a uniform thickness, the lower curve can be found by projecting downward the distance H from the upper points.

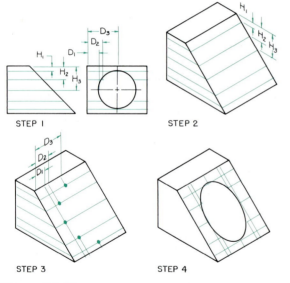

Figure 26.46 Construction of ellipses on an inclined plane.

Step 1 Coordinates are established in the orthographic views.

Step 2 One set of coordinates is transferred to the isometric view.

Step 3 The second set of coordinates is transferred to the isometric to establish points on the ellipse.

Step 4 The plotted points are connected with an elliptical curve. An ellipse template can usually serve as a guide for connecting the points.

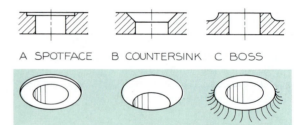

Figure 26.47 Examples of circular features drawn in isometric. These can be drawn by using ellipse templates.

by point-plotting the circular features, using the four-center method, or using an ellipse template, which is the easiest method.

A **threaded shaft** can be drawn in isometric, as Fig. 26.48 shows, by first drawing the cylinder in isometric (Step 1). Next, the major diameters of the crest lines of the thread are drawn separated at a distance of P, the pitch of the thread (Step 2). Ellipses are then drawn by aligning the major diameter of the ellipse template with the perpendiculars to the cylinder's axis (Step 3). Note that the 45° chamfered end is drawn using a smaller ellipse at the end.

A **hexagon-head nut** (Fig. 26.49) is drawn in three steps using an ellipse template. The nut is blocked in, and an ellipse drawn tangent to the

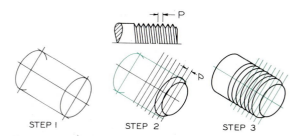

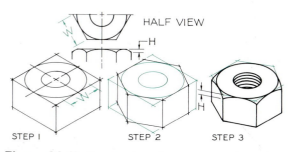

Figure 26.48 Threads in isometric.

Step 1 Draw the cylinder to be threaded by using an ellipse template.

Step 2 Lay off perpendiculars that are spaced by a distance equal to the pitch of the thread, P.

Step 3 Draw a series of ellipses to represent the threads. The chamfered end is drawn by using an ellipse whose major diameter is equal to the root diameter of the threads.

Figure 26.49 Construction of a nut.

Step 1 The overall dimensions of the nut are used to block in the nut.

Step 2 The hexagonal sides are constructed at the top and bottom.

Step 3 The chamfer is drawn with an irregular curve. Threads are drawn to complete the isometric.

rhombus. The hexagon is constructed by locating the distance across a flat, W, parallel to the isometric axes. The other sides of the hexagon are found by drawing lines tangent to the ellipse (Step 2). The distance H is laid off at each corner to establish the amount of chamfer at each corner (Step 3).

A **hexagon-head bolt** is drawn in two positions in Fig. 26.50. The washer face can be seen on the lower side of the head, and the chamfer on the upper side of the head.

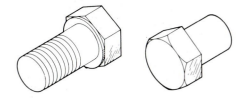

Figure 26.50 Isometric drawings of the upper and lower sides of a hexagon-head bolt.

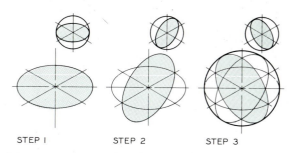

Figure 26.51 An isometric sphere.

Step 1 The three intersecting isometric axes are drawn. The ellipse template is used to draw the horizontal elliptical section.

Step 2 The isometric ellipse template is used to draw one of the vertical elliptical sections.

Step 3 The third vertical elliptical section is drawn, and the center is used to draw a circle tangent to the three ellipses.

To draw a **sphere** in isometric, three ellipses are drawn as isometric planes with a common center (Fig. 26.51). The center is used to construct a circle that will be tangent to each ellipse, as shown in Step 3.

A portion of a sphere is used to draw a **round-head screw** in Fig. 26.52. A hemisphere is constructed in Step 1. The centerline of the slot is located along one of the isometric planes. The thickness of the head is measured at a distance of E from the highest point on the sphere.

26.21 Isometric Sections

A full section can be drawn in isometric to clarify internal details that might otherwise be overlooked (Fig. 26.53). Half sections can also be used, as Fig. 26.54 shows.

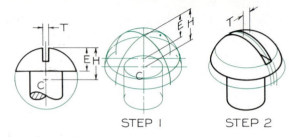

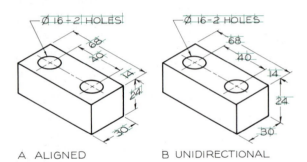

Figure 26.52 Spherical features.

Step 1 An isometric ellipse template is used to draw the elliptical features of a round-head screw.

Step 2 The slot in the head is drawn, and the lines are darkened to complete the isometric of the head.

A ALIGNED

B UNIDIRECTIONAL

Figure 26.55 Dimensions can be placed on isometric drawings by using either technique shown here. Guidelines should always be used for lettering.

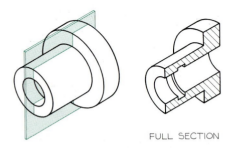

FULL SECTION

Figure 26.53 Parts can be shown in isometric sections to clarify internal features.

26.23 Fillets and Rounds

Fillets and rounds in isometric can be represented by any of the methods shown in Fig. 26.56 to give added realism to a pictorial drawing. Figures 26.56A and 26.56B show how intersecting guidelines are drawn equal in length to the radii of the fillets and rounds, and how arcs are drawn tangent to these lines. These arcs can be drawn freehand or with an ellipse template. The method in Fig. 26.56C uses freehand lines drawn parallel to or concentric with the directions of the fillets and rounds. Figure 26.57 shows an example of these two methods of showing fillets and rounds . The stipple shading was applied by using an adhesive overlay film.

When fillets and rounds are illustrated as shown in Fig. 26.58 and dimensions are applied, it is much easier to understand the features of the part than when it is represented by orthographic views.

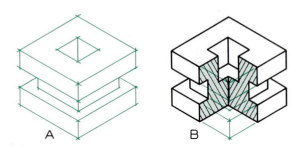

Figure 26.54 An isometric drawing of a half section (B) can be constructed.

26.22 Dimensioned Isometrics

When it is advantageous to dimension and note a part shown in isometric, either the aligned or the unidirectional method can be used to apply the notes (Fig. 26.55). In both cases, notes connected with leaders are positioned horizontally. Always use guidelines for your lettering and numerals.

26.24 Isometric Asemblies

Assemblies are used to explain how parts are assembled. Figure 26.59A shows common mistakes in applying leaders to an assembly; whereas, Fig. 26.59B shows the more acceptable techniques. The numbers in the circles ("balloons") refer to the number given to each part listed in the parts list.

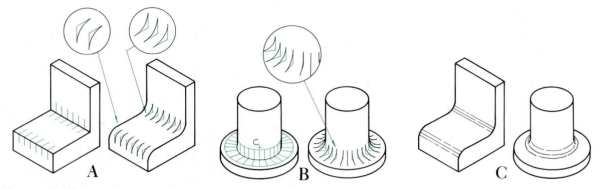

Figure 26.56 Representation of fillets and rounds. (A) Fillets and rounds can be represented by segments of an isometric ellipse if guidelines are constructed at intervals. (B) Fillets and rounds can be represented by elliptical arcs by constructing radial guidelines. (C) Fillets and rounds can be represented by lines that run parallel to the fillets and rounds.

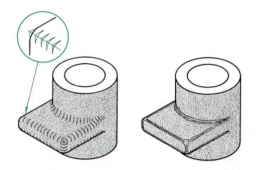

Figure 26.57 Two methods of representing fillets and rounds on a part.

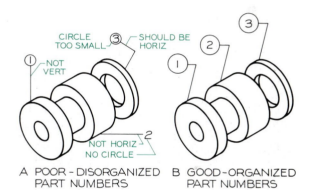

Figure 26.59 (A) Common mistakes in applying part numbers to an assembly. (B) Acceptable techniques of applying part numbers.

Figure 26.60 shows an assembly from a parts manual along with its parts list. This assembly is exploded, which makes it clear how the parts are to be assembled.

26.25 Axonometric Pictorials

An **axonometric pictorial** is a form of orthographic projection in which the pictorial view is projected perpendicularly onto the picture plane with parallel projectors. The object is positioned in an angular position with the picture plane so

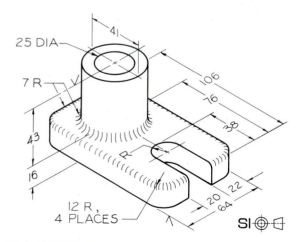

Figure 26.58 An isometric drawing with complete dimensions and fillets and rounds represented.

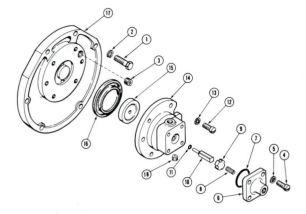

Figure 26.60 An exploded isometric assembly.

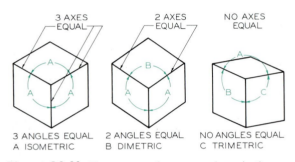

Figure 26.61 Three types of axonometric projection.

that its pictorial projection will be a three-dimensional view. Three types of axonometric pictorials are possible: (1) **isometric,** (2) **dimetric,** or (3) **trimetric.**

The **isometric projection** is the view where the diagonal of a cube is viewed as a point. The planes will be equally foreshortened and the axes equally spaced 120° apart (Fig. 26.61A). The measurements along the three axes will be equal but less than true length since this is true projection.

A **dimetric projection** is the view where two planes are equally foreshortened, and two of the axes are separated by equal angles (Fig. 26.61B). The measurements along two of the axes are equal.

A **trimetric projection** is the view where all three planes are unequally foreshortened, and the angles between the three axes are different (Fig. 26.61C).

26.26 Perspective Pictorials

A **perspective pictorial** is a view normally seen by the eye or a camera; it is the most realistic form of pictorial. In a perspective, all parallel lines converge at infinite vanishing points as they recede from the observer.

The three basic types of perspectives are (1) **one point,** (2) **two point,** and (3) **three point,** depending on the number of vanishing points in their construction (Fig. 26.62).

The **one-point perspective** has one surface of the objective parallel to the picture plane; therefore, it is true shape. The other sides vanish to a single point on the horizon, called a **vanishing point** (Fig. 26.62A).

A **two-point perspective** is a pictorial positioned with two sides at an angle to the picture

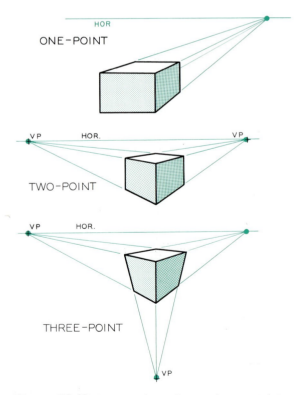

Figure 26.62 A comparison of one-point, two-point, and three-point perspectives.

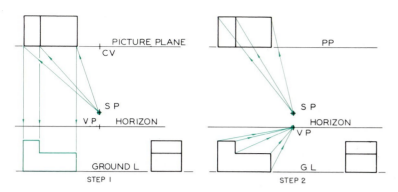
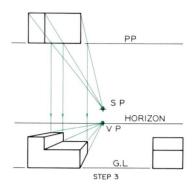

Figure 26.63 Construction of a one-point perspective.

Step 1 Since the object is parallel to the picture plane, there will be only one vanishing point, located on the horizon below the station point. Projections from the top and side views establish the front plane. This surface is true size, since it lies in the picture plane.

Step 2 Draw projectors from the station point to the rear points of the object in the top view and from the front view to the vanishing point on the horizon. In a one-point perspective, the vanishing point is the front view of the station point.

Step 3 Construct vertical projectors from the top view to the front view from the points where the projectors cross the picture plane. These projectors intersect the lines leading to the vanishing point. This is a one-point perspective, since the lines converge at a single VP.

plane; this requires two vanishing points (Fig. 26.62B). All horizontal lines converge at the vanishing points, but vertical lines remain vertical and have no vanishing point.

The **three-point perspective** has three vanishing points since the object is positioned so that all sides of it are at an angle with the picture plane (Fig. 26.62C). The three-point perspective is used in drawing larger objects such as buildings.

26.27   One-Point Perspectives

The steps of drawing a one-point perspective are shown in Fig. 26.63, which shows the top and side views of the object, picture plane, station point, horizon, and ground line.

Picture Plane   is the plane the perspective is projected on. It appears as an edge in the top view.

Station Point (SP)   is the location of the observer's eye in the plan view. The front view of the station point will always lie on the horizon.

Horizon   is a horizontal line in the front view that represents an infinite horizontal, such as the surface of the ocean.

Ground Line   is an infinite horizontal line in the front view that passes through the base of the object being drawn.

Center of Vision (CV)   is a point that lies on the picture plane in the top view and on the horizon in the front view. In both cases, it is on the line from the station point that is perpendicular to the picture plane.

When drawing any perspective, the station point should be far enough away from the object so that the perspective can be contained in a cone of vision not more than 30° (Fig. 26.64). If a larger cone of vision is required, the perspective will be distorted.

26.28   Two-Point Perspectives

If two surfaces of an object are positioned at an angle to the picture plane, two vanishing points will be required to draw it as a perspective. Dif-

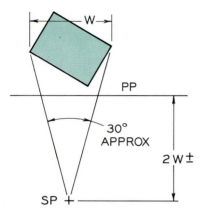

Figure 26.64 The station point (SP) should be far enough away from the object to permit the cone of vision to be less than 30° to reduce distortion.

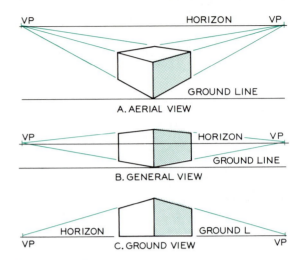

Figure 26.65 Different perspectives can be obtained by locating the horizontal over, under, and through the object.

ferent views can be obtained by changing the relationship between the horizontal and the ground line (Fig. 26.65).

An **aerial view** is obtained when the horizon is placed above the ground line and the top of the

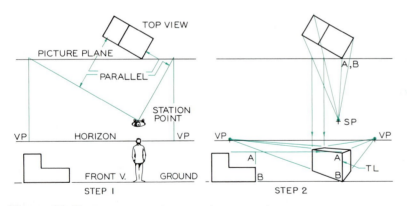

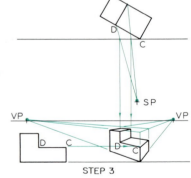

Figure 26.66 Construction of a two-point perspective.

Step 1 Construct projectors that extend from the top view of the station point to the picture plane parallel to the forward edges of the object. Project these points vertically to the horizon in the front view to locate vanishing points. Draw the ground line below the horizon, and construct the side view on the ground line.

Step 2 Since all lines in the picture plane are true length, line *AB* is true length. Thus, line *AB* is projected from the side view to determine its height. Then project each end of *AB* to the vanishing points. Draw projectors from the SP to the exterior edges of the top view. Project the intersections of these projectors with the picture plane to the front view.

Step 3 Determine point *C* in the front view by projecting from the side view to line *AB*. Draw a projector from point *C* to the left vanishing point. Point *D* will lie on this projector beneath the point where a projector from the station point to the top view of point *D* crosses the picture plane. Complete the notch by projecting to the respective vanishing points.

object in the front view. When the ground line and horizon coincide in the front view, a **ground-level view** is obtained. A **general view** is obtained when the horizon is placed above the ground line and through the object, usually equal to the height of a person.

Figure 26.66 shows the steps of constructing a two-point perspective.

Since line *AB* lies in the picture plane, it will be true length in the perspective. All height dimensions must originate at this vertical line because it is the only true-length line in the perspective. Points *C* and *D* are found by projecting to *AB* and then projecting toward the vanishing points.

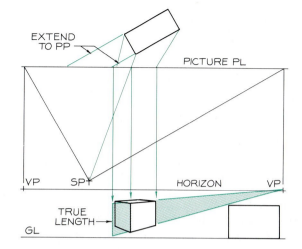

Figure 26.68 A two-point perspective of an object that is not in contact with the picture plane.

drawn to the vanishing point. The corner of the object can be located on this infinite plane by projecting the corner to the picture plane in the top view with a projector from the station point.

Arcs in Perspective

Arcs in two-point perspectives must be found by using coordinates to locate points along the curves in perspective (Fig. 26.69). Points 1

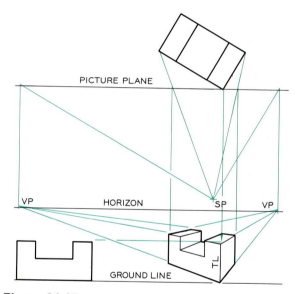

Figure 26.67 A two-point perspective of an object.

Figure 26.67 shows a typical two-point perspective. By referring to Fig. 26.66, you will be able to understand the development of the construction used.

The object in Fig. 26.68 does not contact the picture plane in the top view as in the previous examples. To draw a perspective of this object, the lines of the object must be measured where the extended plane intersects the picture plane. The height is measured, and the infinite plane is

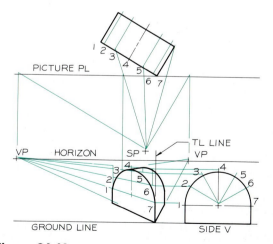

Figure 26.69 A two-point perspective of an object with semicircular features.

through 7 are located along the semicircular arc in the orthographic view. These same points are found in the perspective by projecting coordinates from the top and side views. The points do not form a true ellipse but an egg-shaped oval. An irregular curve is used to connect the points.

26.29 Axonometric Pictorials by Computer

When using AutoCAD, an ISOMETRIC style grid and snap can be used as an aid in making isometric drawings. The STYLE subcommand of the SNAP command can be used to change the rectangular GRID (called STANDARD) (S) to ISOMETRIC (I) where the dots are plotted vertically and at 30° to the horizontal (Fig. 26.70). While in this mode, the cursor's cross hairs, which can be made to snap to the grid points, will align with two axes of the isometric.

Although an isometric drawing can be made by using the grid and snap features, this method is not truly a three-dimensional system; instead, points must be located manually, and elliptical features must be inserted individually, as Fig. 26.71 shows.

Isometric ellipses can be automatically drawn by using the ISOCIRCLE option under the ELLIPSE command. When in this command, you can change the position of the cursor to be aligned with each of the three isometric planes by

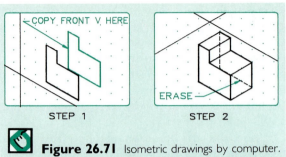

Figure 26.71 Isometric drawings by computer.

Step 1 When the isometric grid is on the screen, the cursor can be made to SNAP to it. The front view is drawn as an isometric plane and duplicated at the back of the object by using the COPY command.

Step 2 The hidden lines are ERASEd to complete the isometric drawing.

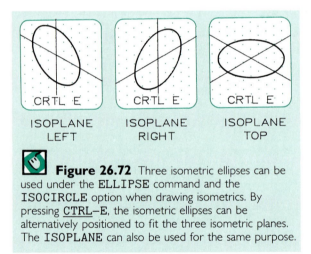

Figure 26.72 Three isometric ellipses can be used under the ELLIPSE command and the ISOCIRCLE option when drawing isometrics. By pressing CTRL–E, the isometric ellipses can be alternatively positioned to fit the three isometric planes. The ISOPLANE can also be used for the same purpose.

pressing CTRL–E on the keyboard (Fig. 26.72). When the cursor is aligned with the proper axes of an isometric plane, you can select the center of the isometric ellipse or its diameter (Fig. 26.73).

The ISOPLANE command can be used to change the position of the cursor in the same manner as CTRL–E. When using ISOPLANE, you will be prompted to select from Left/Top/Right/ ‹Toggle›: options. The Toggle option can be used by pressing (CR) to successively move the cursor position from plane to plane. The other options are self-explanatory.

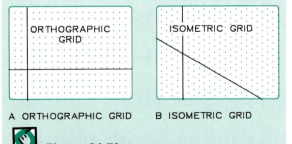

Figure 26.70 The SNAP command permits you to rotate the orthographic grid (Standard) or to select the isometric grid option that can be used as an aid in drawing isometric pictorials.

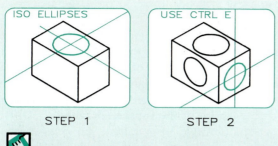

Figure 26.73 Isometric ellipses.

Step 1 While in the Isometric-grid mode, draw isometric ellipses as follows:
```
Command: ELLIPSE (CR)
<Axis endpoint 1>:/Center/Isocircle:
I (CR)
Center of Circle: (Select with cursor.)
<Circle radius>/Diameter: (Select radius
with cursor.)
```
Step 2 Change the orientation of the cursor for drawing isometric ellipses on the other two planes by pressing **CTRL–E**. Repeat the process in Step 1.

26.30 3D Computer Software

A growing number of software programs are available for producing true three-dimensional drawings on the microcomputer that once could only be done by large, mainframe computers. Examples of drawings made by three software programs are covered very briefly in this section. Additional coverage of AutoCAD's 3D options is given in Chapter 38. The programs presented here are (1) **Design Board Professional**®, (2) **DynaPerspective**®, and (3) **AutoCAD**®.

Design Board Professional

Design Board Professional®, developed by MegaCADD, Inc., provides three modes for developing orthographic and perspective drawings of objects: CREATE, VIEW, and MODIFY. The CRE-ATE mode (Fig. 26.74) shows a screen on which a perspective can be created by drawing three-dimensional shapes in their plan views.

The VIEW mode provides a screen on which you may select a viewpoint to look at the orthographic views (Fig. 26.75). The resulting perspec-

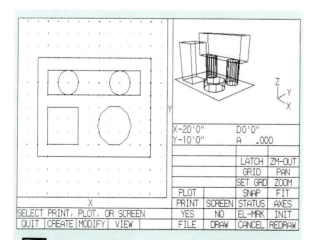

Figure 26.74 The CREATE screen of MegaCADD's Design Board Professional® is used to create the orthographic views of the objects that will be converted into perspective views.

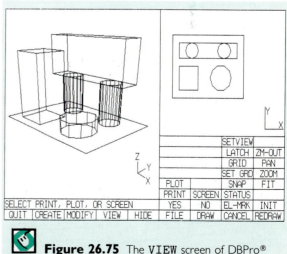

Figure 26.75 The VIEW screen of DBPro® shows the plan view of the model and its corresponding perspective view. The viewpoint can be changed by selecting the observer's position and the target position in the plan view.

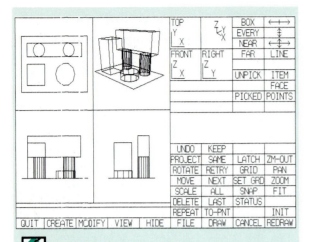

Figure 26.76 The MODIFY screen of DBPro® shows three orthographic views and the perspective of the model created. Modifications of the model can be made in this mode.

tive view will be shown in the large area at the left of the screen.

The MODIFY option (Fig. 26.76) gives three orthographic views of the model as well as its perspective view. Parts of the drawing can be deleted, changed, moved, enlarged, and edited in several other ways. The DRAW command is used to obtain a lot of the finished drawing.

An example of plotted drawing, Mega-CADD's Prohouse, is shown in Fig. 26.77, where hidden lines have been removed from the original wire-diagram drawing. Plots of this type can be scaled to various sizes and tinted with colors. Viewpoints can be selected for different views of the house and many other options. A series of views can be generated to represent a "walk-through" of the house by changing viewpoints and target points within the house.

Another application is the representation of an entire downtown area (Fig. 26.78). Mega-CADD enables you to position yourself at any location and any height to view the city.

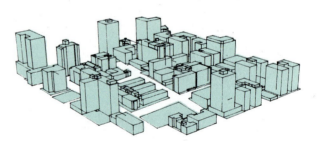

Figure 26.78 An urban area has been plotted as a perspective drawing with hidden lines removed using MegaCADD's Design Board 3D. (Courtesy of MegaCADD, Inc.)

DynaPerspective

DynaPerspective, developed by Dynaware Corporation, is a powerful program that produces orthographic and perspective drawings of three-dimensional drawings. This solid-modeling package can be used to draw, color, and rotate objects about three axes.

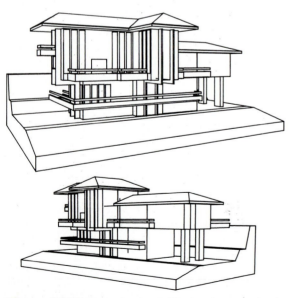

Figure 26.77 An example of different views obtained from the same house developed by DBPro.

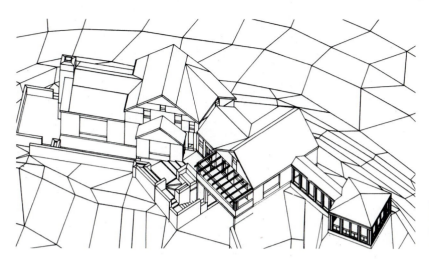

Figure 26.79 This perspective of a house and its site was drawn using DynaPerspective® software as a 3D drawing.

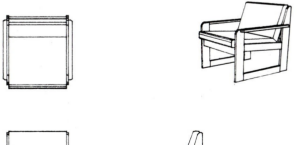

Figure 26.80 DynaPerspective® has many three-dimensional applications for engineers and architects such as this residence.

Figure 26.81 Drawings can be developed simultaneously on the screen to show orthographic and perspective views using DynaPerspective®

A portion of a complex example is the 3D drawing of a house positioned on its irregular site in Fig. 26.79. Because the drawing is truly three-dimensional, the house can be viewed from the outside or "entered" and viewed from within as well. In this case hidden lines have been removed to improve readability.

The house in Fig. 26.80 was drawn with DynaPerspective. Sectional views of its interior can be shown, since it has been created in its entirety inside and out. Furniture, such as the chair in Fig. 26.81, is provided with the software for the architect and designer. The chair can be inserted in the drawing where it is shown in its orthographic views and perspective on the screen.

Once inserted in the drawing, the chair (as well as other furniture) can be rotated about any of the three axes. In Fig. 26.82, the chair has been rotated about its vertical axis. A number a views can be selected in this manner and saved for a slide presentation that enables the designer to show in sequence a number of views of a three-dimensional drawing.

DynaPerspective is a user-friendly, sophisticated program that has many applications. Unfortunately space does not permit an adequate coverage in this book.

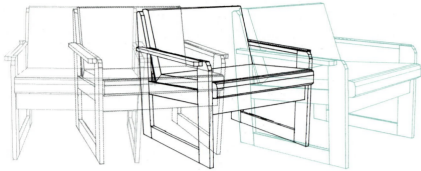

Figure 26.82 DynaPerspective software enables you to rotate objects about three axes. In this example, the chair is rotated about its vertical axis.

Figure 26.83 An example of 3D wire-diagram drawing of a chair by AutoCAD®.

Figure 26.84 The drawing of this airplane was first drawn as a wire diagram and then the invisible lines were hidden by AutoCAD's 3D option.

AutoCAD

AutoCAD's Version 10 offers a program for drawing three-dimenational objects that can be rotated for viewing from different positions. The chair in Fig. 26.83 is an example of a wire-diagram drawing that shows all lines as visible. AutoCAD can hide the invisible lines in order to get a more realistic view of any 3D drawing.

An example of an object with hidden lines omitted is the airplane in Fig. 26.84. The airplane can also be viewed from an infinite number of positions, scaled to an infinite number of sizes, and shaded for improved realism. AutoCAD's 3D features are covered in greater detail in Chapter 38.

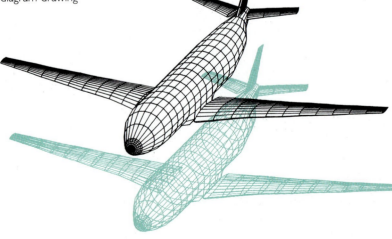

The Future

The future of 3D graphics is truly exciting. What is available today for the microcomputer was not possible on much larger and more expensive computers a few years ago. No doubt the capa-bilities of three-dimensional programs will become more powerful and easier to use with each passing year.

 **Problems**

The following problems (Fig. 26.85) are to be shown on Size A or B sheets as assigned. Select the appropriate scale that will take advantage of the space available on each sheet. By setting each square to 0.20 inch (about 5 mm), two problems can be drawn on each Size A sheet. When each square is set to 0.40 inch (about 10 mm), one problem can be drawn on each Size B sheet.

Obliques

1–24. Construct either cavalier, cabinet, or general obliques of the parts assigned in Fig. 26.85.

Isometrics

1–24. Construct isometric drawings of the objects assigned in Fig. 26.85.

Perspectives

1–24. Lay out perspective views of the assigned parts in Fig. 26.85 on Size B sheets.

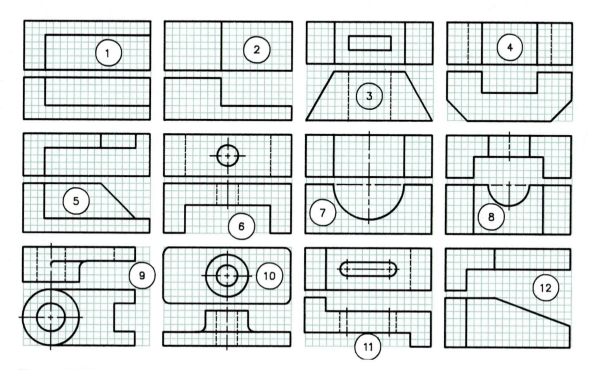

Figure 26.85 Problems 1–24.

(continued)

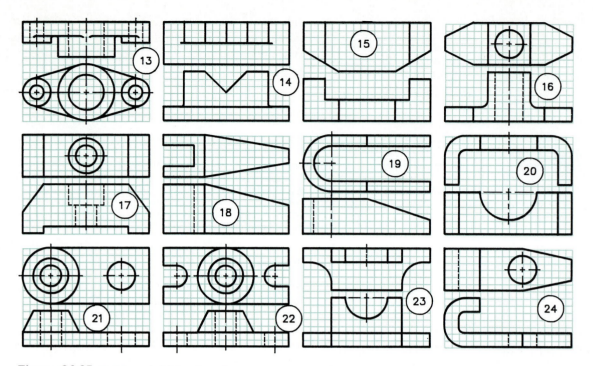

Figure 26.85 Problems 1–24 (continued).

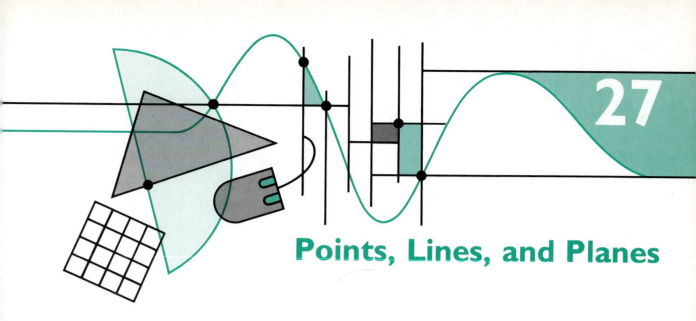

Points, Lines, and Planes

27

27.1 Introduction

Points, lines, and planes are the basic geometric elements used in descriptive geometry. The following rules of solving and labeling descriptive problems are illustrated in Fig. 27.1.

Lettering   should be labeled using $\frac{1}{8}$-in. letters with guidelines. Lines should be labeled at each end and planes at each corner. Either letters or numbers can be used.

Points   should be indicated by two short perpendicular dashes that form a cross, *not* a dot. Each dash should be approximately $\frac{1}{8}$-in. long.

Points on a Line   should be indicated with a short perpendicular dash on the line, *not* a dot. Label the point with a letter or numeral.

Reference Lines   are thin black lines that should be labeled in accordance with the instructions given in Section 27.1.

Object Lines   are used to represent points, lines, and planes. They should be drawn heavier

than reference lines, with an H or F pencil. Hidden lines are drawn thinner than visible lines.

True-Length Lines   should be labeled by the full note, TRUE LENGTH, or by the abbreviation TL.

True-Size Planes   should be labeled by a note, TRUE SIZE, or by the abbreviation TS.

Projection Lines   should be precisely drawn with a 4H pencil. These should be thin, gray lines, just dark enough to be visible. They need not be erased after the problem is completed.

27.2 Orthographic Projection of a Point

A point is a theoretical location in space; it has no dimension. However, a series of points can establish areas, volumes, and lengths.

A point must be projected perpendicularly onto at least two principal planes to establish its position (Fig. 27.2). Note that when the planes of

469

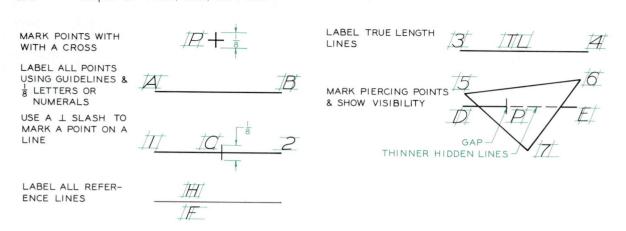

MARK POINTS WITH
WITH A CROSS

LABEL ALL POINTS
USING GUIDELINES &
$\frac{1}{8}$ LETTERS OR
NUMERALS

USE A ⊥ SLASH TO
MARK A POINT ON A
LINE

LABEL ALL REFER-
ENCE LINES

LABEL TRUE LENGTH
LINES

MARK PIERCING POINTS
& SHOW VISIBILITY

GAP
THINNER HIDDEN LINES

Figure 27.1 Standard practices for labeling points, lines, and planes.

the projection box (Fig. 27.2A) are opened into the plane of the drawing surface (Fig. 27.2B), the projectors from each view of point 2 are perpendicular to the reference lines between the planes. The letters *H, F,* and *P* are used to represent the horizontal, frontal, and profile planes, the three principal projection planes.

A point can be located from verbal descriptions with respect to the principal planes. For example, point 2 in Fig. 27.2 can be described as being (1) 4 units left of the profile plane, (2) 3 units below the horizontal plane, and (3) 2 units behind the frontal plane.

When looking at the front view, the horizontal and profile planes appear as edges. The frontal and profile planes appear as edges in the top view, and the frontal and horizontal planes appear as edges in the side view.

27.3 Lines

A line is a straight path between two points in space. A line can appear as (1) a foreshortened line, (2) a true-length line, or (3) a point (Fig. 27.3).

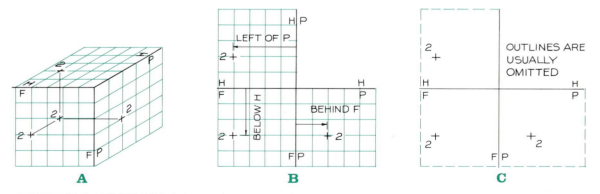

A   **B**   **C**

Figure 27.2 (A) The three projections of point 2 are shown pictorially. (B) The three projections are shown orthographically. The projection planes are opened into the plane of the drawing paper. Point 2 is 4 units to the left of the profile, 3 below the horizontal, and 2 behind the frontal. (C) The outlines of the projection are usually omitted in orthographic projection.

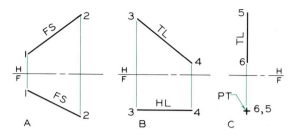

Figure 27.3 A line in orthographic projection can appear as (A) foreshortened (FS), (B) true length (TL), or (C) a point (PT).

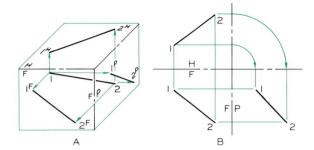

Figure 27.4 (A) A pictorial of the orthographic projection of a line. (B) A standard orthographic projection of a line.

Oblique Lines are neither parallel nor perpendicular to a principal projection plane (Fig. 27.4). When line 1–2 is projected onto the horizontal, frontal, and profile planes, it appears foreshortened in each view.

Principal Lines are parallel to at least one of the principal projection planes. A principal line is true length in the view where the principal plane

to which it is parallel appears true size. The three types of principal lines are **horizontal, frontal,** and **profile lines.**

A **horizontal line** is shown in Fig. 27.5A, where it appears true length in the horizontal view, the top view. It may be shown in an infinite number of positions in the top view and still

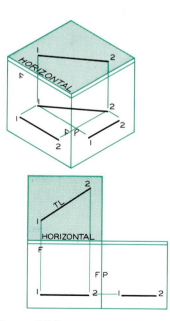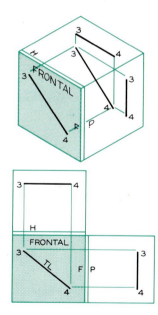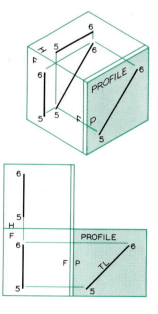

Figure 27.5 Principal lines.

(A) The horizontal line is true length in the horizontal view (the top view). It will appear parallel to the edge view of the horizontal plane in the front and side views.

(B) The frontal line is true length in the front view. It will appear parallel to the edge view of the frontal plane in the top and side views.

(C) The profile line is true length in the profile view (the side view). It will appear parallel to the edge view of the profile plane in the top and front views.

appear true length provided it is parallel to the horizontal plane.

An observer cannot tell whether the line is horizontal when looking at the top view. This must be determined from looking at the front or side views where a horizontal line will be parallel to the edge view of the horizontal, which is the *H-F* fold line. A line that projects as a point in the front view is a combination horizontal and profile line.

A **frontal line** is parallel to the frontal projection plane, and it appears true length in the front view since the observer's line of sight is perpendicular to it in this view. Line 3–4 in Fig. 27.5B is determined to be a frontal line by observing its top and side views where the line is parallel to the edge view of the frontal plane.

A **profile line** is parallel to the profile projection planes and it appears true length in the side views, the profile views. To tell whether or not a line is a profile line, it is necessary to look at a view adjacent to the profile view. In Fig. 27.5C, line 5–6 is parallel to the edge view of the profile plane in the top and side views.

27.4 Location of a Point on a Line

The top and front views of line 1–2 are shown in Fig. 27.6. Point 0 is located on the line in the top view, and it is required that the front view of the point be found.

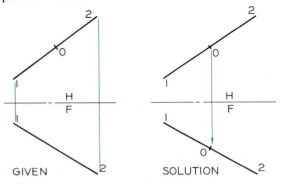

Figure 27.6 A point on a line shown orthographically can be found on the front view by projection. The direction of the projection is perpendicular to the reference line between the two views.

Since the projectors between the views are perpendicular to the *H-F* fold line in orthographic projection, point 0 is found by projecting in this same direction from the top view to the front view of the line.

If a point is to be located at the midpoint of a line, it will be at the line's midpoint whether the line appears true length or foreshortened.

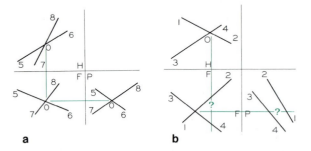

Figure 27.7 (a) The lines intersect because the point of intersection, 0, projects as a point of intersection in all views. (b) The lines cross in the top and front views, but they do not intersect. A common point of intersection does not project from view to view.

27.5 Intersecting and Nonintersecting Lines

Lines that intersect have a point of intersection that lies on both lines. Point 0 in Fig. 27.7a is a point of intersection since it projects to a common crossing point in the three views.

On the other hand, the crossing point of the lines in Fig. 27.7b in the front view is not a point of intersection. Point 0 does not project to a common crossing point in the top view. Therefore the lines do not intersect although they do cross.

27.6 Visibility of Crossing Lines

Lines *AB* and *CD* in Fig. 27.8 do not intersect; however, it is necessary to determine the visibility of the lines by analysis. Select a crossing point in one of the views, the front view in Step 1, and project it to the top view to determine which line is in front of the other. This process of determin-

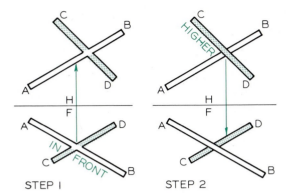

Figure 27.8 Visibility of lines.

Required Find the visibility of the lines in both views.

Step 1 Project the point of crossing from the front to the top view. This projector strikes *AB* before *CD;* therefore, line *AB* is in front and is visible in the front view.

Step 2 Project the point of crossing from the top view to the front view. This projector strikes *CD* before *AB;* therefore, *CD* is above *AB* and is visible in the top view.

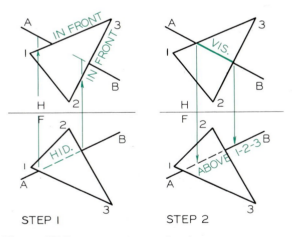

Figure 27.9 Visibility of line and a plane.

Required Find the visibility of the plane and the line in both views.

Step 1 Project the points where *AB* crosses the plane in the front view to the top view. These projectors encounter lines 1–3 and 2–3 of the plane first; therefore, the plane is in front of the line, making the line invisible in the front view.

Step 2 Project the points where *AB* crosses the plane in the top view to the front view. These projectors encounter *AB* first; therefore, the line is higher than the plane, and the line is visible in the top view.

ing visibility is done by analysis rather than by visualization. If only one view were available, it would be impossible to determine visibility.

27.7 Visibility of a Line and a Plane

The principle of visibility of intersecting lines is used in determining the visibility for a line and a plane. In Step 1 of Fig. 27.9, the intersections of *AB* and lines 1–3 and 2–3 are projected to the top view to determine that the lines of the plane are in front of *AB* in the front view. Therefore the line is shown as a dashed line in the front.

Similarly, the two intersections on *AB* in the top view are projected to the front view, where line *AB* is found to be above the two lines of the plane, 2–3 and 1–3. Since it is above the plane, *AB* is drawn as visible in the top view.

27.8 Planes

A plane can be represented by any of the four methods shown in Fig. 27.10. Planes in orthographic projection can appear as (1) an **edge,** (2)

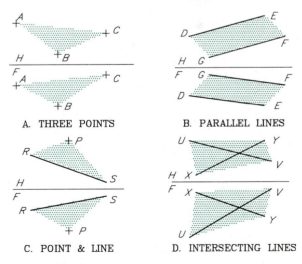

Figure 27.10 Representations of a plane. (A) Three points not on a straight line. (B) Two parallel lines. (C) A line and a point not on the line or its extension. (D) Two intersecting lines.

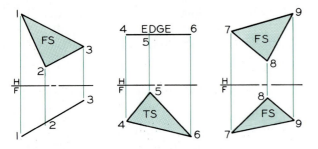

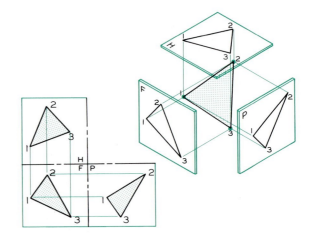

Figure 27.11 A plane in orthographic projection can appear as an edge, true size (TS), or foreshortened (FS). If a plane is foreshortened in all principal views, it is an oblique plane.

true-size plane, (3) a **foreshortened plane** (Fig. 27.11).

Oblique Planes are not parallel to principal projection planes in any view (Fig. 27.12).

Principal Planes are parallel to projection planes (Fig. 27.13). The three types of principal planes are **frontal, horizontal,** and **profile** planes.

A **frontal plane** is parallel to the frontal projection plane, and it appears true size in the front view. To tell that the plane is frontal, you must observe the top or side views where its parallel-

ism to the edge view of the frontal plane can be seen.

A **horizontal plane** is parallel to the horizontal projection plane, and it is true size in the top view. To tell that the plane is horizontal, you must observe the front or side views where its parallelism to the edge view of the horizontal plane can be seen.

Figure 27.12 An oblique plane is one that is neither parallel nor perpendicular to a projection plane; it can be called a general-case plane.

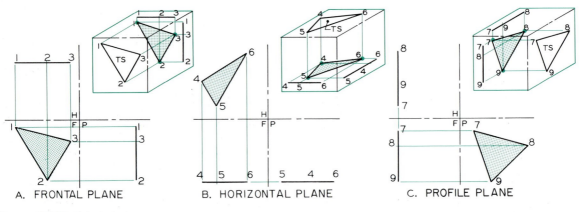

A. FRONTAL PLANE   **B. HORIZONTAL PLANE**   **C. PROFILE PLANE**

Figure 27.13 Principal planes.

(A) The horizontal line is true length in the horizontal view (the top view). It will appear parallel to the edge view of the horizontal plane in the front and side views.

(B) The frontal line is true length in the front view. It will appear parallel to the edge view of the frontal plane in the top and side views.

(C) The profile line is true length in the profile view (the side view). It will appear parallel to the edge view of the profile plane in the top and front views.

A **profile plane** is parallel to the profile projection plane, and it is true size in the side view. To tell that it is a profile plane, you must observe the top or front views where its parallelism to the edge view of the profile plane can be seen.

27.9   A Line on a Plane

Line *AB* is given on the front view of the plane in Fig. 27.14. It is required to find the top view of the line. Points *A* and *B* that lie on lines 1–4 and 2–3 of the plane can be projected to the top view to the same lines of the plane. Points *A* and *B* are found in the top view and are connected to complete the top view of line *AB*. This is an extension of the principle covered in Section 27.4.

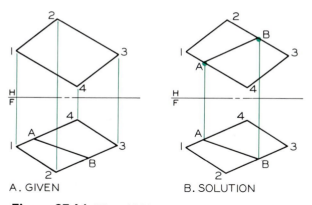

A. GIVEN   **B. SOLUTION**

Figure 27.14 If line *AB* lying on the plane is given, the top view of the line can be found. Points *A* and *B* are projected to lines 1–4 and 2–3, respectively, and are connected to form line *AB*.

27.10   A Point on a Plane

Point 0 is given on the front view of plane 4–5–6 in Fig. 27.15. It is required to locate the point on the plane in the top view. In Step 1, a line in any direction but vertical is drawn through the point to establish a line on the plane. The line is projected to the top view in Step 2, and the point is projected from the front view to the top view of the line.

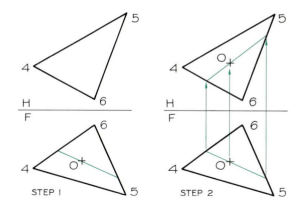

STEP 1   **STEP 2**

Figure 27.15 Location of a point on a plane.
Find the top view of point 0 that lies on the plane.

Step 1 Draw a line through the given view of point 0 in any convenient direction except vertical.

Step 2 Project the ends of the line to the top view, and draw the line. Point 0 is projected to the line.

27.11   Principal Lines on a Plane

Principal lines may be found in any view of a plane when at least two views of the plane are given. An infinite number of principal lines can be drawn on a single plane.

 Horizontal lines parallel to the edge view of the horizontal projection in Fig. 27.16A are drawn in the front view of the plane. These horizontal lines are projected to the top view of the plane, where they are true length.

 Frontal lines are drawn parallel to the frontal projection plane in the top view in Fig. 27.16B. When projected to the front view, the lines are true length.

 Profile lines are drawn parallel to the profile projection plane in the top and front views in Fig. 27.16C. When projected to the side view, the lines will appear true length.

27.12   Parallelism of Lines

If two lines are parallel, they will appear parallel in all views in which they are seen, except where both appear as points.

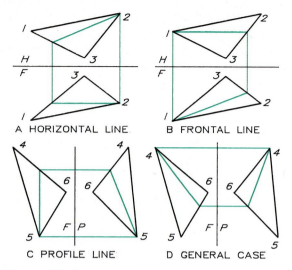

Figure 27.16 Principal lines on a plane. (A) A horizontal line is drawn first in the front view parallel to the edge view of the horizontal plane. It is projected to the top view where it is true length. (B) A frontal line is drawn first in the top view parallel to the edge view of the frontal plane. It is projected to the front view where it is true length. (C) A profile line is drawn first in the front view parallel to the edge view of the profile plane. It is projected to the profile view where it is true length. (D) A general-case line is one that is not parallel to the frontal, horizontal, or profile planes. It is not true length in any principal views.

Parallelism of lines in space cannot be determined if only one view is given; two or more views are required. Using this principle, a line can be drawn parallel to a given line through a specified point (Fig. 27.17).

27.13 Parallelism of a Line and a Plane

> A line is parallel to a plane if it is parallel to any line in the plane.

In Fig. 27.18, it is required that a line with its midpoint at point 0 be drawn parallel to plane 1–2–3. This is done by drawing line *AB* parallel to a line in the plane, line 1–3 in this case, in the top and front views. The line could have been drawn parallel to *any* line in the plane; therefore an infinite number of positions exist.

A similar example is shown in Fig. 27.19, where a line with its midpoint at 0 and parallel to the plane formed by two intersecting lines is required.

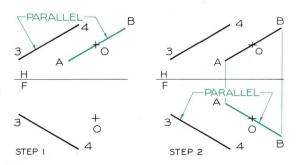

Figure 27.17 A line parallel to a line.

Draw a line through 0 parallel to the given line.

Step 1 Draw line *AB* parallel to the top view of line 3–4 with its midpoint at 0.

Step 2 Draw the front view of line *AB* parallel to the front view of 3–4 through point 0.

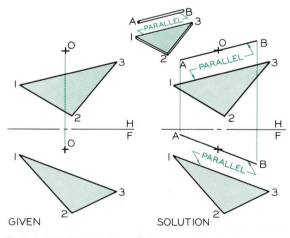

Figure 27.18 A line can be drawn through point 0 that is parallel to the given plane if the line is parallel to any line in the plane. Line *AB* is drawn parallel to line 1–3 in the front and top views, making it parallel to the plane.

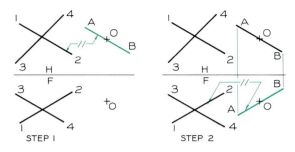

Figure 27.19 A line parallel to plane.

Draw a line parallel to the plane represented by two intersecting lines.

Step I Line *AB* is drawn parallel to line 1–2 through point 0.

Step 2 Line *AB* is drawn parallel to the same line, line 1–2, in the front view, which makes *AB* parallel to the plane.

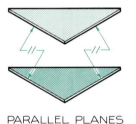

PARALLEL PLANES

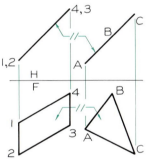

Figure 27.20 Two planes are parallel when intersecting lines in one are parallel to intersecting lines in the other. When parallel planes appear as edges, the edges will be parallel.

27.14 Parallelism of Planes

Two planes are parallel when intersecting lines in one plane are parallel to intersecting lines in the other (Fig. 27.20).

It is easy to determine that planes are parallel when both appear as parallel edges in a view. In Fig. 27.21, it is required that a line be drawn through point 0 and parallel to plane 1–2–3. In Step 1, *EF* is drawn through point 0 and parallel

to line 1–2 in both the top and front views. In Step 2, a second line is drawn through point 0 parallel to line 2–3 of the plane in the front and top views. These two intersecting lines form a plane parallel to plane 1–2–3.

27.15 Perpendicularity of Lines

When two lines are perpendicular, they will project at true 90° angles if **one** or **both** are true length (Fig. 27.22). When two lines are perpendicular but neither is true length, they will not project with a true 90° angle.

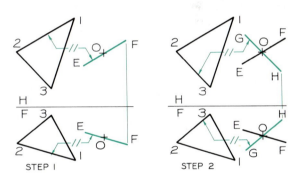

Figure 27.21 A plane through a point parallel to a plane.

Draw a plane through point 0 parallel to the given plane.

Step I Draw line *EF* parallel to any line in the plane, line 1–2 in this case. Show the line in both views.

Step 2 Draw a second line parallel to line 2–3 in the top and front views. These two intersecting lines represent a plane parallel to 1–2–3.

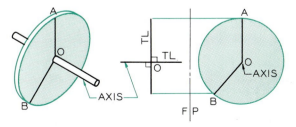

Figure 27.22 Perpendicular lines will intersect at 90° angles in a view where one or both of the lines appear true length.

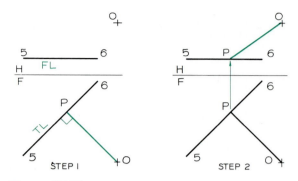

Figure 27.23 A line perpendicular to a frontal line.

Draw a line from point 0 perpendicular to line 5–6.

Step 1 Since line 5–6 is a frontal line and is true length in the front view, a perpendicular from point 0 will make a true 90° angle with it.

Step 2 Project point P to the top view and connect it to point 0. Since neither line is true length, they will not intersect at 90° in the top view.

In Fig. 27.22, it can be seen that the axis is true length in the front view; therefore any spoke will be shown perpendicular to the axis in the front view. Spokes $0A$ and $0B$ are examples where one is true length and the other foreshortened.

27.16 A Line Perpendicular to a Principal Line

In Fig. 27.23, it is required that a line be constructed through point 0 perpendicular to frontal line 5–6, which is true length in the front view. In Step 1, $0P$ is drawn perpendicular to 5–6 since it is true length. In Step 2, point P is projected to the top view of 5–6. Line $0P$ in the top view cannot be drawn as a true 90° angle since neither of the lines is true length in this view.

27.17 A Line Perpendicular to an Oblique Line

In Fig. 27.24, it is required that a line be constructed from point 0 perpendicular to oblique line 1–2. A line could have been found perpendic-

ular to 1–2 in the front view by drawing a frontal line in the top view.

27.18 Perpendicularity Involving Planes

A line can be drawn perpendicular to a plane if it is drawn perpendicular to any two intersecting lines in the plane (Fig. 27.25A). A plane is perpendicular to another plane if a line in one is perpendicular to the other (Fig. 27.25B).

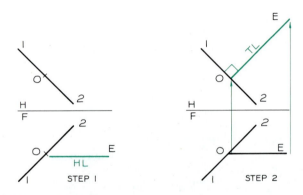

Figure 27.24 A line perpendicular to an oblique line.

Draw a line from point 0 perpendicular to line, 1–2.

Step 1 Draw a horizontal line from point 0 in the front view.

Step 2 Horizontal line $0E$ will be true length in the top view; therefore, it can be drawn perpendicular to line 1–2 in this view.

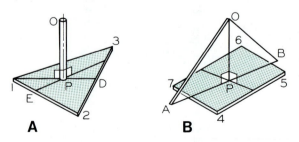

Figure 27.25 (A) A line is perpendicular to a plane if it is perpendicular to two intersecting lines on the plane. (B) A plane is perpendicular to another plane if the plane contains a line perpendicular to the other plane.

27.19 A Line Perpendicular to a Plane

In Fig. 27.26, it is required that a line be drawn from point 0 on the plane perpendicular to the plane.

In Step 1, a frontal line is drawn on the plane in the top view and is projected to the front view, where the line is true length. Line 0P is drawn perpendicular to the true-length line.

In Step 2, a horizontal line is drawn in the front view and then in the top view of the plane through point 0 perpendicular to the true-length line. This results in a line perpendicular to the plane since the line is perpendicular to two intersecting lines in the plane, a horizontal and a frontal line.

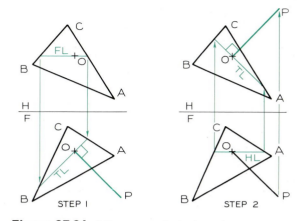

Figure 27.26 A line perpendicular to a plane.

Draw a line from point 0 perpendicular to the plane.

Step 1 Construct a frontal line on the plane through 0 in the top view. This line is true length in the front view; therefore, line 0P can be drawn perpendicular to the true-length line.

Step 2 Construct a horizontal line through point 0 in the front view. This line is true length in the top view; therefore, line 0P can be drawn perpendicular to it.

Problems

Use Size A sheets for the following problems, and lay out the problems using instruments. Each square on the grid is equal to 0.20 inch (about 5 mm). The problems can be laid out on grid paper or plain paper. Label all reference planes and points in each problem with ⅛-inch letters or numbers, using guidelines

1. (Fig. 27.27) A.–D. Draw three views (top, front, and right-side views) of the given point. E and F. Draw the three views of the points and connect them to form lines.

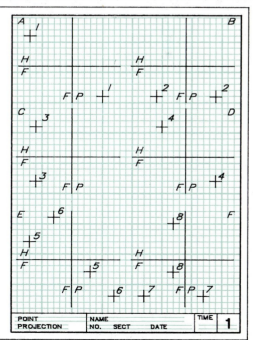

Figure 27.27 Problems 1A–1F.

2. (Fig. 27.28) A–C. Draw three views (top, front, and right-side views) of the partially drawn lines. D–F. Draw the missing views of the lines so that 7–8 is a horizontal line, 1–2 is a frontal line, and 3–4 is a profile line.

3. (Fig. 27.29) A–B. Draw the right-side view of line 1–2 and plane 3–4–5. C–E. Draw the missing views of the planes so that 6–7–8 is a frontal plane, 1–2–3 is a horizontal plane, and 4–5–6 is a profile plane. F. Complete the top and side views of the plane that appears as an edge in the front view.

4. (Fig. 27.30) A. Draw the side view of the plane and locate point A, on the plane in all views. B. Draw the side view and draw two horizontal lines on each view. C. Draw the side view and draw two frontal lines on each view. D. Draw the side view and draw two profile lines on each view. E. Draw a line through B that is parallel to and equal in length to the given line. F. Complete the three views of the intersecting lines.

5. (Fig. 27.31) A–B. Draw 1.50″ lines that pass through points 0 and are parallel to their respective planes. C–D. Draw the top and front views of lines through points 0 that are perpendicular to their respectives line.

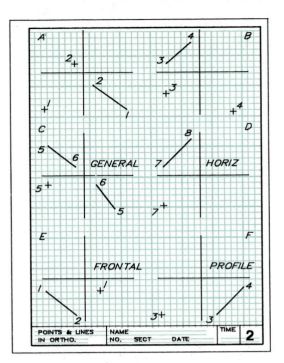

Figure 27.28 Problems 2A–2F.

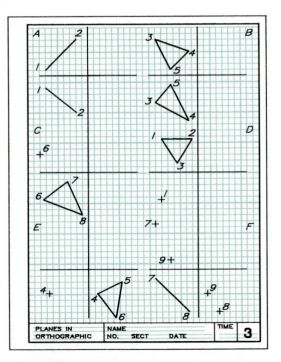

Figure 27.29 Problems 3A–3F.

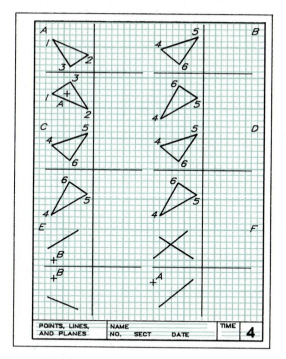

Figure 27.30 Problems 4A–4F.

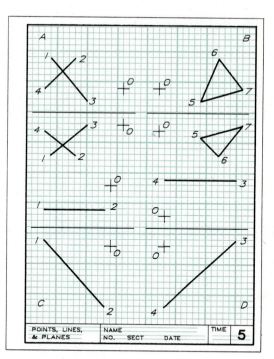

Figure 27.31 Problems 5A–5D.

28

Primary Auxiliary Views in Descriptive Geometry

28.1 Introduction

Descriptive geometry is the projection of three-dimensional figures onto a two-dimensional plane of paper in such manner as to allow geometric manipulations to determine lengths, angles, shapes, and other geometric information by means of graphics. Orthographic projection is the system used for laying out descriptive geometry problems.

A powerful tool of descriptive geometry is the **primary auxiliary view,** which permits the analysis of three-dimensional geometry.

28.2 Descriptive Geometry by Computer

Four useful computer commands for solving descriptive geometry problems are covered in this section and documented in Appendix 52. The commands are PERPLINE, PARALLEL, TRANSFER, COPYDIST, and BISECT.*

---

*These LISP commands were written by Professor Leendert Kersten of the University of Nebraska at Lincoln.

While making a drawing in AutoCAD's Drawing Editor, any of these commands can be executed by typing its name. For example, Command: PERPLINE will enter the command for constructing a line perpendicular to the line of your choice by giving you a series of prompts to follow.

Perpline In Fig. 28.1, line 3–4 is drawn perpendicular to line 1–2 by locating the starting point, and next, the line to which the line is to be perpendicular. The third point is the endpoint in the general area where the line is to end.

Parallel In Fig. 28.2, line 3–4 is drawn parallel to line 1–2 by locating the starting point of the parallel and the general location of its endpoint. Next, select the endpoints of line 1–2 in the same order that you want the parallel to be drawn; in this case, in the direction from 1 to 2.

Transfer The TRANSFER command (Fig. 28.3) is used to transfer distances from reference lines to solve descriptive geometry problems in the same manner that you would transfer the distances with your dividers. First select endpoint 2 in the front view, and then select the reference line. Select endpoint 2 in the top view and the auxiliary reference line, and the endpoint is located with a small circle in the auxiliary view.

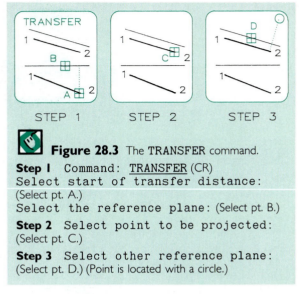

Figure 28.3 The TRANSFER command.

Step 1 Command: <u>TRANSFER</u> (CR)
Select start of transfer distance: (Select pt. A.)
Select the reference plane: (Select pt. B.)
Step 2 Select point to be projected: (Select pt. C.)
Step 3 Select other reference plane: (Select pt. D.) (Point is located with a circle.)

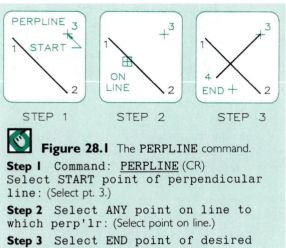

Figure 28.1 The PERPLINE command.

Step 1 Command: <u>PERPLINE</u> (CR)
Select START point of perpendicular line: (Select pt. 3.)
Step 2 Select ANY point on line to which perp'lr: (Select point on line.)
Step 3 Select END point of desired perpendicular (for length only): (Select pt. 4.) (3–4 is drawn.)

Copydist A distance can be copied from a given position to another position as Fig. 28.4 shows. Select the endpoints of the line to be copied, and locate its beginning point in the new position. Next, locate the direction of the copied

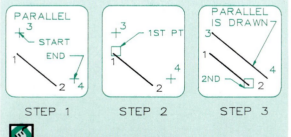

Figure 28.2 The PARALLEL command.

Step 1 Command: <u>PARALLEL</u> (CR)
Select START point of parallel line: (Select pt. 3.)
Select END point of parallel line: (Select pt. 4.)
Step 2 Select 1st point on line for parallelism: (Select pt. 1.)
Step 3 Select 2nd point on line for parallelism: (Select pt. 2.) (3–4 is drawn.)

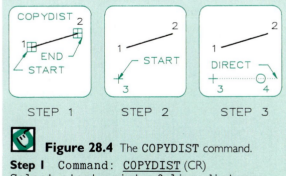

Figure 28.4 The COPYDIST command.

Step 1 Command: <u>COPYDIST</u> (CR)
Select start point of line distance to be copied: (Select end 1.)
End point?: (Select end 2.)
Step 2 Start point of new distance location: (Select pt. 3.)
Step 3 Which direction?: (Select with cursor, and endpoint 4 is located with a circle.)

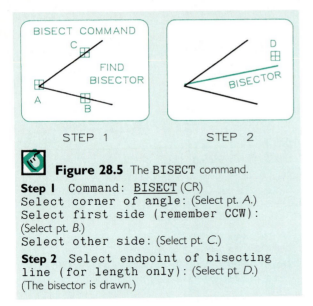

Figure 28.5 The BISECT command.

Step I Command: <u>BISECT</u> (CR)
Select corner of angle: (Select pt. A.)
Select first side (remember CCW):
(Select pt. B.)
Select other side: (Select pt. C.)

Step 2 Select endpoint of bisecting
line (for length only): (Select pt. D.)
(The bisector is drawn.)

line, and the endpoint of the line will be scaled
and located with a small circle.

Bisect   An angle can be bisected (Fig. 28.5) by
selecting the vertex of the angle, the first line,
and the second line. The second line must be
counterclockwise from the first line selected. Locate the endpoint when prompted, and the bisector will be drawn.

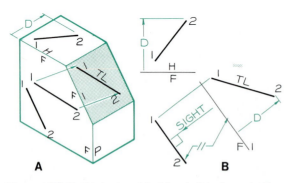

A

B

Figure 28.6 (A) A pictorial of line 1–2 is shown inside a
projection box where a primary auxiliary plane is
established perpendicular to the frontal plane and parallel
to the line. (B) The orthographic arrangement shows the
auxiliary view projected from the front view to find 1–2
true length.

28.3 Primary Auxiliary View of a Line

The top and front views of line 1–2 are shown
pictorially and orthographically in Fig. 28.6.
Since the line is not a principal line, it is not true
length in the principal views. A primary auxiliary
view must be used to find its true-length view.

In Fig. 28.6B, the line of sight is drawn perpendicular to the front view of the line, and reference line F-1 is drawn parallel to its frontal
view. You can see in the pictorial that the auxiliary plane is parallel to the line and perpendicular
to the frontal plane, which accounts for its being
labeled as F-1.

The auxiliary view is found by projecting
parallel to the line of sight and perpendicular to
the F-1 reference line. Point 2 is found by transferring distance D with your dividers to the auxiliary view, since the frontal plane appears as an
edge in both the top and auxiliary views. Point 1

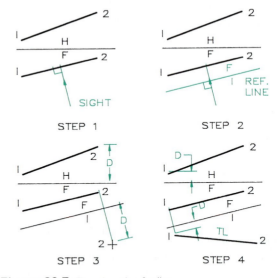

Figure 28.7 True length of a line.

Step I To find 1–2 true length, it must the viewed with a
line of sight perpendicular to one of its views.

Step 2 The F-1 reference line is drawn parallel to the line
and perpendicular to the line of sight.

Step 3 Project point 2 perpendicularly from the front
view. Distance D, from the top view, locates point 2.

Step 4 Point 1 is similarly located. Line 1–2 is true length
in the auxiliary view.

is located in the same manner, and the points are connected to find the true-length view of the line.

Figure 28.7 separates the sequential steps required to find the true length of an oblique line. It is best to letter all reference planes using the notation suggested in the various steps with the exception of dimensions such as *D*, which are transferred from one view to another with your dividers.

A primary auxiliary view can result in a point view of the line if projected from a true-length view of the line in a principal view (Fig. 28.8). The auxiliary view projected from the front view of the line does not give a point view since the line is foreshortened in the front view.

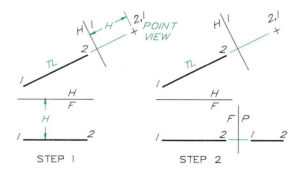

Figure 28.8 Point view of a line.

Step 1 The point view of a line can be found in a primary auxiliary view that is projected from the true-length view of the line.

Step 2 An auxiliary view projected from a foreshortened view of a line will result in a foreshortened view of the line, not a point view.

Computer Method Figure 28.9 illustrates how the PARALLEL and TRANSFER commands can be used to find a true-length view of a line by an auxiliary view. A line is drawn parallel to the top view of line 1–2 in Step 1. The endpoints of the line are TRANSFERred in Steps 2 and 3. To draw the ends of line 1–2 from the centers of the circles (Step 4), use the CENTER option of the OS-NAP command. The circles can be erased afterward if you wish to remove them.

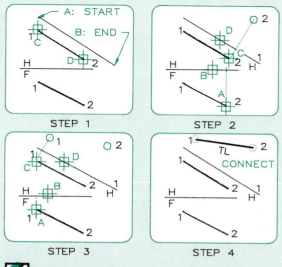

Figure 28.9 True length by computer.

Step 1 Command: <u>PARALLEL</u> (CR) (By following the prompts as covered in Section 28.2, draw the reference line parallel to *CD*.)

Step 2 Command: <u>TRANSFER</u> (CR) (Follow the prompts as covered in Section 28.2 to locate point 2 in the auxiliary view.)

Step 3 Locate point 1 in the auxiliary view using TRANSFER.

Step 4 Draw a line from the centers of the circles found in Steps 2 and 3 by using the CENTER option of the OSNAP command. The circles can be erased after this step.

28.4 True Length by Analytical Geometry

You can see in Fig. 28.10 that the length of a frontal line can be found in the front view by the application of analytical geometry (mathematics) and the Pythagorean theorem, which states that the hypotenuse of a right triangle is equal to the square root of the sum of the squares of the other two sides. The length of the line in Fig. 28.10 can be measured in the front view since it is true length in this view.

The line shown pictorially in Fig. 28.11A can be found true length by analytical geometry by

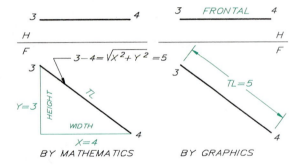

Figure 28.10 A line that appears true length in a view (the front view in this case) can have its length calculated by application of the Pythagorean theorem. Since the line is a frontal line and is true length in the front view, its length can be measured graphically.

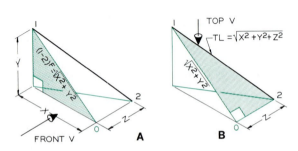

Figure 28.11 A three-dimensional line that is foreshortened in a principal view can be found true length by the Pythagorean theorem in two steps. (A) The frontal projection, 1–0, is found using the X- and Y-coordinates. (B) The hypotenuse of the right triangle 1–0–2 is found using X-, Y-, and Z-coordinates.

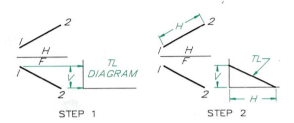

Figure 28.12 True-length diagram.

Step 1 Transfer the vertical distance between the ends of 1–2 to the vertical leg of the TL diagram.

Step 2 Transfer the horizontal length of the line in the top view to the horizontal leg of the TL diagram.

determining the length of the front view where the X- and Y-coordinates form a right triangle. In Fig. 28.11B, a second right triangle, 1–0–2, is solved to find its hypotenuse, which is the true length of 1–2. The true length of an oblique line is the square root of the sum of the squares of the X-, Y-, and Z-coordinates that correspond to width, height, and depth.

28.5 The True-Length Diagram

A true-length diagram is constructed with two perpendicular lines to find a line of true length (Fig. 28.12). This method does not give a direction for the line but merely its true length.

The two measurements laid out on the true-length diagram can be transferred from any two adjacent orthographic views. One measurement is the distance between the endpoints in one of the views. The other measurement, taken from the adjacent view, is measured between the endpoints in a direction perpendicular to the reference line between the two views.

28.6 Angles Between Lines and Principal Planes

To measure the angle between a line and a plane, the line must appear true length in the view where the plane appears as an edge.

Since a principal plane will appear as an edge in a primary auxiliary view, the angle a line makes with this principal plane can be measured if the line is found true length in this view (Fig. 28.13).

28.7 Slope of a Line

Slope is the angle a line makes with the horizontal plane.

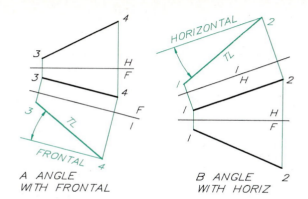

A ANGLE
WITH FRONTAL

B ANGLE
WITH HORIZ

Figure 28.13 Angles between lines and principal planes. (A) When an auxiliary view is projected from the front view, the frontal plane appears as an edge and the line is true length. The angle with the frontal plane can be measured in this view. (B) An auxiliary view projected from the top view shows the horizontal plane as an edge and the line true length. The angle with the horizontal can be measured here.

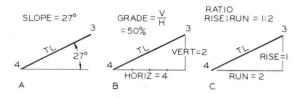

Figure 28.14 The inclination of a line with the horizontal can be measured and expressed by any of three methods. (A) Slope angle. (B) Percent grade. (C) Slope ratio.

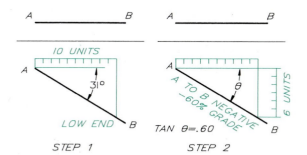

Figure 28.15 Percent grade of a line.

Step 1 The percent grade of a line can be measured in the view where the horizontal appears as an edge and the line is true length (the front view here). Ten units are laid off parallel to the horizontal from the end of the line.

Step 2 A vertical distance from A to the line is measured to be 6 units. The percent grade is 6 divided by 10 or 60%. This is negative when the direction is from A to B. The tangent of this slope angle is $\frac{6}{10}$ or 0.60.

Slope may be specified by any of the three methods shown in Fig. 28.14: **slope angle, percent grade,** or **slope ratio.**

Slope Angle

The slope of a line can be measured in a view where the line is true length and the horizontal plane appears as an edge. Thus, the slope of *AB* in Fig. 28.15 can be measured in the front view, where Θ is found to be 31°. Using the trigonometric tables, this angle can also be found by converting its tangent of 0.60 to 31°.

Percent Grade

The percent grade of a line is found in the view where the line is true length and the horizontal plane appears as an edge.

> **Grade** is the ratio of the vertical **(rise)** divided by the horizontal **(run)** between the ends of a line expressed as a percentage.

The percent grade of *AB* is found in Fig. 28.15 by using a combination of mathematics and graphics. The grade is negative from *A* to *B* since this is downhill; it would be positive from *B* to *A* if it were uphill. Line *AB* has a -60% grade from *A* to *B* in Fig. 28.16, in Step 2.

Slope Ratio

The first number of the slope ratio is the rise, and the second number is the run (see Fig. 28.14). In the slope ratio, the rise is always 1 (for example, 1:10, 1:200).

The graphical method of finding the slope ratio is shown in Fig. 28.16, Step 2, where the rise of 1 is laid off on the true-length view of *EF*. The corresponding horizontal is found to be 2, which results in a slope ratio of 1:2.

Oblique Lines When a line is oblique and does not appear true length in the front view, it must be found true length by an auxiliary view pro-

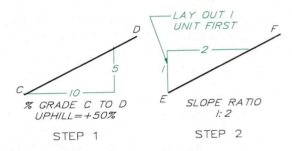

Figure 28.16 Percent grade of a line.

Step 1 The percent grade of a line is positive if uphill, and negative if downhill.

Step 2 The slope ratio is expressed as 1:*XX*, where 1 is the rise and *XX* is the horizontal distance. The 1 unit must be drawn before the horizontal distance.

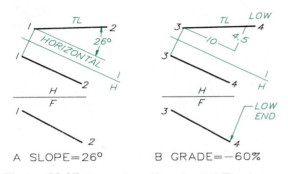

Figure 28.17 Slope of an oblique line. (A) The slope angle can be measured in a view where the horizontal appears as an edge and the line is true length. The slope of 26° is found in an auxiliary view projected from the top view. (B) The percent grade can be measured in a true-length view of the line projected from the top view. Line 3–4 has a −45% grade from 3 to the low end at 4.

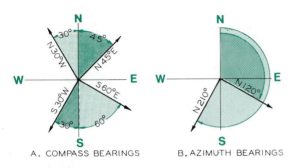

Figure 28.18 (A) Compass bearings are measured with respect to north and south directions on the compass. (B) Azimuth bearings are measured with respect to north in a clockwise direction up to 360°.

jected from the **top view.** This auxiliary view shows the horizontal as an edge and the line true length, making it possible to measure the slope angle (Fig. 28.17A).

Similarly, an auxiliary view projected from the top view must be used to find the percent grade of an oblique line (Fig. 28.17B). Ten units are laid off horizontally, parallel to the H–1 reference line, and the vertical distance is found to be 4.5, or a −45% grade downhill from 3 to 4.

28.8 Compass Bearing of a Line

Two types of bearings of a line's direction are **compass bearings** and **azimuth bearings** (Fig. 28.18).

> Compass bearings always begin with the north or south directions, and the angles with north and south are measured toward east or west.

The line in Fig. 28.18A that makes 30° with north has a bearing of N 30° W. A line making 60° with south toward the east has a compass bearing of south 60° east, or S 60° E. Since a compass can be read only when held horizontally, the compass bearings of a line can be determined only in the top view, or the horizontal view.

> An **azimuth bearing** is measured from north in clockwise direction to 360° (Fig. 28.18B).

Azimuth bearings of a line are written N 120°, N 210°, and so forth, with this notation indicating that the measurements are made from the north.

The compass bearing (direction) of a line is assumed to be toward the low end of the line unless otherwise specified. For example, line 2–3 in Fig. 28.19 has a bearing of N 45° E since the line's low end is point 3. It can be seen in the front view that point 3 is the lower end.

The compass bearing and slope of a line are found in Fig. 28.20. This information can be used to verbally describe the line as having a compass

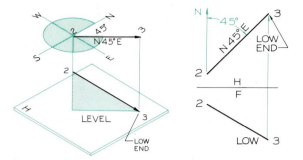

Figure 28.19 The compass bearing of a line is measured in the top view toward its low end (unless specified toward the high end). Line 2–3 has a bearing of N 45° E toward the low end at 3.

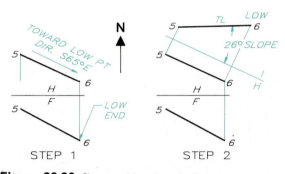

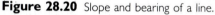

Figure 28.20 Slope and bearing of a line.

Step 1 Slope bearing can be found in the top view toward its low end. Direction of slope is S 65° E.

Step 2 The slope angle of 26° is found in an auxiliary view projected from the top view where the line is found true length.

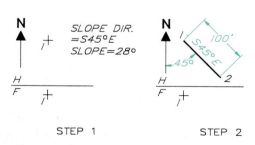

Figure 28.21 A line from slope specifications.

Step 1 It is required to draw a line through point 1 that bears S 45° E for 100′ horizontally and slopes 28°.

Step 2 The bearing and the horizontal distance are drawn in the top view.

bearing of S 60° E and a slope of 26° from 5 to 6. When given verbal information and one point in the top and front views, a line can be drawn as shown in Fig. 28.21.

 Computer Method A general case plot plan (Fig. 28.22) shows a drawing in which the north arrow is at a slight angle to the right on the drawing. AutoCAD provides an option under the UNITS command for aligning a plot plan with the north arrow. When using this command, you are prompted

```
System of angle                (Example)
measure:
   1. Decimal degrees          45.0000
   2. Degree/
      minutes/seconds          45d0'0"
   3. Grads                    50.0000g
   4. Radians                  0.7854r
   5. Surveyor's               N45d0'0"E
      units
Enter choice, 1 to 5 <default>: 5
```

By entering 5, you obtain surveyor's units that will give directional angles with respect to north or south in an east or west direction, such as N 30d45'10", where d = degrees, ′ = minutes, and ″ = seconds. Interior angles measured with the DIM command will give the angles in the same form but without reference to points on the compass such as 152d34'17".

Next, the UNITS command will give a prompt for setting the direction of east to establish the relationship of the drawing to the directional north

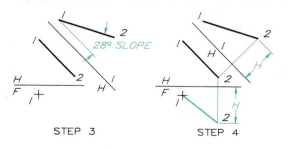

Step 3 An auxiliary view is projected from the top view and the line is drawn at a slope of 28°.

Step 4 The front view of 1–2 is found by locating point 2 in the front view.

Figure 28.22 Most plot plans are drawn where the north arrow is not vertical on the drawing sheet but at an angle since the plot must be positioned to fit the drawing sheet.

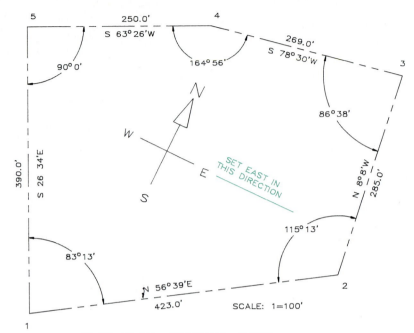

PLOT PLAN: GENERAL CASE
AREA=1,627.0 SQ FT PERIMETER=1,617.0'

arrow:

```
Direction for angle 0:
    East        3 o'clock = 0
    North      12 o'clock = 90
    West        9 o'clock = 180
    South       6 o'clock = 270
Enter direction for angle 0
<current>: (F1 and align.)
```

When you press F1, your screen returns to the drawing you were working on. With the cursor, locate a point on the east line of the compass arrow, and then locate a second point on the line toward the east (Fig. 28.23). By picking these two points, you have indexed the geometry of your drawing to the compass directions being used. The last prompt reads

```
Do you want angles measured
clockwise? <N>: N (CR)
```

By selecting a counterclockwise direction, angles are measured in this direction, the standard direction. By using the DIST command, you can find the lengths and compass directions of lines by selecting points at the ends of the lines in a counterclockwise direction about the traverse (Fig. 28.23).

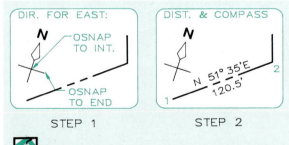

STEP 1   STEP 2

 Figure 28.23 Setting compass direction.

Step 1 First, draw the compass arrow on the plot plan. When in the UNITS command, select 5. Surveyor's units, and you will be prompted for the angle of east. Press F1, and then with the cursor, select two points on the east line (west to east) to calibrate the compass direction.

Step 2 Using the DIST command, select the endpoints of each side in a counterclockwise direction about the traverse. Label the sides with direction inside and lengths outside the plot.

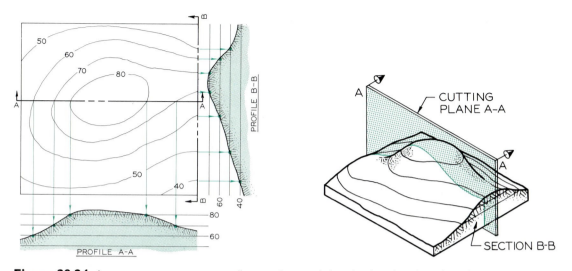

Figure 28.24 A contour map uses contour lines to show variations in elevation on an irregular surface. Vertical sections taken through a contour map are called profiles.

28.9 Contour Maps and Profiles

The **contour map** uses contour lines to show variations in elevation on irregular surfaces of the earth. A pictorial view of a sectional plane and a portion of the earth are shown in Fig. 28.24, along with the conventional orthographic views of the contour map and profiles.

Contour Lines   are horizontal lines that represent constant elevations from a horizontal datum such as sea level. Contour lines can be thought of as the intersections of horizontal planes with the surface of the earth. The vertical interval of spacing between the contours in Fig. 28.24 is 10 ft.

Contour Maps   contain contour lines that are drawn to represent irregularities of the surface (Fig. 28.24). The closer the contour lines are to each other, the steeper the terrain.

Profiles   are vertical sections through a contour map that are used to show the earth's surface at any desired location (Fig. 28.24). Contour lines represent edge views of equally spaced horizontal planes in profiles. The true representation of a profile is drawn with the vertical scale equal to the scale of the contour map; however, this vertical scale is often increased to emphasize changes in elevation that would otherwise not be apparent.

Contoured Surfaces   are also depicted on the drawing board by using contour lines. When applied to objects other than the earth's surface—such as airfoils, automobile bodies, ship hulls, and household appliances—this technique of representing contours is called **lofting**.

Station Numbers   are used to locate distances on a contour map. Since the civil engineer uses a chain (metal tape) 100 feet long, stations are located 100 feet apart (Fig. 28.25). Station 7 is 700 ft from the beginning point, station 0. A point 32 ft from station 7 toward station 8 is labeled station 7 + 32.

28.10 Vertical Sections

In Fig. 28.26, a vertical section (a **profile**) is passed through the top view of an underground pipe that is known to have an elevation of 90 ft

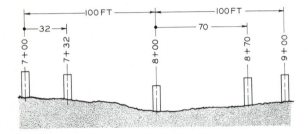

Figure 28.25 Station points are located 100 ft apart. For example, station 7 is 700 ft from the beginning point. A point 32 ft beyond station 7 is labeled station 7 + 32. A point 870 ft from the origin is labeled station 8 + 70.

at point 1 and 60 ft at point 2. An auxiliary view is projected perpendicularly from the top view, contour lines are located, and the top of the earth over the pipe is found in a profile.

To measure the true lengths and angles of slope in the profile, the same scale used to draw the contour map was used to draw the profile. The percent grade and compass bearing of the line are labeled on the contour map.

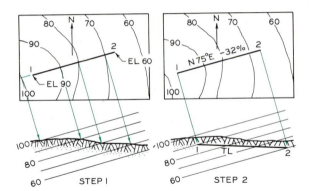

Figure 28.26 Vertical sections (profiles).

Step 1 An underground pipe is known to have elevations of 90 ft and 60 ft at each end. An auxiliary view is projected perpendicularly from the top view, and contours are drawn at 10-ft intervals to correspond to the plan view. The top of the ground is found by projecting from the contour lines in the plan view.

Step 2 Points 1 and 2 are located at elevations of 90 ft and 60 ft in the vertical section (profile). Since 1–2 is TL in the section, its slope or percent grade can be measured. The compass direction and percent grade are labeled in the top view of the line.

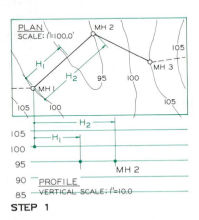

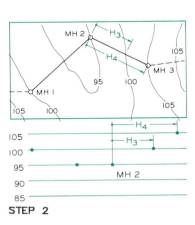

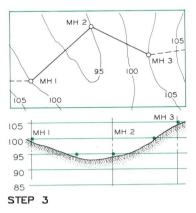

Figure 28.27 Plan-profile.

Required Find the profile of the earth over the drainage system.

Step 1 Distances H_1 and H_2 from manhole 1 are transferred to their respective elevations in the profile. This is not an orthographic projection.

Step 2 Distances H_3 and H_4 are measured from manhole 2 in the plan and are transferred to their respective elevations in the profile. These points represent elevations of points on the earth above the pipe.

Step 3 The five points are connected with a freehand line and the drawing is crosshatched to represent the earth's surface. Centerlines are drawn to show the locations of the three manholes.

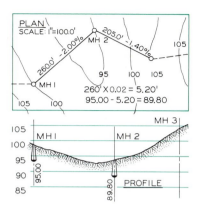
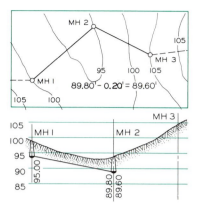
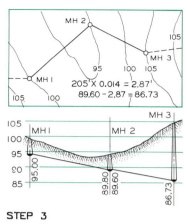

STEP 1

STEP 2

STEP 3

Figure 28.28 Plan-profile—manhole location.

Step 1 The horizontal distance from MH1 to MH2 is multiplied by the percent grade. The elevation of the bottom of manhole 2 is calculated by subtracting from the elevation of manhole 1.

Step 2 The lower side of manhole 2 is 0.20′ lower than the inlet side to compensate for loss of head (pressure) due to the turn in the pipeline. The lower side is found to be 89.60′ and is labeled.

Step 3 The elevation of manhole 3 is calculated to be 86.73′ since the grade is 1.40% from manhole 2 to manhole 3. The flow line of the pipeline is drawn from manhole to manhole, and the elevations are labeled.

28.11 Plan-Profiles

A **plan-profile** is a drawing that includes a plan with contours and a vertical section called a profile. A plan-profile is used to show an underground drainage system from manhole 1 to manhole 3 in Figs. 28.27 and 28.28.

The profile is drawn with an exaggerated vertical scale to emphasize the variations in the earth's surface and the grade of the pipe. Although the vertical scale is usually increased, it can be drawn at the same scale used in the plan if desired.

Manhole 1 is projected to the profile using orthographic projection, but the other points are not orthographic projections (Fig. 28.27). The points where the contour lines cross the top view of the pipe are transferred to their respective elevations in the profile with your dividers. These points are connected to show the surface of the earth over the pipe and the location of manhole centerlines (Fig. 28.28). The drop from manhole 1 to manhole 2 is found to be 5.20 ft by multiplying the horizontal distance of 260.00 ft by a −2.00% grade (Fig. 28.28).

Since the pipes intersect at manhole 2 at an angle, the flow of the drainage is disrupted at the turn; thus, a drop of 0.20 ft is given from the inlet across the floor of the manhole to compensate for the loss of pressure (head) through the manhole.

The true lengths of the pipes cannot be accurately measured in the profile when the vertical scale is different from the horizontal scale. For this computation, trigonometry must be used.

28.12 Edge View of a Plane

> The edge view of a plane can be found in a view where any line on the plane appears as a point.

A line can be found as a point by projecting from a true-length view of the line. (Fig. 28.29).

A true-length line can be found on any plane by drawing the line parallel to one of the principal planes and projecting it to the adjacent view,

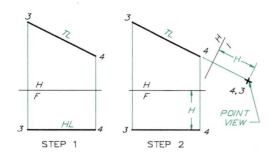

Figure 28.29 Point view of a line.

Step 1 Line 3–4 is horizontal in the front view and is therefore true length in the top view.

Step 2 The point view of 3–4 is found by projecting parallel lines from the top view to the auxiliary view.

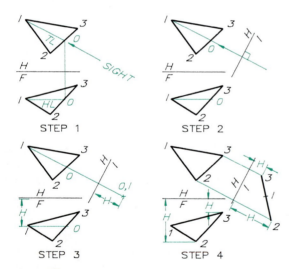

Figure 28.30 Edge view of a plane.

Step 1 To find the edge view of 1–2–3, horizontal line 1–0 is drawn on the front view of the plane, and it is projected to the top view, where it is true length.

Step 2 A line of sight is drawn parallel to the true-length line. The H-1 is drawn perpendicular to the line of sight.

Step 3 The point view of 1–0 is found in the auxiliary view by transferring the height distance.

Step 4 Points 2 and 3 are located in the same manner. These points will align in a line that represents the edge view of the plane.

as shown in Fig. 28.30 (Step 1). Since line 3–4 is true length in the top view, its point view may be found, and the plane will appear as an edge in this auxiliary view.

28.13 Dihedral Angles

> The angle between two planes, called a **dihedral angle,** can be found in the view where the line of intersection between two planes appears as a point.

The line of intersection, 1–2, between the two planes in Fig. 28.31 is true length in the top view. This makes it possible to find the point view of line 1–2 and the edge view of both planes in a primary auxiliary view.

28.14 Piercing Points by Projection

Figure 28.32 shows the steps necessary to find the piercing point of line 1–2 that passes through the plane. Cutting planes are passed through the

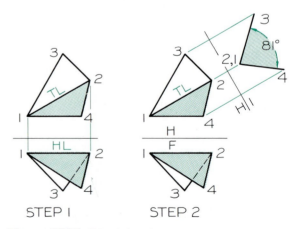

Figure 28.31 Dihedral angle.

Step 1 The line of intersection between the planes, 1–2, is true length in the top view.

Step 2 The angle between the planes (the dihedral angle) can be found in the auxiliary view where the line of intersection appears as a point.

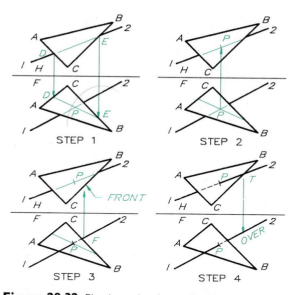

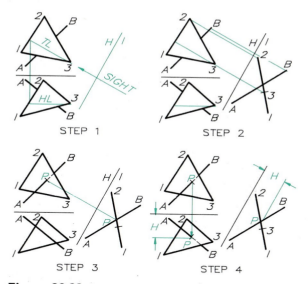

Figure 28.32 Piercing points by projection.

Step 1 Pass a vertical cutting plane through the top view of 1–2, which cuts the plane along line *DE*. Project line *DE* to the front view to locate piercing point *P*.

Step 2 Project point *P* to the top view of line 1–2.

Step 3 Visibility in the front view is found by projecting the crossing point of *CB* and 1–2 to the top view. *CB* is encountered first, which means it is in front of 1–2, making *PF* hidden in the front.

Step 4 Visibility in the top view is found by projecting the crossing point of *CB* and 1–2 to the front view. Line 1–2 is encountered first, which means it is above *CB*, making *PT* visible in the top view.

Figure 28.33 Piercing points by auxiliary view.

Step 1 Draw a horizontal line on the plane in the front view and project it to the top view to find a TL line on the plane.

Step 2 Find the edge view of the plane in an auxiliary view and project *AB* to this view. Point *P* is the piercing point.

Step 3 Project *P* to *AB* in the top view. Since *AP* is nearest the H-1 reference line, it is the highest end of the line and is visible in the top view.

Step 4 Project *P* to *AB* in the front view. *AP* is visible in the front view since *AP* is front of 1–2.

line and plane in the top view. The trace of this cutting plane, line, *DE,* is then projected to the front view, where piercing point *P* is found (Step 1). The top view of *P* is located (Step 2), and the visibility of the line is found (Steps 3 and 4).

28.15 Piercing Points by Auxiliary Views

The piercing point of a line and a plane can be found by an auxiliary view (Fig. 28.33). Piercing point *P* can be seen in Step 2, where the plane is found as an edge. Point *P* is projected back to the line in the top and front views from the auxiliary

view in Steps 3 and 4. The location of *P* in the front view is checked by transferring dimension *H* from the auxiliary view with your dividers.

Visibility is easily determined for the top view since it can be seen that *AP* is higher than the plane in the auxiliary view and is therefore visible in the top view. Analysis of the top view shows the endpoint *A* is the most forward point; therefore *AP* is visible in the front view.

28.16 Perpendicular to a Plane

In Fig. 28.34, it is required that a line be drawn from point *O* perpendicular to the plane.

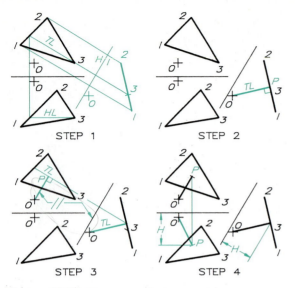

Figure 28.34 Line perpendicular to a plane.

Step I Find the edge view of the plane by finding the point view of a line on it. Project point *O* to this view, also.

Step 2 Draw *OP* perpendicular to the edge view of the plane. *OP* is TL in this view.

Step 3 Since *OP* is TL in the auxiliary view, it must be parallel to the H-1 reference line in the previous view. *OP* is visible in the top view since it is seen that OP is above the edge view in the auxiliary.

Step 4 Project point *P* to the front view and locate its distance from the horizontal with dimension H.

A perpendicular line will appear true length and perpendicular to a plane where the plane appears as an edge.

The true-length perpendicular is drawn in Step 2 to locate piercing point *P*. Point *P* is found in the top view by drawing line *OP* parallel to the H-1 reference line. It must lie in this direction since *OP* is true length in the auxiliary view. Line *OP* will also be perpendicular to a true-length line in the top view of the plane. The front view of point *P*, along with its visibility, is found by projection in Step 4.

28.17 Intersections by Auxiliary View

The intersection between planes can be found by finding the edge view of one of the planes, as shown in Fig. 28.35 (Step 1). Piercing points *L* and *M* are projected from the auxiliary view to their respective lines, 5–6 and 4–6, in the top view (Step 2).

The visibility of plane 4–5–6 in the top view is apparent by inspection of the auxiliary view, where sight line S_1 has an unobstructed view of

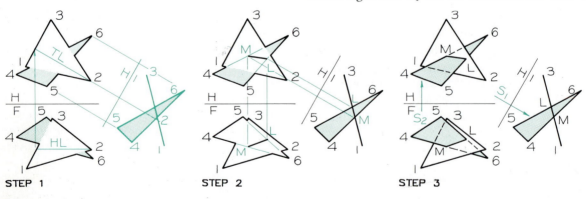

Figure 28.35 Intersection by auxiliary view.

Step I Find the edge view of one of the planes, and project the other plane to this view also.

Step 2 Piercing points *L* and *M* can be seen on the edge view of the plane. *LM* is projected to the top and front views.

Step 3 The line of sight from the top view strikes 1–5 first in the auxiliary view, which makes 1–5 visible in the top view. Line 4–5 is farthest forward in the top view and is visible in the front view.

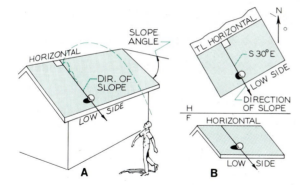

Figure 28.36 (A) The direction of slope of a plane is the compass bearing of a line on the plane. (B) This is measured in the top view toward the low side of the plane.

the 4–L–M portion of the plane. Plane 4–5–L–M is visible in the front view, since sight line S_2 has an unobstructed view of the top view of this portion of the plane.

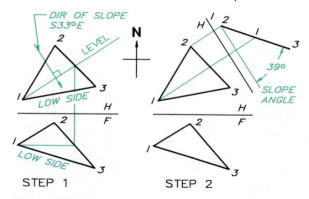

Figure 28.37 Slope and direction of slope of a plane.

Step I Slope direction can be found as perpendicular to a true-length level line in the top view toward the low side of the plane, S 33° E in this case.

Step 2 Slope is measured in an auxiliary view where the horizontal is an edge and the plane is an edge, 39° in this case.

28.18 Slope of a Plane

Planes can be established by using verbal specifications of slope and direction of slope of a plane.

Slope is the angle the plane's edge view makes with the edge of the horizontal plane.

Direction of Slope is the compass bearing of a line perpendicular to a true-length line in the top view of a plane toward its low side. This is the direction in which a ball would roll on the plane. It can be seen in Fig. 28.36A that a ball would roll perpendicular to all horizontal lines of the roof toward the low side.

Figure 28.37 gives the steps of finding the direction of slope and the slope angle of a plane. An understanding of these terms enables you to verbally describe a sloping plane.

A three-dimensional plane can be established by working from slope and direction specifications (Fig. 28.38). The direction of slope is drawn in the given top view, which locates the direction of a true-length horizontal line on the plane (Step 1). The edge view of the plane is then found by

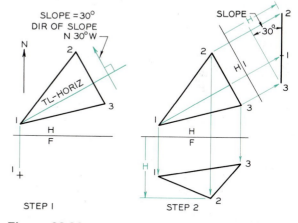

Figure 28.38 Construction from slope specifications.

Step I If the top view, front view of 1, and slope specifications are given, the front view can be completed. Draw the direction of slope in the top view. A true-length horizontal line on the plane is drawn perpendicular to the slope direction.

Step 2 A point view of the TL line is found in the auxiliary view to locate point 1. The edge view of the plane is drawn through 1 at a slope of 30°, according to the specifications. The front view is completed by transferring height dimensions from the auxiliary view to the front view.

Figure 28.39 The road across the top of this dam was built by applying the principles of cut and fill. (Courtesy of the Bureau of Reclamation, U.S. Department of the Interior.)

locating point 1 and constructing a slope of 30° (Step 2). Points 3 and 2 are transferred to the front view to complete that view.

28.19 Cut and Fill

A level roadway routed through irregular terrain or the embankment of a fill used to build a dam involves the principles of cut and fill (Fig. 28.39). **Cut and fill** is the process of cutting away equal amounts of the high ground to fill the lower areas.

In Fig. 28.40, it is required that a level roadway of an elevation of 60 feet be constructed about the given centerline in the contour map using the specified angles of cut and fill.

In Step 1, the roadway is drawn in the top view, and the contour lines in the profile view

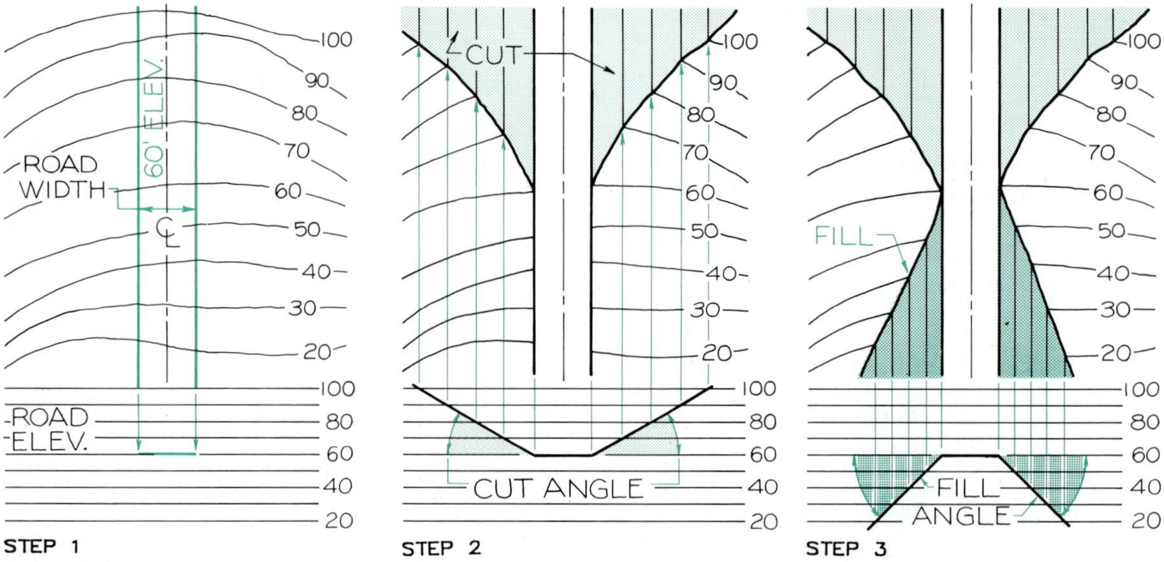

Figure 28.40 Cut and fill of a level roadway.

Step 1 Draw a series of elevation planes in the front view at the same scale as the map, and label them to correspond to the contours on the map. Draw the width of the roadway in the top view and in the front view at the given elevation, 60′ in this case.

Step 2 Draw the cut angles on the upper sides of the road in the front view according to the given specifications. The points of intersection between the cut angles and the contour planes in the front view are projected to their respective contour lines in the top view to determine the limits of cut.

Step 3 Draw the fill angles on the lower sides of the road in the front view. The points in the front view where the fill angles cross the contour lines are projected to their respective contour lines in the top view to give the limits of the fill. Contour lines are changed in the cut-and-fill areas to indicate the new contours parallel to the roadway.

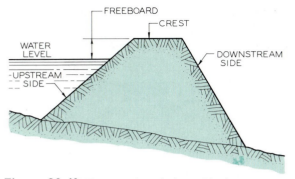

Figure 28.41 Terms and symbols used in the construction of a dam.

28.20 Design of a Dam

Some of the terms used in the drawing of a dam are (1) **crest,** the top of the dam; (2) **water level;** and (3) **freeboard,** the height of the crest above the water level (Fig. 28.41).

An earthen dam is located on the contour map in Fig. 28.42. It makes an arc with its center at point *C*. The top of the dam is to be level to provide a roadway. The method of drawing the top view of the dam and indicating the level of the water held by the dam are shown in Fig. 28.42.

These same principles were used in the design and construction of the 726-foot Hoover Dam, built in the 1930s. Since this dam was made of concrete instead of earth, the dam was built in the shape of an arch bowed toward the water to take advantage of the compressive strength of concrete (Fig. 28.43).

are drawn 10 feet apart, since the contours in the top view are this far apart. In Step 2, the cut angles are measured and drawn on both sides of the roadway on the upper side. The points on the elevation lines crossed by the cut embankments are projected to the top view to find the limits of cut in this view.

In Step 3, the fill angles are laid off in the profile from given specifications. The crossing points on the profile view of the elevation lines are projected to the top view to find the limits of fill. New contour lines are drawn inside the areas of cut and fill to indicate they have been changed.

28.21 Strike and Dip

Strike and **dip** are terms used in geological and mining engineering to describe strata of ore under the surface of the earth.

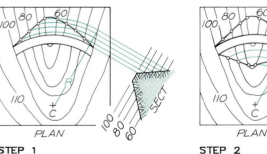

STEP 1

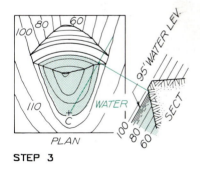

STEP 2   STEP 3

Figure 28.42 Graphical design of a dam.

Step 1 A dam in the shape of an arc with its center at *C* has an elevation of 100'. Draw radius *R* from *C* and project perpendicularly from this line, and draw a section through the dam from specifications. The downstream side of the dam is projected to radial line *R*. Using the radii from *C*, locate points on their respective contours.

Step 2 The elevations of the dam on the upstream side of the section are projected to the radial line, *R*. Using center *C* and your compass, locate points on their respective contour lines in the plan view as they are projected from the section.

Step 3 The elevation of the water level is 95' and is drawn in the section. The point where the water intersects the dam is projected to the radial line in the plan view and is drawn as an arc using center *C*. The limit of the water is drawn between the 90' and 100' contour lines in the top view.

Figure 28.43 The Hoover Dam and Lake Mead, which were built from 1931 to 1935. (Courtesy of the Bureau of Reclamation, U.S. Department of the Interior.)

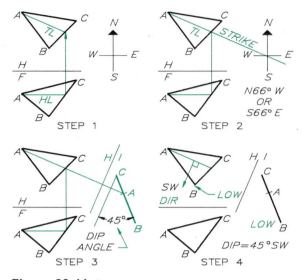

Figure 28.44 Strike and dip of a plane.

Step 1 Draw a horizontal line on the plane in the front view, project it to the top view where it is found TL.

Step 2 Strike is the compass direction of a level line on the plane in the top view. The strike of the plane is either N 66° W or S 66° E.

Step 3 Find the edge view of the plane in an auxiliary view. The dip angle of 45° is the angle between the H-1 and the edge view of the plane.

Step 4 Dip direction is the general compass direction of the dip measured toward the low side, SW in this case. Dip direction is perpendicular to a TL line in the top view and is written 45°SW

Strike is the compass bearing of a level line in the top view of a plane. It has two possible compass bearings since it is the direction of a level line.

Dip is the angle the edge view of a plane makes with the horizontal plus its general compass direction, such as NW or SW. The dip angle is found in the primary auxiliary view projected from the top view, and its dip direction is measured perpendicular to a level line in the top view toward the low side.

The steps of finding the strike and dip of a plane are given in Fig. 28.44. Strike can be measured in the top view by finding a true-length line on the plane in this view. The dip angle is found in an auxiliary view projected from the top

view that shows the horizontal and the plane as edges.

A plane can be constructed from strike and dip specifications (Fig. 28.45). The strike is drawn to represent a true-length horizontal line on the plane. Dip direction is perpendicular to the strike (Step 1). The edge view of the plane is then drawn through point 1 at a dip of 300° (Step 2). Points 2 and 3 are located in the front view.

28.22   Distances From a Point to a Plane

Descriptive geometry principles can be used to find various distances from a point to a plane. An example is shown in Fig. 28.46, where the dis-

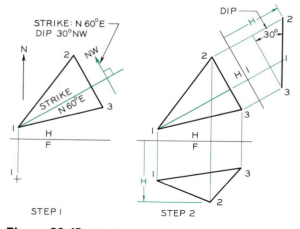

Figure 28.45 Working from strike and dip specifications.

Step 1 Strike is drawn on the top view of the plane, which is a TL horizontal line. The direction of dip is drawn perpendicular to the strike toward the NW, as specified.

Step 2 A point view of the strike line is found in the auxiliary view to locate point 1. The edge view of the plane is drawn through 1 at a dip of 30°, according to the specifications. The front view is completed by transferring height dimensions from the auxiliary to the front view.

tance from point 0 on the ground to an underground ore vein is found.

Three points are located on the top plane of an ore vein. Point 0 is the point on the earth from which the tunnels are to be drilled to the vein. Point 4 is a point on the lower plane of the vein.

The edge view of plane 1–2–3 is found by projecting from the top view. The lower plane is drawn parallel to the upper plane through point 4. The horizontal distance from point 0 to the plane is drawn parallel to the H-1 reference line. The vertical distance is perpendicular to the H-1 line. The shortest distance to the plane is perpendicular to the plane. Each of these lines is true length in this view where the ore vein appears as an edge.

The process of finding the distance from a point to a plane or a vein is a technique often used in solving mining and geological problems; for example, test wells are drilled into coal zones to learn more about them (Fig. 28.47).

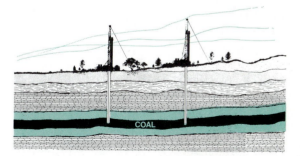

Figure 28.47 Test wells are drilled into coal zones to determine which coal seams will contribute significantly to the production of gas. (Courtesy of Texas Eastern News.)

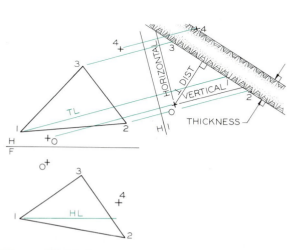

Figure 28.46 The vertical, horizontal, and perpendicular distances from a point to an ore vein can be found in an auxiliary view projected from the top view. The thickness of an ore vein is perpendicular to the upper and lower planes of the vein.

28.23 Outcrop

Strata of ore formations usually approximate planes of a uniform thickness. This assumption is employed in analyzing the orientation of underground ore veins. A vein of ore may be inclined to the surface of the earth and may outcrop on its surface. Outcrops permit open-surface mining operations at a minimum of expense.

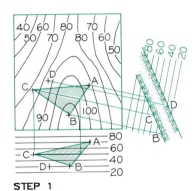

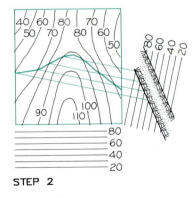

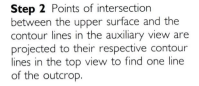

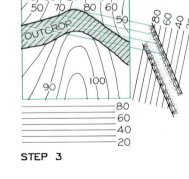

STEP 1

STEP 2

STEP 3

Figure 28.48 Ore vein outcrop.

Step 1 Using points A, B, and C on the upper surface of the plane, find its edge view by projecting an auxiliary off the top view. The lower surface of the plane is drawn parallel to the upper surface through point D, a point on the lower surface.

Step 2 Points of intersection between the upper surface and the contour lines in the auxiliary view are projected to their respective contour lines in the top view to find one line of the outcrop.

Step 3 Points from the lower surface in the auxiliary view are projected to their respective contour lines in the top view to find the second line of outcrop. Crosshatch this area to indicate the outcrop of the vein.

The steps of finding the outcrop of an ore vein are given in Fig. 28.48. The locations of sample drillings, A, B, and C, are shown on the contour map, and their elevations are located on the contours of the profile. These points are known to lie on the upper surface of the vein; point D is known to lie on the lower plane of the vein.

The edge view of the ore vein can be found in an auxiliary view projected from the top view (Step 1). The points on the upper surface are projected back to their respective contour lines in the top view (Step 2). The points on the lower surface of the vein are then projected to the top view (Step 3). If the ore vein continues uniformly at its angle of inclination to the surface, the space be-

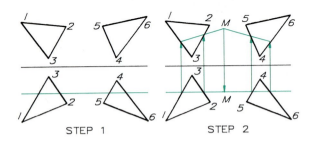

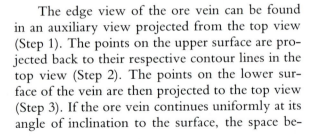

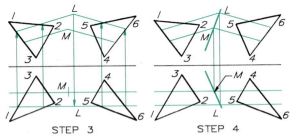

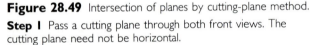

STEP 1

STEP 2

STEP 3

STEP 4

Figure 28.49 Intersection of planes by cutting-plane method.

Step 1 Pass a cutting plane through both front views. The cutting plane need not be horizontal.

Step 2 Project the intersections of the cutting plane to the top views of the planes. Intersection point M is found in the top view and is projected to the front view.

Step 3 A second cutting plane is passed through the front views of the planes and is projected to the top view. Intersection point L is found in the top view and is projected to the front view.

Step 4 Points L and M are connected in the top and front views to represent the line of intersection of the extended planes.

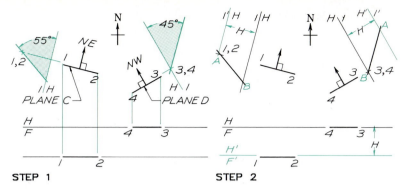
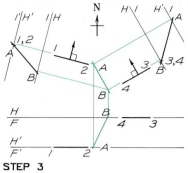

Figure 28.50 Intersection between ore veins by auxiliary view.

Step 1 Lines 1–2 and 3–4 are strike lines and are true length in the horizontal view. The point view of each strike line is found by auxiliary views, using a common reference plane. The edge views of the ore veins can be found by constructing the dip angles with the H-1 line through the point views. The low side is the side of the dip arrow.

Step 2 A supplementary horizontal plane, H'-F', is constructed at a convenient location in the front view. This plane is shown in both auxiliary views located H distance from the H-1 reference line. The H'-1' plane cuts through each ore vein edge in the auxilliary views to locate points A and B on each plane.

Step 3 Points A, on each auxiliary view of the H'-1' plane, are projected to the top view where they intersect at A. Points B on the H-1 plane are projected to their intersection in the top view B. Points A and B are projected to their respective planes in the front view. Line AB is the line of intersection between the two planes.

tween these two lines will be the outcrop of the vein on the surface of the earth.

28.24 Intersection Between Planes—Cutting Plane Method

The top and front views of two planes are given in Fig. 28.49, where it is required to find the line of intersection between them if the planes are infinite in size. Cutting planes are passed through either view at any angle and projected to the adjacent view. The two points, *L* and *M,* found in the top view, form the line of intersection. The compass direction of the line describes its direction of slope toward its low end.

The front view of the line of intersection is found by projecting its end points from the top view to their respective planes in the front view.

28.25 Intersection Between Planes—Auxiliary Method

In Fig. 28.50, two planes have been located and specified using strike and dip. Since the given strike lines are true-length level lines in the top view, the edge view of the planes can be found in the view where the strike appears as a point (Step 1). The edge views are drawn using the given dip angles and directions.

Horizontal datum planes H-F and H'-F' are used to find lines on each plane that will intersect when projected from the auxiliary views to the top view. Points *A* and *B* are connected to determine the line of intersection between the two planes in the top view. These points are projected to the front view to find line *AB*.

Problems

Use Size A sheets for the following problems, and lay out the problems using instruments. Each square on the grid is equal to 0.20 inch (about 5 mm). The problems can be laid out on grid or plain paper. Label all reference planes and points in each problem with ⅛-inch (3 mm) letters or numbers, using guidelines.

1. (Fig. 28.51) A–D. Find the true-length views of the lines by auxiliary view as indicated by the given lines of sight. Alternate method: Find the lines true-length by the Pythagorean theorem.

2. (Fig. 28.52) A–D. Find the true-length views of the lines by auxiliary view as indicated by the lines of sight. Alternate method: Find the lines true length by the Pythagorean theorem.

3. (Fig. 28.53) A–B. Find the lines true length by a true-length diagram. C–D. Find the point views of the lines.

4. (Fig. 28.54) A–D. Find the slope angle, tangent of the slope angle, and the percent grade of the four lines.

5. (Fig. 28.55) A–B. Find the edge views of the planes.

6. (Fig. 28.56) A. Find the angle between the planes. B. By projection, find the projection point between the line and plane and show visibility. C. By the auxiliary-view method, find the point of intersection between the line and plane and show visibility.

7. (Fig. 28.57) A. Construct a 1″ line perpendicular from point O on the plane and show it in all views. B. Construct a line perpendicular to the plane from point O, find the piercing point, and show visibility.

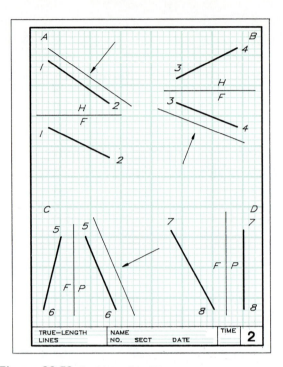

Figure 28.52 Problems 2A–2D.

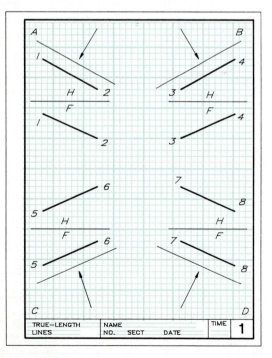

Figure 28.51 Problems 1A–1D.

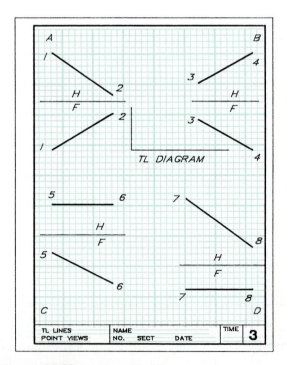

Figure 28.53 Problems 3A–3D.

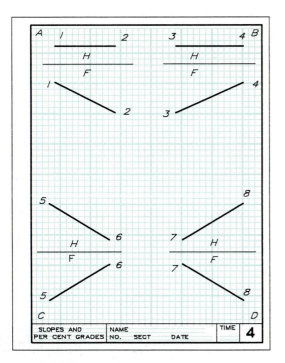

Figure 28.54 Problems 4A–4D.

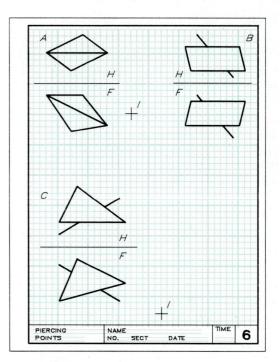

Figure 28.56 Problems 6A–6C.

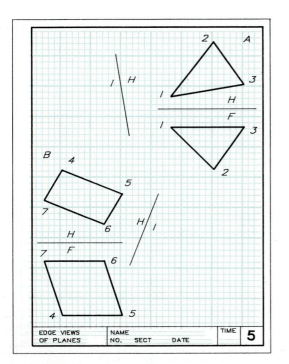

Figure 28.55 Problems 5A–5B.

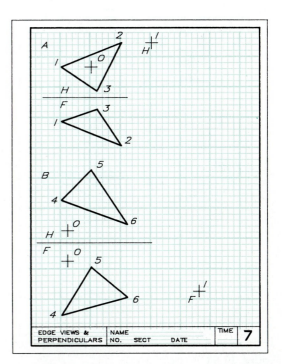

Figure 28.57 Problems 7A–7B.

8. (Fig. 28.58) A. Find the line of intersection between the intersecting planes by an auxiliary view. B. Find the angle between the planes by an auxiliary view.

9. (Fig. 28.59) A–B. Find the direction of slope and the slope angle of the planes. Alternate method: Find the strike and dip of the planes.

10. (Fig. 28.60) Find the shortest distance, the horizontal distance, and the vertical distance from point O to the underground ore vein represented by 1–2–3. Point *B* is on the lower plane of the vein. Find the thickness of the vein.

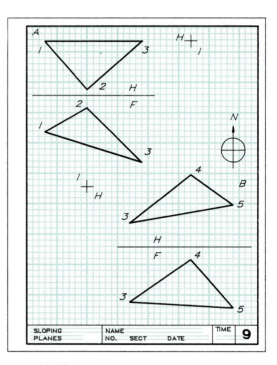

Figure 28.59 Problems 9A–9B.

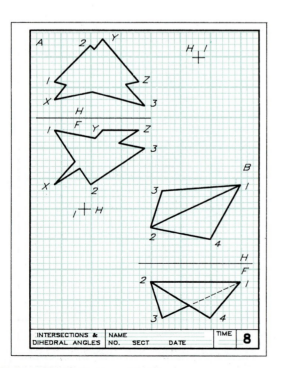

Figure 28.58 Problems 8A–8B.

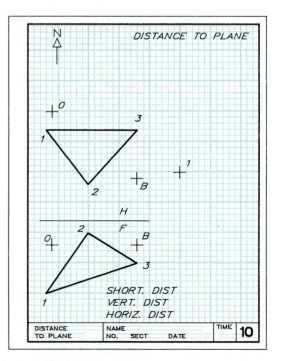

Figure 28.60 Problem 10.

11. (Fig. 28.61) A. Find the line of intersection between the two planes by the cutting-plane method. B. Find the line of intersection between the two planes indicated by strike lines 1–2 and 3–4. The plane with strike 1–2 has a dip of 30° and the one with strike 3–4 has a dip of 55°.

12. (Fig. 28.62) Find the limits of cut and fill in the plan view of the roadway. Use a cut angle of 35° and fill angle of 40°.

13. (Fig. 28.63) Find the outcrop of the ore vien represented by 1–2–3 on the upper surface. Point B is on the lower surface.

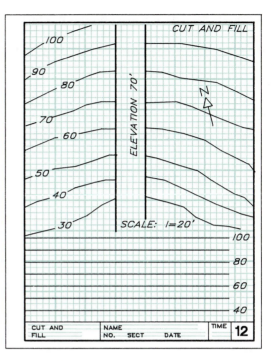

Figure 28.62 Problem 12.

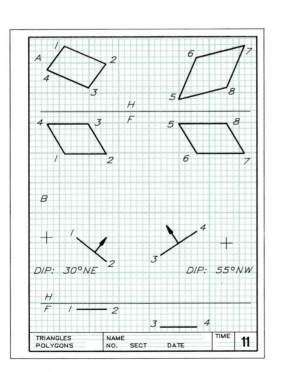

Figure 28.61 Problems 11A–11B.

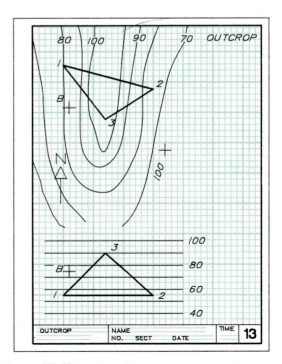

Figure 28.63 Problem 13.

14. (Fig. 28.64) Complete the plan-profile drawing of the drainage system from manhole 1 through manhole 2 to manhole 3, using the grades indicated. Allow a drop of 0.20° across each manhole to compensate for loss of pressure.

15. (Fig. 28.65) Draw the contour map with its contour lines. Give the lengths of the sides, their compass directions, interior angles, scale, and north arrow.

16. (Fig. 28.66) Draw the contour map, and construct the profile (vertical section) as indicated by the cutting-plane line in the plan view. Notice that the profile scale is different from the plan scale.

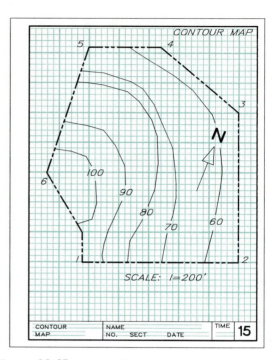

Figure 28.65 Problem 15.

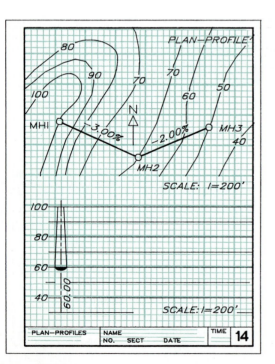

Figure 28.64 Problem 14.

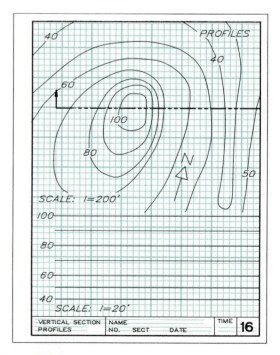

Figure 28.66 Problem 16.

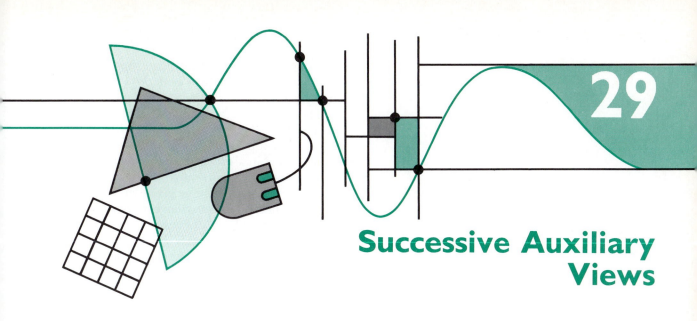

Successive Auxiliary Views

29.1 Introduction

A design cannot be detailed with complete specifications unless its complete geometry has been determined, which usually requires the application of descriptive geometry. The Anechoic Chamber (Fig. 29.1) is an example of a design where various problems of geometry were solved by the use of successive auxiliary views.

> A **secondary auxiliary view** is an auxiliary view projected from a primary auxiliary view. A **successive auxiliary view** is an auxiliary view of a secondary auxiliary view.

Figure 29.1 This internal view of Anechoic Chamber (a testing chamber free from echoes and reverberations) was designed by applying numerous principles of descriptive geometry. (Courtest of NASA.)

29.2 Point View of a Line

When a line appears true length, its point view can be found by projecting an auxiliary view from it. In Fig. 29.2, line 1–2 is true length in the top view since it is horizontal in the front view. Its point view is found in the primary auxiliary view by constructing reference line H-1 perpendicular to the true-length line. The height dimension, *H,* is transferred to the auxiliary view to locate the point view of 1–2.

Since the line in Fig. 29.3 is not true length in either view, the line must be found true length by a primary auxiliary view. By projecting from the front view, the line is found true length. This view could have been projected from the top as

509

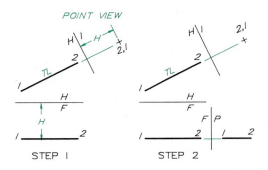

Figure 29.2 The point of view of a line can be found by projecting an auxiliary view from the true-length view of the line.

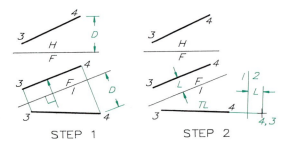

Figure 29.3 Point view of an oblique line.

Step 1 A line of sight is drawn perpendicular to one of the views, the front view in this example. Line 3–4 is found true length by projecting perpendicularly from the front view.

Step 2 A secondary reference line, 1–2, is drawn perpendicular to the true-length view of 3–4. The point view is found by transferring dimension *L* from the front view to the secondary auxiliary view.

well. The point view of the line is found by projecting from the true-length line to a secondary auxiliary view and is labeled 2,1 since point 2 is seen first.

29.3 Angle Between Planes

> The angle between two planes is called a **dihedral angle.**

The dihedral angle can be found in the view where the line of intersection appears as a point.

Since this view results in the point view of a line that lies on both planes, both will appear as edges.

The two planes in Fig. 29.4 represent a special case since the line of intersection, 1–2, is true length in the top view. This permits you to find its point view in a primary auxiliary view where the true angle can be measured.

A more typical case is given in Fig. 29.5, where the line of intersection between the two planes is not true length in either view. The line of intersection, 1–2, is found true length in a primary auxiliary view, and the point view of the line is then found in the secondary auxiliary view, where the dihedral angle is measured.

This principle must be used to determine the angles between side panels of a control tower (Fig. 29.6) so that corner braces can be designed that will hold the structure together.

29.4 True Size of a Plane

> A plane can be found **true size** in a view projected perpendicularly from an edge view of a plane.

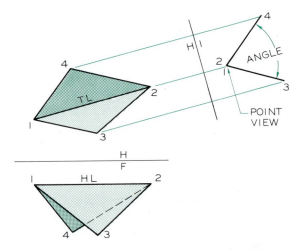

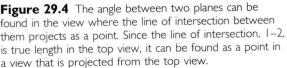

Figure 29.4 The angle between two planes can be found in the view where the line of intersection between them projects as a point. Since the line of intersection, 1–2, is true length in the top view, it can be found as a point in a view that is projected from the top view.

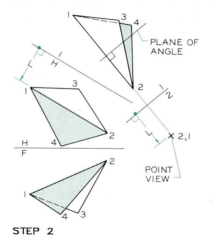

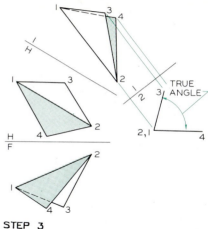

STEP 1 **STEP 2** **STEP 3**

Figure 29.5 Angle between two planes.

Step 1 The angle between two planes can be measured in a view where the line of intersection appears as a point. The line of intersection is first found true length by projecting a primary auxiliary view perpendicularly from the top view, in this case.

Step 2 The point view of the line of intersection is found in the secondary auxiliary view by projecting parallel to the true length view of 1–2. The plane of the dihedral angle is an edge and is perpendicular to the true-length line of intersection.

Step 3 The edge views of the planes are completed in the secondary auxiliary view by locating points 3 and 4. The angle between the planes, the dihedral angle, can be measured in this view.

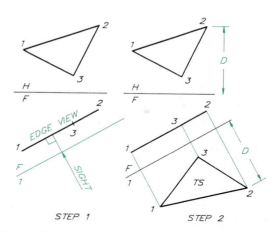

Figure 29.7 True size of a plane.

Step 1 Since plane 1–2–3 appears as an edge in the front view, the line of sight is drawn perpendicular to the edge. The *F*-1 reference line is drawn parallel to the edge.

Step 2 The plane will appear true size in the primary auxiliary by locating the vertex points with depth (D) dimensions.

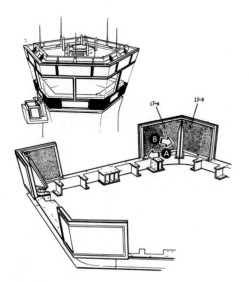

Figure 29.6. The determination of the angle between the planes of the corner panels of the control tower used principles of descriptive geometry. (Courtesy of the Federal Aviation Agency.)

The front view of plane 1–2–3 appears as an edge in the front view (Fig. 29.7). It can be found true size in a primary auxiliary view projected perpendicularly from the edge view.

In Fig. 29.8, the true size of plane 1–2–3 is found by first finding the edge view of the plane

511

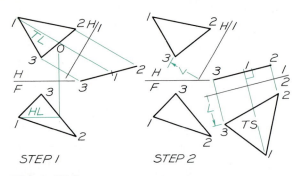

STEP 1   STEP 2

Figure 29.8 True size of a plane.

Step 1 The edge view of plane 1–2–3 is found by finding the point view of TL line, 1–0, in a primary auxiliary view.

Step 2 A true-size view is found by projecting a secondary auxiliary view perpendicularly from the edge view of the plane. Dimension L is shown as a typical measurement used to complete the TS view.

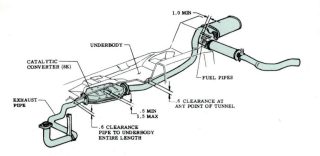

Figure 29.9 The angles of bend in the exhaust pipe were found by applying the principle of finding the angle between two lines. (Courtesy of General Motors Corporations.)

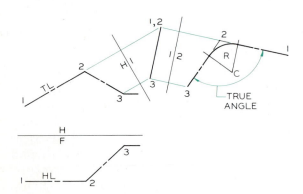

Figure 29.10. The angle between two lines can be found by finding the plane of the lines true size.

(Step 1). The secondary auxiliary view is then projected perpendicularly from the edge view (Step 2) to find a true-size view of the plane where each angle is true size.

This principle can be used to find the angle between lines, such as bends in an exhaust pipe. (Fig. 29.9). This type of problem is shown in Fig. 29.10, where the top and front views of intersecting lines are given. It is required that the angles of bend be determined and a radius of curvature be shown.

Angle 1–2–3 is found as an edge in the primary auxiliary view and true-size in the secondary view, where the angle can be measured and the radius of curvature drawn.

29.5   Shortest Distance From a Point to a Line

> The shortest distance from a point to a line can be measured in the view where the line appears as a point.

The shortest distance from point 3 to line 1–2 is found in a primary auxiliary view in Fig. 29.11

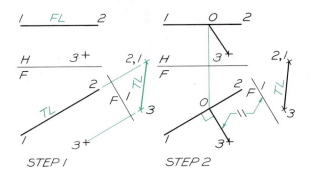

STEP 1   STEP 2

Figure 29.11 Shortest distance from a point to a line.

Step 1 The shortest distance from a point to a line will appear TL where the line appears as a point. The TL line is found in the primary auxiliary view.

Step 2 For the connecting line to be TL in the auxiliary view, it must be parallel to the F-1 reference line in the preceding view, the front view. Line 3–0 is found and projected to the top view.

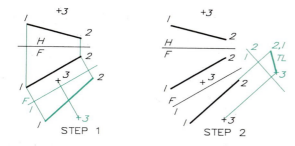

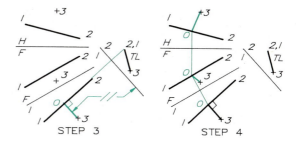

Figure 29.12 Shortest distance from a point to a line.

Step 1 The shortest distance from a point to a line can be found in the view where the line appears as a point. Line 1–2 is found true length by projecting from the front view.

Step 2 Line 1–2 is found as a point in a secondary auxiliary view projected from the true-length view of 1–2. The shortest distance appears true length in this view.

Step 3 Since 3–0 is true length in the secondary auxiliary view, it is parallel to the 1–2 line in the primary auxiliary view and perpendicular to the line.

Step 4 The front and top views of 3–0 are found by projecting from the primary auxiliary view in sequence.

(Step 1). Since the distance from 3 to the line is true length where the line is a point, it must be parallel to reference line F–1 in the front view.

This type of problem is solved in Fig. 29.12 by finding the line 1–2 true length in a primary auxiliary. The point view of the line is found in the secondary auxiliary view, where the distance from point 3 is true length. Since line 0–3 is true length in this view, it must be parallel to the 1–2 reference line in the preceding view, the primary auxiliary view. It is also perpendicular to the true-length view of line 1–2 in the primary auxiliary view.

29.6 Shortest Distance Between Skewed Lines—Line Method

Randomly positioned (non-parallel) lines are called **skewed lines.** The shortest distance between two skewed lines can be measured in the view where one of the lines appears as a point.

The shortest distance between two lines is perpendicular to both lines. The location of the shortest distance is both functional and economical, as demonstrated by the connector between two pipes in Fig. 29.13, since a standard connector is a 90° tee.

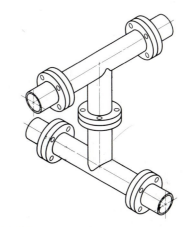

Figure 29.13 The shortest distance between two lines, or planes, is a line perpendicular to both. This is the most economical and functional connection since perpendicular connectors are standard.

A problem of this type is solved by the **line method.** In Fig. 29.14, line 3–4 is found as a point in the secondary auxiliary view, where the shortest distance is drawn perpendicular to line 1–2. Since the distance is true length in the secondary auxiliary view, it must be parallel to the 1–2 reference line in the primary auxiliary view. Point 0 is found by projection, and 0P is drawn perpendicular to line 3–4. The line is projected back to the given principal views.

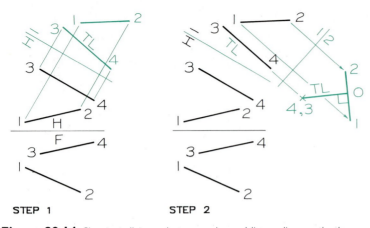

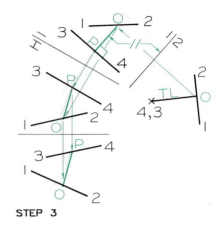

STEP 1   STEP 2   STEP 3

Figure 29.14 Shortest distance between skewed lines—line method.

Step 1 The shortest distance between two skewed lines can be found in the view where one of the lines appears as a point. Line 3–4 is found true length by projecting from the top view along with line 1–2.

Step 2 The point view of line 3–4 is found in a secondary auxiliary view projected from the true-length view of 3–4. The shortest distance between the lines is drawn perpendicular to line 1–2.

Step 3 Since the shortest distance is TL in the secondary auxiliary view, it must be parallel to the reference line in the preceding view. Points 0 and P are projected to the given view.

29.7 Shortest Distance Between Skewed Lines—Plane Method

The distance between skewed lines can be solved using the alternative **plane method** which involves the construction of a plane through one of the lines parallel to the other (Fig. 29.15). The top and front views of 0–2 are drawn parallel to their respective views of line 3–4. Plane 1–2–0 is parallel to line 3–4. Both lines appear parallel in a view where 1–2–0 is an edge.

In Fig. 29.16, plane 3–4–0 is constructed, its edge view is found, and both lines appear parallel (Step 1). A secondary auxiliary view is projected perpendicularly from these parallel lines to find the view where the lines cross (Step 2). This crossing point is the point view of the shortest distance between the lines. It is true length and perpendicular to both lines when projected to the primary auxiliary view, where it is labeled line *LM*. It is projected back to the given views to complete the solution.

This principle of the shortest distance between two skewed lines was applied to the design of the separation of power lines, where clearance is critical (Fig. 29.17).

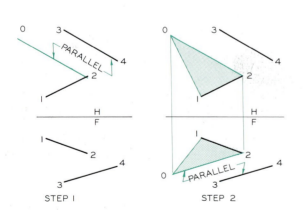

STEP 1   STEP 2

Figure 29.15 A plane can be constructed through a line and parallel to another line.

Step 1 Line 0–2 is drawn parallel to line 3–4 to a convenient length.

Step 2 The front view of line 0–2 is drawn parallel to the front view of line 3–4. The length of the front view of 0–2 is found by projecting from the top view of 0. Plane 1–2–0 is parallel to line 3–4.

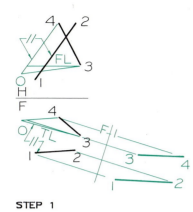
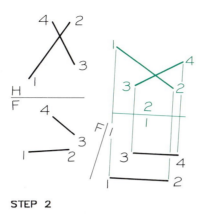
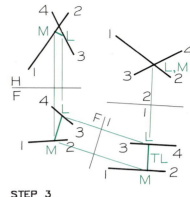

STEP 1   STEP 2   STEP 3

Figure 29.16 Shortest distance between skewed lines—plane method.

Step 1 Construct a plane through line 3–4 that is parallel to line 1–2. When this plane is found as an edge by projecting from the front view, the two lines will appear parallel.

Step 2 The shortest distance will appear TL in the primary auxiliary view, where it will be perpendicular to both lines. The secondary auxiliary view is projected perpendicularly from the lines in the primary auxiliary.

Step 3 The crossing point of the two lines is the point view of perpendicular distance *LM* between them. This distance is projected to the primary auxiliary view, where it is TL, and back to the given views.

Figure 29.17 The determination of clearance between power lines is an electrical engineering problem that is an application of skewed-line principles. (Courtesy of the Tennessee Valley Authority.)

29.8 Shortest Level Distance Between Skewed Lines

To find the shortest level (horizontal) distance between two skewed lines, the **plane method** must be used rather than the line method. Also,

the plane method is used to find a view where the horizontal plane appears as an edge.

In Fig. 29.18, plane 3–4–0 is drawn parallel to line 1–2, and an edge view of the plane is found in the primary auxiliary view. The lines appear parallel in this view, and the horizontal (H-1) appears as an edge. A line of sight is drawn parallel to H-1, and the secondary reference line, 1–2, is drawn perpendicular to H-1. The crossing point of the lines in the secondary auxiliary view locates the point view of the shortest horizontal distance between the lines. This line, *LM,* is true length in the primary auxiliary view and parallel to the H-1 plane. Line *LM* is projected back to the given views. As a check, *LM* must be parallel to the H–F line in the front view since it is a level or horizontal line.

29.9 Shortest Grade Distance Between Skewed Lines

Many lines representing highways, power lines, or conveyors are connected by lines at a specified grade other than horizontal or perpendicular. Conveyors, such as the one shown in Fig. 29.19, are used to transport aggregates or grain.

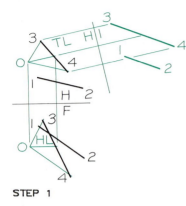

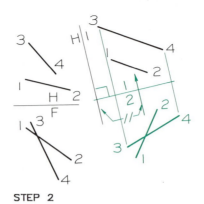

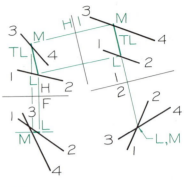

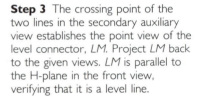

STEP 1   STEP 2   STEP 3

Figure 29.18 Shortest level distance between skewed lines—plane method.

Step 1 Construct plane 0–3–4 parallel to line 1–2 by drawing line 0–4 parallel to 1–2. Find the edge view of 0–3–4 by projecting off the top view. The lines will appear parallel in this view. The auxiliary view must be projected from the top view to find the horizontal plane as an edge.

Step 2 An infinite number of horizontal (level) lines can be drawn parallel to H-1 between the lines in the auxiliary view, but only the shortest level line will appear true length. Construct the secondary auxiliary view by projecting parallel to the horizontal (H-1) to find the point view of the shortest level line.

Step 3 The crossing point of the two lines in the secondary auxiliary view establishes the point view of the level connector, LM. Project LM back to the given views. LM is parallel to the H-plane in the front view, verifying that it is a level line.

Figure 29.19 These conveyors represent the application of skewed-line problems.

If a 50% grade connector between two lines must be found (Fig. 29.20), the plane method is used. A view where the lines appear parallel is constructed, and a 50% grade line is drawn from the edge view of the horizontal (H-1). To have an edge view of the horizontal from which the 50% grade is constructed, the auxiliary views must be projected from the top view.

The grade line can be constructed in two directions from the H-1 line, but the shortest distance will be the one most nearly perpendicular to both lines (Step 2). The secondary auxiliary view is projected parallel to this 50% grade line to find the crossing point of the lines to locate the shortest connector, LM, at a 50% grade. Line LM is projected back to all views. LM is true length in the primary auxiliary view, where the lines appear parallel.

The shortest connector between skewed lines will appear true length in the view where the lines appear parallel. Perpendicular, horizontal, and grade lines are true length in this view.

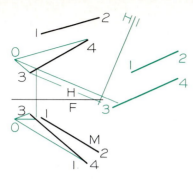
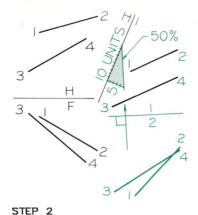
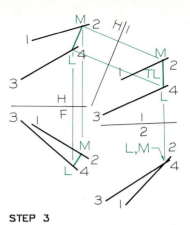

STEP 1   STEP 2   STEP 3

Figure 29.20 Grade distance between skewed lines.

Step 1 To find a level line or a line on a grade between two skewed lines, the primary auxiliary must be projected from the top view. Plane 3–4–0 is constructed parallel to line 1–2. The edge view of the plane is found and the lines appear parallel.

Step 2 Construct a 50% grade line from the edge view of the H-1 line in the primary auxiliary view that is most nearly perpendicular to the lines. Project the secondary auxiliary view parallel to the grade line. The shortest grade distance will appear TL in the primary auxiliary.

Step 3 The point of crossing of the two lines in the secondary auxiliary view establishes the point view of the 50% grade line, *LM*. This line is projected back to the previous views in sequence.

29.10   Angular Distance to a Line

Standard connectors used to connect pipes and structural members are available in two standard angles—90° and 45°. It is far more economical to incorporate these angles into a design rather than to design specially made connectors.

In Fig. 29.21, it is required to locate the point of intersection on line 1–2 of a line drawn from point 0 at a 45° angle to the line. The plane of the line and point, 1–2–0, is found as an edge in the primary auxiliary view and as a true-size plane in the secondary auxiliary view. The angle can be

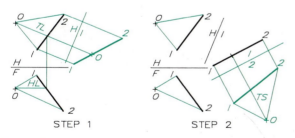
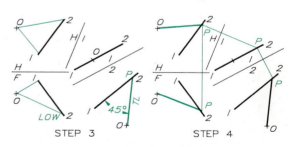

STEP 1   STEP 2   STEP 3   STEP 4

Figure 29.21 Line through a point with a given angle to a line.

Step 1 Connect 0 to each end of the line to form a plane 1–2–0 in both views. Draw a horizontal line in the front view of the plane and project it to the top view, where it is TL. Find the edge view of the plane by obtaining the point view of A0.

Step 2 Find the true size of plane 1–2–0 projecting perpendicularly from the edge view of the plane in the primary auxiliary view. The plane can be omitted in this view and only line 1–2 and point 0 are shown.

Step 3 Line *OP* is constructed at the specific angle with line 1–2, 45°. If the angle is toward point 2, the line slopes downhill; if toward point 1, it slopes uphill.

Step 4 Point *P* is projected back to the other views in sequence.

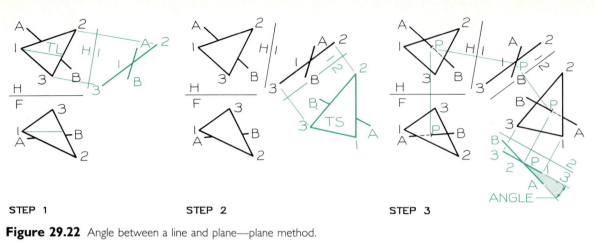

STEP 1   STEP 2   STEP 3

Figure 29.22 Angle between a line and plane—plane method.

Step 1 The angle between a line and a plane can be measured in the view where the plane is an edge and the line is TL. The plane is found as an edge by projecting off the top view. The line is not true length in this view.

Step 2 The plane is found true size by projecting perpendicularly from the edge view of the plane. A view projected in any direction from a TS plane will show the plane as an edge.

Step 3 A third successive auxiliary view is projected perpendicularly from line *AB*. The line appears TL and the plane appears as an edge in this view where the angle is measured. The piercing points and visibility are shown by projecting back in sequence to all views.

measured in this view where the plane of the line and point is true size.

The 45° connector is drawn from 0 to the line toward point 2 if it slopes downhill, or toward point 1 if it slopes uphill. Slope can be determined by referring to the front view where height can be easily seen. Line 0*P* is projected back to the given views.

29.11 Angle Between a Line and a Plane—Plane Method

> The angle between a line and a plane can be measured in the view where the plane appears as an edge and the line appears true length.

In Fig. 29.22, the edge view of the plane is found in a primary auxiliary view projected from any primary view. The plane is then found true size in Step 2, where the line is foreshortened. Line *AB* can be found true length in a third auxiliary view projected perpendicularly from the secondary auxiliary view of *AB*. The line appears true length, and the plane appears as an edge in the third successive auxiliary view.

An alternative method of finding the angle between a line and a plane is the line method in which the line is found as a point, and *TL* in a view where the plane is an edge.

The piercing point is projected back to the views in sequence, and the visibility is determined for each view.

Problems

Use Size A sheets for the following problems, and lay out the problems using instruments. Each square on the grid is equal to 0.20 in. (about 5 mm). The problems can be laid out on grid or plain paper. Label all reference planes and points in each problem with $\frac{1}{8}$-inch letters or numbers, using guidelines.

The crosses marked "1" and "2" are to be used for placing the primary and secondary reference lines.

The primary reference line should pass through "1" and the secondary through "2".

1. (Fig. 29.23) A–B. Find the point views of the lines. C–D. Find the angles between the planes.

2. (Fig. 29.24) A–B. Find the true-size views of the planes.

3. (Fig. 29.25) A–B. Find the angles between the lines.

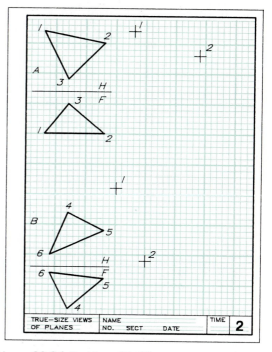

Figure 29.24 Problems 2A–2B.

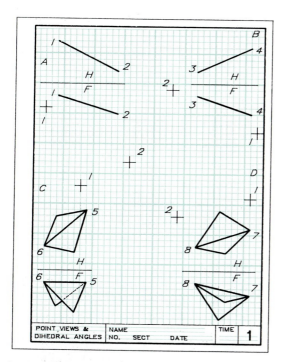

Figure 29.23 Problems 1A–1D.

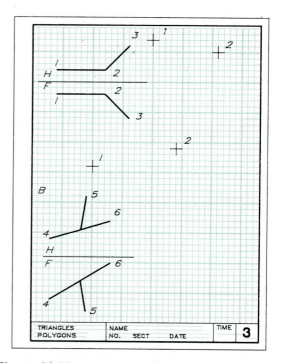

Figure 29.25 Problems 3A–3B.

4. (Fig. 29.26) A–B. Find the shortest distances from the points to the lines. Show this distance in all views.

5. (Fig. 29.27) A–B. Find the shortest distances between the lines by the line method. Show this distance in all views.

6. (Fig. 29.28) Find the shortest distance between the lines by the plane method. Show the line in all views. Alternate problem: Find the shortest horizontal distance between the two lines and show the distance in all views.

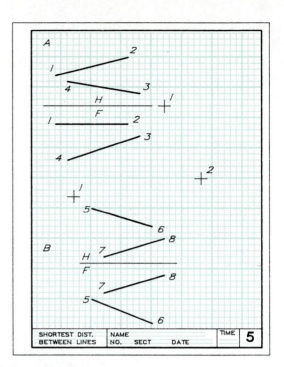

Figure 29.27 Problems 5A–5B.

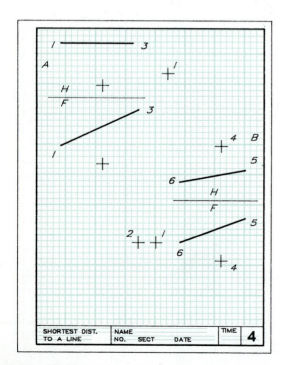

Figure 29.26 Problems 4A–4B.

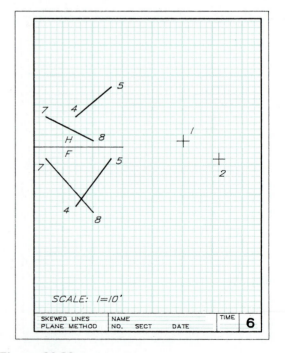

Figure 29.28 Problem 6.

7. (Fig. 29.29) Find the shortest 20% grade distance between the two lines. Show this distance in all views.

8. (Fig. 29.30) Find the connector from point 0 that will intersect line 1–2 at 60°. Show this line in all views. Project from the top view. Scale: full size.

9. (Fig. 29.31) Find the angle between the line and the plane by the plane method. Show visibility in all views.

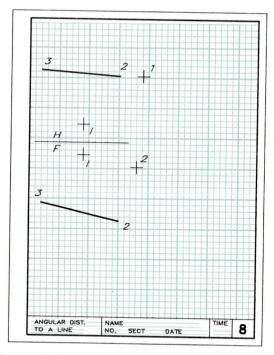

Figure 29.30 Problem 8.

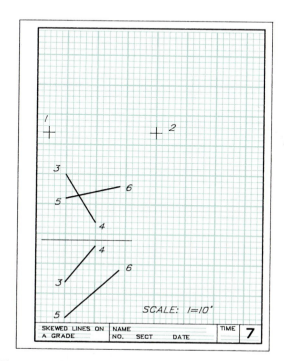

Figure 29.29 Problem 7.

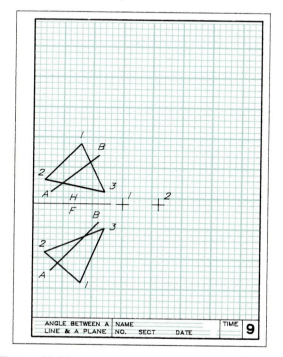

Figure 29.31 Problem 9.

30

Revolution

30.1 Introduction

Figure 30.1 shows the front suspension of an automobile. It was designed to revolve about several axes at each wheel. This is just one of many designs based on the principles of revolution.

Revolution is a technique of revolving an orthographic view into a new position to yield a true-size view of a surface or a line. For example, revolution techniques were used in early descriptive geometry solutions prior to the alternate method of auxiliary views.

30.2 True Length of a Line in the Front View

The object in Fig. 30.2 demonstrates how an inclined surface can be found true size by auxiliary view and revolution. When the auxiliary-view method is used, the observer changes position to an auxiliary vantage point and looks perpendicularly at the inclined surface.

When the revolution method is used, the top view of the object is revolved about the axis until the edge view of the inclined plane is perpendicular to the standard line of sight from the front

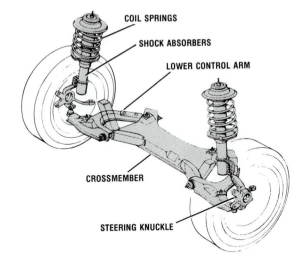

Figure 30.1 An example of the application of the principles of revolution can be seen in this front suspension of an automobile. From the steering wheel to the connecting tires, revolution is a major part of the design. (Courtesy of Chrysler Corporation.)

view. In other words, the observer's line of sight does not change, but the conventional lines of sight between adjacent orthographic views are used.

522

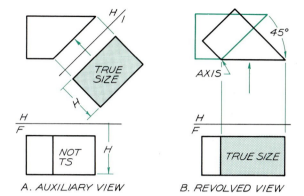

Figure 30.4 illustrates the technique of finding line 1–2 true length in the front view. When in its first position, the observer's line of sight is perpendicular neither to the triangular plane nor to line 1–2. But when it is revolved to be perpendicular to the line of sight, the triangle appears true size, and line 1–2 is true length.

Figure 30.2 (A) The surface is found true size by an auxiliary view. (B) The surface is found true size by revolving the top view.

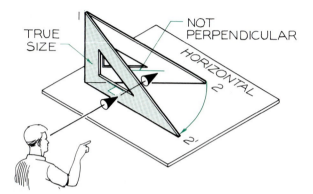

Figure 30.4 Line 1–2 of the triangle does not appear true length in the front view because your line of sight is not perpendicular to it. When the triangle is revolved to a position where your line of sight is perpendicular to it, line 1'–2' can be seen true length.

A single line can be found true length in the front view by revolution, as shown in Fig. 30.3. Line AB is revolved into a position parallel to the frontal plane. The top view represents the circular base of a right cone, and the front view is the triangular view of a cone. Line AB' is the outside element of the cone and is true length.

30.3 True Length of a Line in the Top View

A surface that appears as an edge in the front view can be found true size in the top view by a primary auxiliary view or by a single revolution (Fig. 30.5).

The axis of revolution is located as a point in the front view and is true length in the top view. The edge view of the plane is revolved until it is a horizontal edge in the front view (Fig. 30.5A). It is projected to the top view to find the surface true size. As in the auxiliary-view method, the depth dimension (D) does not change.

Line CD is found true length in the top view by revolving the line into a horizontal position in Fig. 30.6 (Step 2). The arc of revolution in the front view represents the base of a cone of revolution. Line CD' is true length in the top view

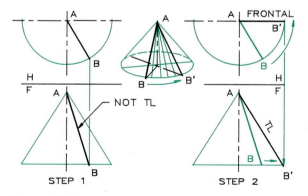

Figure 30.3 True length in the front view.

Step I The top view of line AB is used as a radius to draw the base of a cone with point A as the apex. The front view of the cone is drawn with a horizontal base through point B.

Step 2 The top view of line AB is revolved to be parallel to the frontal plane. When projected to the front view, frontal line AB' is the outside element of the cone and is true length.

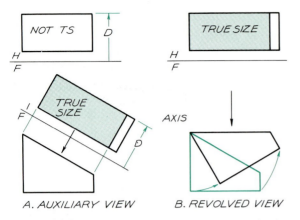

A. AUXILIARY VIEW   B. REVOLVED VIEW

Figure 30.5 (A) The inclined plane is found true size by an auxiliary view with a line of sight perpendicular to the surface. (B) The surface is found true size by revolving the front view until it is perpendicular to the line of sight from the top view.

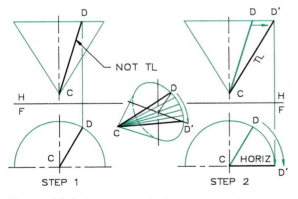

STEP 1   STEP 2

Figure 30.6 True length of a line in the top view.

Step 1 The front view of line *CD* is used as a radius to draw the base of a cone with point *C* as the apex. The top view of the cone is drawn with the base shown as a frontal plane.

Step 2 The front view of line *CD* is revolved into position *CD'* where it is horizontal. When projected to the top view, *CD'* is the outside element of the cone and is true length.

since it is an outside element of the cone. Note that the depth dimension in the top view does not change.

30.4   True Length of a Line in the Profile View

Line *EF* in Fig. 30.7 is found true length by revolving it in the front view until it is parallel to the edge view of the profile plane (Step 1). The circular view of the cone is projected to the side view, where the triangular shape of the cone is seen. Since *EF'* is a profile line in (Step 2), it is true length in the side view, where it is the outside element of the cone.

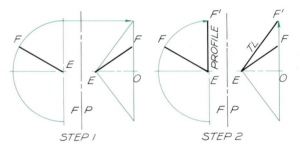

STEP 1   STEP 2

Figure 30.7 True length of a line in the side view.

Given The front and side views of line *EF*.

Required Find the true-length view of line *EF* in the profile review by revolution.

Step 1 The front view of line *EF* is used as a radius to draw the circular view of the base of a cone. The side view of the cone is drawn with a base through point *F* that is a frontal edge.

Step 2 Line *EF* in the frontal view is revolved to position *EF'* where it is a profile line. Line *EF'* in the profile view is true length, since it is a profile line and the outside element of the cone.

In the previous examples, each line has been revolved about one of its ends. However, a line can be revolved about any point on its length. Line 5–6 in Fig. 30.8 is found true length by revolving it about point 0.

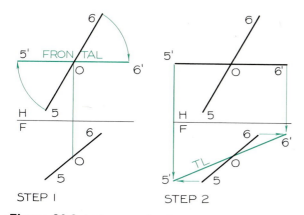

Figure 30.8 In the preceding figures, the lines have been revolved about their ends, but they can be found true length by revolving them about any point on them. Line 5–6 is revolved into a frontal position in the top view and is found true length in the frontal view.

30.5 Angles With a Line and Principal Planes

The angle between a line and plane will appear true size in the view where the plane is an edge and the line is true length. Two principal planes appear as edges in all principal views. Therefore when a line appears true length in a principal view, the angle between the line and the two principal planes can be measured.

The angle between the horizontal and the profile planes can be measured in Fig. 30.9A in the front view. The angle with horizontal and profile planes can be measured in the top view in Fig. 30.9B.

30.6 True Size of a Plane

When a plane appears as an edge in a principal view (Fig. 30.10), it can be revolved to be parallel to the reference line (Step 1). The new front view is true size.

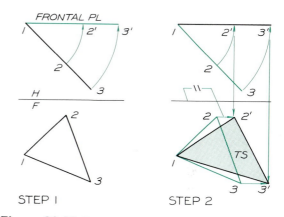

Figure 30.10 True size of a plane.

Step 1 The edge view of a plane is revolved to be parallel to the frontal plane.

Step 2 Points 2' and 3' are projected to the horizontal projectors from points 2 and 3 in the front view.

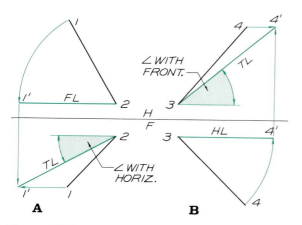

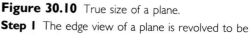

Figure 30.9 Angles with principal planes. (A) The angle with the horizontal plane can be measured in the front view if the line appears true length. (B) The angle with the frontal plane can be measured in the top view if the line appears true length.

The plane in Fig. 30.11 is found true size by the combination of an auxiliary view and a single revolution. The plane is found as an edge projected from a true-length line in the plane. The edge view is revolved to be parallel to the F-1 reference line (Step 3). The true size of the plane is found by projecting the original points, 1, 2, and 3, in the front view parallel to the F-1 line to intersect the projectors from 1' and 2' (Step 4).

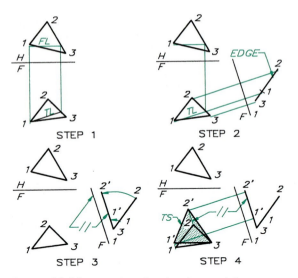

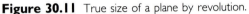

Figure 30.11 True size of a plane by revolution.

Step 1 To find the edge view of the plane by revolution, begin by drawing a frontal line on the plane that is true length in the front view.

Step 2 Find the edge view of the plane by finding the point view of the frontal line.

Step 3 Revolve the edge view of the plane to be parallel to the F–1 reference line.

Step 4 Project the revolved points, 1′ and 2′, to the front view to the projectors from 1 and 2 that are parallel to the F–1 reference line.

The true size of the plane could have been found by projecting from the top view to find the edge view as well.

30.7 True Size of a Plane by Double Revolution

The edge view of a plane can be found by revolution without the use of auxiliary views (Fig. 30.12). A frontal line is drawn on plane 1–2–3, and the line appears true length in the front view. The plane is revolved until the true-length line is vertical in the front view (Step 1). The true-length line will project as a point in the top view; therefore the plane will appear as an edge in this view (Step 2). Projectors from the top view of

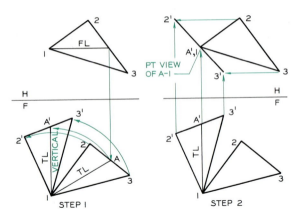

Figure 30.12 Edge view of a plane.

Step 1 A frontal line is found true length on the front view of the plane. The front view is revolved until the true-length line is vertical.

Step 2 Since the TL line (1–A′) is vertical, it will appear as a point in the top view, and the plane will appear as an edge, 1–2′–3′.

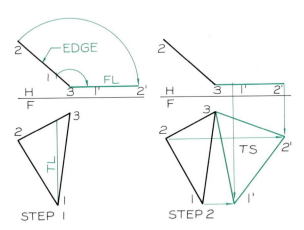

Figure 30.13 True size of a plane.

Step 1 When a plane appears as an edge in the principal view, it can be revolved to a position parallel to a reference line, the frontal line in this case.

Step 2 Points 1′ and 2′ are projected to the front view to intersect with the horizontal projectors from the original points 1 and 2. The plane is true size in this view.

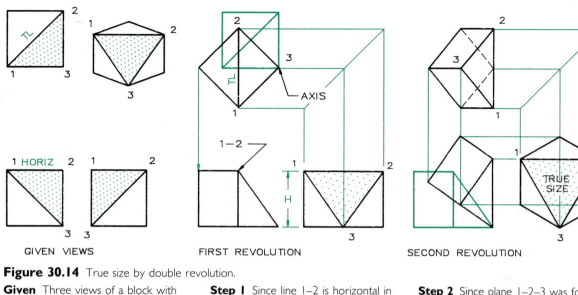

Figure 30.14 True size by double revolution.

Given Three views of a block with an oblique plane across one corner.

Required Find the plane true size by revolution.

Step 1 Since line 1–2 is horizontal in the frontal view, it is true length in the top view. The top view is revolved into a position where line 1–2 can be seen as a point in the front view.

Step 2 Since plane 1–2–3 was found as an edge in Step 1, this plane can be revolved into a vertical position in the front view, to appear true size in the side view. The depth dimension does not change.

points 2 and 3 are parallel to the H-F reference line.

A second revolution, called a **double revolution,** can be made to revolve this edge view of the plane until it is parallel to the frontal plane, as shown in Fig. 30.13. The top views of points 1′ and 2′ are projected to the front view where plane 1–2–3 is true size.

This second revolution could have been performed in Fig. 30.12, but this would have resulted in an overlapping of views, making it difficult to observe the separate steps.

Double revolution is used in Fig. 30.14 to find the oblique plane of the object, plane 1–2–3, true size. In Step 1, the true-length line 1–2 on the plane is revolved in the top view until it is perpendicular to the frontal plane. Thus, line 1–2 appears as a point in the front view, and the plane appears as an edge. This changes the width and depth dimensions, but the height dimension does not change.

In Step 2, the edge view of the plane is revolved into a vertical position parallel to the profile plane. The plane is found true size by projecting to the profile view, where the depth dimension is unchanged and the height dimension has been increased.

30.8 Angle Between Planes

The engine mount frame of a helicopter is an application where the angle between two intersecting planes must be found to provide its design specifications (Fig. 30.15).

In Fig. 30.16, the angle between two planes is found by drawing the edge view of the dihedral angle (the angle between the planes) perpendicular to the line of intersection, and the plane of the angle is projected to the front view (Step 1). The edge view of the angle is revolved until it is a frontal plane in the top view; then, it is

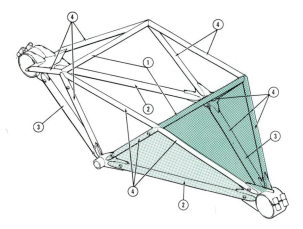

Figure 30.15 By using revolution principles, the angle between any two planes of the helicopter engine mount can be found. (Courtesy of Bell Helicopter Corp.)

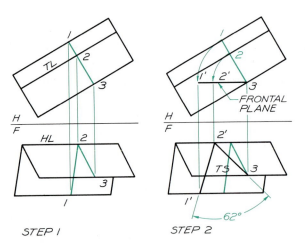

STEP 1   *STEP 2*

Figure 30.16 Angle between planes.

Step 1 A right section is drawn perpendicular to the TL line of intersection between the planes in the top view and is projected to the front view. The section is not true size in the front view.

Step 2 The edge view of the right section is revolved to position 1'–2'–3 in the top view to be parallel to the frontal plane. This section is projected to the frontal view, where it is true size.

projected to the front view where its true-size view is found (Step 2).

A similar problem is solved in Fig. 30.17. In this example, the line of intersection does not appear true length in the given views; therefore an auxiliary view is used to find its true length (Step 1). The plane of the angle between the planes can be drawn as an edge perpendicular to the true-length line of intersection. The foreshortened view of plane 1–2–3 is projected to the top view (Step 1). The edge view of plane 1–2–3 is then revolved in the primary auxiliary view until it is parallel to the H–1 line (Step 2).

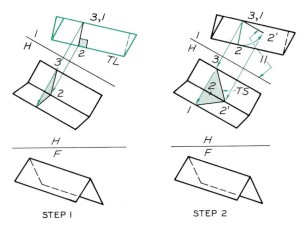

STEP 1   STEP 2

Figure 30.17 Angle between oblique planes.

Step 1 A true-length view of the line of intersection is found in an auxiliary view projected from the top view. The right section is constructed perpendicular to the true length of the line of intersection and is projected to the top view.

Step 2 The edge view of the right section is revolved to be parallel to the H-1 reference line so the plane will appear true size in the top view after being revolved. The angle between the planes is 1–2'–3.

30.9   Location of Directions

To solve more advanced problems of revolution, you must be able to locate the basic directions of up, down, forward, and backward in any given view. In Fig. 30.18A, the directions of backward

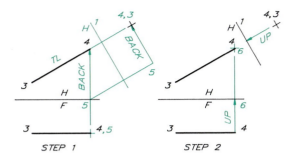

Figure 30.18 To find the direction of back, forward, up, and down in an auxiliary view, construct an arrow pointing in the desired direction in the given principal views, and project this arrow to the auxiliary view. The directions of back (Step 1) and up (Step 2) are shown here.

and up are located by first drawing directional arrows in the given top and front views.

Line 4–5 is drawn pointing backward in the top view, and its front view appears as a point. Arrow 4–5 is projected to the auxiliary view as any other line to locate the direction of backward. By drawing the arrow on the other end of the line, you would find the direction of forward.

The direction of up is located in Fig. 30.18B by drawing line 4–6 in the direction of up in the

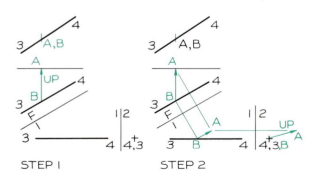

Figure 30.19 Direction in a secondary auxiliary view.

Step 1 To find the direction of up in the secondary auxiliary view, arrow *AB* is drawn pointing upward in the front view. It appears as a point in the top view.

Step 2 Arrow *AB* is projected to the primary and secondary auxiliary views like any other line. The direction of up is located in the secondary auxiliary view.

front view and as a point in the top view. The arrow is found in the primary auxiliary by the usual projection method. The direction of down would be in the opposite direction.

The location of directions in secondary auxiliary views are found in the same manner. The direction of up is found in Fig. 30.19 by beginning with an arrow pointing upward in the front view and appearing as a point in the top view (Step 1). The arrow, *AB,* is projected from the front view to the primary, and then to a secondary auxiliary view to give the direction of up in all views. The other directions can be found in the same manner by beginning with the two given principal views of a known directional arrow.

30.10 Revolution of a Point About an Axis

In Fig. 30.20, it is required that point 0 be revolved about axis 3–4 to its most forward position. The circular path of revolution is drawn in the primary auxiliary view where the axis is a point (Step 1). The direction of forward is drawn (Step 2), and the new location of point 0 is found at 0'. By projecting back through the successive views, point 0' is found in each view. Note that 0' lies on the line in the front view, verifying that 0' is in its most forward position.

The problem in Fig. 30.21 requires an additional auxiliary view since axis 3–4 is not true length in the given views. Therefore the line must be found true length before it can be found as a point where the path of revolution can be drawn as a circle. Point 0 is revolved into its highest position, 0', where the "up" arrow, 3–5, is found in the secondary auxiliary view.

By projecting back to the given views, 0 is located in each view. Its position in the top view is over the axis, which verifies that the point is located at its highest position.

The paths of revolution will appear as edges when their axes are true length, and as ellipses when their axes are not true length. The angle of

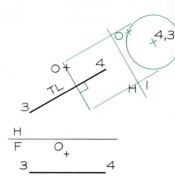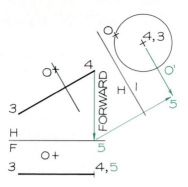

STEP 1   **STEP 2**   **STEP 3**

Figure 30.20 Revolution about an axis.

Step 1 To rotate point 0 about axis 3–4, it is necessary to find the point view of the axis in a primary auxiliary view. The circular path is drawn and the path of revolution is shown in the top view as an edge perpendicular to the axis.

Step 2 If it is required to rotate point 0 to its most forward position, draw an arrow pointing forward in the top view. It will appear as a point in the front view. The arrow, 4–5, is found in the auxiliary view to locate point 0'.

Step 3 Point 0' is projected back to the given views. The path of revolution appears as an ellipse in the front view since the axis is not true length in this view. A 30° ellipse is drawn since this is the angle your line of sight makes with the circular path in the front view.

the ellipse template for drawing the ellipse in the front view is the angle the projectors from the front view make with the edge view of the revolution in the primary auxiliary view. To find the ellipse in the top view, an auxiliary view must be used to find the path of revolution as an

edge perpendicular to the true-length axis projected from the top view.

The handcrank of a casement window (Fig. 30.22) is an example of a problem solved using revolution to determine the clearances between the sill and the window frame.

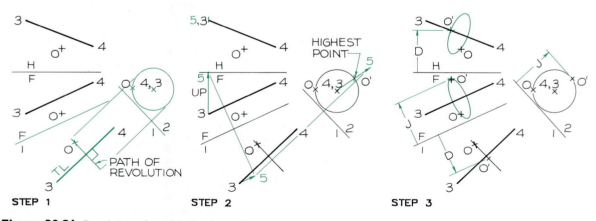

STEP 1   **STEP 2**   **STEP 3**

Figure 30.21 Revolution of a point about an axis.

Step 1 To rotate 0 about axis 3–4, the axis is found as a point in a secondary auxiliary view where the circular path is drawn. The path appears as an edge in the primary auxiliary view where the axis is true length.

Step 2 If it is required to rotate 0 to its highest position, construct arrow 3–5 in the front and top views that points upward. The direction of 3–5 in the secondary auxiliary view locates the highest position, 0'.

Step 3 Point 0' is projected back to the given views by transferring the dimensions J and D using your dividers. The highest point lies over the line in the top view to verify its position. The path of revolution is elliptical wherever the axis is not TL.

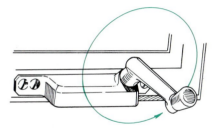

Figure 30.22 The handcrank on a casement window is an example of a problem solved by using revolution principles. The handle must be properly positioned so not to interfere with the windowsill or wall.

30.11 Revolution of a Right Prism About Its Axis

A coal chute between two buildings (Fig. 30.23) is used to convey coal at a continuous rate. The sides of the enclosed chute must be vertical and the bottom of the chute's right section must be horizontal.

In Fig. 30.24, it is required that the right section be positioned about centerline *AB* so that two of its sides will be vertical. This is done by finding the point view of the axis (Step 1), and the direction of up is projected to this view. The

right section is drawn about the axis (Step 2) so that two of its sides are parallel to the upward arrow. The right section is found in the other views. The sides of the chute are then constructed parallel to the axis, (Step 3). The bottom of the chute's right section will be horizontal and properly positioned for conveying coal.

30.12 A Line at a Specified Angle With Two Principal Planes

In Fig. 30.25, it is required that a line be drawn through point 0 that will make angles of 35° with the frontal plane and 44° with the horizontal plane, and that will slope forward and downward.

The cone containing elements making 35° with the frontal plane is drawn (Step 1). The cone with elements making 44° with the horizontal plane is drawn (Step 2). The length of the elements of both cones must be equal, so they will intersect with equal elements. Lines 0–1 and 0–2, which are elements that lie on each cone and make the specified angles with the principal planes, are then found (Step 3).

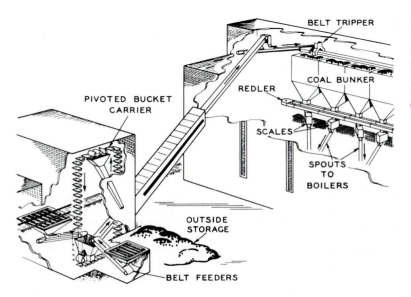

Figure 30.23 A conveyor chute must be installed so that two edges of its right section are vertical for the conveyors to function properly. This requires the application of the revolution of a prism about its axis. (Courtesy of Stephens-Adamson Manufacturing Co.)

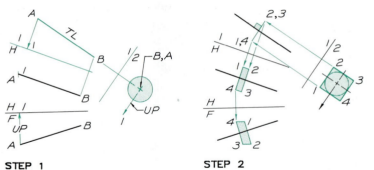

| STEP 1 | STEP 2 | STEP 3 |

Figure 30.24 Revolution of a prism about its axis.

Step I Locate the point view of centerline *AB* in the secondary auxiliary view by drawing a circle about the axis with a diameter equal to one side of the square right section. Draw a vertical arrow in the front and top views, and project it to the secondary auxiliary view to indicate the direction of vertical.

Step 2 Draw the right section, 1–2–3–4, in the secondary auxiliary view with two sides parallel to the vertical directional arrow. Project this section back to the successive views by transferring measurements with dividers. The edge view of the section could have been located in any where along centerline *AB* in the primary auxiliary view.

Step 3 Draw the lateral edges of the prism through the corners of the right section parallel to the centerline in all views. Terminate the ends of the prism in the primary auxiliary view where they appear as edges perpendicular to the centerline. Project the corner points of the ends to the top and front views to establish the ends in these views.

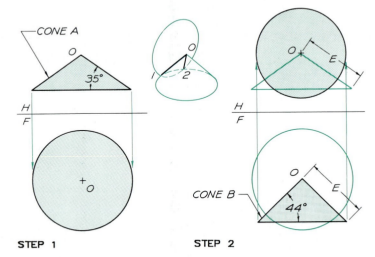
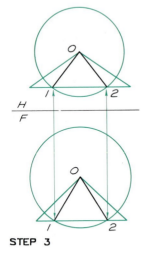

| STEP 1 | STEP 2 | STEP 3 |

Figure 30.25 A line at specified angles.

Step I Draw a triangular view of a cone in the top view such that the extreme elements make an angle of 35° with the frontal plane. Construct the circular view of the cone in the front view, using point 0 as the apex. All elements of this cone make an angle of 35° with the frontal plane.

Step 2 Draw a triangular view of a cone in the front view such that the elements make an angle of 44° with the horizontal plane. Draw the elements of this cone equal in length to element *E* of cone *A*. All elements of cone *B* make an angle of 44° with the horizontal plane.

Step 3 Since elements *A* and *B* are equal in length, there will be two common elements that lie on the surface of each cone, 0–1 and 0–2. Locate points I and 2 at the point where the bases of the cones intersect in both views. Either of these lines will satisfy the problem requirements.

Problems

Use Size A sheets for the following problems, and lay out the problems using instruments. Each square on the grid is equal to 0.20 in. (about 5mm). The problems can be laid out on grid or plain paper. Label all reference planes and points in each problem with $\frac{1}{8}$ inch letters or numbers, using guidelines.

The crosses marked "1" and "2" are to be used for placing primary and secondary reference lines. The primary reference line should pass through "1" and the secondary through "2".

1. (Fig. 30.26) A. and B. Find the true-length views of the lines in their front views by revolution. C. and D. Find the true-length views of the lines in their top views by revolution

2. (Fig. 30.27) A. and B. By revolution, find the true-size views of 1–2–3 in the front view and 4–5–6 in the top view. C. By using an auxiliary view projected from the top view and one revolution, find the true-size view of the plane.

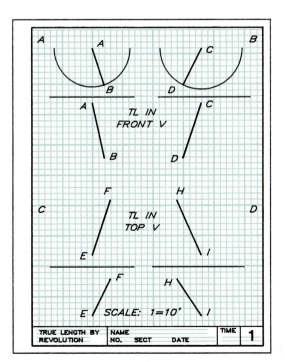

Figure 30.26 Problems IA–D.

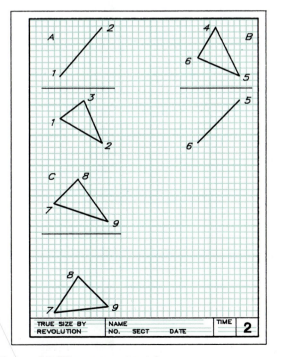

Figure 30.27 Problems 2A–2C.

3. (Fig. 30.28) A. and B. Find the angles between the planes by revolution. Show construction.

4. (Fig. 30.29) A. Perform the construction necessary to show point 0 revolved into its highest position. B. Same as A, but show point 0 in its most forward position.

5. (Fig. 30.30) Construct a chute from *A* to *B* that has the cross section shown. The longer sides are to be vertical sides.

6. (Fig. 30.31) Draw the views of the line that is 3.2″ long that makes 30° with the frontal plane and 52° with the horizontal plane.

7. (Fig. 30.32) A. and B. By revolution find edge views of 1–2–3 in the top view and 4–5–6 in the front view. C. and D. By revolution, find the angle that *AB* makes with the horizontal plane and *CD* makes with the frontal plane.

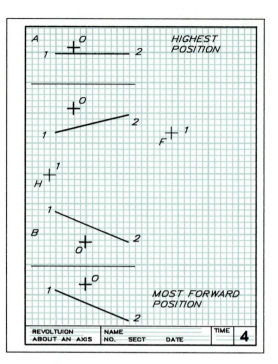

Figure 30.29 Problems 4A–4B.

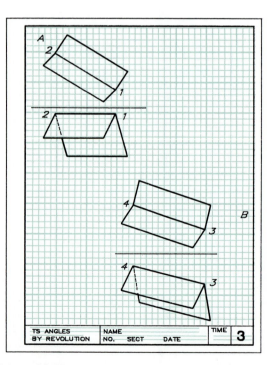

Figure 30.28 Problems 3A–3B.

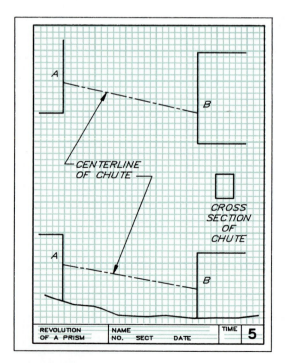

Figure 30.30 Problem 5.

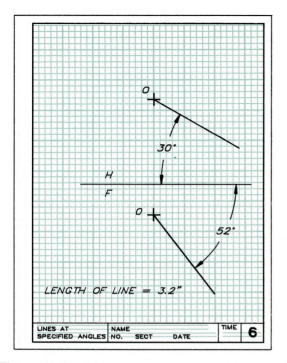

Figure 30.31 Problem 6.

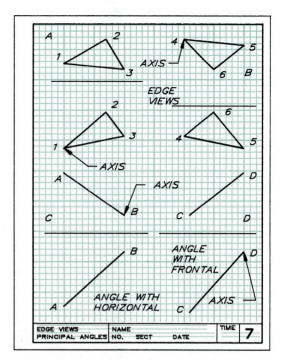

Figure 30.32 Problems 7A–7D.

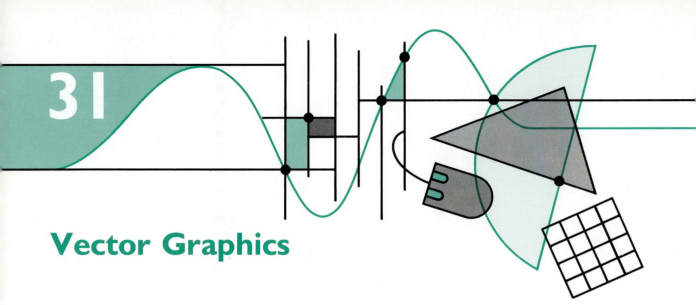

31

Vector Graphics

31.1 Introduction

A system cannot be analyzed for strength without considering the forces of tension and compression within the system. These forces are represented by **vectors**. Other quantities, such as distance, velocity, and electrical properties, may also be represented by vectors.

Graphical methods are useful in the solution of vector problems as alternative methods to conventional trigonometric and algebraic methods.

31.2 Basic Definitions

To understand the techniques of problem solving with vectors, it is necessary to know the terminology of graphical vectors.

Force is a push or pull that tends to produce motion. All forces have (1) magnitude, (2) direction, and (3) a point of application. A force is represented by the rope being pulled in Fig. 31.1A.

Vector is a graphical representation of a quantity of force; it is drawn to scale to indicate magnitude, direction, and point of application. The vector shown in Fig. 31.1B represents the force of the rope pulling the weight, W.

Magnitude is the amount of push or pull; it is represented by the length of the vector line. Magnitude is usually measured in pounds or kilograms of force.

Direction is the inclination of a force (with respect to a reference coordinate system); it is indicated by a line with an arrow at one end.

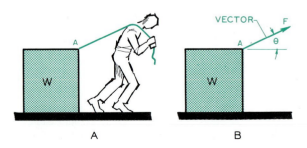

Figure 31.1 Representation of a force (A) by a vector (B).

Point of application is the point through which the force is applied on the object or member (point A in Fig. 31.1A).

Compression is the state created in a member by subjecting it to opposite pushing forces. A member tends to be shortened by the compression (Fig. 31.2A). Compression is represented by a plus sign (+) or the letter C.

Tension is the state created in a member by subjecting it to opposite pulling forces. A member tends to be stretched by tension, as shown in Fig. 31.2B. Tension is represented by a minus sign (−) or the letter T.

Force system is the combination of all forces acting on a given object. Figure 31.3 shows a force system.

Resultant is a single force that can replace all the forces of a force system and have the same effect as the combined forces.

Equilibrant is the opposite of a resultant; it is the single force that can be used to counterbalance all forces of a force system.

Components are any individual forces that, if combined, would result in a given single force. For example, forces A and B are components of resultant R_1 in Step 1 of Fig. 31.3.

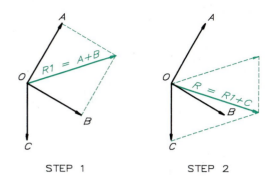

Figure 31.3 Resultant by the parallelogram method.
Step 1 Draw a parallelogram with its sides parallel to vectors A and B. The diagonal R_1, drawn from point 0, is the resultant of forces A and B.

Step 2 Draw a parallelogram using vectors R_1 and C to find diagonal R from P to Q. This is the resultant that can replace forces A, B, and C.

Space diagram is a diagram depicting the physical relationship between structural members. The force system in Fig. 31.3 is given as a space diagram.

Vector diagram is a diagram composed of vectors scaled to their appropriate lengths to represent the forces within a given system. The vector diagram is used to solve for unknowns.

Statics is the study of forces and force systems in equilibrium.

Metric units are standard units for indicating weights and measures. The kilogram (kg) is the standard unit for indicating mass (loads). A comparison of kilograms with pounds is shown in Fig. 31.4. The metric ton is 1000 kilograms. One kilogram = 2.2 pounds.

31.3 Coplanar, Concurrent Force Systems

When several forces, represented by vectors, act through a common point of application, the system is said to be **concurrent**. Vectors A, B, and

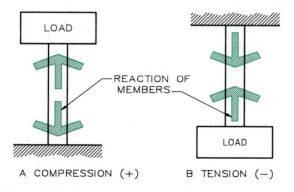

Figure 31.2 Comparison of (A) compression and (B) tension in a member.

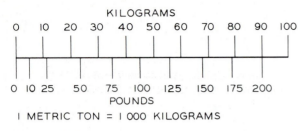

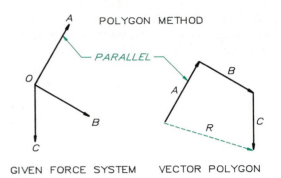

KILOGRAMS

0   10   20   30   40   50   60   70   80   90   100

0 10 25   50   75 100 125 150 175 200
POUNDS

I METRIC TON = 1 000 KILOGRAMS

Figure 31.4 The kilogram is the standard metric unit for measuring forces that are represented by pounds in the English system: 1 kilogram = 2.2 pounds.

POLYGON METHOD

PARALLEL

GIVEN FORCE SYSTEM   VECTOR POLYGON

Figure 31.5 Resultant of a coplanar, concurrent system as determined by the polygon method, in which the vectors are drawn head-to-tail.

C act through a single point in Fig. 31.3; therefore this is a concurrent system. When only one view is necessary to show the true length of all vectors, as in Fig. 31.3, the system is **coplanar**.

The **resultant** represents the composite effect of all forces on the point of application. The resultant is found graphically by (1) the parallelogram method and (2) the polygon method.

31.4 Resultant of a Coplanar, Concurrent System—Parallelogram Method

In Fig. 31.3, all vectors lie in the same plane and act through a common point. The vectors are scaled to known magnitudes.

To apply the parallelogram method to determine the resultant, the vectors for a force system must be known and drawn to scale. Two vectors are used to find a parallelogram; the diagonal of the parallelogram is the resultant of these two vectors and has its point of origin at point O (Fig. 31.3). Resultant R_1 can be called the **vector sum** of vectors A and B.

Since vectors A and B have been replaced by R_1, they can be disregarded in the next step of the solution. Again, resultant R_1 and vector C are resolved by completing a parallelogram (i.e., by drawing a line parallel to each vector). The diagonal of this parallelogram is the resultant of the entire system and is the vector sum of R_1 and C. Resultant R can be analyzed as though it were the only force acting on the point, thereby simplifying the process.

31.5 Resultant of a Coplanar, Concurrent System—Polygon Method

The system of forces shown in Fig 31.3 is shown again in Fig. 31.5, but in this case the resultant is found by the polygon method. The forces are drawn to scale and in their true directions, with each force being drawn head-to-tail to form the polygon. The vectors are drawn in a clockwise sequence, beginning with vector A. The polygon does not close, which means the system is not in **equilibrium**; in other words, it would tend to be in motion, since the forces are not balanced. The resultant R is drawn from the tail of vector A to the head of vector C to close the polygon.

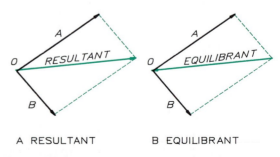

A RESULTANT   B EQUILIBRANT

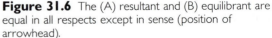

Figure 31.6 The (A) resultant and (B) equilibrant are equal in all respects except in sense (position of arrowhead).

An **equilibrant** has the same magnitude, orientation, and point of application as the **resultant** in a system of forces, but an opposite direction.

The resultant of the system of forces shown in Fig. 31.6 is found by the parallelogram method. The equilibrant can be applied at point 0 to balance the forces A and B and thereby cause the system to be in equilibrium.

31.6 Resultant of Noncoplanar, Concurrent Forces—Parallelogram Method

When vectors lie in more than one plane of projection, they are said to be **noncoplanar**; therefore more than one view is necessary to analyze their spatial relationships. The resultant of a system of noncoplanar forces can be found by the parallelogram method if their true projections are given in two adjacent orthographic views.

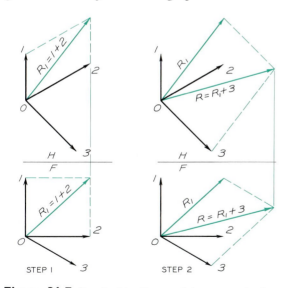

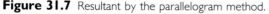

Figure 31.7 Resultant by the parallelogram method.

Step 1 Vectors 1 and 2 are used to construct a parallelogram in the top and front views. The diagonal, R_1, is the resultant of these two vectors.

Step 2 Vectors 3 and R_1 are used to construct a second parallelogram to find the views of the overall resultant, R.

Vectors 1 and 2 in Fig. 31.7 are used to construct the top and front views of a parallelogram. The diagonal of the parallelogram, R_1, is found in both views. As a check, the front view of R_1 must be an orthographic projection of its top view.

In Step 2, resultant R_1 and vector 3 are resolved to form resultant R_2 in both views. The top and front views of R_2 must project orthographically. Resultant R_2 can be used to replace vectors 1, 2, and 3. Since R_2 is an oblique line, its true length must be found by auxiliary view, as shown in Fig. 31.8, or by revolution.

31.7 Resultant of Noncoplanar, Concurrent Forces—Polygon Method

The same system of forces given in Fig. 31.7 is solved in Fig. 31.8 for the resultant of the system by the **polygon method.**

Each vector is laid head-to-tail in a clockwise direction, beginning with vector 1 (Step 1). In each view, the vectors are drawn to be orthographic projections (Step 2). Since the vector polygon did not close, the system is not in equilibrium. The resultant, R, is constructed from the tail of vector 1 to the head of vector 3 in both views. Resultant R is an oblique line and requires an auxiliary view to find its true length.

31.8 Forces in Equilibrium

An example of a coplanar, concurrent structure in equilibrium can be seen in the loading cranes in Fig. 31.9.

The coplanar, concurrent structure given in Fig. 31.10 is designed to support a load of $W = 1000$ kg. The maximum loading in each structural member determines the type and size of the members used in the design.

In Step 1, the only known force, $W = 1000$ kg, is laid off parallel to the given direction. Unknown forces A and B are drawn head-to-tail as vectors to close the force polygon.

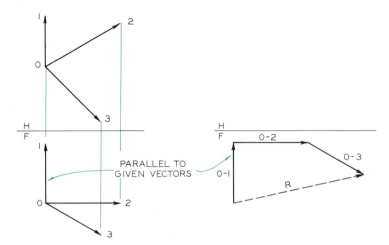

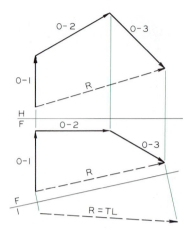

Figure 31.8 Resultant by the polygon method.

Required Find the resultant of this system of concurrent, noncoplanar forces by the polygon method.

Step 1 Each vector is laid off head-to-tail in the front view. The front view of the resultant is the vector found.

Step 2 The same vectors are drawn head-to-tail in the top view to complete the 3-D polygon. The resultant is found true length by an auxiliary view.

In Step 2, vectors *A* and *B* are analyzed to determine whether they are in tension or compression. Vector *B* points upward to the left, which is toward point *O* when transferred to the structural diagram. A vector that acts toward a point of application is in **compression**. Vector *A* points away from point *A* when transferred to

the structural diagram and is therefore in **tension**.

In Fig. 31.11, a similar example of a force system involving a pulley is solved to determine the loads in the structural members caused by the weight of 100 lb. The only difference between this solution and the previous one is the construction of two equal vectors to represent the loads in the cable on both sides of the pulley.

31.9   Truss Analysis

Vector polygons can be used to analyze structural trusses to determine the loads in each member by two graphical methods: (1) joint-by-joint analysis and (2) Maxwell diagrams.

Joint-by-Joint Analysis

The Fink truss in Fig. 31.12 is loaded with 3000 lb. forces concentrated at joints of the structural members. A method of designating forces, called **Bow's notation**, is used. The exterior forces applied to the truss are labeled with letters placed

Figure 31.9 The cargo cranes on this ship are examples of coplanar, concurrent force systems designed to remain in equilibrium. (Courtesy of Exxon Corp.)

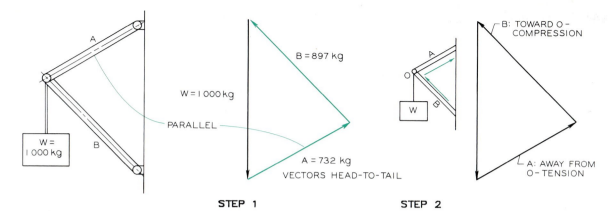

STEP 1   **STEP 2**

Figure 31.10 Coplanar forces in equilibrium.

Required Find the forces in the two structural members caused by the load of 1000 kg.

Step 1 Draw the known load of 1000 kg as a vector. Draw the vectors A and B parallel to their directions. Arrowheads are drawn head-to-tail.

Step 2 Vector A points away from point 0 when transferred to the structural diagram and is in tension. Vector B points toward point 0 and is in compression.

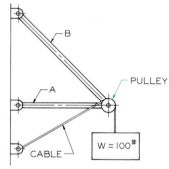 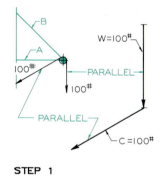 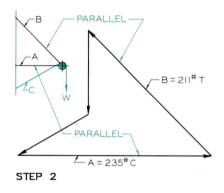

STEP 1   **STEP 2**

Figure 31.11 Determination of forces in equilibrium.

Required Find the forces in the members caused by the load of 100 lb (denoted by #) supported by the pulley.

Step 1 The force in the cable is equal to 100 lb on both sides of the pulley. These two forces are drawn as vectors head-to-tail parallel to their directions in the space diagram.

Step 2 A and B are drawn head-to-tail to close the polygon. The direction of A is toward the point of application and is in compression; B is away from the point and is in tension.

between the forces, and numerals are placed between the interior members.

Each vector is referred to by the number on each of its sides by reading in a clockwise direction. For example, the first vertical load at the left is called AB, with A at the tail and B at the head of the vector.

We first analyze the joint at the left where the reaction of 4500 lb is known. This force, reading in a clockwise direction about the joint, is called EA with an upward direction. The tail is labeled E and the head A. Continuing in a clockwise direction, the next force is A–1, and the next 1–E, which closes the polygon and ends with the beginning letter, E. The arrows are placed, beginning with the known vector EA, in a head-to-tail arrangement.

541

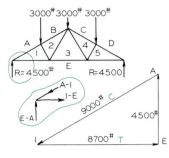

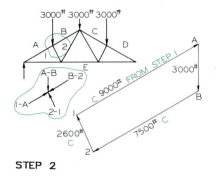

STEP 1

STEP 2

STEP 3

Figure 31.12 Joint analysis of a truss.

Step 1 The truss is labeled using Bow's notation, with letters between the exterior loads and numbers between interior members. The lower left joint can be analyzed since it has only two unknowns, *A*–1 and 1–*E*. These vectors are found by drawing them parallel to their directions from both ends of the 4500 lb reaction in a head-to-tail order.

Step 2 Using vector 1–*A* found in Step 1 and load *AB*, the two unknowns *B*–2 and 2–1 can be found. The known vectors are laid out beginning with vector 1–*A* and moving clockwise about the joint. Vectors *B*–2 and 2–1 close the polygon. If a vector points toward the point of application, it is in compression; if away from the point, it is in tension.

Step 3 The third joint can be analyzed by laying out the vectors *E*–1 and 1–2 from the previous steps. Vectors 2–3 and 3–*E* close the polygon and are parallel to their directions in the space diagram. The directions of 2–3 and 3–*E* are away from the point of application and are in tension.

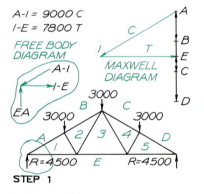

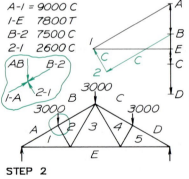

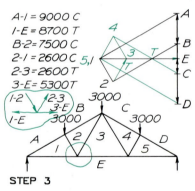

STEP 1

STEP 2

STEP 3

Figure 31.13 Truss analysis.

Step 1 Label the spaces between the outer forces of the truss with letters and the internal spaces with numbers, using Bow's notation. Add the given load vectors in a Maxwell diagram, and sketch a free-body diagram of the first joint. Using vectors *EA*, *A*–1, and 1–*E* drawn head-to-tail, draw a vector diagram to find their magnitudes. Vector *A*–1 is in compression (+) because it points toward the joint, and 1–*E* is in tension (−) because it points away from the joint.

Step 2 Draw a sketch of the next joint to be analyzed. Since *AB* and *A*–1 are known, we have to determine only two unknowns, 2–1 and *B*–2. Draw these parallel to their direction, head-to-tail, in the Maxwell diagram using the existing vectors found in Step 1. Vectors *B*–2 and 2–1 are in compression since each points toward the joint. Note that vector *A*–1 becomes 1–*A* when read in a clockwise direction.

Step 3 Sketch a free-body diagram of the next joint to be analyzed. The unknowns in this case are 2–3 and 3–*E*. Determine the true length of these members in the Maxwell diagram by drawing vectors parallel to given members to find point 3. Vectors 2–3 and 3–*E* are in tension because they act away from the joint. This process is repeated to find the loads of the members on the opposite side.

Tension and compression can be determined by relating the direction of each vector to the original joint. For example, A–1 points toward the joint and is in compression, whereas 1–E points away and is in tension.

Since the truss is symmetrical and equally loaded, the loads in the members on the right will be equal to those on the left.

The other joints are analyzed in the same manner in Steps 2 and 3. The direction of the vectors is opposite at each end. Vector A–1 is toward the left in Step 1, and toward the right in Step 2.

Maxwell Diagrams

The Maxwell diagram is identical to the joint-by-joint analysis except that the polygons overlap, with some vectors common to more than one polygon. Again, Bow's notation is used to good advantage.

The first step (Fig. 31.13) is to lay out the exterior loads beginning clockwise about the truss—AB, BC, CD, DE, and EA—head-to-tail. A letter is placed at each end of the vectors. Since they are parallel, this polygon will be a straight line.

The structural analysis begins at the joint through which reaction EA acts. For easier analysis, a free-body diagram is drawn to isolate this joint. The two unknowns, members A–1 and 1–E, are drawn parallel to their direction in the truss in Step 1, with A–1 beginning at point A and 1–E beginning at point E. To locate point 1, these directions are extended. Because resultant EA points upward, vector A–1 must have its tail at A, giving it a direction toward point 1. By referring to the free-body diagram, we can see that the direction is toward the point of application, which means A–1 is in compression. Vector 1–E points away from the joint, which means it is in tension. The vectors are coplanar and can be scaled to determine their loads.

In Step 2, vectors 1–A and AB are known, whereas vectors B–2 and 2–1 are unknown, making it possible to solve for them. Vector B–2 is drawn parallel to the structural member through point B in the Maxwell diagram, and the line of vector 2–1 is extended from point 1 until it inter-

sects with B–2 at point 2. The arrows of each vector are found by laying off each vector head-to-tail. Both vectors B–2 and 2–1 point toward the joint in the free-body diagram; therefore they are in compression.

In Step 3, the next joint is analyzed in sequence to find the stresses in 2–3 and 3–E. The truss will have equal forces on each side, since it is symmetrical and is loaded symmetrically. The total Maxwell diagram is drawn in Step 3.

If all the polygons in the series do not close at every point with perfect symmetry, there is an error in construction. A slight error of closure can be disregarded, since safety factors are generally applied in derivation of working stresses of structural systems to assure safe construction. Arrowheads are omitted on Maxwell diagrams, since each vector will have opposite directions when applied to different joints.

31.10 Noncoplanar Structural Analysis—Special Case

Three-dimensional structure systems require the use of descriptive geometry, since it is necessary to analyze the system in more than one plane. The manned flying system (MFS) in Fig. 31.14

Figure 31.14 The structural members of this tripod support for a moon vehicle can be analyzed graphically to determine design loads. (Courtesy of NASA.)

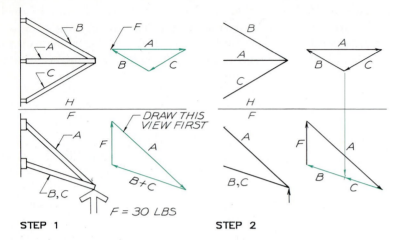

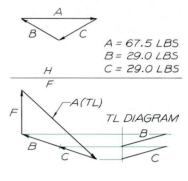

STEP 1   STEP 2   STEP 3

Figure 31.15 Noncoplanar structural analysis—special case.

Step 1 Forces, *B* and *C*, coincide in the front view, resulting in only two unknowns in this view. Vector *F* (30 lb) and the two unknowns are drawn parallel to their front view to complete the front view of the vector polygon. The top view of *A* can be found by projection, from which vectors *B* and *C* can be found.

Step 2 The point of intersection of vectors *B* and *C* in the top view is projected to the front view to separate these vectors. All vectors are drawn head-to-tail. Vectors *B* and *C* are in tension because they act away from the point in the space diagram, whereas *A* is in compression.

Step 3 The completed top and front views found in Step 2 do not give the true lengths of vectors *B* and *C* since they are oblique. The true lengths of these lines are determined by a true-length diagram where they are scaled to find the forces in each member.

can be analyzed to determine the forces in the support members (Fig. 31.15). Weight on the moon can be found by multiplying earth weight by a factor of 0.165. A tripod that must support 182 lb on earth has to support only 30 lb on the moon.

This example in Fig. 31.15 is a special case, since members *B* and *C* lie in the same plane that appears as an edge in the front view. A vector polygon is constructed in the front view in Step 1 by drawing force *F* as a vector and using the other vectors as the other sides of the polygon. One of these vectors is actually a summation of vectors *B* and *C*. The top view is drawn using vectors *B* and *C* to close the polygon from each end of vector *A*. In Step 2, the front view of vectors *B* and *C* is found.

The true lengths of the vectors are found in a true-length diagram in Step 3. The vectors are measured to determine their loads. Vector *A* is found to be in compression because it points toward the point of concurrency. Vectors *B* and *C*

are in tension since they point away from point of concurrency.

31.11 Noncoplanar Structural Analysis—General Case

The structural frame shown in Fig. 31.16 is attached to a vertical wall to support a load of *W* = 600 lb. Since there are three unknowns in each of the views, an auxiliary view must be drawn that will give the edge view of a plane containing two of the vectors, thereby reducing the number of unknowns to two. We no longer need to refer to the front view.

A vector polygon is drawn by constructing vectors parallel to the members in the auxiliary view (Step 1). An adjacent orthographic view of the vector polygon is also drawn by constructing its vectors parallel to the members in the top view (Step 2). A true-length diagram is used to find the true length of the vectors to determine their magnitudes (Step 3).

Figure 31.16 Noncoplanar structural analysis—general case.

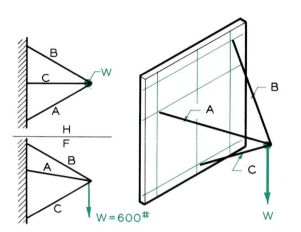

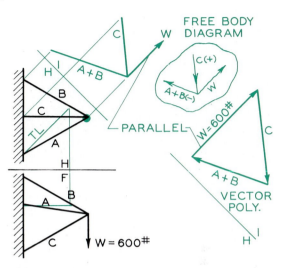

Given The top and front views of a three-member frame which is attached to a vertical wall and supports a weight of 600 lb.

Required Find the loads in the structural members.

Step 1 To limit the unknowns to two, construct an auxiliary view to find two vectors lying in the edge of a plane. Use the auxiliary view and top view in the remainder of the problem. Draw a vector polygon parallel to the members in the auxiliary view in which $W = 600$ lb is the only known vector. Sketch a free-body diagram for preliminary analysis.

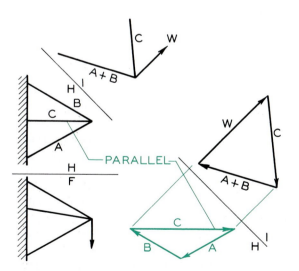

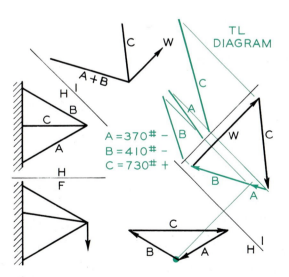

Step 2 Construct an orthographic projection of the view of the vector polygon found in Step 1 so that its vectors are parallel to the members in the top view. The reference plane between the two views is parallel to the H–1 plane. This portion of the problem is closely related to the problem in Fig. 31.15.

Step 3 Project the intersection of vectors A and B in the horizontal view of the vector polygon to the auxiliary view polygon to establish the lengths of vectors A and B. Determine the true lengths of all vectors in a true-length diagram to determine their magnitudes. Analyze for tension or compression.

Figure 31.17 Tractor sidebooms represent noncoplanar, concurrent systems of forces that can be solved graphically. (Courtesy of Trunkline Gas Co.)

A three-dimensional vector system is the side-boom tractors used for lowering pipe into a ditch during pipeline construction (Fig. 31.17).

31.12 Resultant of Parallel, Nonconcurrent Forces

The beam in Fig. 31.18 is on a rotational crane used to move building materials. The magnitude of the weight W is unknown, but the counterbalance weight is 2000 lb; column R supports the beam as shown. Assuming the support cables have been omitted, we desire to find the weight W that would balance the beam.

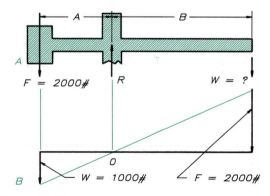

Figure 31.18 Determining the resultant of parallel, nonconcurrent forces.

The graphical solution (Fig. 31.18B) is found by constructing a line to represent the total distance between the forces F and W. Point 0, the point of balance where the summation of the moments will be equal to zero, is projected from the space diagram to this line. Vectors F and W are drawn to scale at each end of the line by transposing them to the opposite ends of the beam. A line is drawn from the end of vector F through point 0 and extended to intersect the direction of vector W. This point represents the end of vector W, which can be scaled to have a magnitude of 1000 lb.

31.13 Resultant of Parallel, Nonconcurrent Forces on a Beam

The beam given in Fig. 31.19 is supported at each end and must carry three given loads. We are required to determine the magnitude of each support, R_1 and R_2, the resultant of the loads, and its location. In Step 1, the spaces between all vectors are labeled in a clockwise direction with Bow's notation, and a force diagram is drawn.

In Step 2, the lines of force in the space diagram are extended, and the strings from the vector diagram are drawn in their respective spaces, parallel to their original direction. For example, string $0a$ is drawn parallel to string $0A$ in space A between forces EA and AB, and string $0b$ is drawn in space B beginning at the intersection of $0a$ with vector AB. The last string, $0e$, is drawn to close the **funicular** diagram. The direction of string $0e$ is transferred to the force diagram, where it is laid off through point 0 to intersect the load line at point E. Vector DE represents support R_2 (refer to Bow's notation as it was applied in Step 1), and vector EA represents support R_1.

In Step 3, the magnitude of the resultant of the loads is the summation of the vertical downward forces, or the distance from A to D. The location of the resultant is found by extending the extreme outside strings in the funicular diagram, $0a$ and $0d$, to their point of intersection. The resultant is found to have a magnitude of 500 lb, a vertical downward direction, and a point of application established by $\overline{X}$.

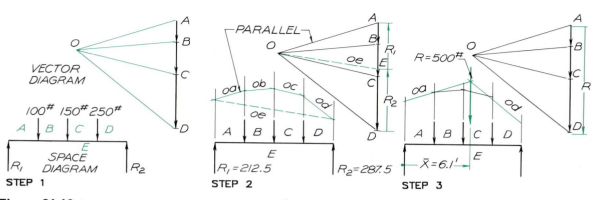

Figure 31.19 Beam analysis with parallel loads.

Step 1 Letter the spaces between the loads with Bow's notation. Find the graphical summation of the vectors by drawing them head-to-tail in a vector diagram at a convenient scale. Locate pole point 0 at a convenient location and draw strings from point 0 to each end of the vectors.

Step 2 Extend the lines of force and draw a funicular diagram with string 0a in the A-space, 0b in the B-space, 0c in the C-space, etc. The last string, which is drawn to close the diagram, is 0e. Transfer 0e to the vector polygon to locate point E, thus establishing the lengths of R_1 and R_2, which are EA and DE, respectively.

Step 3 The resultant of the three downward forces will be equal to their graphical summation, line AD. Locate the resultant by extending strings 0a and 0d in the funicular diagram to a point of intersection. The resultant R = 500 lb will act through this point in a downward direction. $\overline{X}$ is a locating dimension.

 **Problems**

Problems should be presented in instrument drawings on Size A paper, grid or plain. Each grid square represents 0.20 in. All notes, sketches, drawings, and graphical work should be neatly prepared in keeping with good design practices. Written matter should be legibly lettered using $\frac{1}{8}$-inch guidelines.

1. (Fig. 31.20) A. and B. Find the resultants of the force systems by the parallelogram and polygon methods. Scale: 1″ = 100 lb.

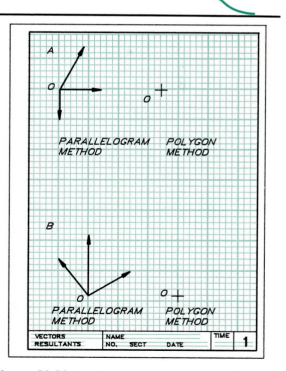

Figure 31.20 Problems 1A and 1B. Resultant of concurrent, coplanar vectors.

2. (Fig. 31.21) A. and B. Find the resultants of the force systems by the parallelogram and polygon methods. Scale: $1'' = 100$ lb.

3. (Fig. 31.22) A. and B. Find the forces in the coplanar force systems. Label the members and assign the forces in each of them and indicate compression or tension.

4. (Fig. 31.23) Find the forces in each member of the truss by using a Maxwell diagram. Make a table of forces and indicate compression and tension.

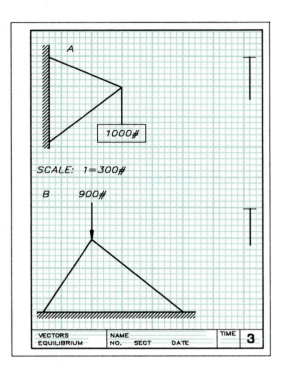

Figure 31.22 Problems 3A and 3B. Coplanar, concurrent forces in equilibrium.

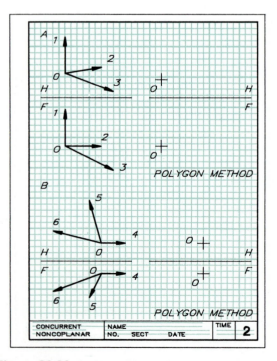

Figure 31.21 Problems 2A and 2B. Resultant of concurrent, noncoplanar vectors.

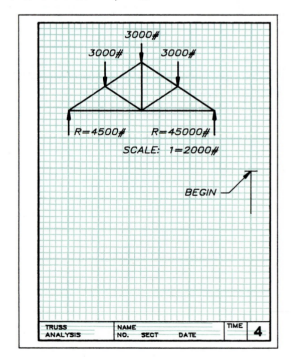

Figure 31.23 Problem 4. Truss analysis.

5. (Fig.31.24) Find the forces in the members of the concurrent noncoplanar force system. Make a table of forces and indicate compression and tension.

6. (Fig. 31.25) Find the forces in the members of the concurrent noncoplanar force system. Make a table of forces and indicate compression and tension.

7. (Fig. 31.26) A. Find the forces in reactions R1 and R2 necessary to equalize the loads applied to the beam. B. Find the value of and the location of the single support that could replace both R1 and R2.

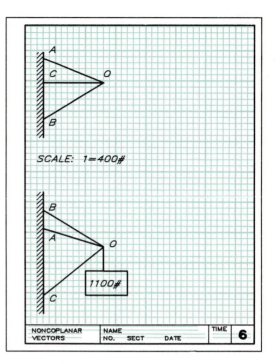

Figure 31.25 Problem 6. Noncoplanar, concurrent forces in equilibrium.

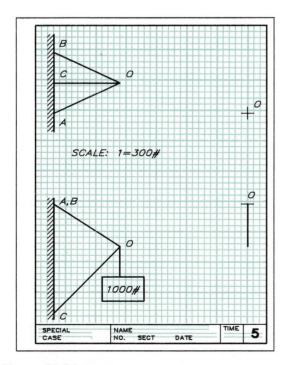

Figure 31.24 Problem 5. Noncoplanar, concurrent forces in equilibrium.

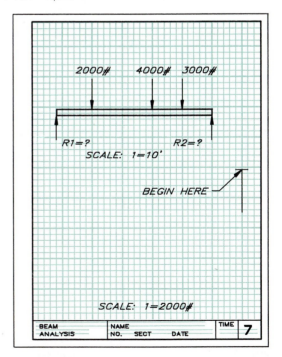

Figure 31.26 Problems 7A and 7B. Coplanar, nonconcurrent forces.

8. (Fig. 31.27) Same as Problem 7.

9. (Fig. 31.28) Find the forces in the support members. Make a table of forces and indicate compression and tension.

10. (Fig. 31.29) Find the forces in the coplanar force system. Label the members, make a table of forces, and indicate compression and tension.

11. Same as Problem 4, but use the joint-by-joint analysis instead of the Maxwell-diagram method.

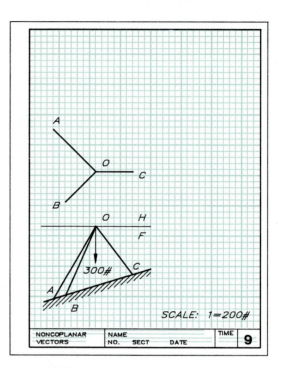

Figure 31.28 Problem 9. Beam analysis.

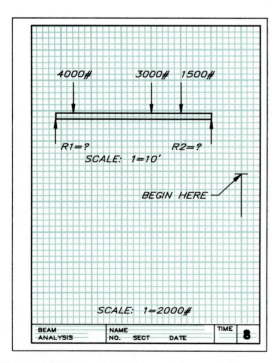

Figure 31.27 Problems 8A and 8B. Beam analysis.

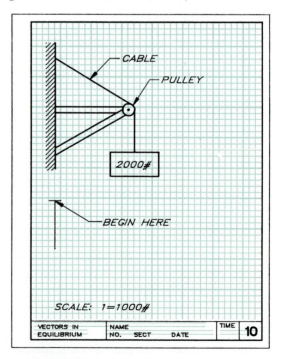

Figure 31.29 Problem 10. Noncoplanar, concurrent forces in equilibrium.

Intersections and Developments

32.1 Introduction

This chapter deals with the methods of finding lines of **intersections** between parts that join. Usually these parts are made of sheet metal or plywood, if used to form concrete.

Once the intersections have been determined, **developments** can be found. These flat patterns can be laid out on the sheet metal and cut to conform to the desired shape. You can see many examples of intersections and developments in Fig. 32.1, in this storage tank facility.

32.2 Intersections of Lines and Planes

The steps of finding the intersection between a line and a plane are illustrated in Fig. 32.2. This is a special case where the point of intersection can be easily seen since the plane appears as an edge (Step 1). It is projected to the front view, and the visibility of the line is found (Step 2).

This principle is used in Fig. 32.3 to find the line of intersection between two planes. The intersection was found by locating the piercing points of lines *AB* and *DC* and connecting these points.

The intersection of a plane at a corner of a prism results in a line of intersection that bends around the corner (Fig. 32.4). Piercing points 2' and 1' are found in Step 1.

Corner point 3 is seen in the side view of Step 2, where the vertical corner pierces the

Figure 32.1 This storage tank facility involves many applications and principles of intersections and developments. (Courtesy of Phillips Petroleum Company.)

551

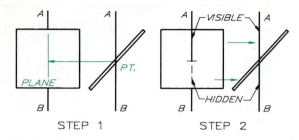

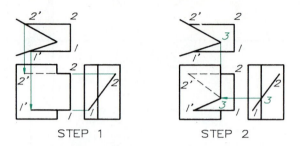

Figure 32.2 Intersection of a line and a plane.

Step 1 The point of intersection can be found in the view where the plane appears as an edge, the side view in this example.

Step 2 Visibility in the front view is determined by looking from the front view to the right-side view.

Figure 32.4 Intersection of a plane at a corner.

Step 1 The intersecting plane appears as an edge in the side view. Intersection points 1′ and 2′ are projected from the top and side views to the front view.

Step 2 The line of intersection from 1′ to 2′ must bend around the vertical corner at 3′ in the top and side views. Point 3′ is projected to the front view to locate line 1′–3′–2′.

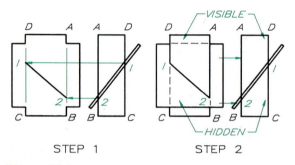

Figure 32.3 Intersection between planes.

Step 1 The points where lines AB and DC intersect the plane are found where the plane appears as an edge. These points are projected to the front view.

Step 2 Line 1–2 is the line of intersection. Visibility is determined by looking from the front view to the right-side view.

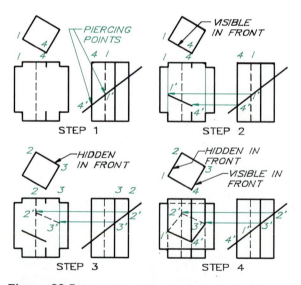

Figure 32.5 Intersections of a plane and a prism.

Step 1 Vertical corners 1 and 4 intersect the edge view of the plane in the side view at 1′ and 4′.

Step 2 Points 1′ and 4′ are projected from the side view to lines 1 and 4 in the front view. They are connected with a visible line as a line of intersection.

Step 3 Vertical corners 2 and 3 intersect the edge view of the plane at 2′ and 3′ in the side view. Points 2′ and 3′ are projected to the front view to form a hidden line of intersection.

Step 4 Points 1′ and 2′ and points 3′ and 4′ are connected and visibility is determined by analyzing the top and side views.

plane. Point 3 is projected to the corner in the front view. Point 2′ is hidden in the front view since it is on the back side.

The intersection of a plane and a prism is found in Fig. 32.5, where the plane appears as an edge. The points of intersection are found for each corner line and are connected; visibility is shown to complete the line of intersection.

An intersection between a plane and a prism is shown in Fig. 32.6. Vertical cutting planes are passed through the planes of the prism in the top view to find the piercing points of the corners in the front view. The points are connected and the visibility is determined to complete the solution.

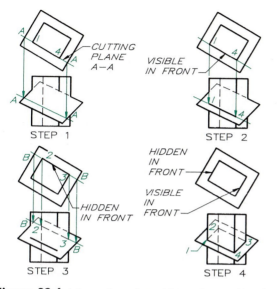

Figure 32.6 Intersection of an oblique plane and a prism.

Step I Vertical cutting plane *A–A* is passed through corners I and 4 in the top view and is projected to the front view.

Step 2 Piercing points I and 4 are found in the front view where the line *A–A* crosses lines I and 4.

Step 3 Vertical cutting plane *B–B* is passed through the corners of 2 and 3 in the top view and is projected to the front view to locate piercing points 2 and 3.

Step 4 The four piercing points are connected and visibility is determined by analysis of the top view.

The intersection between a foreshortened plane and an oblique prism is found in Fig. 32.7. The plane is found as an edge in a primary auxiliary view. The piercing points of the corners of the prism are located in the auxiliary view and are projected back to the given views.

Points 1, 2, and 3 are projected from the auxiliary view to the given views as examples. Visibility is determined by analysis of crossing lines.

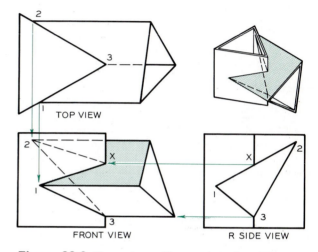

Figure 32.8 Three views of intersecting prisms. The points of intersection can be seen where intersecting planes appear as edges.

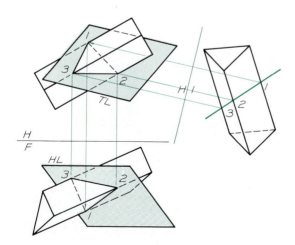

Figure 32.7 The intersection between a plane and a prism can be found by constructing a view in which the plane appears as an edge.

32.3 Intersections Between Prisms

The same principles used to find the intersection between a plane and a line are used to find the intersection between two prisms in Fig. 32.8. Piercing points 1, 2, and 3 are found in the front view by projecting from the side and top views. Point *X* is located in the side view where line of intersection 1–2 bends around the vertical corner of the other prism. Points 1, *X*, and 2 are connected, and visibility is determined.

In Fig. 32.9, an inclined prism intersects a vertical prism. The end view of the inclined prism is found by an auxiliary view (Step 1). In the auxiliary view, you can see where plane 2–3 bends around corner *AB* at point *X* (Step 2). Points of intersection 1′ and 2′ are projected from

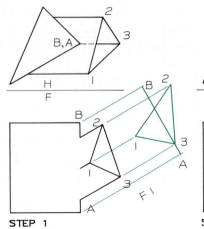

STEP 1

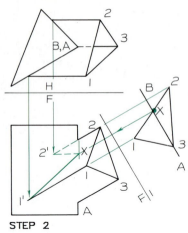

STEP 2

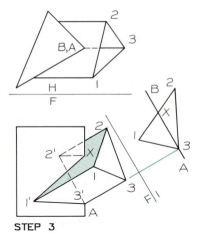

STEP 3

Figure 32.9 Intersection between two prisms.

Step 1 Construct the end view of the inclined prism by projecting an auxiliary view from the front view. Show only line *AB* of the vertical prism in the auxiliary view.

Step 2 Locate piercing points 1′ and 2′ in the top and front views. Intersection line 1′–2′ will bend around corner *AB* at point *X*, which is projected from the auxiliary view.

Step 3 Intersection lines from 2′ and 1′ to 3′ do not bend around the corner. Therefore these are drawn as straight lines. Line 1′–3′ is visible, and line 2′–3′ is invisible.

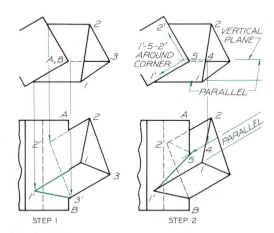

STEP 1   STEP 2

Figure 32.10 Intersection between prisms.

Step 1 The piercing points of lines 1, 2, and 3 are found in the top view and are projected to the front view where piercing points 1′, 2′, and 3′ are found.

Step 2 A cutting plane is passed through corner *AB* in the top view to locate point 5, where line of intersection 1′–2′ bends around the vertical prism. Point 5 is found in the front view, and 1′–5–2′ is drawn.

the top to the front view. The line of intersection 2′–*X*–3′ can be drawn to complete this portion of the line of intersection. The remaining lines, 1′–3′ and 2′–3′, are connected to complete the solution (Step 3).

An alternative method of solving a problem of this type is shown in Fig. 32.10. Piercing

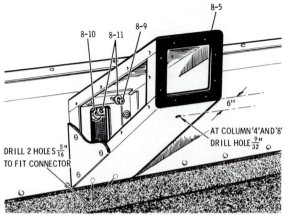

Figure 32.11 This conduit connector was designed using the principles of the intersection of a plane and prism. (Courtesy of the Federal Aviation Administration.)

points 1' and 2' are found in the front view by projecting from the top view (Step 1). Point 5, the point where line 1'–5–2' bends around vertical corner *AB*, is then found (Step 2). A cutting plane is passed through corner *AB* in the top view. The front view of point 5 is found, and the lines of intersection are completed.

The conduit connector in Fig. 32.11 is an example of intersecting planes and prisms.

Figure 32.12 This mid-fuselage section of an aircraft was designed by applying many principles of intersections and developments. (Courtesy of Lockheed-Georgia Company.)

32.4 Intersection of a Plane and Cylinder

The intersections of the components of the mid-fuselage section of an aircraft shown in Fig. 32.12 offer numerous applications of the principles of intersections.

The intersection between a plane and a cylinder is found in Fig. 32.13. Cutting planes are passed vertically through the top view of the cylinder to establish elements on the cylinder and their piercing points. The piercing points are pro-

jected to each view to find the line of intersection, which is an ellipse.

A more general problem is solved in Fig. 32.14, where the cylinder is vertical but the plane is oblique. Vertical cutting planes are passed through the cylinder and the plane in the top view to find piercing points of the cylinder's ele-

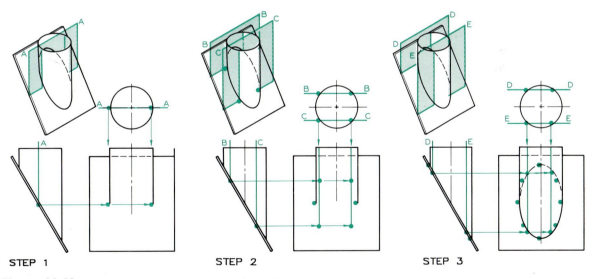

STEP 1　　　　　　　STEP 2　　　　　　　STEP 3

Figure 32.13 Intersection between a cylinder and a plane.

Step 1 A vertical cutting plane, *A–A* is passed through the cylinder parallel to its axis to find two points of intersection.

Step 2 Cutting planes, *B–B* and *C–C*, are used to find four additional points in the top and left side views; these points are projected to the front view.

Step 3 Additional cutting planes are used to find more points. These points are connected to give an elliptical line of intersection.

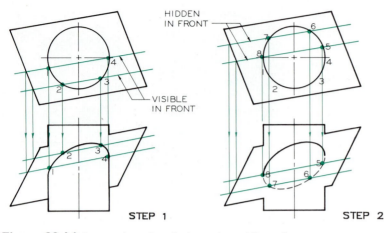

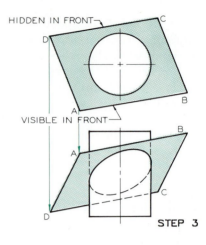

Figure 32.14 Intersection of a cylinder and an oblique plane.

Step 1 Vertical cutting planes are passed through the cylinder in the top view to establish elements on its surface and lines on the oblique plane. Piercing points 1, 2, 3, and 4 are projected to the front view of their respective lines and are connected with a visible line.

Step 2 Additional cutting planes are used to find other piercing points—5, 6, 7, and 8—which are projected to the front view of their respective lines on the oblique plane. These are connected with a hidden line.

Step 3 Visibility of the plane and cylinder is completed in the front view. Line AB is found to be visible by inspection of the top view, and line CD is found to be hidden.

ments on the plane. These points are projected to the front view to complete the line of intersection, an ellipse. The more cutting planes used, the more accurate the line of intersection will be.

The general case of the intersection between a plane and cylinder is solved in Fig. 32.15, where both are oblique in the given views. The

edge view of the plane is found in an auxiliary view. Cutting planes are passed through the cylinder parallel to the cylinder's axis in the auxiliary view to find the piercing points. The piercing points of the elements are connected to give elliptical lines of intersection in the given views.

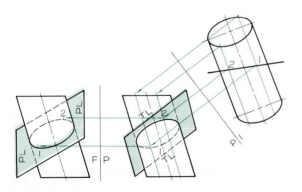

Figure 32.15 The intersection between an oblique cylinder and an oblique plane can be found by constructing a view that shows the plane as an edge.

32.5 Intersections Between Cylinders and Prisms

A series of vertical cutting planes is used in Fig. 32.16 to establish lines that lie on the surfaces of the cylinder and prism. A primary auxiliary view is drawn to show the end view of the inclined prism. Also shown in this view are the vertical cutting planes, which are spaced the same distance apart as in the top view (Step 1).

The line of intersection from 1 to 3 is projected from the auxiliary view to the front view (Step 2), where the intersection is an elliptical curve. The change of visibility of this line is found at point X in the top and auxiliary views,

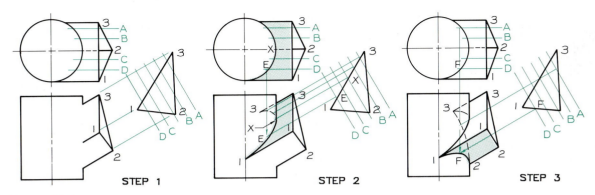

Figure 32.16 Intersection between a cylinder and a prism.

Step I Project an auxiliary view of the triangular prism from the front view to show three of its surfaces as edges. Pass frontal cutting planes through the top view of the cylinder and project them to the auxiliary view. The spacing between the planes is equal in both views.

Step 2 Locate points along the line of intersection 1–3 in the top view and project them to the front view. *Example:* Point *E* on cutting plane *D* is found in the top and auxiliary views and projected to the front view where the projectors intersect. Visibility changes in the front view at point *X*.

Step 3 Determine the remaining points of intersection by using the same cutting planes. Point *F* is shown in the top and auxiliary views and is projected to the front view of 1–2. Connect the points and determine visibility. Space the cutting planes so that they will produce the most accurate line of intersection.

and it is projected to the front view. The process is continued to find the lines of intersection of the other two planes of the prism. (Step 3).

32.6 Intersections Between Two Cylinders

The line of intersection between two perpendicular cylinders can be found by passing cutting planes through the cylinders parallel to the centerlines of each (Fig. 32.17). The points are connected and visibility is determined to complete the solution.

The intersection between nonperpendicular cylinders is found in Fig. 32.18 by a series of vertical cutting planes. Each cutting plane is passed through the cylinders parallel to the centerline of each. As examples of points on the line of intersection, points 1 and 2 are labeled on cutting plane *D*. Other points are found in the same manner. Although the auxiliary view is not required for the solution, it assists you in visualizing the problem. Points 1 and 2 are shown on cutting plane *D* in the auxiliary view, where they

can be projected to the front view as a check on the solution found when projecting from the top view.

32.7 Intersections Between Planes and Cones

To find points of intersection on a cone, cutting planes can be used that are (1) perpendicular to the cone's axis or (2) parallel to the cone's axis. Vertical cutting planes are shown in Fig. 32.19A, where they cut radial lines on the cone. The horizontal planes in Fig. 32.19B cut circular sections that appear true size in the top view.

A series of radial cutting planes is used to find elements on the cone in Fig. 32.20. These elements cross the edge view of the plane in the front view to locate piercing points that are projected to the top view of the same elements to form the line of intersection.

A cone and an oblique plane intersect in Fig. 32.21, and the line of intersection is found by using a series of horizontal cutting planes. The sections cut by these imaginary planes will be circles

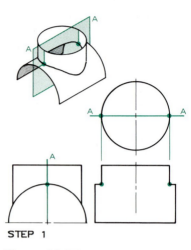

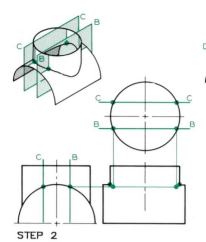

Figure 32.17 Intersection between two cylinders.

Step 1 Cutting plane A–A is passed through the cylinders parallel to the axes of both. Two points of intersection are found.

Step 2 Cutting planes C–C and B–B are used to find four additional points of intersection.

Step 3 Cutting planes D–D and E–E locate four more points. Points found in this manner give the line of intersection.

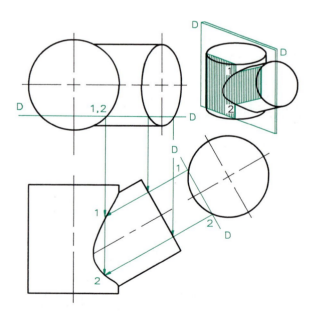

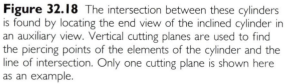

Figure 32.18 The intersection between these cylinders is found by locating the end view of the inclined cylinder in an auxiliary view. Vertical cutting planes are used to find the piercing points of the elements of the cylinder and the line of intersection. Only one cutting plane is shown here as an example.

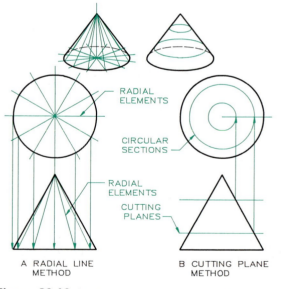

Figure 32.19 (A) Intersections on conical surfaces can be found with radial cutting planes that pass through the cone's centerline and perpendicular to its base. (B) A second method shows cutting planes that are parallel to the cone's base.

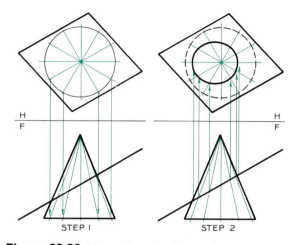

Figure 32.20 Intersection of a plane and a cone.

Step 1 Divide the base into even divisions in the top view, and connect these points with the apex to establish elements on the cone. Project these elements to the front view.

Step 2 The piercing point of each element on the edge view is projected to the top view to the same elements, where they are connected to form the line of intersection.

in the top view. In addition, the cutting planes locate lines on the oblique plane that intersect the same circular sections cut by each respective cutting plane. The points of intersection are found in the top view and are projected to the front view.

The horizontal cutting-plane method also could have been used to solve the example in Fig. 32.20.

32.8 Intersections Between Cones and Prisms

A primary auxiliary view is used to find the end view of the inclined prism that intersects the cone in Fig. 32.22 (Step 1). Cutting planes that radiate from the apex of the cone in the top view are drawn in the auxiliary view to locate elements on the cone's surface that intersect the prism. These elements are drawn in the front view by projection.

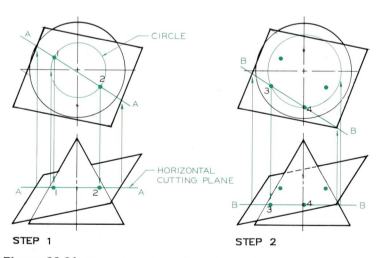

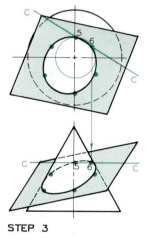

Figure 32.21 Intersection of an oblique plane and a cone.

Step 1 A horizontal cutting plane is passed through the front view to establish a circular section on the cone and a line on the plane in the top view. The piercing point of this line lies on the circular section. Piercing points 1 and 2 are projected to the front view.

Step 2 Horizontal cutting plane B–B is passed through the front view in the same manner to locate piercing points 3 and 4 in the top view. These points are projected to the horizontal plane in the front view from the top view.

Step 3 Additional horizontal planes are used to find sufficient points to complete the line of intersection. Determination of visibility completes the solution.

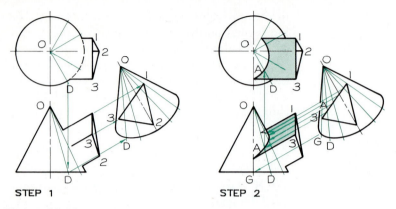

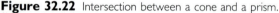

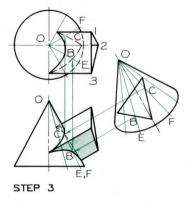

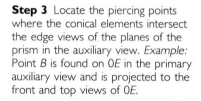

STEP 1 STEP 2 STEP 3

Figure 32.22 Intersection between a cone and a prism.

Step 1 Construct an auxiliary view to obtain the edge views of the lateral surfaces of the prism. In the auxiliary view, pass cutting planes through the cone that radiates from the apex to establish elements on the cone. Project the elements to the front and auxiliary views.

Step 2 Locate the piercing points of the cone's elements with the edge view of plane 1–3 in the primary view and project them to the front and top views. *Example:* Point A lies on element 0D in the primary auxiliary view, so it is projected to the front and top views of 0D.

Step 3 Locate the piercing points where the conical elements intersect the edge views of the planes of the prism in the auxiliary view. *Example:* Point B is found on 0E in the primary auxiliary view and is projected to the front and top views of 0E.

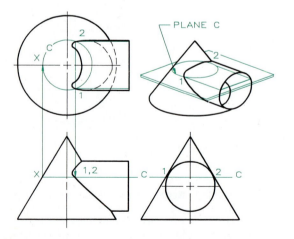

Figure 32.23 Horizontal cutting planes are used to find the intersection between the cone and the cylinder. The cutting planes form circles in the top view. Only one cutting plane is shown here as an example.

Wherever the edge view of plane 1–3 intersects an element in the auxiliary view, the piercing points are projected to the same element in the front and top views (Step 2). An extra cutting plane is passed through point 3 in the auxiliary view to locate an element that is projected to the front and top views. Piercing point 3 is projected to this element in sequence from the auxiliary view to the top view.

This same procedure is used to find the piercing points of the other two planes of the prism (Step 3). All projections of points of intersection originate in the auxiliary view, where the planes of the prism appear as edges.

In Fig. 32.23, horizontal cutting planes are passed through the cone and the intersecting perpendicular cylinder to locate the line of intersection. A series of circular sections are found in the top view. Points 1 and 2 are found on cutting plane C in the top view as examples and are projected to the front view. Other points are found in this same manner.

This method is feasible only when the centerline of the cylinder is perpendicular to the axis of the cone, so that circular sections can be found in the top view, rather than elliptical sections that would be difficult to draw.

The distributor housing in Fig. 32.24 is an example of an intersection between cylinders and a cone.

32.9 Intersections Between Pyramids and Prisms

The intersection between an inclined prism and a pyramid is solved in Fig. 32.25. The end view of the inclined prism is found in a primary auxiliary view; the pyramid is shown in this view also (Step 1). Radial lines 0B and 0A are passed through corners 1 and 3 in the auxiliary view (Step 2). The radial lines are projected from the auxiliary view to the front and top views. Intersecting points 1 and 3 are located on 0B and 0A in each of these views by projection. Point 2 is the point where line 1–3 bends around corner 0C. Lines of intersection 1–4 and 4–3 are then found

Figure 32.24 This electrically operated distributor is an application of intersections between a cone and a series of cylinders. (Courtesy of GATX).

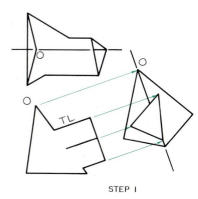

STEP 1

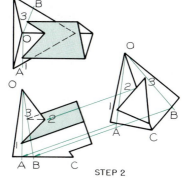

STEP 2

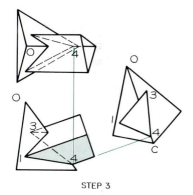

STEP 3

Figure 32.25 Intersection between a prism and a pyramid.

Step 1 Find the edge view of the surfaces of the prism by projecting an auxiliary view from the front view. Project the pyramid into this view also. Only the visible surfaces need be shown in this view.

Step 2 Pass planes A and B through apex 0 and points 1 and 3 in the auxiliary view. Project lines 0A and 0B to the front and top views. Project points 1 and 3 to 0A and 0B in the principal views. Point 2 lies on line 0C. Connect points 1, 2, and 3 to give the intersection of the upper plane.

Step 3 Point 4 lies on line 0C in the auxiliary view. Project this point to the principal views. Connect point 4 to points 3 and 1 to complete the intersections. Visibility is indicated. These geometric shapes are assumed to be hollow, as though constructed of sheet metal.

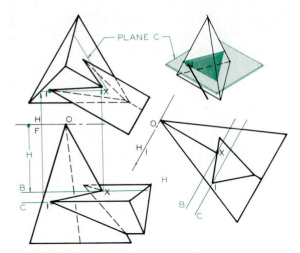

Figure 32.26 The intersection between this pyramid and prism is found by locating the end view of the prism in an auxiliary view. Horizontal cutting planes are passed through the fold lines of the prism to find the piercing points and the line of intersection.

(Step 3). The visibility is determined and the solution is completed.

A prism parallel to the base of a pyramid is shown in Fig. 32.26. Its lines of intersection are found by using a series of horizontal cutting planes that pass through the pyramid parallel to its base to form triangular sections in the top view.

The same cutting planes are passed through the corner lines of the prism in the front and auxiliary views. Each corner edge is extended in the top view to intersect the triangular section formed by the cutting plane passing through it. Point 1 is given as an example.

Corner point *X* is found by passing cutting plane *B* through it in the auxiliary view where it crosses the corner line. This is where the line of intersection of this plane bends around the corner.

32.10 Intersections Between Spheres and Planes

A general case of the intersection between a plane and a sphere is given in Fig. 32.27, where three points, 1, 2, and 3, are located on the

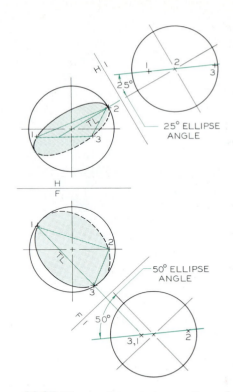

Figure 32.27 Three points are given on the surface of the sphere through which a circle passes. This plane is found as an edge by projecting from the top and front views. Ellipse angles of 25° and 50° are found for drawing the top and front views of the elliptical intersections.

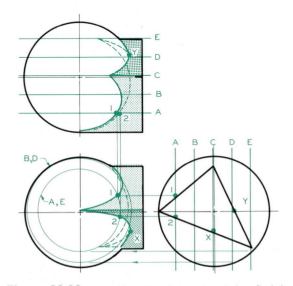

Figure 32.28 Vertical cutting planes are used to find the intersection between a prism and a sphere.

sphere's surface. A circle that lies on the surface of the sphere is to be drawn through these three points.

The edge view of the plane is found by projecting from the top view to the primary auxiliary view. The circle on the sphere passing through 1, 2, and 3 will have an elliptical line of intersection in the top and front views. The ellipse template angle for the top view is the angle between the edge of the plane and the projector from the top view, 25°. The major diameter is drawn parallel to the true-length lines on plane 1–2–3 in the top view.

The ellipse for the front view of the intersection is found in the same manner by finding the edge view of the plane in an auxiliary view projected from the front view. The ellipse template angle is 50°.

32.11 Intersections Between Spheres and Prisms

The intersection between a sphere and a prism is found in Fig. 32.28 by drawing a series of vertical cutting planes in the top and side views. The planes form circular sections in the front view.

The intersections of the edges with the cutting planes in the side view are projected to their respective circles in the front view. Points 1 and 2 are located on cutting plane A in the side view and on the circular path of A in the front view. Point X in the side view locates the point where the visibility changes in the front view. Point Y in the side view is the point where the visibility of the intersection changes in the top view. Both points lie on the centerlines of the sphere on the side view.

32.12 Principles of Developments

The processing plant shown in Fig. 32.29 illustrates examples of sheet metal shapes designed using the principles of developments. In other words, their patterns were laid out on a flat stock

Figure 32.29 Almost all the surfaces shown in this refinery were made from flat stock fabricated to form these irregular shapes. These flat patterns are called developments.

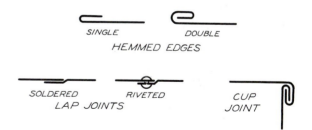

Figure 32.30 Examples of the types of seams used to join developments.

and then formed to the proper shape by bending and seaming the joints.

Examples of standard hemmed edges and joints are shown in Fig. 32.30. The application of the sheet metal design will determine the best method of connecting the seams.

The development of the surfaces of three typical shapes into a flat pattern is shown in Fig. 32.31. The sides of a box are imagined to be unfolded into a common plane. The cylinder is rolled out for a distance equal to its circumference. The pattern of a right cone is developed using the length of an element as a radius.

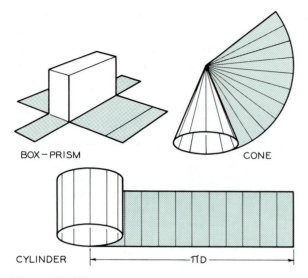

BOX – PRISM

CONE

CYLINDER

Figure 32.31 Three standard types of developments: The box, cylinder, and cone.

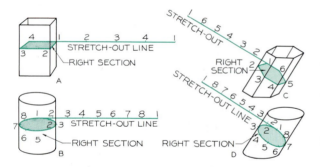

Figure 32.32 (a) and (b) The developments of right prisms and cylinders are found by rolling out the right sections along a stretch-out line. (c) and (d) When these figures are oblique, the right sections are found to be perpendicular to the sides of the prism and cylinder. The development is laid out along the stretch-out line parallel to the edge view of the right section.

Patterns of shapes with parallel elements, such as the prisms and cylinders shown in Figs. 32.32a and b, are begun by constructing stretch-out lines parallel to the edge view of the right section of the parts. The distance around the right section is laid off along the **stretch-out line.** The prism and cylinder in Figs. 32.32 c and d are inclined; thus, the right sections must be drawn perpendicular to their sides, not parallel to their bases.

An **inside pattern** (development) is preferred over an outside pattern because most bending machines are designed to fold metal inward and because markings and scribings will be hidden.

The method of denoting a pattern is labeled by a series of lettered or numbered points about its layout. All lines on a development must be true length.

Seam lines (lines where the pattern is joined) should be the **shortest lines** so that the expense of riveting or welding the pattern is the least possible.

32.13 Development of Prisms

A flat pattern for a prism is developed in Fig. 32.33. Since the edges of the prism are vertical in the front view, its right section is perpendicular to these sides. The top view shows the right section true size. The stretch-out line is drawn parallel to the edge view of the right section, beginning with point 1.

If an inside pattern is drawn and it is to be laid out to the right, point 2 will be to the right of point 1. This is determined by looking from the inside of the top view, where 2 is seen to the right of 1.

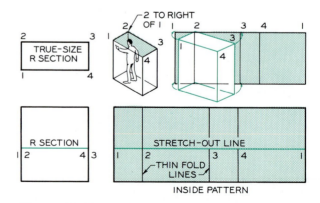

Figure 32.33 The development of a rectangular prism to give an inside pattern. The stretch-out line is parallel to the edge view of the right section.

To locate the fold lines on the pattern, lines 2–3, 3–4, and 4–1 are transferred with your dividers from the right section in the top view to the stretch-out line. The length of each fold line is found by projecting its true length from the front view. The ends of the fold lines are connected to form the limits of the developed surface. Fold lines are drawn as thin lines, and the outside lines are drawn as visible object lines.

The sheet metal assembly in Fig. 32.34 is composed of many developments involving pyramids and prisms.

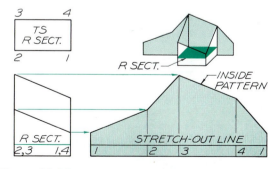

Figure 32.34 This spacecraft was designed and constructed by using principles of intersections and developments. (Courtesy of NASA.)

Figure 32.35 The development of a rectangular prism with a beveled end to give an inside pattern. The stretch-out line is parallel to the right section.

The development of the prism in Fig. 32.35 is similar to the example in Fig. 32.33 except that one end is beveled rather than square. The stretch-out line is drawn parallel to the edge view of the right section in the front view. The true-length distances around the right section are laid off along the stretch-out line, and the fold lines are located. The lengths of the fold lines are found by projecting from the front view of these lines.

32.14 Development of Oblique Prisms

The prism in Fig. 32.36 is inclined to the horizontal plane, but its fold lines are true length in the front view. The right section is drawn as an edge perpendicular to these fold lines, and the stretch-out line is drawn (Step 1). A true-size view of the right section is found in the auxiliary view.

In Step 2, the distances between the fold lines are transferred from the true-size right section to the stretch-out line. The lengths of the fold lines are found by projecting from the front view. In Step 3, the ends of the prism are found and attached to the pattern so they can be folded into position.

In Fig. 32.37, the fold lines are true length in the top view; this enables you to draw the edge view of the right section perpendicular to the fold lines in the top view. The stretch-out line is drawn parallel to the edge view of the section, and the true size of the right section is found in an auxiliary view projected from the top view. The distances about the right section are transferred to the stretch-out line to locate the fold lines. The lengths of the fold lines are found by projecting from the top view. The end portions of the pattern are attached to the pattern to complete the construction.

A prism that does not project true length in either view can be developed as shown in Fig. 32.38. The fold lines are found true length in an auxiliary view projected from the front view. The right section will appear as an edge perpendicular to the fold lines in the auxiliary view. The

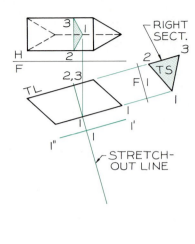

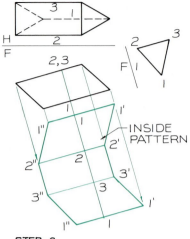

STEP 1

STEP 2

STEP 3

Figure 32.36 Development of an oblique prism

Step 1 The edge view of the right section is perpendicular to the true-length axis of the prism in the front view. Determine the true-size view of the right section by an auxiliary view. Draw the stretch-out line parallel to the edge view of the right section. Line 1'–1" is the first line of the development.

Step 2 Since the pattern is developed toward the right, from line 1'–1", the next point is line 2'–2" by referring to the auxiliary view. Transfer true-length lines 1–2, 2–3, and 3–1 from the right section to the stretch-out line to locate the elements. Determine the lengths of the bend lines by projection.

Step 3 Find the true-size views of the end pieces by projecting auxiliary views from the front view. Connect these surfaces to the development of the lateral sides to form the completed pattern. Fold lines are drawn with thin lines, and outside lines are drawn as object lines.

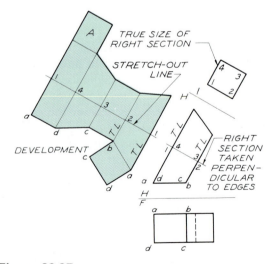

Figure 32.37 The development of this oblique chute is found by locating the right section true size in the auxiliary view. The stretch-out line is drawn parallel to the right section.

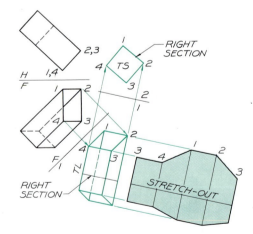

Figure 32.38 The development of an oblique cylinder is found by locating an auxiliary view in which the fold lines are true length and a secondary auxiliary view in which the right section appears true size. The stretch-out line is drawn parallel to the right section.

true size of the right section is found in a secondary auxiliary view.

The stretch-out line is drawn parallel to the edge view on the right section. The fold lines are located on the stretch-out line by measuring around the right section in the secondary auxiliary view. The lengths of the fold lines are then projected to the development from the primary auxiliary view.

32.15 Development of Cylinders

The development of a cylinder is found in Fig. 32.39. Since the elements of the cylinder are true length in the front view, the right section will appear as an edge in this view and true size in the top view. The stretch-out line is drawn parallel to the edge view of the right section, and point 1 is chosen as the beginning point since it is the shortest element. To draw an inside pattern, assume you are standing on the inside looking at point 1 and you will see that point 2 is to the right of point 1; therefore the pattern is laid out with point 2 to the right of point 1.

The spacing between the elements in the top view can be conveniently done by drawing radial lines at 15° or 30° intervals. Using this technique, the elements will be equally spaced, making it convenient to lay them out along the stretch-out line. The lengths of the elements are found by

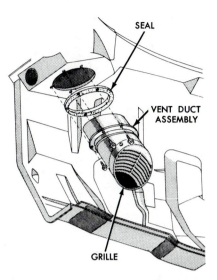

Figure 32.40 This ventilator air duct was designed using development principles. (Courtesy of Ford Motor Co.)

projecting from the front view to complete the pattern.

An application of a developed cylinder with a beveled end is the air-conditioning duct from an automobile shown in Fig. 32.40.

32.16 Development of Oblique Cylinders

The pattern for an oblique cylinder (Fig. 32.41) is found in the same manner as the previous examples but with the addition of one preliminary step: The right section must be found true-size in an auxiliary view. In Step 1, a series of equally spaced elements is located around the right section in the auxiliary view and is projected back to the true-length view. The stretch-out line is drawn parallel to the edge view of the right section in the front view.

In Step 2, the spacing between the elements is laid out along the stretch-out line, and the elements are drawn through these points perpendicular to the stretch-out line. The lengths of the elements are found by projecting from the front view.

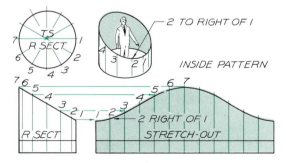

Figure 32.39 The development of a right cylinder's inside pattern. The stretch-out line is parallel to the right section. Point 2 is to the right of point 1 for an inside pattern.

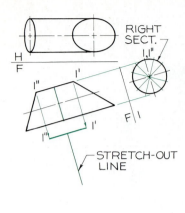

STEP 1

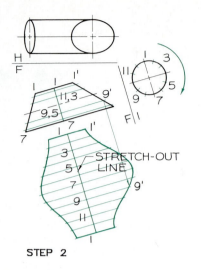

STEP 2

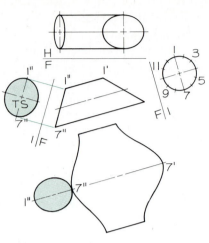

STEP 3

Figure 32.41 Development of an oblique cylinder.

Step 1 The right section is an edge in the front view, and perpendicular to the true-length axis. Construct an auxiliary view to find the true size of the right section. Draw a stretch-out line parallel to the edge view of the right section. Locate element 1'–1". Divide the right section into equal divisions.

Step 2 Project these elements to the front view from the right section. Transfer measurements between the points in the auxiliary view to the stretch-out line to locate the elements in the development. Determine the lengths of the elements by projection.

Step 3 The development of the end pieces will require auxiliary views that project these surfaces as ellipses, as shown for the left end. Attach this true-size ellipse to the pattern. Note that the beginning line for the pattern was line 1'–1", the shortest element, for economy.

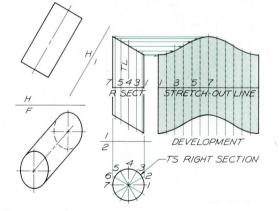

Figure 32.42 The development of an oblique cylinder is found by constructing an auxiliary view in which the elements appear true length. The right section is found true size in a secondary auxiliary view.

The ends of the cylinder are found in Step 3 to complete the pattern. Only one end pattern is shown as an example.

A more general case is the oblique cylinder in Fig. 32.42, where the elements are not true length in the given views. A primary auxiliary view is used to find a view where the elements are true length, and a secondary auxiliary view is drawn to find the true-size view of the right section. The stretch-out line is drawn parallel to the edge view of the right section in the primary auxiliary view, and the elements are located along this line by transferring their distances apart from the true-size section.

The elements are drawn perpendicular to the stretch-out line. The length of each element is found by projecting from the primary auxiliary view. The endpoints are connected with a smooth curve to complete the pattern.

32.17 Development of Pyramids

All lines used to draw a pattern must be true length. Pyramids have only a few lines that are true length in the given views; for this reason, the sloping corner lines must be found true length at the outset.

The corner lines of a pyramid can be found by revolution, as shown in Fig. 32.43. Line 0–5 is revolved in the frontal position of 0–5' in the

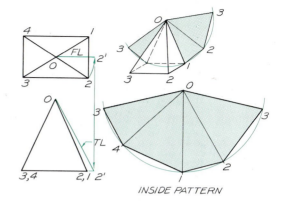

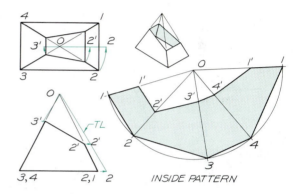

Figure 32.44 Development of a right pyramid.

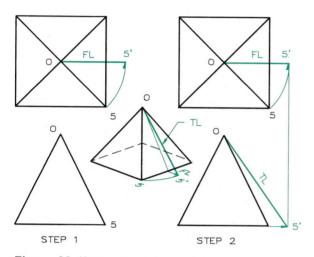

Figure 32.43 True length by revolution.

Step 1 Corner 0–5 of a pyramid is found true length by revolving it into the frontal plane in the top view, to 0–5'.

Step 2 Point 5' is projected to the front view where 0–5' is true length.

Figure 32.45 The development of an inside pattern of a truncated pyramid. The corner lines are found true length by revolution.

top view (Step 1). Since 0–5' is a frontal line, it will be true length in the front view (Step 2).

The development of a pyramid is given in Fig. 32.44. Line 0–2 is revolved into the frontal plane in the top view to find its true length in the front view. Since this is a right pyramid, all bends are equal in length. Line 0–2' is used as a radius to construct the base circle for drawing the development. Distance 3–2 is transferred from the base in the top view to the development, where it forms a chord on the base circle. Lines 2–1, 1–4, and 4–3 are found in the same manner

and in sequence. The bend lines are drawn as thin lines from the base to the apex, point 0.

A variation of this problem is given in Fig. 32.45, in which the pyramid has been truncated or cut at an angle to its axis. The development of the inside pattern is found in the same manner as the previous example; however, to establish the upper lines of the development, an additional step is required. The true-length lines from the apex to points 1', 2', 3', and 4' are found by revolution. These distances are located on their respective lines of the pattern to find the upper limits of the pattern.

The mounting pads in Fig. 32.46 are sections of pyramids that intersect an engine body.

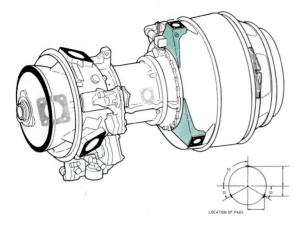

Figure 32.46 Examples of pyramid shapes in the design of mounting pads for an engine. (Courtesy of Avco Lycoming.)

32.18   Development of Cones

All elements of a right cone are equal in length, as illustrated in Fig. 32.47, where 0–6 is found true length by revolution. When revolved to 0–6′ position, it is a frontal line and is therefore true length in the front view where it is an outside element of the cone. Point 7′ is found by projecting horizontally to element 0–6′.

The right cone in Fig. 32.48 is developed by dividing the base into equally spaced elements in the top view and by projecting them to the base in the front view. These elements radiate to the apex at 0. The outside elements in the front view, 0–10 and 0–4, are true length.

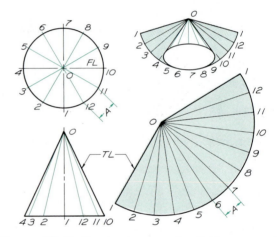

Figure 32.48 The development of an inside pattern of a right cone. In this case, all elements are equal in length.

Using element 0–10 as a radius, draw the base circle of the development. The elements are located along the base circle equal to the chordal distances between them around the base in the top view. You can see this is an inside pattern by inspecting the top view, where point 2 is to the right of point 1 when viewed from the inside.

The sheet metal conical vessel in Fig. 32.49 is an example of a large vessel that was designed using principles of development.

The development of a truncated cone is shown in Fig. 32.50. The pattern is found by lay-

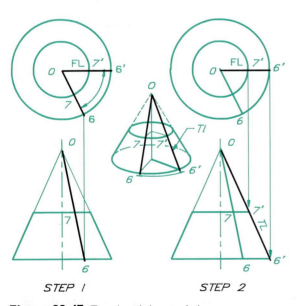

Figure 32.47 True length by revolution.

Step 1 An element of a cone, 0–6, is revolved into a frontal plane in the top view.

Step 2 Point 6′ is projected to the front view where it is the outside element of the cone and is true length. Line 0–7′ is found TL by projecting to the outside element in the front view.

ing out the total cone, ignoring the portion removed from it. The removed upper portion can be found by using true-length line 0–7′ as the radius in the pattern view. The hyperbolic sections through the front view of the cone can be found on their respective elements in the top and front views. Lines 0–2′ and 0–3′ are projected horizontally to the true-length element 0–1 in the front view, where they will appear true length. These distances and others are measured off along their respective elements in the development to establish a smooth curve on the development.

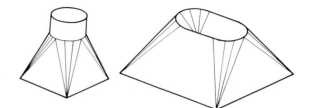

Figure 32.51 Examples of transition pieces that join parts having different cross sections.

Figure 32.49 An example of a conical shape formed from metal panels by applying principles of developments. (Courtesy of NASA.)

Figure 32.52 Transition-piece developments are used to join a circular shape with a rectangular section. (Courtesy of Western Precipitation Group, Joy Manufacturing Co.)

32.19 Development of Transition Pieces

A transition piece is a form that transforms the section at one end to a different shape at the other (Fig. 32.51). Huge transition pieces can be seen in the industrial installation in Fig. 32.52.

The development of a transition piece is shown in Fig. 32.53. In Step 1, radial elements are drawn from each corner to the equally spaced points on the circular end of the piece. Each of these lines is found true length by revolution.

In Step 2, the true-length lines are used with the true-length chordal distance in the top view

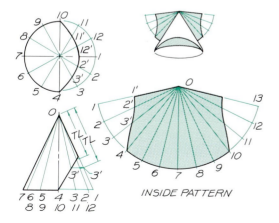

INSIDE PATTERN

Figure 32.50 The development of a conical surface with a side opening.

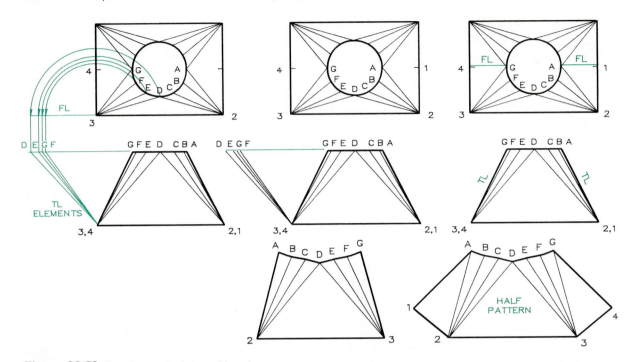

Figure 32.53 Development of a transition piece.

Step 1 Divide the circular edge of the surface into equal parts in the top view. Connect these points with bend lines to the corner points, 2 and 3. Find the true length (TL) of these lines by revolving them into a frontal plane and projecting them to the front view. These lines represent elements on the surface of a cone.

Step 2 Using the TL lines found in the TL diagram and the lines on the circular edge in the top view, draw a series of triangles, which are joined together at common sides. *Examples:* Arcs 2D and 2C are drawn from point 2. Point C is found by drawing arc DC from point D. Line DC is TL in the top view.

Step 3 Construct the remaining planes, A–1–2 and G–3–4, by triangulation to complete the inside half-pattern of the transition piece. Draw the fold lines as thin lines at the places where the surface is to be bent slightly. The line of departure for the pattern is chosen along A–1, the shortest line.

to lay out a series of adjacent triangles to form the pattern beginning with element A–2.

In Step 3, the triangles A–1–2 and G–3–4 are added at each end of the pattern to complete the development of a half-pattern.

32.20 Development of Spheres— Zone Method

In Fig. 32.54, a development of a sphere is found by the zone method. A series of parallels, called **latitudes** in cartography, are drawn in the front view. Each is spaced an equal distance, *D*, apart

along the surface in the front view. Distance *D* can be found mathematically to improve the accuracy of this step.

Cones are passed through the sphere's surface so that they pass through two parallels at the outer surface of the sphere. The largest cone with element *R*1 is found by extending it through where the equator and the next parallel intersect on the sphere's surface in the front view until *R*1 intersects the extended centerline of the sphere. Elements *R*2, *R*3, and *R*4 are found by repeating this process.

The development is begun by laying out the largest zone, using *R*1 as the radius, on the arc

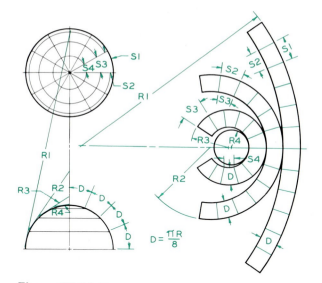

Figure 32.54 The zone method of finding the inside development of a spherical pattern.

that represents the base of an imaginary cone. The breadth of the zone is found by laying off distance D from the front view to the development and drawing the upper portion of the zone with a radius equal to $R1-D$ using the same center. No consideration is given to finding the arc lengths at this time.

The next zone is drawn using radius $R2$ with its center located on a line through the center of arc $R1$. The center of $R2$ is positioned along this line so that the arc drawn will be tangent to the preceding arc, which was drawn with radius $R1-D$. The upper arc of this second zone is drawn with a radius of $R2-D$. The remaining zones are constructed successively in this manner. The last zone will appear as a circle with $R4$ as its radius.

The lengths of the arcs can be established by dividing the top view with vertical cutting planes that radiate through the poles. These lines, which lie on the surface of the sphere, are called **longitudes** in cartography. Arc distances $S1$, $S2$, $S3$, and $S4$ are found on each parallel in the top view. These distances are measured off on the constructed arcs in the development. In this case, there are 12 divisions; smaller divisions would provide a more accurate measurement.

32.21 Development of Spheres— Gore Method

Figure 32.55 illustrates an alternative method of developing a flat pattern for a sphere using a series of spherical elements called **gores.** Equally spaced vertical cutting planes are passed through the poles in the top view. Parallels are located in the front view by dividing the surface into equal zones of dimension D. The gores are projected to the front view.

A true-size view of one of the gores is projected from the top view. Dimensions can be checked mathematically for all points. A series of these gores is laid out in sequence to complete the pattern.

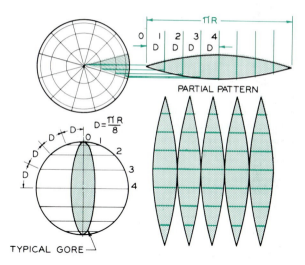

Figure 32.55 The gore method of developing an inside pattern of a sphere.

32.22 Development of Straps

Figure 32.56 illustrates the steps of finding the development of a strap that has been bent to serve as a support bracket, between point A on a vertical surface and point B on horizontal surface. In Step 1, the strap is drawn in the side view where the strap appears as an edge using the specified radius to show the bend. The bend is di-

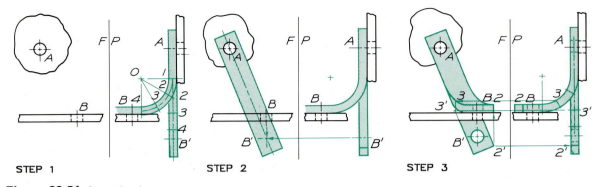

STEP 1 **STEP 2** **STEP 3**

Figure 32.56 Strap development.

Step 1 Construct the edge view of the strap in the side view. Locate points 1, 2, 3, and 4 on the neutral axis at the bend. Revolve this portion of the strap into the vertical plane and measure the distances along this view of the neutral axis. The hole is located at B' in this view.

Step 2 Construct the front view of B' by revolving point B parallel to the profile plane until it intersects the projector from B' in the side view. Draw the centerline of the true-size strap from A to B' in the front view. Add the outline of the strap around this centerline and around the holes.

Step 3 Determine the projection of the strap in the front view by projecting points from the given views. Points 3 and 2 are shown in the views to illustrate the system of projection used. The ends of the strap are drawn in each view to form true projections.

vided into equal arcs and is developed into a flat piece in the vertical plane through point A. Point B' is located in this view.

In Step 2, the location of the hole at B' is found in the front view by projection. The true-size development of the strap is drawn in this view, and it is projected back to the side view to complete that view of the strap.

In Step 3, the projected view of the strap in its bent position in the front view can be found by using projectors from the side view and the true pattern of the strap.

Problems

These problems are scaled to fit two problems per size A sheet when each grid is assigned a value of 0.20″ or 5 mm. When the grid is assigned a value of 0.40″ or 10 mm, one problem per size A sheet can be drawn.

Intersections

1–24. (Fig. 32.57) Lay out the problems assigned and find the intersections that are necessary to complete the views.

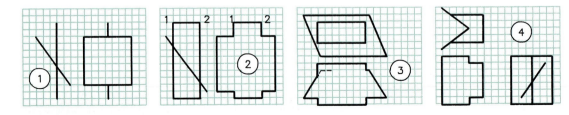

Figure 32.57 Intersections. Problems 1–24.

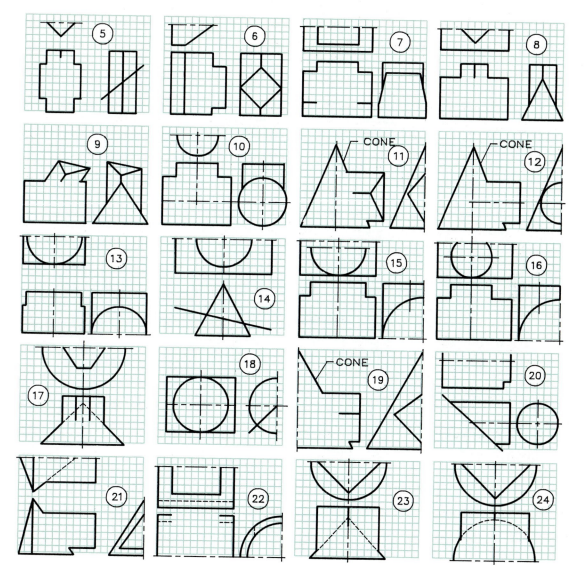

Developments

25–48. (Fig. 32.58) Lay out the problems assigned on a size A sheet. Place the long side of sheet in a horizontal direction to permit room at the right of the given views for the development to be drawn.

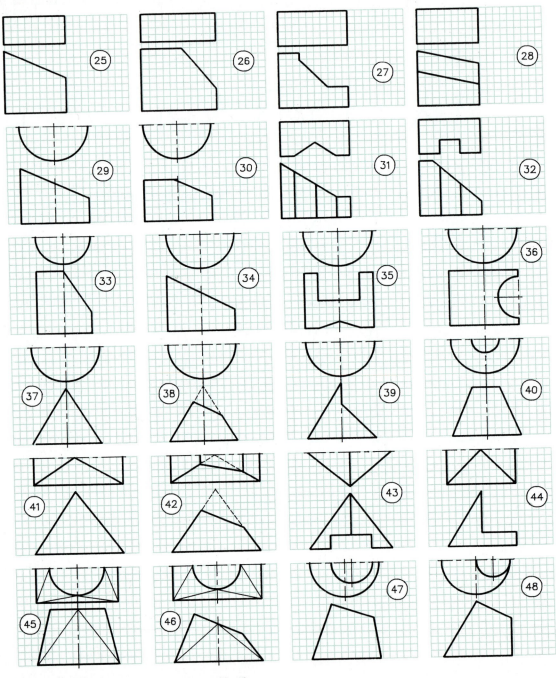

Figure 32.58 Developments. Problems 25–48.

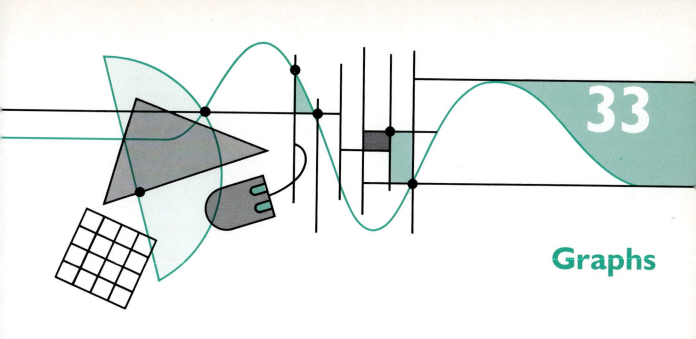

33

Graphs

33.1 Introduction

Data and information expressed as numbers and words are often difficult to analyze or evaluate unless transcribed into graphical form. **Graphs** (charts is an acceptable term but is more appropriate when referring to maps, a specialized form of graphs) are a popular means of briefing other people on trends that might otherwise be difficult to communicate (Fig. 33.1).

The trends of a plotted curve on a graph can be compared to the expressions on a person's

Figure 33.1 Graphs are helpful in presenting technical data to one's associates.

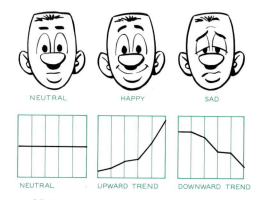

Figure 33.2 Curves on a graph are similar to expressions on a face.

face, which is a graph of sorts that reveals a person's feelings. For example, a flat curve shows no change, whereas an upwardly inclined curve indicates a positive increase. A downward curve, on the other hand, represents a negative result (Fig. 33.2).

In this chapter, we deal with the more commonly used graphs. The basic types are

1. Pie graphs
2. Bar graphs
3. Linear coordinate graphs

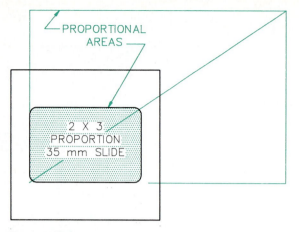

Figure 33.3 This diagonal-line method can be used for construction areas proportional to those of a 35-mm slide.

4. Logarithmic coordinate graphs
5. Semilogarithmic coordinate graphs
6. Polar graphs
7. Schematics and diagrams

33.2 Size Proportions of Graphs

Graphs may be used to illustrate technical reports reproduced in quantity, as well as for projection by slide or overhead projectors. In all cases, the proportion of the graph must be determined so that it will match the proportion of the space or the format of the visual aid.

If a graph is to be photographed by a 35-mm camera, the graph must conform to the standard size of the 35-mm film used. This proportion is approximately 2 × 3 (Fig. 33.3). The proportions of the area in which the graph is to be drawn can be enlarged or reduced by using the diagonal-line method.

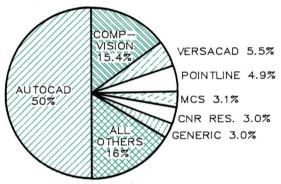

DESKTOP CAD SOFTWARE: 1986 REVENUES

Figure 33.4 A pie graph shows the relationship of the parts to a whole. It is effective when there are only a few parts.

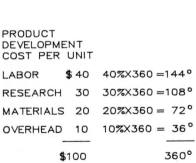

STEP 1

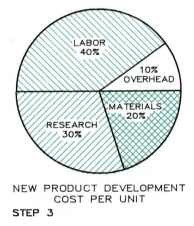

STEP 2   STEP 3

Figure 33.5 Drawing pie graphs.

Step 1 The total sum of the parts is found, and the percentage of each is found. Each percentage is multiplied by 360° to find the angle of each sector of the pie graph.

Step 2 The circle is drawn to represent the pie graph. Each sector is drawn using the degrees found in Step 1. The smaller sectors should be placed as nearly horizontal as possible.

Step 3 The sectors are labeled with their proper names and percentages. In some cases, it might be desirable to include the exact numbers in each sector as well.

33.3 Pie Graphs

> Pie graphs compare the relationship of parts to a whole when there are only a few parts.

Figure 33.4 shows the types of computer graphics software used in industry.

Figure 33.5 shows the method of drawing a pie graph. The tabular data does not give as good an impression of the comparisons as does the pie graph even though the data is quite simple.

To facilitate lettering within narrow spaces, the thin sectors should be placed as nearly horizontal as possible to provide more room for the label. The actual percentage should be given, and it may also be desirable to give the actual numbers or values in each sector.

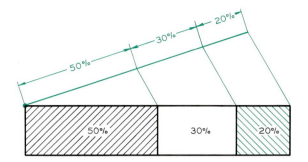

Figure 33.7 The method of constructing a single bar where the sum of all the parts will be 100%.

and the bar is divided into lengths proportional to the percentages represented by each of the three parts of the bar.

Figure 33.8 shows the method of constructing a bar graph. In this case, the title of the graph is placed inside the graph where space is available. Titles are often placed under or over the graph.

> The data should be sorted by arranging the bars in **ascending** or **descending** order since it is desirable to know how the data represented by the bars rank from category to category (Fig. 33.9B).

An alphabetical or numerical arrangement of bars results in a graph more difficult to evaluate (Fig. 33.9A).

If the data are sequential and involve time, such as sales per month, it would be less effective to rank the data in ascending order because it is important to see variations in the data over time.

Bars in a bar graph may be horizontal (Fig. 33.10) or vertical (Fig. 33.11). To show a true comparison of data, it is desirable that the bars begin at zero.

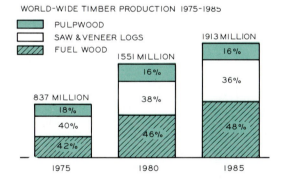

Figure 33.6 In this example, each bar represents 100% of the total amount, and each bar represents different totals.

33.4 Bar Graphs

Since they are well understood by the general public, bar graphs are effective to compare values (Fig. 33.6). In this example, the bars show not only the overall production of timber (the total lengths of the bars) but also the portions of the total devoted to the three uses of the timber.

A bar graph can be composed of a single bar (Fig. 33.7). The total length of the bar is 100%,

33.5 Linear Coordinate Graphs

Figure 33.12 shows a typical coordinate graph, with notes explaining its important features. The axes are linear if the divisions along the axes are equally spaced.

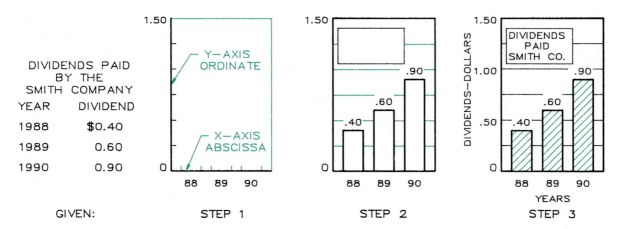

GIVEN: STEP 1   STEP 2   STEP 3

Figure 33.8 Construction of a bar graph.

Given These data are to be plotted as a bar graph.

Step 1 Lay off the vertical and horizontal axes so that the data will fit on the grid. Make the bars begin at zero.

Step 2 Construct and label the bars. The width of the bars should be different from the space between them. Grid lines should not pass through them.

Step 3 Strengthen lines, title the graph, label the axes, and cross-hatch the bars.

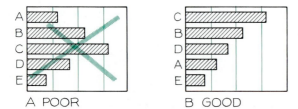

A POOR   B GOOD

Figure 33.9 The bars at A are arranged alphabetically. The resulting graph is not as easy to evaluate as the one at B, where the bars have been arranged in descending order.

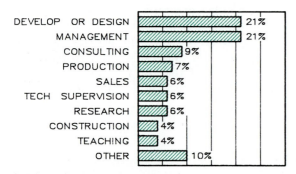

EMPLOYMENT FUNCTIONS OF ENGINEERS

Figure 33.10 A horizontal bar graph that is arranged in descending order to show where engineers are employed.

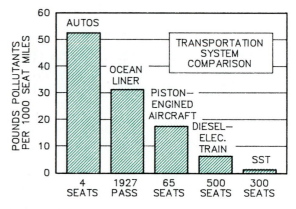

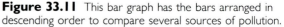

Figure 33.11 This bar graph has the bars arranged in descending order to compare several sources of pollution.

Points are plotted on the grid by using two measurements, called **coordinates**, made along each axis. The plotted points are indicated by using symbols, such as circles, that can be drawn with a template.

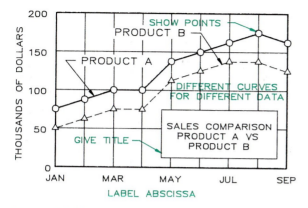

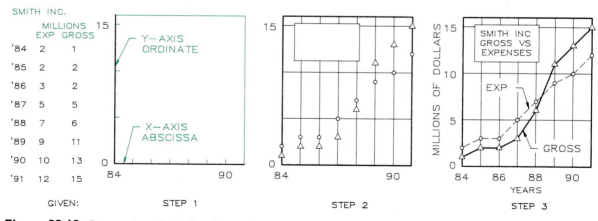

whether it is a smooth or broken line.) The curve should not close up the plotted points; rather, they should be left as open circles or symbols.

The curve must be drawn as a heavy prominent line since it is the most important part of the graph. In Fig. 33.12, there are two curves; therefore, drawing them as different types of lines and labeling them with notes and leaders is helpful. The title of the graph is placed in a box inside the graph.

Units are given along the X and Y axis with labels that designate the units of the graph.

Figure 33.12 The basic linear coordinate graph with the important features identified.

The horizontal scale of the graph is called the **abscissa** or X axis, and the vertical scale is called the **ordinate** or Y axis.

Once the points have been plotted, the curve is drawn from point to point. (The line drawn to represent the plotted points is called a curve

Broken-Line Graphs

Figure 33.13 shows the steps involved in drawing a linear coordinate graph. Because the data points are one year apart on the X axis, it is impossible to assume the change in the data is a smooth, continual progression from point to point. Therefore, the points are connected with a **broken-line curve.**

For the best appearance, the plotted points should not be crossed by the curve or grid lines

Figure 33.13 Construction of a broken-line graph.

Given A record of the Smith Company's gross and expenses.

Step 1 The vertical and horizontal axes are laid off to provide space for the largest values.

Step 2 The points are plotted over the respective years. Different symbols are used for each curve.

Step 3 The data points are connected with straight lines, the axes are labeled, the graph is titled, and the lines are strengthened.

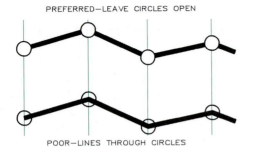

PREFERRED—LEAVE CIRCLES OPEN

POOR—LINES THROUGH CIRCLES

Figure 33.14 The curve of a graph should be drawn from point to point, but it should not close up the symbols used to locate the plotted points.

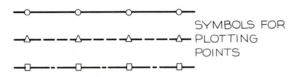

SYMBOLS FOR PLOTTING POINTS

Figure 33.15 Any of these symbols or lines can be effectively used to represent different curves on a single graph. The symbols are about $\frac{1}{8}''$ (3 mm) in diameter.

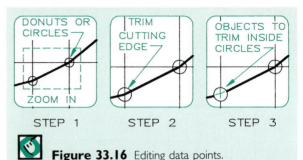

STEP 1   STEP 2   STEP 3

Figure 33.16 Editing data points.

Step 1   Open data points can be plotted on a graph as CIRCLEs, DONUTs, or POLYGONs. To remove lines from inside the open points, ZOOM in on several points.

Step 2 Command: TRIM (CR)
Select cutting edges (s):...
Select objects: (Select the circle.) (CR)

Step 3 Select object to trim: (Select the lines inside the circle, and they will be removed.) Continue this process for all points.

of the graph (Fig. 33.14). Each circle or symbol used to plot points should be about $\frac{1}{8}$ in. (5 mm) in diameter. Figure 33.15 shows several approved symbols and lines.

Computer Method The points on a graph can be drawn as CIRCLEs, DONUTs (open and closed), or POLYGONs. The grid lines and curves that pass through the open symbols can be easily removed by using the TRIM command as shown in Fig. 33.16.

The title of a graph can be placed in any of the positions shown in Fig. 33.17. The title should never be as meaningless as "graph" or "coordinate graph." Instead, it should explain the graph by giving the important information such as the company, date, source of data, and general comparisons being shown.

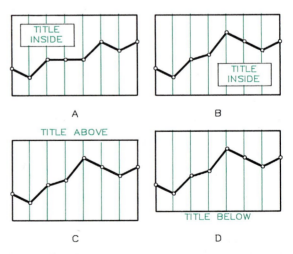

Figure 33.17 Title placement on a graph. (A and B) The title of a graph can be placed inside a box within the area of the graph. Perimeter lines of the box should not coincide with grid lines. (C) The title can be placed over the graph. (D) The title can be placed under the graph.

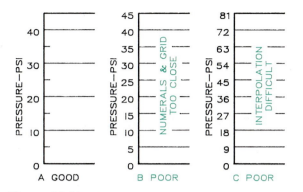

Figure 33.18 The scale at A is the best. It has about the right number of grid lines and divisions, and the numbers are given in well-spaced, easy-to-interpolate form. The numbers at B are too close, and there are too many grid lines. The units at C make interpolation difficult by eye.

The proper calibration and labeling of axes is important to the appearance and use of a graph. Figure 33.18A shows a properly executed axis. The axis in Fig. 33.18B has too many grid lines and too many units labeled along the axis. The units selected in Fig. 33.18C make it difficult to interpolate between the labeled values; for example, it is more difficult to locate a value such as 22 by eye on this scale than on the one in Fig. 33.18A.

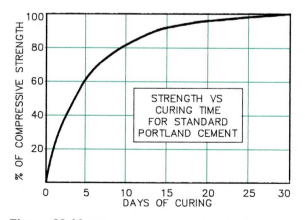

Figure 33.19 When the process graphed involves gradual, continuous changes in relationships, the curve should be drawn as a smooth line.

Smooth-Line Graphs

The strength of cement as related to curing time is plotted in Fig. 33.19. Since you know the strength of cement changes gradually in relation to curing time, the data points are connected with a **smooth curve** rather than a broken-line curve. Even if the data points do not lie on the curve, you can be reasonably certain the deviation is due to errors of measurement or the methods used in collecting the data.

Similarly, the strength of clay tile, as related to its absorption characteristics, is an example of data that yield a smooth curve (Fig. 33.20). Since you know this relationship should be represented by a smooth curve, the **best curve** is drawn to interpret the data to give an average representation of the points.

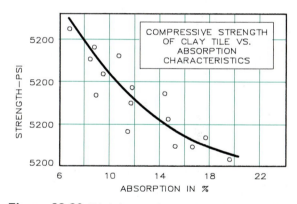

Figure 33.20 If it is known that a relationship plotted in a graph should yield a smooth gradual curve, a smooth-line best curve is drawn to represent the average of the plotted points. You must use your judgment and knowledge of the data in cases of this type.

There is a smooth-line curve relationship between miles per gallon and the speed at which a car is driven. In Fig. 33.21, two engines are compared with two smooth-line curves. Figure 33.22 compares the effect of speed on several automotive characteristics.

When a smooth-line curve is used to connect data points, the implication is that you can **interpolate** between the plotted points to estimate other values. Points connected by a broken-line curve imply you **cannot interpolate** between the plotted points.

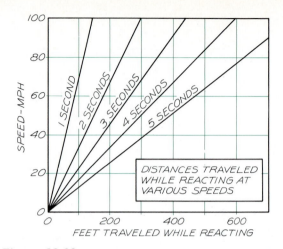

Figure 33.23 A graph can be used to determine a third value when two variables are known. Taking this information from a graph is easier than computing each answer separately.

Straight-Line Graphs

Some graphs have neither broken-line curves nor smooth-line curves but straight-line curves as Fig. 33.23 shows. Using this graph, you can determine a third value from the two given values. For example, if you are driving 70 miles per hour and it takes 5 seconds to react and apply your brakes, you will have traveled 500 feet in this time.

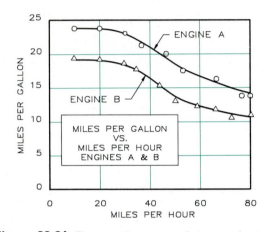

Figure 33.21 These are best curves that approximate the data without necessarily passing through each data point. Inspecting the data tells you this curve should be a smooth-line curve rather than a broken-line curve.

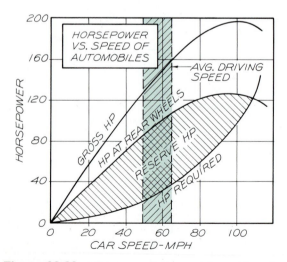

Figure 33.22 A linear coordinate graph is used here to analyze data affecting the design of an automobile's power system.

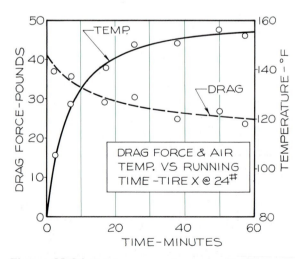

Figure 33.24 This is a composite graph with different scales along each Y axis. The curves are labeled so that reference can be made to the applicable scale.

Two-Scale Coordinate Graphs

Graphs can be drawn with different scales in combination, such as the one shown in Fig. 33.24. The vertical scale at the left is in units of pounds, and the one at the right is in degrees of temperature. Both curves are drawn using their respective Y axes, and each curve is labeled.

With graphs of this type, care must be taken to avoid confusing the reader. These graphs are effective when comparing related variables, such as the drag force and air temperature of a tire, as shown in this example.

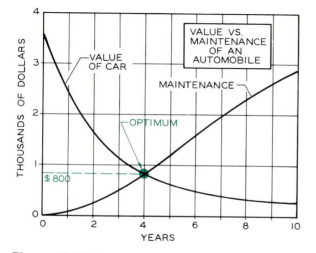

Figure 33.25 This graph shows the optimum time to sell a car based on the intersection of two curves that represent the depreciation of the car's value and its increasing maintenance costs.

Optimization Graphs

Figure 33.25 shows the optimization of the depreciation of an automobile and its increase in maintenance costs. These two sets of data are plotted, and the curves cross at an X axis value of four years. At this time, the cost of maintenance is equal to the value of the car, indicating this might be a desirable time to exchange it for a new one.

Another optimization graph is constructed in Fig. 33.26. The manufacturing cost per unit is reduced as more units are made, but the warehous-

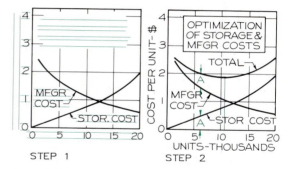

Figure 33.26 Optimization graphs.

Step 1 Lay out the graph, and plot the given curves.

Step 2 Add the two curves to find a third curve. Distance A is shown transferred to locate a point on the third curve. The lowest point of the "total" curve is the optimum point of 11,000 units.

ing cost increases. By adding the two curves, a third curve is found in Step 2. The "total" curve tells you the optimum number to manufacture at a time is about 11,000 units. When more or fewer units are manufactured, the expense per unit is greater.

Composite Graphs

The graph in Fig. 33.27 is a composite (or combination) of an area graph and a coordinate

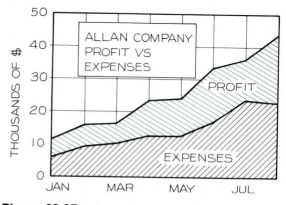

Figure 33.27 This graph is a combination of a coordinate graph and an area graph. The upper curve represents the total of two values plotted, one above the other.

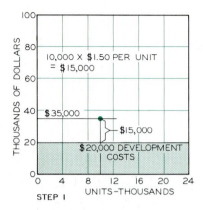

STEP 1

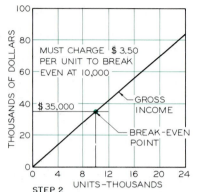

STEP 2

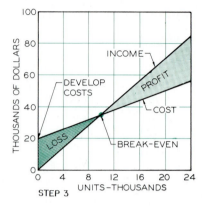

STEP 3

Figure 33.28 Break-even graph.

Step 1 The graph is drawn to show the cost ($20,000 in this case) of developing the product. Each unit would cost $1.50 to manufacture. This is a total investment of $35,000 for 10,000 units.

Step 2 To break even at 10,000, the units must be sold for $3.50 each. Draw a line from zero through the break-even point for $35,000.

Step 3 The loss is $20,000 at zero units and becomes progressively less until the break-even point is reached. The profit is the difference between the cost and income to the right of the break-even point.

graph. The lower curve is plotted first. The upper curve is found by adding the values to the lower curve so that the two areas represent the data. The upper curve is equal to the sum of the two Y values.

Break-Even Graphs

Break-even graphs help evaluate the marketing and manufacturing costs used to determine the selling cost of a product. The break-even graph in Fig. 33.28 reveals that 10,000 units must be sold at $3.50 each to cover the costs of manufacturing and development. Sales in excess of 10,000 result in profit.

A second type of break-even graph (Fig. 33.29) uses the cost of manufacturing per unit versus the number of units produced. In this example, the development costs must be incorporated into the unit costs. The manufacturer can determine how many units must be sold to break even at a given price or the price per unit if a given number is selected. In this example, a sales price of $0.80 requires that 8400 units be sold to break even.

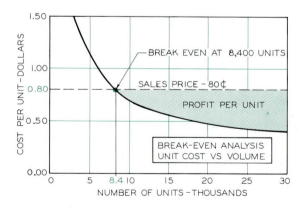

Figure 33.29 The break-even point can be found on a graph that shows the relationship between the cost per unit, which includes the development cost, and the number of units produced. The sales price is a fixed price. The break-even point is reached when 8400 units have been sold at $0.80 each.

33.6 Semilogarithmic Coordinate Graphs

Semilogarithmic graphs are called **ratio graphs** because they give graphical representations of ratios.

One scale, usually the vertical scale, is logarithmic, and the other is linear (divided into equal divisions). Parallel curves on a semilogarithmic graph have equal percentage increases.

Figure 33.30 shows the same data plotted on a linear grid and on a semilogarithmic grid. The semilogarithmic graph reveals that the percent of change from 0 to 5 is greater for curve B than for curve A since curve B is steeper. This comparison was not apparent in the plot on the linear grid.

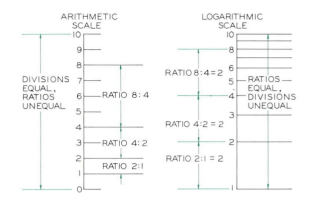

Figure 33.31 The spacings on an arithmetic scale are equal, with unequal ratios between points. The spacings on logarithmic scales are unequal, but equal spaces represent equal ratios.

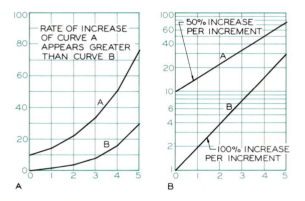

Figure 33.30 When plotted on a standard grid, curve A appears to be increasing at a greater rate than curve B. However, the true rate of change can be seen when the same data are plotted on a semilogarithmic graph in part B.

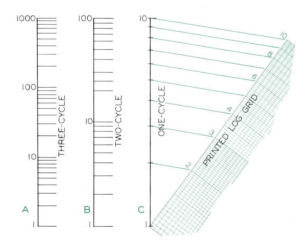

Figure 33.32 Logarithmic paper can be either bought or drawn using several cycles. (A) Three-, (B) two-, and (C) one-cycle scales are shown here. Calibrations can be drawn on a scale of any length by projecting from a printed scale, as shown in part C.

Figure 33.31 shows the relationship between the linear scale and the logarithmic scale. Equal divisions along the linear scale have unequal ratios, and equal divisions along the log scale have equal ratios.

Log scales can be drawn to have one or many cycles. Each cycle increases by a factor of 10. For example, the scale in Fig. 33.32A is a three-cycle scale, and the one in Fig. 33.32B is a two-cycle scale. When scales must be drawn to a special length, commercially printed log scales can be used to graphically transfer the calibrations to the scale being drawn (Fig. 33.32C).

In Fig. 33.33, the calibrations along the log scale are separated by the difference in their logarithms. The logarithms are laid off using a scale

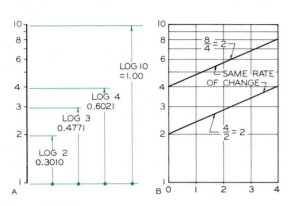

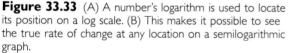

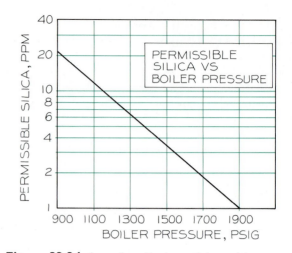

Figure 33.33 (A) A number's logarithm is used to locate its position on a log scale. (B) This makes it possible to see the true rate of change at any location on a semilogarithmic graph.

Figure 33.34 A semilogarithmic graph is used to compare the permissible silica (parts per million) in relation to the boiler pressure.

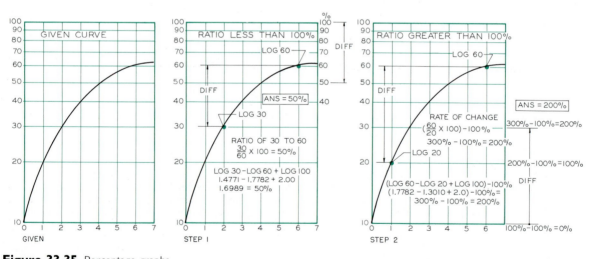

Figure 33.35 Percentage graphs.

Given The data are plotted on a semilogarithmic graph to enable you to determine percentages and ratios in much the same manner that you use a slide rule.

Step I In finding the percent that a smaller number is of a larger number, you know that the percent will be less than 100%. The log of 30 is subtracted from the log of 60 with dividers and is transferred to the percent scale at the right, where 30 is found to be 50% of 60.

Step 2 To find the percent of increase, a smaller number is divided into a larger number to give a value greater than 100%. The difference between the logs of 60 and 20 is found with dividers and is measured upward from 100% to find that the percent of increase is 200%.

calibrated in decimal divisions. It can be seen in Fig. 33.33B that parallel straight-line curves yield equal ratios of increase. Figure 33.34 is an example of a semilogarithmic graph used to present industrial data.

Semilog graphs are sometimes misunderstood by people who do not realize they are different from linear coordinate graphs; also, zero values cannot be shown on log scales.

Percentage Graphs

The percent that one number is of another, or the percent increase of one number that is greater than the other, can be determined by using a semilogarithmic graph (Fig. 33.35).

Data plotted in Step 1 are used to find the percent that 30 is of 60, two points on the curve. The vertical distance between them is equal to the difference of their logarithms. This distance is subtracted from the log of 100 at the right of the graph to give a value of 50% as a direct reading.

In Step 2, the percent of increase between two points is transferred from the grid to the lower end of the log scale and measured upward since the increase is greater than zero. These methods can be used to find percent increases or decreases of any set of points on the grid.

33.7 Polar Graphs

Polar graphs are drawn with a series of concentric circles with the origin at the center. Lines are drawn from the center toward the perimeter of the graph, where the data can be plotted through 360° by measuring values from the origin. For example, the illumination of a lamp is shown in Fig. 33.36, where the maximum lighting of the lamp is 550 lumens at 35° from the vertical.

This type of graph is used to plot the areas of illumination of all types of lighting fixtures and other applications. Polar graph paper is available commercially for drawing graphs of this type.

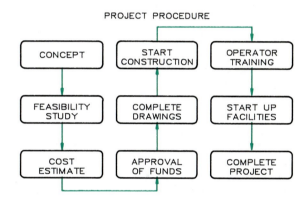

Figure 33.37 This schematic shows a block diagram of the steps required to complete a project.

33.8 Schematics

The **block diagram** in Fig. 33.37 shows the steps required to complete a construction project. Each step is blocked in and connected with arrows to explain the sequence of events.

The organization of a company or group of people can be depicted in an **organizational chart** like the one shown in Fig. 33.38. The offices represented by the blocks in the lower part of the graph are responsible to the offices represented by the blocks above them. The lines of authority connecting the blocks suggest the routes for communication from one office to another in an upward or downward direction.

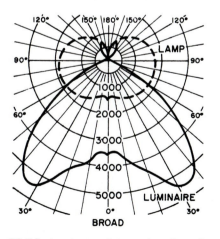

Figure 33.36 A polar graph is used to show the illumination characteristics of luminaires.

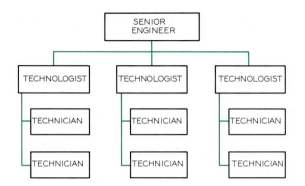

Figure 33.38 This schematic shows the organization of a design team in a block diagram.

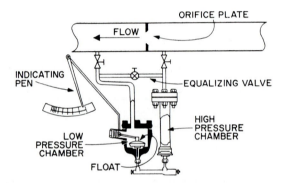

Figure 33.39 A schematic showing the components of a gauge that measures the flow in a pipeline. (Courtesy of Plant Engineering.)

The drawing in Fig. 33.39 is not a graph, nor is it a true view of the apparatus; instead, it is a **schematic** that effectively shows how the parts and their functions relate to one another.

Geographical graphs are used to combine maps and other relationships such as weather (Fig. 33.40). Different symbols represent the annual rainfall in various areas of the nation.

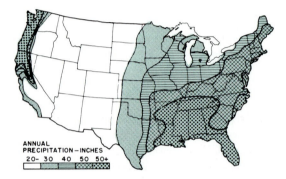

Figure 33.40 A map chart that shows the weather characteristics of various geographical areas. (Courtesy of the Structural Clay Products Institute.)

Problems

The problems below are to be drawn on Size A sheets in pencil or ink, as specified. Follow the techniques covered in this chapter and the examples given as you solve the problems.

Pie Graphs

1. Draw a pie graph that compares the employment of male youth between the ages of 16 and 21: Operators—25%; Craftsmen—9%; Professionals, technicians, and managers—6%; Clerical and sales—17%; Service—11%; Farm workers—13%; and Laborers—19%.

2. Draw a pie graph that shows the relationship between the following members of the technological team: Engineers—985,000; Technicians—932,000; Scientists—410,000.

3. Construct a pie graph of the following percentages of the employment status of graduates of two-year technician programs one year after graduation: Employed—63%, Continuing full-time study—23%; Considering job offers—6%; Military—6%; Other—2%.

4. Construct a pie graph that shows the relationship between the types of degrees held by engineers in aeronautical engineering: Bachelor—65%; Master—29%; Ph.D.—6%.

Bar Graphs

5. Draw a bar graph that depicts the unemployment rate of high school graduates and dropouts in various age categories given in the following table:

	Percent of Labor Force	
Ages	Graduates	Dropouts
16–17	18	22
18–19	12.5	17.5
20–21	8	13
22–24	5	9

6. Draw a single bar that represents 100% of a die casting alloy. The proportional parts of the alloy are as follows: Tin—16%;   Lead—24%;   Zinc—38.8%;   Aluminum—16.4%; Copper—4.8%.

7. Draw a bar graph that compares the number of skilled workers employed in various occupations. Arrange the graph for ease of interpretation and comparison of occupations. Use the following data: Carpenters—82,000;   All-round   machinists—310,000; Plumbers—350,000; Bricklayers—200,000; Appliance servicers—185,000; Automotive mechanics—760,000; Electricians—380,000; Painters—400,000.

8. Draw a bar graph that represents the flow of a river in cubic feet per second (cfs) as shown in the following table. Show bars that represent the data for ten days in the first month. Omit the second month.

Day of Month	Rate of Flow (in 100 cfs)	
	1st Month	2nd Month
1	19	19
2	130	70
3	228	79
4	129	33
5	65	19
6	32	14
7	17	15
8	13	11
9	22	19
10	32	27

9. Draw a bar graph that compares the corrosion resistance of the materials listed in the table below.

	Loss in Weight (%)	
	In Atmosphere	In Sea Water
Common steel	100	100
10% nickel steel	70	80
25% nickel steel	20	55

10. Draw a bar graph using the data in Problem 1.

11. Draw a bar graph using the data in Problem 2.

12. Draw a bar graph using the data in Problem 3.

13. Construct a bar grid graph to show the accident experience of Company A. Plot the numbers of disabling accidents per million person-hours of work on the Y-axis. Years will be plotted on the X-axis. Data: 1973—0.63;   1974—0.76;   1975—0.99;   1976—0.95· 1977—0.55;   1978—0.76;   1979—0.68;   1980—0.55, 1981—0.73;   1982—0.52;   1983—0.46;   1984—0.53; 1985—0.49;   1986—0.55.

Linear coordinate graphs

14. Using the data given in Table 33.1, draw a linear coordinate graph that compares the supply and demand of water in the United States from 1890 to 1990 in billions of gallons of water per day.

15. Present the data in Table 33.2 in a linear coordinate graph to decide which lamps should be selected to provide economical lighting for an industrial plant. The table gives the candlepower directly under the lamps (0°) and at various angles from the vertical when the lamps are mounted at a height of 25 ft.

16. Construct a linear coordinate graph that shows the relationship in energy costs (mills per kilowatt-hour) and the percent capacity of two types of power plants. Plot energy costs along the Y axis, and the capacity factor along the X axis. The plotted curve will compare the costs of a nuclear plant with a gas- or oil-fired plant. Gas-fired plant data: 17 mills, 10%; 12 mills, 20%; 8 mills, 40%; 7 mills, 60%; 6 mills, 80%; 5.8 mills, 100%. Nuclear-plant data: 24 mills, 10%; 14 mills, 20%; 7 mills, 40%; 5 mills, 60%; 4.2 mills, 80%; 3.7 mills, 100%.

17. Plot the data from Problem 13 as a linear coordinate graph.

Table 33.1											
	1890	1900	1910	1920	1930	1940	1950	1960	1970	1980	1990
Supply	80	90	110	135	155	240	270	315	380	450	460
Demand	35	35	60	80	110	125	200	320	410	550	570

Table 33.2										
Angle with vertical	0°	10°	20°	30°	40°	50°	60°	70°	80°	90°
Candlepower (thous.) 2–400W	37	34	25	12	5.5	2.5	2	0.5	0.5	0.5
Candlepower (thous.) 1–1000W	22	21	19	16	12.3	7	3	2	0.5	0.5

18. Construct a linear coordinate graph to show the relationship between the transverse resilience in inch-pounds (Y axis) and the single-blow impact in foot-pounds (X axis) of gray iron. Data: 21 fp, 375 ip; 22 fp, 350 ip; 23 fp, 380 ip; 30 fp, 400 ip; 32 fp, 420 ip; 33 fp, 410 ip; 38 fp, 510 ip; 45 fp, 615 ip; 50 fp, 585 ip; 60 fp, 785 ip; 70 fp, 900 ip; 75 fp, 920 ip.

19. Draw a linear coordinate graph to compare the two sets of data in the following table: capacity vs. diameter, and capacity vs. weight of a brine cooler. The horizontal scale is to be tons of capacity, and the vertical scales are to be outside diameter on the left and weight (cwt) on the right.

Tons Refrigerating Capacity	Outside Diameter (in)	Weight (cwt)
15	22	25
30	28	46
50	34	73
85	42	116
100	46	136
130	50	164
160	58	215
210	60	263

Use 20 × 20 graph paper $8\frac{1}{2}'' \times 11''$. Horizontal scale of $1'' = 40$ tons. Vertical scales of $1'' = 10''$ of outside DIA and $1'' = 40$ cwt (hundred weight).

20. Draw a linear coordinate graph that shows the voltage characteristics for a generator as given in the following table of values—abscissa-armature current in amperes (I_a); ordinate-terminal voltage in volts (E_t):

I_a	E_t	I_a	E_t	I_a	E_t
0	288	31.1	181.8	41.5	68
5.4	275	35.4	156	40.5	42.5
11.8	257	39.7	108	39.5	26.5
15.6	247	40.5	97	37.8	16
22.2	224.5	40.7	90	13.0	0
26.2	217	41.4	77.5		

21. Draw a linear coordinate graph for the centrifugal pump test data in the table below. The units along the X axis are to be gallons per minute. There will be four curves to represent the variables given.

Gallons per Minute	Discharge Pressure	Water HP	Electric HP	Efficiency (%)
0	19.0	0.00	1.36	0.0
75	17.5	0.72	2.25	32.0
115	15.0	1.00	2.54	39.4
154	10.0	1.00	2.74	36.5
185	5.0	0.74	2.80	26.5
200	3.0	0.63	2.83	22.2

22. Draw a linear coordinate graph that compares two of the values shown in Table 33.3—ultimate strength and elastic limit—with degrees of temperature labeled along the X axis.

Table 33.3

°F	Ultimate Strength	Elastic Limit	Elongation (%)	Reduction of Area (%)	Brinell Hardness No.
400	257,500	208,000	10.8	31.3	500
500	247,000	224,500	12.5	39.5	483
600	232,500	214,000	13.3	42.0	453
700	207,500	193,500	15.0	47.5	410
800	180,500	169,000	17.0	52.5	358
900	159,500	146,500	18.5	56.5	313
1000	142,500	128,500	20.3	59.2	285
1100	126,500	114,000	23.0	60.8	255
1200	114,500	96,500	26.3	67.8	230
1300	108,000	85,500	25.8	58.3	235

23. Draw a linear coordinate graph that compares two of the values shown in the table in Problem 22—percent of elongation and percent of reduction of area of the cross section—with the degrees of temperature that will be represented along the X axis.

Break-Even Graphs

24. Construct a break-even graph that shows the earnings for a new product that has a development cost of $12,000. The first 8000 will cost $0.50 each to manufacture, and you wish to break even at this quantity. What would be the profit at volumes of 20,000 and 25,000?

25. Same as Problem 24 except that the development costs are $80,000, the manufacturing cost of the first 10,000 is $2.30 each, and the desired break-even point is 10,000. What would be the profit at volumes of 20,000 and 30,000?

26. A manufacturer has incorporated the manufacturing and development costs into a cost-per-unit estimate. He wishes to sell the product at $1.50 each. On the Y axis, plot cost per unit in dollars; on the X axis, plot number of units in thousands. Data: 1000, $2.55; 2000, $2.01; 3000, $1.55; 4000, $1.20; 5000, $0.98; 6000, $0.81; 7000, $0.80; 8000, $0.75; 9000, $0.73; 10,000, $0.70. How many must be sold to break even? What will be the total profit when 9000 are sold?

27. The cost per unit to produce a product by a manufacturing plant is given below. Construct a break-even graph with the cost per unit plotted on the Y axis and the number of units on the X axis. Data: 1000, $5.90; 2000, $4.50; 3000, $3.80; 4000, $3.20; 5000, $2.85; 6000, $2.55; 7000, $2.30; 8000, $2.17; 9000, $2.00; 10,000, $0.95.

Logarithmic Graphs

28. Using the data given in Table 33.4, construct a logarithmic graph where the vibration amplitude (A) is plotted as the ordinate, and the vibration frequency (F) is plotted as the abscissa. The data for curve 1 represent the maximum limits of machinery in good condition with no danger from vibration. The data for curve 2 are the lower limits of machinery that is being

Table 33.4

F	100	200	500	1000	2000	5000	10,000
A(1)	0.0028	0.002	0.0015	0.001	0.0006	0.0003	0.00013
A(2)	0.06	0.05	0.04	0.03	0.018	0.005	0.001

vibrated excessively to the danger point. The vertical scale should be three cycles, and the horizontal scale should be two cycles.

29. Plot the data below on a two-cycle log graph to show the current in amperes (Y axis) versus the voltage in volts (X axis) of precision temperature-sensing resistors. Data: 1 volt, 1.9 amps; 2 volts, 4 amps; 4 volts, 8 amps; 8 volts, 17 amps; 10 volts, 20 amps; 20 volts, 30 amps; 40 volts, 36 amps; 80 volts, 31 amps; 100 volts, 30 amps.

30. Plot the data from Problem 14 as a logarithmic graph.

31. Plot the data from Problem 20 as a logarithmic graph.

Semilogarithmic graphs

32. Construct a semilogarithmic graph with the Y axis a two-cycle log scale from 1 to 100 and the X axis a linear scale from 1 to 7. Plot the data below to show the survivability of a shelter at varying distances from a one-megaton bomb exploding in air. The data consists of overpressure in psi along the Y axis, and distance from ground zero in miles along the X axis. The data points represent an 80% chance of survival of the shelter. Data: 1 mile, 55 psi; 2 miles, 11 psi; 3 miles, 4.5 psi; 4 miles, 2.5 psi; 5 miles, 2.0 psi; 6 miles, 1.3 psi.

33. The growth of two divisions of a company, Division A and Division B, is given in the data below. Plot the data on a rectilinear graph and on a semilog graph. The semilog graph should have a one-cycle log scale on the Y axis for sales in thousands of dollars, and a linear scale on the X axis showing years for a six-year period. Data in dollars: 1 yr, A = \$11,700 and B = \$44,000; 2 yr, A = \$19,500 and B = \$50,000; 3 yr, A = \$25,000 and B = \$55,000; 4 yr, A = \$32,000 and B = \$64,000; 5 yr, A = \$42,000 and B = \$66,000; 6 yr, A = \$48,000 and B = \$75,000. Which division has the better growth rate?

34. Draw a semilog chart showing probable engineering progress. Use the following indices: 40,000 B.C. = 21; 30,000 B.C. = 21.5; 20,000 B.C. = 22; 16,000 B.C. = 23; 10,000 B.C. = 27; 6000 B.C. = 34; 4000 B.C. = 39; 2000 B.C. = 49; 500 B.C. = 60; A.D. 1900 = 100. Horizontal scale 1″ = 10,000 years. Height of cycle = about 5″. Two-cycle printed paper may be used if available.

35. Plot the data from Problem 20 as a semilogarithmic graph.

36. Plot the data from Problem 22 as a semilogarithmic graph.

Percentage Graphs

37. Plot the data given in Problem 14 on a semilog graph to determine the percentages and ratios of the data. What is the percent of increase in the demand for water from 1890 to 1920? What percent of demand is the supply for 1900, 1930, and 1970?

38. Using the graph plotted in Problem 33, determine the percent of increase of Division A and Division B from Year 1 to Year 4. What percent of sales of Division A are the sales of Division B at the end of Year 2? At the end of Year 6?

39. Plot two values from Problem 22—water horsepower and electric horsepower—on semilog paper compared with gallons per minute along the X axis. What percent of electric horsepower is water horsepower when 75 gallons per minute are being pumped? What is the percent increase of the electric horsepower from 0 to 185 gallons per minute?

Organizational Charts

40. Draw an organizational chart for a city government organized as follows: The electorate elects school board, city council, and municipal court officers; the city council is responsible for the civil service commission, city manager, and city planning board; the city manager's duties cover finance, public safety, public works, health and welfare, and law.

41. Draw an organizational chart for a manufacturing plant. The sales manager, chief engineer, treasurer, and general manager are responsible to the president. The general manager has three department heads: master mechanic, plant superintendent, and purchasing agent. The plant superintendent has charge of the shop foremen, under whom are the working forces, and direct charge of the shipping, tool and die, inspection, order, and stores and supplies departments.

Polar Graphs

42. Construct a polar graph of the data given in Problem 15.

43. Construct a polar graph of the following illumination, in lumens at various angles, emitted from a luminaire. The zero-degrees position is vertically under the overhead lamp. Data: 0°, 12,000; 10°, 15,000; 20°, 10,000; 30°, 8000; 40°, 4200; 50°, 2500; 60°, 1000; 70°, 0. The illumination is symmetrical about the vertical.

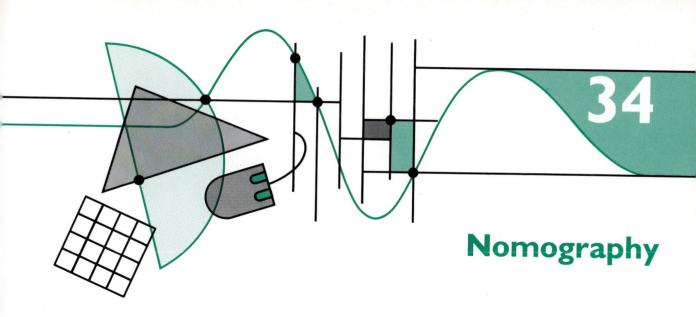

Nomography

34.1 Nomography

An additional aid in analyzing data is a graphical computer called a **nomogram** or **nomograph.** Basically, a nomogram, or "number chart," is any graphical arrangement of calibrated scales and lines that may be used to facilitate calculations, usually those of a repetitive nature.

The term "nomogram" frequently denotes a specific type of scale arrangement called an **alignment chart.** Typical examples of alignment charts are shown in Fig. 34.1. Many other types are also used that have curved scales or other scale arrangements for more complex problems. The discussion of nomograms in this chapter will be limited to the simpler conversion, parallel-scale, and N-type graphs and their variations.

Using an Alignment Graph

An alignment graph is usually constructed to solve for one or more unknowns in a formula or empirical relationship between two or more quantities; for example, it can be used to convert degrees Celsius to degrees Fahrenheit or to find the size of a structural member to sustain a certain load. An alignment chart is read by placing a straightedge, or by drawing a line called an **isopleth,** across the scales of the chart and reading corresponding values from the scale on this line. The example in Fig. 34.2 shows readings for the formula $U + V = W$.

34.2 Alignment-Graph Scales

To construct any alignment graph, you must first determine the graduations of the scales that will give the desired relationships. Alignment-graph scales are called **functional scales.** A functional scale is graduated according to values of some function of a variable. A functional scale for $F(U) = U^2$ is illustrated in Fig. 34.3. It can be seen in this example that if a value of $U = 2$ was substituted into the equation, the position of U on the functional scale would be 4 units from zero, or $2^2 = 4$. This procedure can be repeated with all values of U by substitution.

The Scale Modulus

Since the graduations on a functional scale are spaced in proportion to values of the function, a proportionality, or scaling factor, is needed. This

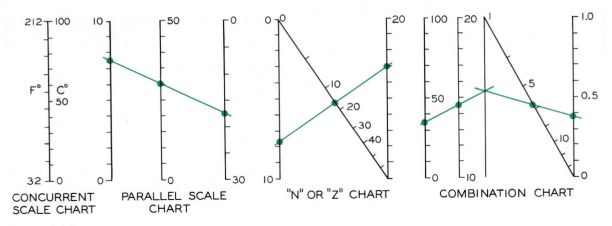

CONCURRENT SCALE CHART PARALLEL SCALE CHART "N" OR "Z" CHART COMBINATION CHART

Figure 34.1 Typical examples of types of alignment charts.

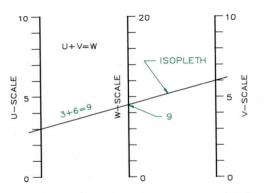

Figure 34.2 Use of an isopleth to solve graphically for unknowns in the given equation.

constant of proportionality is called the **scale modulus,** and it is given by the equation

$$m = \frac{L}{F(U_2) - F(U_1)}, \qquad (1)$$

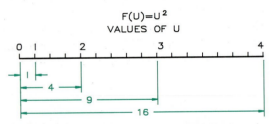

$F(U) = U^2$
VALUES OF U

Figure 34.3 Functional scale for units of measurement proportional to $F(U) = U^2$.

where

m = scale modulus, in inches per functional unit,

L = desired length of the scale, in inches,

$F(U_2)$ = functional value at the end of the scale,

$F(U_1)$ = functional value at the start of the scale,

For example, suppose we are to construct a functional scale for $F(U) = \sin U$ with $0° \leq U \leq 45°$ and a scale 6″ in length. Thus, L = 6″, $F(U_2) = \sin 45° = 0.707$, $F(U_1) = \sin 0° = 0$. Therefore Eq. (1) can be written in the following form by substitution:

$$m = \frac{6}{0.707 - 0} = 8.49 \text{ inches per (sine) unit.}$$

The Scale Equation

Graduation and calibration of a functional scale are made possible by a **scale equation.** The general form of this equation may be written as a variation of Eq. (1) in the following form:

$$X = m[F(U) - F(U_1)], \qquad (2)$$

where

X = distance from the measuring point of the scale to any graduation point.

Table 34.1									
U	0°	5°	10°	15°	20°	25°	30°	35°	40°	45°
X	0	0.74	1.47	2.19	2.90	3.58	4.24	4.86	5.45	6.00

m = scale modulus,

$F(U)$ = functional value at the graduation point,

$F(U_1)$ = functional value at the measuring point of the scale.

For example, a functional scale is constructed for the previous equation, $F(U) = \sin U$ $(0° \leq U \leq 45°)$. It has been determined that $m = 8.49$, $F(U) = \sin U$, and $F(U_1) = \sin 0° = 0$. Thus, by substitution (2) becomes

$$X = 8.49\,(\sin U - 0) = 8.49 \sin U.$$

Using the equation, we can substitute values of U and construct a table of positions. In this case, the scale is calibrated at 5° intervals, as reflected in Table 34.1.

The values of X from the table give the positions, in inches, for the corresponding graduations, measured from the start of the scale ($U = 0°$); see Fig. 34.4. Note that the measuring point does *not* need to be at one end of the scale, but it is usually the most convient point, especially if the functional value is zero at that point.

A graphical method of locating the functional values along a scale can be found as shown in Fig.

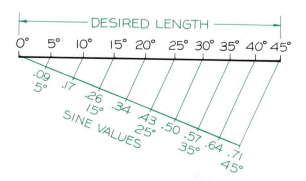

Figure 34.5 A functional scale that shows the sine of the angles from 0° to 45° can be drawn graphically by the proportional-line method. The scale is drawn to a desired length and the sine values of angles at 5° intervals are laid off along a construction line that passes through the 0° end of the scale. These values are projected back to the scale.

34.5 by the proportional-line method. The sine functions are measured off along a line at 5° intervals, with the end of the line passing through the 0° end of the scale. The functions are transferred from the inclined line with parallel lines back to the scale where the functions are represented and labeled.

34.3 Concurrent Scales

Concurrent scales are useful in the rapid conversion of one value into terms of a second system of measurement. Formulas of the type $F_1 = F_2$, which relate two variables, can be adapted to the concurrent-scale format. Typical examples might be the Fahrenheit-Celcius temperature relation,

$$°F = \frac{9}{5}\,°C + 32,$$

or the area of a circle,

$$A = \pi r^2$$

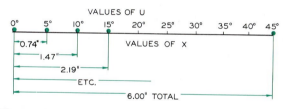

Figure 34.4 Construction of a functional scale using values from Table 34.1, which were derived from the scale equation.

Table 34.2

r	1	2	3	4	5	6	7	8	9	10
X_r	0	0.15	0.40	0.76	1.21	1.77	2.42	3.18	4.04	5.00

Design a concurrent-scale chart involves the construction of a functional scale for each side of the mathematical formula in such a manner that the **position** and **lengths** of each scale coincide. For example, to design a conversion chart 5 in. long that will give the areas of circles whose radii range from 1 to 10, we first write $F_1(A) = A$ $F_2(r) = \pi r^2$, and $r_1 = 1$, $r_2 = 10$. The scale modulus for r is

$$m_r = \frac{L}{F_2(r_2) - F_2(r_1)}$$

$$= \frac{5}{\pi(10)^2 - \pi(1)^2} = 0.0161.$$

Thus, the scale equation for r becomes

$$X_r = m_r [F_2(r) - F_2(r_1)]$$
$$= 0.0161[\pi r^2 - \pi(1)^2]$$
$$= 0.0161\pi(r^2 - 1)$$
$$= 0.0505(r^2 - 1).$$

A table of values for X_r and r may be completed as shown in Table 34.2. The r-scale can be drawn from this table, as shown in Fig. 34.6. From the original formula, $A = \pi r^2$, the limits of A are found to be $A_1 = \pi = 3.14$, and $A_2 = 100\pi = 314$. The scale modulus for concurrent scales is always the same for equal-length scales; therefore $m_A = m_r = 0.0161$, and the scale equation for A becomes

$$X_A = m_A[F_1(A) - F_1(A_1)]$$
$$= 0.0161 (A - 3.14).$$

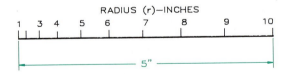

Figure 34.6   Calibration of one scale of a concurrent scale chart using values from Table 34.2.

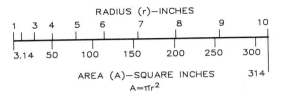

Figure 34.7   The completed concurrent scale chart for the formula $A = \pi r^2$. Values for the A-scale are taken from Table 34.3.

The corresponding table of values is then computed for selected values of A, as shown in Table 34.3.

The A-scale is now superimposed on the r-scale; its calibrations have been placed on the other side of the line to facilitate reading (Fig. 34.7). If it is desired to expand or contract one of the scales, an alternative arrangement may be used, as shown in Fig. 34.8. The two scales are drawn parallel at any convenient distance, and calibrated in **opposite** directions. A different scale modulus and corresponding scale equation must be calculated for each scale if they are *not* the same length.

Table 34.3

A	(3.14)	50	100	150	200	250	300	(314)
X_A	0	0.76	1.56	2.36	3.16	3.96	4.76	5.00

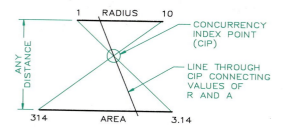

Figure 34.8 Concurrent scale chart with unequal scales.

A graphical method can be used to construct concurrent scales, as shown in Fig. 34.9, by using the proportional-line method. Since there are 101.6 mm in 4 inches, the units of millimeters can be located on the upper side of the inch scale by projecting to the scale with a series of parallel projectors.

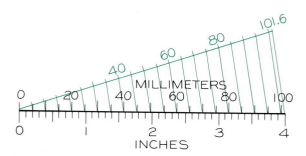

Figure 34.9 The proportional-line method can be used to construct an alignment graph that converts inches to millimeters. This requires that the units at each end of the scales be known. For example, there are 101.6 millimeters in 4 inches.

34.4 Construction of Alignment Graphs with Three Variables

For a formula of three functions (of one variable each), the general approach is to select the lengths and positions of **two** scales according to the range of variables and size of the chart desired. These are then calibrated by means of the scale equations, as shown in the preceding section. Although definite mathematical relationships exist that may be used to locate the third scale, graph-

ical constructions are simpler and usually less subject to error. Examples of the various forms are presented in the following sections.

34.5 Parallel-Scale Nomographs

Any formula of the type $F_1 + F_2 = F_3$ may be represented as a parallel-scale alignment chart, as shown in Fig. 34.10A. Note that all scales increase (functionally) in the same direction and that the function of the middle scale represents the **sum** of the other two. Reversing the direction of any scale changes the sign of its function in the formula, as for $F_1 - F_2 = F_3$ in Fig. 34.10B.

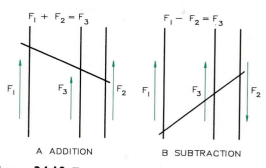

Figure 34.10 Two common forms of parallel-scale alignment nomographs.

To illustrate this type of alignment graph, we will use the formula $Z = X + Y$, as shown in Fig. 34.11. The outer scales for X and Y are drawn and calibrated. They can be drawn to any length and positioned any distance apart, as shown in Fig. 34.12. Two sets of data that yield a Z of 8 in Fig. 34.11 (Step 1) are used to locate the parallel Z-scale. The Z-scale is drawn and divided into 16 equal units (Step 2). The finished nomograph (Step 3) can be used to add various values of X and Y to find their sums along the Z-scale.

A more complex alignment graph is illustrated in Fig. 34.13, where the formula $U + 2V = 3W$ is expressed in the form of a nomograph. First it is necessary to determine and calibrate the two outer scales for U and V; we can

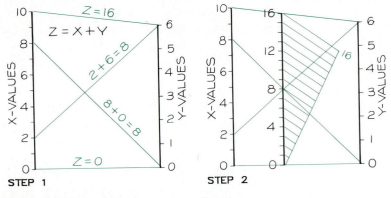

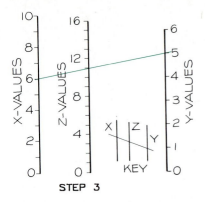

Figure 34.11 Parallel-scale nomogram (linear).

Step 1 Two parallel scales are drawn at any length and calibrated. The location of the parallel Z-scale is found by selecting two sets of values that will give the same value of Z, 8 in this example. The ends of the Z-scale will have values of 0 and 16, the sum of the end values of X and Y.

Step 2 The Z-scale is drawn through the point located in Step 1 parallel to the other scales. The scale is calibrated from 0 to 16 by using the proportional-line method. Note that the two sets of X- and Y-values cross at 8, the sum of each set.

Step 3 The Z-scale is calibrated and labeled. A key is drawn to show how the nomograph is used. If the Y-scale were calibrated with 0 at the upper end instead of the bottom, a different Z-scale could be computed and the nomograph could be used for $Z = X - Y$.

make them any convenient length and position them any convenient distance apart, as shown in Fig. 34.12. These scales are the basis for the construction shown in Fig. 34.13.

The limits of calibration for the middle scale are found by connecting the endpoints of the outer scales and substituting these values into the formula. Here, W is found to be 0 and 10 at the

extreme ends (Step 1). Two pairs of corresponding values of U and V are selected that will give the same value of W. For example, values of $U = 0$ and $V = 7.5$ give a value of 5 for W. We also find that $W = 5$ when $U = 14$ and $V = 0.5$. To verify this, we connect these corresponding pairs of values with isopleths to locate their intersection, which establishes the position of the W-scale.

Since the W-scale is linear ($3W$ is a linear function), it may be subdivided into uniform intervals by equal parts (Step 2). For a nonlinear scale, the scale modulus (and the scale equation) may be found in Step 2 by substituting length and its two end values into Eq. (1). The scales can be used to determine an infinite number of problem solutions (Step 3).

Parallel-Scale Graph with Logarithmic Scales

Problems involving formulas of the type $F_1 \times F_2 = F_3$ can be solved in a manner similar to the example given in Fig. 34.1 when logarithmic scales are used.

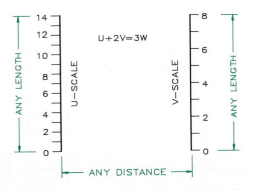

Figure 34.12 Calibration of the outer scales for the formula $U + 2V = 3W$, where $0 \leq U \leq 14$ and $0 \leq V \leq 8$.

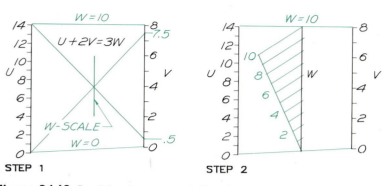
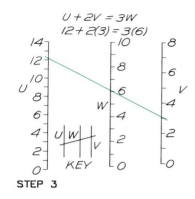

Figure 34.13 Parallel-scale nomograph (linear).

Step 1 Substitute the end values of the U- and V-scales into the formula to find the end values of the W-scale: $W = 10$ and $W = 0$. Select two sets of U and V that will give the same value of W. *Example:* When $U = 0$ and $V = 7.5$, W will equal 5, and when $U = 14$ and $V = 0.5$, W will equal 5. Connect these sets of values; the intersection locates the W-scale.

Step 2 Draw the W-scale parallel to the outer scales; its length is controlled by the previously established lines of $W = 10$ and $W = 0$. Since this scale is 10 linear divisions long, divide it graphically into ten units as shown. This will be a linear scale constructed as shown in Fig. 34.11.

Step 3 The nomogram can be used as illustrated by selecting any two known variables and connecting them with an isopleth to determine the third unknown. A key is included to illustrate how the nomogram is to be used. An example of $U = 12$ and $V = 3$ is shown to verify the accuracy of the graph.

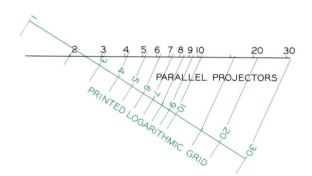

Figure 34.14 Graphical calibration of a scale using logarithmic paper.

The first step in drawing a nomograph with logarithmic scales is learning how to transfer logarithmic functions to the scale. The graphical method is shown in Fig. 34.14, where units along the scale are found by projecting from a printed logarithmic scale with parallel lines.

The formula $Z = XY$ is converted into a nomograph in Fig. 34.15. The X- and Y-scales

are drawn within the desired limits from 1 to 10 on each (Step 1). Sets of values of X and Y that yield the same value of Z, 10 in this case, are used to locate the Z-axis. The limits of the Z-axis are 1 and 100.

The Z-axis is drawn and calibrated as a two-cycle log scale (Step 2). A key is drawn to explain how an isopleth is used to add the logarithms of X and Y to give the log of Z (Step 3). When logarithms are added the result is multiplication. Had the Y-axis been calibrated in the opposite direction with 1 at the upper end and 10 at the lower end, a new Z-axis could have been calibrated, and the nomograph used for the formula, $Z = Y/X$, since it would be subtracting logarithms.

34.6 N- or Z-Graphs

Whenever F_2 and F_3 are linear functions, we can partially avoid using logarithmic scales for for-

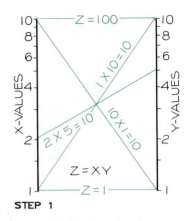

STEP 1

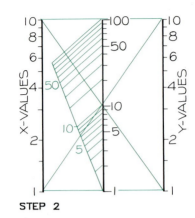

STEP 2

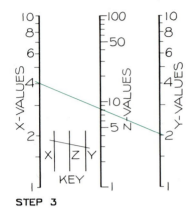

STEP 3

Figure 34.15 Parallel-scale nomogram (lograithmic).

Step 1 For the equation $Z = XY$, parallel log scales are drawn. Sets of X and Y points that yield the same value of Z, 10 in this example, are drawn. Their intersection locates the Z-scale with end values of 1 and 100.

Step 2 The Z-axis is graphically calibrated as a two-cycle logarithmic scale from 1 to 100. This scale is parallel to the X- and Y-scales.

Step 3 A key is drawn. An isopleth is drawn to show that $4 \times 2 = 8$. By reversing the Y-value scale from 1 downward to 10 and computing a different Z-scale, the nomogram could be used for the equation $Z = X/Y$.

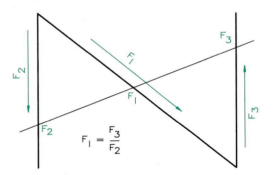

Figure 34.16 An N-graph for solving an equation of the form $F_1 = F_2/F_3$.

mulas of the type

$$F_1 = \frac{F_2}{F_3}.$$

Instead, we use an N-graph, as shown in Fig. 34.16. The outer scales, or "legs," of the N are functional scales and will therefore be linear if F_2 and F_3 are linear, whereas if the same formula were drawn as a parallel-scale graph, all scales would have to be logarithmic.

Some main features of the N-chart are

1. The outer scales are parallel functional scales of F_2 and F_3.
2. They increase (functionally) in **opposite** directions.
3. The diagonal scale connects the (functional) **zeros** of the outer scale.
4. In general, the diagonal scale is not a functional scale for the function F_1 and is nonlinear.

Construction of an N-graph is simplified because locating the middle (diagonal) scale is usually less of a problem than it is for a parallel-scale graph. Calibration of the diagonal scale is most easily accomplished by graphical methods.

The steps in constructing a basic N-graph of the equation $Z = Y/X$ are shown in Fig. 34.17. The diagonal is drawn to connect the zero ends of the scales (Step 1). Whole values are located along the diagonal by using combinations of X- and Y-values. It is important that the units located along the diagonal are whole values that are easy to interpolate between (Step 2). The diagonal is labeled, and a key is given explaining how

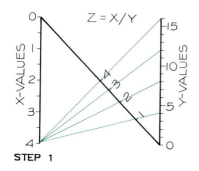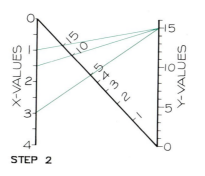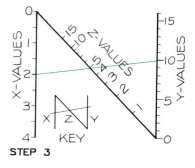

Figure 34.17 An N-chart nomograph.

Step 1 A Z-nomograph for the equation of $Z = Y/X$ can be drawn with two parallel scales. The zero ends of each scale are connected with a diagonal scale. Isopleths are drawn to locate units along the diagonal.

Step 2 Additional isopleths are drawn to locate other units along the diagonal. It is important that the units labeled on the diagonal be whole units that are easy to interpolate between.

Step 3 The diagonal scale is labeled and a key is drawn. An isopleth is drawn to show that $10/2 = 5$. The accuracy of the N-chart is greater at the 0 end of the diagonal. It approaches infinity at the other end.

to use the nomograph (Step 3). A sample isopleth is also given verifying the correctness of the graphical relationship between the scales.

A more advanced N-graph is constructed for the equation

$$A = \frac{B + 2}{C + 5},$$

where $0 \le B \le 8$ and $0 \le C \le 15$. This equation follows the form of

$$F_1 = \frac{F_2}{F_3},$$

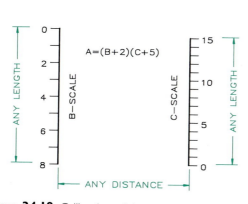

Figure 34.18 Calibration of the outer scales of an N-graph for the equation $A = (B + 2)(C + 5)$.

where $F_1(A) = A$, $F_2(B) = B + 2$, and $F_3(C) = C + 5$. Thus the outer scales will be for $B + 2$ and $C + 5$, and the diagonal scale will be for A.

The construction is begun in the same manner as for a parallel-scale graph by selecting the layout of the outer scales (Fig. 34.18). The limits of the diagonal scale are determined by connecting the endpoints on the outer scales, giving $A = 0.1$ for $B = 0$; $C = 15$ and $A = 2.0$ for $B = 8$; and $C = 0$, as shown in Fig. 34.19, which also gives the remainder of the construction.

The diagonal scale is located by finding the **function zeros** of the outer scales (i.e., the points where $B + 2 = 0$ or $B = -2$, and $C + 5 = 0$ or $C = -5$). The diagonal scale may then be drawn by connecting these points (Step 1). Calibration of the diagonal scale is most easily accomplished by substituting into the formula. Select the upper limit of an outer scale, for example, $B = 8$. This gives the formula

$$A = \frac{10}{C + 5}.$$

Solve this equation for the other outer scale variable,

$$C = \frac{10}{A} - 5.$$

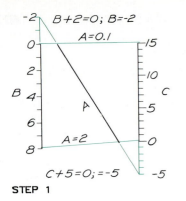

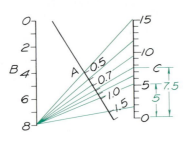

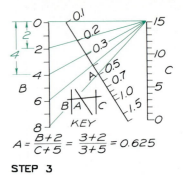

$$A = \frac{B+2}{C+5} = \frac{3+2}{3+5} = 0.625$$

STEP 1   STEP 2   STEP 3

Figure 34.19 Construction of an N-graph.

Step 1 Locate the diagonal scale by finding the functional zeros of the outer scales. This is done by setting $B + 2 = 0$ and $C + 5 = 0$, which gives a zero value for A.

Step 2 Select the upper limit of one of the outer scales, $B = 8$ in this case, and substitute it into the given equation to find a series of values of C for the desired values of A, as shown in Table 34.4. Draw isopleths from $B = 8$ to the values of C to calibrate the A-scale.

Step 3 Calibrate the remainder of the A-scale in the same manner by substituting $C = 15$ into the equation to determine a series of values on the B-scale for desired values on the A-scale, as listed in Table 34.5. Draw isopleths from $C = 15$ to calibrate the A-scale as shown.

Table 34.4							
A	2.0	1.5	1.0	0.9	0.8	0.7	0.6	0.5
C	0	1.67	5.0	6.11	7.50	9.28	11.7	15.0

Using this as a "scale equation," make a table of values for the desired values of A and corresponding values of C (up to the limit of C in the chart), as shown in Table 34.4. Connect isopleths from $B = 8$ to the tabulated values of C. Their intersections with the diagonal scale give the required calibrations for approximately half the diagonal scale (Step 2).

The remainder of the diagonal scale is calibrated by substituting the end value of the other outer scale ($C = 15$) into the formula, giving

$$A = \frac{B + 2}{20}.$$

Solving this for B yields

$$B = 20A - 2.$$

A table for the desired values of A can be constructed as shown in Table 34.5. Isopleths connecting $C = 15$ with the tabulated values of B will locate the remaining calibrations on the A-scale (Step 3).

Table 34.5					
A	0.5	0.4	0.3	0.2	0.1
B	8.0	6.0	4.0	2.0	0

Problems

The following problems are to be solved on Size A sheets using the principles covered in this chapter. Problems involving geometric construction and mathematical calculations should show the construction and calculations as part of the solutions. If the calculations are extensive, include them on a separate sheet.

Concurrent scales

Construct concurrent scales for converting the following relationships of one type of unit to another. The range of units for the scales is given for each relationship.

1. Kilometers and miles:

1.609 km = 1 mile, from 10 to 100 miles.

2. Liters and U.S. gallons:

1 liter = 0.2692 U.S. gallons, from 1 to 10 liters.

3. Knots and miles per hour:

1 knot = 1.15 miles per hour, from 0 to 45 knots.

4. Horsepower and British thermal units:

1 horsepower = 42.4 Btu, from 0 to 1200 hp.

5. Centigrade and Fahrenheit:

$$°F = \frac{9}{5}°C + 32, \text{ from } 32°F \text{ to } 212°F.$$

6. Radius and area of a circle:

Area = πr^2, from $r = 0$ to 10.

7. Inches and millimeters:

1 inch = 25.4 millimeters, from 0 to 5 inches.

8. Numbers and their logarithms:

(use logarithm tables), numbers from 1 to 10.

Addition and Subtraction Nomographs

Construct parallel-scale nomographs to solve the following addition and subtraction problems.

9. $A = B + C$, where $B = 0$ to 10, and $C = 0$ to 5.

10. $Z = X + Y$, where $X = 0$ to 8, and $Y = 0$ to 12.

11. $Z = Y - X$, where $X = 0$ to 6, and $Y = 0$ to 24.

12. $A = C - B$, where $C = 0$ to 30, and $B = 0$ to 6.

13. $W = 2V + U$, where $U = 0$ to 12, and $V = 0$ to 9.

14. $W = 3U + V$, where $U = 0$ to 10, and $V = 0$ to 10.

15. Electrical current at a circuit junction:

$$I = I_1 + I_2$$

where

I = current entering the junction in amperes,
I_1 = current leaving the junction, varying from 2 to 15 amps,
I_2 = current leaving junction, varying from 7 to 36 amps.

16. Pressure change in fluid flowing in a pipe:

$$\Delta P = P_2 - P_1,$$

where

ΔP = pressure change between two points in pounds per square inch,
P_1 = pressure upstream, varying from 3 psi to 12 psi,
P_2 = pressure downstream, varying from 10 psi to 15 psi.

Multiplication and Division— Parallel Scales

Construct parallel-scale nomographs with logarithmic scales that will perform the following multiplication and division operations.

17. Area of a rectangle: Area = Height × Width, where $H = 1$ to 10, and $W = 1$ to 12.

18. Area of a triangle: A = Base × Height/2, where $B = 1$ to 10, and $H = 1$ to 5.

19. Electrical potential between terminals of a conductor:

$$E = IR,$$

where

E = electrical potential in volts,
I = current, varying from 1 to 10 amperes,
R = resistance, varying from 5 to 30 ohms.

20. Pythagorean theorem:

$$C^2 = A^2 + B^2,$$

where

C = hypotenuse of a right triangle in centimeters,
A = one leg of the right triangle, varying from 5 to 50 cm,
B = second leg of the right triangle, varying from 20 to 80 cm.

21. Allowable pressure on a shaft bearing:

$$P = \frac{ZN}{100},$$

where

P = pressure in pounds per square inch,
Z = viscosity of lubricant from 15 to 50 cp (centipoises),
N = angular velocity of shaft from 10 to 1000 rpm.

22. Miles per gallon an automobile travels: mpg = miles/gallon. Miles vary from 1 to 500, and gallons vary from 1 to 24.

23. Cost per mile (cpm) of an automobile: cpm = cost/miles. Miles vary from 1 to 500, and cost varies from $1 to $28.

24. Angular velocity of a rotating body:

$$W = \frac{V}{R},$$

where

W = angular velocity, in radians per second,
V = peripheral velocity, varying from 1 to 100 meters per second,
R = radius, varying from 0.1 to 1 meter.

N-Graphs

Construct N-graphs that will solve the following equations.

25. Stress = P/A, where P varies from 0 to 1000 psi, and A varies from 0 to 15 square inches.

26. Volume of a cylinder:

$$V = \pi r^2 h,$$

where

V = volume in cubic inches,
r = radius, varying from 5 to 10 feet,
h = height, varying from 2 to 20 inches.

27. Same as Problem 17.

28. Same as Problem 18.

29. Same as Problem 19.

30. Same as Problem 20.

31. Same as Problem 21.

32. Same as Problem 22.

33. Same as Problem 23.

34. Same as Problem 24.

Empirical Equations and Calculus

35.1 Empirical Data

Data gathered from laboratory experiments and tests or from actual field tests are called **empirical data.** Often empirical data can be trans-

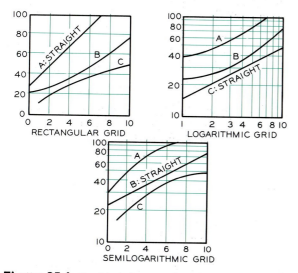

Figure 35.1 Empirical data are plotted on each of these types of grids to determine which will render a straight-line plot. If the data can be plotted as a straight line on one of these grids, their equation can be found.

formed to equation form by means of one of three types of equations to be covered in this chapter.

The analysis of empirical data begins by plotting the data on rectangular grids, logarithmic grids, or semilogarithmic grids. Curves are then drawn through each point to determine which of the grids renders a straight-line relationship (Fig. 35.1). When the data plots as a straight line, its equation may be determined.

35.2 Selection of Points on a Curve

Two methods of finding the equation of a curve are (1) the selected-points method and (2) the slope-intercept method. These are compared on a linear graph in Fig. 35.2.

Selected-Points Method

Two widely separated points, such as (1, 30) and (4, 60), can be selected on the curve. These points are substituted in the equation below.

$$\frac{Y - 30}{X - 1} = \frac{60 - 30}{4 - 1}$$

607

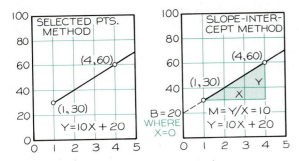

Figure 35.2 The equation of a straight line on a grid can be determined by selecting any two points on the line. The slope-intercept method requires that the intercept be found where $X = 0$ on a semilog grid. This requires the extension of the curve to the Y-axis.

The resulting data for the equation is

$$Y = 10X + 20.$$

Slope-Intercept Method

To apply the slope-intercept method, the intercept on the Y-axis where $X = 0$, must be known. If the X-axis is logarithmic, the log of $X = 1$ is 0, and the intercept must be found above the value of $X = 1$.

In Fig. 35.2, the data do not intercept the Y-axis; therefore the curve must be extended to find the intercept $B = 20$. The slope of the curve is found ($\Delta Y/\Delta X$), and it is substituted into the slope-intercept form to give the equation $Y = 10X + 20$.

The two methods illustrated here make the best use of the graphical process and are the most direct methods of introducing these concepts.

35.3 The Linear Equation: $Y = MX + B$

The curve fitting the experimental data plotted in Fig. 35.3 is a straight line; therefore these data are linear, meaning each measurement along the Y-axis is directly proportional to the X-axis units. We may use either the slope-intercept method or the selected-points method to find the equation for the data.

In the slope-intercept method, two known points are selected along the curve. The vertical and horizontal differences between the coordinates of each of these points are determined to establish the right triangle shown in Fig. 35.3A. In the slope-intercept equation, $Y = MX + B$, M is the tangent of the angle between the curve and the horizontal, B is the intercept of the curve with the Y-axis where $X = 0$, and X and Y are variables. In this example, $M = \frac{30}{5} = 6$, and the intercept is 20.

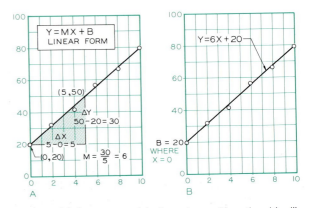

Figure 35.3 (A) A straight line on an arithmetic grid will have an equation in the form $Y = MX + B$. The slope, M, is found to be 6. (B) The intercept, B, is found to be 20. The equation is written as $Y = 6X + 20$.

If the curve had sloped downward to the right, the slope would have been negative. By substituting this information into the slope-intercept equation, we obtain $Y = 6X + 20$, from which we can determine values of Y by substituting any value of X into the equation.

The selected-points method could also have been used to arrive at the same equation if the intercept were not known. By selecting two widely separated points, such as (2, 32) and (10, 80), one can write the equation in this form:

$$\frac{Y - 32}{X - 2} = \frac{80 - 32}{10 - 2}, \qquad \therefore Y = 6X + 20,$$

which results in the same equation as was found by the slope-intercept method ($Y = MX + B$).

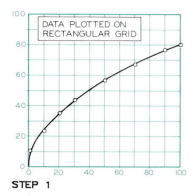
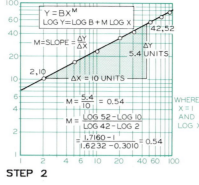
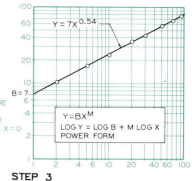

STEP 1   STEP 2   STEP 3

Figure 35.4 The power equation: $Y = BX^M$.

Given The data plotted on the rectangular grid give an approximation of a parabola. Since the data do not form a straight line on the rectangular grid, the equation will not be linear.

Step 1 The data forms a straight line on a logarithmic grid. The slope, M, can be found graphically with an engineer's scale, setting dX at 10 units and measuring the slope (dY) using the same scale.

Step 2 The intercept $B = 7$ is found where $X = 1$. The slope and intercept are substituted into the equation, which becomes $Y = 7X^{0.54}$.

35.4 The Power Equation: $Y = BX^M$

Data plotted on a logarithmic grid, as in Fig. 35.4, are found to form a straight line (Step 1). Therefore we express the data in the form of a power equation in which Y is a function of X raised to a given power, or $Y = BX^M$. The equation of the data is obtained by using the point where the curve intersects the Y-axis where $X = 0$, and letting M equal the slope of the curve. Two known points are selected on the curve to form the slope triangle. The engineers' scale can then be used, when the cycles along the X- and Y-axes are equal, to measure the slope between the coordinates of the two points.

If the horizontal distance of the right triangle is drawn to be 1 or a multiple of 10, the vertical distance can be read directly. The slope M (tangent of the angle) is found to be 0.54 (Step 2). The intercept B is 7; thus, the equation is $Y = 7X^{0.54}$, which can be evaluated for each value of X by converting this power equation into the logarithmic form of log Y:

$$\log Y = \log B + M \log X,$$
$$\log Y = \log 7 + 0.54 \log X.$$

When the slope-intercept method is used, the intercept is found on the Y-axis where $X = 1$. In Fig. 35.5, the Y-axis at the left of the graph has an X value of 0.1; thus, the intercept is located midway across the graph where $X = 1$. This is

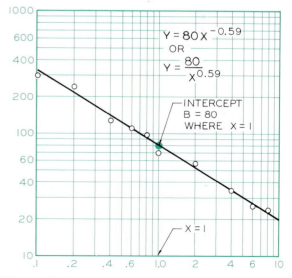

Figure 35.5 When the slope-intercept equation is used, the intercept can be found only where $X = 1$. Therefore in this example, the intercept is found at the middle of the graph.

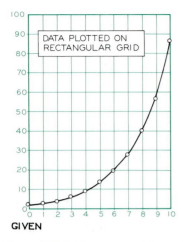

GIVEN

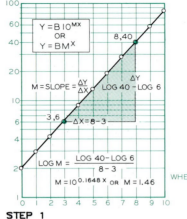

STEP 1

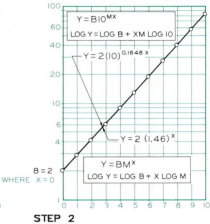

STEP 2

Figure 35.6 The exponential equation: $Y = BM^X$.

Given These data give a straight line on a semilogarithmic grid. Therefore the data fit the equation form, $Y = BM^X$.

Step 1 The slope must be found by mathematical calculations; it cannot be found graphically. Slopes may be written in either of the forms shown here.

Step 2 The intercept $B = 2$ is found where $X = 0$. Values, M, and B, are substituted into the equation to give $Y = 2(10)^{0.1648X}$ or $Y = 2(1.46)^X$.

analogous to the linear form of the equation, since the log of 1 is 0. The curve slopes downward to the right; thus the slope, M, is negative.

Base-10 logarithms are used in these examples, but natural logs could be used with e (2.718) as the base.

The value of M can be substituted in the equation in following manner:

$$Y = BM^X \qquad \text{or} \qquad Y = 2(1.46)^X,$$
$$Y = B(10)^{MX} \qquad \text{or} \qquad Y = 2(10)^{0.1648X},$$

where X is a variable that can be substituted into the equation to give an infinite number of values for Y. We can write this equation in its logarith-

35.5 The Exponential Equation: $Y = BM^X$

When data plotted on a semilogarithmic grid (Fig. 35.6, Step 1) approximate a straight line, we can write the equation $Y = BM^X$, where B is the Y-intercept of the curve, and M is the slope of the curve. To derive the equation, two points are selected along the curve so that a right triangle can be drawn to represent the differences between the coordinates of the points selected (Step 2). The slope of the curve is found to be

$$\log M = \frac{\log 40 - \log 6}{8 - 3} = 0.1648,$$

or

$$M = (10)^{0.1648} = 1.46.$$

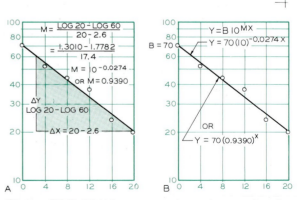

Figure 35.7 (A) When a curve slopes downward to the right, its slope is negative, as calculated here.
(B) Two forms of a final equation are shown here by substitution.

mic form, which enables us to solve it for the unknown value of Y for any given value of X. The equation can be written as

$$\log Y = \log B + X \log M,$$

or

$$\log Y = \log 2 + X \log 1.46.$$

To find the slope of a curve with a negative slope, the same methods are used. The curve of the data in Fig. 35.7 slopes downward to the right; therefore the slope is negative. Two points are selected to find the slope, M, which is the antilog of -0.0274. The intercept, 70, can be combined with the slope, M, to find the final equations, (Fig. 35.7B).

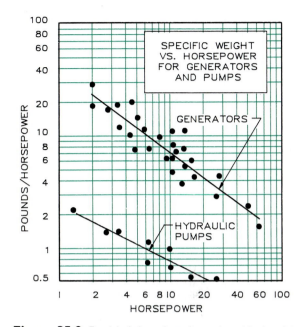

Figure 35.9 Empirical data plotted on a logarithmic grid, showing the specific weight versus horsepower of electric generators and hydraulic pumps. The curve is the average of points plotted.

equation

$$Y = MX + B.$$

Figure 35.9 is an example of how empirical data can be plotted to compare the specific weight (pounds per horsepower) of generators and hydraulic pumps versus horsepower. The weight of these units decreases linearly as the horsepower increases. Therefore these data can be written in the form of the power equation

$$Y = BX^{M}.$$

The half-life decay of radioactivity is plotted in Fig. 35.10 to show the relationship of decay to time. Since the half-life of different isotopes varies, different units would have to be assigned to time along the X-axis; however, the curve would be a straight line for all isotopes. The exponential form of the equation discussed in Section 35.5 can be applied to find the equation for these data in the form of

$$Y = BM^{X}.$$

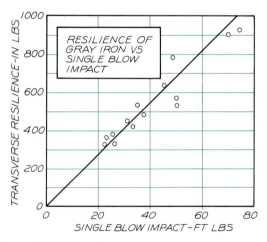

Figure 35.8 The relationship between the transverse strength of gray iron and impact resistance results in a straight line with an equation of the form $Y = MX + B$.

35.6 Applications of Empirical Graphs

Figure 35.8 is an example of how empirical data can be plotted to compare the transverse strength and impact resistance of gray iron. Although the data are scattered, the best curve is drawn. Since the curve is a straight line on a linear graph, the equation of these data can be found by the

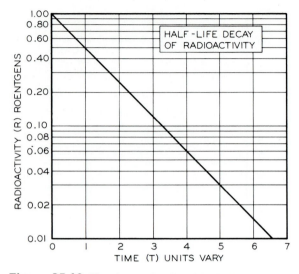

Figure 35.10 The decay of radioactivity is plotted as a straight line on this semilog graph, making it possible for its equation to be found in the form $Y = BM^X$.

area under a given curve, (the product of the two variables plotted on the X- and Y-axes). The area under a curve is approximated by dividing one of the variables into a number of very small intervals, which become small rectangular areas under the curve (Fig. 35.11B). The bars are extended so that as much of the square end of the bar is below the curve as above it and the average height of the bar is, near its midpoint.

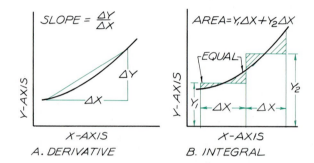

Figure 35.11 (A) The derivative of a curve is the change at any point that is the slope of the curve, Y/X. (B) The integral of a curve is the cumulative area enclosed by the curve, which is the summation of the products of the areas.

35.7 Introduction to Graphical Calculus

If the equation of the curve is known, calculus will solve the problem. However, many experimental data cannot be converted to standard equations. In these cases, it is desirable to use the graphical method of calculus, which provides solutions to irregular problems.

The two basic forms of calculus are (1) **differential calculus** and (2) **integral calculus.** Differential calculus is used to determine the rate of change of one variable with respect to another, as shown in Fig. 35.11A. The rate of change at any instant along the curve is the slope of a line tangent to the curve at that point. This slope can be approximated by constructing a chord at a given interval (Fig. 35.11A). The slope of this chord can be measured by finding the tangent of $\Delta Y/\Delta X$, which represents miles per hour, weight versus length, or a number of other rates important to the analysis of data.

Integral calculus is the reverse of differential calculus. Integration is the process of finding the

35.8 Graphical Differentiation

Graphical differentiation is the determination of the rate of change of two variables with respect to each other at any given point. Figure 35.12 illustrates the preliminary construction of the derivative scale that would be used to plot a continuous derivative.

The steps in completing the graphical differentiation are given in Fig. 35.13. The maximum slope of the data curve is estimated to be slightly less than 12. A scale is selected that will provide an ordinate that will accommodate the maximum slope. A line is drawn from point 12 on the ordinate axis of the derivative grid parallel to the known slope on the given curve grid and to the extension of the X-axis, to locate the pole point.

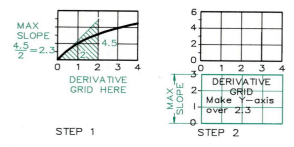

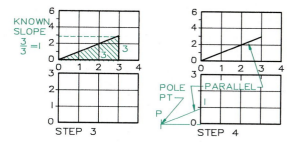

Figure 35.12 Scales for graphical differentiation.

Step 1 The maximum slope, 2.3, of the curve is found by constructing a line tangent to the curve where it is steepest.

Step 2 The derivative grid is drawn with a maximum ordinate of 3.0 to accommodate the slope of 2.3.

Step 3 A known slope of 1 is found on the given grid. The slope has no relationship to the curve.

Step 4 Draw a line from 1 on the Y-axis of the derivative grid parallel to the slope of the triangle drawn in the given grid. This line locates the pole point on the extension of the X-axis.

A series of chords is constructed on the given curve. The interval between 0 and 1, where the curve is sharpest, was divided in half to provide a more accurate plot. A smooth curve is constructed through the top of these bars so that the area above the horizontal top of the bar is the same as that below it. The rate of change, $\Delta Y / \Delta X$, can be found at any interval of the variable X by reading directly from the derivative graph at the value of X in question.

35.9 Applications of Graphical Differentiation

The mechanical handling shuttle shown in Fig. 35.14 is used to convert rotational motion into linear motion.

The linkage is drawn to show the end positions of point P, which will be used as the zero point for plotting the travel versus the degrees of revolution. Since rotation is constant at one revolution per three seconds, the degrees of revolution can be converted to time, as shown in the data curve at the top of Fig. 35.15. The drive crank, R_1, is revolved at 30° intervals, and the distance that point P travels from its end position is plotted on the graph to give the distance-versus-time relationship.

We determine the ordinate scale of the derivative grid by estimating the maximum slope of the given data curve, which is a little less than 100 in./sec. A convenient scale is chosen that will have a maximum limit of 100 units. A slope of 40 is drawn on the given data curve to pole P in the derivative grid. From point 40 on the derivative ordinate scale, we draw a line parallel to the known slope, which is found on the given grid. Point P is the point where this line intersects the extension of the X-axis.

A series of chords are drawn on the given curve to approximate the slope at various points. Lines are constructed through point P of the derivative scale parallel to the chord lines and extended to the ordinate scale. These points are then projected across to their respective intervals to form vertical bars. A smooth curve is drawn through the tops of the bars to give an average of the bars. This curve can be used to find the velocity of the shuttle in inches per second at any time interval.

The construction of the second derivative curve, the acceleration, is similar to that of the first derivative. We estimate the maximum slope to be 200 in./sec/sec by inspecting the first derivative. An easily measured scale is established for the ordinate scale of the second derivative curve. Point P is found in the same manner.

Figure 35.13 Graphical differentiation.

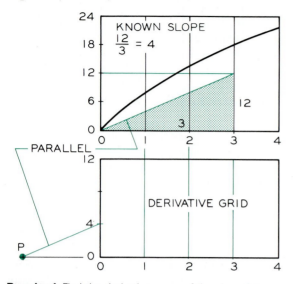

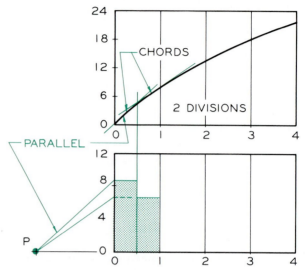

Required Find the derivative curve of the given data.

Step 1 Find the derivative grid and the pole point using the construction illustrated in Fig. 35.12.

Step 2 Construct chords between intervals on the given curve and draw lines parallel to them through point P on the derivative grid. These lines locate the heights of bars in their respective intervals. The first interval is divided in half since the curve is changing sharply in this interval.

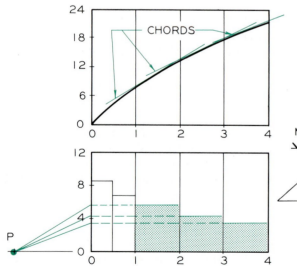

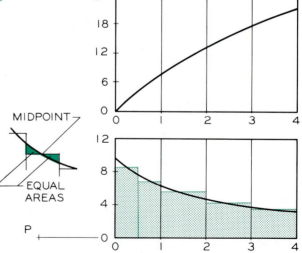

Step 3 Additional chords are drawn in the last two intervals. Lines parallel to these chords are drawn from the pole point to the Y-axis to find additional bars in their respective intervals.

Step 4 The vertical bars represent the slopes of the curve at different intervals. The derivative curve is drawn through the midpoints of the bars so that the areas under and above the bars are approximately equal.

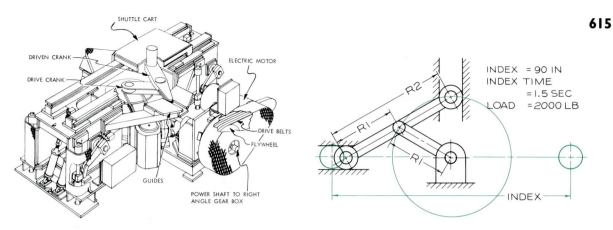

Figure 35.14 A pictorial and scale drawing of an electrically powered mechanical handling shuttle used to move automobile parts on an assembly line. (Courtesy of General Motors Corp.)

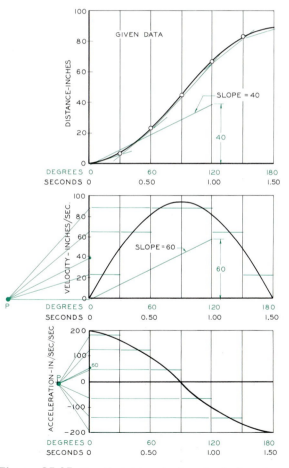

Figure 35.15 Graphical determination of velocity and acceleration of the mechanical handling shuttle by differential calculus.

Chords are drawn at intervals on the first derivative curve. Lines are drawn parallel to these chords from point P in the second derivative curve to the Y-axis, where they are projected horizontally to their respective intervals to form a series of bars. A smooth curve is drawn through the tops of the bars to give a close approximation of the average areas of the bars. A minus scale is given to indicate deceleration.

From the velocity and acceleration plots, it can be seen that the parts being handled by the shuttle are accelerated at a rapid rate until the maximum velocity is attained at 90°, at which time deceleration begins and continues until the parts come to rest.

35.10 Graphical Integration

Integration is the process of determining the area (product of two variables) under a given curve. For example, if the Y-axis were pounds and the X-axis were feet, the integral curve would give the product of the variables, foot-pounds, at any interval of feet along the X-axis. Figure 35.16 depicts the method of constructing scales for graphical integration.

This technique is used in Fig. 35.17 to integrate the equation of the given curve, $Y = 2X^2$. The total area under the curve can be estimated to be less than 40 units. This value becomes the

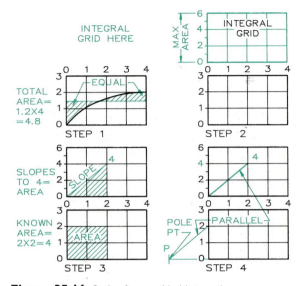

Figure 35.16 Scales for graphical integration.

Step 1 To determine the maximum value on the Y-axis, a line is drawn to approximate the area under the given curve, 4.8 in this case.

Step 2 The integral is drawn with a Y-axis of 6 as the maximum value to represent the maximum area.

Step 3 A known area of 4 is found on the given grid. A slope from 0 to 4 is drawn on the integral grid directly above the known area to establish the integral for this model.

Step 4 A line is drawn from 2 on the Y-axis of the given grid parallel to the slope line in the integral grid. The pole point is located on the extension of the X-axis.

maximum height of the Y-axis on the integral curve. A convenient scale is selected for the ordinate, and the pole point, P, is found.

A series of vertical bars is constructed to approximate the areas under the curve at these intervals. The narrower the bars, the more accurate will be the resulting calculations. The interval between 1 and 2 was divided in half to provide a more accurate plot. The top lines of the bars are extended horizontally to the Y-axis, where the points are then connected by lines to point P. Lines are drawn parallel to AP, BP, CP, DP, and EP in the integral grid to correspond to the respective intervals in the given grid. The intersection points of the chords are connected by a smooth curve—the integral curve to give the cu-

mulative product of the X- and Y-variables along the X-axis. The area under the curve at $X = 3$ can be read as 18.

35.11 Applications of Graphical Integration

An example is shown in Fig. 35.18, where a truck exerts a total force of 36,000 lb on a beam of a bridge. From the load diagram shown in Fig. 35.19 we can, by integration, find the shear diagram, which indicates the points in the beam where failure due to crushing is most critical. In the shear diagram, the left-end resultant of 15.9 kips is drawn to scale from the axis. The first load of 4 kips acting in a downward direction is subtracted from this value directly over its point of application. The second load of 16 kips exerts a downward force and is subtracted from the 11.9 kips (15.9 − 4). The third load of 16 kips is also subtracted, and the right-end resultant will bring the shear diagram back to the X-axis.

The moment diagram is used to evaluate the bending characteristics of the applied loads in foot-pounds at any interval along the beam. The ordinate of any X-value in the moment diagram must represent the cumulative foot-pounds in the shear diagram as measured from either end of the beam.

Pole point P is located in the shear diagram by applying the method described in Fig. 35.19. A rectangular area of 200 ft-kips is found in the shear diagram. We estimate the total area to be less than 600 ft-kips, so we select a scale that will allow an ordinate scale of 600 units for the moment diagram. We draw a known area of 200 (10 × 20) on the shear diagram. A diagonal line in the moment diagram is drawn that slopes upward from 0 to 200, where $X = 20$. The diagonal, 0K, is transferred to the shear diagram, where it is drawn from the ordinate of the given rectangle to point P on the extension of the X-axis. Rays AP, BP, CP, and DP are found in the shear diagram by projecting horizontally from the various values of shear. In the moment diagram, these rays are then drawn in their re-

Figure 35.17 Graphical integration.

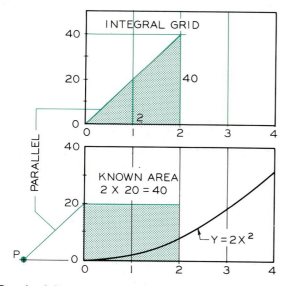

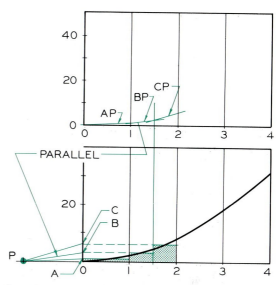

Required Plot the integral curve of the given data.

Step 1 Find the pole point, *P*, using the technique illustrated in Fig. 35.16.

Step 2 Construct bars to approximate the areas under the curve. The interval from 1 to 2 was divided in half to improve the approximation. The heights of the bars are projected to the *Y*-axis, and lines are drawn to the pole point. Sloping lines *AP*, *BP*, and *CP* are drawn in their respective intervals parallel to the lines drawn to the pole, *P*.

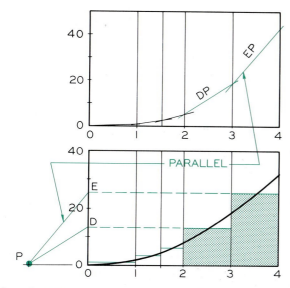

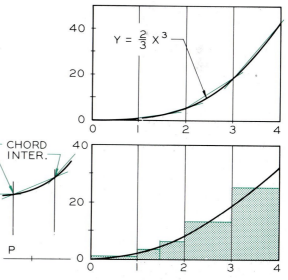

Step 3 Additional bars are drawn from 2 to 4 on the *X*-axis. The heights of the bars are projected to the *Y*-axis, and rays are drawn to the pole point, *P*. Lines *DP* and *DE* are drawn in their respective intervals and parallel to their rays in the integral grid.

Step 4 The straight lines connected in the integral grid represent chords of the integral curve. Construct the integral curve to pass through the points where the chords intersect. Ordinate value on the integral curve represents the cumulative area under the given curve from zero to that point on the *X*-axis.

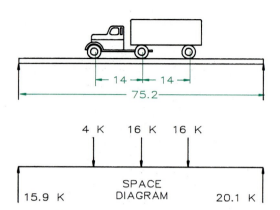

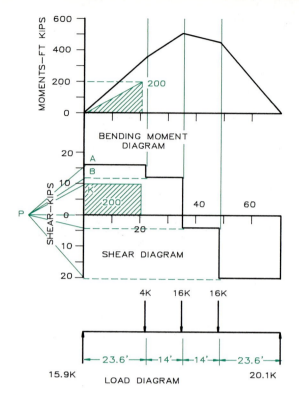

Figure 35.18 The forces on a beam of a bridge loaded with a truck.

Figure 35.19 Determination of shear and bending moment by graphical integration.

spective intervals to form a straight-line curve that represents the cumulative area of the shear diagram, in units of ft-kips. The maximum bending is scaled to be about 560 ft-kips at the center of the beam. This beam must be capable of withstanding a shear of 20.1 kips and a bending moment of 560 ft-kips.

Problems

The following problems are to be solved on Size A sheets. Solutions involving mathematical calculations should show these calculations on separate sheets if space is not available on the sheet where the graphical solution is drawn.

Empirical Equations—Logarithmic

1. Find the equation of the data shown in the following table. The empirical data compare input voltage (*V*) with input current amperes (*I*) to a heat pump.

Y-axis	V	0.8	1.3	1.75	1.85
Y-axis	I	20	30	40	45

2. Find the equation of the data in the following table. The empirical data give the relationship between peak allowable current in amperes (*I*) with the overload operating time in cycles at 60 cycles per second (*C*). Place *I* on the Y-axis.

Y-axis	I	2000	1840	1640	1480	1300	1200	1000
X-axis	C	1	2	5	10	20	50	100

3. Find the equation of the data in the following table. The empirical data for a low-voltage circuit breaker used on a welding machine give the maximum loading during weld in amperes (rms) for the percent of duty (pdc). Place rms along the Y-axis and pdc along the X-axis.

Y-axis	rms	7500	5200	4400	3400	2300	1700
X-axis	pdc	3	6	9	15	30	60

4. Construct a three-cycle × three-cycle logarithmic graph to find the equation of a machine's vibration displacement in mills along the *Y*-axis and vibration frequency in cycles per minute (cpm) along the *X*-axis. Data: 100 cpm, 0.80 mills; 400 cpm, 0.22 mills; 1000 cpm, 0.09 mills; 10,000 cpm, 0.009 mills; 50,000 cpm, 0.0017 mills.

5. Find the equation of the data in the following table that compares the velocities of air moving over a plane surface in feet per second *(v)* at different heights in inches *(y)* above the surface. Plot *Y*- values on the *Y*-axis.

Y	0.1	0.2	0.3	0.4	0.6	0.8	1.2	1.6	2.4	3.2
V	18.8	21.0	22.6	24.1	26.0	27.3	29.2	30.6	32.4	33.7

6. Find the equation of the data in the following table that shows the distance traveled in feet *(s)* at various times in seconds *(t)* of a test vehicle. Plot *s* and the *Y*-axis and then *t* on the *X*-axis.

t	1	2	3	4	5	6
s	15.8	63.3	146.0	264.0	420.0	580.0

Empirical Equations—Linear

7. Construct a linear graph to determine the equation for the yearly cost of a compressor in relationship to the compressor's size in horsepower. The yearly cost should be plotted on the *Y*-axis and the compressor's size in horsepower on the *X*-axis. Data: 0 hp, $0; 50 hp, $2100; 100 hp, $4500; 150 hp, $6700; 200 hp, $9000; 250 hp, $11,400. What is the equation of these data?

8. Construct a linear graph to determine the equation for the cost of soil investigation by boring to determine the proper foundation design for varying sizes of buildings. Plot the cost of borings in dollars along the *Y*-axis and the building area in sq ft along the *X*-axis. Data 0 sq ft, $0; 25,000 sq ft, $35,000; 50,000 sq ft, $70,000; 750,000 sq ft, $100,000; 1,000,000 sq ft, $130,000.

9. Find the equation of the empirical data plotted in Fig. 35.8.

10. Plot the data in the table below on a linear graph and determine its equation. The empirical data show the deflection in centimeters of a spring *(d)* when it is loaded with different weights in kilograms *(w)*. Plot *w* along the *X*-axis and *d* along the *Y*-axis.

w	0	1	2	3	4	5
d	0.45	1.10	1.45	2.03	2.38	3.09

11. Plot the data in the table below on a linear graph and determine its equation. The empirical data show the temperatures read from a Fahrenheit thermometer at B and a centigrade thermometer at A. Plot the *A*-values along the *X*-axis and the *B*-values along the *Y*-axis.

°A	−6.8	6.0	16.0	32.2	52.0	76.0
°B	20.0	43.0	60.8	90.0	125.8	169.0

Empirical Equations—Semilogarithmic

12. Construct a semilog graph of the following data to determine their equation. The *Y*-axis should be a two-cycle log scale and the *X*-axis a 10-unit linear scale. Plot the voltage *(E)* along the *Y*-axis and time *(T)* in sixteenths of a second along the *X*-axis to represent resistor voltage during capacitor charging. Data: 0 sec., 10 volts; 2 sec., 6 volts; 4 sec., 3.6 volts; 6 sec., 2.2 volts; 8 sec., 1.4 volts; 10 sec., 0.8 volts.

13. Find the equation of the data plotted in Fig. 35.10.

14. Construct a semilog graph of the following data to determine their equation. The *Y*-axis should be a three-cycle log scale and the *X*-axis a linear scale from 0 to 250. These data give a comparison of the reduction factor, *R* (*Y*-axis), with the mass thickness per square foot (*X*-axis), of a nuclear protection barrier. Data: 0 MT, 1.0*R*; 100 MT, 0.9*R*; 150 MT, 0.028*R*; 200 MT, 0.009*R*; 300 MT, 0.0011*R*.

15. An engineering firm is considering its expansion by reviewing its past sales as shown in the table below. Their years of operation are represented by *x,* and *N* is their annual income in tens of thousands. Plot *x* along the *X*-axis and *N* along the *Y*-axis and determine the equation of their progress.

x	1	2	3	4	5	6	7	8	9	10
N	0.05	0.08	0.12	0.20	0.32	0.51	0.80	1.30	2.05	3.25

Empirical Equations—General Types

16–21. Plot the experimental data shown in Table 35.1 on the grid where the data will appear as straight-line curves. Determine the equations of the data.

A	X	0	40	80	120	160	200	240	280	
	Y	4.0	7.0	9.8	12.5	15.3	17.2	21.0	24.0	
B	X	1	2	5	10	20	50	100	200	500	1000
	Y	1.5	2.4	3.3	6.0	9.0	15.0	23.0	24.0	60.0	85.0
C	X	1	5	10	50	100	500	1000			
	Y	3	10	19	70	110	400	700			
D	X	2	4	6	8	10	12	14			
	Y	6.5	14.0	32.0	75.0	115.0	320.0	710.0			
E	X	0	2	4	6	8	10	12	14		
	Y	20	34	53	96	115	270	430	730		
F	X	0	1	2	3	4	5	6	7	8	9	10
	Y	1.8	2.1	2.2	2.5	2.7	3.0	3.4	3.7	4.1	4.5	5.0

Table 35.1

Calculus—Differentiation

22. Plot the equation $Y = X^3/6$ as a rectangular graph. Graphically differentiate the curve to determine the first and second derivatives.

23. Plot the following data on a graph, and find the derivative curve of the data on a graph placed below the first: $Y = 2X^2$.

24. Plot the following equation on a graph, and find the derivative curve of the data on a graph placed below the first: $4Y = 8 - X^2$.

25. Plot the following data on a graph, and find the derivative curve of the data on a graph placed below the first: $3Y = X^2 + 16$.

26. Plot the following data on a graph, and find the derivative curve of the data on a graph placed below the first: $X = 3Y^2 - 5$.

Calculus—Integration

27. Plot the following equation on a graph, and find the integral curve of the data on a graph placed above the first: $Y = X^2$.

28. Plot the following equation on a graph, and find the integral curve of the data on a graph placed above the first: $Y = 9 - X^2$.

29. Plot the following equation on a graph, and find the integral curve of the data on a graph placed above the first: $Y = X$.

30. A plot plan shows that a tract of land is bounded by a lake front (Fig. 35.20). By graphical integration, determine a graph that will represent the cumulative area of the land from points A to E. What is the total area? What is the area of each lot?

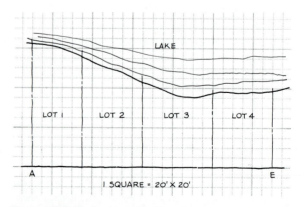

Figure 35.20 Problem 30. Plot plan of a tract bounded by a lake front.

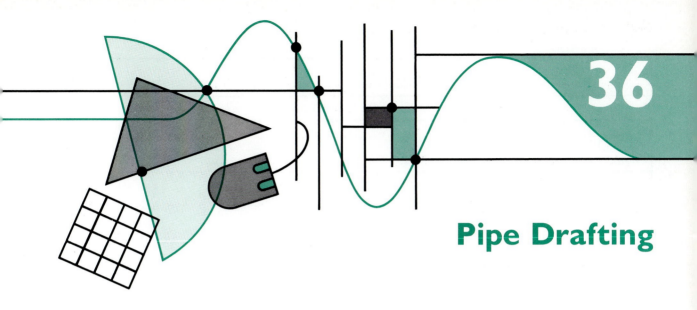

36

Pipe Drafting

36.1 Introduction

To understand pipe drafting, you first must become familiar with the types of pipe available. Commonly used types of pipe are (1) steel pipe, (2) cast-iron pipe, (3) copper, brass, and bronze pipe and tubing, and (4) plastic pipe.

To ensure uniformity of size and strength of interchangeable components, the standards for the grades and weights of pipe and pipe fittings are specified by several organizations, including the American National Standards Institute (ANSI), the American Society for Testing Materials (ASTM), the American Petroleum Institute (API), and the Manufacturers Standardization Society (MSS).

36.2 Welded and Seamless Steel Pipe

Traditionally, steel pipe has been specified in three weights: **standard** (STD), **extra strong** (XS), and **double extra strong** (XXS). These designations and their specifications are listed in the ANSI B 36.10–1979 standards. However, additional pipe designations, called **schedules,** have

been introduced to provide the pipe designer with a wider selection of pipe.

The ten schedules are Schedule 10, Schedule 20, Schedule 30, Schedule 40, Schedule 60, Schedule 80, Schedule 100, Schedule 120, Schedule 140, and Schedule 160. The wall thicknesses of the pipes vary from the thinnest, in Schedule 10, to the thickest, in Schedule 160. The outside diameters are of a constant size for pipes of the same nominal size in all schedules.

Schedule designations correspond to STD, XS, and XXS specifications in some cases, as Table 36.1 partly shows. This table has been abbreviated from the ANSI B 36.10–1979 tables by omitting several pipe sizes and schedules. The most often used schedules are 40, 80, and 120.

Pipes from the smallest size up to and including 12-in. pipes are specified by their inside diameter (ID), which means the outside diameter (OD) is larger than the specified size. The inside diameters are the same size as the nominal sizes of the pipe for STD weight pipe. For XS and XXS pipe, the size of the inside diameters is slightly different from the nominal size. Beginning with the 14-in. diameter pipes, the nominal sizes represent the outside diameters of the pipe.

Table 36.1
Dimensions and weights of welded and seamless steel pipe (ANSI B36.10–1979)★

Inch Nominal Size	OD (in.)	Wall Thk. (in.)	Weight (lb/ft)	STD† XS XXS	Sch. No.	OD (mm)	Wall Thk. (mm)	Weight (kg/m)
½	0.84	0.11	0.85	STD	40	21.3	2.8	1.3
1	1.32	0.13	1.68	STD	40	33.4	3.4	2.5
1	1.3	0.18	2.17	XS	80	33.4	4.6	3.2
1	1.3	0.36	3.66	XXS		33.4	9.1	5.5
2	2.38	0.22	3.65	STD	40	60.3	3.9	5.4
2	2.38	0.22	5.02	XS	80	60.3	5.5	7.5
2	2.38	0.44	9.03	XXS		60.3	11.1	13.4
4	4.50	0.23	10.79	STD	40	114.3	6.0	16.1
4	4.50	0.34	14.98	XS	80	114.3	8.6	42.6
4	4.50	0.67	27.54	XXS		114.3	17.1	41.0
8	8.63	0.32	28.55	STD	40	219.1	8.2	42.6
8	8.63	0.50	43.39	XS	80	219.1	12.7	64.6
8	8.63	0.88	74.40	XXS		219.1	22.2	107.9
12	12.75	0.38	49.56	STD		323.0	9.5	67.9
12	12.75	0.50	65.42	XS	120	323.0	12.7	97.5
12	12.75	1.00	125.4	XXS	30	133.9	25.4	187.0
14	14.00‡	0.38	54.57	STD		355.6	9.5	87.3
14	14.00	0.50	72.08	XS		355.6	12.7	107.4
18	18.00	0.38	70.59	STD		457	9.5	106.2
18	18.00	0.50	93.45	XS	20	457	12.7	139.2
24	24.00	0.38	94.62	STD		610	9.5	141.1
24	24.00	0.50	125.49	XS		610	12.7	187.1
30	30.00	0.38	118.65	STD	20	762	9.5	176.8
30	30.00	0.50	157.53	XS		762	12.7	234.7
40	40.00	0.38	158.70	STD		1016	9.5	236.5
40	40.00	0.50	210.90	XS		1016	12.7	314.2

★This table has been compressed by omitting many of the available pipe sizes. The nominal sizes of pipes listed in the complete table are ⅛", ¼", ⅜", ½", ¾", 1", 1¼", 1½", 2", 2½", 3", 3½", 4", 5", 6", 8", 10", 12", 14", 16", 18", 20", 22", . . . (at 2" increments up to 60").

†Standard (STD)
X-strong (XS)
XX-strong (XXS)

‡Beginning with 14-in. DIA pipe, the nominal size represents the outside diameter (OD).

The standard lengths for steel pipe are 20 ft and 40 ft. Seamless steel (SMLS STL) pipe is a smooth pipe with no weld seams along its length. Welded pipe is formed into a cylinder and butt-welded (BW) at the seam, or it is joined with an electric resistance weld (ERW).

36.3 Cast-Iron Pipe

Cast-iron pipe is used for transporting liquids, water, gas, and sewage. When used as a sewerage pipe, cast-iron pipe is called soil pipe. Cast-iron pipe is available in diameters of from 3 in. to 60 in.

The standard lengths of cast-iron pipe are 5 ft and 10 ft. Cast iron is more brittle and more subject to cracking when it is loaded than is steel pipe, so it should not be used where high pressures or weights will be applied to it.

36.4 Copper, Brass, and Bronze Pipe

Copper, brass, and bronze are used to manufacture piping and tubing for applications where there must be a high resistance to corrosive elements, such as acidic soils and chemicals, that are transmitted through the pipes. Copper pipe is used when the pipes are placed within or under concrete slab foundations of buildings. These nonferrous materials ensure the pipes will not have to be replaced because of corrosion. The standard length of these pipes is 12 ft.

Tubing is a smaller-size pipe that can be easily bent when it is made of copper, brass, or bronze. The term **piping** applies to rigid pipes usually larger than 2 in. in diameter.

36.5 Miscellaneous Pipes

Other materials used to manufacture pipes are aluminum, asbestos-cement, concrete, polyvinyl chloride (PVC), and various other plastics. The method the pipedrafter uses to design and detail piping systems is essentially the same regardless of the piping material used.

36.6 Pipe Joints

The basic connection in a pipe system is the joint where two straight sections of pipe fit together. Figure 36.1 shows three types of joints: **screwed, welded,** and **flanged.**

Screwed Joints are joined by pipe threads of the type covered in Chapter 18 and Appendix 9. Pipe threads are tapered at a ratio of 1 to 16 along

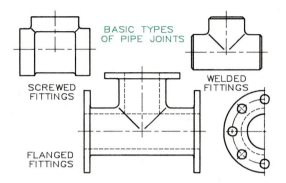

Figure 36.1 The three basic types of joints are screwed, welded, and flanged joints.

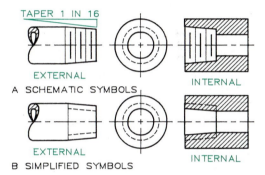

Figure 36.2 A screwed pipe uses a pipe thread that binds tightly as pipes are screwed together.

the outside diameter (Fig. 36.2). As the pipes are screwed together, the threads bind to form a snug, locking fit.

Welded Joints are joined by welded seams around the perimeter of the pipe to form butt welds. Welded joints are used extensively in "big inch" pipelines used in cross-country transporting of petroleum products.

Flanged Joints (Fig. 36.3) are welded to the straight sections of pipe, which are then bolted together around the perimeter of the flanges. Flanged joints form strong, rigid joints that can withstand high pressure and permit disassembly. Figure 36.4 and Appendix 12 show types of flange faces.

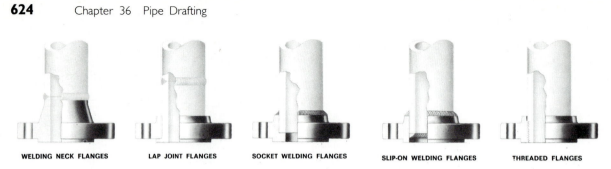

WELDING NECK FLANGES   LAP JOINT FLANGES   SOCKET WELDING FLANGES   SLIP-ON WELDING FLANGES   THREADED FLANGES

Figure 36.3 Types of flanged joints and the methods of attaching the flanges to the pipes. (Courtesy of Vogt Machine Co.)

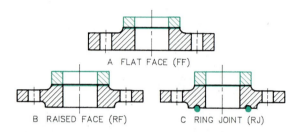

A FLAT FACE (FF)

B RAISED FACE (RF)   C RING JOINT (RJ)

Figure 36.4 Three types of flange faces are the raised face (RF), flat face (FF), and ring joint (RJ).

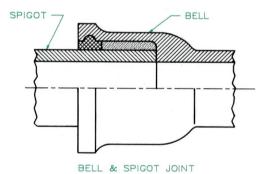

BELL & SPIGOT JOINT

Figure 36.5 A bell and spigot joint (B&S) is used to connect cast-iron pipes.

Bell and Spigot (B&S) Joints are used to join cast-iron pipes (Fig. 36.5). The spigot is placed inside the bell, and the two are sealed with molten lead or a sealing ring that snaps into position to form a sealed joint.

Soldering is used to connect smaller pipes and tubular connections. Soldering is usually limited to nonferrous tubing. Screwed fittings are also available to connect tubing (Fig. 36.6).

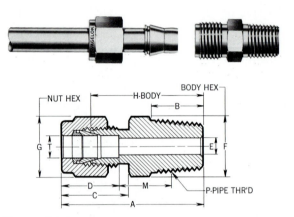

Figure 36.6 This fitting is used to attach small tubing. (Courtesy of Crawford Fitting Co.)

36.7 Screwed Fittings

Figure 36.7 shows several standard screwed fittings. The two types of graphical symbols used to represent fittings and pipe are **double-line symbols** and **single-line symbols.**

Double-line symbols are more descriptive of the fittings and pipes since they are drawn to scale with double lines. Single-line symbols are more symbolic since the pipe and fittings are drawn with single lines, and the fittings are drawn as single-line schematic symbols.

Fittings are available in three weights: standard (STD), extra strong (XS), and double extra strong (XXS), matching the weights of the pipes they will be connected with.

A piping system of screwed fittings is shown in Fig. 36.8, with double-line symbols in a sin-

REDUCER   HALF COUPLING   PIPE CAP   SQUARE HEAD PLUG   HEX. HD. PLUG   ROUND HEAD PLUG   HEXAGON BUSHING   FLUSH BUSHING

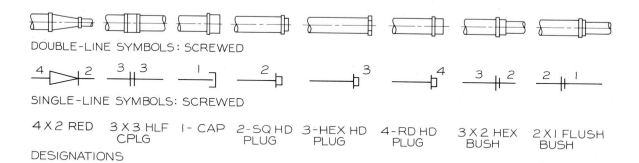

DOUBLE-LINE SYMBOLS: SCREWED

SINGLE-LINE SYMBOLS: SCREWED

4 X 2 RED   3 X 3 HLF CPLG   1- CAP   2-SQ HD PLUG   3-HEX HD PLUG   4-RD HD PLUG   3 X 2 HEX BUSH   2 X I FLUSH BUSH

DESIGNATIONS

90° ELBOW   TEE   45° ELBOW   CROSS   STREET ELBOW   LATERAL   COUPLING

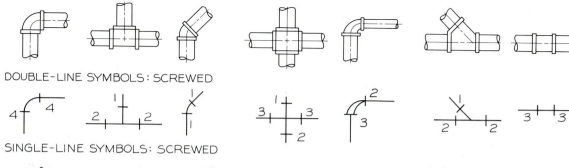

DOUBLE-LINE SYMBOLS: SCREWED

SINGLE-LINE SYMBOLS: SCREWED

4 X 90° ELL   2 X 2 X I TEE   I X 45° ELL   3 X 3 X 2 X I CROSS   3 X 2 RED ELL   2 X 2 X I LAT   3 COUPL

DESIGNATIONS

Figure 36.7 Examples of standard screwed fittings along with the single-line and double-line symbols that represent them. Nominal pipe sizes can be indicated by numbers placed near the joints. The major flow direction is labeled first, and the branches are labeled second. The large openings are labeled to precede the smaller openings.

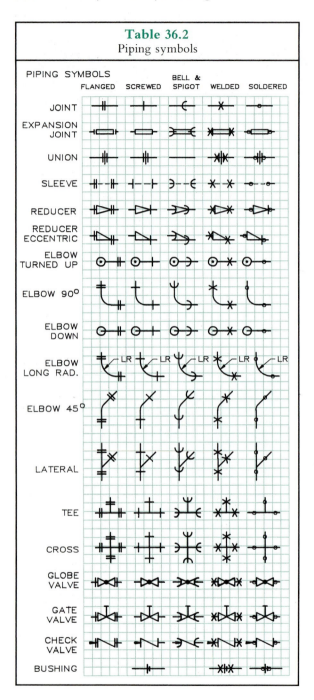

Table 36.2
Piping symbols

PIPING SYMBOLS	FLANGED	SCREWED	BELL & SPIGOT	WELDED	SOLDERED
JOINT					
EXPANSION JOINT					
UNION					
SLEEVE					
REDUCER					
REDUCER ECCENTRIC					
ELBOW TURNED UP					
ELBOW 90°					
ELBOW DOWN					
ELBOW LONG RAD.					
ELBOW 45°					
LATERAL					
TEE					
CROSS					
GLOBE VALVE					
GATE VALVE					
CHECK VALVE					
BUSHING					

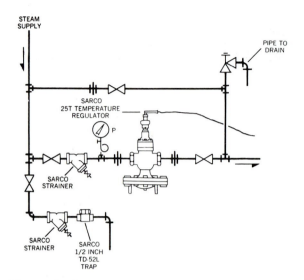

Figure 36.8 A single-line piping system with the major valves represented as double-line symbols. (Courtesy of Sarco, Inc.)

gle-line system to call attention to them. These could just as well have been drawn using the single-line symbols.

Table 36.2 shows the most common symbols for representing fittings; these have been extracted from the ANSI Z 32.2.3 standards.

36.8   Flanged Fittings

Flanges are used to connect fittings into a piping system when heavy loads are supported in large pipes and where pressures are great. Flanges are welded to straight pipe sections so that they can be bolted together.

Examples of several fittings are drawn as double-line and single-line symbols in Fig. 36.9. The elbow, commonly referred to as an **ell,** is available in angles of turn of 90° and 45° in both long and short radii. The long-radius (LR) ells have radii approximately 1.5 times the nominal diameter of the large end of the ell. The radius of a short-radius ell equals the diameter of the larger end.

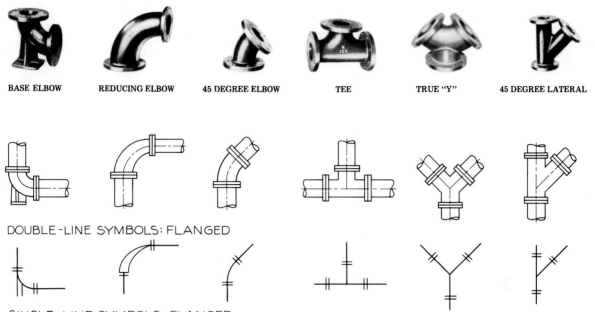

BASE ELBOW   REDUCING ELBOW   45 DEGREE ELBOW   TEE   TRUE "Y"   45 DEGREE LATERAL

DOUBLE-LINE SYMBOLS: FLANGED

SINGLE-LINE SYMBOLS: FLANGED

Figure 36.9 Examples of standard flanged fittings along with the single-line and double-line symbols that represent them.

Appendixes 10–13 give a table of dimensions for 125-lb and 250-lb cast-iron fittings.

36.9 Welded Fittings

Welding is a method of joining pipes and fittings for permanent, pressure-resistant joints. Figure 36.10 shows examples of double-line and single-line fittings connected by welding. Fittings are available with beveled edges prepared for welding.

A piping layout in Fig. 36.11 illustrates a series of welded joints with a double-line drawing. The location of the welded joints has been dimensioned.

36.10 Valves

Valves are used to regulate the flow within a pipeline or to turn off the flow. Types of valves are **gate, globe, angle, check, safety, dia-**

phragm, float, and **relief,** to name a few. The four basic types of valves—gate, globe, angle, and check—are shown in Fig. 36.12 using single-line symbols. Figure 36.13 shows the symbols for the other types of valves

Gate Valves are used to turn the flow within a pipe on or off with the least restriction of flow through the valve. These valves are not meant to be used to regulate the degree of flow.

Globe Valves are used to turn the flow on and off and to regulate the flow to a desired level.

Angle Valves are types of globe valves that turn at 90° angles at bends in the piping system. They have the same controlling features as the straight globe valves.

Check Valves restrict the flow in the pipe to only one direction. A backward flow is checked by either a movable piston or a swinging washer.

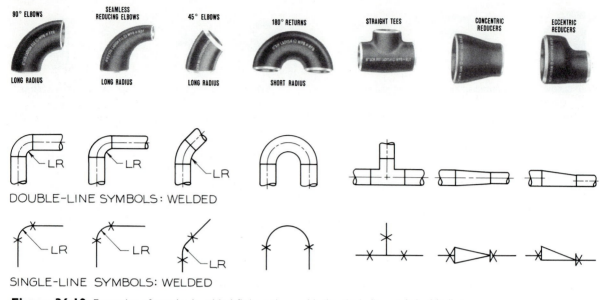

DOUBLE-LINE SYMBOLS: WELDED

SINGLE-LINE SYMBOLS: WELDED

Figure 36.10 Examples of standard welded fittings along with the single-line and double-line symbols that represent them.

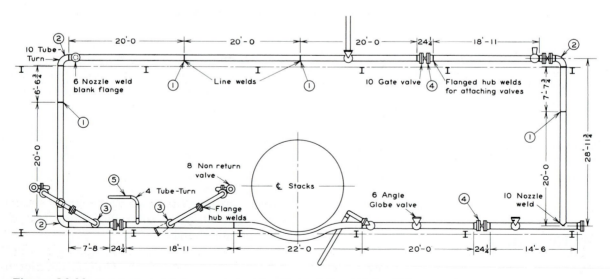

Figure 36.11 A piping layout that uses a series of welded and flanged joints. This system is drawn using double-line symbols.

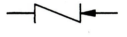

GATE VALVE   GLOBE VALVE   ANGLE VALVE   CHECK VALVE

SINGLE—LINE PIPING SYMBOLS

Figure 36.12 The four basic types of valves are gate, globe, angle, and check valves. (Courtesy of Vogt Machine Co.)

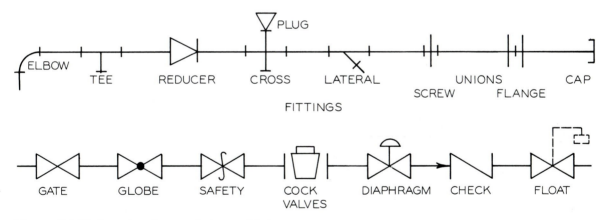

Figure 36.13 fittings labels: ELBOW, TEE, REDUCER, CROSS, PLUG, LATERAL, UNIONS, SCREW, FLANGE, CAP — FITTINGS

GATE, GLOBE, SAFETY, COCK, DIAPHRAGM, CHECK, FLOAT — VALVES

Figure 36.13 Examples of types of valves and fittings.

36.11 Fittings in Orthographic Views

Figure 36.14 shows orthographic views of typical fittings and valves. These fittings are drawn as single-line screwed fittings, but the same general principles can be used to represent other types of joints as double-line drawings. Observe the var-

ious views of the fittings, and notice how the direction of an elbow can be shown by a slight variation in the different views.

A piping system is shown in a single orthographic view in Fig. 36.15, with a combination of double-line and single-line symbols. Arrows are used to give the direction of flow in the sys-

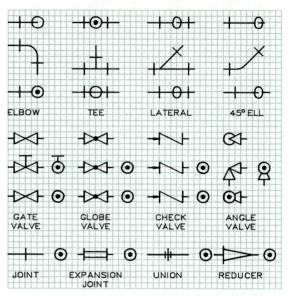

Figure 36.14 Fittings and valves drawn with single-line symbols are drawn differently in these various orthographic views.

Figure 36.15 This piping system is drawn using a combination of single-line and double-line symbols. The connections are shown as screwed, welded, and flanged. (Courtesy of Bechtel Corp.)

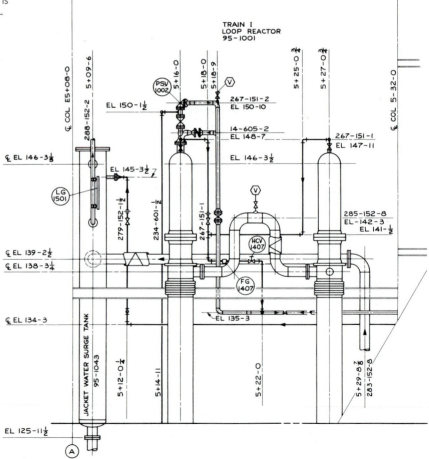

tem. Different joints are screwed, welded, and flanged. Horizontal elevation lines give the heights of each horizontal pipe. Station 5 + 12 − 0-$\frac{1}{4}$″ represents a distance of 500′ plus 12′ − 0-$\frac{1}{4}$″, or 512′ − 0-$\frac{1}{4}$″ from the beginning station point of 0 + 00.

The dimensions in Fig. 36.15 are measured from the centerlines of the pipes, indicated by the CL symbols. In some cases, the elevations of the pipes are dimensioned to the bottom of the pipe, abbreviated **BOP** (see Fig. 36.20).

36.12 Piping Systems in Pictorial

Isometric and axonometric drawings of piping systems are called **spool drawings;** they can be drawn using either single-line or double-line symbols.

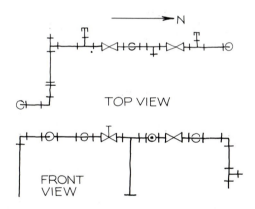

Figure 36.16 Top and front views are used to represent this three-dimensional piping system with single-line symbols.

A three-dimensional piping system is sketched orthographically in Fig. 36.16 with top and front views. Although this is a relatively simple three-dimensional system, a thorough understanding of orthographic projection is needed to read the drawing.

In Fig. 36.17, the piping system is drawn with all the pipes revolved into the same horizontal plane. The vertical pipes and their fittings are drawn true size in the top view. This is called a

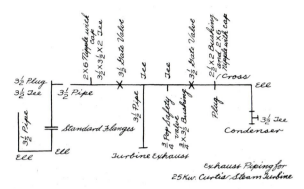

Figure 36.17 The vertical pipes shown in Fig. 36.16 are revolved into the horizontal plane to form a developed drawing. The fittings and valves are noted on the sketch.

developed pipe drawing. The fittings and pipe sizes are noted on this preliminary sketch, from which the finished drawing will be made in Fig. 36.18.

An axonometric schematic (Fig. 36.19) explains the three-dimensional relationship of the parts of the system. The rounded bends in the elbows in an isometric drawing can be constructed with ellipses using the isometric ellipse template, or the corners can be drawn square to reduce the time and effort needed.

Piping layouts are often drawn as isometrics to clarify complex systems that are hard to interpret when shown in orthographic views. Some of the more often used piping connectors are shown in Fig. 36.20 as both orthographic and isometric views.

Part of a pipe system is shown in Fig. 36.21 as an isometric drawing.

A north arrow is drawn on the plan view of the piping system in Fig. 36.16 to orient the isometric pictorial. North is not necessarily related to compass north; rather, it is a direction parallel to a major set of pipes within the system.

36.13 Dimensioned Isometrics

The spool drawing in Fig. 36.22 shows the specifications for the pipe, fittings, flanges, and valves noted on the drawing and itemized in the bill of materials.

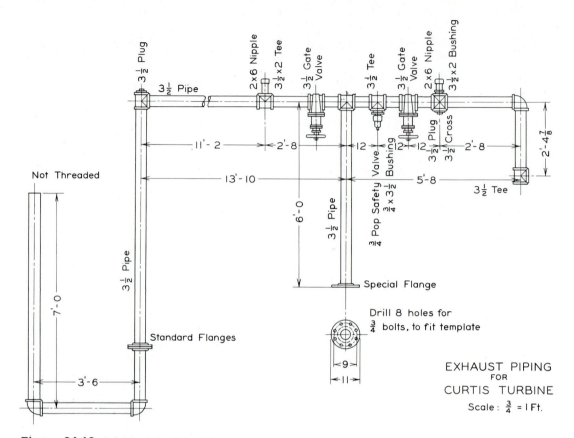

Figure 36.18 A finished developed drawing that shows all the components in the system true size with double-line symbols.

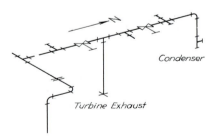

Figure 36.19 An axonometric pictorial of the pipe system shown in Fig. 36.18 is drawn to give a three-dimensional picture of the system.

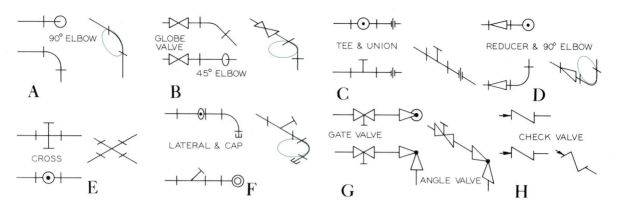

Figure 36.20 A comparison of orthographic views and isometric pictorials of single-line representations of piping symbols. The isometric template is used for constructing rounded corners in the isometric drawings.

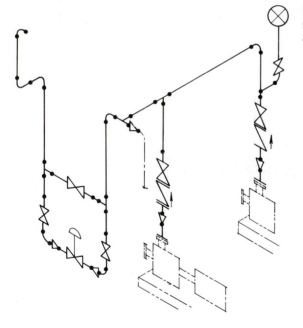

Figure 36.21 A typical piping system drawn in isometric using the appropriate symbols.

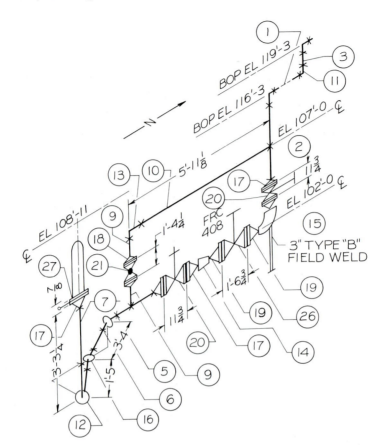

Figure 36.22 This dimensioned isometric pictorial is called a spool drawing. Because it is sufficiently complete, it can serve as a working drawing when used with the bill of materials.

NO	QTY	DESCRIPTION	MATL
		PIPE	
1	1	8" X 18'–2-5/8 SCH 40 SMLS STL OH	A–53
2	1	8" X 10'–7-1/4" SCH 40 SMLS STL OH	A–53
3	1	8" X 1'–0 SCH 40 SMLS STL OH	A–53
4	1	8" X 0'–6-7/8 SCH 40 SMLS STL OH	A–53
5	1	8" X 0'–7 SCH SMLS STL OH	A–53
6	1	8" X 2'–6 SCH 40 SMLS STL OH	A–53
7	1	8" X 2'–1-1/2 SCH 40 SMLS STL OH	A–53
8	1	6" X 7'–5-7/8 SCH 40 SMLS STL OH	A–53
9	2	6" X 1'–1-7/8 SCH 40 SMLS STL OH	A–53
10	1	6" X 5'–2-1/8 SCH 40 SMLS STL OH	A–53
		FITTINGS	
11	3	8"–90 DEG LR ELL STD WT BW SMLS	A–53
12	1	8"–90 DEG LR ELL, LONG TANGENT	A–53
13	1	6"–90 DEG SR ELL STD BW SMLS	A–53
14	1	8" X 6" CONCENTRIC RED STD BW SMLS	A–53
15	1	8" X 6" RED ELL STD BW SMLS	A–53

NO	QTY	DESCRIPTION	MATL
16	2	8"–45 DEG LR ELL STD BW SMLS	A–53
		FLANGES	
17	5	8"–150 LBS RF FS WN	A–181
18	2	6"–150 LBS RF FS WN	A–181
19	2	6"–300 LBS RF FS WN	A–181
		VALVES	
20	2	8"–150 LBS CS FLG RF	47X
21	1	6"–150 LBS CS FLG RF GLOBE	143X
		OTHER	
22	48	3/4 DIA ASTM ALLOY STL STUD BOLTS	A–193
23	48	ASTM HVY HEX NUT, EACH BOLT	A–194
24	24	3/4 DIA ASTM ALLOY STL STUD BOLTS	A–193
25	24	ASTM HVY HEX NUT, EACH BOLT	A–194
26	1	FLUID RECORDER CONTROLLER	
27	1	8" SPEC BLIND	
28	5	8" – 150 LBS SPIRAL WOUND 1/8" THK GASKET	304SS
29	4	6"–150 LBS SPIRAL WOUND 1/8" THK GASKET	304SS

Table 36.3
Standard abbreviations associated with pipe specifications

AVG	average	FS	forged steel	SPEC	specification
BC	bolt circle	FSS	forged stainless steel	SR	short radius
BE	beveled ends	FW	field weld	SS	stainless steel
BF	blind flange	GALV	galvanized	STD	standard
BM	bill of materials	GR	grade	STL	steel
BOP	bottom of pope	ID	inside diameter	STM	steam
B&S	bell & spigot	INS	insulate	SW	socket weld
BWG	Birmingham wire gauge	IPS	iron pipe size	SWP	standard working pressure
CAS	cast alloy steel	LR	long radius	TC	test connection
CI	cast iron	LW	lap weld	TE	threaded end
CO	clean out	MI	malleable iron	TEMP	temperature
CONC	concentric	MFG	manufacture	T&G	tongue & groove
CPLG	coupling	OD	outside diameter	TOS	top of steel
CS	carbon steel, cast steel	OH	open hearth	TYP	typical
DWG	drawing	PE	plain end—not beveled	VC	vitrified clay
ECC	eccentric	PR	pair	WE	weld end
EF	electric furnace	RED	reducer	WN	weld neck
EFW	electric fusion weld	RF	raised face	WB	welded bonnet
ELEV	elevation	RTG or RJ	right type joint	WT	weight
ERW	electric resistance weld	SCH	schedule	XS	extra strong
FF	flat face	SCRD	screwed	XXS	double extra strong
FLG	flange	SMLS	seamless		
FOB	flat on bottom	SO	slip-on		

Several abbreviations are used to specify piping components and fittings. Part number 1, for example, is an 8-in. diameter pipe of a Schedule 40 weight that is made of seamless steel by the open hearth (OH) process. Instead of OH, EF may be used, which is the abbreviation for electric furnace. Table 36.3 gives standard abbreviations associated with pipe drawings.

Under the column mat (materials), you will notice a code beginning with the letter *A*. The letter *A* represents a grade of carbon steel listed in Table A of ANSI B31.3, *Petroleum Refinery Piping Standards*. The codes for fittings, flanges, and valves are taken from the manufacturers' catalogues of these products.

Figure 36.23 shows a suggested format for spool drawings. This format is used by the Bechtel Corporation, a major construction company, in designing and constructing pipelines and refineries.

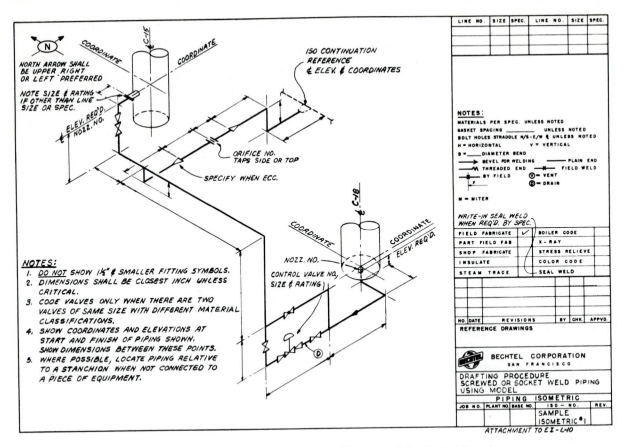

Figure 36.23 A suggested format for preparing spool drawings. (Courtesy of the Bechtel Corp.)

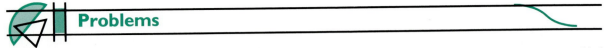

Problems

1. On a Size A sheet, draw five orthographic views of the fittings listed below. The views should include the front view, top view, bottom view, and left and right views. Draw two fittings per page. Refer to Fig. 36.14 and Table 36.2 as guides in making these drawings. Use single-line symbols to draw the following screwed fittings: 90° ell, 45° ell, tee, lateral, cap reducing ell, cross, check valve, union, globe valve, gate valve, and bushing.

2. Same as Problem 1, but draw the fittings as flanged fittings.

3. Same as Problem 1, but draw the fittings as welded fittings.

4. Same as Problem 1, but draw the fittings as double-line screwed fittings.

5. Same as Problem 1, but draw the fittings as double-line flanged fittings.

6. Same as Problem 1, but draw the fittings as double-line welded fittings.

7. Convert the single-line sketch in Fig. 36.24 into a double-line system that will fit on a Size A sheet.

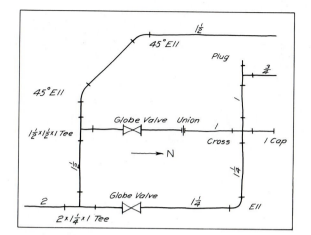

Figure 36.24 Problem 7.

8. Convert the pipe system shown in isometric in Fig. 36.17 into a double-line isometric drawing that will fit on a Size B sheet.

9. Convert the pipe system shown in Fig. 36.17 into a two-view orthographic drawing using single-line symbols.

10. Convert the isometric drawing of the pipe system shown in Fig. 36.23 into a two-view orthographic drawing that will fit on a Size B sheet. Take the measurements from the given drawing, and select a convenient scale.

37

Electric and Electronics Drafting

37.1 Introduction

Electric and electronics drafting is a specialty area in the field of drafting technology. **Electrical drafting** is related to the transmission of electrical power used in many homes and industries for lighting, heating, and equipment operation. **Electronics drafting** deals with circuits in which integrated circuits or transistors are used and where power is used in small quantities, such as radios, televisions, computers, and video cassette recorders.

Electronics drafters are responsible for the preparation of drawings that will be used in fabricating the circuit and thereby bringing the product into being. They work from sketches and specifications developed by the engineer or electronics technologist.

Most of this chapter has been extracted from ANSI Y14.15, *Electrical and Electronics Diagrams,* the standards that regulate the drafting techniques used in this area. The symbols used were taken from ANSI Y32.2, *Graphic Symbols for Electrical and Electronics Diagrams.*

37.2 Types of Diagrams

Electronic circuits are classified and drawn in the format of one of the following types of diagrams:

1. Single-line diagrams
2. Schematic diagrams
3. Connection diagrams

Figure 37.1 shows the suggested line weights for drawing these diagrams.

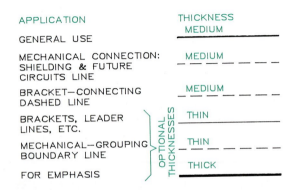

Figure 37.1 The recommended line weights for drawing electronics diagrams.

Single-Line Diagrams

Single-line diagrams use single lines and graphic symbols to show an electric circuit or system of circuits and the parts and devices within it. A single line is used to represent both AC and DC systems, as Fig. 37.2 shows. Figure 37.3 shows a single-line diagram of an audio system. Primary circuits are indicated by thick connecting lines, and medium lines represent connections to the current and potential sources.

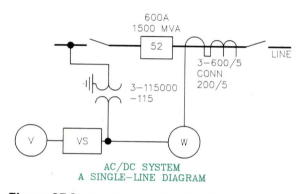

AC/DC SYSTEM
A SINGLE-LINE DIAGRAM

Figure 37.2 A portion of a single-line diagram where heavy lines represent the primary circuits, and medium lines represent the connections to the current and potential sources.

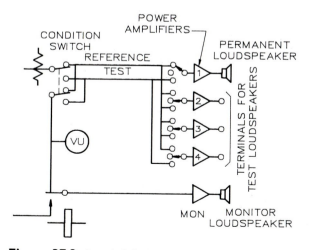

Figure 37.3 A typical single-line diagram for illustrating electronics and communications circuits.

Single-line diagrams show the connections of meters, major equipment, and instruments. In addition, ratings are often given to provide such information as kilowatts, voltages, cycles and revolutions per minute, and generator ratings.

Schematic Diagrams

Schematic diagrams use graphic symbols to show the electrical connections and functions of a specific circuit arrangement. Although the schematic diagram allows one to trace the circuit and its functions, the physical size, shapes, and locations at various components are not given. Figure 37.4 on page 640 shows a schematic diagram.

Connection Diagrams

Connection diagrams show the connections and installations of the parts and devices of the system. Besides showing the internal connections, external connections, or both, they show the physical arrangement of the parts. Figure 37.5 shows a three-dimensional connection diagram.

37.3 Schematic Diagram Connecting Symbols

The most basic symbols of a circuit are the connections of parts within the circuit. Using dots to show connections is optional; it is preferable to omit them if clarity is not sacrificed by doing so. Connections, or junctions, are indicated by using small black dots (Fig. 37.6A). The dots distinguish between connecting lines and those that simply pass over each other (Fig. 37.6B). It is preferred that connecting wires have single junctions wherever possible. When the layout of a circuit does not permit using single junctions, and lines within the circuit must cross, dots must be used to distinguish between crossing and connecting lines (Figs. 37.6A and 37.6B).

Interrupted Paths are breaks in lines within a schematic diagram; they are used to conserve space when this can be done without confusion.

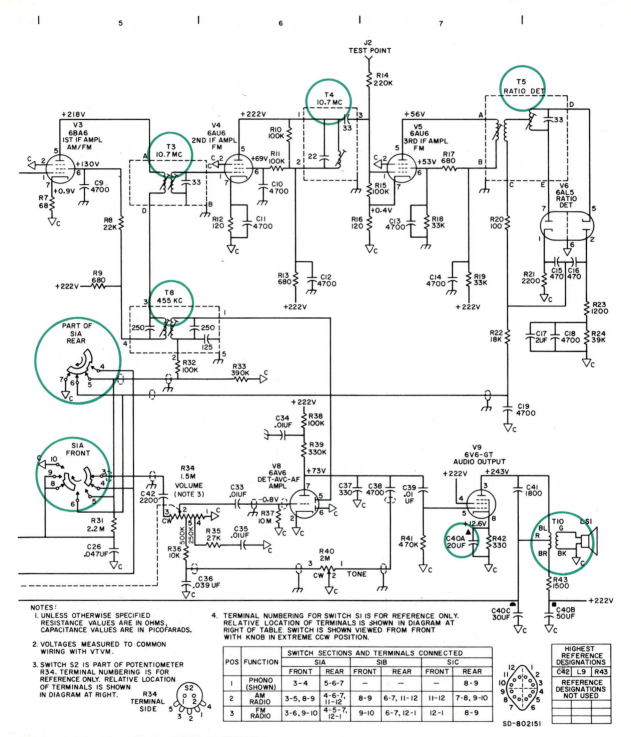

Figure 37.4 A portion of a typical schematic diagram of an AM/FM radio circuit with all parts of the system labeled. The title of a drawing of this type should specify it is a schematic diagram as well as the type of circuit. (Courtesy of ANSI.)

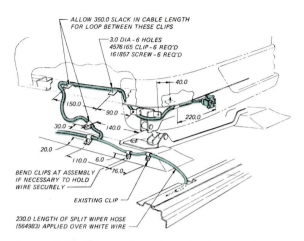

Figure 37.5 This is a three-dimensional connection diagram that shows the circuit and its components with the dimensions necessary to explain how it is connected or installed. (Courtesy of the General Motors Corp.)

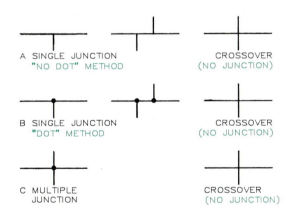

Figure 37.6 (A) Connections should be shown with single-point junctions. (B) Dots may be used to call attention to connectors. (C) Dots must be used when there are multiples of the type shown here.

For example, the circuit in Fig. 37.7 has been interrupted; instead, of connecting the left and right sides of the circuit, the lines are labeled to correspond to the matching notes at the other side of the interrupted circuit.

Occasionally, sets of lines in a horizontal or vertical direction will be interrupted (Fig. 37.8). Brackets will be used to interrupt the circuit, and notes will be placed outside the brackets to indicate the destinations of the wires or their connections.

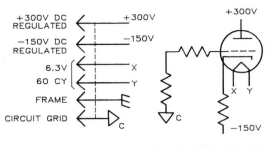

CONNECTOR INPUT   CIRCUIT ARRANGEMENT

Figure 37.7 Circuits may be interrupted and connections not shown by lines if they are properly labeled to clarify their relationship to the removed part of the circuit. The connections above are labeled to match those on the left and right sides of the illustration.

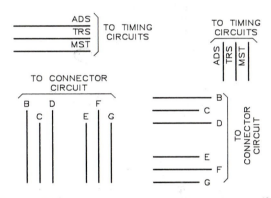

Figure 37.8 Brackets and notes may be used to specify the destinations of interrupted circuits.

In some cases, a dashed line is used to connect brackets that interrupt circuits (Fig. 37.9). The dashed line should be drawn so that it will not be mistaken as a continuation of one of the lines within the bracket.

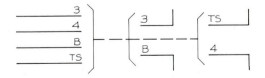

Figure 37.9 The connections of interrupted circuits can be indicated by using brackets and a dashed line in addition to labeling the lines. The dashed line should not be drawn to appear as an extension of one of the lines in the circuit.

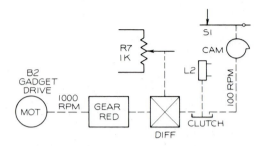

Figure 37.10 If mechanical functions are closely related to electrical functions, it may be desirable to link the mechanical components within the schematic diagram.

Mechanical Linkages that are closely related to electronic functions may be shown as part of a schematic diagram (Fig. 37.10).

37.4 Graphic Symbols

The electronics drafter must be familiar with the basic graphic symbols used to represent the parts and devices within electrical and electronics circuits.

The symbols we cover here are extracted from the ANSI Y32.2 standards and are adequate for practically all diagrams. But when a highly specialized part needs to be shown and a symbol for it is not provided in these standards, it is permissible for the drafter to devise a symbol provided it is properly labeled and its meaning clearly conveyed.

The symbols presented in Figs. 37.11–37.14 are drawn on a grid of 5 mm (0.20 in.) squares that has been reduced. The actual dimensions of the symbols can be approximated by using the grid and equating each square to its full-size measurement of 5 mm. Symbols may be drawn larger or smaller to fit the size of your layout. Nearly 600 different variations of the basic electrical and electronics symbols are listed in the ANSI standards.

Electronics symbols can be drawn with conventional drawing equipment or with drafting templates of electronics symbols (Fig. 37.15).

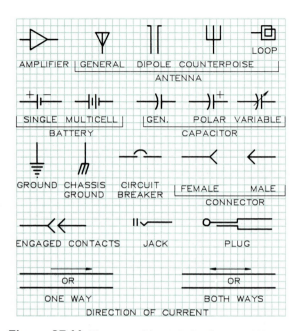

Figure 37.11 These graphic symbols of parts within a schematic diagram are drawn on a 5 mm (0.20 in.) grid. The suggested sizes of the symbols can be found by taking the dimensions from the grid to draw the symbols full size.

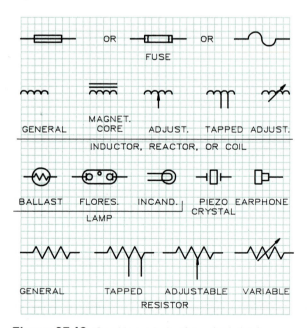

Figure 37.12 Graphic symbols of standard circuit components.

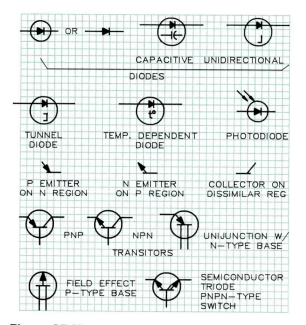

Figure 37.13 Graphic symbols for semiconductors. The arrows in the middle of the figure illustrate the meanings of the arrows used in the transistor symbols.

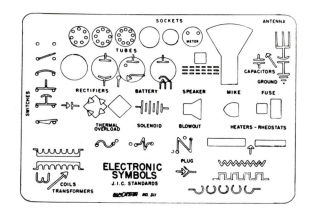

Figure 37.15 Many types of templates are available for drawing the graphics symbols in schematic drawings. (Courtesy of Frederick Post Co.)

37.5 Terminals

Terminals are the end, or devices attached in a circuit with connecting wires. Examples of devices with terminals are switches, relays, and transformers. The graphic symbol for a terminal is an open circle drawn the same size as the solid circle used to indicate a connection.

Switches are used to turn a circuit on or off or to actuate a certain part of it while turning another part off. The schematic diagram in Fig. 37.4 shows examples of labeling switches. In this case, a table clarifies the switching connections of the terminals.

When a group of parts is enclosed or shielded (drawn enclosed with dashed lines) and the terminal circles have been omitted, the terminal markings should be placed immediately outside the enclosure, as shown in Fig. 37.4 at T2, T3, T4, T5, and T8. Terminal identifications should be added to the graphic symbols that correspond to the actual physical markings that appear on or near the terminals of the part, such as 10.7 MC for transformer T3 in Fig. 37.4. Figures 37.16–37.18 show several examples of notes and symbols that explain the parts of a diagram.

Colored wires or symbols are often used to identify the various leads that connect to termi-

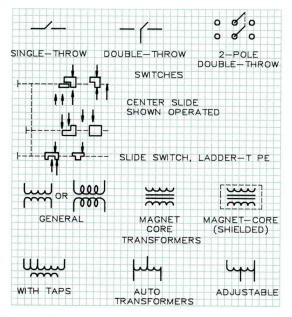

Figure 37.14 Graphic symbols for representing switches and transformers.

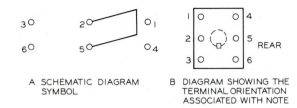

A SCHEMATIC DIAGRAM SYMBOL

B DIAGRAM SHOWING THE TERMINAL ORIENTATION ASSOCIATED WITH NOTE

Figure 37.16 (A) An example of a method of labeling the terminals of a toggle switch on a schematic diagram. (B) A diagram that illustrates the toggle switch when viewed from its rear.

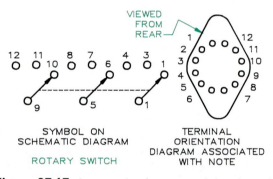

SYMBOL ON SCHEMATIC DIAGRAM

TERMINAL ORIENTATION DIAGRAM ASSOCIATED WITH NOTE

ROTARY SWITCH

Figure 37.17 An example of a rotary switch as it would appear on a schematic diagram, and a diagram that shows the numbered terminals of the switch when viewed from its rear.

nals. When colored wires need to be identified on a diagram, the colors are lettered on the drawing, as was done for transformer T10 in Fig. 37.4. Figure 37.4 also shows an example of the identification of a capacitor, C40, marked with geometric symbols.

Rotary Terminals are used to regulate the resistance in some circuits, and the direction of rotation of the dial is indicated on the schematic diagram. The abbreviations CW (clockwise) or CCW (counterclockwise) are placed adjacent to the movable contact when it is in its extreme clockwise position, as shown in Fig. 37.19A. The movable contact has an arrow at its end.

If the device terminals are not marked, numbers may be used with the resistor symbol and the number 2 assigned to the adjustable contact (Fig. 37.19B). Other fixed types may be sequentially numbered and added (Fig. 37.19C).

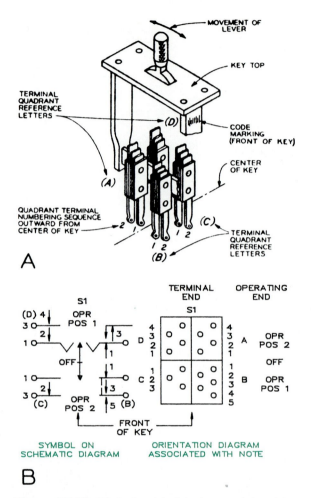

SYMBOL ON SCHEMATIC DIAGRAM

ORIENTATION DIAGRAM ASSOCIATED WITH NOTE

Figure 37.18 (A) A pictorial of the lever switch and its four quadrants. (Courtesy of ANSI.) (B) An example of a typical lever switch as it would appear on a schematic diagram, and an orientation diagram that shows the numbered terminals of the switch when viewed from its operating end.

The position of a switch as it relates to the function of a circuit should be indicated on a schematic diagram. Figure 37.20 illustrates a method of showing functions of a variable switch. The arrow represents the movable end of the switch that can be positioned to connect with several circuits. The functional positions of the rotary switch are shown both by symbol and table.

Another method of representing a rotary switch is shown in Fig. 37.21 by symbol and ta-

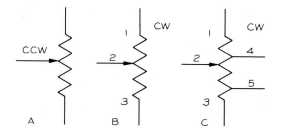

Figure 37.19 (A) To indicate the direction of rotation of rotary switches on a schematic diagram, the abbreviations CW (clockwise) and CCW (counterclockwise) are placed near the movable contact. (B) If the device terminals are not marked, numbers may be used with the resistor symbols and the number 2 assigned to the adjustable contact. (C) Additional contacts may be labeled as shown.

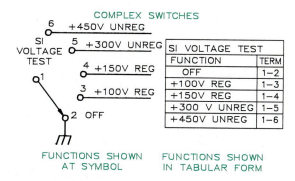

COMPLEX SWITCHES

S1 VOLTAGE TEST	
FUNCTION	TERM
OFF	1–2
+100V REG	1–3
+150V REG	1–4
+300 V UNREG	1–5
+450V UNREG	1–6

FUNCTIONS SHOWN AT SYMBOL   FUNCTIONS SHOWN IN TABULAR FORM

Figure 37.20 For more complex switches, position-to-position function relations may be shown using symbols on the schematic diagram, or they may be shown by a table of values located elsewhere on the diagram.

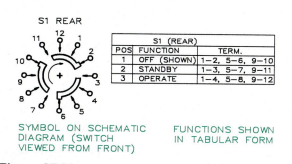

S1 (REAR)		
POS	FUNCTION	TERM.
1	OFF (SHOWN)	1–2, 5–6, 9–10
2	STANDBY	1–3, 5–7, 9–11
3	OPERATE	1–4, 5–8, 9–12

SYMBOL ON SCHEMATIC DIAGRAM (SWITCH VIEWED FROM FRONT)   FUNCTIONS SHOWN IN TABULAR FORM

Figure 37.21 A rotary switch may be shown on a schematic diagram with its terminals labeled as shown at the left, or its functions can be given in a table placed elsewhere on the drawing as shown at the right. Dashes are used to indicate the linkage of the numbered terminals. For example, 1–2 means terminals 1 and 2 are connected in the off position.

ble. Because of the complexity of this particular switch, the table representation is preferred. The dashes between the numbers in the table indicate the numbers have been connected. For example, when the switch is in position 2, the following terminals are connected: 1 and 3, 5 and 7, and 9 and 11. A table of this type is used at the bottom of Fig. 37.4.

Electron tubes have pins that fit into sockets, which have terminals that connect with circuits. Pins are labeled with numbers placed outside the symbol used to represent the tube (Fig. 37.22) and, with the tube viewed from its bottom, are numbered in a clockwise direction.

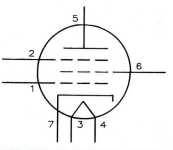

PLACEMENT OF TUBE PIN NUMBERS

Figure 37.22 Tube pin numbers should be placed outside the tube envelope and adjacent to the connecting lines.

37.6 Separation of Parts

In complex circuits, it is often advantageous to separate elements of a multielement part with portions of the graphic symbols drawn in different locations on the drawing. An example of this method of separation is the switch labeled S1A, S1B, S1C, and so on in Fig. 37.4. The switch is labeled S1, and the letters that follow, called suffixes, designate different parts of the switch. Suffix letters may also be used to label subdivisions of an enclosed unit made up of a series of internal parts, such as the crystal unit shown in Fig. 37.23. These crystals are referred to as Y1A and Y1B.

Rotary switches of the type shown in Fig. 37.24 are designated S1A, S1B, and so on. The

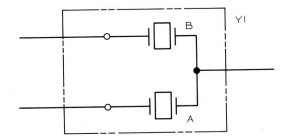

Figure 37.23 As subdivisions within the complete part, crystals A and B are referred to as Y1A and Y1B.

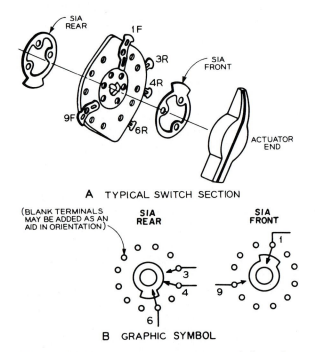

A   TYPICAL SWITCH SECTION

(BLANK TERMINALS MAY BE ADDED AS AN AID IN ORIENTATION)

B   GRAPHIC SYMBOL

Figure 37.24 Parts of rotary switches are designated with suffix letters A, B, C, etc. and are referred to as S1A, S1B, S1C, etc. The words FRONT and REAR are added to these designations when both sides of the switch are used. (Courtesy of ANSI.)

suffix letters are labeled in sequence beginning with the knob and working away from it. Each end of the sections of the switch should be viewed from the same end. When the rear and front of the switches are used, the words FRONT and REAR are added to the designations.

Parts of items such as terminal boards, connectors, or rotary switches may be separated on a diagram. The words PART OF may precede the identification of the portion of the circuit it is a part of, as Fig. 37.25A shows. A second method of showing a part of a system is by using conventional break lines, which makes the note PART OF unnecessary.

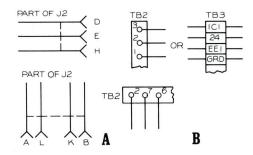

Figure 37.25 (A) The portions of connectors or terminal boards are functionally separated on a diagram; the words PART OF may precede the reference designation of the entire portion. (B) Or conventional breaks can be used to graphically indicate that the part drawn is only a portion of the whole.

37.7   Reference Designations

A combination of letters and numbers that identify items on a schematic diagram are called reference designations. These designations identify the components not only on the drawing but also in the documents referring to them. Reference designations should be placed close to the symbols that represent the replaceable items of a circuit on a drawing. Items not separately replaceable may be identified if necessary. Mounting devices for electron tubes, lamps, fuses, and so forth are seldom identified on schematic diagrams.

Standard practice is to begin each reference designation with an uppercase letter that may be followed by a number with no hyphen between them. The number usually represents a portion of the part. The lowest number of a designation should be assigned to begin at the upper left of the schematic diagram and proceed consecutively from left to right and top to bottom throughout the drawing.

Some of the standard abbreviations that designate parts of an assembly are amplifier, A; battery, BT; capacitor, C; connector, J; piezoelectric crystal, Y; fuse, F; electron tube, V; generator, G; rectifier, CR; resistor, R; transformer, T; and transistor, Q.

As the circuit is being designed, some of the numbered elements may be deleted from the drawing. The remaining numbered elements should not be renumbered even though there is a missing element within the sequence of numbers. Instead, a table like the one shown in Fig. 37.26

HIGHEST REFERENCE DESIGNATIONS	
R72	C40
REFERENCE DESIGNATIONS NOT USED	
R8, R10, R61 R64, R70	C12, C15, C17 C20, C22

Figure 37.26 Reference designations are used to identify parts of a circuit. They are labeled in a numerical sequence from left to right beginning at the upper left of the diagram. If parts are later deleted from the system, the ones deleted should be listed in a table along with the highest reference number designations.

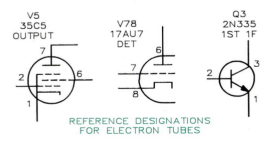

Figure 37.27 Three lines of notes can be used with electron tubes or semiconductors to specify reference designations, type designations, and function.

can be used to list the parts omitted from the circuit. The highest designations are also given in the table to be sure that all parts were considered.

Electron tubes and transitors are labeled not only with reference designations but also with type designation and circuit function, as Fig. 37.27 shows. This information is labeled on three lines, such as

V5

35C5

OUTPUT

which are placed adjacent to the symbol.

37.8 Numerical Units of Function

Functional units such as the values of resistance, capacitance, inductance, and voltage should be specified with the fewest number of zeros by using the multipliers in Fig. 37.28A as prefixes. Examples of this are shown in Figs. 37.28B and 37.28C, where units of resistance and capacitance are given. When four-digit numbers are given, the commas should be omitted (for example, one thousand should be written 1000, not 1,000). You should recognize and use the lowercase or uppercase prefixes indicated in Fig. 37.28A.

Where certain units are repeated, general notes can be used to reduce time and effort:

UNLESS OTHERWISE SPECIFIED:
RESISTANCE
VALUES ARE IN OHMS.

CAPACITANCE VALUES ARE IN
MICROFARADS.

CAPACITANCE VALUES ARE IN
PICOFARADS.

A note for specifying capacitance values is

CAPACITANCE VALUES SHOWN AS NUMBERS EQUAL TO OR GREATER THAN UNITS ARE IN pF, AND NUMBERS LESS THAN UNITY ARE IN μF.

A MULTIPLIERS		SYMBOL	
		METHOD 1	METHOD 2
MULTIPLIER	PREFIX		
10^{12}	TERA	T	T
10^{9}	GIGA	G	G
10^{6} (1,000,000)	MEGA	M	M
10^{3} (1,000)	KILO	k	K
10^{-3} (0.001)	MILLI	m	MILLI
10^{-6} (0.000,001)	MICRO	μ	U
10^{-9}	NANO	n	N
10^{-12}	PICO	p	P
10^{-13}	FEMTO	f	F
10^{-16}	ATTO	a	A

B RESISTANCE

RANGE IN OHMS	EXPRESS AS	EXAMPLE
LESS THAN 1000	OHMS	0.031 470
1000 TO 99,999	OHMS OR KILOHMS	1800 15,853 10k 82k
100,000 to 999,999	KILOHMS OR MEGOHMS	220k 0.22M
1,000,000 OR MORE	MEGOHMS	3.3M

C CAPACITANCE

RANGE IN PICOFARADS	EXPRESS AS	EXAMPLE
LESS THAN 10,000	PICOFARADS	152.4pF 4700pF
10,000 OR MORE	MICROFARADS	0.015μF 30μF

Figure 37.28 (A) Multipliers should be used to reduce the number of zeros in a number. (B and C) Examples of expressing units of capacitance and resistance.

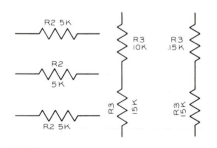

Figure 37.29 Methods of labeling the units of resistance on a schematic diagram.

Figure 37.29 shows examples of the placement of the reference designations and the numerical values of resistors.

37.9 Functional Identification of Parts

The readability of a circuit is improved if parts are labeled to indicate their functions. Test points are labeled on drawings with the letters *Tp* and their suffix numbers. The sequence of the suffix numbers should be the same as the sequence of troubleshooting the circuit when it is defective. Alternatively, the test function can be indicated on the diagram below the reference designation.

Additional information may be included on the schematic diagram to aid in maintaining the system.

- DC resistance of windings and coils
- Critical input and output impedance values
- Wave shapes (voltage or current) at significant points
- Wiring requirements for critical ground points, shielding, pairing, and so on
- Power or voltage ratings of parts
- Caution notation for electrical hazards at maintenance points
- Circuit voltage values at significant points (tube pins, test points, terminal boards, and so on)
- Zones (grid system) on complex schematics
- Signal flow direction in main signal paths

37.10 Printed Circuits

Printed circuits are universally used for miniature electronic components and computer systems.

The drawings of printed circuits are drawn four or more times the size of the circuit ultimately printed. The drawings are usually drawn in black India ink on acetate film and are then photographically reduced to the desired size. The circuit is "printed" onto an insulated board made of plastic or ceramics, and the devices within the circuit are connected (Fig. 37.30).

Figure 37.30 This is a magnified view of a printed circuit board where the circuit has been printed and etched on the board and the devices have been soldered into position. (Courtesy of Bishop Industries Corp.)

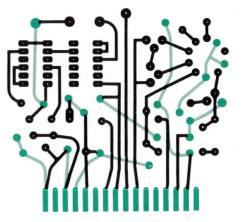

Figure 37.32 By using two colors, such as blue and red, one circuit drawing can be made. From the same drawing, two negatives can be made by using camera filters that screen out one of the colors with each shot. The circuits are then printed on each side of the board. (Courtesy of Bishop Industries Corp.)

Figure 37.31 A printed circuit attached to two sides of the circuit board requires two circuit drawings, one for each side. The drawings are photographically converted to reduced negatives for printing. (Courtesy of Bishop Industries Corp.)

37.11 Shortcut Symbols

Several manufacturers produce preprinted symbols that can be used for "drawing" electronic and printed circuits. The symbols are available on sheets or tapes that can be burnished onto the surface of the drawing to form a permanent schematic diagram (Fig. 37.33). The symbols can be connected with matching tape to represent wires between them instead of drawing the lines.

Some printed circuits are printed on both sides of the circuit board, which requires two photographic negatives (Fig. 37.31) made from positive drawings (black lines on a white background). Each drawing for each side can be made on separate sheets of acetate that are laid over each other when the second diagram is drawn. However, a more efficient method of making a single drawing, from which two negatives are photographically made, uses red and blue tape (Fig. 37.32). A filter on the process camera drops out the red for one negative, and a different filter drops out the blue for the second negative.

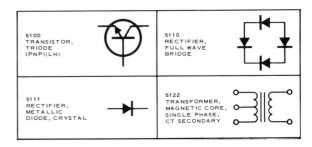

Figure 37.33 Stick-on symbols are available for laying out schematic diagrams. The resulting layouts have a contrast and sharpness better than that found in drawn schematic diagrams. (Courtesy of Bishop Industries Corp.)

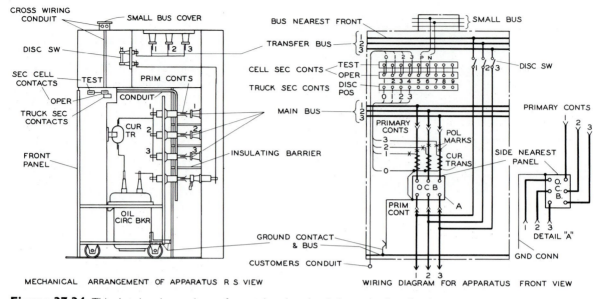

Figure 37.34 This drawing shows views of a metal-enclosed switchgear to describe the arrangement of the apparatus; it also gives the wiring diagram for the unit.

37.12 Installation Drawings

Many types of electric and electronics drawings are used to produce the finished installation, from the designer who visualized the system to the contractor who builds it. Drawings are used to design the circuit, detail its parts for fabrication, specify the arrangement of the devices within the system, and instruct the contractor on how to install the project.

A combination arrangement and wiring-diagram drawing is shown in Fig. 37.34, where the system is shown in a front and right-side view. The wiring diagram explains how the wires and components within the system are connected for the metal-encased switchgear. Busses are conductors for the primary circuits.

Problems

1. (Fig. 37.35) On a size A sheet, make a schematic diagram of the circuit.

2. (Fig. 37.36) On a Size A sheet, make a schematic diagram of the circuit.

3. (Fig. 37.37) On a Size A sheet, make a schematic diagram of the circuit.

4. (Fig. 37.38) On a Size A sheet, make a schematic diagram of the circuit.

5. (Fig. 37.39) On a Size A sheet, make a schematic diagram of the circuit.

6. (Fig. 37.40) On a Size B sheet, make a schematic diagram of the circuit.

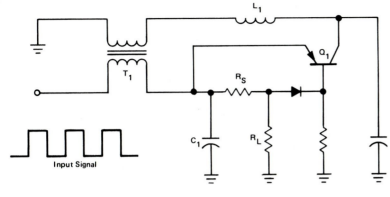

Input Signal

Figure 37.35 Problem 1. A low-pass inductive-input filter. (Courtesy of NASA.)

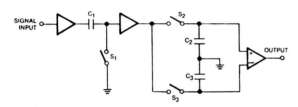

Figure 37.36 Problem 2. A quadruple-sampling processor. (Courtesy of NASA.)

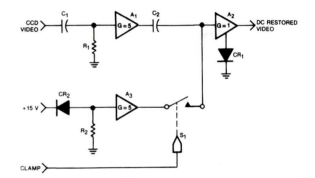

Figure 37.37 Problem 3. A temperature-compensating DC restorer circuit designed to condition the video circuit of a Fairchild area-image sensor. (Courtesy of NASA.)

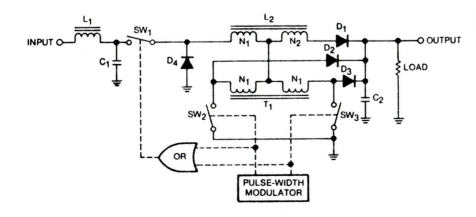

Figure 37.38 Problem 4. A "buck/boost" voltage regulator. (Courtesy of NASA.)

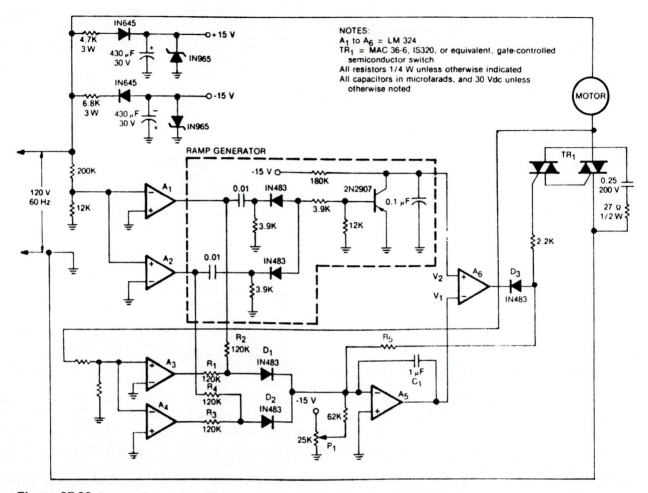

Figure 37.39 Problem 5. An improved power-factor controller. (Courtesy of NASA.)

Figure 37.40 Problem 6. A magnetic-amplifier DC transducer. (Courtesy of NASA.)

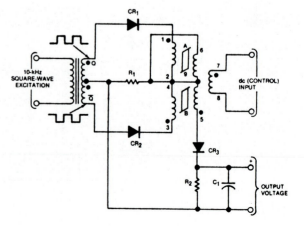

AutoCAD Computer Graphics

38.1 Introduction

This chapter is devoted to the coverage of computer graphics using AutoCAD software (Release 10) on an IBM AT or compatible computer with a minimum of 640 KB RAM, a 20 MB hard disk, a mouse (or tablet), and an A-B plotter (Fig. 38.1). AutoCAD is used as a basis for presenting

Figure 38.1 A view of one of the Engineering Design Graphics computer laboratories at Texas A&M University.

microcomputer graphics because it is the most widely used software for the microcomputer. Although CAD software differs, much of what is learned on one system is transferrable to another system.

The coverage of computer graphics has been introduced in previous chapters to demonstrate how it can be applied to specific types of problems. However, this chapter has been placed at the end of the book so that any last-minute updates from AutoCAD can be included before the book goes to press.

Although commands and data can be input through the keyboard when using AutoCAD, the drafter will be more productive if a mouse or tablet is used. AutoCAD operates on a two-button mouse (Fig. 38.2), where the left button is used to select entities, and the right button acts as a carriage return, which is abbreviated throughout the text as (CR).

Most of AutoCAD's principles and commands are covered in this chapter and previous chapters. However, a great many more sophisticated applications are possible through the application of specialized commands and LISP

653

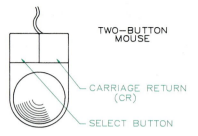

Figure 38.2 The left button on the mouse is the select button, and the right button is the carriage return (CR) button.

programming. Several AutoCAD manuals accompany the software that can be used as references for these specialized applications.

Release 10

The AutoCAD Release 10 covered in this chapter is a three-dimensional extension of Release 9. It can be used in the same two-dimensional manner as Release 9 was used by setting the system variable, FLATLAND, to 1 in the following manner:

```
Command: SETVAR
Variable name or ?: FLATLAND
New value of FLATLAND <0>: 1
```

By using the same procedure, FLATLAND can be set to zero to activate the three-dimensional aspects of AutoCAD. The two-dimensional commands of Release 9 are almost identical to those in Release 10 except in a few instances. Consequently, files created by Release 9 can be used in Release 10.

38.2 Starting Up

Begin AutoCAD by booting up the system and typing ACAD. The Main Menu will appear on the screen:

```
0. Exit AutoCAD
1. Begin a NEW drawing
2. Edit an EXISTING drawing
3. Plot a drawing
4. Printer plot a drawing
5. Configure AutoCAD
```

```
6. File Utilities
7. Compile shape/font description
   file
8. Convert old drawing file
```

```
Enter Selection: 1 (CR)*
Enter NAME of drawing: A:DRAW1 (CR)
```

The number 1 was entered to specify that a new drawing, called DRAW1, is to be drawn. Names of drawings can be no longer than eight characters with no blank spaces, exclusive of drive prefix (A:, for example). AutoCAD automatically adds a file type of .DWG following the assigned name, which will be filed as A:DRAW1.DWG. The Main Menu leaves the screen, and the drawing area with the Root Menu ready for making a drawing is displayed.

Had you wished to edit a previous drawing, you could have responded:

```
Enter selection: 2 (Edit an existing draw-
ing.) (CR)
Enter NAME of drawing: A:DRAW2 (CR)
```

The drawing, A:DRAW2, would then be displayed on the screen as it appeared when last exited from. You may now edit (change) the drawing as a continuation of the last drawing session.

If you have forgotten the names of the drawing files in memory, enter selection 6 from the Main Menu and enter 1 for a listing of them. Type 0 to return to the Main Menu, and enter 2 to name the drawing to edit.

38.3 Experimenting

The beginner who has not used any version of computer graphics, or AutoCAD in particular, can benefit from turning the machine on, loading the system by typing ACAD, and responding to the screen prompts to see how much could be understood from the prompts alone. For example, to draw a line, you would select DRAW from the Root Menu, and LINE from the submenu. By experimenting, you will become familiar with

---

*(CR) will be used to indicate a carriage return or the ENTER key.

the system, its prompts and menus. But you will soon find the need for specific instructions.

38.4 Introduction to Plotting

You may wish to make a plot of your drawing before ending your first session on the computer. To do so, see that a pen is inserted properly in your plotter and that a sheet of paper is loaded. (You may need help from your instructor on how your plotter works.) Type the command PLOT, and press (CR). You will be prompted at the command line as follows:

```
What to plot—Display, Extents, Lim-
its, View, or Window <D>: E (E for Ex-
tents.)
```

The graphics on the screen will be replaced with text that gives the current plotting specifications. For now, look at the current PLOT ORIGIN and the SCALE. If they are not 0,0 and F, respectively, type Y in response to the last question:

```
Do you want to change anything? <N>:
Y
```

The current settings will be shown, one at a time on the screen, and each setting can be changed by typing new values after each prompt. But for now, change only the ORIGIN to 0,0 and SCALE to F for FIT, which means the plot will be sized to fit the extents of the paper size. Continue by pressing (CR) at the other prompts until the plotter begins to plot your drawing.

There is more to know about plotting, but this will introduce you to how your plotter works. In Section 38.73, we give a more detailed coverage of plotting.

38.5 Shutting Down

The first step in turning off the machine is to select the UTILITY command from the Root Menu, which gives the following options: END, QUIT, and SAVE.

End By selecting END, your drawing is saved under the name it was given at the start of the session; your drawing leaves the screen, and the Main Menu reappears.

Save The SAVE command asks for the name to save the drawing under. You may give it a new name and keep the current drawing on the screen in the Drawing Editor. SAVE can also be used to protect against a power failure by periodically saving the drawing under its original name. For example, place an A: in front of the name (A:DRAW3) if you want to save the drawing to drive A. Now enter QUIT.

Quit If you are changing a drawing or looking at a previously made drawing and wish to return to the Main Menu and discard all changes made during the current session, use the QUIT command. You will receive the prompt Do you really want to discard all changes to drawing? By responding Yes or Y, you will be returned to the Main Menu, and the drawing will be left unchanged.

Exit Type 0 to exit AutoCAD and return to DOS (disk operating system). Turn the monitor and computer off to end the session.

38.6 Drawing Layers

AutoCAD provides an infinite number of layers to make a drawing on. Each layer is assigned its own name, color, and line type. For example, you may have a layer name HIDDEN that appears on the screen in yellow, and the line types drawn on it will be dashed lines to represent hidden lines.

Layers can be used to reduce the duplication of effort. For example, architects commonly use the same floor plan for several different applications: one for dimensions, one for furniture arrangement, one for floor finishes, one for electrical details, and so forth. The same basic plan is used for all these applications by turning on the needed layers and turning off others.

Setting Layers   The following layers are sufficient for most working drawings:

Layer Name	State	Color	LineType
0	ON★OFF	7 (White)	CONTINUOUS
VISIBLE	ON★OFF	1 (Red)	CONTINUOUS
HIDDEN	ON★OFF	2 (Yellow)	HIDDEN
CENTER	ON★OFF	3 (Green)	CENTER
HATCH	ON★OFF	4 (Cyan)	CONTINUOUS
DIMEN	ON★OFF	5 (Blue)	CONTINUOUS
CUT	ON★OFF	6 (Magenta)	PHANTOM

Layers are given names that correspond with their line types, and each is assigned a different color to distinguish it on a color monitor. The 0 (zero) layer is the system default layer, which can be turned off but not deleted.

Layers   By selecting LAYERs from the Root Menu, you will receive a subcommand of LAYER, which is selected from the keyboard or by mouse. The following is AutoCAD's dialogue with you while you are setting layers:

```
Command: LAYER (CR)
?/Set/New/On/Off/Color/Ltype/
Freeze/Thaw: New or N (CR)
Layer name(s): VISIBLE,
HIDDEN,CENTER,HATCH,DIMEN,
CUT (CR)
```

This series of responses has created six new layers by name, and each will have a white default color and a continuous line type.

Color   To set the COLOR of each layer, we must respond in the following manner:

```
?/Set/NEW/ON/OFF/COLOR/Ltype/
Freeze/Thaw: COLOR or C (CR)
Color: RED (or 1) (CR)
Layer name(s) for color 1
(red) <VISIBLE>: VISIBLE (CR)
```

The VISIBLE layer has been assigned the color red, which could have been assigned with the number 1 instead of the word, RED. The colors

of the other layers must be assigned in the same manner, one at a time.

Line Types   are assigned to layers in the following manner:

```
?/Set/New/On/Off/Color/Ltype/
Freeze/Thaw: Ltype (or L) (CR)
Linetype (or ?) <CONTINOUS>:
HIDDEN (CR)
Layer name(s) for linetype
HIDDEN <0>: HIDDEN (CR)
```

The lines on the HIDDEN layer will now be drawn with dashed (hidden) lines. By typing LINETYPE, and pressing (CR) twice, the line types available from AutoCAD will be displayed (Fig. 38.3).

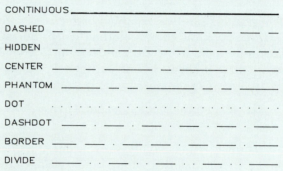

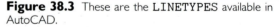

Figure 38.3   These are the LINETYPES available in AutoCAD.

Set   We must SET a layer to make it the current layer in order to draw on it by responding in the following manner:

```
?/Set/New/On/Off/Ltype/Freeze/
Thaw: SET (or S) (CR)
New current layer <0>:
VISIBLE (CR) (CR) (Two carriage returns.)
```

On/Off   Any line drawn will appear on the VISIBLE layer in red with continuous lines. Once we have defined a number of layers and have drawn on each, they can be turned on or off by the ON/OFF command in the following manner:

```
?/Set/New/On/Off/Ltype/Freeze/
Thaw: ON (or OFF)
```

Layer name(s) to turn on:
<u>VISIBLE</u> (or <u>*</u> for all layers)
(Or <u>VISIBLE</u>, <u>HIDDEN</u>, <u>CENTER</u>, to turn
these layers ON or OFF.) (CR)

Even though several layers are on, you can draw on only one layer, the **current layer.** By selecting the question-mark option (?), the screen will display the current listing of the layers, their line types, colors, and on/off status. Press function key F1 to change the screen back to the graphics editor.

Freeze and Thaw   FREEZE and THAW are options under the LAYER command. By FREEZEing a specified layer, it will not be redrawn or plotted until it has been THAWed; thus a drawing can be regenerated much faster on the screen than when OFF is used. This option allows unneeded layers to be turned off much like the layer option, OFF. To save your file of layers and their specifications, use the command END, which will return you to the Main Menu.

But before we END, let's set additional parameters as shown in the next section.

38.7 Setting Screen Parameters

To set screen parameters, you must become familiar with the under SETTINGS of the Root Menu which are LIMITS, GRID, UNITS, SNAP, AXIS, ORTHO, DRAGMODE, BLIPMODE, LTSCALE, and the function keys.

Limits   The LIMITS command is used to establish the size of the screen area that represents the area of the paper a drawing will be plotted on. A full-size drawing that will fit on a size A sheet (11 × 8.5 inches) will have a drawing area of about 10.1 × 7.8 inches for most plotters. If millimeters are used, the limits will be approximately 257, 198. LIMITS are set as follows:

Command: <u>LIMITS</u> (CR)
ON/OFF Lower Left corner:
‹0.00,0.00› : (CR) (To accept default value.)
Upper right corner:
‹36.00,24.00› : <u>10</u>,<u>7.8</u> (CR)

Grid   A grid on the screen can be set in the following manner:

Command: <u>GRID</u> (CR)
On/Off/Value (X)/Aspect
‹0.00› : <u>.2</u> (CR)

This command paints the screen with a square pattern of dots that are each 0.2 in. apart. To make the newly assigned limits fill the working area of the screen, use the ZOOM/ALL command from the DISPLAY command on the Root Menu.

Units   The UNITS used in a drawing must be assigned before the limits can be set. By typing the command <u>UNITS</u>, AutoCAD will give the following listing to select from:

1. Scientific (2.67E+02)
2. Decimal (267.00 inches or millimeters)
3. Engineering (12'−4.50")
4. Architectural (16'−3 1/2")

Enter choice, 1 to 4 ‹2›: <u>2</u> (CR)

Select the type of units, and respond to the following prompts to specify the details desired, such as the number of decimal places. Decimal units should be used for the metric system (millimeters) and the English system (inches). To return to the Drawing Editor, press function key F1.

Snap   The SNAP command can be used to make the cursor on the screen snap to a visible or invisible grid. If your grid is set at spacings of 0.2 in., the cursor can be made to snap to the grid or to an invisible 0.1-in. grid between the visible grid points (Fig. 38.4). Press function key F9 to turn SNAP on or off.

Command: <u>SNAP</u> (CR)
On/Off/Value/Aspect/
Rotate/Style: <u>0.1</u> (CR)

Axis   The AXIS command is used to display ruler lines of a specified spacing on the screen across the bottom and right sides in the following manner:

Command: <u>AXIS</u> (CR)
On/Off/Tick spacing
(X)/Aspect: <u>0.20</u> (CR)

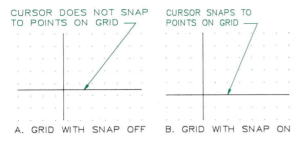

A. GRID WITH SNAP OFF   B. GRID WITH SNAP ON

Figure 38.4 The SNAP command can be used to make the cursor stop on visible or invisible grids on the screen.

This response sets the ruler with tick marks 0.20 units apart. By responding with 5X, the tick marks are spaced at a multiple of five times the snap resolution, which places tick marks at every fifth snap point. The ASPECT option allows you to set different values on each scale of the axis.

Ortho The ORTHO command forces all lines to be either vertical or horizontal, which aids in making orthographic views where most lines meet at right angles.

> Command: ORTHO (Select ON or OFF from the screen menu.) (CR)

ORTHO can be turned off by using Ctrl O or by pressing function key F8 (Fig. 38.5). Function keys can be used to turn screen parameters on and off once they have been set. **Running coordinates** can be obtained at the top of the screen to give the coordinates of the cursor's position by pressing F6. The **status**

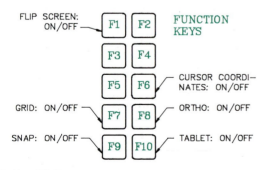

Figure 38.5 The function keys on the IBM (or compatible) keyboard can be used to activate six functions of AutoCAD.

line (Fig. 38.6) at the top of the screen gives the current layer by name; it shows when SNAP and ORTHO are turned on by displaying these words.

Dragmode When an object is moved, it can be dynamically moved across the screen by setting the DRAGMODE command to ON. If this command was set to OFF, the object would be moved, but its image would appear only after its destination point had been selected. By setting DRAGMODE to AUTO, all objects that can be dragged are dragged automatically without prompting.

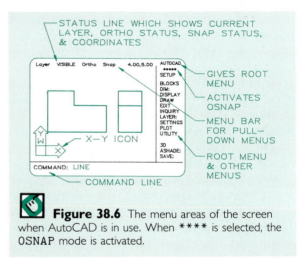

Figure 38.6 The menu areas of the screen when AutoCAD is in use. When **** is selected, the OSNAP mode is activated.

Blipmode By setting BLIPMODE to ON, a temporary blip (+ sign) will appear on the screen when a point is selected with the cursor. Blips are removed by pressing function key F7 twice. If this command is set to OFF, no blips will appear on the screen.

LTscale Hidden lines, centerlines, phantom lines, and other noncontinuous lines can be drawn with varying sizes of dashes and spaces by changing the LTSCALE value to a larger or smaller number.

Format File If the previous parameters were specified, they would be saved when you ENDed the file. This file should be saved with its parameters as an "empty" file with no drawing, and named `A:FORMAT` so that it can be called up as a drawing to begin a new drawing on. Once the drawing is completed in this `FORMAT` file, `SAVE` the drawing, and give it a name different from `A:FORMAT`; then <u>QUIT</u>, and respond <u>YES</u> when asked if you wish to discard all changes in the drawing. In this way, you have returned to the original file, `A:FORMAT`, without disturbing its parameters, allowing it to be used again, and you have saved a drawing that was made from it.

Root Menu By placing the cursor on `AUTO-CAD`, the Root Menu can be recalled. The four stars, `****`, can be selected to activate the `OSNAP` command (Fig. 38.6).

Several submenus are longer than will fit on the screen at one time, so the `NEXT` command is used to obtain them.

Menus Commands can be executed from the **keyboard,** the **root menu,** or the **pull-down menu bar.** The keyboard is used like a typewriter to type in commands that are shown at the **command line** as they are typed (Fig. 38.6). Your mouse or digitizer can be used to select commands from the Root Menu at the right of the screen by moving to the desired command and pressing the left mouse button.

Pull-Down Menus

Several pull-down menus and dialogue boxes can be accessed by placing the cursor on the menu bar at the top of the screen with the mouse as shown in Fig. 38.7. Only one heading can be activated at a time. Subcommands from which to select various options are given in the pull-down menus under the headings (Fig. 38.7).

Several of the options give dialogue boxes and icon menus on the screen from which to select or add information. Figure 38.8 shows a dialogue box that is filed under the **Modes** heading of the **Menu Bar.** The cursor is moved from box to box by pressing carriage return (CR), or by using your mouse, and new information is typed

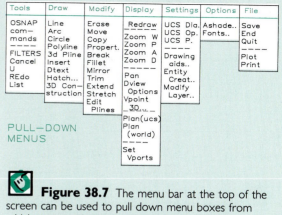

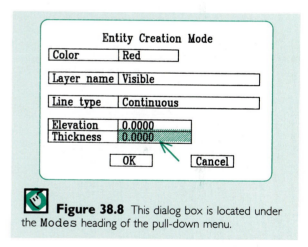

Figure 38.7 The menu bar at the top of the screen can be used to pull down menu boxes from which commands can be selected with your cursor.

Figure 38.8 This dialog box is located under the `Modes` heading of the pull-down menu.

in if desired. Some dialogue boxes have subdialogue boxes. For example, the **Entity Creation** dialogue box has a subdialogue box that shows all the colors that can be selected.

The **Modify Layer** dialogue box (Fig. 38.9) located under the `SETTINGS` heading of the **Menu Bar** can be used change the status of a layer. It can be used to turn layers on or off, freeze or thaw them, assign the current layer, or create a new one.

The **Drawing Aids** dialogue box (Fig. 38.10) is also found under the `SETTINGS` heading

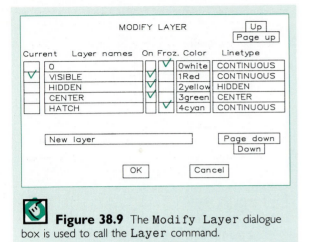

Figure 38.9 The Modify Layer dialogue box is used to call the Layer command.

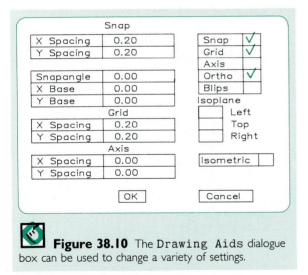

Figure 38.10 The Drawing Aids dialogue box can be used to change a variety of settings.

of the **Menu Bar.** It can also be used to set SNAP specification, turn on or off the GRID, AXIS, OR-THO, BLIPMODE, or isometric features. Other available dialogue boxes will be discussed later.

38.8 Utility Commands

Utility commands can be used to control the operation of AutoCAD and make changes in the files that are developed.

Help The HELP (?) command gives a list of commands on the screen. Press Ctrl C to abort the listing process. Press F1 to return to the graphics mode on the screen. To obtain additional information on a specific command, respond to the prompts as follows:

Command: HELP (or ?) (CR)
Command name (Return for list): LINE
(CR)

The screen will give information about the LINE command.

Rename The RENAME command is used to change the name of a block, layer, linetype, text style, named view, UCS, or viewport. The command prompts you for the Old name and the New name. Continuous lines and layer 0 cannot be RE-NAMEd.

Files The FILES command permits you to list, delete, rename, and copy files from the **Drawing Editor.** By typing FILES, the following listing will appear on the screen:

File Utility Menu
0. Exit file utility menu
1. List drawing files
2. List user specified files
3. Delete files
4. Rename files
5. Copy files

Enter selection: 1 (CR)

By typing 1, you will be prompted for the disk drive to search, Enter prefix. Enter C: if that is the disk drive you wish to have searched.

Option 2, List user-specified files, can be used to display the files that you specify. For example, by responding to ENTER FILE SEARCH SPECIFICATIONS: with C:*.BAK, you will get a listing of all drawing files with a suffix .BAK that are stored on disk drive C.

Option 3, Delete files, allows you to eliminate unneeded files on your disks. You can specify a file such as A:DRAW1.DWG, or you can use wild cards such as A:*.DWG to delete all files with a DWG suffix. When using wild cards, each file that meets the specifications will be listed one at a time, and you are given the option to delete them by entering <u>Y</u> or <u>N</u> as they are listed.

Option 4, Rename files, is used to change the name of a file by responding to the prompts as follows:

```
Enter current file name:
A:DRAW1.DWG (CR)
Enter new file name:
A:DRAW2.DWG (CR)
```

Option 5, Copy files, lets you copy a file with the following prompts:

```
Enter name of Source file
name: B:DRAW1.DWG (CR)
Enter name of Destination file name:
C:DRAW1.DWG (CR)
```

Shell   The SHELL, or SH, command gives access to the DOS operating system while remaining in the **Drawing Editor** by responding as follows:

```
Command: SHELL (CR)
DOS command: DIR A: (Or similar com-
mand.) (CR)
Command: (Reappears on screen.)
```

Purge   The PURGE command can be used when you **first** begin editing an existing drawing in the following manner:

```
Command: PURGE (CR)
Purge unused Blocks/Layers/
LTypes/SHapes/STyles?
All: ALL (CR)
```

The ALL response is used to eliminate any unused objects that are in the drawing file. The other options can be used to purge specific features of a drawing. You will be prompted for each PURGE one at a time.

38.9   Custom-designed Lines

In Section 38.6, we introduced you to Auto-CAD's LINETYPEs. By typing <u>LINETYPE</u>, (CR), <u>?</u>, the types and names of lines available will be displayed on the screen (Fig. 38.3). These lines can be assigned to a layer by selecting the LTYPE option of the LAYER command and typing in the

preferred line type, such as hidden, center, or dashed.

These lines are composed of lines, points, and spaces. You can get long dashes or short dashes by the LTSCALE command and the assignment of an LTSCALE factor to lengthen or shorten the dashes. When this command is used, **all** lines composed of dashes and spaces are changed globally on the screen. Therefore, you may find a need for a line, such as a centerline, with smaller spaces and shorter dashes for small circles.

You can custom design line types by using the LINETYPE command and its CREATE option (Fig. 38.11). In Step 1, you will be prompted for the line's name and the file it is to be stored in. In Step 2, a description can be given by typing in keyboard dashes and dots to represent the line. When asked to enter the line pattern, begin with a dash (a positive length) or a dot, represented by a zero. Spaces are represented by minus values. Spaces, dots, and dashes are separated by commas. Only one typical pattern that will be used

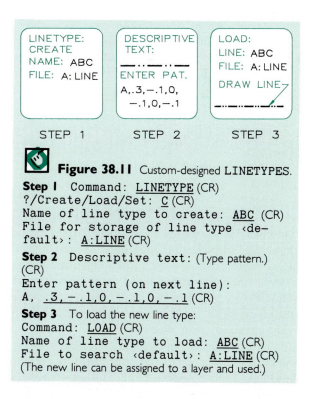

Figure 38.11 Custom-designed LINETYPES.

Step 1   Command: <u>LINETYPE</u> (CR)
?/Create/Load/Set: <u>C</u> (CR)
Name of line type to create: <u>ABC</u> (CR)
File for storage of line type ‹default›: <u>A:LINE</u> (CR)

Step 2   Descriptive text: (Type pattern.) (CR)
Enter pattern (on next line):
A, <u>.3,−.1,0,−.1,0,−.1</u> (CR)

Step 3   To load the new line type:
Command: <u>LOAD</u> (CR)
Name of line type to load: <u>ABC</u> (CR)
File to search ‹default›: <u>A:LINE</u> (CR)
(The new line can be assigned to a layer and used.)

in multiples need be shown, for example, dash, -space, short dash, -space.

Use the LINETYPE command and its LOAD option to gain access to the newly designed line style that can be assigned to a LAYER by the LTYPE option (Step 3).

38.10   Making a Drawing—Lines

Turn your computer on and obtain AutoCAD's **Main Menu** by the method required by your system. It might involve entering a sequence of commands, such as C:CD\ACAD (CR) and then ACAD (CR). Once you have the **Main Menu,** type 1, Begin a NEW drawing, if you are going to set your drawing parameters from scratch. If you wish to use your FORMAT file that was developed in Section 38.7 with its assigned parameters, type 2, Edit an EXISTING drawing. When prompted for a filename, type A:FORMAT, and the file will be loaded into the drawing editor, and the screen will appear ready for a drawing to be made.

If you wish to draw a line, you have three methods that can be used: the **Root Menu,** the keyboard, or the pull-down menu bar. If you select DRAW and then LINE from the Root Menu (Fig. 38.12), the command line on the screen will prompt you as follows:

```
Command: LINE (CR)
From point: (Select a point or give coordi-
nates.)
To point: (Select second point or give coordi-
nates.)
To point: (You can continue in this manner
with a series of connected lines.)
```

The lines are drawn by moving the cursor about the screen and selecting endpoints by pressing the select button of the mouse (the left button). The current line will **rubber band** from its last point (Fig. 38.13), and lines can be drawn in succession until you press (CR) on the keyboard, or the right button of your mouse.

The pull-down menu bar is used by moving the cursor to the top of the screen until the menu headings appear. Select DRAW and its subcom-

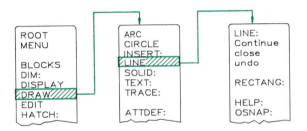

Figure 38.12 You may progress from the Root Menu to levels of submenus by selecting commands with the keyboard or mouse.

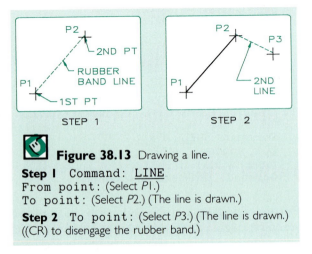

Figure 38.13 Drawing a line.

Step 1 Command: LINE
From point: (Select P1.)
To point: (Select P2.) (The line is drawn.)

Step 2 To point: (Select P3.) (The line is drawn.)
((CR) to disengage the rubber band.)

mand, LINE, and draw the line in the usual manner.

The keyboard input can be used separately or as a supplemental method of drawing lines by one of three methods: **absolute coordinates, relative coordinates,** or **polar coordinates.** By typing LINE, you will be prompted, From points, at which time you can either locate the starting point with your mouse or type in the coordinates. This technique of typing specifications is especially useful to precisely locate points that involve decimal places, such as 4.3521.

Absolute Coordinates can be typed at the keyboard, such as 2.566, 5.903, which gives the coordinates from the origin of 0,0 on the screen. Successive points are entered in the same manner

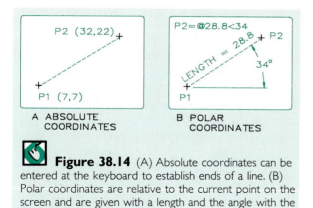

A ABSOLUTE COORDINATES   B POLAR COORDINATES

Figure 38.14 (A) Absolute coordinates can be entered at the keyboard to establish ends of a line. (B) Polar coordinates are relative to the current point on the screen and are given with a length and the angle with the horizontal measured clockwise.

with each point located from the same origin of 0,0 (Fig. 38.14A).

Relative Coordinates are *X* and *Y* values measured from the current point on the screen. These coordinates are preceded by the @ symbol, for example, @4.25,6.78.

Polar Coordinates are given by a distance and an angle measured from the current point. The distance is preceded by the @ symbol followed by the ‹ sign and the angle measured from 0 in a counterclockwise direction. Figure 38.14B gives

an example of a line that is 28.8″ long and makes 34° with the horizontal, @28.8‹34.

You can correct errors when typing commands by one of the following methods:

Ctrl X (Deletes the line.)

Ctrl C (Cancels the current command and returns the Command: prompt.)

Backspace (Deletes one character at a time.)

The status line at the top of the screen shows the length of the line and angle from the last point when you are rubber-banding from point to point. When drawing a continuous series of lines, the CLOSE command can be used to draw the last line to the beginning point, as Fig. 38.15 shows.

38.11 Selection of Entities

One of the most often-occurring prompts is Select objects:, which asks you to select an entity, or set of entities, that are to be ERASED, CHANGED, or modified in other ways. Several methods of SELECTing entities are available. You may select entities one at a time with the cursor; use a window, a crossing window, or a box; type LAST or PREVIOUS; or use the auto option.

Figure 38.16 illustrates how a single entity can be selected with the cursor. Or, when prompted, Select objects:, M can be typed at the keyboard, and several entities can be selected one at a time.

When prompted to select, the window option (W) can be used, after which you will be prompted for the first and second diagonals of a window. Only entities totally within the window will be selected (Fig. 38.16).

The crossing option (C) shown in Fig. 38.16, selects the entities within the window and those crossing the window. When prompted to select objects, BOX can be typed at the keyboard, and a window or a crossing window can be used by selecting the second diagonal to the right or to the left respectively (Fig. 38.16).

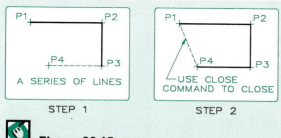

A SERIES OF LINES — STEP 1   USE CLOSE COMMAND TO CLOSE — STEP 2

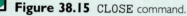
Figure 38.15 CLOSE command.

Step 1 A series of lines are drawn using the LINE command from *P1* through *P3*. (The last line is a rubber-band line until the next point is selected.)

Step 2 After *P4* has been selected, select CLOSE, and the line will be drawn to the first point of the series.

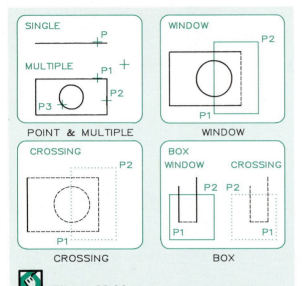

Figure 38.16 Object select options. When prompted by a command to `Select objects:`, you may select them as single or multiple points. A window option (`W`) can be used to select objects that lie completely within the windows, or crossing option (`C`) can be used to select objects within or crossed by the window. The `BOX` option can be used for both the window and crossing options, determined by the sequence in which the diagonal corners are selected.

Selection can be made by last (`L`), which selects the most recently created object.

The previous option (`P`) allows you to recall the last selected set of objects for editing. For example, a number of objects can be selected for `MOVE`ing, then can be `MOVE`d. After entering the `MOVE` command and typing P, the last group of entities are remembered and can be `MOVE`d to a new position.

When selecting a set of objects, you can remove each of them in reverse order one at a time by typing U (undo) as many times as needed.

While in the object selection process, you may remove a selected object by typing R (remove) and selecting the objects to be removed. To add others to the set, type A (add) and select the objects to be added. Once the selection is finished to your satisfaction, give a null reply (press CR) to the `Select/remove object:` prompt.

By setting the selection option to `SI` (single) the object, or sets of objects, will be acted upon without pausing for interaction with the drafter.

38.12 Erasing and Breaking Lines

The `ERASE` command, a subcommand under `EDIT`, (or `MODIFY`), is used to remove parts of a drawing. The `LAST` option calls for the erasure of single entities (lines, text, circles, arcs, and blocks) one at a time working backward from the one most recently drawn. The use of a `WINDOW` is the second option for erasing, where entities that lie completely within the window will be erased as shown in Fig. 38.17.

```
Command: ERASE (CR)
Select objects or Window or
Last: WINDOW (or W) (CR)
First corner: (Select P1.)
Other corner: (Select P2.) 4 found
Select objects: (CR) or Window or
last
(The entities are erased.)
```

A similar method of erasing lines is the `CROSSING` (`C`) option shown in Fig. 38.18. Any entity (line, arc, circle, or text) that is crossed by the window is removed in its entirety. The default of the `ERASE` command is `Select Ob`—

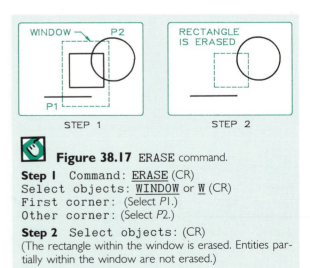

Figure 38.17 ERASE command.

Step 1 Command: `ERASE` (CR)
`Select objects:` `WINDOW` or `W` (CR)
`First corner:` (Select P1.)
`Other corner:` (Select P2.)

Step 2 `Select objects:` (CR)
(The rectangle within the window is erased. Entities partially within the window are not erased.)

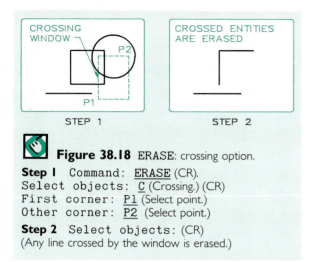

Figure 38.18 ERASE: crossing option.

Step 1 Command: ERASE (CR).
Select objects: C (Crossing.) (CR)
First corner: P1 (Select point.)
Other corner: P2 (Select point.)
Step 2 Select objects: (CR)
(Any line crossed by the window is erased.)

jects, which requires that you point to entities on the screen with the cursor to indicate those to erase (Fig. 38.19). By pressing (CR) the entities are erased. By using the pull-down menu, the entities are erased as they are selected.

Entities lying within the window that should not be erased can be removed by using the RE—MOVE command and selecting the entities with the cursor. When selected, the entities become dashed lines on the screen, indicating they have been marked for erasure. Additional lines can be

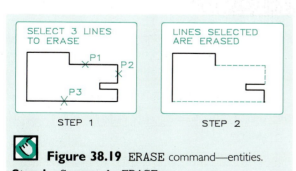

Figure 38.19 ERASE command—entities.

Step 1 Command: ERASE
Select Objects: (Select entities, lines, with P1, P2 and P3.)
Step 2 Press (CR) again, and the lines are erased. Use Command: OOPS (CR) to recall erased entities.

marked for erasure by the ADD command in the same manner as the REMOVE command but with an opposite effect.

Should you mistakenly erase something, the OOPS command can be used to restore the last erasure, but only the last erasure.

The BREAK command is used to remove a part of an entity, such as a line (Fig. 38.20).

Command: BREAK (CR)
Select object: (Select point P1 on the line to be broken.)
Enter second point or F:F (CR)
Enter first point: (Select P2.)
Enter second point: (Select P3.)
(The line is broken from P2 to P3.)

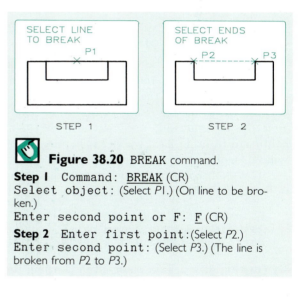

Figure 38.20 BREAK command.

Step 1 Command: BREAK (CR)
Select object: (Select P1.) (On line to be broken.)
Enter second point or F: F (CR)
Step 2 Enter first point:(Select P2.)
Enter second point: (Select P3.) (The line is broken from P2 to P3.)

Had the break not begun and ended at intersections of lines, you could have omitted the step in which the F response was given. But this extra step ensures that the computer understands which line is to be broken.

A portion of a line may be broken away by using the BREAK command (Fig. 38.21). By windowing the line to be broken and selecting two points at the ends of the break, that part of the line is removed.

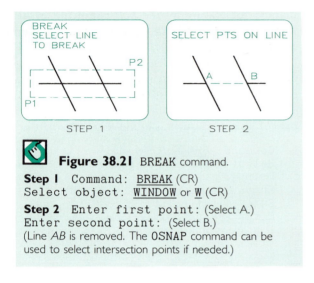

Figure 38.21 BREAK command.

Step 1 Command: <u>BREAK</u> (CR)
Select object: <u>WINDOW</u> or <u>W</u> (CR)

Step 2 Enter first point: (Select A.)
Enter second point: (Select B.)
(Line *AB* is removed. The OSNAP command can be used to select intersection points if needed.)

38.13   UNDO Command

The U command can be used to undo the latest entity placed on the screen. By entering <u>U</u> commands, one after another, the drawing can be erased back to its beginning point. Immediately after the U command, you may use the REDO command to bring back the deleted entity; OOPS will not work.

The UNDO command can be used for several different operations. Its options are as follows:

Command: <u>UNDO</u> (CR)
Auto/Back/Control/End/Group/
Mark/‹Number›: <u>4</u> (CR)

By entering <u>4</u>, it has the same effect as using the U command four separate times.

MARK can be used to mark a point in a drawing and then to add other experimental features that can be disposed of if desired by the BACK option. This will UNDO only that part of the drawing back to what was drawn when you used the MARK option. You will be prompted, This will undo everything. OK? ‹Y›. By responding <u>Y</u>, the mark will be removed also, making it possible for the next U or UNDO command to proceed backward past the mark.

The UNDO GROUP, UNDO END, and AUTO subcommands can be used to remove groups of en-

tities at a time, but these options are meant to be used with menus.

The CONTROL subcommand has three options: ALL, NONE, and ONE. ALL turns on the full features of the UNDO command, and NONE turns off the features entirely. The ONE option limits U and UNDO commands to single operations and requires the least amount of disk space.

38.14   TRACE Command

Wide lines, thicker than the point of a pen, can be drawn using the TRACE command in the following manner:

Command: <u>TRACE</u> (CR)
Width: <u>0.4</u> (CR)
From point: <u>2, 3</u> (CR)
To point: <u>4, 6</u> (CR)
To point: <u>6, 2</u> (CR)

With FILL on, the line will be drawn as shown in Fig. 38.22A. When FILL is off, the line will be drawn as parallel lines (Fig. 38.22B). The plotting of the line being entered will not be plotted until the endpoint of the next line is indicated in order for the program to compute the "miter" angles at each corner.

38.15   POINT Command

The POINT command is used to locate points on a drawing. The point markers can be any of those shown in Fig. 38.23. The marker is selected by using two options under the SETVAR command: PDMODE and PDSIZE. The basic markers can be set in the following manner:

Command: <u>SETVAR</u> (CR) Variable name
or ?:<u>PDMODE</u> (CR)
New value for variable-name ‹0›: <u>3</u> (CR)
(Marker = X.)

The PDSIZE command is used to assign a size to the marker in the following way:

Command: <u>SETVAR</u> (CR) <u>Variable</u> name or
?: <u>PDSIZE</u> (CR)
New value for variable-name ‹0.000›:
<u>0.1</u> (Size of marker.) (CR)

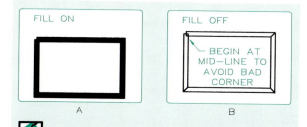

Figure 38.22 TRACE command. (A) When the TRACE command is used with FILL ON, lines are drawn solid to the width specified. (B) When FILL is turned off, parallel lines are drawn.

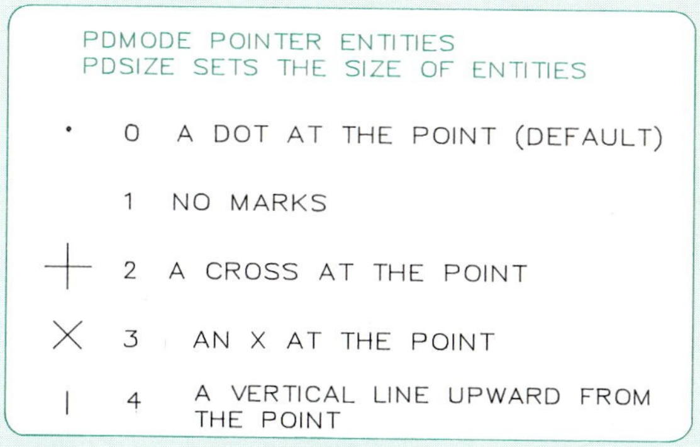

```
PDMODE POINTER ENTITIES
PDSIZE SETS THE SIZE OF ENTITIES

·   0   A DOT AT THE POINT (DEFAULT)

    1   NO MARKS

+   2   A CROSS AT THE POINT

X   3   AN X AT THE POINT

|   4   A VERTICAL LINE UPWARD FROM
        THE POINT
```

Figure 38.23 The pointer entities (markers) used to locate points on a drawing can be selected and sized using the PDMODE and PDSIZE commands.

The PDSIZE variable gives the size of the marker on the drawing. If this value is a negative number, the marker will be sized as a percentage of the screen and will appear the same size regardless of the zooming that may occur. When entered as a positive number, the size of the marker will vary with each zoom.

Additional styles of markers can be obtained by adding 32, 64, and 96 to the basic values of 0 through 4 (Fig. 38.24).

When you are drawing lines and other entities, a plus sign (BLIP) is made temporarily on the screen to mark the points. When the screen is

redrawn by pressing the F7 function key, they are removed since the plus signs are markers and not points on the screen. By turning the BLIPMODE command either ON or OFF, you choose whether or not to have BLIPS shown on the screen. It is recommended that beginners set the BLIPMODE to ON.

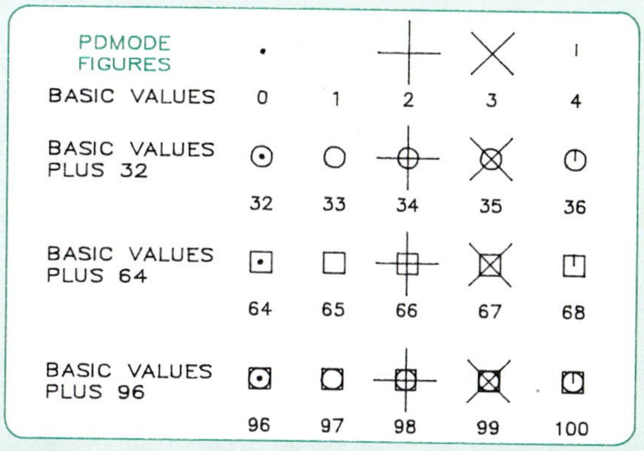

Figure 38.24 The basic pointer symbols can be changed by adding 32, 64, and 96 to the basic symbols, 1 through 4.

38.16 Drawing Circles

The command for drawing circles is found under the DRAW submenu, where you may give the center and radius, the center and diameter, or three points. Used with DRAG, you can move the cursor and see the circle change until it is the size you want it to be (Fig. 38.25). DRAG can be turned ON or OFF by inserting the command, DRAGMODE, followed by ON or OFF.

38.17 Tangent Options of the CIRCLE Command

By using the tangent options of the CIRCLE command, a circle can be drawn tangent to a circle and a line, three lines, three circles, or two lines and a circle. Figure 38.26 shows a circle drawn

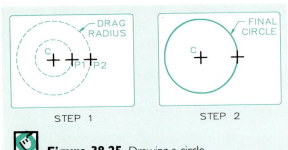

Figure 38.25 Drawing a circle.

Step 1 Command: <u>CIRCLE</u> (CR)
3P/2P/‹Center point›: (Locate center *C* with cursor.)
Diameter/‹Radius›: <u>DRAG</u> (Select center *P1*, move to *P2*, and the circle will dynamically change if DRAGMODE is ON.)

Step 2 The final radius is selected with the select button, and the circle is drawn.

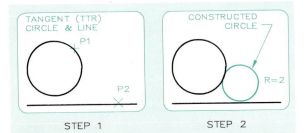

Figure 38.26 Tangent options (TTR).

Step 1 Command <u>CIRCLE</u> (CR)
3P/TTR/‹Center Pt›: <u>TTR</u>
Enter Tangent spec: (Select point on circle.)
Enter second Tangent spec: (Select point on line.)

Step 2 Radius: <u>2</u> (CR)
(The circle with radius = 2 is drawn tangent to the line and circle.)

tangent to a line and a circle. The TTR option is chosen, the radius is given, and points on the circle and line are selected.

An arc is drawn tangent to three lines in Fig. 38.27 by using the 3P option of the CIRCLE command. In this case, the radius cannot be given since it must be computed.

A circle can be drawn tangent to two lines and a circle or tangent to three circles, as Fig. 38.28 shows. These circles are also found by using the 3P option.

38.18 Drawing Arcs

The ARC command is found under the DRAW menu, where arcs can be drawn using nine combinations of variables, including starting point,

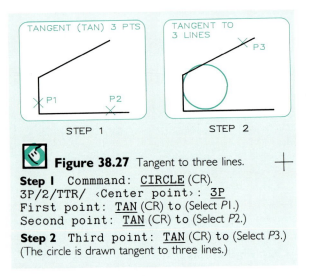

Figure 38.27 Tangent to three lines.

Step 1 Commmand: <u>CIRCLE</u> (CR).
3P/2/TTR/ ‹Center point›: <u>3P</u>
First point: <u>TAN</u> (CR) to (Select *P1*.)
Second point: <u>TAN</u> (CR) to (Select *P2*.)

Step 2 Third point: <u>TAN</u> (CR) to (Select *P3*.)
(The circle is drawn tangent to three lines.)

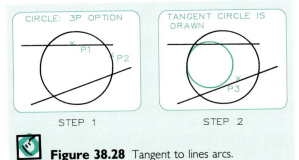

Figure 38.28 Tangent to lines arcs.

Step 1 Command: <u>CIRCLE</u> (CR)
3P/2P/TTR/‹Center point›: <u>3P</u> (CR)
First point: <u>TAN</u> (CR) to (Select *P1* on line.)
Second point: <u>TAN</u> (CR) to (Select *P2* on circle.)

Step 2 Third point: <u>TAN</u> (Select *P3* on line.)
(A circle is drawn tangent to the lines and circle.)

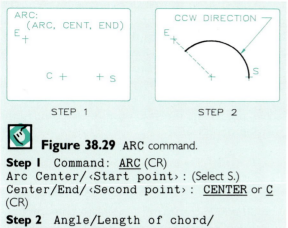

STEP 1   STEP 2

Figure 38.29 ARC command.

Step 1 Command: ARC (CR)
Arc Center/<Start point> : (Select S.)
Center/End/<Second point> : CENTER or C
(CR)

Step 2 Angle/Length of chord/
<Endpoint> : DRAG (Select point E.) (The arc is
drawn to an imaginary line from S to E in a counterclockwise direction.)

center, angle, ending point, length of arc, and radius. For example, the S, C, E version requires
that you locate the starting point (S), the center
(C), and the ending point (E) (Fig. 38.29). The
arc begins at point S, but point E need not lie on
the arc. Arcs are drawn in a counterclockwise direction by default.

To continue a line with an arc drawn from
the last point of the line and tangent to it, re-

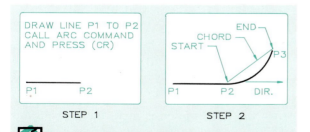

STEP 1   STEP 2

Figure 38.30 ARC tangent to end of line.

Step 1 Command: LINE (CR)
From point: (Select P1.)
To point: (Select P2.) (CR)

Step 2 Command: ARC (CR)
Center/<Start point> : (CR)
End point: (Select P3.)

spond as follows:

    Command: ARC (CR)
    Center/<Start point> : (CR)

An arc may now be drawn tangent to the last
point of the line, which is useful for drawing runouts on a part (Fig. 38.30).

With the only difference being the order of
the commands, the same technique can be used
for drawing a line from a previously drawn arc.

The DRAGMODE command can be turned on to
allow the arcs to be seen before they are selected
for their final positions.

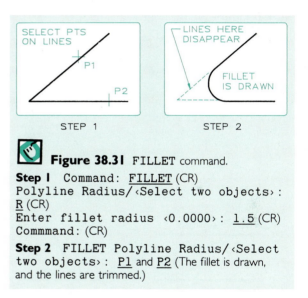

STEP 1   STEP 2

Figure 38.31 FILLET command.

Step 1 Command: FILLET (CR)
Polyline Radius/<Select two objects> :
R (CR)
Enter fillet radius <0.0000> : 1.5 (CR)
Commmand: (CR)

Step 2 FILLET Polyline Radius/<Select
two objects> : P1 and P2 (The fillet is drawn,
and the lines are trimmed.)

38.19   FILLET Command

FILLETs can be drawn to any desired radius between two nonparallel lines, whether or not they
intersect. The fillet is drawn, and the lines are
trimmed as shown in Fig. 38.31.

By entering a fillet radius of 0, noninteresct-
ing lines will be automatically extended to form
a perfect intersection. Once the radius is assigned,
it remains in memory as the default radius for
drawing additional fillets.

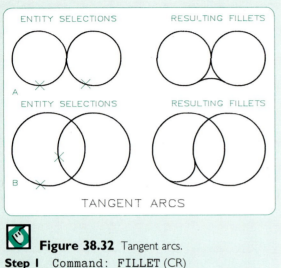

Figure 38.32 Tangent arcs.

Step 1 Command: <u>FILLET</u> (CR)
Polyline/Radius/‹Select two objects›:
<u>R</u> (CR)
Enter fillet radius ‹0.0000›: <u>1.2</u> (Example.) (CR)

Step 2 Command: (CR)
FILLET Polyline/Radius/‹Select two objects›: (Select a point on each circle, and the fillet arc is drawn.)

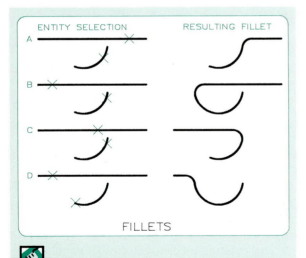

Figure 38.33 When a fillet radius has been selected, points on each line or arc can be selected for filleting. Each fillet is determined by the location of the points selected.

Arcs (fillets) can be drawn tangent to circles or arcs, as shown in Fig. 38.32. The radius must be specified and points on two circles selected. AutoCAD will draw fillet arcs that most nearly approximate the locations selected on the circles.

Figure 38.33 shows examples of fillets that connect lines and arcs. The location of the selected points on the two entities determines the position of the fillet.

38.20 CHAMFER Command

The CHAMFER command is used to construct angular bevels at the intersections of lines or polylines. Select two lines, and the lines are trimmed or extended, and the CHAMFER is drawn. Press (CR) and you are ready to repeat this command using the last values if other corners are to be chamfered (Fig. 38.34). A polyline is chamfered in the same manner.

38.21 POLYGON Command

A many-sided figure composed of equal sides can be drawn using the POLYGON command. The center is located, the number of sides given, and inscribing or circumscribing options entered (Fig. 38.35).

Command: <u>POLYGON</u> (CR)
Number of sides: <u>5</u> (CR)
Edge/‹Center of polygon›: (Locate center.)
Inscribed in circle\Circumscribed about circle (I/C): <u>I</u> (CR)
Radius of circle: (Type length and (CR) or select length with pointer.)

When the EDGE option is used, the next prompt asks

First endpoint of edge: (Select point.)
Second endpoint of edge: (Select point.)

The polygon will then be drawn in a counterclockwise direction about the center point. A maximum of 1024 sides can be drawn with this command.

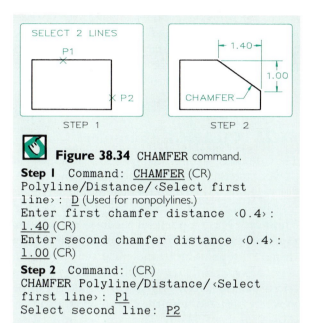

Figure 38.34 CHAMFER command.

Step I Command: <u>CHAMFER</u> (CR)
Polyline/Distance/‹Select first
line›: <u>D</u> (Used for nonpolylines.)
Enter first chamfer distance ‹0.4›:
<u>1.40</u> (CR)
Enter second chamfer distance ‹0.4›:
<u>1.00</u> (CR)
Step 2 Command: (CR)
CHAMFER Polyline/Distance/‹Select
first line›: P1
Select second line: <u>P2</u>

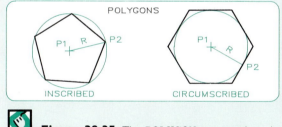

Figure 38.35 The POLYGON command can be used to draw polygons inscribed in or circumscribed about a circle.

38.22 Enlarging, Reducing, and Panning Drawings

Parts of a drawing or the entire drawing can be enlarged by the ZOOM command, a submenu under the DISPLAY menu. A part of the drawing in Fig. 38.36 is too small to read at its present size, but by using a ZOOM and WINDOW, you can select the area that you want enlarged to fill the screen.

Instead of responding with <u>WINDOW</u>, you could have responded with <u>ALL</u>, <u>CENTER</u>, DY-NAMIC, <u>EXTENTS</u>, <u>LEFT</u>, <u>PREVIOUS</u>, or a number to indicate the factor by which you want the present drawing changed in size.

All The ALL response expands the drawing's LIMITS to fill the display screen.

Center The CENTER reply allows you to pick the center of the drawing and degree of magnification or reduction desired.

Dynamic The DYNAMIC option allows you to zoom and pan on the screen by selecting points with the cursor.

Extents The EXTENTS response enlarges the drawing to fill the screen while disregarding its LIMITS if it does not fill them.

Left By responding with <u>L</u> (left), you can pick the lower left corner and the height of the drawing that you wish to have enlarged.

Previous The PREVIOUS reply displays the last view that was used. Views can be ZOOMed an almost infinite number of times.

Pan The PAN command moves a part of the drawing about the display screen, which is useful

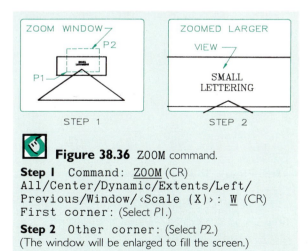

Figure 38.36 ZOOM command.
Step I Command: <u>ZOOM</u> (CR)
All/Center/Dynamic/Extents/Left/
Previous/Window/‹Scale (X)›: <u>W</u> (CR)
First corner: (Select P1.)
Step 2 Other corner: (Select P2.)
(The window will be enlarged to fill the screen.)

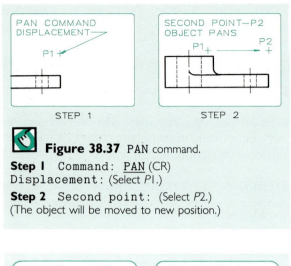

Figure 38.37 PAN command.

Step 1 Command: <u>PAN</u> (CR)
Displacement: (Select *P1*.)

Step 2 Second point: (Select *P2*.)
(The object will be moved to new position.)

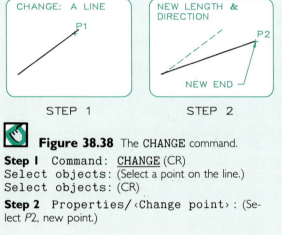

Figure 38.38 The CHANGE command.

Step 1 Command: <u>CHANGE</u> (CR)
Select objects: (Select a point on the line.)
Select objects: (CR)

Step 2 Properties/‹Change point›: (Select *P2*, new point.)

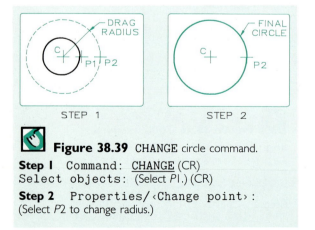

Figure 38.39 CHANGE circle command.

Step 1 Command: <u>CHANGE</u> (CR)
Select objects: (Select *P1*.) (CR)

Step 2 Properties/‹Change point›:
(Select *P2* to change radius.)

when the drawing extends beyond the screen's limits. The drawing in Fig. 38.37 is PANned by entering two points, the displacement point (a point on the screen to be moved) and the second point (the new position of the first point).

38.23 CHANGE Command

The CHANGE command permits changes in LINES, LAYERS, BLOCKS, CIRCLES, and TEXT. For example, to change the length and direction of a line in Fig. 38.38, you may use the CHANGE command, select a point on the line, and move it to its new end point. The size of a circle can be changed by pointing to the circle and selecting a new radius, and the circle will be enlarged or reduced on the screen (Fig. 38.39). Blocks can be moved or rotated with the CHANGE command, as shown in Fig. 38.40. Text can be moved or changed.

If you wish to change part of a drawing to a different layer, execute CHANGE, and select the part to move by pointing to entities, windowing, or using the "last" option. When prompted for

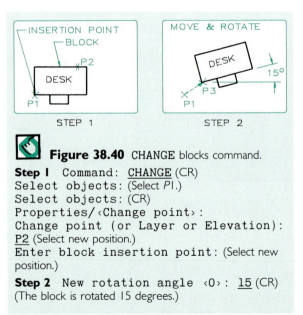

Figure 38.40 CHANGE blocks command.

Step 1 Command: <u>CHANGE</u> (CR)
Select objects: (Select *P1*.)
Select objects: (CR)
Properties/‹Change point›:
Change point (or Layer or Elevation):
<u>P2</u> (Select new position.)
Enter block insertion point: (Select new position.)

Step 2 New rotation angle ‹0›: <u>15</u> (CR)
(The block is rotated 15 degrees.)

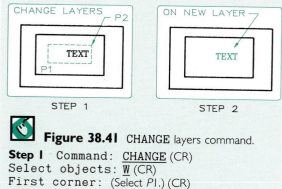

Figure 38.41 CHANGE layers command.

Step 1 Command: CHANGE (CR)
Select objects: W (CR)
First corner: (Select *P1*.) (CR)
Other corner: (Select *P2*.) (CR)

Step 2 Properties ‹ Change point › : L
New layer: VISIBLE (Name of existing layer.) (CR)

Properties Change point, respond with L (layer), and give the name of the layer the drawing will be moved to (Fig. 38.41).

The CHANGE command can be used to change the variables of previously inserted text: **style, height, rotation angle,** and the **text** itself. The command is executed as follows:

```
Command: CHANGE (CR)
Select objects: (Point to text.)
Properties/‹Change point›: (Select in-
sertion point or CR.)
Text style: STANDARD
New style or RETURN for no change:
SIMP (CR)
New height ‹0.10›: .20 (CR)
New rotation angle ‹0›: (Select two
points or CR.)
New text ‹Now is the time›: (Type new
text or CR.)
```

You may press Return to accept the current values at each prompt.

38.24 POLYLINE (PLINE) Command

The PLINE command is used to connect a series of lines and arcs of varying widths to form a PO-LYLINE. The prompt after typing PLINE is

```
Command: PLINE: (CR)
From point: (Select starting point.)
Current line-width is 0.0000
Arc/Close/Halfwidth/Length/
Undo/Width/‹Endpoint of line›:
```

The default response, End point of line, shown in brackets, will result in a line drawn from the last point on the screen at the time you executed PLINE. You will be prompted for the width of the line at its beginning and end. This command requires experimentation to become familiar with its many options.

Straight Lines The PLINE command defaults to the straight-line mode and prompts

```
From point: (Select point.)
Current line width is 0.3:
Arc/Close/Halfwidth/Length/
Undo/Width/‹End point of
line›: W (For width.) (CR)
Starting width ‹0.000›: 0.4
(CR)
Ending width ‹1.000›: 0.6 (CR)
(Select endpoint.)
```

The line is drawn. The longer the line, the thinner and more tapered it becomes.

A value of zero can be entered for the thinnest lines that can be drawn by the system. CLOSE will automatically close to the beginning point of the PLINE and terminate the command. LENGTH lets you continue a PLINE at its last angle by specifying the length of the segment. If the first line was an arc, this command will produce a line tangent to the arc.

The UNDO option erases the last segment of the polyline, and it can be repeated to continue erasing segments of the PLINE. The HALFWIDTH option lets you specify the width of the line from the center of a wide line, as shown in Fig. 38.42.

Arcs When you respond to the PLINE option line with an A (ARC), you can specify arc segments of a PLINE. AutoCAD will give the following command line:

```
Angle/Center/Close/Direction/
Halfwidth/Line/Radius/Second
pt/Undo/Width ‹End point of
arc›:
```

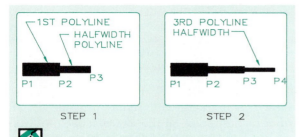

Figure 38.42 PLINE command.

Step 1 Command: <u>PLINE</u> (CR)
From point: <u>P1</u> (Select point.)
Current line width is 0.0000
Arc/Close/Halfwidth/Length/Undo/
Width/‹Endpoint of line›: <u>WIDTH</u> (CR)
Starting width ‹0.0000›: <u>.12</u> (CR)
Ending width ‹.12›: (CR)
Arc/Close/ . . . /‹Endpoint of line›:
<u>P2</u>
Arc/Close/ . . . /‹Endpoint of line›:
<u>Halfwidth</u> (CR)
Starting half-width ‹0.3000›: <u>.3</u> (CR)
Ending half-width ‹0.3000›: (CR)
Arc/Close/ . . . /‹Endpoint of line›:
<u>P3</u>

Step 2 Arc/Close/ . . . /‹Endpoint of
line›: <u>Halfwidth</u> (CR)
Starting half-width ‹0.2›: <u>.1</u> (CR)
Ending half-width ‹0.1000›: (CR)
Arc/Close/ . . . ?‹Endpoint of line›:
<u>P4</u>

The default assumes an arc will be drawn tangent to the last line drawn and will pass through the next point selected.

By using the ANGLE option, you may give the Included angle: as a positive or negative value, and the next prompt will ask for Center/Radius/‹End point›:. AutoCAD then draws an arc tangent to the previous line segment. If you select Center, you will be asked to give the center of the next arc segment. The next prompt asks for Angle/Length/‹End point›:, where Angle refers to the included angle, and Length is length of the arc's chord.

The CLOSE option causes the PLINE to be closed with an arc segment to the beginning point. DIRECTION allows you to override the de-

fault, which draws the next arc tangent to the last PLINE segment. AutoCAD prompts Direction from starting point:, and you can point to the desired beginning point and respond to the next prompt, Endpoint, to give the direction of the arc.

The LINE option switches the PLINE command back to the straight-line mode. The RADIUS option gives a prompt, Radius:, that allows you to specify the radius of the next arc. The following prompt, Angle/Length/‹End point›:, lets you specify the included angle or the length of the arc's chord.

SECOND PT is used to select the second point and the endpoint of a three point arc. The two prompts are Second point: and Endpoint:.

38.25 PEDIT Command

The PEDIT command is used to edit polylines drawn with the PLINE command. The prompts for this command are

Select polyline: (Select line with cursor.)
Entity selected is not a
polyline
Do you want it turned into
one? <u>Y</u>
Close/Join/Width/Edit vertex/
Spline/Fit curve/Decurve/Undo
/exit‹X›: <u>CLOSE</u>
(Used to close PLINE.)

If the PLINE is already closed, the CLOSE command will be replaced by the OPEN option.

Join The JOIN option lets you respond to the prompt Select objects Window or Last: by selecting segments that are to be joined to the polyline. Once chosen, these segments become part of the polyline. Segments must have exact meeting points and must not meet with an overlapping intersection for joining to take place.

Width The WIDTH option gives the prompt Enter new width for all segments:. Enter the new width from the keyboard, and the PLINE is redrawn to this new thickness.

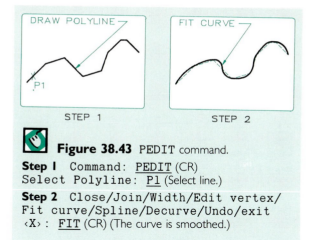

Figure 38.43 PEDIT command.
Step I Command: <u>PEDIT</u> (CR)
Select Polyline: <u>P1</u> (Select line.)
Step 2 Close/Join/Width/Edit vertex/
Fit curve/Spline/Decurve/Undo/exit
‹X›: <u>FIT</u> (CR) (The curve is smoothed.)

Fit Curve The FIT CURVE (F) option constructs a smooth curve that passes through all vertices of the PLINE with pairs of arcs that join sequential vertices (Fig. 38.43). To change the resulting curve to better suit your needs, use the EDIT VERTEX command discussed below.

Spline The SPLINE (S) option draws a cubic curve that passes through the first and last points, but not necessarily through the other points (Fig. 38.44). The SPLINE is useful for drawing curves representing data plotted on graphs.

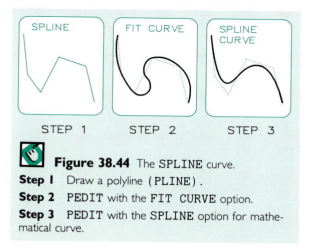

Figure 38.44 The SPLINE curve.
Step I Draw a polyline (PLINE).
Step 2 PEDIT with the FIT CURVE option.
Step 3 PEDIT with the SPLINE option for mathematical curve.

Decurve DECURVE (D) removes the arcs inserted by the FIT CURVE option and returns the PLINE to its straight-line form.

Undo (U) The UNDO (U) option undoes the most recently done PEDIT 399 editing step.

Edit Vertex EDIT VERTEX (E) allows you to select a single vertex of the PLINE and edit it. When this option is used, the first vertex of the PLINE will be marked with an X on the screen. An arrow will be shown if you have specified a tangent direction for the vertex, and you will receive the following prompt:

Next/Previous/Break/Insert/
Move/Regen/Straighten/Tangent/
Width/eXit/‹N›:

Next and Previous The NEXT (N) and PRE-VIOUS (P) options move the X marker to the next or previous vertex. To move to a vertex several vertices away, select NEXT or PREVIOUS, and press (CR) until the vertex is reached.

Break When the BREAK (B) option is selected, the location of the position of the X is shown, and the following prompt appears:

Next/Previous/Go/eXit ‹N›:

By using NEXT or PREVIOUS you can select a second point and enter <u>GO</u>, and the line between the two points will be erased (Fig. 38.45). Enter <u>EXIT</u>, and the BREAK will be canceled, and you will return to EDIT VERTEX.

Insert The INSERT option gives the following prompt:

Enter location of new vertex:

This lets you add a new vertex to a polyline (Fig. 38.46). A vertex can also be moved by the MOVE (M) option (Fig. 38.47).

Straighten A polyline can be straightened by the STRAIGHTEN (S) option, which saves the current location of the vertex specified by an X and gives the following prompt:

    Next/Previous/Go/eXit/‹N›:

By moving the X to a new vertex on the line and specifying <u>GO</u>, the line will be straightened be-

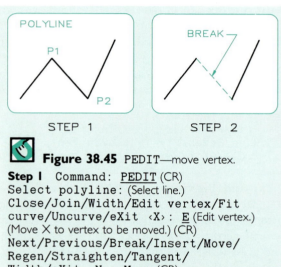

STEP 1   STEP 2

Figure 38.45 PEDIT—move vertex.

Step 1 Command: <u>PEDIT</u> (CR)
Select polyline: (Select line.)
Close/Join/Width/Edit vertex/Fit curve/Uncurve/eXit ‹X›: <u>E</u> (Edit vertex.)
(Move X to vertex to be moved.) (CR)
Next/Previous/Break/Insert/Move/ Regen/Straighten/Tangent/ Width/eXit ‹N›: <u>Move</u> (CR)
Enter location of new vertex: (Select P1.)

Step 2 Press (CR), and the polyline will be changed to pass through the moved vertex.

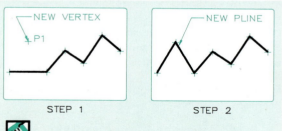

STEP 1   STEP 2

Figure 38.46 PEDIT—add vertex.

Step 1 Use **EDIT VERTEX** option and place X on line before the new vertex.
Next/Previous/Break/Insert/Move/ Regen/Straighten/Tangent/Width/eXit ‹N›: <u>Insert</u> (CR)
Enter location of new vertex: (Select P1.)

Step 2 Press (CR), and the new vertex will be inserted, and the polyline will pass through it.

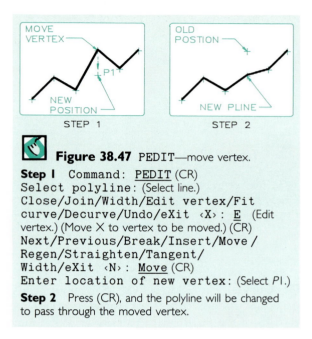

STEP 1   STEP 2

Figure 38.47 PEDIT—move vertex.

Step 1 Command: <u>PEDIT</u> (CR)
Select polyline: (Select line.)
Close/Join/Width/Edit vertex/Fit curve/Decurve/Undo/eXit ‹X›: <u>E</u> (Edit vertex.) (Move X to vertex to be moved.) (CR)
Next/Previous/Break/Insert/Move / Regen/Straighten/Tangent/ Width/eXit ‹N›: <u>Move</u> (CR)
Enter location of new vertex: (Select P1.)

Step 2 Press (CR), and the polyline will be changed to pass through the moved vertex.

tween the two vertices (Fig. 38.48). Enter <u>X</u> for <u>eXit</u> if you change your mind, and you will be returned to the EDIT VERTEX prompt.

The TANGENT (T) suboption lets you indicate a tangent direction at the vertex marked by

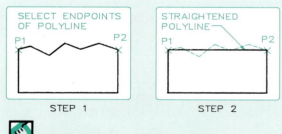

STEP 1   STEP 2

Figure 38.48 PEDIT—straighten line.

Step 1 Use **EDIT VERTEX** option of PEDIT, and place X at the vertex at the beginning of the line to be straightened, P1.
Next/Previous/Break/Insert/Move/ Regen/Straighten/Tangent/Width/eXit ‹N›: <u>Straighten</u> (CR)

Step 2 Next/Previous/Go/eXit ‹N›: <u>Next</u> (CR) (Move to P2.)
Next/Previous/Go/eXit ‹N›: <u>Go</u> (CR) (Line P1–P2 is straightened.)

the X for use in curve fitting when it is used next. The prompt is

    Direction of tangent:

Enter the angle from the keyboard, or select a point with the cursor on the screen from the current point.

The WIDTH (W) suboption lets you change the beginning and ending widths of an existing line segment from the X-marked vertex. Use the NEXT and PREVIOUS options to confirm which direction the line will be drawn. To draw the changed polyline on the screen, use the REGEN (R) option.

Exit The eXit option is used to exit from the PEDIT command and return to the command: prompt.

38.26 HATCH Command

The HATCH command is used to cross-hatch an area that has been sectioned (Fig. 38.49). The prompts are

    Command: HATCH (CR)
    Pattern (? or name/U, style):
    ‹default›: ANSI31 (CR)

By responding with ?, you will be given a list of the standard patterns in ACAD.PAT. A response of U is a user-defined pattern.

Several dialogue boxes that show the hatching patterns can be displayed on the screen by using the HATCH option under the DRAW heading on the **Menu Bar.** A pattern can be selected by the cursor.

    Angle   for   crosshatch   lines
    ‹0›: 45 (Or show with the pointer.) (CR)
    Spacing between lines ‹0.1›:
    0.2 (Or show with the pointer.) (CR)
    Double hatch area? ‹N› (CR)
    Select Objects or Window or Last:

By responding as shown above, equally spaced hatch lines will be drawn 0.2 units apart and at 45° with the horizontal.

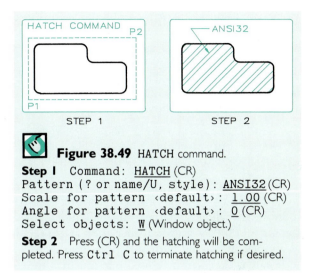

STEP 1   STEP 2

Figure 38.49 HATCH command.

Step I   Command: HATCH (CR)
Pattern (? or name/U, style): ANSI32 (CR)
Scale for pattern ‹default›: 1.00 (CR)
Angle for pattern ‹default›: 0 (CR)
Select objects: W (Window object.)

Step 2   Press (CR) and the hatching will be completed. Press Ctrl C to terminate hatching if desired.

The DRAW heading of the pull-down menu of Release 10 can be selected to gain access to the HATCH option. Drawings (icons) of the various hatch patterns are shown in several screens from which to make pattern selections by using the cursor arrow. Access to icons makes it unnecessary to memorize the patterns by their numbers.

The letters N, O, or I can be added to your response to the PATTERN prompt. For example, PATTERN: ANSI32,N. These letters will cause the crosshatching to be given in the style shown in Fig. 38.50 and as defined below:

> N–Normal (Hatches alternate areas beginning with the outermost.)

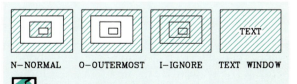

N-NORMAL   O-OUTERMOST   I-IGNORE   TEXT WINDOW

Figure 38.50 Whenever you specify a hatching pattern, enter a comma and the letter N, O, or I after the pattern, and the hatching will be applied as shown above; for example, Pattern: ANSI31,N, to hatch the outside and alternate layers inside the figure. Hatching automatically leaves a window around text within the area.

0–Outermost (Hatches only the outermost areas.)

I–Ignore (Hatches all inside areas ignoring contents.)

If a pattern is selected from the ACAD.PAT file, you will receive the following prompt:

```
Scale for pattern <1.00>: .75 (CR)
Angle for pattern <0>: 0 (CR)
Select Objects or Window or Last:
```

Both responses can be made at the keyboard or by selecting two points on the screen for each to indicate the angle and the scale. All hatch lines can be ERASED as a group since they are entered as a block. If entered with an * preceding the pattern name, they can be edited one line at a time. The area to be hatched is selected one line at a time or WINDOWed as shown in Fig. 38.50.

38.27 Text and Numerals

The TEXT command, a subcommand under DRAW can be used for inserting plotted text in a drawing from the keyboard at any size and angle (Fig. 38.51).

```
Command: TEXT (CR)
Start point or Align/Center/
Fit/Middle/Right/Style: (Select
point with cursor.)
Height <0.18>: 0.5 (CR)
Rotation angle <0>: (CR)
Text: TEXT (CR) (The word, TEXT, is written
beginning with the point selected and is left-justi-
fied.)
```

Other responses are A (aligned), which changes the text to fit exactly between two end-points (Fig. 38.51A); C centers the text about a point; R (right) justifies the text to a point located at the right; and S (style) selects a text style from those available.

Figure 38.52 shows the FIT and MIDDLE options. When using the FIT option, the height of the text does not change, but the width of the line is stretched or compressed to fit between the two selected points.

Figure 38.51 (A) The TEXT command defaults to left-justified text, and the point indicated represents the lower left corner of capital letters.
(B) Text can also be centered, right justified, or aligned so that the first and last letters of a word or line span the distance between two selected points.

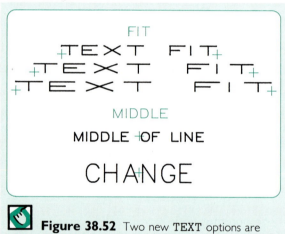

Figure 38.52 Two new TEXT options are FIT and MIDDLE. When using FIT, a string of text is stretched to fit between two points without changing its height. Using MIDDLE places the center of the string and letters at the point selected.

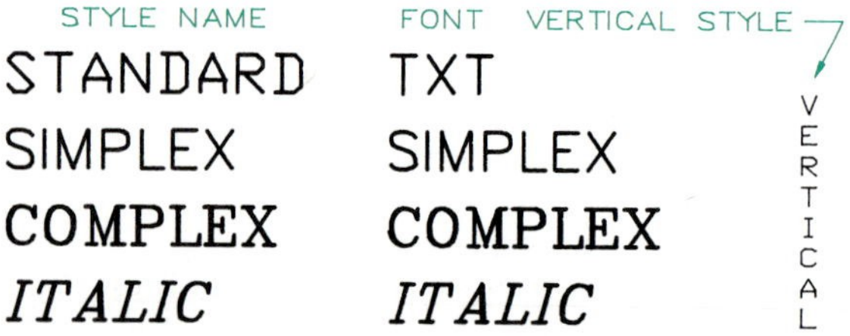

Figure 38.53 Five fonts of text available from AutoCAD: Txt, Simplex, Complex, Italic, and Vertical.

The MIDDLE option centers a word or a line about a point in the center of the height of the letters used and centered about the width of the word or line.

38.28   The STYLE Command

The five text fonts that were available with AutoCAD's previous version are TXT, SIMPLEX, COMPLEX, ITALIC, and VERTICAL (Fig. 38.53). The default style is STANDARD, which uses TXT font. Figure 38.54 shows additional fonts in Release 10.

The STYLE command is used to create variations of any of the fonts in the following manner:

```
Command: STYLE (CR)
Text style name (or?): PRETTY (CR)
Font file ‹TXT›: COMPLEX (CR)
Height (0.20): 0 (CR)
Width factor ‹default›: 1.00 (CR)
Obliquing angle ‹45›: 0 (CR)
Backwards? (Y/N): N (CR)
Upside–down? ‹Y/N›: N (CR)
```

Responding to the first prompt with ? will give a list of the defined text styles, or the pull-down menu under the heading DRAW can be used to access the TEXT command that shows examples of

SIMPLEX FONTS

ROMANS	ROMAN SIMPLEX LETTERING, 1234
SCRIPTS	*Simplex Script Text, 1234567890*
GREEKS	ΓΡΕΕΚ ΣΙΜΠΛΕΞ ΤΕΞΤ, 123456789

DUPLEX FONT

| ROMAND | **ROMAN DOUBLE LINE, SANS SERIF** |

COMPLEX FONTS

ROMANC	NOW IS THE TIME FOR ALL GOOD
ITALICC	*NOW IS THE TIME FOR ALL*
SCRIPTC	*NOW IS THE TIME FOR ALL GOOD*
GREEKC	ΝΟΩ ΙΣ ΤΗΕ ΤΙΜΕ ΦΟΡ ΑΛΛ ΓΟΟΔ

TRIPLEX FONTS

| ROMANT | **NOW IS THE TIME FOR ALL GOOD** |
| ITALICT | *NOW IS THE TIME FOR ALL* |

GOTHIC FONTS

GOTHICE	ENGLISH GOTHIC TEXT
GOTHICG	GERMAN GOTHIC LETTERING
GOTHICI	ITALIAN GOTHIC TEXT

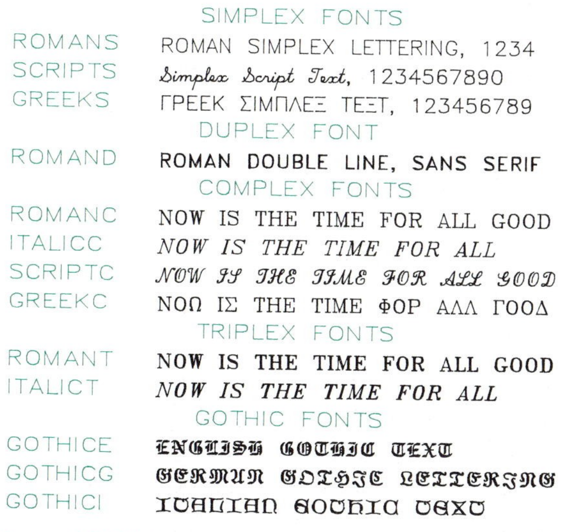

Figure 38.54 Additional text fonts available in AutoCAD Release 10.

SYMBOL FONTS

(LETTERS A THRU Z)

SYASTRO	
SYMAP	
SYMATH	
SYMETEO	
SYMUSIC	

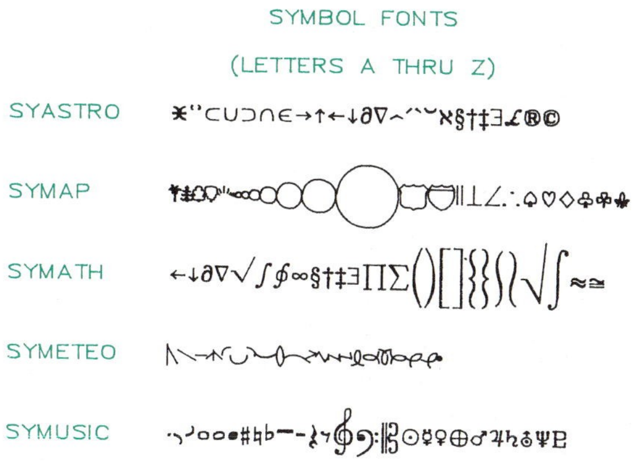

Figure 38.55 Symbols available in AutoCAD Release 10.

the fonts that can be selected. Figure 38.55 shows examples of symbol fonts. The text style created above is named PRETTY. Each time PRETTY is specified, it will be unnecessary to reassign the values given above unless you wish to change them. However, if you select a new font file such as SIMPLEX for PRETTY, previously drawn text will be redrawn when regenerated. By giving a height of 0, you will be prompted for the height of the lettering desired whenever PRETTY is specified as the style of text.

The font, VERTICAL, can be used for drawing vertical lines of text with letters stacked one under the other. Text must be entered with a rotation angle of 270° when using this font.

The QTEXT command can be used to reduce the time required to display a drawing on the screen when it is redrawn. QTEXT (quick text) can be set to ON, and the text will be drawn on the screen as a series of boxes representing the space required for the text (Fig. 38.56). When QTEXT is turned OFF, the full text will be plotted on the screen.

Special characters must be entered from the keyboard when using the TEXT command. By preceding the codes with a double percent sign, %%, the following characters will be entered:

%%O	Toggle overscore mode on/off
%%U	Toggle underscore mode on/off
%%d	Draw degrees symbol
%%P	Draw plus/minus tolerance symbol

%%c	Draw circle diameter dimension-ing symbol
%%%	Force a single percent sign
%%nnn	Draw special character number "nnn" (ASCII)

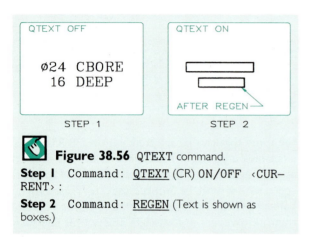

Figure 38.56 QTEXT command.

Step 1 Command: <u>QTEXT</u> (CR) ON/OFF ‹CUR—RENT›:

Step 2 Command: <u>REGEN</u> (Text is shown as boxes.)

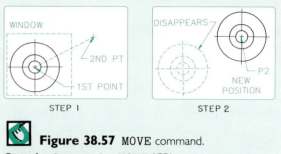

Figure 38.57 MOVE command.

Step 1 Command: <u>MOVE</u> (CR)
Select objects: <u>W</u> (CR)
First corner: (Select point.)
Other corner: (Select point.) 5 found.
Select objects: (CR)
Base point or displacement: (Select 1st point.)

Step 2 Second point of displacement: (Select 2nd point, *P2*.) (The object is drawn at its new position, *P2*, and the original drawing disappears.)

38.29 Moving and Copying Drawings

A drawing can be moved to a new position by the MOVE command (Fig. 38.57), or it can be duplicated by the COPY command. When copied, the original drawing is left in its original position, and a copy of it is located where specified.

The COPY procedure is the same as MOVE except the command is COPY instead of MOVE. The COPY command has a MULTIPLE prompt that can be used for copying multiples of drawings by moving your cursor about the screen and pressing the select button.

A drawing can be moved or copied by dragging it into position by the DRAG option (Fig. 38.58). DRAG is activated after responding to Base point or displacement with <u>DRAG</u>. When the cursor is moved about the screen, the drawing is dynamically moved until it is set by pressing the select button of the mouse.

Besides a window, you may use the default, which is set to select the entities to be moved with the cursor, one at a time. You may

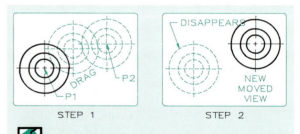

Figure 38.58 DRAG mode.

Step 1 Command: <u>MOVE</u> (CR)
Select objects: <u>W</u> (CR) (Window object.)
Base point or displacement: <u>DRAG</u> (CR)
Base point or displacement: (Select *P1*.) (Or X,Y distance.)
Second point of displacement: (Select new location.) (As the cursor is moved, the drawing is DRAGged about the screen.)

Step 2 When moved to the desired location, press the Select button to draw the object in its final position. (The original drawing disappears.) If DRAGMODE is ON or AUTO, dragging will be automatic.

also select LAST, which will move the last entity, such as a LINE, ARC, CIRCLE, TEXT, or BLOCK.

38.30 Mirroring Drawings

Symmetrical objects can be drawn by drawing a portion of the figure and then mirroring the drawing about one or more axes. The schematic threads in Fig. 38.59 are drawn using the MIRROR command. If a line coincides with the MIRROR line, such as P1–P2 in Fig. 38.59, the line will be drawn twice when mirrored. Therefore, a line of this type should be drawn after the view has been mirrored.

38.31 Mirrored Text (MIRRTEXT)

MIRRTEXT is a system variable that is a subcommand of the SETVAR command, which is used for mirroring a drawing that has text. By setting MIRRTEXT to 0, the text will not be mirrored, but the drawing will be (Fig. 38.60). If MIRRTEXT is set to 1, the text will be mirrored along with the drawing.

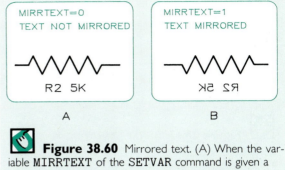

Figure 38.60 Mirrored text. (A) When the variable MIRRTEXT of the SETVAR command is given a value of 0, the text will not be mirrored. (B) When MIRRTEXT is given a value of 1, the text will be mirrored.

38.32 Snapping to Objects (OSNAP)

On many occasions, you will want to draw lines that SNAP to features of other objects within drawings rather than SNAP to a grid. This type of snap is called an OSNAP, which is short for object snap. The line of stars ***** in the **Root Menu** is selected to activate the OSNAP command. The TOOLS heading of the pull-down menu also gives OSNAP options. For example, Fig. 38.61 shows

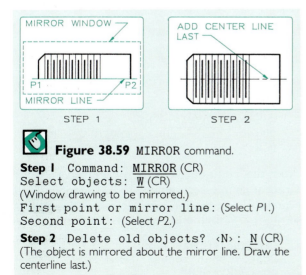

Figure 38.59 MIRROR command.
Step 1 Command: MIRROR (CR)
Select objects: W (CR)
(Window drawing to be mirrored.)
First point or mirror line: (Select P1.)
Second point: (Select P2.)
Step 2 Delete old objects? ‹N›: N (CR)
(The object is mirrored about the mirror line. Draw the centerline last.)

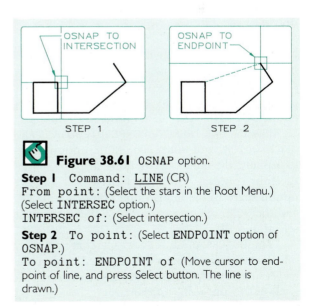

Figure 38.61 OSNAP option.
Step 1 Command: LINE (CR)
From point: (Select the stars in the Root Menu.)
(Select INTERSEC option.)
INTERSEC of: (Select intersection.)
Step 2 To point: (Select ENDPOINT option of OSNAP.)
To point: ENDPOINT of (Move cursor to endpoint of line, and press Select button. The line is drawn.)

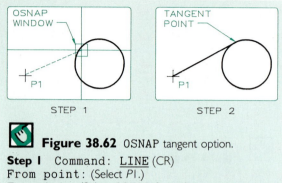

Figure 38.62 OSNAP tangent option.

Step I Command: <u>LINE</u> (CR)
From point: (Select *P1*.)
To point: (Select OSNAP from the menu, and select TANGENT.)

Step 2 To point: (Select point on circle.) (The line is drawn from *P1* tangent to the circle on the side where the tangent point was selected.)

the steps in drawing a line from an intersection of lines to the endpoint of a line.

From *P*1 in Fig. 38.62 a line is drawn tangent to the circle. The same procedure as shown above is used, but instead of ENDPOINT for the second point, TANGENT is selected on the side of the circle where the tangent point is to be located. The tangent point is found automatically, and the line is drawn.

By using the TANGENT option of OSNAP, a line can be drawn tangent to two arcs as shown in Fig. 38.63. The tangent line is automatically trimmed to end at its two tangent points.

The various options of OSNAP are NEAREST, ENDPOINT, MIDPOINT, CENTER, NODE, QUADRANT, INTERSECTION, INSERT, PERPENDICULAR, TANGENT, QUICK, and NONE. The NODE option snaps to a point, the QUADRANT options snaps to one of the four compass points on a circle, the INSERT option snaps to the intersection point of a BLOCK; and the NONE option turns off OSNAP for the next selection.

The QUICK option reduces searching time by selecting the first object encountered within the target area, rather than searching for the one that is closest to the center of the target.

The APERTURE command is used to assign a size to the target box that appears at the cursor when OSNAP options are in use. The operator can vary from 1 to 50 pixels square.

Other OSNAP options permit you to snap to centers of arcs and circles and to insert points of blocks, midpoints of lines, nearest points of an object, points (nodes) perpendicular to lines, tangent to arcs and circles, and quadrant points of a circle.

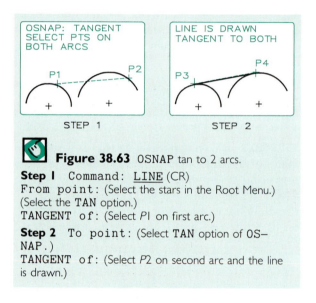

Figure 38.63 OSNAP tan to 2 arcs.

Step I Command: <u>LINE</u> (CR)
From point: (Select the stars in the Root Menu.) (Select the TAN option.)
TANGENT of: (Select *P1* on first arc.)

Step 2 To point: (Select TAN option of OSNAP.)
TANGENT of: (Select *P2* on second arc and the line is drawn.)

One or more OSNAP settings can be made when the same options are to be used many times. For example, if you wish to repetitively snap to endpoints and circle centers, do the following:

Command: <u>OSNAP</u> (CR)
Object snap modes: <u>ENDPOINT, CENTER</u> (CR)

From this point on, your cursor will have a select target at its intersection for picking endpoints and centers of arcs and circles. Remove this OSNAP setting in the following manner:

Command: <u>OSNAP</u> (CR)
Object snap modes: (Select NONE, OFF, or press (CR).) (CR)

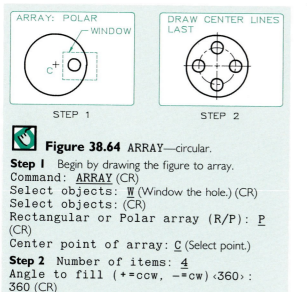

Figure 38.64 ARRAY—circular.

Step 1 Begin by drawing the figure to array.
```
Command: ARRAY (CR)
Select objects: W (Window the hole.) (CR)
Select objects: (CR)
Rectangular or Polar array (R/P): P
(CR)
Center point of array: C (Select point.)
```

Step 2 Number of items: 4
```
Angle to fill (+=ccw, -=cw) <360>:
360 (CR)
Rotate objects as they are copied?
<Y> (CR)
(The array is drawn.)
```

38.33 ARRAY Command

The ARRAY command is used to draw repetitive shapes in circular and rectangular patterns. For example, a series of holes can be located on a bolt circle by drawing the first hole in its desired position and then activating the ARRAY command, as shown in Fig. 38.64.

A rectangular ARRAY is begun by drawing the first part in the lower left corner of the array. Once drawn, follow the commands of the rectangular ARRAY shown in Fig. 38.65. Termination of the array can be caused by Ctrl C.

Rectangular arrays may be drawn at angles (Fig. 38.66) if the SNAP mode has been rotated to some angle other than zero. The first object is then drawn with the DRAW commands in the lower left corner of the ARRAY. The ARRAY command is activated, and the number of rows, number of columns, and cell distances are given in response to the command prompts. The array will be drawn at the angle of the SNAP mode.

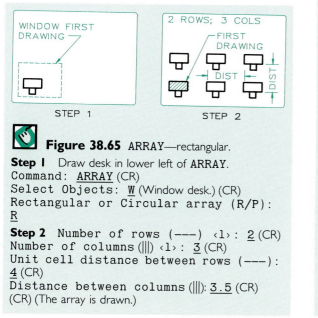

Figure 38.65 ARRAY—rectangular.

Step 1 Draw desk in lower left of ARRAY.
```
Command: ARRAY (CR)
Select Objects: W (Window desk.) (CR)
Rectangular or Circular array (R/P):
R
```

Step 2 Number of rows (---) <1>: 2 (CR)
```
Number of columns (|||) <1>: 3 (CR)
Unit cell distance between rows (---):
4 (CR)
Distance between columns (|||): 3.5 (CR)
(CR) (The array is drawn.)
```

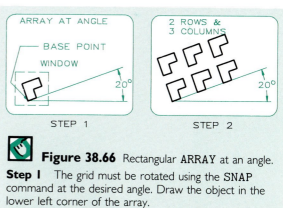

Figure 38.66 Rectangular ARRAY at an angle.

Step 1 The grid must be rotated using the SNAP command at the desired angle. Draw the object in the lower left corner of the array.
```
Command: ARRAY (CR)
Select objects. (Window the object.)
Rectangular or Polar array <R/P>: R
(CR)
```

Step 2 Number of rows (---) <1>: 2 (CR)
```
Number of columns (|||) <1>: 3 (CR)
Unit cell or distance between rows
(---): 2 (CR)
Distance between columns (|||): 3 (CR)
```

38.34   DONUT Command

A "donut" can be drawn by using the DONUT command in which the inside diameter, outside diameter, and center point are given (Fig. 38.67). By setting the inside diameter to 0, the DONUT will draw a solid circle (Fig. 38.68).

38.35   SCALE Command

The SCALE command permits you to make drawn objects larger or smaller. For example, the desk in Fig. 38.69 is enlarged by windowing the

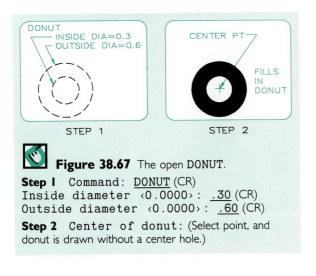

Figure 38.67 The open DONUT.

Step 1   Command: <u>DONUT</u> (CR)
Inside diameter ‹0.0000› : <u>.30</u> (CR)
Outside diameter ‹0.0000› : <u>.60</u> (CR)

Step 2   Center of donut: (Select point, and donut is drawn without a center hole.)

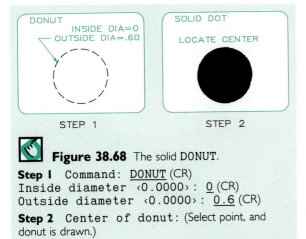

Figure 38.68 The solid DONUT.

Step 1   Command: <u>DONUT</u> (CR)
Inside diameter ‹0.0000› : <u>0</u> (CR)
Outside diameter ‹0.0000› : <u>0.6</u> (CR)

Step 2   Center of donut: (Select point, and donut is drawn.)

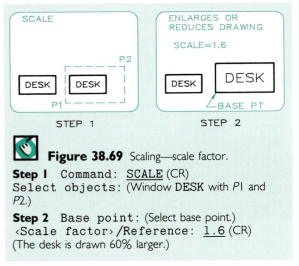

Figure 38.69 Scaling—scale factor.

Step 1   Command: <u>SCALE</u> (CR)
Select objects: (Window DESK with *P1* and *P2*.)

Step 2   Base point: (Select base point.)
‹Scale factor›/Reference: <u>1.6</u> (CR)
(The desk is drawn 60% larger.)

desk, selecting a base point, and giving it a scale factor of 1.6. The text and the drawing are enlarged in the *X* and *Y* directions.

A second option of the SCALE command enables you to select a length of a given figure (Fig. 38.70), specify its present length, and then specify the new length, which is a ratio of the first specified dimension. The lengths can be given by using the cursor or by typing them in at the keyboard as numeric values.

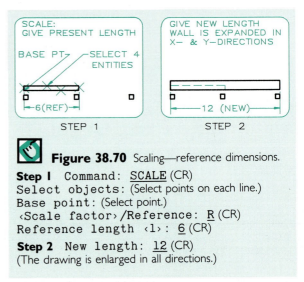

Figure 38.70 Scaling—reference dimensions.

Step 1   Command: <u>SCALE</u> (CR)
Select objects: (Select points on each line.)
Base point: (Select point.)
‹Scale factor›/Reference: <u>R</u> (CR)
Reference length ‹1› : <u>6</u> (CR)

Step 2   New length: <u>12</u> (CR)
(The drawing is enlarged in all directions.)

38.36 STRETCH Command

The STRETCH command is used to move a portion of a drawing while retaining the connections it has with other lines and entities. The window in a floor plan in Fig. 38.71 is moved to a new position by the STRETCH command while remaining attached to the line of the wall. A CROSSING window is used to select the line that will be stretched.

38.37 ROTATE Command

An object drawn on the screen can be rotated about a base point by using the ROTATE command (Fig. 38.72). The object, which may comprise a number of lines, is windowed, a base point is selected, and a rotation angle is typed in at the keyboard, or the cursor is used to show the angle of rotation on the screen. Lines from several layers can be rotated if they are included within the window used to select the object.

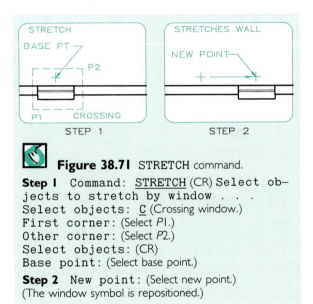

Figure 38.71 STRETCH command.
Step 1 Command: STRETCH (CR) Select objects to stretch by window . . .
Select objects: C (Crossing window.)
First corner: (Select P1.)
Other corner: (Select P2.)
Select objects: (CR)
Base point: (Select base point.)

Step 2 New point: (Select new point.)
(The window symbol is repositioned.)

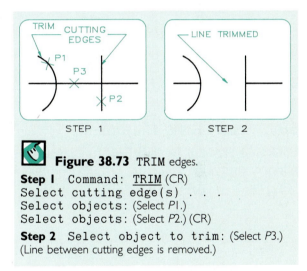

Figure 38.73 TRIM edges.
Step 1 Command: TRIM (CR)
Select cutting edge(s) . . .
Select objects: (Select P1.)
Select objects: (Select P2.) (CR)

Step 2 Select object to trim: (Select P3.)
(Line between cutting edges is removed.)

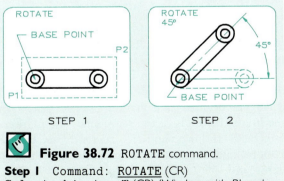

Figure 38.72 ROTATE command.
Step 1 Command: ROTATE (CR)
Select objects: W (CR) (Window with P1 and P2.)

Step 2 Base point: (Select point.)
‹Rotation angle›/Reference: 45 (CR)
(Object is rotated 45° CCW.)

38.38 TRIM Command

By using the TRIM command, edges of entities can be used as cutting edges to trim a line, as shown in Fig. 38.73. You are prompted to select the entities that will serve as cutting edges, and the objects between the cutting edges are selected. The cutting edges trim the lines, giving perfect intersections at the crossing points.

Using the CROSSING option of the TRIM command, a window is placed around a set of intersecting lines (Fig. 38.74), and lines selected

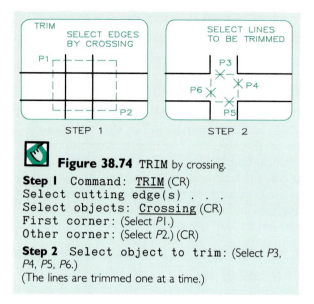

Figure 38.74 TRIM by crossing.

Step 1 Command: <u>TRIM</u> (CR)
Select cutting edge(s) . . .
Select objects: <u>Crossing</u> (CR)
First corner: (Select P1.)
Other corner: (Select P2.) (CR)

Step 2 Select object to trim: (Select P3, P4, P5, P6.)
(The lines are trimmed one at a time.)

to be the cutting edges are crossed. The portions of the lines between the cutting edges are selected one at a time and are then removed or trimmed.

38.39 EXTEND Command

Lines and arcs can be lengthened to intersect a selected entity by using the EXTEND command (Fig. 38.75). You are first prompted to select the

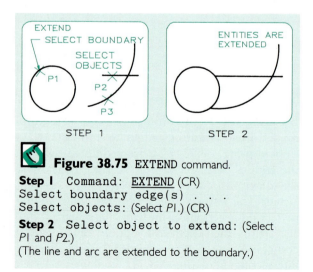

Figure 38.75 EXTEND command.

Step 1 Command: <u>EXTEND</u> (CR)
Select boundary edge(s) . . .
Select objects: (Select P1.) (CR)

Step 2 Select object to extend: (Select P1 and P2.)
(The line and arc are extended to the boundary.)

boundary entity and then the line or arc to be extended. More than one entity can be extended at a time to join the previously selected boundary entity.

A polyline can be extended with the EXTEND command to a selected boundary, as shown in Fig. 38.76. The boundary is selected, the ends of the PLINES are selected, and both PLINES are extended. The EXTEND command will not work on "closed" PLINES, such as a polygon.

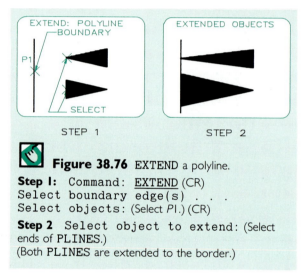

Figure 38.76 EXTEND a polyline.

Step 1: Command: <u>EXTEND</u> (CR)
Select boundary edge(s) . . .
Select objects: (Select P1.) (CR)

Step 2 Select object to extend: (Select ends of PLINES.)
(Both PLINES are extended to the border.)

38.40 DIVIDE Command

An entity can be divided into a specified number of equal segments by using the DIVIDE command (Fig. 38.77). The entity is selected by locating a point on it with the cursor. You will be prompted for the number of segments into which it is to be divided, and markers will be placed on the line at the ends of the segments. The markers will be of the type and size currently set by the PDMODE and PDSIZE variables under the SETVAR command.

The BLOCK option under the DIVIDE command allows you to select a previously saved block to mark the ends of the segments of a divided line (Fig. 38.78). The blocks can be either ALIGNED or NOT ALIGNED as shown. In this ex-

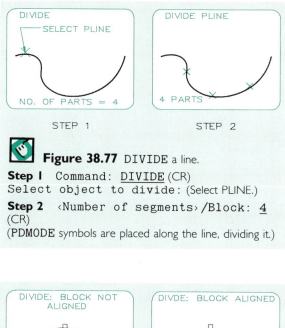

Figure 38.77 DIVIDE a line.

Step I Command: <u>DIVIDE</u> (CR)
Select object to divide: (Select PLINE.)

Step 2 ‹Number of segments›/Block: <u>4</u>
(CR)
(**PDMODE** symbols are placed along the line, dividing it.)

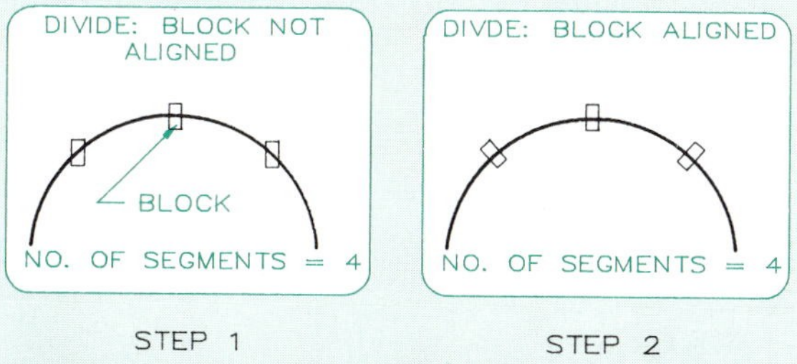

Figure 38.78 DIVIDE an arc.

Step I Command: <u>DIVIDE</u> (CR)
Select object to divide: (Select arc.)
‹Number of segments›/Block: <u>B</u> (CR)
Block name to insert: <u>RECT</u> (CR)
Align block with object? ‹Y› <u>N</u> (CR)
Number of segments: <u>4</u> (CR)

Step 2 Align block with object? ‹Y›
(CR)
(The blocks will radiate from the arc's center.)

ample, the BLOCKS are rectangles, but they could have been drawn in any shape.

38.41 MEASURE Command

Markers can be placed along an arc, circle, polyline, or line at a specified distance apart (Fig. 38.79) by using the MEASURE command. The seg-

ment length option asks you for the entity to be segmented, which you must select; then it asks for the segment length. Markers will be placed along the line (or entity) beginning with the end nearest to the location of the point selected. The last segment probably will not be equal to the specified segment length.

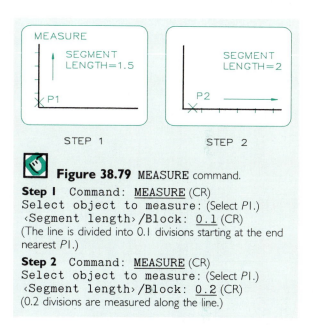

Figure 38.79 MEASURE command.

Step I Command: <u>MEASURE</u> (CR)
Select object to measure: (Select *PI*.)
‹Segment length›/Block: <u>0.1</u> (CR)
(The line is divided into 0.1 divisions starting at the end nearest *PI*.)

Step 2 Command: <u>MEASURE</u> (CR)
Select object to measure: (Select *PI*.)
‹Segment length›/Block: <u>0.2</u> (CR)
(0.2 divisions are measured along the line.)

38.42 OFFSET Command

An entity can be drawn parallel to and offset from another entity by the OFFSET command (Fig. 38.80). In this example, a polyline is drawn offset from a given polyline. You will be prompted for the offset distance or the point the offset drawing is to pass through. Next, you will be prompted for the entity (a PLINE, in this case) to be used as the pattern for the offset PLINE.

The OFFSET command is an excellent aid when drawing parallel lines, as architects would do when drawing floor plans (Fig. 38.81). The offset distance can be set to a precise value, and lines can be repetitively selected, and the side to offset is indicated with the cursor. The FILLET command, with a radius set to 0, can be used to

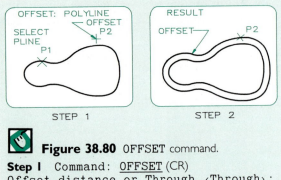

Figure 38.80 OFFSET command.

Step 1 Command: <u>OFFSET</u> (CR)
Offset distance or Through ‹Through›:
<u>T</u> (CR)
Select object to offset: (Select *P1*.)
Through point: (Select *P2*.)

Step 2 An enlarged **PLINE** is drawn that passes through *P2*.

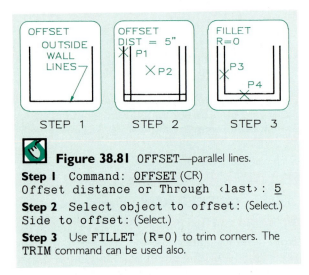

Figure 38.81 OFFSET—parallel lines.

Step 1 Command: <u>OFFSET</u> (CR)
Offset distance or Through ‹last›: <u>5</u>

Step 2 Select object to offset: (Select.)
Side to offset: (Select.)

Step 3 Use **FILLET** (R=0) to trim corners. The **TRIM** command can be used also.

trim the corners to perfect intersections. The TRIM command can be used as an alternate method for squaring corners.

38.43 BLOCKS

One of the more powerful features of computer graphics is the option of building a file of drawings or symbols to be used repetitively on drawings. In AutoCAD, these drawing files are called BLOCKS.

BLOCKS, such as the SI symbol in Fig. 38.82, are drawn in the conventional manner and are blocked as shown below.

Command: <u>BLOCK</u> (CR)
BLOCK name (or ?): <u>SI</u> (CR)
Insertion base point: (Select insert point.)
Select objects: <u>WINDOW</u> or <u>W</u> (CR)
First corner: (Select point.)
Other corner: (Select point.)
Select objects: (CR)
(The BLOCK is filed into memory and disappears from screen.)

The BLOCK is inserted into the drawing by the steps shown in Fig. 38.82. BLOCKS are inserted as entities, which means they cannot be edited by erasing parts of them or breaking lines within them.

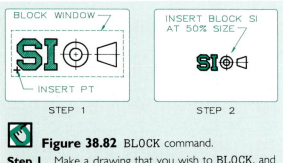

Figure 38.82 BLOCK command.

Step 1 Make a drawing that you wish to **BLOCK**, and respond to the **BLOCK** command as follows:
BLOCK name (or ?): <u>SI</u> (CR)
Insertion base point: (Select insert point.)
Select objects: <u>WINDOW</u> or <u>W</u> (CR)
(Window the drawing, and the object disappears into memory.)

Step 2 To Insert:
Command: <u>INSERT</u> (CR)
Block name (or ?): <u>SI</u> (CR)
Insertion point: (Select point with cursor.)
X-scale factor ‹1›/Corner/XYZ: <u>0.5</u>
(CR)
Y-scale factor ‹default=X›: (CR)
Rotation angle ‹0›: (CR)
(Block SI is inserted at 50% size.)

When an attempt is made to erase a portion of a BLOCK, the whole BLOCK is erased. However, BLOCKS can be edited if a star is inserted in front of the block name, for example, Block name (or ?): *SI (CR).

BLOCKS can be used only on the current drawing file unless they are converted to WBLOCKS, Write Blocks, which become permanent. This conversion is performed as follows:

Command: WBLOCK (CR)
File Name: A:SI (CR) (This assigns the name of the WBLOCK to the drive A.)
Block Name: SI (CR) (This is the name of the BLOCK that is being changed to a WBLOCK.)

A library of WBLOCKS that can be inserted in different files can relieve drafters of making drawings and will greatly improve their productivity.

The EXPLODE command is used to separate a BLOCK into individual entities that can be erased one at a time. Type EXPLODE, and select any point on the BLOCK.

38.44 Transparent Commands

Transparent commands can be used when another command is in progress. For example, if you are dimensioning a part and wish to use the HELP command, type 'HELP to use this command, and (CR) to complete the dimensioning command. Transparent commands are typed with apostrophes in front of them. Available transparent commands are 'GRAPHSCR, 'HELP, 'PAN, 'REDRAW, 'RESUME, 'SETVAR, 'TEXTSCR, 'VIEW, and 'ZOOM.

38.45 VIEW Command

A drawing can be saved as several separate named views with the VIEW command. For example, a large drawing such as a house is too large to see in its entirety. Therefore, it is desirable to show it as a series of views—one for the kitchen, one for the living room, and so forth.

The method of creating a VIEW is illustrated in Fig. 38.83, where the command prompts are

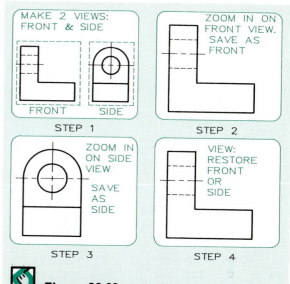

Figure 38.83 VIEW command.

Step 1 The two-view drawing can be saved as separate VIEWs.

Step 2 ZOOM the front view to fill the screen.
Command: VIEW (CR)
?/Delete/Restore/Save/Window: S (CR)
View name to save: FRONT (CR)

Step 3 ZOOM the side view to fill the screen.
Command: VIEW (CR)
?/Delete/Restore/Save/Window: S (CR)
View name to save: SIDE (CR)

Step 4 To display a view:
Command: VIEW (CR)
?/Delete/Restore/Save/Window: R (CR)
View name to save: FRONT (CR) (View is displayed on screen.)

given as

Command: VIEW (CR) ?/Delete/
Restore/Save/Window: (Select one.)
View name: (NAME) (CR)

The exact image on the screen can be saved as it is by selecting the Save option and giving it a name when prompted. The Window option is used to select a portion of a drawing currently displayed on the screen to become the VIEW. Type Restore, give the VIEW name, and it will be displayed. The Delete subcommand removes the VIEW from the list. To review the list of VIEWS that have been made, use the ? option.

38.46 Inquiry Commands

The INQUIRY commands are used to determine relationships among entities and information about the file being used. INQUIRY commands are LIST, DBLIST, STATUS, TIME, HELP, DIST, ID, and AREA.

Use the LIST command in the following manner:

Command: <u>LIST</u> (CR)
Select objects: (Select entities.) (CR)

The screen will replace the graphics with a listing of text information about the entity or entities selected, such as coordinates, lengths, angles, areas and so on.

DBLIST gives a similar listing of all the entities of a drawing. Use Ctrl S to stop and start scrolling on the screen. Ctrl C will abort the command.

DIST command is used to find the distance between two points. You will be prompted for the first and second points that can be selected with the cursor or by coordinates typed at the

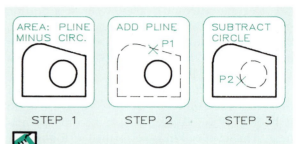

STEP 1   STEP 2   STEP 3

Figure 38.84 AREAS.

Step 1   The object drawn as a continuous **PLINE** and a circle can have its area determined with the **AREA** command.

Step 2   Command: <u>AREA</u> (CR)
‹First point›/Entity/Add/Subtract: <u>A</u>(CR)
‹First point›/Entity/Subtract: <u>E</u>(CR)
(ADD mode) Select circle or polyline:
(Point to polyline, and area is given.) (CR)

Step 3   (ADD mode) Select circle or polyline: (CR)
‹First point›/Entity/Subtract: <u>S</u>(CR)
‹First point›/Entity/Add: <u>E</u>(CR)
(SUBTRACT mode) Select circle or polyline: (Point to circle and its area is given along with the total area of the part.) (CR)

```
   133 entities in B:G-25-18
Limits are        X:    0.0000    3.4000   Off
                  Y:    0.0000    3.0000
Drawing uses      X:    0.1000    3.1000
                  Y:   -0.1750    3.4000  **Over
Display shows     X:    0.1000    5.3022
                  Y:   -0.1750    3.4000
Insertion base is X:    0.0000   Y:    0.0000   Z:    0.0000
Snap resolution is X:   0.0500   Y:    0.0500
Grid spacing is   X:    0.1000   Y:    0.1000

Current layer:       2DIMEN (Off)
Current color:       BYLAYER -- 5 (blue)
Current linetype:    BYLAYER -- CONTINUOUS
Current elevation:       0.0000 thickness:    0.0000
Axis off Fill on Grid off Ortho on Qtext off Snap on Tablet off
Object snap modes: None
Free RAM: 12250 bytes    Free disk: 764416 bytes
I/O page space: 64K bytes
```

Figure 38.85 The STATUS command will give this information about your drawing.

```
TIME
Current time:            27 Apr 1989 at 19:09:49.270
Drawing created:         12 Apr 1989 at 20:11:37.950
Drawing last updated:    19 Apr 1989 at 19:24:42.360
Time in drawing editor:  0 days 00:06:18.660
Elapsed timer:           0 days 00:06:18.660
```

Figure 38.86 The TIME command is used to record dates and times of drawings on file.

keyboard. At the command line, you will be given the distance, its angle, the delta *X* and the delta *Y,* where delta means the change in values from the first point.

ID is used to pick a point on the screen and obtain its *X, Y,* and *Z* coordinates.

AREA is used to find the area and perimeter of a space on the screen. You will be prompted, First point, Next point:, Next point:, and so on until you finish picking the points about the area and press (CR). An area can be calculated by having areas added or subtracted as shown in Fig. 38.84.

The STATUS command can be used by typing, Command: STATUS (CR). The data shown in Fig. 38.85 will be displayed on the screen, temporarily replacing the graphics display. This screen gives you information about the settings, layers, coordinates, and disk space.

The TIME command gives text display of information about the time spent on a drawing as shown in Fig. 38.86. The timer can be RESET and turned ON to measure the time for a drawing session, but the cumulative time cannot be erased without destroying the drawing file. The DIS-PLAY option displays the time opposite the heading, Elapsed timer:, to show how long the current session has taken after RESETting.

By typing FILES when in the Drawing Editor, the File Utility Menu (Fig. 38.87) will be displayed on the screen. You may select an option by number, use it, and return to Drawing Editor by pressing 0 and the F1 key.

```
File Utility Menu

  0. Exit File Utility Menu
  1. List Drawing files
  2. List user specified files
  3. Delete files
  4. Rename files
  5. Copy file

Enter selection (0 TO 5) <0>:
```

Figure 38.87 The FILES command will enable you to have access to other files on your disk without leaving the Drawing Editor.

38.47 Dimensioning Principles

Figure 38.88 shows the types of dimensions that can be used with AutoCAD. These options can be found in the submenu of the DIM: command.

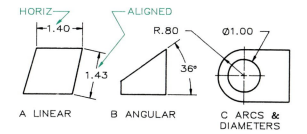

Figure 38.88 The types of dimensions that appear on a drawing.

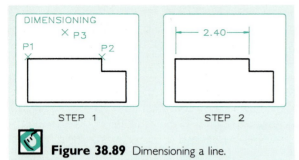

Figure 38.89 Dimensioning a line.

Step 1 Command: <u>DIM</u> (CR)
Dim: <u>VER</u>tical or <u>HOR</u>izontal (CR)
First extension line origin or (CR)
to select: (Select P1.)
Second extension line origin: (Select
P2.)
Dimension line location: (Select P3.)

Step 2 Press (CR), and the measurement appears at the command line, 2.40. Press (CR) to accept this value (or give a different value), and the dimension line is drawn.

```
DIM VARS  DEFAULT
DIMALT    OFF    Alternate units selected
DIMALTD   2      Alternate unit decimal places
DIMALTF   25.4   Alternate unit scale facator
DIMAPOST         Default suffix for alternate text
DIMASO    OFF    Create associative dimensions
DIMASZ    .12    Arrow size
DIMBLK           Arrow block name
DIMBLK1          First arrow block name
DIMBLK2          Second arrow block name
DIMCEN    −.05   Center mark size
DIMDLE    .00    Dimension line extension
DIMDLI    .38    Dim. line increment for continuation
DIMEXE    .12    Extension beyond dimension line
DIMEXO    .06    Extension line offset
DIMLFAC   1.00   Length factor
DIMLIM    OFF    Generates dimen tolerance limits
DIMPOST          Character suffix after dimensions
DIMRND    .00    Rounding value for distances
DIMSAH    OFF    Separate arrow heads at each end
DIMSCALE  1.00   Scale factor for all dim. variables
DIMSE1    OFF    Suppress first extension line
DIMSE2    OFF    Suppress second extension line
DIMSHO    OFF    Shows new dimens. while dragging
DIMSOXD   OFF    Suppresses outside dimen. lines
DIMTAD    OFF    Text placed above dimension lines
DIMTIH    ON     Text inside ext. lines is horizontal
DIMTIX    OFF    Text forced inside extension lines
DIMTM     .00    Minus tolerance value
DIMTOFL   OFF    Forces dim. line inside/text outside
DIMTOH    ON     Text outside ext lines is horizontal
DIMTOL    OFF    Applies tolerances to dimensions
DIMTP     .00    Plus tolerance value
DIMTSZ    .00    Assigns sizes to ticks
DIMTVP    .00    Locates text over or under dim. line
DIMTXT    .12    Text height
DIMZIN    0      Suppress 0 inches in feet & inches
```

Figure 38.90 A listing of the DIM VARS that can be selected and changed when dimensioning. Type DIM and STATUS to get this listing.

When in the **LINEAR** mode, you are prompted to specify if the dimension line is to be horizontal, aligned, or rotated. In Fig. 38.89, a horizontal line is dimensioned by selecting its two endpoints (*P*1 and *P*2) and locating the dimension line with *P*3. The dimension of 2.40 in., which is measured by the program, is accepted by (CR) and the dimension is shown on the drawing (Step 2).

> All drawings should be drawn **full size** when using computer graphics. The scaling process will take place at the time of plotting the drawings.

38.48 Dimensioning Variables—Introduction

Before becoming proficient with dimensioning, you must learn to control dimensioning variables, Dim Vars, shown in Fig. 38.90. In this section, we cover the basic variables necessary to begin dimensioning; the more advanced variables we cover in Section 38.51.

A listing of the current Dim Vars and their assigned values (Fig. 38.90) can be displayed on the screen by using the STATUS option under the DIM: command. The values shown in this figure are the default values set by AutoCAD. For an average drawing, you should reset the variables as shown in Fig. 38.91 based on the letter height,

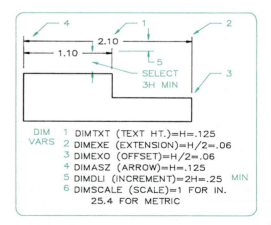

Figure 38.91 Dimensioning variables are based on the height of the lettering, usually about $\frac{1}{8}$ inch.

.125″, for example. You can change each variable by selecting the DIM: heading in the Root Menu, and then select the Dim Vars command, which will list the names of the variables. You will need to set DIMTXT, DIMASZ, DIMEXE, DIMEXO, DIM-TAD, and DIMDLI to the values shown in Fig. 38.91. Set the DIMSCALE variable to 1.

These variables will be retained in memory throughout the drawing session. They can be saved as part of your FORMAT file by the command SAVE, which will add them to this file for future use.

The UNITs command must be used to assign the number of decimals that the dimension units are to have. Two decimal places are used for inches, and none for millimeters. Architectural units are feet and inches.

Dimensions can be placed over or within the dimension lines by turning the DIMTAD command on or off (Fig. 38.92).

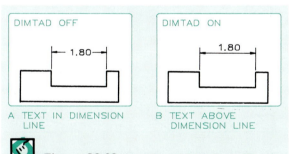

Figure 38.92 (A) When DIMTAD is OFF, the text is inserted within the dimension line. (B) When DIMTAD is ON, the text is placed above the dimension line.

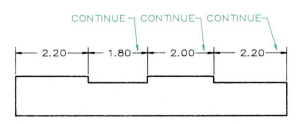

Figure 38.93 When dimensions are placed end to end, the option, CONTINUE, can be used to specify the Second extension line origin after the first dimension line has been drawn.

When a series of horizontal dimensions link together end-to-end, the CONTINUE command can be used to join successive dimension lines in a line, as shown in Fig. 38.93.

BASELINE dimensioning can be used to give a series of dimensions that originate from the same baseline. The DIMDLI variable (dimension line increment for continuation) automatically separates the parallel dimension lines (Fig. 38.94).

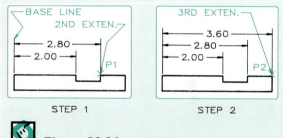

Figure 38.94 BASELINE option.

Step 1 Command: DIM (CR)
Dimension the first line as shown in Fig. 25.93. When prompted for Dim: for the next dimension, respond with BASELINE (CR) and you will be asked for the Second extension line origin: (Select P1.)

Step 2 The second dimension line will be drawn. Respond to Dim: with BASELINE (CR) again, and when asked for Second extension line origin:, select P2. Continue in this manner for any number of dimensions using the same baseline.

38.49 Dimensioning Arcs and Circles

Dimensions of circles will be given, as shown in Fig. 38.95, based on the size of the circle unless you override the defaults in the system. The computer follows the same decision process in selecting the style of dimensions as you would when using a pencil. Figure 38.96 shows the process of dimensioning a circle. A point on the circumference is selected with the cursor, and the diameter is automatically computed and placed across the circle.

Arcs are dimensioned with an R placed in front of the dimension of the radius (Fig. 38.97).

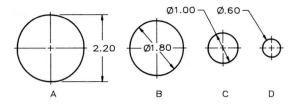

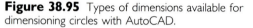

Figure 38.95 Types of dimensions available for dimensioning circles with AutoCAD.

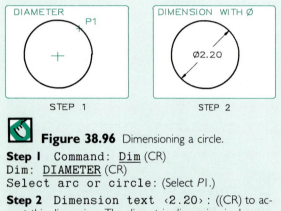

Figure 38.96 Dimensioning a circle.

Step I Command: <u>Dim</u> (CR)
Dim: <u>DIAMETER</u> (CR)
Select arc or circle: (Select *P*1.)

Step 2 Dimension text ‹2.20›: ((CR) to accept this dimension. The diametric dimension is drawn from *P*1 through the center.)

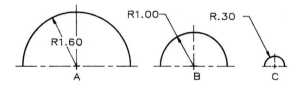

Figure 38.97 Arcs will be dimensioned by one of the formats given here depending on the size of the radius.

LEADERS can be used to position diametric and radial dimensions at locations of your choosing; but <u>R</u> and <u>0</u> will need to be entered at the keyboard in front of the measurements. Figure 38.98 shows the steps of dimensioning an arc.

38.50 Dimensioning Angles

Figure 38.99 shows variations in dimensioned angles, which occur because of inadequate space. Begin by selecting two lines of the angle, and

then select the location for the dimension line arc. Where room permits, the dimension value will be centered in the arc between the arrows. Figure 38.100 shows the commands for dimensioning an angle.

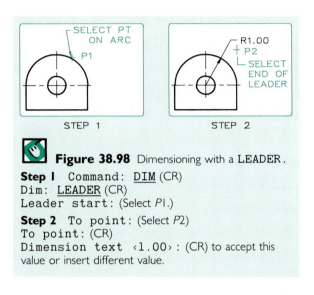

Figure 38.98 Dimensioning with a LEADER.

Step I Command: <u>DIM</u> (CR)
Dim: <u>LEADER</u> (CR)
Leader start: (Select *P*1.)

Step 2 To point: (Select *P*2)
To point: (CR)
Dimension text ‹1.00›: (CR) to accept this value or insert different value.

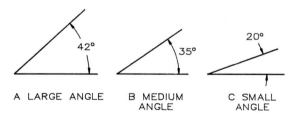

Figure 38.99 Angles will be dimensioned in any of these three formats using AutoCAD.

38.51 Dimensioning Variables—Advanced

Several advanced dimensioning variables given in Fig. 38.90 are explained below.

DIMTSZ gives the length of a slash (tick) that is used at the ends of dimension lines rather than arrows. When set to 0, arrows will be used.

DIMDLE is used to assign the length of the dimension line extension beyond the slash (tick) at

694

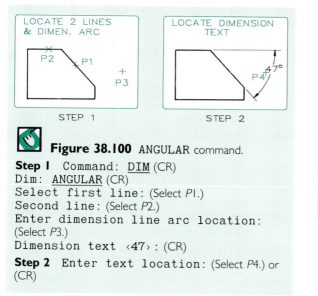

Figure 38.100 ANGULAR command.

Step 1 Command: DIM (CR)
Dim: ANGULAR (CR)
Select first line: (Select P1.)
Second line: (Select P2.)
Enter dimension line arc location:
(Select P3.)
Dimension text ‹47›: (CR)

Step 2 Enter text location: (Select P4.) or
(CR)

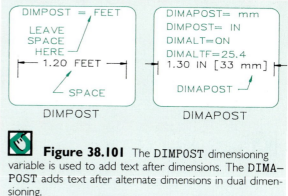

Figure 38.101 The DIMPOST dimensioning variable is used to add text after dimensions. The DIMA-POST adds text after alternate dimensions in dual dimensioning.

the ends of dimension lines. This variable will be inactive when DIMTSZ is set to 0.

DIMRND sets the rounding factor for dimensions. The number of decimal places in the rounding values is determined by the UNIT command.

DIMTIH, if ON, positions text inside extension lines horizontally; if OFF, the text is aligned with the dimension lines.

DIMTOH, if ON, positions text outside the extension lines horizontally; if OFF, the text is aligned with the dimension lines.

DIMSE1 suppresses the first extension lines.

DIMSE2 suppresses the second extension lines.

DIMALT, if ON, gives dimensions as dual dimensions (in two values such as inches and millimeters). These values are taken from the DIMALTF and DIMALTD settings.

DIMALTF sets alternate units scale factor for dual dimensions. The default value is 25.4, which gives the metric conversion for measurements made in inches.

DIMALTD sets the decimal places for the alternate units from 0 to 4 decimal places.

DIMPOST is used to assign a character string to follow all dimensions such as millimeters, feet, and miles. Type a period when prompted for this value to disable this variable.

DIMAPOST is used to assign a character string to follow all alternate dimensions (in dual dimensioning). Type a period to disable this variable (Fig. 38.101).

DIMLFAC sets the length factor that is used to scale all dimensions by the factor specified before the dimensions are inserted on the drawing. The default value is 1.00.

DIMZIN is used with architectural units of feet and inches to control the usage of zero values that precedes feet and inches as shown in Fig. 38.102.

DIMASO (associative dimensions) creates dimensions that are associated with lengths of the parts being dimensioned. When DIMASO is being

DIMZIN		EXAMPLES			
0	OMIT ZERO & INCHES	3/4"	9"	2'	2'–0 1/2"
1	ZERO FT & ZERO IN	0'–0 1/4"	0'–9"	2'–0"	2'–0 1/2"
2	ZERO FEET	0'–0 1/2"	0'–9"	2'	2'–0 1/2"
3	ZERO IN	3/4"	9"	2'–0"	2'–0 1/2"

Figure 38.102 The DIMZIN dimensioning variable controls the use of zeros when architectural units are being used. Options 0 and 1 are the most commonly used responses.

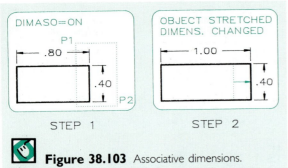

Figure 38.103 Associative dimensions.

Step 1   Set DIMASO to ON and apply dimensions to the part. Use the STRETCH command and WINDOW the end of the part with the C option.

Step 2   Select a new endpoint for the part and it will be lengthened and new dimensions will be calculated.

used, the STRETCH command can be used to stretch a part (Fig. 38.103), and the dimension lines will be recalculated and replaced on the drawing to correspond to the new stretched lengths.

DIMSHO works the same as DIMASO, but the dimension values are dynamically recalculated on the screen as the dimensioned feature is stretched. The default value is OFF.

DIMBLK (dimensioning block) is the name you assign to a block that you have created for specially designed arrows at the ends of dimension lines. If you set DIMBLK to DOT, you will get a dimension line with a dot located at the intersection of the dimension and extension lines.

To create your own arrow (a dot in this case), draw a right-end dot with a segment of the dimension line attached and an extension line beyond the dot (Fig. 38.104). The total length (including the dimension line segment, but not the extended portion) should be one drawing unit long. Make a BLOCK of the dot and dimension line, set the insertion point at the center of the dot where it will join the extension line, enter the dimensioning command, DIM, and type DIMBLK. When prompted, assign the dot BLOCK to DIMBLK. Now, when you dimension a linear distance, your special arrowheads (dots) will be inserted at the ends of all dimension lines.

To stop using the special dot, change the DIMBLK variable to a period (.).

Three new dimensioning subcommands, UPDATE, HOMETEXT, and NEWTEXT are available when DIMASO or DIMSHO are ON. The commands can be applied to entities that have been selected or by a window that encompasses the entire drawing.

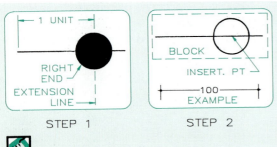

Figure 38.104 Custom-designed arrowhead.

Step 1   An arrowhead, a DONUT in this case, is drawn to be one unit long.

Step 2   The terminator is BLOCKed with the insertion point at the extension line. Use the DIMBLK option of the DIM VARS command to assign the BLOCK by its name. When dimensioning, the BLOCK will become the new arrowhead.

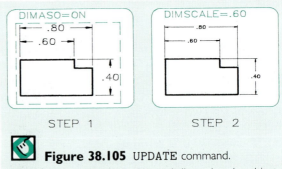

Figure 38.105 UPDATE command.

Step 1   Turn DIMASO ON, and dimension the object. Change any dimensioning variable you wish.
Dim: UPDATE (CR)
Select objects: (Window drawing.)

Step 2   If the DIMSCALE variable was changed, all dimensioning factors will be changed to their new values.

UPDATE is used to reassign dimension variables to the current dimension variables, UNITS, and text STYLE. By changing dimensioning variables and the text STYLE, these values can be updated on the current drawing as follows:

DIM: UPDATE (CR)
Select objects: (Window the drawing for selecting individual dimension lines.)

Figure 38.105 shows how the DIMSCALE variables are updated.

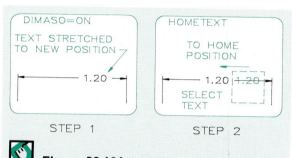

Figure **38.106** HOMETEXT command.
Step 1 Turn DIMASO ON, and dimension the object. If the object and dimensions are STRETCHed, the dimension numerals will not be centered.
Step 2 Dim: HOMETEXT (CR)
Select objects: (Select text.) (The text will automatically center itself in the dimension line.)

HOMETEXT is used to reposition text to its standard position at the center of the dimension line after the dimension line has been lengthened by the STRETCH command (Fig. 38.106).

Dim: HOMETEXT (CR)
Select objects: (Select the dimension entities.)

NEWTEXT is used to change a text string to different values (Fig. 38.107).

Dim: NEWTEXT (CR)
Enter new dimension text: (Enter value.) (CR)
Select objects: (Select dimension text to be changed.) (CR)

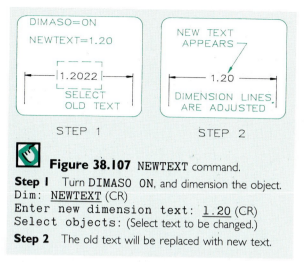

Figure **38.107** NEWTEXT command.
Step 1 Turn DIMASO ON, and dimension the object.
Dim: NEWTEXT (CR)
Enter new dimension text: 1.20 (CR)
Select objects: (Select text to be changed.)
Step 2 The old text will be replaced with new text.

If you give a null reply or enter ‹› for new text, the actual dimension measurements will be used as the text.

38.52 New Dimensioning Variables: Release 10

AutoCAD's Release 10 has added five new dimensioning variables: DIMTOFL, DIMTIX, DIMSOXD, DIMSAH, and DIMTVP. The first three of these variables are illustrated in Fig. 38.108.

The DIMTOFL variable places the text outside the extension lines and forces a line between them.

The DIMTIX variable forces the text between the extension lines.

The DIMSOXD variable suppresses dimension lines that would normally be drawn outside of extension lines.

The DIMSAH variable can be turned ON to give separate types of arrowheads at each end of a dimension line as shown in Fig. 38.109. The arrowheads are defined as DIMBLK1 and DIMBLK2 for the first and second types. When the dimension is inserted, the two types of arrowheads will be inserted. For this option to work, DIMTSZ must be set to zero, or ticks will be drawn instead.

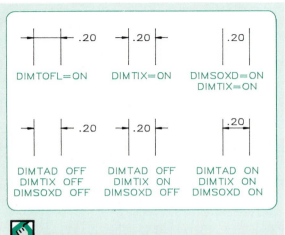

Figure 38.108 Examples of the application of dimensioning variables, `DIMTOFL`, `DIMTIX`, and `DIMSOXD` in conjunction with `DIMTAD`.

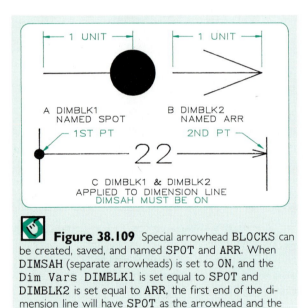

Figure 38.109 Special arrowhead BLOCKS can be created, saved, and named `SPOT` and `ARR`. When `DIMSAH` (separate arrowheads) is set to `ON`, and the Dim Vars `DIMBLK1` is set equal to `SPOT` and `DIMBLK2` is set equal to `ARR`, the first end of the dimension line will have `SPOT` as the arrowhead and the second end will have `ARR` as the arrowhead.

The `DIMTVP` variable is used with `DIMTAD` turned `OFF` to position text vertically; either above or below a dimension line. Text will be above the dimension line if the number is positive and below the dimension line if negative. The number that is inserted is used as a multiplier of the text height for the vertical space (`DIMTVP X DIMTXT`).

38.53 Toleranced Dimensions

Dimensions can be toleranced automatically, using any of the forms shown in Fig. 38.110, by setting the `Dim Vars`: `DIMTP` (plus tolerance), `DIMTM` (minus tolerance)—and by setting the `DIMTOL` to `ON`. As long as `DIMTOL` is on, all dimensions will have tolerances applied to them. By turning `ON` the `DIMLIM`, AutoCAD will compute the upper and lower limits of the dimension and apply them to the dimension line. Figure 38.111

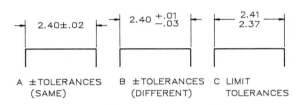

Figure 38.110 Dimensions can be toleranced in any of these three formats.

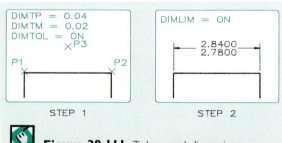

Figure 38.111 Toleranced dimensions.

Step I To tolerance a dimension, set `Dim Vars`, `DIMTOL` to `ON`. Set `DIMTP` (plus tolerance) and `DIMTM` (minus tolerance) to the desired values. Set `DIMLIM` to `ON` to convert the tolerances into the limit form.

Step 2 Dimension the line by using *P1*, *P2*, and *P3* to specify the first extension line, second extension line, and the dimension location. (The toleranced dimension is drawn.)

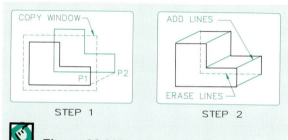

Figure 38.112 An oblique pictorial.

Step 1   Draw the frontal surface of the oblique, and COPY this view at *P2*, which is the desired angle and distance from *P1*.

Step 2   Connect the corner points, and erase the invisible lines to complete the oblique.

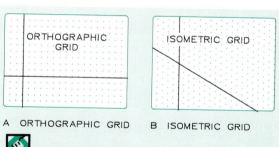

Figure 38.113 (A) The orthographic grid is called the STANDARD style of the SNAP mode. (B) The ISOMETRIC style of the SNAP mode.

shows the steps of giving tolerance limits on a dimension.

38.54   Oblique Pictorials

An oblique pictorial can be constructed as shown in Fig. 38.112. The front orthographic view is constructed and then COPYed behind the first view at the angle desired for the receding axis. Using OSNAP, the visible endpoints are connected, and invisible lines are erased. Circles can be drawn as true circles on the true-size front surface, but circular features should be avoided on the receding planes since their construction is complex.

38.55   Isometric Pictorials

The STYLE subcommand of the SNAP command can be used to change the rectangular GRID (called STANDARD) to ISOMETRIC (I) where the dots are plotted vertically and at 30° with the horizontal (Fig. 38.113). The lines of the cursor on the screen align with two of the isometric axes. The axes can be rotated by using Ctrl E or by activating the ISOPLANE command from the screen menu. When ORTHO is ON, lines are forced to be drawn parallel to the isometric axes. Figure

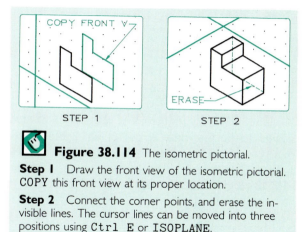

Figure 38.114 The isometric pictorial.

Step 1   Draw the front view of the isometric pictorial. COPY this front view at its proper location.

Step 2   Connect the corner points, and erase the invisible lines. The cursor lines can be moved into three positions using Ctrl E or ISOPLANE.

38.114 shows the steps for constructing an isometric using the grid.

To depict circular features in isometric, a four-center ellipse one unit across its diameter is constructed (Fig. 38.115). This ellipse should be made into a BLOCK and inserted where needed. It can be scaled to fit different drawings by giving it an X–VALUE factor at the time of insertion, and it can be rotated to fit each isometric plane.

Figure 38.116 shows an isometric pictorial with elliptical features. The ellipse was inserted using the four-center ellipse BLOCK developed in Fig. 38.115. The size of the block is changed by entering the size factor when prompted for the

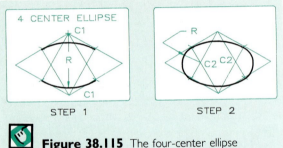

Figure 38.115 The four-center ellipse block.

Step 1 Construct a rhombus that is one unit along both axes. Using the four-center ellipse method, construct an ellipse using four centers and four arcs. (The rhombus should be drawn on a construction layer that can be turned off so that the lines will not show.)

Step 2 Make a WBLOCK of this ellipse using the center as the insertion point. This block can be INSERTed as a * BLOCK so that it can be broken when needed.

X-value. It is advantageous if the BLOCK is designed to be one unit across (a UNIT block) so that its size can be scaled with ease. For example, a UNIT BLOCK will be multiplied by an X-value of 2.48 to give a BLOCK that is 2.48 in.

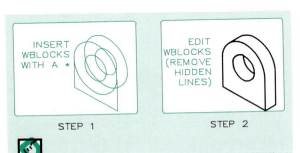

Figure 38.116 An isometric drawing with elliptical features.

Step 1 The UNIT BLOCK of the ellipse developed in Fig. 38.115 is inserted and sized (X-scale factor) to match the drawing's dimensions. The ellipse can be rotated as a BLOCK to fit any of the three isometric surfaces.

Step 2 The back side of the isometric is drawn. The hidden lines are removed with the BREAK command.

38.56 ELLIPSE Command

The ELLIPSE command can be used to draw ellipses by several methods. In Fig. 38.117, an ellipse is drawn by selecting the endpoints of the major axis and the third point, P3, which gives the minor radius of the ellipse.

A second method (Fig. 38.118) is begun by locating the center of the ellipse, P1, then the axis

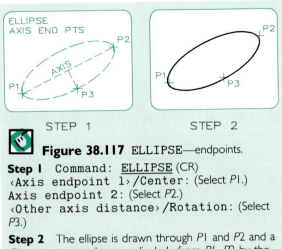

Figure 38.117 ELLIPSE—endpoints.

Step 1 Command: ELLIPSE (CR)
‹Axis endpoint 1›/Center: (Select P1.)
Axis endpoint 2: (Select P2.)
‹Other axis distance›/Rotation: (Select P3.)

Step 2 The ellipse is drawn through P1 and P2 and a distance measured perpendicularly from P1–P2 by the location of P3.

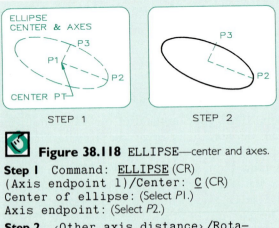

Figure 38.118 ELLIPSE—center and axes.

Step 1 Command: ELLIPSE (CR)
(Axis endpoint 1)/Center: C (CR)
Center of ellipse: (Select P1.)
Axis endpoint: (Select P2.)

Step 2 ‹Other axis distance›/Rotation: (Select P3.)
(The ellipse is drawn.)

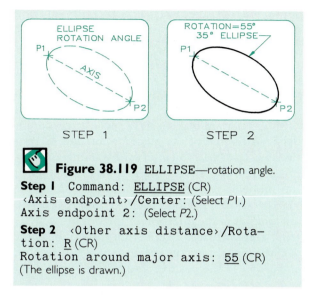

STEP 1   STEP 2

Figure 38.119 ELLIPSE—rotation angle.

Step 1 Command: <u>ELLIPSE</u> (CR)
‹Axis endpoint›/Center: (Select *P1*.)
Axis endpoint 2: (Select *P2*.)

Step 2 ‹Other axis distance›/Rotation: <u>R</u> (CR)
Rotation around major axis: <u>55</u> (CR)
(The ellipse is drawn.)

endpoint, *P2*, and the length of the other axis, *P3*. The ellipse will be drawn to pass through points *P2* and a point on the minor diameter specified by *P3*.

A third method (Fig. 38.119) requires the location of the endpoints of the axis and the specification of a rotation angle about the axis. When the rotation angle is 0°, the ellipse is a full circle; when the rotation is 90°, the ellipse is an edge.

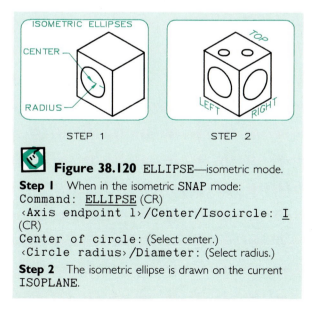

STEP 1   STEP 2

Figure 38.120 ELLIPSE—isometric mode.

Step 1 When in the isometric SNAP mode:
Command: <u>ELLIPSE</u> (CR)
‹Axis endpoint 1›/Center/Isocircle: <u>I</u> (CR)
Center of circle: (Select center.)
‹Circle radius›/Diameter: (Select radius.)

Step 2 The isometric ellipse is drawn on the current ISOPLANE.

When in the isometric SNAP mode, the EL-LIPSE command will be

‹Axis endpoint 1›/
Center/Isocircle: <u>I</u> (CR)

By selecting the ISOCIRCLE option, ellipses can be automatically drawn in the correct orientation on each of the three ISOPLANEs. The isometric ellipses can be selected using the previous techniques (Fig. 38.120).

Portions of ellipses drawn with these new commands can be removed by the BREAK command. Since ellipses are drawn with connected curves, some peculiarities can occur when the BREAK command is used.

38.57 Introduction to 3D Pictorials

To take full advantage of the three-dimensional aspects of Release 10, set the FLATLAND system variable to 0 (OFF) in the following manner:

Command: <u>SETVAR</u>
Variable name or ?: <u>FLATLAND</u>
New value of FLATLAND? ‹0› <u>0</u>

The commands ELEV, VPOINT, and HIDE can be used as introductions to three-dimensional drawing. ELEVation is under the SETTINGS menu, and VPOINT and HIDE commands are under the DIS-PLAY menu (VIEWPOINT option).

Command: <u>ELEV</u> (CR)
New current elevation ‹0›: (CR) (To accept 0.)
New current thickness ‹0›: <u>3</u> (To assign a height of 3″ units.)

These responses have set the base to 0 and the height to 3 units. Select VPOINT and its PLAN option from the **Root Menu.** VPOINT gives a plan view (top view) of the drawing area on the screen. Use the regular DRAW commands such as LINE, CIRCLE, TEXT, and ARC to complete the top view of the drawing. All entities (except TEXT) that are drawn have a height of 3 in. in the *Z* direction.

Obtain the viewpoint globe by typing VPOINT and striking (CR) twice; a set of axes will

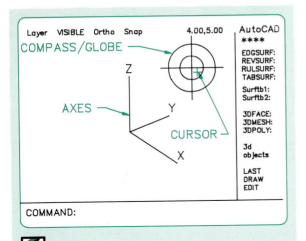

Figure 38.121 When **VPOINT** command is selected, and (CR) is pressed twice, a globe and a set of axes will appear on the screen for selecting a viewpoint.

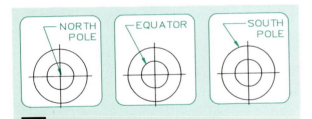

Figure 38.122 By positioning the cursor within the small circle of the globe, the viewpoint will be looking down on the upper hemisphere of the globe. Any point on the small circle gives a horizontal (equatorial) view. A point between the small and large circles is a view of the lower hemisphere.

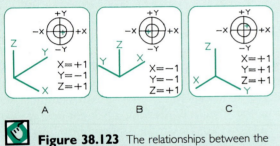

Figure 38.123 The relationships between the points on the **VPOINT** globe and the **VPOINT** values selected from the keyboard.

appear on the screen (Fig. 38.121). By moving your cursor about the globe, you can position the axes for your desired **VPOINT**, and the three-dimension solid will appear on the screen in wire-diagram form. Using this same method, the viewpoint can be changed to see the object from different directions (Fig. 38.122).

Observation of Fig. 38.123 will help you understand the use of the globe option and its relationship to **VPOINTS** specified by numerals. Figure 38.124 also shows the graphical meaning of the numerals that are used for establishing the **VPOINT**. A **VPOINT** of 1, −1, 1 means that you are looking in the direction of the origin (0,0,0) from a point that is 1 unit from the origin in the X-direction, 1 unit in the negative Y-direction, and 1 unit in the positive Z-direction.

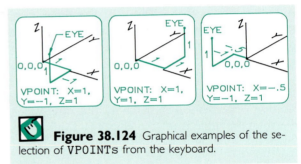

Figure 38.124 Graphical examples of the selection of **VPOINTs** from the keyboard.

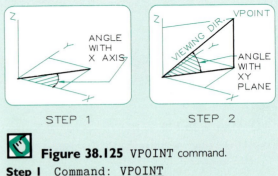

Figure 38.125 VPOINT command.

Step 1 Command: <u>VPOINT</u>
Rotate/‹View Point› ‹current›: <u>R</u>
Enter angle in X–Y plane from X axis
‹current›: <u>45</u>

Step 2 Enter angle from X–Y plane
‹current›: <u>30</u>

A pull-down menu under the DISPLAY heading, VPOINT 3D, can be used to select viewpoints, including the globe and HIDE options.

The VPOINT command offers the ROTATE option for specifying the point of view as shown in Fig. 38.125. In this case the angle with the X-axis and the XY-plane are given. Figure 38.126 shows an example 3D drawing.

The HIDE command can be turned on to remove hidden lines, which results in an empty box without a top (Fig. 38.127). TEXT in 3D will not have an extruded thickness but will lie on the ELEV plane. Circles will appear as vertical cylinders.

The six principal orthographic views can be found by typing the following values when in the VPOINT command:

Top view	0,0,1
Front view	0,−1,0
Right-side view	1,0,0
Left-side view	−1,0,0
Rear view	0,1,0
Bottom view	0,0,−1

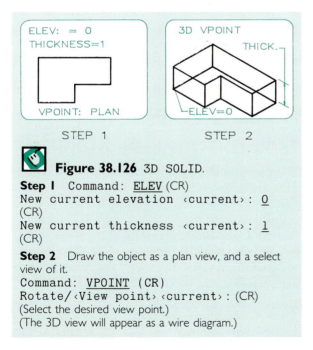

STEP 1   STEP 2

Figure 38.126 3D SOLID.

Step 1 Command: <u>ELEV</u> (CR)
New current elevation ‹current›: <u>0</u> (CR)
New current thickness ‹current›: <u>1</u> (CR)

Step 2 Draw the object as a plan view, and a select view of it.
Command: <u>VPOINT</u> (CR)
Rotate/‹View point› ‹current›: (CR)
(Select the desired view point.)
(The 3D view will appear as a wire diagram.)

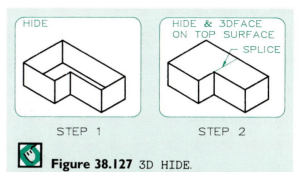

STEP 1   STEP 2

Figure 38.127 3D HIDE.

Step 1 Once a desired 3D view has been obtained, select HIDE from the menu, and the object will appear as an open box.

Step 2 By using the SOLID or 3DFACE command at the elevation of the top plane, the upper surface can be filled in to make it appear as solid when HIDE is applied. Only rectangular or triangular areas can be filled. Splice lines between these solid areas will show.

38.58 Fundamentals of 3D Drawing

With system variable, FLATLAND, turned OFF (0) and UCSION turned on, you are ready to experiment with 3D drawing. When prompted for a point, you establish it with three coordinates such as 1, 2, 4, which means that the point is located 1 unit in the X-direction from the origin (0, 0, 0), 2 units in the Y-direction from the first point, and 4 units in the Z-direction from the second point.

Most of your drawings will be made on the screen where the positive X-direction is horizontally to the right, the positive Y-direction is vertically to the top, and Z-direction is a point imagined to be projecting out of the screen. The right-hand rule (Fig. 38.128), can be used to establish the 3 axes by pointing your thumb toward the X-axis and your index finger in the direction of the Y-axis. Your middle finger will be perpendicular to both axes and in the direction of the positive Z-axis.

When making a drawing, an X-Y icon can be made to appear in the lower left corner of the screen to show the direction of the X- and Y-axes; the Z-axis is not represented by an icon. Examples of directional icons are shown in Fig.

38.129, The *X-Y* icon with a W represents the direction of the **World Coordinate System** (WCS). One without the W represents the directions of the **User Coordinate System** (UCS).

The broken-pencil icon gives a warning that a projection plane appears as an edge, making drawing and other commands hard to perform in

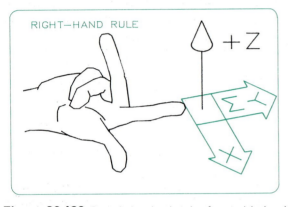

Figure 38.128 By pointing the thumb of your right hand in the positive *X* direction, and your index finger in the positive *Y* direction, your middle finger will point in the positive *Z* direction.

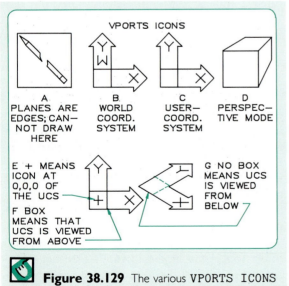

Figure 38.129 The various VPORTS ICONS that appear on the screen to show the *X*- and *Y*-axes.

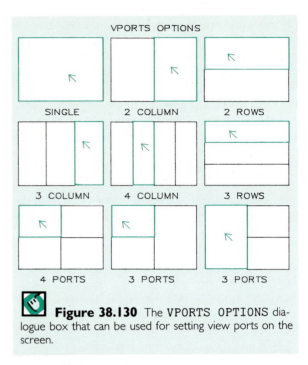

Figure 38.130 The VPORTS OPTIONS dialogue box that can be used for setting view ports on the screen.

this view. The perspective view of the box indicates that the current viewpoint of a drawing gives a perspective.

A maximum of four VPORTS can be selected from a pulldown menu, VPORTS OPTION, shown in Fig. 38.130. By using these VPORTS, you can make multiple views of the part being drawn, view all of them at one time, and change them individually. VPORTS can be SAVEd by name.

38.59 The Coordinate Systems

Two coordinates systems can be used in 2D and 3D drawings: the World Coordinate System (WCS) and the User Coordinate System (UCS).

World Coordinate System The WCS has an origin where *X*, *Y*, and *Z* are 0. The normal view of the WCS is one where the *X*- and *Y*-axes are perpendicular and true length on the screen.

User Coordinate System The UCS can be located within the WCS with its *X*- and *Y*-axes positioned in any direction with an origin located at any point of your choosing.

Establish a User Coordinate System in the following manner:

```
Command: UCS (CR)
ORigin/ZAxis/3point/Entity/
View/X/Y/Z/Prev/Restore/Save/
Del/?/<World> : O (CR)
Origin point <0,0,0> : (Select desired
origin.) (CR)
```

Show the UCS icon at the UCS origin in the following manner:

```
Command: UCSICON (CR)
ON/OFF/All/Noorgin/ORigin
<ON> : OR (CR)
```

An *X-Y* icon is drawn at the UCS origin.

An easy way of setting the UCS to a previously drawn object is the 3point option of the UCS command as shown in Fig. 38.131. By using this option points are located at the origin, on the *X*-axis, and the *Y*-axis.

An explanation of the other options of the UCS command are given below.

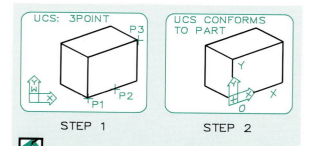

Figure 38.131 The 3Point option.

Step 1 Command: UCS (Select 3point option.) Origin point <0, 0, 0> : P1 (Select origin.) Point on positive portion of the X axis: P2 (On X-axis.) Point on positive Y portion of UCS X–Y plane: P3 (On Y-axis.)

Step 2 The UCS icon will be transferred to the origin. The plus sign at the corner box indicates that it is at the origin.

Zaxis With the ZAXIS option, by selecting the origin and a point on the positive portion of the *Z*-axis, *X*- and *Y*-axes are located.

Entity With the ENTITY option, you can point to an entity (other than 3D Polyline or a polygon mesh) and the UCS will have the same positive *Z*-axis as the selected entity.

View The VIEW option establishes a coordinate system with an *X-Y* plane that is parallel to the screen. This command allows you to apply non-pictorial text to a 3D drawing.

X/Y/Z By specifying *X* with the X/Y/Z option, you may rotate the UCS about the *X*-, *Y*-, or *Z*-axes.

Previous The Previous option restores the previous UCS.

Restore The Restore option prompts you for the name of the saved UCS and makes it the current UCS.

Save The Save option is used to name and save a UCS.

Delete The Delete option removes a saved UCS from memory.

? The ? option gives a listing of the current and saved coordinate systems. If the current UCS is unnamed, it is listed as *WORLD or *NO NAME.

World The World option sets the UCS to the same as the WCS.

The options of the UCSICON command are explained below.

On The ON option turns the icon on.

Off The OFF option turns the icon off.

All The ALL option applies the icon to all active viewports when more than one is being used.

Noorigin The NOORIGIN option displays the icon at the lower left corner regardless of the location of the UCS origin.

Origin The ORIGIN option places the icon at the origin of the current coordinate system if space is available; otherwise, the icon is displayed at the lower left of the viewport.

The pull-down dialogue box, Modify UCS (Fig. 38.132A), lists the defined coordinate systems with a check indicating the current system. The

List button displays information about the X-, Y-, and Z-axes of each UCS origin. By selecting Define New UCS, a new dialogue box is called that can be used to define a new UCS.

Under the Settings heading of the **Menu Bar,** the UCS dialogue box entitled User Coordinate System Options can be pulled down for selecting views of a 3D drawing (Fig. 38.132B) if you are in the WCS. For example, when a 3D view of an object is on the screen, you can select the icon for the top view and you will be prompted for the origin (0,0,0). Select a point on the screen and points will appear representing the new UCS for X- and Y-axes for a top view. Use the PLAN command, and the top view will appear on the screen.

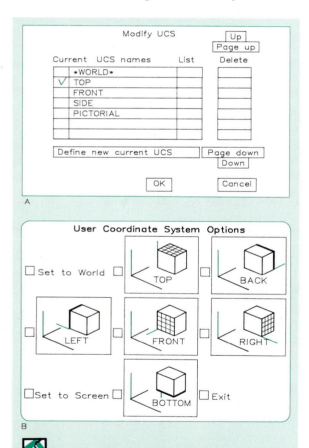

A

B

Figure 38.132 (A) This dialogue box can be pulled down from the Settings heading of the Menu Bar. (B) When a 3D view of an object is on the screen you can select an icon for the desired viewpoint. You will be prompted for origin of the selected UCS. Use the PLAN command and the designated view will appear on the screen. This works only while in the WCS.

38.60 The DVIEW Command

The DVIEW command enables you to dynamically select a viewpoint for looking at a three-dimensional drawing. In addition to obtaining axonometric drawings (parallel projections), perspectives can be drawn where parallel lines tend to converge toward vanishing points, yielding highly realistic pictorials (Fig. 38.133).

The DVIEW commands gives the following options:

```
Command: DVIEW (CR)
CAmera/TArget?Distance/POints/
PAn/Zoom/TWist/CLip/Hide
/Off/Undo/
<eXit>:
```

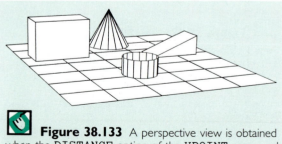

Figure 38.133 A perspective view is obtained when the DISTANCE option of the VPOINT command is selected.

With the CAmera (Camera) option your viewpoint is rotated about the object as if you were moving about it with a camera. When prompted, you specify the viewpoint at the keyboard with angles, or you can use slider bars. A slider bar appears at the right of the screen (Fig. 38.134) with an angle from −90° to +90° for looking at the object from bottom to top.

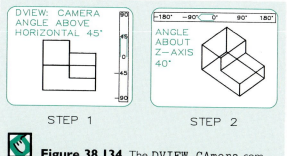

STEP 1   STEP 2

Figure 38.134 The DVIEW CAmera command.

Step 1 Command: DVIEW (Select CAmera option.) Select objects: (Window object.) (CR) (A vertical slider bar appears at right. Select viewpoint elevation with cursor.)

Step 2 A horizontal slider bar appears at top of screen. Select desired viewpoint with cursor.

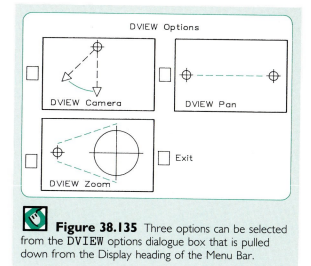

Figure 38.135 Three options can be selected from the DVIEW options dialogue box that is pulled down from the Display heading of the Menu Bar.

Once the vertical viewpoint is selected, a second slider bar appears at the top of the screen that allows you to select a viewpoint from −180° to +180° about the object.

Three of the DVIEW options, including the CAmera option, can be selected from the pull-down menu under the DISPLAY heading of the menu bar (Fig. 38.135). Figure 38.136 gives an

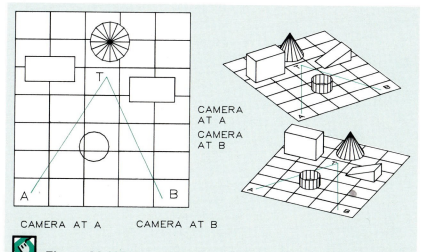

CAMERA AT A   CAMERA AT B

Figure 38.136 The CAmera option of the DVIEW command can be used to obtain different views of a target point by moving the position of the camera.

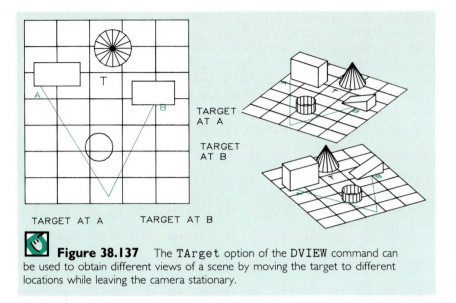

Figure 38.137 The TArget option of the DVIEW command can be used to obtain different views of a scene by moving the target to different locations while leaving the camera stationary.

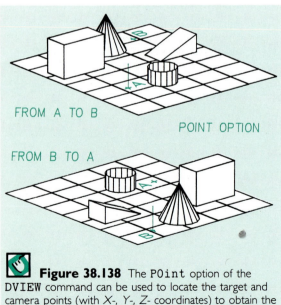

Figure 38.138 The POint option of the DVIEW command can be used to locate the target and camera points (with X-, Y-, Z- coordinates) to obtain the desired view. Both perspective and axonometric views can be found with this option.

example of how the camera viewpoint can be changed for viewing a target.

TArget is used to rotate a specified target point about the camera as shown in Fig. 38.137. The prompts are identical to the CAmera option, but the target position is being rotated about the camera instead of the camera about the object.

DISTANCE places the camera with respect to the object and automatically turns on the **perspective option.** The X-Y icon is replaced with the perspective-box icon. You are prompted for a distance to the target, which can be entered as a number at the keyboard or specified by a slider bar from 0X to 16X. 1X is the current distance to the target. The DVIEW, ZOOM option can be used to magnify a drawing without turning the perspective option on.

POints is used to locate the target point first and the camera position second for viewing a drawing as shown in Fig. 38.138.

PAn is used to move a drawing on the screen without changing viewpoint or magnification. When in the perspective mode, the pointing device must be used; but when perspective is OFF, coordinates can be typed at the keyboard for panning coordinates.

Zoom is used to change the magnification of a drawing on the screen in the same manner as the Zoom/Center command when perspective options are not being used. If the perspective option is ON, Zoom varies the magnification of the drawing as if you were changing the lens of a camera from a wide-angle to a telescopic lens. Therefore, the view of the object is dynamically distorted.

TWist is used to rotate the 3D drawing on the screen. For example, you may wish to plot a drawing with a vertical format on a sheet with a horizontal format. TWist can be used for this operation like the ROTATE command is used for two-dimensional drawings.

CLip is used to place planes perpendicular to the line sight in a drawing (perspective and parallel projections) to remove all of the drawing either in front of or in back of the plane. When the CLip option is selected, the prompts Back/Front/‹Off›: appear.

By using the Back option, the clipping plane is located on the drawing for removing all that is behind it (Fig. 38.139).

By using the Front option, the clipping plane is located for removing portions of the drawing that are in front of it (Fig. 38.139).

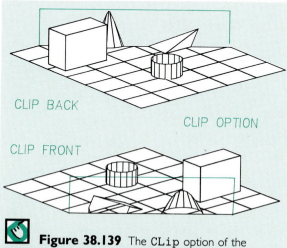

Figure 38.139 The CLip option of the DVIEW command is used to remove Back or Front portions of a drawing. The clipping plane is assumed to be parallel to the drawing screen.

The Off option turns clipping off. When in the perspective mode, clipping cannot be turned off, but the clipping plane is located at the camera position for front clipping.

Hide suppresses the invisible lines of a 3D drawing.

OFF turns off the perspective mode that was enabled by the Distance option.

Undo reverses the previous operations performed under the DVIEW commands, one at a time.

eXit ends the DVIEW command.

When in the perspective mode, some operations cannot be performed, and a prompt will be given asking if you wish to continue with perspective off. A negative response will leave the perspective ON and a positive response will turn it OFF.

38.61 Basic 3D Forms

A pull-down menu entitled 3D Construction can be found under the **Draw** heading of the **Menu Bar.** From this menu, you can select boxes, pyramids, domes, spheres, dishes, wedges, cones, and toruses (toriods). Also, included in this menu are options for surface revolution, ruled surface, edge surface, tabulated surface, and meshes. Boxes may be selected for setting values for surface tabulation 1 and surface tabulation 2.

Box Figure 38.140 illustrates how, with the Box command, a cube or box can be drawn by selecting a corner, measuring the length and height, and selecting the angle of rotation about the Z-axis.

Wedge Figure 38.141 shows how a wedge can be drawn in the exact manner as the box was drawn, using the WEDGE command.

Pyramid By using the PYRAMID command, a pyramid can be drawn that extends to its apex, as illustrated in Fig. 38.142. An option is available for drawing a truncated pyramid with its apex removed.

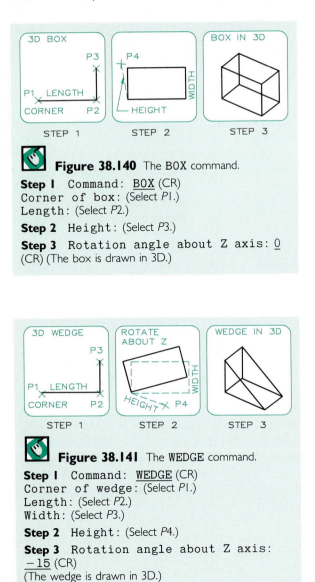

Figure 38.140 The BOX command.

Step 1 Command: <u>BOX</u> (CR)
Corner of box: (Select *P1*.)
Length: (Select *P2*.)

Step 2 Height: (Select *P3*.)

Step 3 Rotation angle about Z axis: <u>0</u>
(CR) (The box is drawn in 3D.)

Figure 38.141 The WEDGE command.

Step 1 Command: <u>WEDGE</u> (CR)
Corner of wedge: (Select *P1*.)
Length: (Select *P2*.)
Width: (Select *P3*.)

Step 2 Height: (Select *P4*.)

Step 3 Rotation angle about Z axis:
<u>−15</u> (CR)
(The wedge is drawn in 3D.)

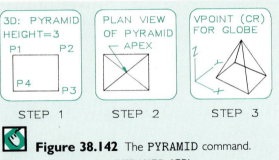

Figure 38.142 The PYRAMID command.

Step 1 Command: <u>PYRAMID</u> (CR)
First base point: (Select *P1*.)
Second base point: (Select *P2*.)
Third base point: (Select *P3*.)
Tetrahedron/‹Fourth base point› : (Select
P4.)

Step 2 Ridge/Top. ‹Apex point› : <u>.XY</u>
of (need Z): <u>2</u> (CR)

Step 3 The pyramid is drawn in 3D.

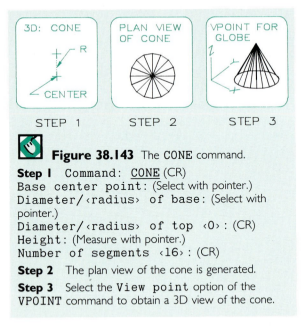

Figure 38.143 The CONE command.

Step 1 Command: <u>CONE</u> (CR)
Base center point: (Select with pointer.)
Diameter/‹radius› of base: (Select with
pointer.)
Diameter/‹radius› of top ‹0› : (CR)
Height: (Measure with pointer.)
Number of segments ‹16› : (CR)

Step 2 The plan view of the cone is generated.

Step 3 Select the View point option of the
VPOINT command to obtain a 3D view of the cone.

Cone Figure 38.143 shows how the CONE command is used to draw a cone as a basic 3D form. A truncated cone can also be drawn by following the given prompts.

Sphere The SPHERE command is used to draw a sphere by selecting its center and its radius (Fig. 38.144).

Dish Figure 38.145 shows how a dish, the lower hemisphere of a sphere, is drawn by selecting its center and its radius using the DISH command.

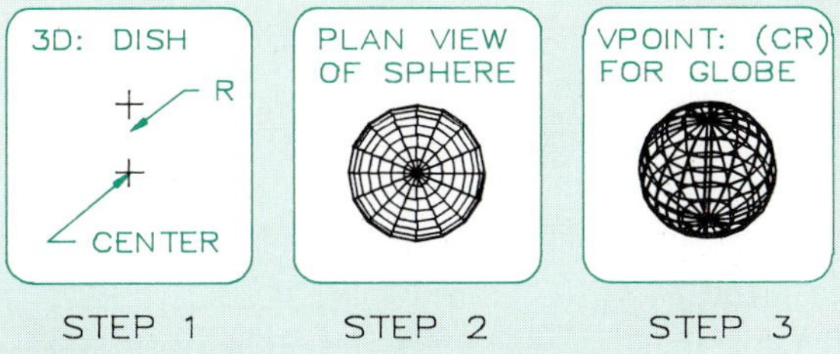

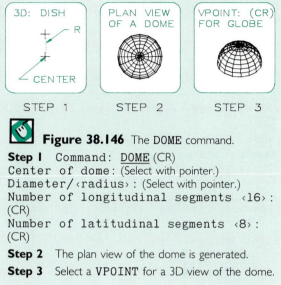

Figure 38.144 The SPHERE command.

Step 1 Command: <u>SPHERE</u> (CR)
Center of sphere: (Select with pointer.)
Diameter/<radius> : (Select with pointer.)
Number of longitudinal segments<16> :
(CR)
Number of latitudinal segments <16> :
(CR)

Step 2 The plan view of the sphere will be generated.

Step 3 Select a VPOINT to obtain a 3D view of the sphere.

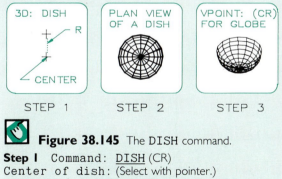

STEP 1   STEP 2   STEP 3

Figure 38.145 The DISH command.

Step 1 Command: <u>DISH</u> (CR)
Center of dish: (Select with pointer.)
Diameter/<radius> : (Select with pointer.)
Number of longitudinal segments <16> :
(CR)
Number of latitudinal segments <8> :
(CR)

Step 2 The plan view of the dish is generated.

Step 3 Select a VPOINT for a 3D view of the dish.

Figure 38.146 The DOME command.

Step 1 Command: <u>DOME</u> (CR)
Center of dome: (Select with pointer.)
Diameter/<radius> : (Select with pointer.)
Number of longitudinal segments <16> :
(CR)
Number of latitudinal segments <8> :
(CR)

Step 2 The plan view of the dome is generated.

Step 3 Select a VPOINT for a 3D view of the dome.

Dome Using the DOME command, Fig. 38.146 illustrates the steps of drawing a dome, the upper hemisphere of a sphere, by selecting the center and its radius.

Torus Figure 38.147 shows the steps of drawing a donut shape called a torus or toriod using the TORUS command.

Hide The HIDE command can be used to remove the hidden (invisible) lines from these 3D forms to give them a more realistic three-dimensional appearance.

38.62 3D Polygon Meshes

Three-dimensional meshes are available for showing the surfaces of an object. Also, these meshes can be edited by the PEDIT command to change the form of the meshes. The commands available for drawing meshes are: 3DMESH, RULE–SURF, TABSURF, REVSURF, and EDGESURF.

3DMESH is used to define a 3D polygon mesh as shown in Fig. 38.148. In this example a plan

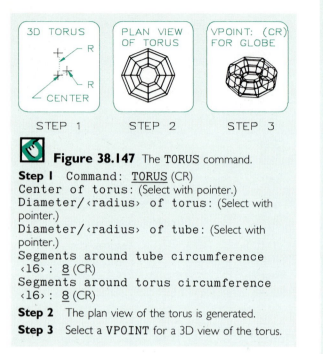

Figure 38.147 The TORUS command.

Step 1 Command: <u>TORUS</u> (CR)
Center of torus: (Select with pointer.)
Diameter/‹radius› of torus: (Select with pointer.)
Diameter/‹radius› of tube: (Select with pointer.)
Segments around tube circumference ‹16›: <u>8</u> (CR)
Segments around torus circumference ‹16›: <u>8</u> (CR)

Step 2 The plan view of the torus is generated.

Step 3 Select a VPOINT for a 3D view of the torus.

view of the mesh is shown where vertices are specified in the M- and N-directions with 256 in each direction being the maximum number permitted. Once the M- and N-values of the mesh have been specified, you will be prompted for the X-, Y-, and Z-values for each vertex, starting from the origin. The values can be used to establish the vertices in any position, not just in a rectangular format.

When all the points have been provided the mesh will appear on the screen (Fig. 38.149). It can be viewed as a three-dimensional surface from any angle of your choosing.

PEDIT is a command that can be used to change a mesh:

```
Command: PEDIT (CR)
Edit vertex/Smooth surface/
Desmooth/Mclose/Nclose/Undo/
eXit ‹X›:
```

The Smooth option can be used to form the mesh into a smooth three-dimensional shape. Desmooth is used to discard the smoothed surface and reverts it to its previous form. The Undo

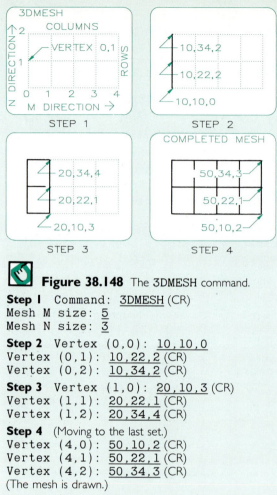

Figure 38.148 The 3DMESH command.

Step 1 Command: <u>3DMESH</u> (CR)
Mesh M size: <u>5</u>
Mesh N size: <u>3</u>

Step 2 Vertex (0,0): <u>10,10,0</u>
Vertex (0,1): <u>10,22,2</u> (CR)
Vertex (0,2): <u>10,34,2</u> (CR)

Step 3 Vertex (1,0): <u>20,10,3</u> (CR)
Vertex (1,1): <u>20,22,1</u> (CR)
Vertex (1,2): <u>20,34,4</u> (CR)

Step 4 (Moving to the last set.)
Vertex (4,0): <u>50,10,2</u> (CR)
Vertex (4,1): <u>50,22,1</u> (CR)
Vertex (4,2): <u>50,34,3</u> (CR)
(The mesh is drawn.)

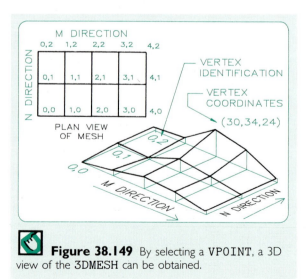

Figure 38.149 By selecting a VPOINT, a 3D view of the 3DMESH can be obtained.

command is used to undo PEDIT options that have been performed. If the mesh is closed, Mo-pen and Nopen will replace the commands Mclose and Nclose as options. The command eXit is used to exit from the PEDIT command.

Edit vertex enables you to edit individual vertices within the mesh:

```
Vertex (M,N): Next/Previous/
Left/Right/Up/Down/Move/REgen/
eXit <N>
```

An X appears at the origin for selecting the vertex to be changed. The Next and Previous options are used to move forward and backward from the origin toward the last vertex. The Left and Right options moves the pointer in the *N*-direction. The Up and Down options move the pointer in the *M*-direction to make locating vertices faster.

Once a vertex is selected for changing, select the Move option, and you will be prompted, Enter new location. At this time, type the *X*-, *Y*-, and *Z*-coordinates of the new position. The revised mesh can be displayed by the REgen option.

38.63   The RULESURF Command

The RULESURF command is used to connect two entities with ruled lines. The entities can be curves, arcs, polylines, lines, or points (but only one of the boundaries can be a point). Examples of ruled surfaces are shown in Fig. 38.150.

The system variable, SURFTAB1, is used to assign the number of vertices that are to be placed along each entity. Figure 38.151 demonstrates the results of varying the SURFTAB1 system variable.

38.64   The TABSURF Command

The TABSURF command is used to tabulate a surface from a path curve and a direction vector. In Fig. 38.152, a circle and a vector are given. By selecting the circle first, and the endpoint of the vector, a cylinder is tabulated with its elements equal in length to the vector. The SURFTAB1 sys-

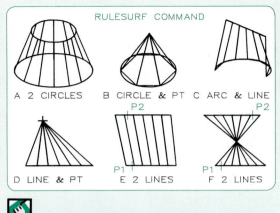

Figure 38.150 The RULESURF command connects entities with a series of straight lines. You can see at E and F that the positions of the selected points determine the direction of the ruled lines.

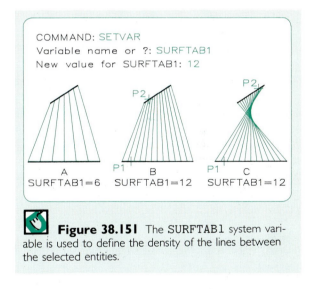

Figure 38.151 The SURFTAB1 system variable is used to define the density of the lines between the selected entities.

tem variable is used to control the density of the tabulated surface.

38.65   The REVSURF Command

A surface can be revolved about an axis with the REVSURF command as shown in Fig. 38.153. A line, polyline, arc, or circle can be revolved about

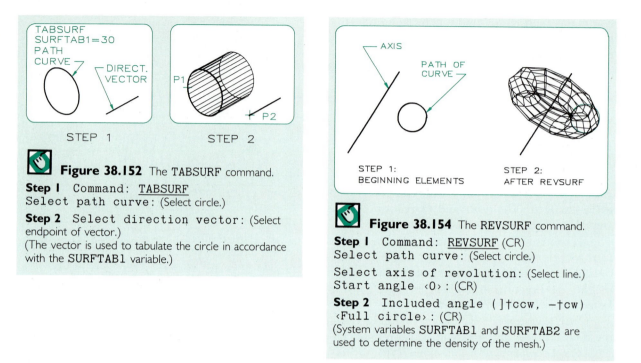

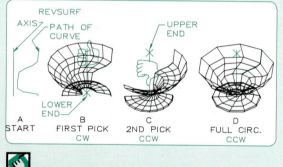

Figure 38.152 The TABSURF command.

Step 1 Command: <u>TABSURF</u>
Select path curve: (Select circle.)

Step 2 Select direction vector: (Select endpoint of vector.)
(The vector is used to tabulate the circle in accordance with the **SURFTAB1** variable.)

Figure 38.154 The REVSURF command.

Step 1 Command: <u>REVSURF</u> (CR)
Select path curve: (Select circle.)

Select axis of revolution: (Select line.)
Start angle ‹0›: (CR)

Step 2 Included angle (]↑ccw, −↑cw)
‹Full circle›: (CR)
(System variables **SURFTAB1** and **SURFTAB2** are used to determine the density of the mesh.)

Figure 38.153 The REVSURF command.

Step 1 An axis and a polyline defining the path curve are given.

Step 2 Command: <u>REVSURF</u>
Select path of curve: (Select with pointer.)
Select axis of revolution: (Select lower end of axis.)
Start angle ‹0›: (CR)
Included angle (]↑ccw,−↑cw) ‹Full circle›: <u>180</u> (Rotated counterclockwise.) (CR)

Step 3 By selecting a point on the upper end of the axis, the path curve is rotated clockwise.

Step 4 A full-circle revolution.

an axis to form a surface of revolution. System variable **SURFTAB1** controls the density of the revolved surface.

If a circle or a closed polyline is to be revolved about an axis, as shown in Fig. 38.154, the system variable **SURFTAB2** is used to control the density of the path curve. System variable **SURFTAB1** controls the density in the direction of the revolution.

38.66 The EDGESURF Command

Four edges that intersect at their corners can be connected with a mesh by using the EDGESURF command as shown in Fig. 38.155. System variables **SURFTAB1** and **SURFTAB2** are used to control density in the *M*- and *N*-directions, respectively. The first line selected determines the *M*-direction.

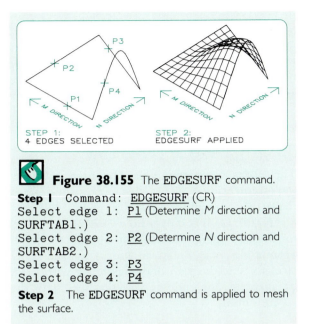

Figure 38.155 The EDGESURF command.

Step 1 Command: <u>EDGESURF</u> (CR)
Select edge 1: <u>P1</u> (Determine *M* direction and SURFTAB1.)
Select edge 2: <u>P2</u> (Determine *N* direction and SURFTAB2.)
Select edge 3: <u>P3</u>
Select edge 4: <u>P4</u>

Step 2 The EDGESURF command is applied to mesh the surface.

38.67 3DPOLY and 3DLINE Commands

The 3DPOLY and 3DLINE commands are used in the same manner as the PLINE and LINE commands are used in 2D drawing, but the *Z*-coordinates must be given in addition to *X*- and *Y*-coordinates. 3DPOLY lines can be exploded to convert them to 3DLINES. Fig. 38.156 shows 3DPOLY lines drawn with absolute coordinates from the origin of 0,0,0 to 3,0,0 to 3,2,1.5 to 1,2,1.5 and back to the origin. POINTS in 3D space can be located in the same manner.

Figure 38.157 shows how the ends of three-dimensional lines are located by typing relative coordinates from point to point. Each relative coordinate is written in the form, @X,Y,Z or @ 3,0,0, which locates an endpoint with respect to the last point.

Three-dimensional shapes drawn with the ELEV command and THICKNESS option, as discussed in Sections 38.58, can be rotated with the VPOINT command and visibility determined with the HIDE command. The SOLID command can be

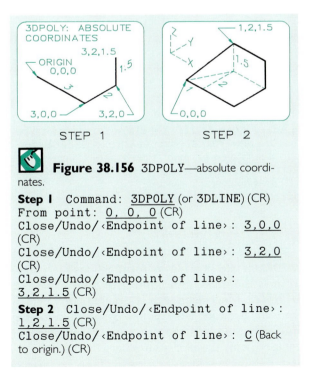

STEP 1   STEP 2

Figure 38.156 3DPOLY—absolute coordinates.

Step 1 Command: <u>3DPOLY</u> (or 3DLINE) (CR)
From point: <u>0, 0, 0</u> (CR)
Close/Undo/‹Endpoint of line›: <u>3,0,0</u> (CR)
Close/Undo/‹Endpoint of line›: <u>3,2,0</u> (CR)
Close/Undo/‹Endpoint of line›: <u>3,2,1.5</u> (CR)

Step 2 Close/Undo/‹Endpoint of line›: <u>1,2,1.5</u> (CR)
Close/Undo/‹Endpoint of line›: <u>C</u> (Back to origin.) (CR)

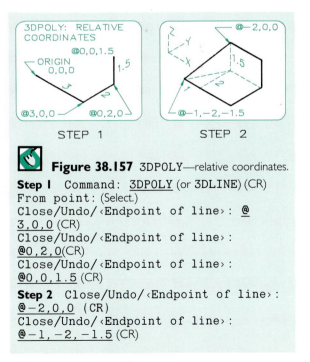

STEP 1   STEP 2

Figure 38.157 3DPOLY—relative coordinates.

Step 1 Command: <u>3DPOLY</u> (or 3DLINE) (CR)
From point: (Select.)
Close/Undo/‹Endpoint of line›: <u>@ 3,0,0</u> (CR)
Close/Undo/‹Endpoint of line›: <u>@0,2,0</u> (CR)
Close/Undo/‹Endpoint of line›: <u>@0,0,1.5</u> (CR)

Step 2 Close/Undo/‹Endpoint of line›: <u>@−2,0,0</u> (CR)
Close/Undo/‹Endpoint of line›: <u>@−1,−2,−1.5</u> (CR)

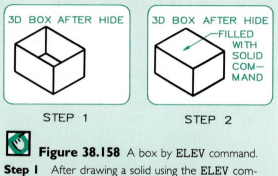

STEP 1   STEP 2

Figure 38.158 A box by **ELEV** command.

Step I After drawing a solid using the **ELEV** command and the **THICKNESS** option, the solid's visibility can be shown by using the **HIDE** command.

Step 2 By using the **SOLID** or **3DFACE** command, the top can be filled so it will appear opaque when the **HIDE** command is used.

used to fill the top surface to make it appear opaque rather than as an open box (Fig. 38.158).

The **ENDPOINT** and **MIDPOINT** options of the **OSNAP** command can be used to locate the endpoints of lines drawn when using the **LINE** command. Three-dimensional forms can also be stretched in the *X* and *Y* directions but not in the *Z* direction. However, the heights in the *Z* direction can be changed by using the **ELEV** command.

3DPOLY lines drawn as three-dimensional lines can be edited with the **PEDIT** command to form a spline curve (a helix in this case) as shown

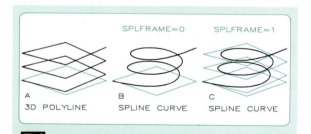

A   B   C
3D POLYLINE   SPLINE CURVE   SPLINE CURVE

Figure 38.159 The **PEDIT** command and the **SPLFRAME** system variable set to zero can be used to spline a three-dimensional curve. By setting the `SPLFRAME=1`, both the spline curve and the defining polyline are displayed.

in Fig. 38.159. System variable **SPLINEGS** controls the accuracy of the spline curve and the system variable **SPLFRAME**, which controls the display of the control points. If set to 0 (zero) only the spline curve is shown; if set to a 1, the curve and its defining polyline are shown.

38.68 3DFACE command

The **3DFACE** command can be used to draw three-dimensional planes that are opaque when the **HIDE** command is applied to them.

Figure 38.160 shows how **3DFACE** was used to draw sloping planes that appear as solid, opaque planes. **3DFACE** can also be used to apply faces to wire-diagram drawings made with **LINE** by snapping the endpoints of the lines with the **OSNAP** option.

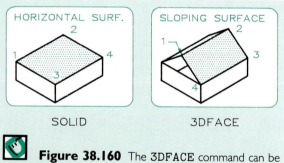

SOLID   3DFACE

Figure 38.160 The **3DFACE** command can be used to draw objects with opaque planes. **3DFACE** can be used to draw planes, or it can be used to apply faces to previously drawn lines by snapping to the endpoints of the lines outlining the planes.

Figure 38.161 illustrates how four corners of a **3DFACE** are selected to form an opaque plane. Four is the maximum number of points that can be selected before an area is defined, after which you are prompted for points 3 and 4, with the understanding that the previous two points will be used with the next two points that will be selected. The successive areas will be connected

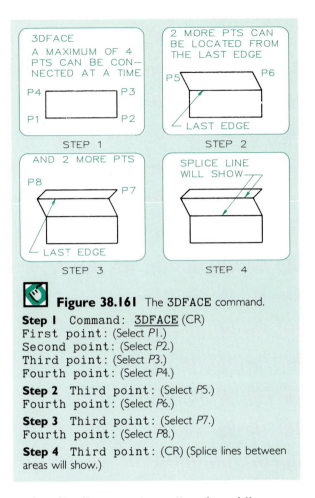

Figure 38.161 The 3DFACE command.

Step I Command: <u>3DFACE</u> (CR)
First point: (Select *P1*.)
Second point: (Select *P2*.)
Third point: (Select *P3*.)
Fourth point: (Select *P4*.)

Step 2 Third point: (Select *P5*.)
Fourth point: (Select *P6*.)

Step 3 Third point: (Select *P7*.)
Fourth point: (Select *P8*.)

Step 4 Third point: (CR) (Splice lines between areas will show.)

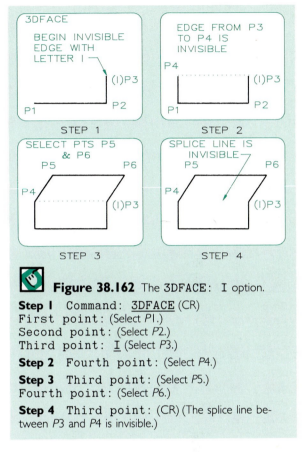

Figure 38.162 The 3DFACE: I option.

Step I Command: <u>3DFACE</u> (CR)
First point: (Select *P1*.)
Second point: (Select *P2*.)
Third point: <u>I</u> (Select *P3*.)

Step 2 Fourth point: (Select *P4*.)

Step 3 Third point: (Select *P5*.)
Fourth point: (Select *P6*.)

Step 4 Third point: (CR) (The splice line between *P3* and *P4* is invisible.)

with splice lines creating a "patchwork" appearance that may detract from the surface.

By inserting the letter I prior to selecting a perimeter line of a 3DFACE, the splice line will be omitted, thereby eliminating the "patchwork" appearance as shown in Fig. 38.162.

When 3DFACE is used to make surfaces opaque, holes are filled in so it is impossible to see through them when the HIDE command is used. The SLOT command, written in LISP, can be used with 3DFACE to opaque the surface up to the edge of the hole as shown in Fig. 38.163.

The SLOT command is loaded by typing (LOAD 'SLOT') and SLOT is typed again to use it. A slot or a hole can be drawn by responding to the prompts. The depth of the hole, or slot, is extruded upward from the point selected. By set-

ting the system variable, SPLFRAME, to 1, it is turned on and the corner lines surrounding the hole or slot are visible.

By using the 3DFACE, with the "I" (invisible) option, the trapezoidal shapes can be made opaque. When SPLFRAME is set to 0 (off) and HIDE is applied, the surface around the hole becomes opaque, and the hole can be seen through.

38.69 XYZ Filters

Filters are options under 3DLINE and 3DFACE commands that adopt the coordinates of three-dimensional points of a drawing. Figure 38.164 illustrates how filters are applied to a two-dimen-

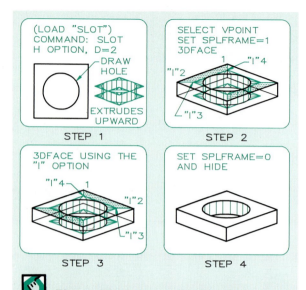

Figure 38.163 The SLOT command.

Step 1 Command: (LOAD 'SLOT')
Command: SLOT
Hole or Slot? H/S ‹S›: H
First center point of slot: (Select.)
Slot radius: 4 Depth: 2 (Depth is extruded upward.)

Step 2 Command: VPOINT
Rotate/‹View point› ‹0,00,0,00,1.00›:
1,−1,.5
(obtain 3D view of box)
Set Var: SPLFRAME (Set to I and REGEN.)
(Use 3DFACE with "I" option to connect corner points.)

Step 3 Continue using 3DFACE and the "I" option to face the other two sides.

Step 4 Set SPLFRAME=0 and apply the HIDE command. The surface around the holes becomes opaque.

sional drawing. To locate the front view of a point projected from the left-side view and top view, select the .Y of the left point and the .X of the top point. The position of the front view is automatically snapped to a point located by these two coordinates.

A three-dimensional drawing can be made using filters as shown in Fig. 38.165, where two

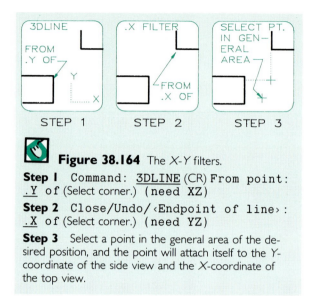

Figure 38.164 The X-Y filters.

Step 1 Command: 3DLINE (CR) From point: .Y of (Select corner.) (need XZ)

Step 2 Close/Undo/‹Endpoint of line›: .X of (Select corner.) (need YZ)

Step 3 Select a point in the general area of the desired position, and the point will attach itself to the Y-coordinate of the side view and the X-coordinate of the top view.

coordinates for a given point are selected using the filter option, and the third one is specified when prompted. For example, you can select the coordinates of a point in the XZ plane, and spec-

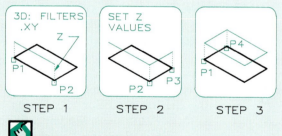

Figure 38.165 3D filters.

Step 1 Command: 3DPOLY (or 3DLINE) (CR)
From point: .XY of (Select P1.)
of (need Z) 2 (CR)
Close/Undo/‹Endpoint of line›: .XY of (Select P2.) (need Z) 2 (CR)

Step 2 Close/Undo/‹Endpoint of line›: .XY of (Select P3.) (Need Z) 2 (CR)

Step 3 Close/Undo/‹Endpoint of line›: .XY of (Select P4.)
(Need Z) 2 (CR)
Close/Undo/‹Endpoint of line›: .XY of(Select I.) (need Z) 2 (CR)

ify the distance of the point from it in the *Y* direction.

38.70 New Drawing in 3D

The steps of constructing a three-dimensional object are shown in Fig. 38.166. The plan view of the object is drawn in Step 1 and moved to origin, 0,0,0. In Step 2, an isometric viewpoint is selected for showing a three-dimensional view of the object's surface.

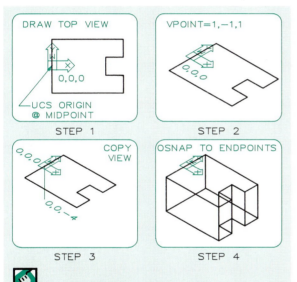

DRAW TOP VIEW   STEP 1
UCS ORIGIN @ MIDPOINT
0,0,0

VPOINT=1,−1,1   STEP 2
0,0,0

COPY VIEW   STEP 3
0,0,0
0,0,−4

OSNAP TO ENDPOINTS   STEP 4

Figure 38.166 A drawing in 3D.

Step 1 Draw the top view of the part. Use the UCS command to set the origin at the midpoint of the line on the view. Use the UCSICON command to locate the icon at the UCS origin.

Step 2 Command: <u>VPOINT</u> (CR)
Rotate/‹View point› ‹0,0,0›: <u>1, −1, 1</u>
(A 3D view is displayed.)

Step 3 Command: <u>COPY</u> (CR)
Select objects: <u>W</u> (Window 3D view.)
‹Base point or displacement›
/Multiple: <u>0,0,−4</u> (CR)
Second point of displacement: (CR)

Step 4 Set OSNAP to END and connect the corner points with the LINE command.

In Step 3, the first view is copied and located at 0,0,−4 to establish the lower plane of the object and a height of 4 units. By using the END option of OSNAP, the corners of each plane are connected to form a wire diagram of the object.

In Fig. 38.167 (Step 1), a four-window VPORTS is selected for drawing the wire diagram in each port. In Step 2, the PLAN option of the UCS command is used to show the top view of the object. This UCS is SAVEd with that option of the UCS command.

In Step 3, the UCS is rotated 90 degrees about the X-axis, and the front view is found by using the PLAN command. This UCS is SAVEd, as FRONT. In Step 4, the *X*-axis is rotated 90° about the *Y*-axis, and the PLAN command is used to find the side view. Its UCS is SAVEd as SIDE.

This process has resulted in top, front, right-side views, and a pictorial view of the wire-frame drawing. Additional lines can be added in any of the four views, and the other views will be updated simultaneously.

The system variable, UCSFOLLOW, can be set to ON (1) by using the command, SETVAR. When turned on, the UCSFOLLOW variable will automatically augment the PLAN option when a UCS is RE-STORED. Therefore the *X*- and *Y*-axes will automatically appear true size in the restored view. If UCSFOLLOW is set prior to selecting VPORTS, the setting will apply to all VIEWPORTS. Once VIEW-PORTS have been set, each viewport can have a UCSFOLLOW variable assigned. For example, UCSFOLLOW can be active in two viewports and inactive in two others, and so forth.

Note: While working in VPORTS, the CHANGE command may not always work properly. Therefore, use the CHPROP command as an alternative method of changing layers and so forth.

38.71 Object With an Inclined Surface

The part shown in Fig. 38.168 has an inclined plane that requires an additional rotation for representation in 3D.

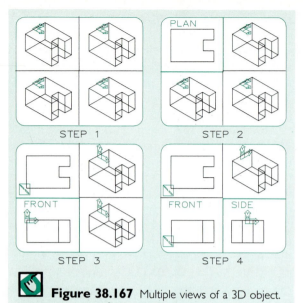

STEP 1 STEP 2

STEP 3 STEP 4

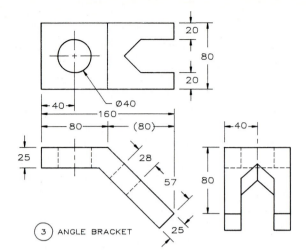

Figure 38.168 The angle bracket below with an inclined surface is drawn as a 3D object in the following figures.

③ ANGLE BRACKET

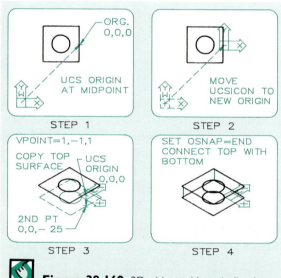

Figure 38.167 Multiple views of a 3D object.

Step 1 Command: <u>VPORTS</u> (CR)
Save/Restore/Delete/Join/SIngle/?/2/
‹3›/4: <u>4</u> (CR)
(The 3D drawing appears in four ports.)

Step 2 (Select the upper left port.)
Command: <u>PLAN</u> (CR)
‹Current UCS/UCS/World: (CR)
Command: <u>ZOOM</u>
All/Center/.... ‹Scale(X)›: <u>.6X</u> (CR)
Command: <u>UCS</u> (CR)
Origin/Xaxis/...‹World›: <u>S</u> (CR)
?/Name of UCS: <u>TOP</u> (CR)

Step 3 (Select lower left port.)
Command: <u>UCS</u> (CR)
Origin/Xaxis/...‹World›: <u>X</u> (CR)
Rotation angle about X axis: <u>90</u> (CR)
Command: <u>PLAN</u> (CR)
‹Current UCS›/UCS/World: (CR)
(Get front orthographic view.)
UCS (Command repeated.)
Origin/Xaxis...‹World›: <u>S</u> (CR)
?/Name of UCS: <u>FRONT</u> (CR)

Step 4 (Select lower right port.)
Use the UCS command to rotate the UCS 90° about the Y-axis to obtain a right-side view. Obtain its orthographic view with the **PLAN** command and save the UCS as **SIDE**, as in Step 3.

Figure 38.169 3D object with an inclined surface.

Step 1 The top view of the object is drawn and the UCS origin is set at the midpoint of the line on the object.

Step 2 The UCSICON command is used to place the icon at the new origin.

Step 3 The VPOINT command is used to select a view point of 1,−1,1 of the surface. The surface is copied to a position of 0,0,−25 from the UCS origin.

Step 4 Set the OSNAP to the END option and connect the two planes.

In Step 1 of Fig. 38.169, the top view of the object is drawn and the UCS origin is moved to the midpoint of one of its lines. In Step 2, the UCSICON is moved to the UCS origin. An isometric viewpoint is selected in Step 3, and the surface is copied at a location 0.25 inches below the upper surface. The corner points are connected in Step 4 with the END option of OSNAP.

In Fig. 38.170 (Step 1), a four-panel VPORTS is selected for displaying the views of the object. In Step 2, the UCS of the upper right port is rotated 90° about the X-axis for obtaining a front view. The lower left port is activated, and the UCS is SAVEd in Step 3. The PLAN command is used to obtain and orthographic front view in Step 4.

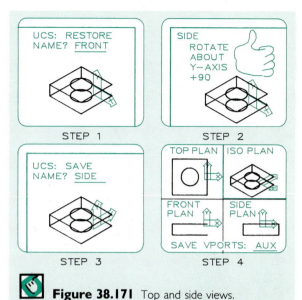

Figure 38.171 Top and side views.

Step 1   When desired, you may **RESTORE** the **FRONT** view by the UCS command, but it is unnecessary here.

Step 2   Select the lower right port. Command: UCS (CR) (Select Y and rotate the UCS 90° about the Y-axis.)

Step 3   Command: UCS (CR) (SAVE) ?/Name of UCS: SIDE (CR)

Step 4   Command: PLAN (CR) (Obtain an orthographic side view.) Use the same steps to obtain and save a top view in the upper left corner. Leave the 3D view in the upper right corner.

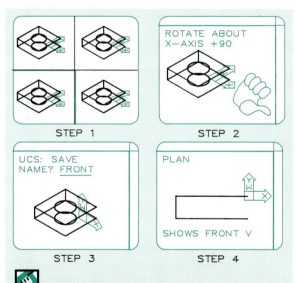

Figure 38.170 The 4-port construction.

Step 1   Command: VPORTS (Select 4 ports.) (The 3D drawing is shown in all 4 ports.)

Step 2   (Select lower left port.) Command: UCS (CR) (Select X option.) Rotate UCS 90° about the X-axis for a front-view UCS.

Step 3   Command: UCS (Enter SAVE.) ?/Name of UCS: FRONT (CR)

Step 4   Command: PLAN (CR) (Display an orthographic front view. Use ZOOM to obtain desired size.)

In Step 1 of Fig. 38.171, the pictorial view is RESTOREd, and the UCS is rotated 90° about the Y-axis to place the X- and Y-axes in the profile plane of the object. The lower-right panel is activated, the UCS is SAVEd, and the PLAN command is used to give an orthographic right-side view.

The top view, upper right panel is found in the manner as the previous two orthographic were found.

In Step 1 of Fig. 38.172, the lower right port is activated and the UCS is RESTOREd. Draw a line from the origin to 0, −80, 80 to establish the sloping surface (Step 2). The UCS is rotated −45° about the X-axis to make it line in the plane of

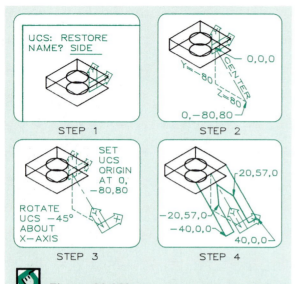

Figure 38.172 Drawing the inclined surface.

Step 1 Select the side-view port. `Command: UCS` (CR) (Select `RESTORE`.) `?/Name of UCS: SIDE` (CR) `Command: VIEWPORTS` (CR) (`SAVE` this view port configuration as `AUX1`.) `Command: VPORTS` (CR)(Select `SIngle` to obtain a full-screen image.)

Step 2 `Command: UCS Origin/Zaxis/ ...` `<World>: OR` (Select origin.) `Origin point:` `0, -80, 80`

Step 3 `Command: UCSICON` (CR) (Select `ORigin` option to place icon at a new origin.) `Command:` `UCS` (CR) (Select X and rotate UCS − 45° about the X-axis.) Use `UCS` command to `SAVE` as `INCLINED`.

Step 4 Use the `LINE` command and dimensions from the given view of the object to specify the corners of the V-notch.

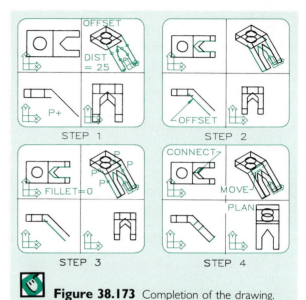

Figure 38.173 Completion of the drawing.

Step 1 `Command: VPORTS` (`RESTORE` the 4-port configuration, `AUX1`, to restore the four views.) Use the `OFFSET` command and select point, on the V-notch in 3D view with an offset distance of 25, and locate the side of the offset in the fronts view port.

Step 2 The lower surface of the inclined plane is displayed.

Step 3 The `FILLET` command is used to trim the ends of the intersecting lines. The lines are picked in the 3D view.

Step 4 With `OSNAP` set to `END`, end points of lines are connected. The horizontal intersection line is `MOVE`d to its new position. Notice that you can use any of the views for changing or developing the drawing.

the inclined surface (Step 3). In Step 4, the coordinates of the V-notch (taken from the given view of the object) are used to construct the upper inclined surface of the part.

In Step 1 of Fig. 38.173, the `OFFSET` command is used to select the points of the 3D pictorial of the inclined surface, the lower left port is activated, a distance of 0.25 inches assigned, and *P* locates the side of the offset. In Step 2, the offset surface is drawn in all views.

In Step 3, the `FILLET` command is applied to the line in upper right port to make them join precisely. In Step 4, the `END` option of the `OSNAP`

command is used to connect the corner points to complete the drawing.

38.72 Drawing With Meshes

An object can be represented by applying the previously covered commands `RULESURF`, `TABSURF`, `REVSURF`, and `EDGESURF`.

By referring to Part A of Fig. 38.174, you can see where circles, arcs, and lines have been used to form a wire diagram of a part and the surface has been tabulated between the circular

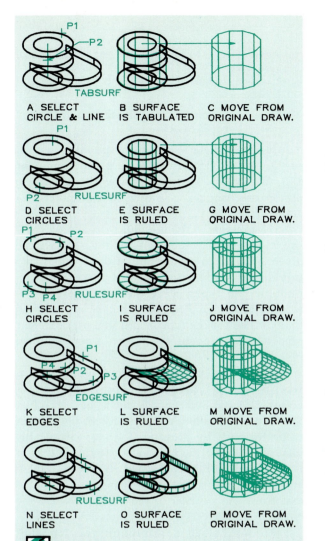

A SELECT CIRCLE & LINE

B SURFACE IS TABULATED

C MOVE FROM ORIGINAL DRAW.

D SELECT CIRCLES

E SURFACE IS RULED

G MOVE FROM ORIGINAL DRAW.

H SELECT CIRCLES

I SURFACE IS RULED

J MOVE FROM ORIGINAL DRAW.

K SELECT EDGES

L SURFACE IS RULED

M MOVE FROM ORIGINAL DRAW.

N SELECT LINES

O SURFACE IS RULED

P MOVE FROM ORIGINAL DRAW.

Figure 38.174 3D object with meshes applied.

A, B, C A wire diagram of a 3D part is given. TAB–SURF is used to tabulate the upper circles. The resulting lines are moved from the original drawing.

D, E, F RULESURF is used to connect the small circles. The ruled lines are moved to the same SNAP point as the tabulated lines.

H, I, J The upper and lower donut areas are connected with RULESURF. The ruled surfaces are moved to the location of the other meshes.

K, L, M The EDGESURF command is used to apply a mesh to the upper and lower surfaces of the lug, and to move the mesh to the other meshes.

N, O, P Use RULESURF to apply a mesh to the vertical edges of the lug. Transfer the mesh to the other meshes with the MOVE command.

ends with the TABSURF command at Part B, and the ruled lines are moved to the side of the original drawing with the MOVE command and LAST option.

The RULESURF command is used at Parts D and E to rule the hole. The ruled lines are MOVEd to the same location as in the previous step. RU–LESURF is used to rule the ends of the cylinders, which are also moved (Parts H, I, and J).

In Parts K, L, and M, the upper surface of the lug is meshed using the EDGESURF command. This mesh is also moved to the right of the original drawing.

In Parts N, O, and P, the RULESURF command is used to apply mesh to the thin edges of the lug, which are moved to the right to add to the other meshed surfaces. This method can be continued to create meshes and transfer them to a removed position.

Once completed, the visibility of the object can be shown by using the HIDE command (Fig. 38.175).

38.73 Plotting a Drawing

A hard copy of a drawing can be plotted with either a pen plotter or a printer plotter (a typewriter printer with graphics capability). You may

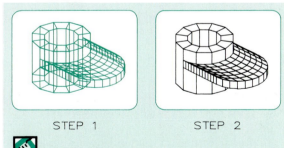

STEP 1

STEP 2

Figure 38.175 Visibility of meshed surfaces.

Step 1 The 3D drawing developed in the last example is a hollow shell enclosed in meshes with no wire-frame outlines.

Step 2 The HIDE command is applied to remove the hidden lines to form a more realistic three-dimensional view of the part.

initiate the commands for plotting from the **Main Menu** or the **Root Menu** when the drawing to be plotted appears on the display. The command for activating the pen plotter is PLOT, and PRPLOT is used for the printer plotter.

You must tell AutoCAD which part of the drawing to plot by responding to the following prompt:

```
What to plot Display, Extents,
Limits, View, or Window <D>: L (CR)
```

The meanings of these options are

DISPLAY (D) plots the view currently visible on the display screen, or the last view that was displayed before SAVE, or END.

EXTENTS (E) plots all entities even though they may exceed the limits.

LIMITS (L) plots the drawing within the area defined by the limits.

VIEW (V) plots a named view that was saved by the VIEW command. You will be prompted for the VIEW name.

WINDOW (W) plots a selected part of the drawing that is enclosed in a window specified by two diagonal points. When PLOT is begun from the Drawing Editor, you can point to the corners of the window, but you must use coordinates of the window entered at the keyboard when beginning at the Main Menu.

The plotter origin for a Hewlett-Packard 7475 is located at 0.42 in., 0.35 in. from the lower left corner of an 11- × -8½-in. sheet, which gives an effective plotting area of 10.15 × 7.8 in. when you instruct the plotter to place the origin at 0,0.

After selecting the part of the drawing to be plotted, you will be given the following specifications on the display screen:

```
Sizes are in Inches
Plot origin is at (0.00, 0.00)
Plotting area is 10.50 wide by 8.00
high (A size)
Plot is NOT rotated 90 degrees
Pen width is 0.010
```

```
Area fill will NOT be adjusted for
pen width
Hidden lines will NOT be removed
Scale is 1 = 1

Do you want to change anything?
<N>: N or Y (CR)
```

You can use these values by responding with N or NO, or you can change them by entering Y or YES. When YES is entered you will be given the option of changing as shown in Computer Example 38.1.

By pressing (CR) for NO, you accept these specifications. If you enter Y for YES, you will be given a chance to change any of these values. Do **not** change the line types if your plotter supports multiple pen types; leave the line type set at 0 (continuous line). Pen speed can be set to yield the best line for the type of pen you use. PEN NO. gives the location of the pen in a multiple-pen plotter.

After Y has been entered to indicate that you wish to change parameters, AutoCAD will list Layer Color, Pen No., Line type, and Pen Speed one at a time. You may accept the current value for each by pressing (CR), or change the values by typing the new value after each default is displayed. When completed, type S to obtain an updated display of your changes. Type X to exit from this portion of the program when the changes are correct.

AutoCAD will now prompt you for additional plot specifications.

```
Write the plot to a file? <N>: N or
(CR)
Size units (Inches or Millimeters)
<I>: I (I for inches and M for millimeters.)
(CR)
Plot origin in units <0.00, 0.00>:
2,1 (CR) (Enter coordinates of the orgin from the
"home" position of your plotter, using millime-
ters or inches depending on the units you have
specified.)
```

AutoCAD will list the standard plotting sizes:

Standard values for plotting size

Size	Width	Height
A	10.50	8.00
MAX	16.00	10.00

Layer Color	Pen No.	Line Type	Pen Speed	Layer Color	Pen No.	Line Type	Pen Speed
1 (Red)	1	0	9	9	3	0	2
2 (Yellow)	2	0	9	10	4	0	36
3 (Green)	2	0	9	11	5	0	36
4 (Cyan)	2	0	9	12	6	0	36
5 (Blue)	2	0	9	13	1	0	36
6 (Magenta)	1	0	9	14	2	0	36
7 (White)	2	0	9	15	3	0	36
8	2	0	36				

Computer Example 38.1

Line types 0 = continuous line
1 = ...
2 = ---- ---- ---- ----
3 = ----- ----- ----- -----
4 = ------. ------. ------. ------.
5 = ---- - ---- - ---- - ---- -
6 = --- - - --- - - --- - - --- - -

Do you want to change any of these parameters? <N>

Enter the Size or Width, Height (in units) ‹B›: A (CR)
(Size A sheet 11 × 8.5 inches or MAX for the largest size the plotter will accept, perhaps a Size C sheet.)

Instead of responding with A or MAX for sheet size, you may type in the desired limits of the sheet, for example, 10.15,7.8. The values will be listed as a third option, U (for User), that can be recalled during future plots.

Rotate 2D plots 90 degrees clockwise? ‹N›: (CR)
Pen width ‹0.10›: (CR) (to accept)
Adjust area fill boundaries for pen width?‹N›: (CR)
(By responding Y the boundaries of filled areas will be moved inward one-half pen width for a higher degree of accuracy.)
Remove hidden lines?‹N› (CR)
Specify scale by entering:
Plotted units=Drawing units or Fit or? ‹Fit›:

If your drawing was to be plotted in inches, enter 1 = 1 for a full-size drawing where a plotted inch was equal to 1 in. on the screen. For an architectural
scale such as $\frac{1}{2}'' = 1'-0''$, the unit at the left of the equal sign should be converted to 1. This conver-

sion results in an equality of $\frac{1}{2}$ in. = 12 in. or 1 in. = 24 in. If your drawing was drawn in millimeters, when prompted, Size units (Inches or Millimeters) ‹I›:, enter M, and the drawing will be plotted in metric units.

Specify scale by entering:
Plotted Millimeters=Drawing units
or Fit or ? ‹F›: 1=1 (CR)

By responding with F, the drawing will be scaled to fill the available space, but the scale will be a nonstandard scale. Responding with ? will give a list of the various scaling options and their descriptions.

Plot specifications are saved by AutoCAD so that you will not have to change plot parameters until you wish. AutoCAD will display the following message:

Effective plotting area: 10.50 wide by 8.00 high
Position the paper in plotter (Pause for this step.)
Press RETURN to continue or S to Stop for hardware setup
Press RETURN to begin plotting

Some plotters have other features that can be adjusted according to manufacturer's specifications. However, the preceding steps are typical of the ones used by plotters and printers.

38.74 Attributes

Attributes are combinations of drawings saved as BLOCKS and text. A drawing can be made with attributes defined using the ATTDEF command, saved as a BLOCK, and then inserted repetitively. With each insertion, you will be prompted for attribute values that will be written on the drawing or left on file as an invisible value for listing in an attribute report.

An attribute drawing is shown in Fig. 38.176, where a title block is drawn using the DRAW command. To define the attributes, use the ATTDEF command.

```
Command: ATTDEF (CR)
Attribute modes—Invisible: N Con-
stant:N Verify:N Preset:N
Enter (ICV) to change. RETURN when
done: V (CR)
```

By entering I, C, or V, one at a time, you can change the settings of INVISIBLE, CONSTANT, and VERIFY modes of the attributes. If the IN-VISIBLE mode is on, the attribute values will not be shown on the drawing to prevent clutter. If the CONSTANT mode is on, a fixed attribute value will be assigned to each BLOCK insertion. If the VERIFY mode is on, you will be able to verify the attribute values at insertion time.

The next prompts are

```
Attribute tag: DATE (CR) (No blanks al-
lowed.)
Attribute prompt: DATE? (CR)
Default Attribute value: (Blank if none.)
(CR or enter value.)
Attribute value: (CR)
Start point or Align/Center/Fit/
Middle/Right/Style: C (CR)
Height <0.100>: .125
Rotation angle <0>: (CR)
(The attribute tag appears on the drawing.)
```

Multiple attributes can be added to each drawing.

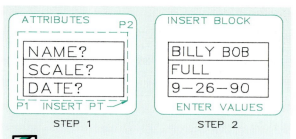

Figure 38.176 Attribute definitions.

Step 1 Command: ATTDEF (CR)
Attribute modes––Invisible:N con-stant:N Verify:N Preset:N
Enter (ICVP) to change, RETURN when done: V (CR)
Attribute modes––Invisible:N Con-stant:N Verify:Y Preset:N
Enter (ICVP) to change, RETURN when done: (CR)
Attribute tag: NAME (CR)
Attribute prompt: NAME? (CR)
Default attribute value: (Blank)
Start point or
Align/Center/Fit/Middle/Right/Style:
(Locate tag on title block.)
Height <0.2000>: .125 (CR)
Rotation angle <0>: (CR)

Step 2 BLOCK the title block with a *P1* and *P2* window.

Command: INSERT
Block name <or?>: TITLE
Insertion point: (Select point.)
X scale factor <1>/Corner/XYZ: (CR)
Y scale factor <default=X>: (CR)
Rotation angle <0>: (CR)
(Respond to the attribute prompts to complete the title block.)

Once all attributes have been added, the drawing is windowed as a BLOCK in the usual manner.

When the BLOCK is inserted, you are prompted for attribute values, and default values will be shown on the screen.

```
Enter attribute values
Name?: <NONE>: BILLY BOB (CR)
Scale? <full size>: (CR) (Accepts full size
default.)
Date: <NONE>: 9-26-89 (CR)
```

A listing of these attribute values will be given as

you (CR) through the list, enabling you to verify the correctness of your entries. When the end of the list is reached, the values will be plotted on the drawing in the places previously assigned (Fig. 38.177).

The visibility of the attributes can be changed by the ATTDISP command, which may be different from the ATTDEF command specifications.

```
Command: ATTDISP (CR)
Normal/On/Off <current value>:
N (CR)
```

By entering N, the attributes specified either hidden or visible will appear that way on the drawing. The ON option will display all values on the drawing, whether hidden or visible. The OFF option does not show any of the values on the drawing. For the values to be redisplayed, the drawing must be REGENerated.

The ATTEDIT command is used to edit attributes one at a time or globally, where all are changed at one time. The command is used as follows:

```
Command: ATTEDIT (CR)
Edit attributes one by one? <Y>  Y
(CR)
```

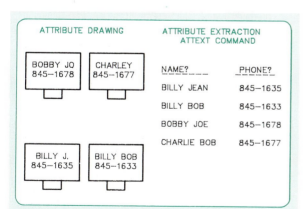

Figure 38.177 A listing of the attributes on a drawing can be made using the **ATTEXT** command in conjunction with BASIC. More elaborate listings can be obtained when attributes are used with a database program.

By responding yes (Y), you may edit any of the attributes currently visible on the screen. By responding no (N), you may edit all attributes with global editing. You can select the attributes to edit by responding to the following prompts:

```
Block name specification <*>:
Attribute tag specification <*>:
Attribute value specification <*>:
```

If you press (CR) after the BLOCK name specification, attributes of BLOCKS with all names will be selected for editing. If you specified the BLOCK name but entered (CR) after the tag specification, attributes for all tags will be selected. However, if you responded with NAME for the tag specification, only the attributes of the tag NAME will be selected. By entering a \, you will be able to edit null-value attributes.

You will now be prompted

```
Select ATTRIBUTES:
```

Select the attributes by pointing to them on the drawing or by windowing the group to be edited. Each attribute will be marked with an X on the drawing to indicate those eligible for editing. The next prompt is

```
Value/Position/Height/Angle/
Style/Layer/Color/Next <N>:
```

By selecting the first letter of these commands, you may change the attributes in many ways by responding to AutoCAD's prompts.

Global editing may be done by using the following sequence of commands:

```
Command: ATTEDIT
Edit attributes one by one?
<Y>  N (Global option.)
Global edit of Attribute values.
Edit only Attributes visible  on
screen? <Y>  N
(Screen goes to Text mode if N.)
(Drawing must be regenerated afterward.)
```

The next steps involve selecting the attributes in the same manner as when they were done one at a time. In this case, changes are applied uniformly to all selected blocks and attributes.

38.75 Attribute Listing (ATTEXT)

The ATTEXT command of Release 9 (not Release 10) can be used for producing tabular listings—such as parts lists, inventories, and bills of materials—using the method below. When using Release 10, a template file must be created instead of using the method below.

A BLOCK with attributes is INSERTed four times, as shown in Fig. 38.177, and the drawing is saved under the name A:OFFICE. Begin by loading AutoCAD, and the A:OFFICE file, and then enter the ATTEXT command.

```
Command: ATTEXT (CR)
CDF, SDF, or DXF Attribute extract
(or Entities)?<C> D (CR)
(This file can be read by other programs.)
Extract file name <OFFICE>: (CR)
??? entities in extract file
```

Use the SHELL command to go to BASIC.

```
Command: SHELL
DOS command: BASICA
??? bytes free
OK
Load "ATTEXT"
OK
RUN
Extract file name: A:OFFICE
```

Your BASIC program will print a bill of materials, as shown in Fig. 38.177. Return to DOS by typing SYSTEM and you will get AutoCAD's prompt, Command:.

38.76 Grid Rotation

To draw lines parallel or perpendicular to a given line, the grid on the screen can be rotated to align with existing lines. The ROTATE command is an option under the SNAP command. The prompts are as follows:

```
Command: SNAP (CR)
ON/OFF/Value/Aspect/Rotate/
Style:R (Rotate) (CR)
Base point <0, 0>: (Select a point.)
Rotation angle <0>: (Type the angle or
specify the angle on the screen.)
```

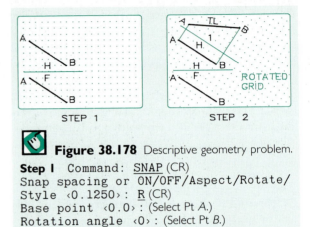

Figure 38.178 Descriptive geometry problem.

Step 1 Command: SNAP (CR)
Snap spacing or ON/OFF/Aspect/Rotate/
Style <0.1250>: R (CR)
Base point <0,0>: (Select Pt A.)
Rotation angle <0>: (Select Pt B.)

Step 2 Draw H-1 reference line. OSNAP from points A and B perpendicular to H-1. Extend the projectors to locate the true-length view.

The grid will be plotted on the screen in alignment with the selected points. Figure 38.178 shows an example of a descriptive geometry problem with a true-length auxiliary view of line AB.

The grid is returned to its original position by selecting the SNAP command and the ROTATE option prompts.

```
Base point <2, 3>: (CR)
Rotation angle <37>: 0 (CR) (Realigns
grid to its original position.)
```

38.77 Digitizing With the Tablet

Drawings on paper can be digitized into the computer point by point when a tablet is available. Tablets vary in size from 11″ × 8.5″ to several square feet.

To calibrate a drawing and tablet for digitizing, tape the drawing to the tablet and use the following steps:

```
Command: TABLET
Option (ON/OFF/CAL/CFG): CAL
(CR) Calibrate tablet for use
Digitize first known point:
(Digitize point.)
```

```
Enter coordinates for first point:
1,1 (CR)
Digitize second known point:
(Digitize point.)
Enter coordinates for second point:
10,1
```

You should digitize points from left to right or from bottom to top of the drawing. Further, your limits must be large enough to contain the limits on the tablet. You may now use the AutoCAD **Root Menu** for copying the drawing. For example, activate the `LINE` command, select a beginning point on the tablet with your pointer, and select other points in sequence.

Use the `ON` or `OFF` commands to turn the tablet mode on or off. When turned off, your pointer can be used to access the **Root Menu** at the right of the tablet. Function key F10 on the IBM XT/AT can also be used for this purpose.

38.78 SKETCH Command

The `SKETCH` command can be used with the tablet for tracing drawings composed of irregular lines (Fig. 38.179). Begin by attaching the draw-

ing on the tablet, and calibrate the tablet as discussed above. Enter the `SKETCH` command.

```
Command: SKETCH (CR)
Record increment <0.1>: 0.01 (CR)
Sketch. Pen eXit Quit Record Erase
Connect.
```

The record increment specifies the distances between the endpoints of the connecting lines that represent the irregular lines you sketch. The other command options are defined below.

`Pen`	Raises or lowers pen
`eXit`	Records lines and exits
`Quit`	Discards temporary lines and exits
`Record`	Records temporary lines
`Erase`	Erases selected lines
`Connect`	Connects current line to last endpoint
`.(period)`	Line from endpoint of last line to the current location of the pointer

Begin sketching by moving your pointer to the first point with the pen up, lower the pen (`P`), and trace over the line to be traced with the pointer. An irregular line is drawn on the screen. Erase by raising the pen (`P`), enter `Erase (E)`, move the pointer backward from the current point, and select the point where you want the erasure to stop. All drawn lines are temporary until you select `Record (R)` or `eXit (X)`. After recording the lines, you can begin new lines by repeating these steps. The mouse is unsatisfactory for sketching; for this type of digitizing only a tablet can be effectively used.

The `SKPOLY` system variable can be set by typing `SETVAR`, entering `SKPOLY` and typing 1 (on). When you enter the `SKETCH` command, the lines that you draw will be continuous polylines as an alternative to `SKETCH`ed lines, which connected but are separate segments.

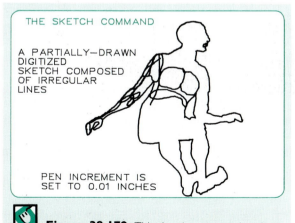

Figure 38.179 This drawing was made using the SKETCH command at a tablet instead of a mouse. An original drawing was taped to the tablet, and the cursor was used to trace over it using increments of 0.01″

38.79 Slide Shows

Several commands can be used to create a slide show on the computer screen: `SCRIPT`, `MSLIDE`, `RSCRIPT`, `DELAY`, `RESUME`, and `VSLIDE`.

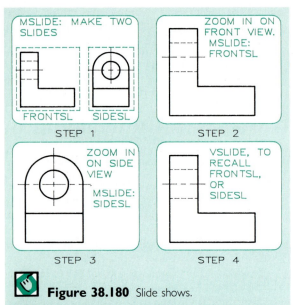

Figure 38.180 Slide shows.

Step 1 A drawing is made that is to be saved as two separate slides.

Step 2 ZOOM in on the front view.
Command: <u>MSLIDE</u> (CR) Slide file ‹current›: <u>FRONTSL</u> (CR)

Step 3 ZOOM in on the side view.
Command: <u>MSLIDE</u> (CR) Slide file ‹current›: <u>SIDESL</u> (CR)

Step 4 Command: <u>VSLIDE</u> (CR) Slide file <u>FRONTSL</u> (CR)
(The slide is displayed on the screen.)

Enter the Drawing Editor, and make a drawing that you wish to use as a slide (Fig. 38.180). Enter the MSLIDE command and respond to the prompt with a slide name, <u>A:FRONTSL</u> for example. Erase the drawing and make a second drawing, enter <u>MSLIDE</u>, and name it <u>A:SIDESL</u>. You may make as many slides as you wish by following these steps. When you are finished, erase the last drawing from the screen.

Enter the command VSLIDE and type <u>A:FRONTSL</u>, and this image will be recalled and displayed on the screen. Type <u>REDRAW</u> to remove A:SIDESL from the screen. Recall SLIDE2 with VSLIDE in the same manner. Now that you have checked your slides, QUIT, and EXIT AutoCAD (or use the SHELL command to gain access to DOS).

To create a SCRIPT file, type <u>SH</u> to obtain the DOS prompt, C›, and start a script file with the EDLIN command that will be called A:DISPLAY.SCR in the following manner:

```
C› EDLIN A:DISPLAY.SCR
* I
```

Type the following script file. Notice that the disk drive, A:, is placed in front of both slide name.

```
1: VSLIDE A:FRONTSL (CR)
2: DELAY 2000 (CR)
3: VSLIDE A:SIDESL (CR)
4: DELAY 1000 (CR)
5: RSCRIPT (CR)
6: RESUME (CR)
7: (Ctrl Break)
*E (Saves file and ends session.)
```

The DELAY command is specified in milliseconds; this value will be an approximation because this speed varies with different types of computers.

To show the slides according to the SCRIPT, load AutoCAD's **Drawing Editor** by calling up any new drawing. Type <u>SCRIPT</u> and <u>A:DISPLAY</u> (the name of SCRIPT) in response to the filename prompt. The slides will be shown in sequence with the delays between each slide; then the sequence will recycle, beginning with the first slide. This sequence can be stopped by pressing two keys—<u>Ctrl C</u> or Backspace.

If the RSCRIPT command had been omitted from the SCRIPT file, the two slides would have been shown, and the sequence would have stopped. To repeat the show, you would then type <u>RESUME</u>. When you are through, type <u>QUIT</u> to return to the **Main Menu.**

Many more advanced slides can be developed using methods covered in AutoCAD's manual. For example, the SCRIPT file can include such variables as LIMITS, SNAP, GRID, UNITS, and TEXT, which are activated by SCRIPT each time it is called up. By using your imagination, a series of slides can be made to give animated action.

The EDLIN command is an MS-DOS line editor that can be used to write the SCRIPT file. Some of the commands used with EDLIN are as follows:

‹line No.›   Edits a specified line

A   Appends lines to a file

D   Deletes lines

E   Ends session and writes file to disk

I   Inserts line into file

L   Lists lines

Q   Quits editing without saving file

S   Searches for a specified string

W   Writes a specified number of lines to disk

Ctrl Break   Ends insertion

38.80 SETVAR Command

Many modes, sizes, limits, and variables remaining in AutoCAD's memory under the SETVAR command are available. These variables can be inspected and changed by means of the SETVAR command, unless they are read-only commands. Some variables are saved in AutoCAD's configuration file, and others are saved in drawings that are made.

To review system variables, use the SETVAR command.

```
Command: SETVAR (CR)
Variable name or ?: ? (CR)
```

A complete listing of the current variables will be given on the screen. To change one or more variables respond as follows:

```
Commmand: SETVAR (CR)
Variable name or ?: TEXTSIZE (CR)
New value for TEXTSIZE ‹current›:
0.125 (CR)
```

The default value, given in brackets, can be changed by typing a new value, 0.125. By entering the SETVAR command with an apostrophe in front of it, you can use it transparently while another command is in progress.

A listing of the SETVAR options are given in Appendixes 38.1 and 38.2

38.81 Command Tree

The branching diagram in Fig. 38.181 (pages 732–733) shows the relationships of the various commands and subcommands of AutoCAD to one another. Reviewing this diagram before working on the computer will help you learn the commands and become proficient with AutoCAD software.

Problems

The problems at the ends of the previous chapters can be drawn and plotted using computer-graphics techniques as covered in this chapter and previous chapters.

Additional problems are given in Figures 38.182–38.195 that can be used to develop computer graphics skills. Each can be drawn for plotting on an A-Size sheet (11 × 8.5).

AutoCAD® Release 10

(Menu Bar and Pull—Down Menus)

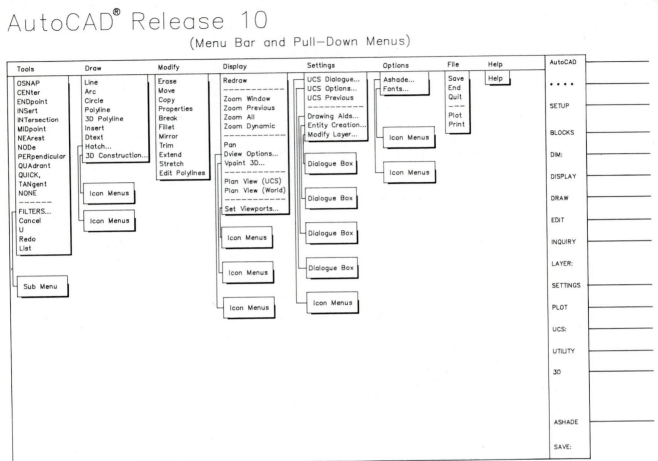

Figure 38.181 This menu tree shows the locations of various commands and their relationships to their sequential subcommands. A review of these commands will help you become familar with AutoCAD. (Courtesy of AutoDesk, Inc.)

(Primary Screen Menu Hierarchy)

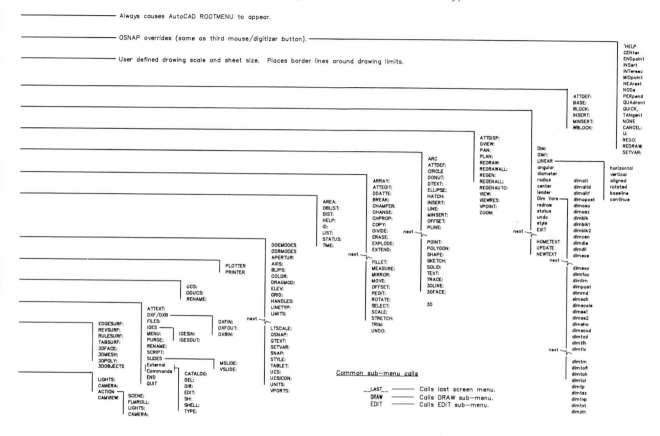

Always causes AutoCAD ROOTMENU to appear.

OSNAP overrides (same as third mouse/digitizer button).

User defined drawing scale and sheet size. Places border lines around drawing limits.

'HELP
CENter
ENDpoint
INSert
INTersec
MIDpoint
NEArest
NODe
PERpend
QUAdrant
QUICK,
TANgent
NONE
CANCEL:
U:
REDO:
REDRAW:
SETVAR:

ATTDEF:
BASE:
BLOCK:
INSERT:
MINSERT:
WBLOCK:

ATTDISP:
DVIEW:
PAN:
PLAN:
REDRAW:
REDRAWALL:
REGEN:
REGENALL:
REGENAUTO:
VIEW:
VIEWRES:
VPOINT:
ZOOM:

ARC
ATTDEF:
CIRCLE
DONUT:
DTEXT:
ELLIPSE:
HATCH:
INSERT:
LINE:
MINSERT:
OFFSET:
PLINE:

POINT:
POLYGON:
SHAPE:
SKETCH:
SOLID:
TEXT:
TRACE:
3DLINE:
3DFACE:

30

DIM:
DIM1:
LINEAR
angular
diameter
radius
center
leader
Dim Vars
redraw
status
undo
style
EXIT
HOMETEXT
UPDATE
NEWTEXT
next

horizontal
vertical
aligned
rotated
baseline
continue

dimalt
dimaltd
dimaltf
dimapost
dimaso
dimasz
dimblk
dimblk1
dimblk2
dimcen
dimdle
dimdli
dimexe
dimexo
dimlfac
dimlim
dimpost
dimrnd
dimsah
dimscale
dimse1
dimse2
dimsho
dimsoxd
dimtod
dimtih
dimtix

dimtm
dimtofl
dimtoh
dimtol
dimtp
dimtsz
dimtvp
dimtxt
dimzin

AREA:
DBLIST:
DIST:
HELP:
ID:
LIST:
STATUS:
TIME:

ARRAY:
ATTEDIT:
DDATTE:
BREAK:
CHAMFER:
CHANGE:
CHPROP:
COPY:
DIVIDE:
ERASE:
EXPLODE:
EXTEND:

FILLET:
MEASURE:
MIRROR:
MOVE:
OFFSET:
PEDIT:
ROTATE:
SELECT:
SCALE:
STRETCH:
TRIM:
UNDO:

DDEMODES
DDRMODES
APERTUR:
AXIS:
BLIPS:
COLOR:
DRAGMOD:
ELEV:
GRID:
HANDLES:
LINETYP:
LIMITS:

PLOTTER
PRINTER

UCS:
DDUCS:
RENAME:

LTSCALE:
OSNAP:
QTEXT:
SETVAR:
SNAP:
STYLE:
TABLET:
UCS:
UCSICON:
UNITS:
VPORTS:

DXFIN:
DXFOUT:
DXBIN:

ATTEXT:
DXF/DXB
FILES:
IGES
MENU:
PURGE:
RENAME:
SCRIPT:
SLIDES
External
Commands
END
QUIT

IGESIN:
IGESOUT:

MSLIDE:
VSLIDE:

CATALOG:
DEL:
DIR:
EDIT:
SH:
SHELL:
TYPE:

EDGESURF:
REVSURF:
RULESURF:
TABSURF:
3DFACE:
3DMESH:
3DPOLY:
3DOBJECTS

LIGHTS:
CAMERA:
ACTION
CAMVIEW:

SCENE:
FLMROLL:
LIGHTS:
CAMERA:

Common sub-menu calls

__LAST__ ——— Calls last screen menu.
DRAW ——— Calls DRAW sub-menu.
EDIT ——— Calls EDIT sub-menu.

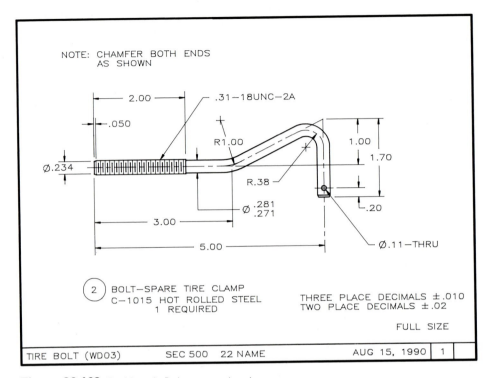

Figure 38.182 Problem 1. Bolt—spare tire clamp.

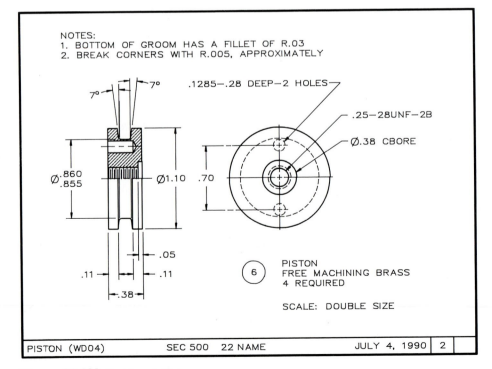

Figure 38.183 Problem 2. Piston.

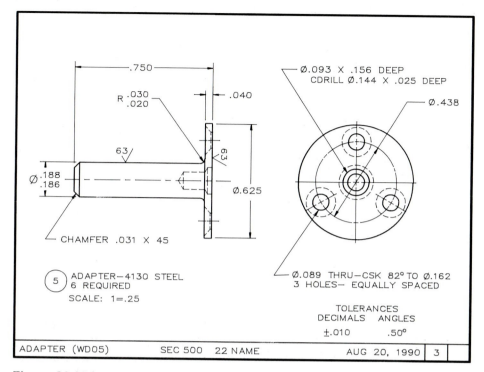

Figure 38.184 Problem 3. Adapter.

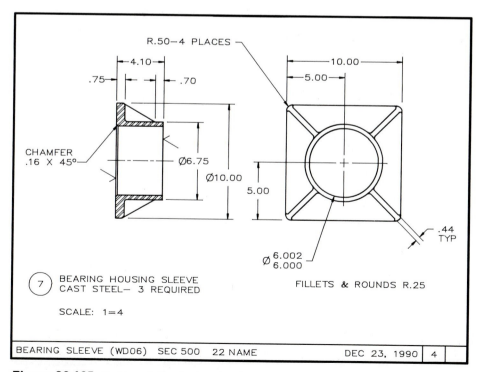

Figure 38.185 Problem 4. Bearing housing sleeve.

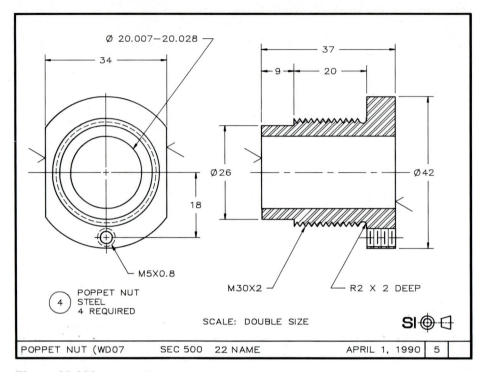

Figure 38.186 Problem 5. Poppet nut.

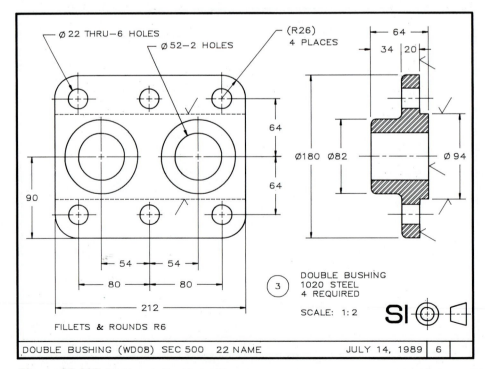

Figure 38.187 Problem 6. Double bushing.

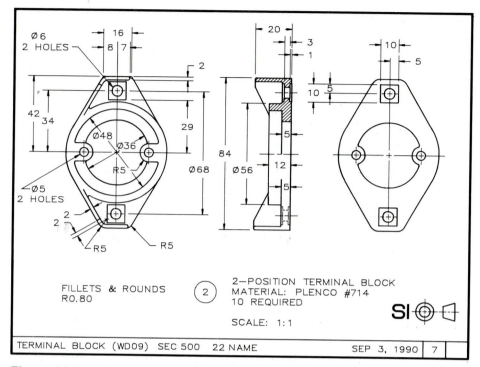

Figure 38.188 Problem 7. Terminal block.

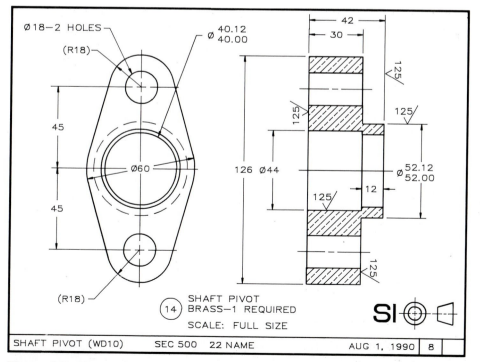

Figure 38.189 Problem 8. Shaft pivot.

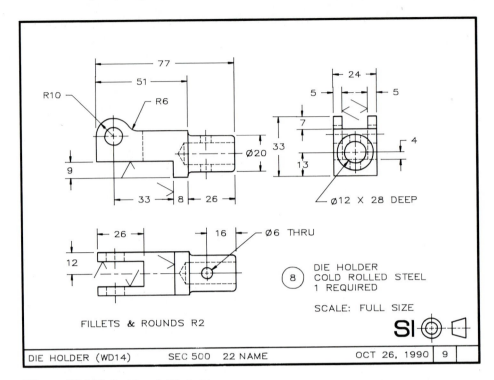

Figure 38.190 Problem 9. Die holder.

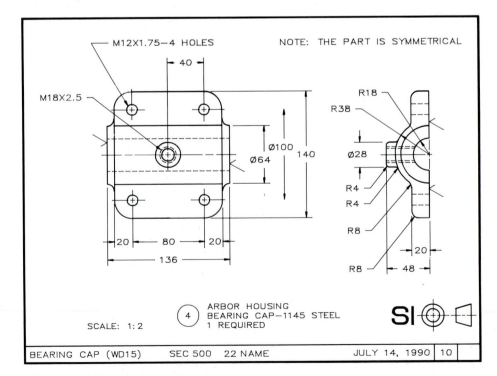

Figure 38.191 Problem 10. Bearing cap.

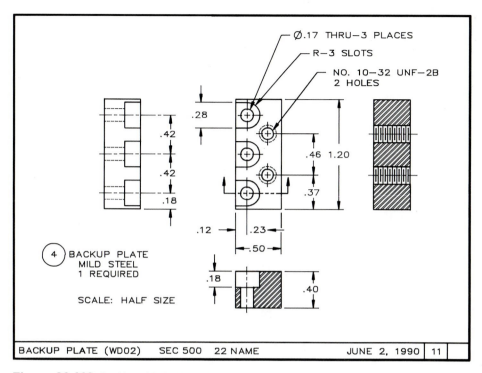

Figure 38.192 Problem 11. Backup plate.

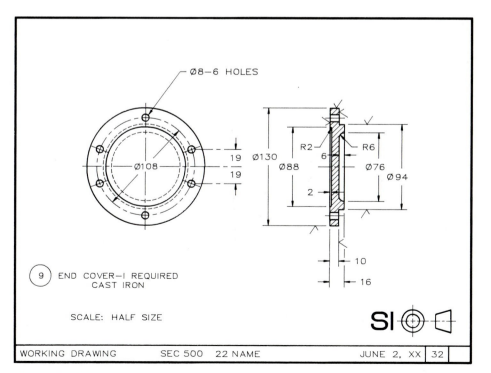

Figure 38.193 Problem 12. End cover.

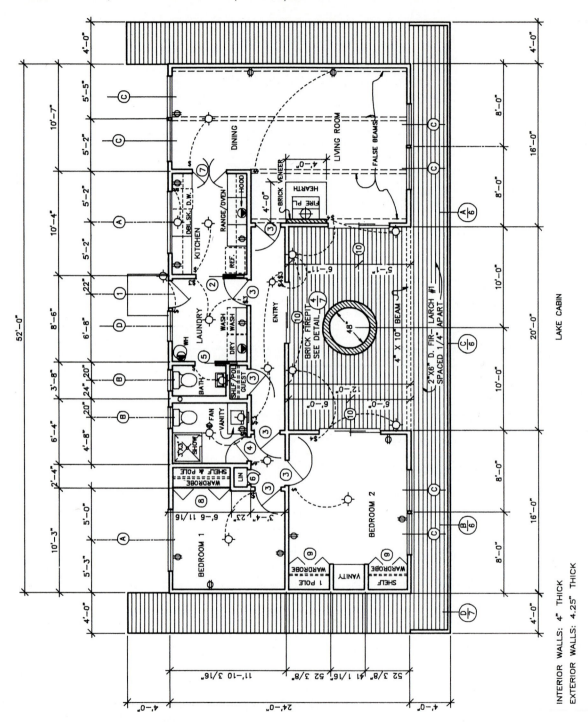

INTERIOR WALLS: 4" THICK
EXTERIOR WALLS: 4.25" THICK

LAKE CABIN

Figure 38.194 Architectural Plan. Draw the plan of the house on a C-Size sheet at a scale of 1/4" = 1'-0". Include a door schedule, window schedule, and legend on the same sheet (Figure 25.195).

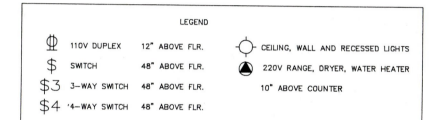

LEGEND

⊕	110V DUPLEX	12" ABOVE FLR.
$	SWITCH	48" ABOVE FLR.
$3	3-WAY SWITCH	48" ABOVE FLR.
$4	'4-WAY SWITCH	48" ABOVE FLR.

○ CEILING, WALL AND RECESSED LIGHTS

▲ 220V RANGE, DRYER, WATER HEATER

10" ABOVE COUNTER

WINDOW SCHEDULE					
MARK	NO.	STOCK	TYPE	MFGR.	ROUGH OPENING
A	2	6030	HORIZONTAL SLIDING	ALENCO	72 3/4" X 36 1/2"
B	2	2030	HORIZONTAL SLIDING	ALENCO	24 3/4" X 36 1/2"
C	6	4040	HORIZONTAL SLIDING	ALENCO	48 3/4" X 48 1/2"
D	1	3030	HORIZONTAL SLIDING	ALENCO	36 3/4" X 36 1/2"

DOOR SCHEDULE				
MARK	NO.	TYPE	MATERIAL	SIZE
1	1	1 LT. PANEL SASH DOOR	BIRCH	2'-8" X 6'-8" X 1 3/4"
2	1	FLUSH HOLLOW CORE – INTERIOR	BIRCH	2'-8" X 6'-8" X 1 3/8"
3	6	FLUSH HOLLOW CORE – INTERIOR	BIRCH	2'-6" X 6'-8" X 1 3/8"
4	1	FLUSH HOLLOW CORE – INTERIOR	BIRCH	2'-4" X 6'-8" X 1 3/8"
5	1	FLUSH HOLLOW CORE – INTERIOR	BIRCH	2'-0" X 6'-8" X 1 3/8"
6	1	FLUSH HOLLOW CORE – INTERIOR	BIRCH	1'-6" X 6'-8" X 1 3/8"
7	1 PR.	SWING. CAFE DOORS	BIRCH	1'-8" X 4'-0" X 1 1/8"
8	1	FLUSH HOLLOW CORE – FOLDING	BIRCH	6'-0" X 6'-8" X 1 3/8"
9	2	FLUSH HOLLOW CORE – INTERIOR	BIRCH	4'-0" X 6'-8" X 1 3/8"
10	3	SLIDING GLASS (6069 – 2V)	GLASS	6'-0" X 6'-8 1/2"

Figure 38.195 Schedules and legends of the architectural plan.

Appendix 38.1
AutoCAD Command Reference

Command	Description		Options
APERTURE	Controls the size of the object snap target box		
AREA	Computes the area of a polygon, polyline, or circle	A	Sets "add" mode
		S	Sets "subtract" mode
		E	Computes area of a selected circle or polyline
ATTEDIT	Permits editing of Attributes		

(CONTINUED)

Command	Description	Options	
ATTEXT	Extracts Attribute data from a drawing	C	CDF comma-delimited format extract
		D	DXF format extract
		S	SDF format extract
		E	Extract selected objects only
AXIS	Displays a "ruler line" on the graphics monitor	ON	Turn axis (ruler line) on
		OFF	Turn axis off
		S	Lock tick spacing to Snap resolution
		A	Set aspect (differing X-Y spacings)
		number	Set tick spacing (0 = use snap spacing)
		numberX	Set spacing to multiple of snap spacing
BLIPMODE	Controls display of marker blips for point selection	ON	Enable temporary marker blips
		OFF	Disable temporary marker blips
CHANGE	Alters the location, size, orientation, or other properties of selected objects. Especially useful for Text entries.	P	Change common properties of objects
		C	Color
		E	Elevation (to be dropped in the next major update)
		LA	Layer
		LT	Linetype
		T	Thickness
CHPROP	Modifies properties of selection objects	C	Color
		LA	Layer
		LT	Linetype
		T	Thickness
COLOR	Establishes the color for subsequently drawn objects	number	Set entity color number
		name	Set entity color to standard color name
		BYBLOCK	Set "floating" entity color
		BYLAYER	Use layer's color for entities
DBLIST	Lists database information for every entity in the drawing		
DRAGMODE	Allows control of the dynamic specification ("dragging") feature for all appropriate commands	ON	Honor "DRAG" requests when applicable
		OFF	Ignore "DRAG" requests
		A	Set "Auto" mode: drag whenever possible
DVIEW	Defines parallel or visual perspective views dynamically	CA	Select the camera angle relative to the target
		CL	Set front and back clipping planes
		D	Set camera-to-target distance, turn on perspective
		H	Remove hidden lines on the selection set
		OFF	Turn perspective off
		PA	Pan drawing across the screen
		PO	Specify the camera and target points
		TA	Rotate the target point about the camera
		TW	Twist the view around your line of sight
		U	Undo a DVIEW subcommand
		X	Exit the DVIEW command
		Z	Zoom in/out, or set lens length

Command	Description	Options	
END	Exits the Drawing Editor after saving the updated drawing		
EXPLODE	Shatters a Block or Polyline into its constituent parts		
EXTEND	Lengthens a Line, Arc, or Polyline to meet another object		
FILES	Performs disk file utility tasks		
FILL	Controls whether Solids, Traces, and wide Polylines are automatically filled on the screen and the plot output	ON OFF	Solids, Traces, and wide Polylines filled Solids, Traces, and wide Polylines outlined
FILMROLL	Generates a file for rendering by AutoShade		
HIDE	Regenerates a 3D visualization with "hidden" lines removed		
ID	Displays the coordinates of a specified point		
LAYER	Creates named drawing layers and assigns color and linetype properties to those layers	Cc Fa,b Lt Ma Na,b ONa,b OFFa,b Sa Ta,b ?	Set specified layers to color "c" Freeze layers "a" and "b" Set specified layers to linetype "t" Make "a" the current layer, creating it if necessary Create new layers "a" and "b" Turn on layers "a" and "b" Turn off layers "a" and "b" Set current layer to existing layer "a" Thaw layers "a" and "b" List layers and their associated colors & linetypes
LINETYPE	Defines linetypes (sequences of alternating line segments and spaces), loads them from libraries, and sets the linetype for subsequently drawn objects	? C L S "Set" suboptions: name BYBLOCK BYLAYER ?	List a linetype library Create a linetype definition Load a linetype definition Set current entity linetype Set entity linetype name Set "floating" entity linetype Use layer's linetype for entities List loaded linetypes
LOAD	Loads a file of user-defined Shapes to be used with the SHAPE command	?	List the names of loaded Shape files
LTSCALE	Sets scale factor to be applied to all linetypes within the drawing		
MULTIPLE	Causes the next command to repeat until canceled		

(CONTINUED)

Command	Description	Options	
OSNAP	Enables points to be precisely located on reference points of existing objects	CEN ENDP INS INT MID NEA NOD NON PER QUA QUI TAN	Center of Arc or Circle Closest endpoint of Arc or line Insertion point of Text/Block/Shape Intersection of Line/Arc/Circle Midpoint of Arc or Line Nearest point of Arc/Circle/Line/Point Node (point) None (off) Perpendicular to Arc/Line/Circle Quadrant point of Arc or Circle Quick mode (first find, not closest) Tangent to Arc or Circle
PEDIT (3D)	Permits editing	During vertex editing:	
		B G I M N P R S X	Set first vertex for Break Go (perform Break or Straighten operation) Insert new vertex after current one Move current vertex Make next vertex current Make previous vertex current Regenerate the Polyline Set first vertex for Straighten Exit vertex editing, or cancel Break/Straighten
PEDIT (Mesh)	Permits editing of 3D polygon meshes:	D E M N S U X	Desmooth-restore original mesh Edit mesh vertices (see below for suboptions) Open (or close) the mesh in the M direction Open (or close) the mesh in the N direction Fit a smooth surface as defined by SURFTYPE Undo one editing operation Exit PEDIT command
		During vertex editing:	
		D L M N P R RE U X	Move "down" to previous vertex in M direction Move "left" to previous vertex in N direction Reposition the marked vertex Move to next vertex Move to previous vertex Move "right" to next vertex in N direction Redisplay the polygon mesh Move "up" to next vertex in M direction Exit vertex editing

Command	Description	Options	
PLAN	Puts the display in PLAN view (VPOINT 0,0,1) relative to either the current UCS, a specified UCS, or the World Coordinate System	C U W	Establishes a plan view of the current UCS Establishes a plan view of the specified UCS Establishes a plan view of the World Coordinate System
PRPLOT	Plots a drawing on a printer plotter		
PURGE	Removes unused Blocks, text styles, layers, or linetypes from the drawing	A B LA LT SH ST	Purge all unused named objects Purge unused Blocks Purge unused layers Purge unused linetypes Purge unused shape files Purge unused text styles
REDO	Reverses the previous command if it was U or UNDO		
'REDRAW	Refreshens or cleans up the current viewport		
'REDRAWALL	Redraws all viewpoints		
REGEN	Regenerates the current viewport		
REGENALL	Regenerates all viewports		
REGENAUTO	Controls automatic regeneration performed by other commands	ON OFF	Allow automatic regens Prevent automatic regens
RENAME	Changes the names associated with text styles, layers, linetypes, Blocks, views, User Coordinate Systems, and viewport configurations	B LA KT S U VI VP	Rename Block Rename layer Rename linetype Rename text style Rename UCS Rename view Rename viewport configuration
'RESUME	Resumes an interrupted command script		
RSCRIPT	Restarts a command script from the beginning		
SCALE	Alters the size of existing objects	R	Resize with respect to reference size
SCRIPT	Executes a command script		
SELECT	Groups objects into a selection-set for use in subsequent commands		
SETVAR	Allows you to display or change the value of system variables		
SH	Allows access to internal PC-DOS/MS-DOS commands		
SHAPE	Draws pre-defined shapes	?	List available Shape Names
SHELL	Allows access to other programs while running AutoCAD		

(CONTINUED)

Command	Description	Options	
SNAP	Specifies a "round-off" interval for digitizer point entry so entities can be placed at precise locations easily	number ON OFF A R S	Set snap alignment resolution Align designated points Do not align designated points Set aspect (differing X-Y spacings) Rotate snap grid Select style, standard or isometric
STATUS	Displays drawing statistics & modes		
STYLE	Creates named text styles, with user-selected combinations of font, mirroring, obliquing, and horizontal scaling	?	List currently defined text styles
TABLET	Aligns the digitizing tablet with coordinates of a paper drawing to accurately copy it with AutoCAD	ON OFF CAL CFG	Turn tablet mode on Turn tablet mode off Calibrate tablet Configure tablet menus, pointing area
TABSURF	Creates a polygon mesh approximating a general tabulated surface defined by a path and a direction vector		
'TEXTSCR	Flips to the text display on single-screen systems. Used in command scripts and menus		
TIME	Displays drawing creation and update times, and permits control of an elapsed timer	D ON OFF R	Display current times Start user elapsed timer Stop user elapsed timer Reset user elapses timer
U	Reverses the effect of the previous command		
UCS	Defines or modifies the current User Coordinate System	D E O P R S V W X Y Z ZA 3 ?	Delete one or more saved coordinate systems Set a UCS with the same extrusion direction as that of the selected entity Shift the origin of the current coordinate system Restore the previous UCS Restore a previously-saved UCS Save the current UCS Establish a new UCS whose Z axis is parallel to the current viewing direction Set the current UCS equal to the World Coordinate System Rotate the current UCS around the X axis Rotate the current UCS around the Y axis Rotate the current UCS around the Z axis Define a UCS using an origin point and a point on the positive portion of the Z axis

Command	Description		Options
			Define a UCS using an origin point, a point on the positive portion of the X axis, and a point on the positive Y-portion of the XY plane
			List the saved coordinate systems
UCSICON	Controls visibility and placement of the User Coordinate System icon, which indicates the origin and orientation current UCS. The options normally affect only the current viewport.	A	Change settings in all active viewports
		N	Display the icon at the lower left corner of the viewport
		O	Display the icon at the origin of the current UCS if possible
		OFF	Disable the coordinate system icon
		ON	Enable the coordinate system icon
UNDO	Reverses the effect of multiple commands, and provides control over the "undo" facility	number	Undoes the number of most recent commands
		A	Auto: controls treatment of menu items as Undo Groups
		B	Back: undoes back to previous Undo Mark
		C	Control: enables/disables the Undo feature
		E	End: terminates an UNDO Group
		G	Group: begins sequence to be treated as one command
		M	Mark: places marker in Undo file (for Back)
'VIEW	Saves the current graphics display as a Named View, or restores a saved view to the display	D	Delete named view
		R	Restore named view to screen
		S	Save current display as named view
		W	Save specified window as named view
		?	List named views
VIEWPORTS or VPORTS	Divides AutoCAD's graphics display into multiple viewports, each of which may contain a different view of the current drawing	D	Delete a saved viewport configuration
		J	Join (merge) two viewports
		R	Restore a saved viewport configuration
		S	Save the current viewport configuration
		SI	Display a single viewport filling the entire graphics area
		2	Divide the current viewport into 2 viewports
		3	Divide the current viewport into 3 viewports
		4	Divide the current viewport into 4 viewports
		?	List the current and saved viewport configurations
VIEWRES	Allows you to control the precision and speed of Circle and Arc drawing on the monitor by specifying the number of sides in a Circle		

(CONTINUED)

Command	Description	Options	
VPOINT	Selects the viewpoint for a 3D visualization	R RETURN x,y,z	Select view point via two rotation angles Select view point via compass & axes tripod Specifies view point
VSLIDE	Displays a previously-created slide file	file *file	View slide Preload slide, next VSLIDE will view
WBLOCK	Writes selected entities to a disk file	name = * RETURN	Write specified Block Definition Block name same as file name Write entire drawing Write selected objects
3DFACE	Draws three-dimensional plane sections	I	Make the following edge invisible.
3DMESH	Defines a three-dimensional polygon mesh by specifying its size (in terms of M and N) and the location of each vertex in the mesh		
3DPOLY	Creates a three-dimensional polyline	C U RETURN	Close the polyline back to the first point Undo (delete) the the last segment entered Exit 3DPOLY command

Reprinted with permission from the AutoCAD Reference Manual © 1989, Autodesk, Inc.

Appendix 38.2
System Variables

Variable name	Type	Saved in	Meaning
ANGBASE	real	drawing	Angle 0 direction (with respect to the current UCS)
ANGDIR	integer	drawing	1 = clockwise angles, 0 = counterclockwise (with respect to the current UCS)
APERTURE	integer	config	Object snap target height, in pixels
AREA	real		True area computed by AREA, LIST or DBLIST. (read-only)
ATTDIA	integer	drawing	1 to cause the INSERT command to use a dialogue box for entry of Attribute values, 0 to issue prompts.
ATTMODE	integer	drawing	Attribute display mode (0 = off, 1 = normal, 2 = on)
ATTREQ	integer	drawing	0 to assume defaults for the values of all Attributes during INSERT of Blocks, 1 to enable prompts (or dialogue box) for Attributes values, as selected by ATTDIA.
AUNITS	integer	drawing	Angular units mode (0 = decimal degrees, 1 = degrees/minutes/seconds, 2 = grads, 3 = radians, 4 = surveyor's units)
AUPREC	integer	drawing	Angular units decimal places
AXISMODE	integer	drawing	Axis display on if 1, off if 0
AXISUNIT	2D point	drawing	Axis spacing, X and Y
BACKZ	real	drawing	Back clipping plane offset for the current viewport, in drawing units. Meaningful only if the "Back clipping" bit in VIEWMODE is ON. The distance of the back clipping plane from the camera point can be found by subtracting BACKZ from the camera-to-target distance. (read-only)
BLIPMODE	integer	drawing	Marker blips on if 1, off if 0
CDATE	real		Calendar date/time (read-only) (special format; see below)
CHAMFERA	real	drawing	First chamfer distance
CHAMFERB	real	drawing	Second chamfer distance
CMDECHO	integer		When AutoLISP's (command) function is used, prompts and input are echoed if this variable is 1, but not if it is 0
COORDS	integer	drawing	If zero, coordinate display is updated on point picks only. If 1, display of absolute coordinates is continuously updated. If 2, distance and angle from last point are displayed when a distance or angle is requested.
CVPORT	integer	drawing	The identification number of the current viewport
DATE	real		Julian date/time (read-only) (special format; see below)
DISTANCE	real		Distance computed by DIST command (read-only)
DRAGMODE	integer	drawing	0 = no dragging, 1 = on if requested, 2 = auto
DWGNAME	string		Drawing name as entered by the user. If the user specified a drive/directory prefix, it is included as well. (read-only)
ELEVATION	real	drawing	Current 3D elevation, relative to the current UCS. (To be dropped in the next major update.)
FILLETRAD	real	drawing	Fillet radius
FILLMODE	integer	drawing	Fill mode on if 1, off if 0
FLATLAND	integer	drawing	This is a temporary 3D conversion aid, to be removed in the next major update. If set to 1, object snap, DXF, and AutoLISP operate

(CONTINUED)

Variable name	Type	Saved in	Meaning
			as they did prior to Release 10; if set to 0, these functions take full advantage of Release 10's new capabilities. For details, see Appendix D. Default = 0 for new drawings, 1 for drawings created with versions prior to Release 10.
FRONTZ	real	drawing	Front clipping plane offset for the current viewport, in drawing units. Meaningful only if the "Front clipping" bit in VIEWMODE is ON and the "Front clip not at eye" bit is also ON. The distance of the front clipping plane from the camera point can be found by subtracting FRONTZ from the camera-to-target distance. (read-only)
GRIDMORE	integer	drawing	1 = grid on for current viewport; 0 = grid off
GRIDUNIT	2D point	drawing	Grid spacing for current viewport, X and Y
HANDLES	integer	drawing	If 0, entity handles are disabled. If 1, handles are on (read-only)
HIGHLIGHT	integer		Object selection highlighting on if 1, off if 0
INBASE	3D point	drawing	Insertion base point (set by BASE command) expressed in UCS coordinates
LENSLENGTH	real	drawing	Length of the lens (in millimeters) used in perspective viewing, for the current viewport (read-only)
LIMMAX	2D point	drawing	Upper right drawing limits expressed in World coordinates
LIMMIN	2D point	drawing	Lower left drawing limits expressed in World coordinates
LTSCALE	real	drawing	Global linetype scale factor
MIRRTEXT	integer	drawing	MIRROR reflects text if nonzero, retains text direction if zero.
ORTHOMODE	integer	drawing	Ortho node on if 1, off if 0
PMODE	integer	drawing	Point entity display mode
PDSIZE	real	drawing	Point entity display size
PERIMETER	real		Perimeter computed by AREA, LIST, or DBLIST. (read-only)
POPUPS	integer		1 if the currently configured disply driver supports dialogue boxes, the menu bar, pull-down menus, and icon menus. 0 if these advanced user interface features are not available (read-only)
QTEXTMODE	integer	drawing	Quick textmode on if 1, off if 0
SCREENSIZE	2D point		Current viewport size in pixels, X and Y (read-only)
SKETCHINC	real	drawing	Sketch record increment
SKPOLY	integer	drawing	SKETCH generates lines if 0, Polylines if 1
SNAPANG	real	drawing	Snap / grid rotation angle (UCS-relative) for the current viewport
SNAPBASE	2D point	drawing	Snap / grid origin point for the current viewport (in UCS XY coordinates)
SNAPISOPAIR	integer	drawing	Current isometric plane (0 = left, 1 = top, 2 = right) for the current viewport
SNAPMODE	integer	drawing	1 = snap on for current viewport; 0 = snap off
SNAPSTYL	integer	drawing	Snap style for current viewport (0 = standard, 1 = isometric)
SNAPUNIT	2D point	drawing	Snap spacing for current viewport, X and Y
SPLFRAME	integer	drawing	If = 1: - the control polygon for spline fit Polylines is to be displayed - only the defining mesh of a surface fit polygon mesh is to be displayed (the fit surface is not displayed) - "invisible" edges of 3D Faces are displayed

Variable name	Type	Saved in	Meaning
			If = 0: - do not display the control polygon for spline fit Polylines - display the fit surface of a polygon mesh, not the defining mesh - do not display the "invisible" edges of 3D Faces
SPLINESEGS	integer	drawing	The number of line segments to be generated for each spline patch.
SPLINETYPE	integer	drawing	Type of spline curve to be generated by "PEDIT Spline". The valid values are: 5 = quadratic B-spline 6 = cubic B-spline
SURFTAB1	integer	drawing	Number of tabulations to be generated for RULESURF and TABSURF. Also mesh density in the M Direction for REVSURF and EDGESURF
SURFTAB2	integer	drawing	Mesh density in the N direction for REVSURF and EDGESURF
SURFTYPE	integer	drawing	Type of surface fitting to be performed by "PEDIT Smooth". The valid values are: 5 = quadratic B-spline surface 6 = cubic B-spline surface 8 = Bezier surface
SURFU	integer	drawing	Surface density in the M direction
SURFV	integer	drawing	Surface density in the N direction
TARGET	3D point	drawing	Location (in UCS coordinates) of the target ("look-at") point for the current viewport (read-only)
TEXTEVAL	integer		If = 0, all responses to prompts for text strings and attribute values are taken literally. If = 1, text starting with "(" or "!" is evaluated as an AutoLISP expression, as for nontextual input. NOTE: The DTEXT command takes all input literally, regardless of the setting of TEXTEVAL.
TEXTSIZE	real	drawing	The default height for new Text entities drawn with the current text style (meaningless if the style has a fixed height). (In previous versions, this variable was nearly useless, since it controlled only the initial setting of the default height when creating a new text style.)
TEXTSTYLE	string	drawing	This variable contains the name of the current text style. (read-only)
THICKNESS	real	drawing	Current 3D thickness
TRACEWID	real	drawing	Default trace width
UCSFOLLOW	integer	drawing	If = 1, any UCS change causes an automatic change to plan view of the new UCS (in the current viewport). If = 0, a UCS change doesn't affect the view.
UCSNAME	string	drawing	Name of the current coordinate system. Returns a null string if the current UCS is unnamed (read-only)
UCSORG	3D point	drawing	The origin point of the current coordinate system. This value is always returned in World coordinates (read-only)
WORLDUCS	integer		If = 1, the current UCS is the same as the World Coordinate System. If = 0, it is not. (read-only)
WORLDVIEW	integer	drawing	DVIEW and VPOINT command input is relative to the current UCS. If this variable is set to 1, the current UCS is changed to the WCS for the duration of a DVIEW or VPOINT command. Default value = 0.

Appendix Contents

Appendix 1
Abbreviations (ANSI Z 32.13)

Word	Abbreviation	Word	Abbreviation	Word	Abbreviation
Abbreviate	ABBR	Bench mark	BM	Centigrade	C
Absolute	ABS	Between	BET.	Centigram	CG
Account	ACCT	Between centers	BC	Centimeter	cm
Actual	ACT.	Between		Chain	CH
Adapter	ADPT	perpendiculars	BP	Chamfer	CHAM
Addendum	ADD.	Bevel	BEV	Change notice	CN
Adjust	ADJ	Bill of material	B/M	Change order	CO
Advance	ADV	Birmingham wire gage	BWG	Channel	CHAN
After	AFT.	Blueprint	BP	Check	CHK
Aggregate	AGGR	Board	BD	Check valve	CV
Air condition	AIR COND	Boiler	BLR	Chemical	CHEM
Airport	AP	Bolt circle	BC	Chord	CHD
Airplane	APL	Both sides	BS	Circle	CIR
Allowance	ALLOW	Bottom	BOT	Circuit	CKT
Alloy	ALY	Bottom chord	BC	Circular	CIR
Alteration	ALT	Boundary	BDY	Circular pitch	CP
Alternate	ALT	Bracket	BRKT	Circumference	CIRC
Alternating current	AC	Brake horsepower	BHP	Clockwise	CW
Altitude	ALT	Brass	BRS	Coated	CTD
Aluminum	AL	Brazing	BRZG	Cold drawn	CD
American National		Break	BRK	Cold drawn steel	CDS
Standard	AMER NATL STD	Breaker	BKR	Cold finish	CF
American wire gage	AWG	Bridge	BRDG	Cold punched	CP
Ammeter	AM	Brinnell hardness	BH	Cold rolled	CR
Amount	AMT	British Standard	BR STD	Cold rolled steel	CRS
Ampere	AMP	British Thermal Units	BTU	Column	COL
Anneal	ANL	Broach	BRO	Combination	COMB.
Antenna	ANT.	Bronze	BRZ	Combustion	COMB
Apparatus	APP	Brown & Sharp (Wire gage,		Commutator	COMM
Appendix	APPX	same as AWG)	B&S	Company	CO
Approved	APPD	Building	BLDG	Concentric	CONC
Approximate	APPROX	Bulkhead	BHD	Concrete	CONC
Arc weld	ARC/W	Bureau	BU	Condition	COND
Area	A	Bureau of Standards	BU STD	Connect	CONN
Armature	ARM.	Bushing	BUSH.	Contact	CONT
Asbestos	ASB	Button	BUT.	Cord	CD
Asphalt	ASPH	Buzzer	BUZ	Corporation	CORP
Assembly	ASSY	By-pass	BYP	Corrugate	CORR
Association	ASSN			Cotter	COT
Atomic	AT	Cabinet	CAB.	Counterclockwise	CCW
Authorized	AUTH	Cadmium plate	CD PL	Counterbore	CBORE
Auxiliary	AUX	Calculate	CALC	Counterdrill	CDRILL
Avenue	AVE	Calibrate	CAL	Counterpunch	CPUNCH
Average	AVG	Calorie	CAL	Countersink	CSK
Avoirdupois	AVDP	Capacitor	CAP	Coupling	CPLG
Azimuth	AZ	Cap screw	CAP SCR	Crank	CRK
		Case harden	CH	Cross section	XSECT
Babbitt	BAB	Cast iron	CI	Cubic	CU
Back pressure	BP	Cast steel	CS	Cubic centimeter	cc
Balance	BAL	Casting	CSTG	Cubic feet per minute	CFM
Ball bearing	BB	Castle nut	CAS NUT	Cubic feet per second	CFS
Barometer	BAR	Catalogue	CAT.	Cubic foot	CU FT
Barrel	BBL	Cement	CEM	Cubic inch	CU IN.
Base line	BL	Center	CTR	Cubic meter	CU M
Base plate	BP	Centerline	CL	Cubic yard	CU YD
Battery	BAT.	Center of gravity	CG	Current	CUR
Bearing	BRG	Center to center	C to C	Cylinder	CYL

Cont.

Appendix 1
Abbreviations (ANSI Z 32.13) (cont.)

Word	Abbreviation	Word	Abbreviation	Word	Abbreviation
Decimal	DEC	Fillister	FIL	Illustrate	ILLUS
Dedendum	DED	Filter	FLT	Inch	(") IN.
Degree	(°) DEG	Finish	FIN.	Inches per second	IPS
Department	DEPT	Finish all over	FAO	Include	INCL
Design	DSGN	Flange	FLG	Industrial	IND
Detail	DET	Flat head	FH	Information	INFO
Develop	DEV	Fluid	FL	Inside diameter	ID
Diagonal	DIAG	Focus	FOC	Instrument	INST
Diagram	DIAG	Foot	(') FT	Insulate	INS
Diameter	DIA	Forging	FORG	Interior	INT
Diametrical pitch	DP	Forward	FWD	Internal	INT
Dimension	DIM.	Foundation	FDN	Intersect	INT
Direct current	DC	Foundry	FDRY	Iron	I
Discharge	DISCH	Frequency	FREQ	Irregular	IRREG
Distance	DIST	Front	FR		
District	DIST			Jack	J
Ditto	DO			Joint	JT
Dovetail	DVTL	Gage	GA	Junction	JCT
Dowel	DWL	Gallon	GAL	Junction box	JB
Down	DN	Galvanize	GALV		
Dozen	DOZ	Galvanized iron	GI	Key	K
Drafting	DFTG	Galvanized steel	GS	Keyseat	KST
Draftsman	DFTSMN	Gasket	GSKT	Keyway	KWY
Drawing	DWG	General	GEN	Kiln-dried	KD
Drill	DR	Government	GOVT	Kip (1000 lb)	K
Drive	DR	Governor	GOV	Knots	KN
Drop forge	DF	Grade	GR		
		Grade line	GL	Laboratory	LAB
		Gram	G	Lateral	LAT
Each	EA	Gravity	G	Latitude	LAT
East	E	Grind	GRD	Left	L
Eccentric	ECC	Groove	GRV	Left hand	LH
Effective	EFF	Ground	GRD	Length	LG
Elbow	ELL	Gypsum	GYP	Letter	LTR
Electric	ELEC			Light	LT
Elevation	ELEV	Half-round	½ RD	Line	L
Engineer	ENGR	Handle	HDL	Logarithm	LOG.
Equal	EQ	Hanger	HGR	Lubricate	LUB
Equipment	EQUIP.	Hard	H	Lumber	LBR
Equivalent	EQUIV	Hard-drawn	HD		
Estimate	EST	Harden	HDN	Machine	MACH
Exterior	EXT	Hardware	HDW	Malleable	MALL
Extra heavy	X HVY	Head	HD	Malleable iron	MI
Extra strong	X STR	Headless	HDLS	Manhole	MH
		Headquarters	HQ	Manual	MAN.
Fabricate	FAB	Heat	HT	Manufacture	MFR
Face to face	F to F	Heat treat	HT TR	Material	MATL
Fahrenheit	F	Hexagon	HEX	Maximum	MAX
Fairing	FAIR.	High-pressure	HP	Mechanical	MECH
Far side	FS	High-speed	HS	Mechanism	MECH
Federal	FED.	Horizontal	HOR	Median	MED
Feet	(') FT	Horsepower	HP	Metal	MET.
Feet per minute	FPM	Hot rolled	HR	Meter (Instrument or	
Feet per second	FPS	Hot rolled steel	HRS	measure of length)	M
Field	FLD	Hour	HR	Miles	MI
Figure	FIG.	Hundredweight	CWT	Miles per gallon	MPG
Fillet	FIL	Hydraulic	HYD	Miles per hour	MPH

Appendix 1
Abbreviations (ANSI Z 32.13) (cont.)

Word	Abbreviation	Word	Abbreviation	Word	Abbreviation
Millimeter	MM	Precast	PRCST	Shaft	SFT
Minimum	MIN	Prefabricated	PREFAB	Sketch	SK
Minute	(') MIN	Preferred	PFD	Sleeve	SLV
Miscellaneous	MISC	Prepare	PREP	Slide	SL
Mixture	MIX.	Pressure	PRESS.	Slotted	SLOT.
Model	MOD	Pressure angle	PA.	Socket	SOC
Month	MO	Process	PROC	Solder	SLD
Morse taper	MOR T	Production	PROD	South	S
Multiple	MULT	Profile	PF	Space	SP
		Project	PROJ	Special	SPL
National	NATL	Proof	PRF	Specific gravity	SP GR
Near side	NS			Spherical	SPHER
Negative	NEG	Quadrant	QUAD	Spot faced	SF
Neutral	NEUT	Quart	QT	Spring	SPG
Nipple	NIP.	Quarter	QTR	Square	SQ
Nominal	NOM	Quarter-round	¼ RD	Stainless	STN
Normal	NOR			Stainless steel	SST
North	N	Radial	RAD	Standard	STD
Not to scale	NTS	Radius	R	Station	STA
Number	NO.	Railroad	RR	Steel	STL
		Ream	RM	Stock	STK
Obsolete	OBS	Received	RECD	Straight	STR
Octagon	OCT	Record	REC	Structural	STR
Ohm	Ω	Rectangle	RECT	Substitute	SUB
On center	OC	Reference	REF	Summary	SUM.
Opposite	OPP	Reference line	REF L	Supply	SUP
Optical	OPT	Relief	REL	Surface	SUR
Original	ORIG	Remove	REM	Symbol	SYM
Ounce	OZ	Require	REQ	Symmetrical	SYM
Outlet	OUT.	Required	REQD	System	SYS
Outside diameter	OD	Return	RET.		
Outside face	OF	Revise	REV	Tangent	TAN.
Outside radius	OR	Revolution	REV	Taper	TPR
Overall	OA	Revolutions per minute	RPM	Technical	TECH
		Rheostat	RHEO	Temperature	TEMP
Pack	PK	Right	R	Template	TEMP
Packing	PKG	Right hand	RH	Tensile strength	TS
Parallel	PAR.	Rivet	RIV	Tension	TENS.
Part	PT	Rockwell hardness	RH	Thick	THK
Patent	PAT.	Roller bearing	RB	Thousand	M
Permanent	PERM	Room	RM	Thousand pound	KIP
Perpendicular	PERP	Root diameter	RD	Thread	THD
Photograph	PHOTO	Root mean square	RMS	Tolerance	TOL
Piece	PC	Rough	RGH	Tongue & groove	T&G
Pint	PT	Round	RD	Tool steel	TS
Pitch	P	Rubber	RUB.	Tooth	T
Pitch circle	PC			Total	TOT
Pitch diameter	PD	Safety	SAF	Transfer	TRANS
Plastic	PLSTC	Sand blast	SD BL	Typical	TYP
Plate	PL	Schedule	SCH		
Point	PT	Screen	SCRN	Ultimate	ULT
Polish	POL	Screw	SCR	Unit	U
Position	POS	Sea level	SL	Universal	UNIV
Positive	POS	Second	SEC		
Pound	LB	Section	SECT	Vacuum	VAC
Pounds per square inch	PSI	Separate	SEP	Valve	V
Power	PWR	Set screw	SS	Variable	VAR
					Cont.

Appendix 1
Abbreviations (ANSI Z 32.13) (cont.)

Word	Abbreviation	Word	Abbreviation	ABBREVIATIONS FOR COLORS	
Vertical	VERT	West	W	Amber	AMB
Volt	V	Width	W	Black	BLK
Voltmeter	VM	Wood	WD	Blue	BLU
Volume	VOL	Woodruff	WDF	Brown	BRN
		Wrought iron	WI	Green	GRN
Washer	WASH.			Orange	ORN
Watt	W	Yard	YD	White	WHT
Weight	WT	Year	YR	Yellow	YEL

Appendix 2
Conversion tables

Length conversions

Angstrom units	$\times\ 1 \times 10^{-10}$	= meters
	$\times\ 1 \times 10^{-4}$	= microns
	$\times\ 1.650\ 763\ 73 \times 10^{-4}$	= wavelengths of orange-red line of krypton 86
Cables	$\times\ 120$	= fathoms
	$\times\ 720$	= feet
	$\times\ 219.456$	= meters
Fathoms	$\times\ 6$	= feet
	$\times\ 1.828\ 8$	= meters
Feet	$\times\ 12$	= inches
	$\times\ 0.3048$	= meters
Furlongs	$\times\ 660$	= feet
	$\times\ 201.168$	= meters
	$\times\ 220$	= yards
Inches	$\times\ 2.54 \times 10^{8}$	= Angstroms
	$\times\ 25.4$	= millimeters
	$\times\ 8.333\ 33 \times 10^{-2}$	= feet
Kilometers	$\times\ 3.280\ 839 \times 10^{3}$	= feet
	$\times\ 0.62$	= miles
	$\times\ 0.539\ 956$	= nautical miles
	$\times\ 0.621\ 371$	= statute miles
	$\times\ 1.093\ 613 \times 10^{3}$	= yards
Light-years	$\times\ 9.460\ 55 \times 10^{12}$	= kilometers
	$\times\ 5.878\ 51 \times 10^{12}$	= statute miles
Meters	$\times\ 1 \times 10^{10}$	= Angstroms
	$\times\ 3.280\ 839\ 9$	= feet
	$\times\ 39.370\ 079$	= inches
	$\times\ 1.093\ 61$	= yards
Microns	$\times\ 10^{4}$	= Angstroms
	$\times\ 10^{-4}$	= centimeters
	$\times\ 10^{-6}$	= meters
Nautical Miles (International)	$\times\ 8.439\ 049$	= cables
	$\times\ 6.076\ 115\ 49 \times 10^{3}$	= feet
	$\times\ 1.852 \times 10^{3}$	= meters
	$\times\ 1.150\ 77$	= statute miles

Appendix 2
Conversion tables (cont.)

Length conversions

Statute Miles	$\times$ 5.280 $\times$ 10^3	= feet
	$\times$ 8	= furlongs
	$\times$ 6.336 0 $\times$ 10^4	= inches
	$\times$ 1.609 34	= kilometers
	$\times$ 8.689 7 $\times$ 10^{-1}	= nautical miles
Miles	$\times$ 10^{-3}	= inches
	$\times$ 2.54 $\times$ 10^{-2}	= millimeters
	$\times$ 25.4	= micrometers
	$\times$ 0.61	= kilometers
Yards	$\times$ 3	= feet
	$\times$ 9.144 $\times$ 10^{-1}	= meters
Feet/hour	$\times$ 3.048 $\times$ 10^{-4}	= kilometers/hour
	$\times$ 1.645 788 $\times$ 10^{-4}	= knots
Feet/minute	$\times$ 0.3048	= meters/minute
	$\times$ 5.08 $\times$ 10^{-3}	= meters/second
Feet/second	$\times$ 1.097 28	= kilometers/hour
	$\times$ 18.288	= meters/minute
Kilometers/hour	$\times$ 3.280 839 $\times$ 10^3	= feet/hour
	$\times$ 54.680 66	= feet/minute
	$\times$ 0.277 777	= meters/second
	$\times$ 0.621 371	= miles/hour
Kilometers/minute	$\times$ 3.280 839 $\times$ 10^3	= feet/minute
	$\times$ 37.282 27	= miles/hour
Knots	$\times$ 6.076 115 $\times$ 10^3	= feet/hour
	$\times$ 101.268 5	= feet/minute
	$\times$ 1.687 809	= feet/second
	$\times$ 1.852	= kilometers/hour
	$\times$ 30.866	= meters/minute
	$\times$ 0.514 4	= meters/second
	$\times$ 1.150 77	= statute miles/hour
Meters/hour	$\times$ 3.280 839	= feet/hour
	$\times$ 88	= feet/minute
	$\times$ 1.466	= feet/second
	$\times$ 1 $\times$ 10^{-3}	= kilometers/hour
	$\times$ 1.667 $\times$ 10^{-2}	= meters/minute
Feet/second^2	$\times$ 1.097 28	= kilometers/hour/second
	$\times$ 0.304 8	= meters/second^2

Area conversions

Acres	$\times$ 4.046 85 $\times$ 10^{-3}	= square kilometers
	$\times$ 4.046 856 $\times$ 10^3	= square meters
	$\times$ 4.356 0 $\times$ 10^4	= square feet
Ares	$\times$ 2.471 053 8 $\times$ 10^{-2}	= acres
	$\times$ 1	= square dekameters
	$\times$ 10^2	= square meters
Barns	$\times$ 1 $\times$ 10^{-28}	= square meters
Circular mils	$\times$ 1 $\times$ 10^{-6}	= circular inches
	$\times$ 5.067 074 8 $\times$ 10^{-4}	= square millimeters
	$\times$ 0.785 398 1	= square mils

Cont.

Appendix 2
Conversion tables (cont.)

Area conversions

Hectares	$\times$ 2.471 05	= acres
	$\times\ 10^2$	= ares
	$\times\ 10^4$	= square meters
Square feet	$\times$ 2.295 684 $\times\ 10^{-5}$	= acres
	$\times$ 9.290 3 $\times\ 10^{-4}$	= ares
	$\times$ 144	= square inches
	$\times$ 9.290 304 $\times\ 10^{-2}$	= square meters
Square inches	$\times$ 1.273 239 5 $\times\ 10^6$	= circular mils
	$\times$ 6.944 4 $\times\ 10^{-3}$	= square feet
	$\times$ 6.451 6 $\times\ 10^{-4}$	= square meters
Square kilometers	$\times$ 247.105 38	= acres
	$\times$ 1.076 391 0 $\times\ 10^7$	= square feet
	$\times$ 1.000	= cubic meters
	$\times$ 1.307 950 6	= cubic yards
	$\times$ 219.969	= imperial gallons

Volume conversions

Liters	$\times\ 10^3$	= cubic centimeters
	$\times$ 1.000 $\times\ 10^6$	= cubic millimeters
	$\times$ 1.000 $\times\ 10^{-3}$	= cubic meters
	$\times$ 61.023 74	= cubic inches
	$\times$ 3.531 5 $\times\ 10^{-2}$	= cubic feet
	$\times$ 1.307 95 $\times\ 10^{-3}$	= cubic yards
	$\times$ 0.22	= gallons
	$\times$ 0.219 969	= imperial gallons
	$\times$ 0.879 877	= imperial quarts
Imperial pints	$\times$ 0.125	= imperial gallons
	$\times$ 0.568 261	= liters
	$\times$ 20	= imperial fluid ounces
	$\times$ 0.5	= imperial quarts
	$\times$ 568.260 9	= cubic centimeters
Imperial quarts	$\times$ 1.136 52 $\times\ 10^3$	= cubic centimeters
	$\times$ 69.354 8	= cubic inches
	$\times$ 1.136 522 8	= liters

Power conversions

British Thermal Units/hour	$\times$ 2.928 7 $\times\ 10^{-4}$	= kilowatts
	$\times$ 0.292 875	= watts
BTU/minute	$\times$ 1.757 25 $\times\ 10^{-2}$	= kilowatts
BTU/pound	$\times$ 2.324 4	= joules/gram
BTU/second	$\times$ 1.413 91	= horsepower
	$\times$ 107.514	= kilogrammeters/second
	$\times$ 1.054 35	= kilowatts
	$\times$ 1.054 35 $\times\ 10^3$	= watts
Foot-pound-force/hour	$\times$ 5.050 $\times\ 10^{-7}$	= horsepower
	$\times$ 3.766 16 $\times\ 10^{-7}$	= kilowatts
Foot-pound-force/ minute	$\times$ 3.030 303 $\times\ 10^{-5}$	= horsepower
	$\times$ 2.259 70 $\times\ 10^{-2}$	= joules/second
	$\times$ 2.259 70 $\times\ 10^{-5}$	= kilowatts

Appendix 2
Conversion tables (cont.)

Power conversions

Horsepower	$\times$ 42.435 6	= BTU/minute
	$\times$ 550	= footpounds/second
	$\times$ 0.746	= kilowatts
	$\times$ 746	= joules/second
Kilogrammeters/second	$\times$ 9.806 65	= watts
Kilowatts	$\times$ 3.414 43 $\times$ 10^3	= BTU/hour
	$\times$ 2.655 22 $\times$ 10^6	= footpounds/hour
	$\times$ 4.425 37 $\times$ 10^4	= footpounds/minute
	$\times$ 737.562	= footpounds/second
	$\times$ 1.019 726 $\times$ 10^7	= gramcentimeters/second
	$\times$ 1.341 02	= horsepower
	$\times$ 3.6 $\times$ 10^6	= joules/hour
	$\times$ 10^3	= joules/second
	$\times$ 3.671 01 $\times$ 10^5	= kilogrammeters/hour
	$\times$ 999.835	= international watt
Watts	$\times$ 44.253 7	= footpounds/minute
	$\times$ 1.341 02 $\times$ 10^{-3}	= horsepower
	$\times$ 1	= joules/second

Time conversions

(No attempt has been made in this brief treatment to correlate solar, mean solar, sidereal, and mean sidereal days.)

Mean solar days	$\times$ 24	= mean solar hours
Mean solar hours	$\times$ 3.600 $\times$ 10^3	= mean solar seconds
	$\times$ 60	= mean solar minutes

Angle conversions

Degrees	$\times$ 60	= minutes
	$\times$ 1.745 329 3 $\times$ 10^{-2}	= radians
Degrees/foot	$\times$ 5.726 145 $\times$ 10^{-4}	= radians/centimeter
Degrees/minute	$\times$ 2.908 8 $\times$ 10^{-4}	= radians/second
	$\times$ 4.629 629 $\times$ 10^{-5}	= revolutions/second
Degrees/second	$\times$ 1.745 329 3 $\times$ 10^{-2}	= radians/second
	$\times$ 0.166	= revolutions/minute
	$\times$ 2.77 $\times$ 10^{-3}	= revolutions/second
Minutes	$\times$ 1.667 $\times$ 10^{-2}	= degrees
	$\times$ 2.908 8 $\times$ 10^{-4}	= radians
	$\times$ 60	= seconds
Radians	$\times$ 0.159 154	= circumferences
	$\times$ 57.295 77	= degrees
	$\times$ 3.437 746 $\times$ 10^3	= minutes
Seconds	$\times$ 2.777 $\times$ 10^{-4}	= degrees
	$\times$ 1.667 $\times$ 10^{-2}	= minutes
	$\times$ 4.848 136 8 $\times$ 10^{-6}	= radians
Steradians	$\times$ 0.159 154 9	= hemispheres
	$\times$ 7.957 74 $\times$ 10^{-2}	= spheres
	$\times$ 0.636 619 7	= spherical right angles

Cont.

Appendix 2
Conversion tables (cont.)

Mass conversions

Grains	$\times\ 6.479\ 8 \times 10^{-2}$	= grams
	$\times\ 2.285\ 71 \times 10^{-3}$	= avoirdupois ounces
Grams	$\times\ 15.432\ 358$	= grains
	$\times\ 3.527\ 396 \times 10^{-2}$	= avoirdupois ounces
	$\times\ 2.204\ 62 \times 10^{-3}$	= avoirdupois pounds
Kilograms	$\times\ 564.383\ 4$	= avoirdupois drams
	$\times\ 2.204\ 622\ 6$	= avoirdupois pounds
	$\times\ 2.2$	= pounds
	$\times\ 9.842\ 065 \times 10^{-4}$	= long tons
	$\times\ 10^{-3}$	= metric tons
	$\times\ 1.102\ 31 \times 10^{-3}$	= short tons
Avoirdupois ounces	$\times\ 28.349\ 5$	= grams
	$\times\ 6.25 \times 10^{-2}$	= avoirdupois pounds
	$\times\ 0.911\ 458$	= troy ounces
Avoirdupois pounds	$\times\ 256$	= drams
	$\times\ 4.535\ 923\ 7 \times 10^{2}$	= grams
	$\times\ 0.453\ 592\ 4$	= kilograms
	$\times\ 16$	= ounces
Long tons	$\times\ 2.24 \times 10^{3}$	= avoirdupois pounds
	$\times\ 1.106\ 046\ 9$	= metric tons
	$\times\ 1.12$	= short tons
Metric tons	$\times\ 10^{3}$	= kilograms
	$\times\ 2.204\ 622 \times 10^{3}$	= avoirdupois pounds
Short tons	$\times\ 2 \times 10^{3}$	= avoirdupois pounds
	$\times\ 907.184\ 74$	= kilograms

Force conversions

Dynes	$\times\ 10^{-5}$	= newtons
Newtons	$\times\ 10^{5}$	= dynes
	$\times\ 0.224\ 808$	= pounds-force
Pounds	$\times\ 4.448\ 22$	= newtons

Energy conversions

British Thermal Units (thermochemical)	$\times\ 1.054\ 35 \times 10^{3}$	= joules
	$\times\ 2.928\ 27 \times 10^{-4}$	= kilowatthours
	$\times\ 1.054\ 35 \times 10^{3}$	= wattseconds
Foot-pound-force	$\times\ 1.355\ 818\ 0$	= joules
	$\times\ 0.138\ 255$	= kilogramforce-meters
	$\times\ 3.766\ 16 \times 10^{-7}$	= kilowatthours
	$\times\ 1.355\ 818\ 0$	= newtonmeters
Joules	$\times\ 9.484\ 5 \times 10^{-4}$	= British Thermal Units
	$\times\ 0.737\ 562$	= foot-pounds-force
	$\times\ 0.101\ 971\ 6$	= kilogramforce-meters
	$\times\ 2.777\ 7 \times 10^{-7}$	= kilowatthours
	$\times\ 1$	= wattseconds

Appendix 2
Conversion tables (cont.)

Energy conversions

Kilogramforce-meters	$\times\ 9.287\ 7\ \times\ 10^{-3}$	= British Thermal Units
	$\times\ 7.233\ 01$	= foot-pounds-force
	$\times\ 9.806\ 65$	= joules
	$\times\ 9.806\ 65$	= newtonmeters
	$\times\ 2.724\ 0\ \times\ 10^{-3}$	= watthours
Kilowatthours	$\times\ 3.409\ 52\ \times\ 10^{3}$	= British Thermal Units
	$\times\ 2.655\ 22\ \times\ 10^{6}$	= foot-pounds-force
	$\times\ 1.341\ 02$	= horsepowerhours
	$\times\ 3.6\ \times\ 10^{6}$	= joules
	$\times\ 3.670\ 98\ \times\ 10^{5}$	= kilogramforce-meters
Newtonmeters	$\times\ 0.101\ 971$	= kilogramforce-meters
	$\times\ 0.737\ 562$	= poundforce-feet
Watthours	$\times\ 3.414\ 43$	= British Thermal Units
	$\times\ 2.655\ 22\ \times\ 10^{3}$	= foot-pounds-force
	$\times\ 3.6\ \times\ 10^{3}$	= joules
	$\times\ 3.670\ 98\ \times\ 10^{2}$	= kilogramforce-meters

Pressure conversions

Atmospheres	$\times\ 1.013\ 25$	= bars
	$\times\ 1.033\ 23\ \times\ 10^{3}$	= grams/square centimeter
	$\times\ 1.033\ 23\ \times\ 10^{7}$	= grams/square meter
	$\times\ 14.696\ 0$	= pounds/square inch
	$\times\ 760$	= torrs
	$\times\ 101$	= kilopascals
Bars	$\times\ 0.986\ 923$	= atmospheres
	$\times\ 10^{6}$	= baryes
	$\times\ 1.019\ 716\ \times\ 10^{7}$	= grams/square meter
	$\times\ 1.019\ 716\ \times\ 10^{4}$	= kilogramsforce/square meter
	$\times\ 14.503\ 8$	= poundsforce/square inch
Baryes	$\times\ 10^{-6}$	= bars
Inches of mercury	$\times\ 3.386\ 4\ \times\ 10^{-2}$	= bars
	$\times\ 345.316$	= kilogramsforce/square meter
	$\times\ 70.726\ 2$	= poundsforce/square foot
Pascal	$\times\ 1$	= newton/square meter

Appendix 3
Logarithms of numbers

N	0	1	2	3	4	5	6	7	8	9
1.0	.0000	.0043	.0086	.0128	.0170	.0212	.0253	.0294	.0334	.0374
1.1	.0414	.0453	.0492	.0531	.0569	.0607	.0645	.0682	.0719	.0755
1.2	.0792	.0828	.0864	.0899	.0934	.0969	.1004	.1038	.1072	.1106
1.3	.1139	.1173	.1206	.1239	.1271	.1303	.1335	.1367	.1399	.1430
1.4	.1461	.1492	.1523	.1553	.1584	.1614	.1644	.1673	.1703	.1732
1.5	.1761	.1790	.1818	.1847	.1875	.1903	.1931	.1959	.1987	.2014
1.6	.2041	.2068	.2095	.2122	.2148	.2175	.2201	.2227	.2253	.2279
1.7	.2304	.2330	.2355	.2380	.2405	.2430	.2455	.2480	.2504	.2529
1.8	.2553	.2577	.2601	.2625	.2648	.2672	.2695	.2718	.2742	.2765
1.9	.2788	.2810	.2833	.2856	.2878	.2900	.2923	.2945	.2967	.2989
2.0	.3010	.3032	.3054	.3075	.3096	.3118	.3139	.3160	.3181	.3201
2.1	.3222	.3243	.3263	.3284	.3304	.3324	.3345	.3365	.3385	.3404
2.2	.3424	.3444	.3464	.3483	.3502	.3522	.3541	.3560	.3579	.3598
2.3	.3617	.3636	.3655	.3674	.3692	.3711	.3729	.3747	.3766	.3784
2.4	.3802	.3820	.3838	.3856	.3874	.3892	.3909	.3927	.3945	.3962
2.5	.3979	.3997	.4014	.4031	.4048	.4065	.4082	.4099	.4116	.4133
2.6	.4150	.4166	.4183	.4200	.4216	.4232	.4249	.4265	.4281	.4298
2.7	.4314	.4330	.4346	.4362	.4378	.4393	.4409	.4425	.4440	.4456
2.8	.4472	.4487	.4502	.4518	.4533	.4548	.4564	.4579	.4594	.4609
2.9	.4624	.4639	.4654	.4669	.4683	.4698	.4713	.4728	.4742	.4757
3.0	.4771	.4786	.4800	.4814	.4829	.4843	.4857	.4871	.4886	.4900
3.1	.4914	.4928	.4942	.4955	.4969	.4983	.4997	.5011	.5024	.5038
3.2	.5051	.5065	.5079	.5092	.5105	.5119	.5132	.5145	.5159	.5172
3.3	.5185	.5198	.5211	.5224	.5237	.5250	.5263	.5276	.5289	.5302
3.4	.5315	.5328	.5340	.5353	.5366	.5378	.5391	.5403	.5416	.5428
3.5	.5441	.5453	.5465	.5478	.5490	.5502	.5514	.5527	.5539	.5551
3.6	.5563	.5575	.5587	.5599	.5611	.5623	.5635	.5647	.5658	.5670
3.7	.5682	.5694	.5705	.5717	.5729	.5740	.5752	.5763	.5775	.5786
3.8	.5798	.5809	.5821	.5832	.5843	.5855	.5866	.5877	.5888	.5899
3.9	.5911	.5922	.5933	.5944	.5955	.5966	.5977	.5988	.5999	.6010
4.0	.6021	.6031	.6042	.6053	.6064	.6075	.6085	.6096	.6107	.6117
4.1	.6128	.6138	.6149	.6160	.6170	.6180	.6191	.6201	.6212	.6222
4.2	.6232	.6243	.6253	.6263	.6274	.6284	.6294	.6304	.6314	.6325
4.3	.6335	.6345	.6355	.6365	.6375	.6385	.6395	.6405	.6415	.6425
4.4	.6435	.6444	.6454	.6464	.6474	.6484	.6493	.6503	.6513	.6522
4.5	.6532	.6542	.6551	.6561	.6571	.6580	.6590	.6599	.6609	.6618
4.6	.6628	.6637	.6646	.6656	.6665	.6675	.6684	.6693	.6702	.6712
4.7	.6721	.6730	.6739	.6749	.6758	.6767	.6776	.6785	.6794	.6803
4.8	.6812	.6821	.6830	.6839	.6848	.6857	.6866	.6875	.6884	.6893
4.9	.6902	.6911	.6920	.6928	.6937	.6946	.6955	.6964	.6972	.6981
5.0	.6990	.6998	.7007	.7016	.7024	.7033	.7042	.7050	.7059	.7067
5.1	.7076	.7084	.7093	.7101	.7110	.7118	.7126	.7135	.7143	.7152
5.2	.7160	.7168	.7177	.7185	.7193	.7202	.7210	.7218	.7226	.7235
5.3	.7243	.7251	.7259	.7267	.7275	.7284	.7292	.7300	.7308	7316
5.4	.7324	.7332	.7340	.7348	.7356	.7364	.7372	.7380	.7388	.7396
N	0	1	2	3	4	5	6	7	8	9

Appendix 3
Logarithms of numbers (cont.)

N	0	1	2	3	4	5	6	7	8	9
5.5	.7404	.7412	.7419	.7427	.7435	.7443	.7451	.7459	.7466	.7474
5.6	.7482	.7490	.7497	.7505	.7513	.7520	.7528	.7536	.7543	.7551
5.7	.7559	.7566	.7574	.7582	.7589	.7597	.7604	.7612	.7619	.7627
5.8	.7634	.7642	.7649	.7657	.7664	.7672	.7679	.7686	.7694	.7701
5.9	.7709	.7716	.7723	.7731	.7738	.7745	.7752	.7760	.7767	.7774
6.0	.7782	.7789	.7796	.7803	.7810	.7818	.7825	.7832	.7839	.7846
6.1	.7853	.7860	.7868	.7875	.7882	.7889	.7896	.7903	.7910	.7917
6.2	.7924	.7931	.7938	.7945	.7952	.7959	.7966	.7973	.7980	.7987
6.3	.7993	.8000	.8007	.8014	.8021	.8028	.8035	.8041	.8048	.8055
6.4	.8062	.8069	.8075	.8082	.8089	.8096	.8102	.8109	.8116	.8122
6.5	.8129	.8136	.8142	.8149	.8156	.8162	.8169	.8176	.8182	.8189
6.6	.8195	.8202	.8209	.8215	.8222	.8228	.8235	.8241	.8248	.8254
6.7	.8261	.8267	.8274	.8280	.8287	.8293	.8299	.8306	.8312	.8319
6.8	.8325	.8331	.8338	.8344	.8351	.8357	.8363	.8370	.8376	.8382
6.9	.8388	.8395	.8401	.8407	.8414	.8420	.8426	.8432	.8439	.8445
7.0	.8451	.8457	.8463	.8470	.8476	.8482	.8488	.8494	.8500	.8506
7.1	.8513	.8519	.8525	.8531	.8537	.8543	.8549	.8555	.8561	.8567
7.2	.8573	.8579	.8585	.8591	.8597	.8603	.8609	.8615	.8621	.8627
7.3	.8633	.8639	.8645	.8651	.8657	.8663	.8669	.8675	.8681	.8686
7.4	.8692	.8698	.8704	.8710	.8716	.8722	.8727	.8733	.8739	.8745
7.5	.8751	.8756	.8762	.8768	.8774	.8779	.8785	.8791	.8797	.8802
7.6	.8808	.8814	.8820	.8825	.8831	.8837	.8842	.8848	.8854	.8859
7.7	.8865	.8871	.8876	.8882	.8887	.8893	.8899	.8904	.8910	.8915
7.8	.8921	.8927	.8932	.8938	.8943	.8949	.8954	.8960	.8965	.8971
7.9	.8976	.8982	.8987	.8993	.8998	.9004	.9009	.9015	.9020	.9025
8.0	.9031	.9036	.9042	.9047	.9053	.9058	.9063	.9069	.9074	.9079
8.1	.9085	.9090	.9096	.9101	.9106	.9112	.9117	.9122	.9128	.9133
8.2	.9138	.9143	.9149	.9154	.9159	.9165	.9170	.9175	.9180	.9186
8.3	.9191	.9196	.9201	.9206	.9212	.9217	.9222	.9227	.9232	.9238
8.4	.9243	.9248	.9253	.9258	.9263	.9269	.9274	.9279	.9284	.9289
8.5	.9294	.9299	.9304	.9309	.9315	.9320	.9325	.9330	.9335	.9340
8.6	.9345	.9350	.9355	.9360	.9365	.9370	.9375	.9380	.9385	.9390
8.7	.9395	.9400	.9405	.9410	.9415	.9420	.9425	.9430	.9435	.9440
8.8	.9445	.9450	.9455	.9460	.9465	.9469	.9474	.9479	.9484	.9489
8.9	.9494	.9499	.9504	.9509	.9513	.9518	.9523	.9528	.9533	.9538
9.0	.9542	.9547	.9552	.9557	.9562	.9566	.9571	.9576	.9581	.9586
9.1	.9590	.9595	.9600	.9605	.9609	.9614	.9619	.9624	.9628	.9633
9.2	.9638	.9643	.9647	.9652	.9657	.9661	.9666	.9671	.9675	.9680
9.3	.9685	.9689	.9694	.9699	.9703	.9708	.9713	.9717	.9722	.9727
9.4	.9731	.9736	.9741	.9745	.9750	.9754	.9759	.9763	.9768	.9773
9.5	.9777	.9782	.9786	.9791	.9795	.9800	.9805	.9809	.9814	.9818
9.6	.9823	.9827	.9832	.9836	.9841	.9845	.9850	.9854	.9859	.9863
9.7	.9868	.9872	.9877	.9881	.9886	.9890	.9894	.9899	.9903	.9908
9.8	.9912	.9917	.9921	.9926	.9930	.9934	.9939	.9943	.9948	.9952
9.9	.9956	.9961	.9965	.9969	.9974	.9978	.9983	.9987	.9991	.9996
N	0	1	2	3	4	5	6	7	8	9

Appendix 4
Values of trigonometric functions

Degrees	Radians	Sine	Tangent	Cotangent	Cosine		
0° 00′	.0000	.0000	.0000		1.0000	1.5708	90° 00′
10′	.0029	.0029	.0029	343.77	1.0000	1.5679	50′
20′	.0058	.0058	.0058	171.89	1.0000	1.5650	40′
30′	.0087	.0087	.0087	114.59	1.0000	1.5621	30′
40′	.0116	.0116	.0116	85.940	.9999	1.5592	20′
50′	.0145	.0145	.0145	68.750	.9999	1.5563	10′
1° 00′	.0175	.0175	.0175	57.290	.9998	1.5533	89° 00′
10′	.0204	.0204	.0204	49.104	.9998	1.5504	50′
20′	.0233	.0233	.0233	42.964	.9997	1.5475	40′
30′	.0262	.0262	.0262	38.188	.9997	1.5446	30′
40′	.0291	.0291	.0291	34.368	.9996	1.5417	20′
50′	.0320	.0320	.0320	31.242	.9995	1.5388	10′
2° 00′	.0349	.0349	.0349	28.636	.9994	1.5359	88° 00′
10′	.0378	.0378	.0378	26.432	.9993	1.5330	50′
20′	.0407	.0407	.0407	24.542	.9992	1.5301	40′
30′	.0436	.0436	.0437	22.904	.9990	1.5272	30′
40′	.0465	.0465	.0466	21.470	.9989	1.5243	20′
50′	.0495	.0494	.0495	20.206	.9988	1.5213	10′
3° 00′	.0524	.0523	.0524	19.081	.9986	1.5184	87° 00′
10′	.0553	.0552	.0553	18.075	.9985	1.5155	50′
20′	.0582	.0581	.0582	17.169	.9983	1.5126	40′
30′	.0611	.0610	.0612	16.350	.9981	1.5097	30′
40′	.0640	.0640	.0641	15.605	.9980	1.5068	20′
50′	.0669	.0669	.0670	14.924	.9978	1.5039	10′
4° 00′	.0698	.0698	.0699	14.301	.9976	1.5010	86° 00′
10′	.0727	.0727	.0729	13.727	.9974	1.4981	50′
20′	.0756	.0756	.0758	13.197	.9971	1.4952	40′
30′	.0785	.0785	.0787	12.706	.9969	1.4923	30′
40′	.0814	.0814	.0816	12.251	.9967	1.4893	20′
50′	.0844	.0843	.0846	11.826	.9964	1.4864	10′
5° 00′	.0873	.0872	.0875	11.430	.9962	1.4835	85° 00′
10′	.0902	.0901	.0904	11.059	.9959	1.4806	50′
20′	.0931	.0929	.0934	10.712	.9957	1.4777	40′
30′	.0960	.0958	.0963	10.385	.9954	1.4748	30′
40′	.0989	.0987	.0992	10.078	.9951	1.4719	20′
50′	.1018	.1016	.1022	9.7882	.9948	1.4690	10′
6° 00′	.1047	.1045	.1051	9.5144	.9945	1.4661	84° 00′
10′	.1076	.1074	.1080	9.2553	.9942	1.4632	50′
20′	.1105	.1103	.1110	9.0098	.9939	1.4603	40′
30′	.1134	.1132	.1139	8.7769	.9936	1.4573	30′
40′	.1164	.1161	.1169	8.5555	.9932	1.4544	20′
50′	.1193	.1190	.1198	8.3450	.9929	1.4515	10′
7° 00′	.1222	.1219	.1228	8.1443	.9925	1.4486	83° 00′
10′	.1251	.1248	.1257	7.9530	.9922	1.4457	50′
20′	.1280	.1276	.1287	7.7704	.9918	1.4428	40′
30′	.1309	.1305	.1317	7.5958	.9914	1.4399	30′
40′	.1338	.1334	.1346	7.4287	.9911	1.4370	20′
50′	.1367	.1363	.1376	7.2687	.9907	1.4341	10′
8° 00′	.1396	.1392	.1405	7.1154	.9903	1.4312	82° 00′
10′	.1425	.1421	.1435	6.9682	.9899	1.4283	50′
20′	.1454	.1449	.1465	6.8269	.9894	1.4254	40′
30′	.1484	.1478	.1495	6.6912	.9890	1.4224	30′
40′	.1513	.1507	.1524	6.5606	.9886	1.4195	20′
50′	.1542	.1536	.1554	6.4348	.9881	1.4166	10′
9° 00′	.1571	.1564	.1584	6.3138	.9877	1.4137	81° 00′
		Cosine	Cotangent	Tangent	Sine	Radians	Degrees

Appendix 4
Values of trigonometric functions (cont.)

Degrees	Radians	Sine	Tangent	Cotangent	Cosine		
9° 00′	.1571	.1564	.1584	6.3138	.9877	1.4137	81° 00′
10′	.1600	.1593	.1614	6.1970	.9872	1.4108	50′
20′	.1629	.1622	.1644	6.0844	.9868	1.4079	40′
30′	.1658	.1650	.1673	5.9758	.9863	1.4050	30′
40′	.1687	.1679	.1703	5.8708	.9858	1.4021	20′
50′	.1716	.1708	.1733	5.7694	.9853	1.3992	10′
10° 00′	.1745	.1736	.1763	5.6713	.9848	1.3963	80° 00′
10′	.1774	.1765	.1793	5.5764	.9843	1.3934	50′
20′	.1804	.1794	.1823	5.4845	.9838	1.3904	40′
30′	.1833	.1822	.1853	5.3955	.9833	1.3875	30′
40′	.1862	.1851	.1883	5.3093	.9827	1.3846	20′
50′	.1891	.1880	.1914	5.2257	.9822	1.3817	10′
11° 00′	.1920	.1908	.1944	5.1446	.9816	1.3788	79° 00′
10′	.1949	.1937	.1974	5.0658	.9811	1.3759	50′
20′	.1978	.1965	.2004	4.9894	.9805	1.3730	40′
30′	.2007	.1994	.2035	4.9152	.9799	1.3701	30′
40′	.2036	.2022	.2065	4.8430	.9793	1.3672	20′
50′	.2065	.2051	.2095	4.7729	.9787	1.3643	10′
12° 00′	.2094	.2079	.2126	4.7046	.9781	1.3614	78° 00′
10′	.2123	.2108	.2156	4.6382	.9775	1.3584	50′
20′	.2153	.2136	.2186	4.5736	.9769	1.3555	40′
30′	.2182	.2164	.2217	4.5107	.9763	1.3526	30′
40′	.2211	.2193	.2247	4.4494	.9757	1.3497	20′
50′	.2240	.2221	.2278	4.3897	.9750	1.3468	10′
13° 00′	.2269	.2250	.2309	4.3315	.9744	1.3439	77° 00′
10′	.2298	.2278	.2339	4.2747	.9737	1.3410	50′
20′	.2327	.2306	.2370	4.2193	.9730	1.3381	40′
30′	.2356	.2334	.2401	4.1653	.9724	1.3352	30′
40′	.2385	.2363	.2432	4.1126	.9717	1.3323	20′
50′	.2414	.2391	.2462	4.0611	.9710	1.3294	10′
14° 00′	.2443	.2419	.2493	4.0108	.9703	1.3265	76° 00′
10′	.2473	.2447	.2524	3.9617	.9696	1.3235	50′
20′	.2502	.2476	.2555	3.9136	.9689	1.3206	40′
30′	.2531	.2504	.2586	3.8667	.9681	1.3177	30′
40′	.2560	.2532	.2617	3.8208	.9674	1.3148	20′
50′	.2589	.2560	.2648	3.7760	.9667	1.3119	10′
15° 00′	.2618	.2588	.2679	3.7321	.9659	1.3090	75° 00′
10′	.2647	.2616	.2711	3.6891	.9652	1.3061	50′
20′	.2676	.2644	.2742	3.6470	.9644	1.3032	40′
30′	.2705	.2672	.2773	3.6059	.9636	1.3003	30′
40′	.2734	.2700	.2805	3.5656	.9628	1.2974	20′
50′	.2763	.2728	.2836	3.5261	.9621	1.2945	10′
16° 00′	.2793	.2756	.2867	3.4874	.9613	1.2915	74° 00′
10′	.2822	.2784	.2899	3.4495	.9605	1.2886	50′
20′	.2851	.2812	.2931	3.4124	.9596	1.2857	40′
30′	.2880	.2840	.2962	3.3759	.9588	1.2828	30′
40′	.2909	.2868	.2994	3.3402	.9580	1.2799	20′
50′	.2938	.2896	.3026	3.3052	.9572	1.2770	10′
17° 00′	.2967	.2924	.3057	3.2709	.9563	1.2741	73° 00′
10′	.2996	.2952	.3089	3.2371	.9555	1.2712	50′
20′	.3025	.2979	.3121	3.2041	.9546	1.2683	40′
30′	.3054	.3007	.3153	3.1716	.9537	1.2654	30′
40′	.3083	.3035	.3185	3.1397	.9528	1.2625	20′
50′	.3113	.3062	.3217	3.1084	.9520	1.2595	10′
18° 00′	.3142	.3090	.3249	3.0777	.9511	1.2566	72° 00′
		Cosine	Cotangent	Tangent	Sine	Radians	Degrees

Cont.

Appendix 4
Values of trigonometric functions (cont.)

Degrees	Radians	Sine	Tangent	Cotangent	Cosine		
18° 00′	.3142	.3090	.3249	3.0777	.9511	1.2566	72° 00′
10′	.3171	.3118	.3281	3.0475	.9502	1.2537	50′
20′	.3200	.3145	.3314	3.0178	.9492	1.2508	40′
30′	.3229	.3173	.3346	2.9887	.9483	1.2479	30′
40′	.3258	.3201	.3378	2.9600	.9474	1.2450	20′
50′	.3287	.3228	.3411	2.9319	.9465	1.2421	10′
19° 00′	.3316	.3256	.3443	2.9042	.9455	1.2392	71° 00′
10′	.3345	.3283	.3476	2.8770	.9446	1.2363	50′
20′	.3374	.3311	.3508	2.8502	.9436	1.2334	40′
30′	.3403	.3338	.3541	2.8239	.9426	1.2305	30′
40′	.3432	.3365	.3574	2.7980	.9417	1.2275	20′
50′	.3462	.3393	.3607	2.7725	.9407	1.2246	10′
20° 00′	.3491	.3420	.3640	2.7475	.9397	1.2217	70° 00′
10′	.3520	.3448	.3673	2.7228	.9387	1.2188	50′
20′	.3549	.3475	.3706	2.6985	.9377	1.2159	40′
30′	.3578	.3502	.3739	2.6746	.9367	1.2130	30′
40′	.3607	.3529	.3772	2.6511	.9356	1.2101	20′
50′	.3636	.3557	.3805	2.6279	.9346	1.2072	10′
21° 00′	.3665	.3584	.3839	2.6051	.9336	1.2043	69° 00′
10′	.3694	.3611	.3872	2.5826	.9325	1.2014	50′
20′	.3723	.3638	.3906	2.5605	.9315	1.1985	40′
30′	.3752	.3665	.3939	2.5386	.9304	1.1956	30′
40′	.3782	.3692	.3973	2.5172	.9293	1.1926	20′
50′	.3811	.3719	.4006	2.4960	.9283	1.1897	10′
22° 00′	.3840	.3746	.4040	2.4751	.9272	1.1868	68° 00′
10′	.3869	.3773	.4074	2.4545	.9261	1.1839	50′
20′	.3898	.3800	.4108	2.4342	.9250	1.1810	40′
30′	.3927	.3827	.4142	2.4142	.9239	1.1781	30′
40′	.3956	.3854	.4176	2.3945	.9228	1.1752	20′
50′	.3985	.3881	.4210	2.3750	.9216	1.1723	10′
23° 00′	.4014	.3907	.4245	2.3559	.9205	1.1694	67° 00′
10′	.4043	.3934	.4279	2.3369	.9194	1.1665	50′
20′	.4072	.3961	.4314	2.3183	.9182	1.1636	40′
30′	.4102	.3987	.4348	2.2998	.9171	1.1606	30′
40′	.4131	.4014	.4383	2.2817	.9159	1.1577	20′
50′	.4160	.4041	.4417	2.2637	.9147	1.1548	10′
24° 00′	.4189	.4067	.4452	2.2460	.9135	1.1519	66° 00′
10′	.4218	.4094	.4487	2.2286	.9124	1.1490	50′
20′	.4247	.4120	.4522	2.2113	.9112	1.1461	40′
30′	.4276	.4147	.4557	2.1943	.9100	1.1432	30′
40′	.4305	.4173	.4592	2.1775	.9088	1.1403	20′
50′	.4334	.4200	.4628	2.1609	.9075	1.1374	10′
25° 00′	.4363	.4226	.4663	2.1445	.9063	1.1345	65° 00′
10′	.4392	.4253	.4699	2.1283	.9051	1.1316	50′
20′	.4422	.4279	.4734	2.1123	.9038	1.1286	40′
30′	.4451	.4305	.4770	2.0965	.9026	1.1257	30′
40′	.4480	.4331	.4806	2.0809	.9013	1.1228	20′
50′	.4509	.4358	.4841	2.0655	.9001	1.1199	10′
26° 00′	.4538	.4384	.4877	2.0503	.8988	1.1170	64° 00′
10′	.4567	.4410	.4913	2.0353	.8975	1.1141	50′
20′	.4596	.4436	.4950	2.0204	.8962	1.1112	40′
30′	.4625	.4462	.4986	2.0057	.8949	1.1083	30′
40′	.4654	.4488	.5022	1.9912	.8936	1.1054	20′
50′	.4683	.4514	.5059	1.9768	.8923	1.1025	10′
27° 00′	.4712	.4540	.5095	1.9626	.8910	1.0996	63° 00′
		Cosine	Cotangent	Tangent	Sine	Radians	Degrees

Appendix 4
Values of trigonometric functions (cont.)

Degrees	Radians	Sine	Tangent	Cotangent	Cosine		
27° 00′	.4712	.4540	.5095	1.9626	.8910	1.0996	63° 00′
10′	.4741	.4566	.5132	1.9486	.8897	1.0966	50′
20′	.4771	.4592	.5169	1.9347	.8884	1.0937	40′
30′	.4800	.4617	.5206	1.9210	.8870	1.0908	30′
40′	.4829	.4643	.5243	1.9074	.8857	1.0879	20′
50′	.4858	.4669	.5280	1.8940	.8843	1.0850	10′
28° 00′	.4887	.4695	.5317	1.8807	.8829	1.0821	62° 00′
10′	.4916	.4720	.5354	1.8676	.8816	1.0792	50′
20′	.4945	.4746	.5392	1.8546	.8802	1.0763	40′
30′	.4974	.4772	.5430	1.8418	.8788	1.0734	30′
40′	.5003	.4797	.5467	1.8291	.8774	1.0705	20′
50′	.5032	.4823	.5505	1.8165	.8760	1.0676	10′
29° 00′	.5061	.4848	.5543	1.8040	.8746	1.0647	61° 00′
10′	.5091	.4874	.5581	1.7917	.8732	1.0617	50′
20′	.5120	.4899	.5619	1.7796	.8718	1.0588	40′
30′	.5149	.4924	.5658	1.7675	.8704	1.0559	30′
40′	.5178	.4950	.5696	1.7556	.8689	1.0530	20′
50′	.5207	.4975	.5735	1.7437	.8675	1.0501	10′
30° 00′	.5236	.5000	.5774	1.7321	.8660	1.0472	60° 00′
10′	.5265	.5025	.5812	1.7205	.8646	1.0443	50′
20′	.5294	.5050	.5851	1.7090	.8631	1.0414	40′
30′	.5323	.5075	.5890	1.6977	.8616	1.0385	30′
40′	.5352	.5100	.5930	1.6864	.8601	1.0356	20′
50′	.5381	.5125	.5969	1.6753	.8587	1.0327	10′
31° 00′	.5411	.5150	.6009	1.6643	.8572	1.0297	59° 00′
10′	.5440	.5175	.6048	1.6534	.8557	1.0268	50′
20′	.5469	.5200	.6088	1.6426	.8542	1.0239	40′
30′	.5498	.5225	.6128	1.6319	.8526	1.0210	30′
40′	.5527	.5250	.6168	1.6212	.8511	1.0181	20′
50′	.5556	.5275	.6208	1.6107	.8496	1.0152	10′
32° 00′	.5585	.5299	.6249	1.6003	.8480	1.0123	58° 00′
10′	.5614	.5324	.6289	1.5900	.8465	1.0094	50′
20′	.5643	.5348	.6330	1.5798	.8450	1.0065	40′
30′	.5672	.5373	.6371	1.5697	.8434	1.0036	30′
40′	.5701	.5398	.6412	1.5597	.8418	1.0007	20′
50′	.5730	.5422	.6453	1.5497	.8403	.9977	10′
33° 00′	.5760	.5446	.6494	1.5399	.8387	.9948	57° 00′
10′	.5789	.5471	.6536	1.5301	.8371	.9919	50′
20′	.5818	.5495	.6577	1.5204	.8355	.9890	40′
30′	.5847	.5519	.6619	1.5108	.8339	.9861	30′
40′	.5876	.5544	.6661	1.5013	.8323	.9832	20′
50′	.5905	.5568	.6703	1.4919	.8307	.9803	10′
34° 00′	.5934	.5592	.6745	1.4826	.8290	.9774	56° 00′
10′	.5963	.5616	.6787	1.4733	.8274	.9745	50′
20′	.5992	.5640	.6830	1.4641	.8258	.9716	40′
30′	.6021	.5664	.6873	1.4550	.8241	.9687	30′
40′	.6050	.5688	.6916	1.4460	.8225	.9657	20′
50′	.6080	.5712	.6959	1.4370	.8208	.9628	10′
35° 00′	.6109	.5736	.7002	1.4281	.8192	.9599	55° 00′
10′	.6138	.5760	.7046	1.4193	.8175	.9570	50′
20′	.6167	.5783	.7089	1.4106	.8158	.9541	40′
30′	.6196	.5807	.7133	1.4019	.8141	.9512	30′
40′	.6225	.5831	.7177	1.3934	.8124	.9483	20′
50′	.6254	.5854	.7221	1.3848	.8107	.9454	10′
36° 00′	.6283	.5878	.7265	1.3764	.8090	.9425	54° 00′
		Cosine	Cotangent	Tangent	Sine	Radians	Degrees

Cont.

Appendix 4
Values of trigonometric functions (cont.)

Degrees	Radians	Sine	Tangent	Cotangent	Cosine		
36° 00′	.6283	.5878	.7265	1.3764	.8090	.9425	54° 00′
10′	.6312	.5901	.7310	1.3680	.8073	.9396	50′
20′	.6341	.5925	.7355	1.3597	.8056	.9367	40′
30′	.6370	.5948	.7400	1.3514	.8039	.9338	30′
40′	.6400	.5972	.7445	1.3432	.8021	.9308	20′
50′	.6429	.5995	.7490	1.3351	.8004	.9279	10′
37° 00′	.6458	.6018	.7536	1.3270	.7986	.9250	53° 00′
10′	.6487	.6041	.7581	1.3190	.7969	.9221	50′
20′	.6516	.6065	.7627	1.3111	.7951	.9192	40′
30′	.6545	.6088	.7673	1.3032	.7934	.9163	30′
40′	.6574	.6111	.7720	1.2954	.7916	.9134	20′
50′	.6603	.6134	.7766	1.2876	.7898	.9105	10′
38° 00′	.6632	.6157	.7813	1.2799	.7880	.9076	52° 00′
10′	.6661	.6180	.7860	1.2723	.7862	.9047	50′
20′	.6690	.6202	.7907	1.2647	.7844	.9018	40′
30′	.6720	.6225	.7954	1.2572	.7826	.8988	30′
40′	.6749	.6248	.8002	1.2497	.7808	.8959	20′
50′	.6778	.6271	.8050	1.2423	.7790	.8930	10′
39° 00′	.6807	.6293	.8098	1.2349	.7771	.8901	51° 00′
10′	.6836	.6316	.8146	1.2276	.7753	.8872	50′
20′	.6865	.6338	.8195	1.2203	.7735	.8843	40′
30′	.6894	.6361	.8243	1.2131	.7716	.8814	30′
40′	.6923	.6383	.8292	1.2059	.7698	.8785	20′
50′	.6952	.6406	.8342	1.1988	.7679	.8756	10′
40° 00′	.6981	.6428	.8391	1.1918	.7660	.8727	50° 00′
10′	.7010	.6450	.8441	1.1847	.7642	.8698	50′
20′	.7039	.6472	.8491	1.1778	.7623	.8668	40′
30′	.7069	.6494	.8541	1.1708	.7604	.8639	30′
40′	.7098	.6517	.8591	1.1640	.7585	.8610	20′
50′	.7127	.6539	.8642	1.1571	.7566	.8581	10′
41° 00′	.7156	.6561	.8693	1.1504	.7547	.8552	49° 00′
10′	.7185	.6583	.8744	1.1436	.7528	.8523	50′
20′	.7214	.6604	.8796	1.1369	.7509	.8494	40′
30′	.7243	.6626	.8847	1.1303	.7490	.8465	30′
40′	.7272	.6648	.8899	1.1237	.7470	.8436	20′
50′	.7301	.6670	.8952	1.1171	.7451	.8407	10′
42° 00′	.7330	.6691	.9004	1.1106	.7431	.8378	48° 00′
10′	.7359	.6713	.9057	1.1041	.7412	.8348	50′
20′	.7389	.6734	.9110	1.0977	.7392	.8319	40′
30′	.7418	.6756	.9163	1.0913	.7373	.8290	30′
40′	.7447	.6777	.9217	1.0850	.7353	.8261	20′
50′	.7476	.6799	.9271	1.0786	.7333	.8232	10′
43° 00′	.7505	.6820	.9325	1.0724	.7314	.8203	47° 00′
10′	.7534	.6841	.9380	1.0661	.7294	.8174	50′
20′	.7563	.6862	.9435	1.0599	.7274	.8145	40′
30′	.7592	.6884	.9490	1.0538	.7254	.8116	30′
40′	.7621	.6905	.9545	1.0477	.7234	.8087	20′
50′	.7650	.6926	.9601	1.0416	.7214	.8058	10′
44° 00′	.7679	.6947	.9657	1.0355	.7193	.8029	46° 00′
10′	.7709	.6967	.9713	1.0295	.7173	.7999	50′
20′	.7738	.6988	.9770	1.0235	.7153	.7970	40′
30′	.7767	.7009	.9827	1.0176	.7133	.7941	30′
40′	.7796	.7030	.9884	1.0117	.7112	.7912	20′
50′	.7825	.7050	.9942	1.0058	.7092	.7883	10′
45° 00′	.7854	.7071	1.0000	1.0000	.7071	.7854	45° 00′
		Cosine	Cotangent	Tangent	Sine	Radians	Degrees

Appendix 5
Weights and measures

UNITED STATES SYSTEM

LINEAR MEASURE

Inches	Feet	Yards	Rods	Furlongs	Miles
1.0 =	.08333 =	.02778 =	.0050505 =	.00012626 =	.00001578
12.0 =	1.0 =	.33333 =	.0606061 =	.00151515 =	.00018939
36.0 =	3.0 =	1.0 =	.1818182 =	.00454545 =	.00056818
198.0 =	16.5 =	5.5 =	1.0 =	.025 =	.003125
7920.0 =	660.0 =	220.0 =	40.0 =	1.0 =	.125
63360.0 =	5280.0 =	1760.0 =	320.0 =	8.0 =	1.0

SQUARE AND LAND MEASURE

Sq. Inches	Square Feet	Square Yards	Sq. Rods	Acres	Sq. Miles
1.0 =	.006944 =	.000772			
144.0 =	1.0 =	.111111			
1296.0 =	9.0 =	1.0 =	.03306 =	.000207	
39204.0 =	272.25 =	30.25 =	1.0 =	.00625 =	.0000098
	43560.0 =	4840.0 =	160.0 =	1.0 =	.0015625
		3097600.0 =	102400.0 =	640.0 =	1.0

AVOIRDUPOIS WEIGHTS

Grains	Drams	Ounces	Pounds	Tons
1.0 =	.03657 =	.002286 =	.000143 =	.0000000714
27.34375 =	1.0 =	.0625 =	.003906 =	.00000195
437.5 =	16.0 =	1.0 =	.0625 =	.00003125
7000.0 =	256.0 =	16.0 =	1.0 =	.0005
14000000.0 =	512000.0 =	32000.0 =	2000.0 =	1.0

DRY MEASURE

Pints	Quarts	Pecks	Cubic Feet	Bushels
1.0 =	.5 =	.0625 =	.01945 =	.01563
2.0 =	1.0 =	.125 =	.03891 =	.03125
16.0 =	8.0 =	1.0 =	.31112 =	.25
51.42627 =	25.71314 =	3.21414 =	1.0 =	.80354
64.0 =	32.0 =	4.0 =	1.2445 =	1.0

LIQUID MEASURE

Gills	Pints	Quarts	U. S. Gallons	Cubic Feet
1.0 =	.25 =	.125 =	.03125 =	.00418
4.0 =	1.0 =	.5 =	.125 =	.01671
8.0 =	2.0 =	1.0 =	.250 =	.03342
32.0 =	8.0 =	4.0 =	1.0 =	.1337
			7.48052 =	1.0

METRIC SYSTEM

UNITS

Length—Meter   :   Mass—Gram   :   Capacity—Liter
for pure water at 4°C. (39.2°F.)
1 cubic decimeter or 1 liter = 1 kilogram

1000 Milli $\begin{Bmatrix} meters\ (mm) \\ grams\ (mg) \\ liters\ (ml) \end{Bmatrix}$ = 100 Centi $\begin{Bmatrix} meters\ (cm) \\ grams\ (cg) \\ liters\ (cl) \end{Bmatrix}$ = 10 Deci $\begin{Bmatrix} meters\ (dm) \\ grams\ (dg) \\ liters\ (dl) \end{Bmatrix}$ = 1 $\begin{Bmatrix} meter \\ gram \\ liter \end{Bmatrix}$

1000 $\begin{Bmatrix} meters \\ grams \\ liters \end{Bmatrix}$ = 100 Deka $\begin{Bmatrix} meters\ (dkm) \\ grams\ (dkg) \\ liters\ (dkl) \end{Bmatrix}$ = 10 Hecto $\begin{Bmatrix} meters\ (hm) \\ grams\ (hg) \\ liters\ (hl) \end{Bmatrix}$ = 1 Kilo $\begin{Bmatrix} meter\ (km) \\ gram\ (kg) \\ liter\ (kl) \end{Bmatrix}$

1 Metric Ton	= 1000 Kilograms
100 Square Meters	= 1 Are
100 Ares	= 1 Hectare
100 Hectares	= 1 Square Kilometer

Appendix 6
Decimal equivalents and temperature conversions

DECIMAL EQUIVALENTS—INCH-MILLIMETER CONVERSION TABLE

1/2	1/4	1/8	1/16	1/32	1/64	Decimals	Millimeters
					1	.015625	.396875
				1		.031250	.793750
					3	.046875	1.190625
			1			.062500	1.587500
					5	.078125	1.984375
				3		.093750	2.381250
					7	.109375	2.778125
		1				.125000	3.175000
					9	.140625	3.571875
				5		.156250	3.968750
					11	.171875	4.365625
			3			.187500	4.762500
					13	.203125	5.159375
				7		.218750	5.556250
					15	.234375	5.953125
	1					.250000	6.350000
					17	.265625	6.746875
				9		.281250	7.143750
					19	.296875	7.540625
			5			.312500	7.937500
					21	.328125	8.334375
				11		.343750	8.731250
					23	.359375	9.128125
		3				.375000	9.525000
					25	.390625	9.921875
				13		.406250	10.318750
					27	.421875	10.715625
			7			.437500	11.112500
					29	.453125	11.509375
				15		.468750	11.906250
					31	.484375	12.303125
1						.500000	12.700000

1/2	1/4	1/8	1/16	1/32	1/64	Decimals	Millimeters
					33	.515625	13.096875
				17		.531250	13.493750
					35	.546875	13.890625
			9			.562500	14.287500
					37	.578125	14.684375
				19		.593750	15.081250
					39	.609375	15.478125
		5				.625000	15.875000
					41	.640625	16.271875
				21		.656250	16.668750
					43	.671875	17.065625
			11			.687500	17.462500
					45	.703125	17.859375
				23		.718750	18.256250
					47	.734375	18.653125
	3					.750000	19.050000
					49	.765625	19.446875
				25		.781250	19.843750
					51	.796875	20.240625
			13			.812500	20.637500
					53	.828125	21.034375
				27		.843750	21.431250
					55	.859375	21.828125
		7				.875000	22.225000
					57	.890625	22.621875
				29		.906250	23.018750
					59	.921875	23.415625
			15			.937500	23.812500
					61	.953125	24.209375
				31		.968750	24.606250
					63	.984375	25.003125
2	4	8	16	32	64	1.000000	25.400000

Appendix 7
Weights and specific gravities

Substance	Weight Lb. per Cu. Ft.	Specific Gravity	Substance	Weight Lb. per Cu. Ft.	Specific Gravity
METALS, ALLOYS, ORES			**TIMBER, U. S. SEASONED**		
Aluminum, cast, hammered	165	2.55-2.75	**Moisture Content by Weight:**		
Brass, cast, rolled	534	8.4-8.7	Seasoned timber 15 to 20%		
Bronze, 7.9 to 14% Sn	509	7.4-8.9	Green timber up to 50%		
Bronze, aluminum	481	7.7	Ash, white, red	40	0.62-0.65
Copper, cast, rolled	556	8.8-9.0	Cedar, white, red	22	0.32-0.38
Copper ore, pyrites	262	4.1-4.3	Chestnut	41	0.66
Gold, cast, hammered	1205	19.25-19.3	Cypress	30	0.48
Iron, cast, pig	450	7.2	Fir, Douglas spruce	32	0.51
Iron, wrought	485	7.6-7.9	Fir, eastern	25	0.40
Iron, spiegel-eisen	468	7.5	Elm, white	45	0.72
Iron, ferro-silicon	437	6.7-7.3	Hemlock	29	0.42-0.52
Iron ore, hematite	325	5.2	Hickory	49	0.74-0.84
Iron ore, hematite in bank	160-180	....	Locust	46	0.73
Iron ore, hematite loose	130-160	....	Maple, hard	43	0.68
Iron ore, limonite	237	3.6-4.0	Maple, white	33	0.53
Iron ore, magnetite	315	4.9-5.2	Oak, chestnut	54	0.86
Iron slag	172	2.5-3.0	Oak, live	59	0.95
Lead	710	11.37	Oak, red, black	41	0.65
Lead ore, galena	465	7.3-7.6	Oak, white	46	0.74
Magnesium, alloys	112	1.74-1.83	Pine, Oregon	32	0.51
Manganese	475	7.2-8.0	Pine, red	30	0.48
Manganese ore, pyrolusite	259	3.7-4.6	Pine, white	26	0.41
Mercury	849	13.6	Pine, yellow, long-leaf	44	0.70
Monel Metal	556	8.8-9.0	Pine, yellow, short-leaf	38	0.61
Nickel	565	8.9-9.2	Poplar	30	0.48
Platinum, cast, hammered	1330	21.1-21.5	Redwood, California	26	0.42
Silver, cast, hammered	656	10.4-10.6	Spruce, white, black	27	0.40-0.46
Steel, rolled	490	7.85	Walnut, black	38	0.61
Tin, cast, hammered	459	7.2-7.5			
Tin ore, cassiterite	418	6.4-7.0			
Zinc, cast, rolled	440	6.9-7.2			
Zinc ore, blende	253	3.9-4.2	**VARIOUS LIQUIDS**		
			Alcohol, 100%	49	0.79
			Acids, muriatic 40%	75	1.20
VARIOUS SOLIDS			Acids, nitric 91%	94	1.50
			Acids, sulphuric 87%	112	1.80
Cereals, oats....bulk	32	....	Lye, soda 66%	106	1.70
Cereals, barley....bulk	39	....	Oils, vegetable	58	0.91-0.94
Cereals, corn, rye....bulk	48	....	Oils, mineral, lubricants	57	0.90-0.93
Cereals, wheat....bulk	48	....	Water, 4°C. max. density	62.428	1.0
Hay and Straw....bales	20	....	Water, 100°C.	59.830	0.9584
Cotton, Flax, Hemp	93	1.47-1.50	Water, ice	56	0.88-0.92
Fats	58	0.90-0.97	Water, snow, fresh fallen	8	.125
Flour, loose	28	0.40-0.50	Water, sea water	64	1.02-1.03
Flour, pressed	47	0.70-0.80			
Glass, common	156	2.40-2.60			
Glass, plate or crown	161	2.45-2.72	**GASES**		
Glass, crystal	184	2.90-3.00			
Leather	59	0.86-1.02	Air, 0°C. 760 mm.	.08071	1.0
Paper	58	0.70-1.15	Ammonia	.0478	0.5920
Potatoes, piled	42		Carbon dioxide	.1234	1.5291
Rubber, caoutchouc	59	0.92-0.96	Carbon monoxide	.0781	0.9673
Rubber goods	94	1.0-2.0	Gas, illuminating	.028-.036	0.35-0.45
Salt, granulated, piled	48	....	Gas, natural	.038-.039	0.47-0.48
Saltpeter	67	....	Hydrogen	.00559	0.0693
Starch	96	1.53	Nitrogen	.0784	0.9714
Sulphur	125	1.93-2.07	Oxygen	.0892	1.1056
Wool.	82	1.32			

The specific gravities of solids and liquids refer to water at 4°C., those of gases to air at 0°C. and 760 mm. pressure. The weights per cubic foot are derived from average specific gravities, except where stated that weights are for bulk, heaped or loose material, etc.

(Courtesy of the American Institute of Steel Construction.)

Appendix 7
Weights and specific gravities (cont.)

Substance	Weight Lb. per Cu. Ft.	Specific Gravity	Substance	Weight Lb. per Cu. Ft.	Specific Gravity
ASHLAR MASONRY			**MINERALS**		
Granite, syenite, gneiss......	165	2.3-3.0	Asbestos....................	153	2.1-2.8
Limestone, marble............	160	2.3-2.8	Barytes....................	281	4.50
Sandstone, bluestone..........	140	2.1-2.4	Basalt....................	184	2.7-3.2
			Bauxite....................	159	2.55
MORTAR RUBBLE			Borax....................	109	1.7-1.8
MASONRY			Chalk....................	137	1.8-2.6
Granite, syenite, gneiss......	155	2.2-2.8	Clay, marl....................	137	1.8-2.6
Limestone, marble............	150	2.2-2.6	Dolomite....................	181	2.9
Sandstone, bluestone..........	130	2.0-2.2	Feldspar, orthoclase............	159	2.5-2.6
			Gneiss, serpentine............	159	2.4-2.7
DRY RUBBLE MASONRY			Granite, syenite............	175	2.5-3.1
Granite, syenite, gneiss......	130	1.9-2.3	Greenstone, trap............	187	2.8-3.2
Limestone, marble............	125	1.9-2.1	Gypsum, alabaster............	159	2.3-2.8
Sandstone, bluestone..........	110	1.8-1.9	Hornblende....................	187	3.0
			Limestone, marble............	165	2.5-2.8
BRICK MASONRY			Magnesite....................	187	3.0
Pressed brick	140	2.2-2.3	Phosphate rock, apatite......	200	3.2
Common brick................	120	1.8-2.0	Porphyry....................	172	2.6-2.9
Soft brick....................	100	1.5-1.7	Pumice, natural............	40	0.37-0.90
			Quartz, flint....................	165	2.5-2.8
CONCRETE MASONRY			Sandstone, bluestone..........	147	2.2-2.5
Cement, stone, sand......	144	2.2-2.4	Shale, slate....................	175	2.7-2.9
Cement, slag, etc............	130	1.9-2.3	Soapstone, talc............	169	2.6-2.8
Cement, cinder, etc.........	100	1.5-1.7			
			STONE, QUARRIED, PILED		
VARIOUS BUILDING			Basalt, granite, gneiss........	96	....
MATERIALS			Limestone, marble, quartz	95	....
Ashes, cinders....................	40-45	....	Sandstone....................	82	....
Cement, portland, loose......	90		Shale....................	92	....
Cement, portland, set........	183	2.7-3.2	Greenstone, hornblende......	107	....
Lime, gypsum, loose............	53-64				
Mortar, set....................	103	1.4-1.9			
Slags, bank slag....................	67-72	....	**BITUMINOUS SUBSTANCES**		
Slags, bank screenings........	98-117	....	Asphaltum....................	81	1.1-1.5
Slags, machine slag............	96	....	Coal, anthracite............	97	1.4-1.7
Slags, slag sand............	49-55	....	Coal, bituminous............	84	1.2-1.5
			Coal, lignite....................	78	1.1-1.4
EARTH, ETC., EXCAVATED			Coal, peat, turf, dry............	47	0.65-0.85
Clay, dry....................	63	....	Coal, charcoal, pine............	23	0.28-0.44
Clay, damp, plastic............	110	....	Coal, charcoal, oak............	33	0.47-0.57
Clay and gravel, dry............	100	....	Coal, coke....................	75	1.0-1.4
Earth, dry, loose............	76	....	Graphite....................	131	1.9-2.3
Earth, dry, packed............	95	....	Paraffine....................	56	0.87-0.91
Earth, moist, loose............	78	....	Petroleum....................	54	0.87
Earth, moist, packed............	96	....	Petroleum, refined............	50	0.79-0.82
Earth, mud, flowing............	108	....	Petroleum, benzine............	46	0.73-0.75
Earth, mud, packed............	115	....	Petroleum, gasoline............	42	0.66-0.69
Riprap, limestone............	80-85	....	Pitch....................	69	1.07-1.15
Riprap, sandstone............	90	....	Tar, bituminous............	75	1.20
Riprap, shale............	105	....			
Sand, gravel, dry, loose......	90-105	....			
Sand, gravel, dry, packed....	100-120	....			
Sand, gravel, dry, wet............	118-120	....			
EXCAVATIONS IN WATER					
Sand or gravel....................	60	....	**COAL AND COKE, PILED**		
Sand or gravel and clay......	65	....	Coal, anthracite............	47-58	....
Clay....................	80	....	Coal, bituminous, lignite..	40-54	....
River mud....................	90	....	Coal, peat, turf............	20-26	....
Soil....................	70	....	Coal, charcoal............	10-14	....
Stone riprap....................	65	....	Coal, coke....................	23-32	....

The specific gravities of solids and liquids refer to water at 4°C., those of gases to air at 0°C. and 760 mm. pressure. The weights per cubic foot are derived from average specific gravities, except where stated that weights are for bulk, heaped or loose material, etc.

Appendix 8
Wire and sheet metal gages

WIRE AND SHEET METAL GAGES
IN DECIMALS OF AN INCH

Name of Gage	United States Standard Gage*		The United States Steel Wire Gage	American or Brown & Sharpe Wire Gage	New Birmingham Standard Sheet & Hoop Gage	British Imperial or English Legal Standard Wire Gage	Birmingham or Stubs Iron Wire Gage	Name of Gage
Principal Use	Uncoated Steel Sheets and Light Plates		Steel Wire except Music Wire	Non-Ferrous Sheets and Wire	Iron and Steel Sheets and Hoops	Wire	Strips, Bands, Hoops and Wire	Principal Use
Gage No.	Weight Oz. per Sq. Ft.	Approx. Thickness Inches	Thickness, Inches					Gage No.
7/0's			.4900		.6666	.500		7/0's
6/0's			.4615	.5800	.625	.464		6/0's
5/0's			.4305	.5165	.5883	.432	.500	5/0's
4/0's			.3938	.4600	.5416	.400	.454	4/0's
3/0's			.3625	.4096	.500	.372	.425	3/0's
2/0's			.3310	.3648	.4452	.348	.380	2/0's
0			.3065	.3249	.3964	.324	.340	0
1			.2830	.2893	.3532	.300	.300	1
2			.2625	.2576	.3147	.276	.284	2
3	160	.2391	.2437	.2294	.2804	.252	.259	3
4	150	.2242	.2253	.2043	.250	.232	.238	4
5	140	.2092	.2070	.1819	.2225	.212	.220	5
6	130	.1943	.1920	.1620	.1981	.192	.203	6
7	120	.1793	.1770	.1443	.1764	.176	.180	7
8	110	.1644	.1620	.1285	.1570	.160	.165	8
9	100	.1495	.1483	.1144	.1398	.144	.148	9
10	90	.1345	.1350	.1019	.1250	.128	.134	10
11	80	.1196	.1205	.0907	.1113	.116	.120	11
12	70	.1046	.1055	.0808	.0991	.104	.109	12
13	60	.0897	.0915	.0720	.0882	.092	.095	13
14	50	0747	.0800	.0641	.0785	.080	.083	14
15	45	.0673	.0720	.0571	.0699	.072	.072	15
16	40	.0598	.0625	.0508	.0625	.064	.065	16
17	36	.0538	.0540	.0453	.0556	.056	.058	17
18	32	.0478	.0475	.0403	.0495	.048	.049	18
19	28	.0418	.0410	.0359	.0440	.040	.042	19
20	24	.0359	.0348	.0320	.0392	.036	.035	20
21	22	.0329	.0318	.0285	.0349	.032	.032	21
22	20	.0299	.0286	.0253	.0313	.028	.028	22
23	18	.0269	.0258	.0226	.0278	.024	.025	23
24	16	.0239	.0230	.0201	.0248	.022	.022	24
25	14	.0209	.0204	.0179	.0220	.020	.020	25
26	12	.0179	.0181	.0159	.0196	.018	.018	26
27	11	.0164	.0173	.0142	.0175	.0164	.016	27
28	10	.0149	.0162	.0126	.0156	.0148	.014	28
29	9	.0135	.0150	.0113	.0139	.0136	.013	29
30	8	.0120	.0140	.0100	.0123	.0124	.012	30
31	7	.0105	.0132	.0089	.0110	.0116	.010	31
32	6.5	.0097	.0128	.0080	.0098	.0108	.009	32
33	6	.0090	.0118	.0071	.0087	.0100	.008	33
34	5.5	.0082	.0104	.0063	.0077	.0092	.007	34
35	5	.0075	.0095	.0056	.0069	.0084	.005	35
36	4.5	.0067	.0090	.0050	.0061	.0076	.004	36
37	4.25	.0064	.0085	.0045	.0054	.0068		37
38	4	.0060	.0080	.0040	.0048	.0060		38
39			.0075	.0035	.0043	.0052		39
40			.0070	.0031	.0039	.0048		40

* U. S. Standard Gage is officially a weight gage, in oz. per sq. ft. as tabulated. The Approx. Thickness shown is the "Manufacturers' Standard" of the American Iron and Steel Institute, based on steel as weighing 501.81 lbs. per cu. ft. (489.6 true weight plus 2.5 percent for average over-run in area and thickness). The A.I.S.I. standard nomenclature for flat rolled carbon steel is as follows:

Widths, Inches	Thicknesses, Inch						
	0.2500 and thicker	0.2499 to 0.2031	0.2030 to 0.1875	0.1874 to 0.0568	0.0567 to 0.0344	0.0343 to 0.0255	0.0254 to 0.0142	0.0141 and thinner
To 3½ incl.	Bar	Bar	Strip	Strip	Strip	Strip	Sheet	Sheet
Over 3½ to 6 incl.	Bar	Bar	Strip	Strip	Strip	Sheet	Sheet	Sheet
" 6 to 12 "	Plate	Strip	Strip	Strip	Sheet	Sheet	Sheet	Sheet
" 12 to 32 "	Plate	Sheet	Sheet	Sheet	Sheet	Sheet	Sheet	Black Plate
" 32 to 48 "	Plate	Sheet	Sheet	Sheet	Sheet	Sheet	Sheet	Sheet
" 48	Plate	Plate	Plate	Sheet	Sheet	Sheet	Sheet	——

Appendix 9
American standard taper pipe threads, NPT[1]

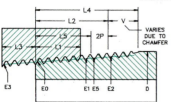

1	2	3	4	5	6	7	8	9	10	11
				Pitch Di-ameter at	Hand-Tight Engagement			Effective Thread, External		
	Outside Diameter of Pipe D	Threads per Inch n	Pitch of Thread p	Beginning of External Thread E_0	Length[2] L_1		Dia E_1	Length L_2		Dia E_2
Nominal Pipe Size					In.	Thds		In.	Thds	In.
$\frac{1}{16}$	0.3125	27	0.03704	0.27118	0.160	4.32	0.28118	0.2611	7.05	0.28750
$\frac{1}{8}$	0.405	27	0.03704	0.36351	0.180	4.86	0.37476	0.2639	7.12	0.38000
$\frac{1}{4}$	0.540	18	0.05556	0.47739	0.200	3.60	0.48989	0.4018	7.23	0.50250
$\frac{3}{8}$	0.675	18	0.05556	0.61201	0.240	4.32	0.62701	0.4078	7.34	0.63750
$\frac{1}{2}$	0.840	14	0.07143	0.75843	0.320	4.48	0.77843	0.5337	7.47	0.79179
$\frac{3}{4}$	1.050	14	0.07143	0.96768	0.339	4.75	0.98887	0.5457	7.64	1.00179
1	1.315	$11\frac{1}{2}$	0.08696	1.21363	0.400	4.60	1.23863	0.6828	7.85	1.25630
$1\frac{1}{4}$	1.660	$11\frac{1}{2}$	0.08696	1.55713	0.420	4.83	1.58338	0.7068	8.13	1.60130
$1\frac{1}{2}$	1.900	$11\frac{1}{2}$	0.08696	1.79609	0.420	4.83	1.82234	0.7235	8.32	1.84130
2	2.375	$11\frac{1}{2}$	0.08696	2.26902	0.436	5.01	2.29627	0.7565	8.70	2.31630
$2\frac{1}{2}$	2.875	8	0.12500	2.71953	0.682	5.46	2.76216	1.1375	9.10	2.79062
3	3.500	8	0.12500	3.34062	0.766	6.13	3.38850	1.2000	9.60	3.41562
$3\frac{1}{2}$	4.000	8	0.12500	3.83750	0.821	6.57	3.88881	1.2500	10.00	3.91562
4	4.500	8	0.12500	4.33438	0.844	6.75	4.38712	1.3000	10.40	4.41562
5	5.563	8	0.12500	5.39073	0.937	7.50	5.44929	1.4063	11.25	5.47862
6	6.625	8	0.12500	6.44609	0.958	7.66	6.50597	1.5125	12.10	6.54062
8	8.625	8	0.12500	8.43359	1.063	8.50	8.50003	1.7125	13.70	8.54062
10	10.750	8	0.12500	10.54531	1.210	9.68	10.62094	1.9250	15.40	10.66562
12	12.750	8	0.12500	12.53281	1.360	10.88	12.61781	2.1250	17.00	12.66562
14 OD	14.000	8	0.12500	13.77500	1.562	12.50	13.87262	2.2500	18.90	13.91562
16 OD	16.000	8	0.12500	15.76250	1.812	14.50	15.87575	2.4500	19.60	15.91562
18 OD	18.000	8	0.12500	17.75000	2.000	16.00	17.87500	2.6500	21.20	17.91562
20 OD	20.000	8	0.12500	19.73750	2.125	17.00	19.87031	2.8500	22.80	19.91562
24 OD	24.000	8	0.12500	23.71250	2.375	19.00	23.86094	3.2500	26.00	23.91562

All dimensions are given in inches.

[1] The basic dimensions of the American Standard Taper Pipe Thread are given in inches to four or five decimal places. While this implies a greater degree of precision than is ordinarily attained, these dimensions are the basis of gage dimensions and are so expressed for the purpose of eliminating errors in computations.

[2] Also length of thin ring gage and length from gaging notch to small end of plug gage.

(Courtesy of ANSI; B2.1–1960.)

Appendix 10
American standard 250-lb cast iron flanged fittings

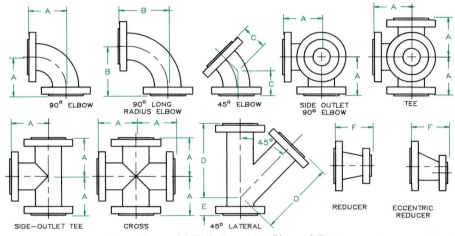

90° ELBOW 90° LONG RADIUS ELBOW 45° ELBOW SIDE OUTLET 90° ELBOW TEE
SIDE—OUTLET TEE CROSS 45° LATERAL REDUCER ECCENTRIC REDUCER

Dimensions of 250-lb Cast Iron Flanged Fittings

Nominal Pipe Size	Flanges			Fittings		Straight					
	Dia of Flange	Thickness of Flange (Min)	Dia of Raised Face	Inside Dia of Fittings (Min)	Wall Thickness	Center to Face 90 Deg Elbow Tees, Crosses and True "Y"	Center to Face 90 Deg Long Radius Elbow	Center to Face 45 Deg Elbow	Center to Face Lateral	Short Center to Face True "Y" and Lateral	Face to Face Reducer
						A	B	C	D	E	F
1	4 7/8	11/16	2 11/16	1	7/16	4	5	2	6 1/2	2	....
1 1/4	5 1/4	3/4	3 1/16	1 1/4	7/16	4 1/4	5 1/2	2 1/2	7 1/4	2 1/4	....
1 1/2	6 1/8	13/16	3 9/16	1 1/2	7/16	4 1/2	6	2 3/4	8 1/2	2 1/2	....
2	6 1/2	7/8	4 3/16	2	7/16	5	6 1/2	3	9	2 1/2	5
2 1/2	7 1/2	1	4 15/16	2 1/2	1/2	5 1/2	7	3 1/2	10 1/2	2 1/2	5 1/2
3	8 1/4	1 1/8	5 11/16	3	9/16	6	7 3/4	3 1/2	11	3	6
3 1/2	9	1 3/16	6 5/16	3 1/2	9/16	6 1/2	8 1/2	4	12 1/2	3	6 1/2
4	10	1 1/4	6 15/16	4	5/8	7	9	4 1/2	13 1/2	3	7
5	11	1 3/8	8 5/16	5	11/16	8	10 1/4	5	15	3 1/2	8
6	12 1/2	1 7/16	9 11/16	6	3/4	8 1/2	11 1/2	5 1/2	17 1/2	4	9
8	15	1 5/8	11 15/16	8	13/16	10	14	6	20 1/2	5	11
10	17 1/2	1 7/8	14 1/16	10	15/16	11 1/2	16 1/2	7	24	5 1/2	12
12	20 1/2	2	16 7/16	12	1	13	19	8	27 1/2	6	14
14	23	2 1/8	18 15/16	13 1/4	1 1/8	15	21 1/2	8 1/2	31	6 1/2	16
16	25 1/2	2 1/4	21 1/16	15 1/4	1 1/4	16 1/2	24	9 1/2	34 1/2	7 1/2	18
18	28	2 3/8	23 5/16	17	1 3/8	18	26 1/2	10	37 1/2	8	19
20	30 1/2	2 1/2	25 9/16	19	1 1/2	19 1/2	29	10 1/2	40 1/2	8 1/2	20
24	36	2 3/4	30 5/16	23	1 5/8	22 1/2	34	12	47 1/2	10	24
30	43	3	37 3/16	29	2	27 1/2	41 1/2	15	....	....	30

All dimensions are given in inches.
(Courtesy of ANSI; B16.1–1967.)

Appendix 11
American standard 126-lb cast iron flanged fittings

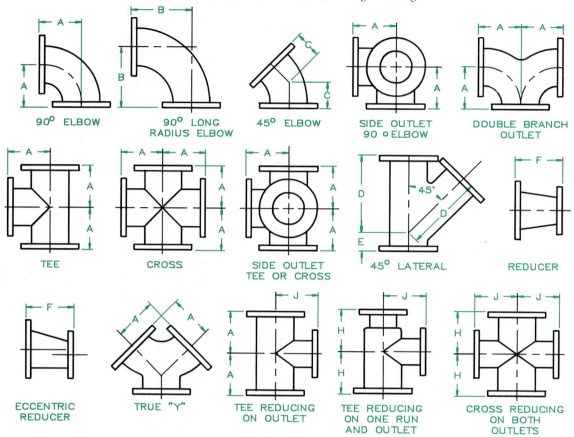

90° ELBOW

90° LONG RADIUS ELBOW

45° ELBOW

SIDE OUTLET 90° ELBOW

DOUBLE BRANCH OUTLET

TEE

CROSS

SIDE OUTLET TEE OR CROSS

45° LATERAL

REDUCER

ECCENTRIC REDUCER

TRUE "Y"

TEE REDUCING ON OUTLET

TEE REDUCING ON ONE RUN AND OUTLET

CROSS REDUCING ON BOTH OUTLETS

Appendix 11

American standard 125-lb cast iron flanged fittings (cont.)

Nominal Pipe Size	Flanges: Dia of Flange	Flanges: Thickness of Flange (Min)	General: Inside Dia of Flange Fittings	General: Wall Thickness	Straight Fittings: Center to Face 90 deg Elbow Tees, Crosses True "Y" and Double Branch Elbow A	Straight Fittings: Center to Face 90 deg Long Radius El-bow B	Straight Fittings: Center to Face 45 deg Elbow C	Straight Fittings: Center to Face Lateral D	Straight Fittings: Short Center to Face True "Y" and Lateral E	Straight Fittings: Face to Face Reducer F	Reducing Fittings (Short Body Patterns), Tees and Crosses: Size of Outlet and Smaller	Reducing Fittings (Short Body Patterns), Tees and Crosses: Center to Face Run H	Reducing Fittings (Short Body Patterns), Tees and Crosses: Center to Face Outlet or Side Outlet J
1	4¼	7/16	1	5/16	3½	5	1¾	5¾	1¾	. . .			
1¼	4⅝	½	1¼	5/16	3¾	5½	2	6¼	1¾	. . .			
1½	5	9/16	1½	5/16	4	6	2¼	7	2				
2	6	⅝	2	5/16	4½	6½	2½	8	2½	5			
2½	7	11/16	2½	5/16	5	7	3	9½	2½	5½			
3	7½	¾	3	⅜	5½	7¾	3	10	3	6			
3½	8½	13/16	3½	7/16	6	8½	3½	11½	3	6½	All reducing tees and crosses, sizes 16 in. and smaller, shall have same center to face dimensions as straight size fittings, corresponding to the size of the largest opening.		
4	9	15/16	4	½	6½	9	4	12	3	7			
5	10	15/16	5	½	7½	10¼	4½	13½	3½	8			
6	11	1	6	9/16	8	11½	5	14½	3½	9			
8	13½	1⅛	8	⅝	9	14	5½	17½	4½	11			
10	16	1 3/16	10	¾	11	16½	6½	20½	5	12			
12	19	1¼	12	13/16	12	19	7½	24½	5½	14			
14	21	1⅜	14	⅞	14	21½	7½	27	6	16			
16	23½	1 7/16	16	1	15	24	8	30	6½	18			
18	25	1 9/16	18	1 1/16	16½	26½	8½	32	7	19	12	13	15½
20	27½	1 11/16	20	1⅛	18	29	9½	35	8	20	14	14	17
24	32	1⅞	24	1¼	22	34	11	40½	9	24	16	15	19
30	38¾	2⅛	30	1 7/16	25	41½	15	49	10	30	20	18	23
36	46	2⅜	36	1⅝	28*	49	18	. . .	. . .	36	24	20	26
42	53	2⅝	42	1 13/16	31*	56½	21	. . .	. . .	42	24	23	30
48	59½	2¾	48	2	34*	64	24	. . .	. . .	48	30	26	34

All dimensions are given in inches.
(Courtesy of ANSI; B16.1–1967.)

Appendix 12

American standard 125-lb cast iron flanges*

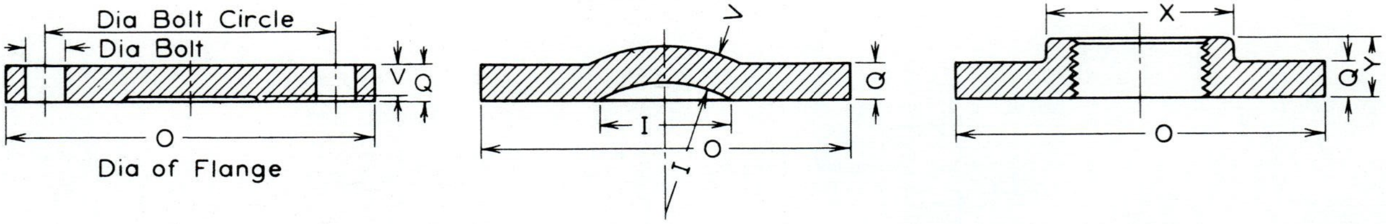

Size *I*	*O*	*Q*	*V*	*X*	*Y*	Dia. Bolt Circle	No. of Bolts	Dia. Bolts	Dia. Bolt Holes	Length of Bolts
1	4 1/4	7/16	—	1 15/16	0.68	3 1/8	4	1/2	5/8	1 3/4
1 1/4	4 5/8	1/2	—	2 5/16	0.76	3 1/2	4	1/2	5/8	2
1 1/2	5	9/16	—	2 9/16	0.87	3 7/8	4	1/2	5/8	2
2	6	5/8	—	3 1/16	1.00	4 3/4	4	5/8	3/4	2 1/4
2 1/2	7	11/16	—	3 9/16	1.14	5 1/2	4	5/8	3/4	2 1/2
3	7 1/2	3/4	—	4 1/4	1.20	6	4	5/8	3/4	2 1/2
3 1/2	8 1/2	13/16	—	4 13/16	1.25	7	8	5/8	3/4	2 3/4
4	9	15/16	—	5 5/16	1.30	7 1/2	8	5/8	3/4	3
5	10	15/16	—	6 7/16	1.41	8 1/2	8	3/4	7/8	3 1/4
6	11	1	—	7 9/16	1.51	9 1/2	8	3/4	7/8	3 1/4
8	13 1/2	1 1/8	—	9 11/16	1.71	11 3/4	8	3/4	7/8	3 1/2
10	16	1 3/16	—	11 15/16	1.93	14 1/4	12	7/8	1	3 3/4
12	19	1 1/4	12/16	14 1/16	2.13	17	12	7/8	1	3 3/4
14 O.D.	21	1 3/8	7/8	15 3/8	2.25	18 3/4	12	1	1 1/8	4 1/4
16 O.D.	23 1/2	1 7/16	1	17 1/2	2.45	21 1/4	16	1 1/8	1 1/8	4 1/2
18 O.D.	25	1 9/16	1 1/16	19 5/8	2.65	22 3/4	16	1 1/8	1 1/4	4 3/4

All dimensions in inches.

* Extracted from American Standards, "Cast-Iron Pipe Flanges and Flanged Fittings" (ANSI B16.1), with the permission of the publisher, The American Society of Mechanical Engineers.

Appendix 13
American national standard 125-lb cast iron screwed fittings*

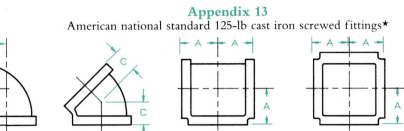

| ELBOW | 45° ELBOW | TEE | CROSS | TYPICAL SECTION |

Nominal Pipe Size	A	C	B Min	E Min	F Min	F Max	G Min	H Min
¼	0.81	0.73	0.32	0.38	0.540	0.584	0.110	0.93
⅜	0.95	0.80	0.36	0.44	0.675	0.719	0.120	1.12
½	1.12	0.88	0.43	0.50	0.840	0.897	0.130	1.34
¾	1.31	0.98	0.50	0.56	1.050	1.107	0.155	1.63
1	1.50	1.12	0.58	0.62	1.315	1.385	0.170	1.95
1¼	1.75	1.29	0.67	0.69	1.660	1.730	0.185	2.39
1½	1.94	1.43	0.70	0.75	1.900	1.970	0.200	2.68
2	2.25	1.68	0.75	0.84	2.375	2.445	0.220	3.28
2½	2.70	1.95	0.92	0.94	2.875	2.975	0.240	3.86
3	3.08	2.17	0.98	1.00	3.500	3.600	0.260	4.62
3½	3.42	2.39	1.03	1.06	4.000	4.100	0.280	5.20
4	3.79	2.61	1.08	1.12	4.500	4.600	0.310	5.79
5	4.50	3.05	1.18	1.18	5.563	5.663	0.380	7.05
6	5.13	3.46	1.28	1.28	6.625	6.725	0.430	8.28
8	6.56	4.28	1.47	1.47	8.625	8.725	0.550	10.63
10	8.08	5.16	1.68	1.68	10.750	10.850	0.690	13.12
12	9.50	5.97	1.88	1.88	12.750	12.850	0.800	15.47
14 O.D.	10.40	—	2.00	2.00	14.000	14.100	0.880	16.94
16 O.D.	11.82	—	2.20	2.20	16.000	16.100	1.000	19.30

All dimensions in inches.
* Extracted from American National Standards, "Cast-Iron Screwed Fittings, 125- and 250-lb" (ANSI B16.4), with the permission of the publisher, The American Society of Mechanical Engineers.

Appendix 14
American national standard unified inch screw threads
(UN and UNR thread form)★

Sizes Primary	Sizes Secondary	Basic Major Diameter	Coarse UNC	Fine UNF	Extra Fine UNEF	4UN	6UN	8UN	12UN	16UN	20UN	28UN	32UN	Sizes
0		0.0600	—	80	—	—	—	—	—	—	—	—	—	0
	1	0.0730	64	72	—	—	—	—	—	—	—	—	—	1
2		0.0860	56	64	—	—	—	—	—	—	—	—	—	2
	3	0.0990	48	56	—	—	—	—	—	—	—	—	—	3
4		0.1120	40	48	—	—	—	—	—	—	—	—	—	4
5		0.1250	40	44	—	—	—	—	—	—	—	—	—	5
6		0.1380	32	40	—	—	—	—	—	—	—	—	UNC	6
8		0.1640	32	36	—	—	—	—	—	—	—	—	UNC	8
10		0.1900	24	32	—	—	—	—	—	—	—	—	UNF	10
	12	0.2160	24	28	32	—	—	—	—	—	—	UNF	UNEF	12
$\frac{1}{4}$		0.2500	20	28	32	—	—	—	—	—	UNC	UNF	UNEF	$\frac{1}{4}$
$\frac{5}{16}$		0.3125	18	24	32	—	—	—	—	—	20	28	UNEF	$\frac{5}{16}$
$\frac{3}{8}$		0.3750	16	24	32	—	—	—	—	UNC	20	28	UNEF	$\frac{3}{8}$
$\frac{7}{16}$		0.4375	14	20	28	—	—	—	—	16	UNF	UNEF	32	$\frac{7}{16}$
$\frac{1}{2}$		0.5000	13	20	28	—	—	—	—	16	UNF	UNEF	32	$\frac{1}{2}$
$\frac{9}{16}$		0.5625	12	18	24	—	—	—	UNC	16	20	28	32	$\frac{9}{16}$
$\frac{5}{8}$		0.6250	11	18	24	—	—	—	12	16	20	28	32	$\frac{5}{8}$
	$\frac{11}{16}$	0.6875	—	—	24	—	—	—	12	16	20	28	32	$\frac{11}{16}$
$\frac{3}{4}$		0.7500	10	16	20	—	—	—	12	UNF	UNEF	28	32	$\frac{3}{4}$
	$\frac{13}{16}$	0.8125	—	—	20	—	—	—	12	16	UNEF	28	32	$\frac{13}{16}$
$\frac{7}{8}$		0.8750	9	14	20	—	—	—	12	16	UNEF	28	32	$\frac{7}{8}$
	$\frac{15}{16}$	0.9375	—	—	20	—	—	—	12	16	UNEF	28	32	$\frac{15}{16}$
1		1.0000	8	12	20	—	—	UNC	UNF	16	UNEF	28	32	1
	$1\frac{1}{16}$	1.0625	—	—	18	—	—	8	12	16	20	28	—	$1\frac{1}{16}$
$1\frac{1}{8}$		1.1250	7	12	18	—	—	8	UNF	16	20	28	—	$1\frac{1}{8}$
	$1\frac{3}{16}$	1.1875	—	—	18	—	—	8	12	16	20	28	—	$1\frac{3}{16}$
$1\frac{1}{4}$		1.2500	7	12	18	—	—	8	UNF	16	20	28	—	$1\frac{1}{4}$
	$1\frac{5}{16}$	1.3125	—	—	18	—	—	8	12	16	20	28	—	$1\frac{5}{16}$
$1\frac{3}{8}$		1.3750	6	12	18	—	UNC	8	UNF	16	20	28	—	$1\frac{3}{8}$
	$1\frac{7}{16}$	1.4375	—	—	18	—	6	8	12	16	20	28	—	$1\frac{7}{16}$
$1\frac{1}{2}$		1.5000	6	12	18	—	UNC	8	UNF	16	20	28	—	$1\frac{1}{2}$
	$1\frac{9}{16}$	1.5625	—	—	18	—	6	8	12	16	20	—	—	$1\frac{9}{16}$
$1\frac{5}{8}$		1.6250	—	—	18	—	6	8	12	16	20	—	—	$1\frac{5}{8}$
	$1\frac{11}{16}$	1.6875	—	—	18	—	6	8	12	16	20	—	—	$1\frac{11}{16}$
$1\frac{3}{4}$		1.7500	5	—	—	—	6	8	12	16	20	—	—	$1\frac{3}{4}$
	$1\frac{13}{16}$	1.8125	—	—	—	—	6	8	12	16	20	—	—	$1\frac{13}{16}$
$1\frac{7}{8}$		1.8750	—	—	—	—	6	8	12	16	20	—	—	$1\frac{7}{8}$
	$1\frac{15}{16}$	1.9375	—	—	—	—	6	8	12	16	20	—	—	$1\frac{15}{16}$
2		2.0000	$4\frac{1}{2}$	—	—	—	6	8	12	16	20	—	—	2
	$2\frac{1}{8}$	2.1250	—	—	—	—	6	8	12	16	20	—	—	$2\frac{1}{8}$
$2\frac{1}{4}$		2.2500	$4\frac{1}{2}$	—	—	—	6	8	12	16	20	—	—	$2\frac{1}{4}$
	$2\frac{3}{8}$	2.3750	—	—	—	—	6	8	12	16	20	—	—	$2\frac{3}{8}$
$2\frac{1}{2}$		2.5000	4	—	—	UNC	6	8	12	16	20	—	—	$2\frac{1}{2}$
	$2\frac{5}{8}$	2.6250	—	—	—	4	6	8	12	16	20	—	—	$2\frac{5}{8}$
$2\frac{3}{4}$		2.7500	4	—	—	UNC	6	8	12	16	20	—	—	$2\frac{3}{4}$
	$2\frac{7}{8}$	2.8750	—	—	—	4	6	8	12	16	20	—	—	$2\frac{7}{8}$

* Series designation shown indicates the UN thread form; however, the UNR thread form may be specified by substituting UNR in place of UN in all designations for external use only.

Cont.

Apppendix 14
American national standard unified inch screw threads
(UN and UNR thread form)★ (cont.)

Sizes		Basic Major Diameter	Threads per Inch											Sizes
			Series with Graded Pitches			Series with Constant Pitches								
Primary	Secondary		Coarse UNC	Fine UNF	Extra Fine UNEF	4UN	6UN	8UN	12UN	16UN	20UN	28UN	32UN	
3		3.0000	4	—	—	UNC	6	8	12	16	20	—	—	3
	$3\frac{1}{8}$	3.1250	—	—	—	4	6	8	12	16	—	—	—	$3\frac{1}{8}$
$3\frac{1}{4}$		3.2500	4	—	—	UNC	6	8	12	16	—	—	—	$3\frac{1}{4}$
	$3\frac{3}{8}$	3.3750	—	—	—	4	6	8	12	16	—	—	—	$3\frac{3}{8}$
$3\frac{1}{2}$		3.5000	4	—	—	UNC	6	8	12	16	—	—	—	$3\frac{1}{2}$
	$3\frac{5}{8}$	3.6250	—	—	—	4	6	8	12	16	—	—	—	$3\frac{5}{8}$
$3\frac{3}{4}$		3.7500	4	—	—	UNC	6	8	12	16	—	—	—	$3\frac{3}{4}$
	$3\frac{7}{8}$	3.8750	—	—	—	4	6	8	12	16	—	—	—	$3\frac{7}{8}$
4		4.0000	4	—	—	UNC	6	8	12	16	—	—	—	4
	$4\frac{1}{8}$	4.1250	—	—	—	4	6	8	12	16	—	—	—	$4\frac{1}{8}$
$4\frac{1}{4}$		4.2500	—	—	—	4	6	8	12	16	—	—	—	$4\frac{1}{4}$
	$4\frac{3}{8}$	4.3750	—	—	—	4	6	8	12	16	—	—	—	$4\frac{3}{8}$
$4\frac{1}{2}$		4.5000	—	—	—	4	6	8	12	16	—	—	—	$4\frac{1}{2}$
	$4\frac{5}{8}$	4.6250	—	—	—	4	6	8	12	16	—	—	—	$4\frac{5}{8}$
$4\frac{3}{4}$		4.7500	—	—	—	4	6	8	12	16	—	—	—	$4\frac{3}{4}$
	$4\frac{7}{8}$	4.8750	—	—	—	4	6	8	12	16	—	—	—	$4\frac{7}{8}$
5		5.0000	—	—	—	4	6	8	12	16	—	—	—	5
	$5\frac{1}{8}$	5.1250	—	—	—	4	6	8	12	16	—	—	—	$5\frac{1}{8}$
$5\frac{1}{4}$		5.2500	—	—	—	4	6	8	12	16	—	—	—	$5\frac{1}{4}$
	$5\frac{3}{8}$	5.3750	—	—	—	4	6	8	12	16	—	—	—	$5\frac{3}{8}$
$5\frac{1}{2}$		5.5000	—	—	—	4	6	8	12	16	—	—	—	$5\frac{1}{2}$
	$5\frac{5}{8}$	5.6250	—	—	—	4	6	8	12	16	—	—	—	$5\frac{5}{8}$
$5\frac{3}{4}$		5.7500	—	—	—	4	6	8	12	16	—	—	—	$5\frac{3}{4}$
	$5\frac{7}{8}$	5.8750	—	—	—	4	6	8	12	16	—	—	—	$5\frac{7}{8}$
6		6.0000	—	—	—	4	6	8	12	16	—	—	—	6

(Courtesy of ANSI; B1.1–1974.)

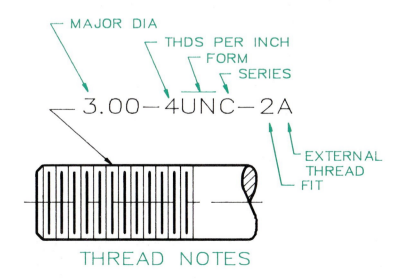

THREAD NOTES

Appendix 15

Tap drill sizes for american national and unified coarse and fine threads

$$p = \text{pitch} = \frac{1}{\text{No. thd per in}}$$

$$d = \text{depth} = p \times 0.650$$

$$f = \text{flat} = \frac{p}{8}$$

$$\text{pitch dia} = d - \frac{0.650}{N}$$

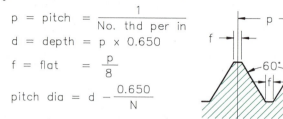

For nos. 575 and 585 screw thread micrometers

Size	Threads per inch NC UNC	NF UNF	Outside Diameter Inches	Pitch Diameter Inches	Root Diameter Inches	Tap Drill Approx. 75% Full Thread	Decimal Equiv. of Tap Drill
0	..	80	.0600	.0519	.0438	3/64	.0469
1	64	..	.0730	.0629	.0527	53	.0595
1	..	72	.0730	.0640	.0550	53	.0595
2	56	..	.0860	.0744	.0628	50	.0700
2	..	64	.0860	.0759	.0657	50	.0700
3	48	..	.0990	.0855	.0719	47	.0785
3	..	56	.0990	.0874	.0758	46	.0810
4	40	..	.1120	.0958	.0795	43	.0890
4	..	48	.1120	.0985	.0849	42	.0935
5	40	..	.1250	.1088	.0925	38	.1015
5	..	44	.1250	.1102	.0955	37	.1040
6	32	..	.1380	.1177	.0974	36	.1065
6	..	40	.1380	.1218	.1055	33	.1130
8	32	..	.1640	.1437	.1234	29	.1360
8	..	36	.1640	.1460	.1279	29	.1360
10	24	..	.1900	.1629	.1359	26	.1470
10	..	32	.1900	.1697	.1494	21	.1590
12	24	..	.2160	.1889	.1619	16	.1770
12	..	28	.2160	.1928	.1696	15	.1800
1/4	20	..	.2500	.2175	.1850	7	.2010
1/4	..	28	.2500	.2268	.2036	3	.2130
5/16	18	..	.3125	.2764	.2403	F	.2570
5/16	..	24	.3125	.2854	.2584	I	.2720
3/8	16	..	.3750	.3344	.2938	5/16	.3125
3/8	..	24	.3750	.3479	.3209	Q	.3320
7/16	14	..	.4375	.3911	.3447	U	.3680
7/16	..	20	.4375	.4050	.3726	25/64	.3906
1/2	13	..	.5000	.4500	.4001	27/64	.4219
1/2	..	20	.5000	.4675	.4351	29/64	.4531
9/16	12	..	.5625	.5084	.4542	31/64	.4844
9/16	..	18	.5625	.5264	.4903	33/64	.5156
5/8	11	..	.6250	.5660	.5069	17/32	.5312
5/8	..	18	.6250	.5889	.5528	37/64	.5781
3/4	10	..	.7500	.6850	.6201	21/32	.6562
3/4	..	16	.7500	.7094	.6688	11/16	.6875
7/8	9	..	.8750	.8028	.7307	49/64	.7656
7/8	..	14	.8750	.8286	.7822	13/16	.8125

Cont.

Appendix 15
Tap drill sizes for american national and unified coarse and fine threads (cont.)

Size	Threads per inch		Outside Diameter Inches	Pitch Diameter Inches	Root Diameter Inches	Tap Drill Approx. 75% Full Thread	Decimal Equiv. of Tap Drill
	NC UNC	NF UNF					
1	8	..	1.0000	.9188	.8376	$\frac{7}{8}$	.8750
1	..	12	1.0000	.9459	.8917	$\frac{59}{64}$	.9219
1⅛	7	..	1.1250	1.0322	.9394	$\frac{63}{64}$	.9844
1⅛	..	12	1.1250	1.0709	1.0168	$1\frac{3}{64}$	1.0469
1¼	7	..	1.2500	1.1572	1.0644	$1\frac{7}{64}$	1.1094
1¼	..	12	1.2500	1.1959	1.1418	$1\frac{11}{64}$	1.1719
1⅜	6	..	1.3750	1.2667	1.1585	$1\frac{7}{32}$	1.2187
1⅜	..	12	1.3750	1.3209	1.2668	$1\frac{19}{64}$	1.2969
1½	6	..	1.5000	1.3917	1.2835	$1\frac{11}{32}$	1.3437
1½	..	12	1.5000	1.4459	1.3918	$1\frac{27}{64}$	1.4219
1¾	5	..	1.7500	1.6201	1.4902	$1\frac{9}{16}$	1.5625
2	4½	..	2.0000	1.8557	1.7113	$1\frac{25}{32}$	1.7812
2¼	4½	..	2.2500	2.1057	1.9613	$2\frac{1}{32}$	2.0313
2½	4	..	2.5000	2.3376	2.1752	2¼	2.2500
2¾	4	..	2.7500	2.5876	2.4252	2½	2.5000
3	4	..	3.0000	3.8376	2.6752	2¾	2.7500
3¼	4	..	3.2500	3.0876	2.9252	3	3.0000
3½	4	..	3.5000	3.3376	3.1752	3¼	3.2500
3¾	4	..	3.7500	3.5876	3.4252	3½	3.5000
4	4	..	4.0000	3.3786	3.6752	3¾	3.7500

(Courtesy of the L. S. Starrett Company.)

FROM TABLE

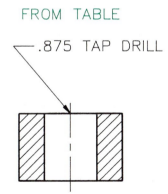

.875 TAP DRILL

HOLE IS THREADED

.875 TAP DRILL
1.00−8UNC−2B

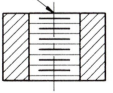

Appendix 16
Length of thread engagement groups

Nominal Size Diam.		Pitch P	Length of Thread Engagement				Nominal Size Diam.		Pitch P	Length of Thread Engagement			
			Group S		Group N		Group L				Group S		Group N		Group L
Over	To and Incl		To and Incl	Over	To and Incl	Over	Over	To and Incl		To and Incl	Over	To and Incl	Over
1.5	2.8	0.2	0.5	0.5	1.5	1.5	22.4	45	1	4	4	12	12
		0.25	0.6	0.6	1.9	1.9			1.5	6.3	6.3	19	19
		0.35	0.8	0.8	2.6	2.6			2	8.5	8.5	25	25
		0.4	1	1	3	3			3	12	12	36	36
		0.45	1.3	1.3	3.8	3.8			3.5	15	15	45	45
2.8	5.6	0.35	1	1	3	3			4	18	18	53	53
		0.5	1.5	1.5	4.5	4.5			4.5	21	21	63	63
		0.6	1.7	1.7	5	5	45	90	1.5	7.5	7.5	22	22
		0.7	2	2	6	6			2	9.5	9.5	28	28
		0.75	2.2	2.2	6.7	6.7			3	15	15	45	45
		0.8	2.5	2.5	7.5	7.5			4	19	19	56	56
5.6	11.2	0.75	2.4	2.4	7.1	7.1			5	24	24	71	71
		1	3	3	9	9			5.5	28	28	85	85
		1.25	4	4	12	12			6	32	32	95	95
		1.5	5	5	15	15	90	180	2	12	12	36	36
11.2	22.4	1	3.8	3.8	11	11			3	18	18	53	53
		1.25	4.5	4.5	13	13			4	24	24	71	71
		1.5	5.6	5.6	16	16			6	36	36	106	106
		1.75	6	6	18	18	180	355	3	20	20	60	60
		2	8	8	24	24			4	26	26	80	80
		2.5	10	10	30	30			6	40	40	118	118

All dimensions are given in millimeters. (Courtesy of ISO Standards.)

Appendix 17
ISO metric screw thread standard series

Nominal Size Dia. (mm) Column[a]			Pitches (mm) Series with Graded Pitches		Series with Constant Pitches												Nominal Size Dia. (mm)
1	2	3	Coarse	Fine	6	4	3	2	1.5	1.25	1	0.75	0.5	0.35	0.25	0.2	
0.25			0.075	—	—	—	—	—	—	—	—	—	—	—	—	—	0.25
0.3			0.08	—	—	—	—	—	—	—	—	—	—	—	—	—	0.3
	0.35		0.09	—	—	—	—	—	—	—	—	—	—	—	—	—	0.35
0.4			0.1	—	—	—	—	—	—	—	—	—	—	—	—	—	0.4
	0.45		0.1	—	—	—	—	—	—	—	—	—	—	—	—	—	0.45
0.5			0.125	—	—	—	—	—	—	—	—	—	—	—	—	—	0.5
	0.55		0.125	—	—	—	—	—	—	—	—	—	—	—	—	—	0.55
0.6			0.15	—	—	—	—	—	—	—	—	—	—	—	—	—	0.6
	0.7		0.175	—	—	—	—	—	—	—	—	—	—	—	—	—	0.7
0.8			0.2	—	—	—	—	—	—	—	—	—	—	—	—	—	0.8
	0.9		0.225	—	—	—	—	—	—	—	—	—	—	—	—	—	0.9
1			0.25	—	—	—	—	—	—	—	—	—	—	—	—	0.2	1
	1.1		0.25	—	—	—	—	—	—	—	—	—	—	—	—	0.2	1.1
1.2			0.25	—	—	—	—	—	—	—	—	—	—	—	—	0.2	1.2
	1.4		0.3	—	—	—	—	—	—	—	—	—	—	—	—	0.2	1.4
1.6			0.35	—	—	—	—	—	—	—	—	—	—	—	—	0.2	1.6
	1.8		0.35	—	—	—	—	—	—	—	—	—	—	—	—	0.2	1.8
2			0.4	—	—	—	—	—	—	—	—	—	—	—	0.25	—	2
	2.2		0.45	—	—	—	—	—	—	—	—	—	—	—	0.25	—	2.2
2.5			0.45	—	—	—	—	—	—	—	—	—	—	0.35	—	—	2.5
3			0.5	—	—	—	—	—	—	—	—	—	—	0.35	—	—	3
	3.5		0.6	—	—	—	—	—	—	—	—	—	—	0.35	—	—	3.5
4			0.7	—	—	—	—	—	—	—	—	—	0.5	—	—	—	4
	4.5		0.75	—	—	—	—	—	—	—	—	—	0.5	—	—	—	4.5
5			0.8	—	—	—	—	—	—	—	—	—	0.5	—	—	—	5
		5.5	—	—	—	—	—	—	—	—	—	—	0.5	—	—	—	5.5
6			1	—	—	—	—	—	—	—	—	0.75	—	—	—	—	6
		7	1	—	—	—	—	—	—	—	—	0.75	—	—	—	—	7
8			1.25	1	—	—	—	—	—	—	1	0.75	—	—	—	—	8
		9	1.25	—	—	—	—	—	—	—	1	0.75	—	—	—	—	9
10			1.5	1.25	—	—	—	—	—	1.25	1	0.75	—	—	—	—	10
		11	1.5	—	—	—	—	—	—	—	1	0.75	—	—	—	—	11
12			1.75	1.25	—	—	—	—	1.5	1.25	1	—	—	—	—	—	12
	14		2	1.5	—	—	—	—	1.5	1.25[b]	1	—	—	—	—	—	14
		15	—	—	—	—	—	—	1.5	—	1	—	—	—	—	—	15
16			2	1.5	—	—	—	—	1.5	—	1	—	—	—	—	—	16
		17	—	—	—	—	—	—	1.5	—	1	—	—	—	—	—	17
	18		2.5	1.5	—	—	—	2	1.5	—	1	—	—	—	—	—	18
20			2.5	1.5	—	—	—	2	1.5	—	1	—	—	—	—	—	20
	22		2.5	1.5	—	—	—	2	1.5	—	1	—	—	—	—	—	22

[a] Thread diameter should be selected from columns 1, 2 or 3, with preference being in that order.

[b] Pitch 1.25 mm in combination with diameter 14 mm has been included for sparkplug applications.

[c] Diameter 35 mm has been included for bearing locknut applications.

The use of pitches shown in parentheses should be avoided wherever possible.

The pitches enclosed in the bold frame, together with the corresponding nominal diameters in columns 1 and 2, are those combinations which have been established by ISO Recommendations as a selected "coarse" and "fine" series for commercial fasteners.

Appendix 17
ISO metric screw thread standard series (cont.)

Nominal Size Dia. (mm)			Pitches (mm)													Nominal Size Dia. (mm)
Column[a]			Series with Graded Pitches		Series with Constant Pitches											
1	2	3	Coarse	Fine	6	4	3	2	1.5	1.25	1	0.75	0.5	0.35	0.25	0.2	
24			3	2	—	—	—	2	1.5	—	1	—	—	—	—	—	24
		25	—	—	—	—	—	2	1.5	—	1	—	—	—	—	—	25
		26	—	—	—	—	—	—	1.5	—	1	—	—	—	—	—	26
	27		3	2	—	—	—	2	1.5	—	1	—	—	—	—	—	27
		28	—	—	—	—	—	2	1.5	—	1	—	—	—	—	—	28
30			3.5	2	—	—	(3)	2	1.5	—	1	—	—	—	—	—	30
		32	—	—	—	—	—	2	1.5	—	—	—	—	—	—	—	32
	33		3.5	2	—	—	(3)	2	1.5	—	—	—	—	—	—	—	33
		35[c]	—	—	—	—	—	—	1.5	—	—	—	—	—	—	—	35[c]
36			4	3	—	—	—	2	1.5	—	—	—	—	—	—	—	36
		38	—	—	—	—	—	—	1.5	—	—	—	—	—	—	—	38
	39		4	3	—	—	—	2	1.5	—	—	—	—	—	—	—	39
		40	—	—	—	—	3	2	1.5	—	—	—	—	—	—	—	40
42			4.5	3	—	4	3	2	1.5	—	—	—	—	—	—	—	42
	45		4.5	3	—	4	3	2	1.5	—	—	—	—	—	—	—	45
48			5	3	—	4	3	2	1.5	—	—	—	—	—	—	—	48
		50	—	—	—	—	3	2	1.5	—	—	—	—	—	—	—	50
	52		5	3	—	4	3	2	1.5	—	—	—	—	—	—	—	52
		55	—	—	—	4	3	2	1.5	—	—	—	—	—	—	—	55
56			5.5	4	—	4	3	2	1.5	—	—	—	—	—	—	—	56
		58	—	—	—	4	3	2	1.5	—	—	—	—	—	—	—	58
	60		5.5	4	—	4	3	2	1.5	—	—	—	—	—	—	—	60
		62	—	—	—	4	3	2	1.5	—	—	—	—	—	—	—	62
64			6	4	—	4	3	2	1.5	—	—	—	—	—	—	—	64
		65	—	—	—	4	3	2	1.5	—	—	—	—	—	—	—	65
	68		6	4	—	4	3	2	1.5	—	—	—	—	—	—	—	68
		70	—	—	6	4	3	2	1.5	—	—	—	—	—	—	—	70
72			—	—	6	4	3	2	1.5	—	—	—	—	—	—	—	72
		75	—	—	—	4	3	2	1.5	—	—	—	—	—	—	—	75
	76		—	—	6	4	3	2	1.5	—	—	—	—	—	—	—	76
		78	—	—	—	—	—	2	—	—	—	—	—	—	—	—	78
80			—	—	6	4	3	2	1.5	—	—	—	—	—	—	—	80
		82	—	—	—	—	—	2	—	—	—	—	—	—	—	—	82
	85		—	—	6	4	3	2	—	—	—	—	—	—	—	—	85
90			—	—	6	4	3	2	—	—	—	—	—	—	—	—	90

Cont.

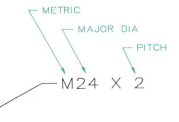

METRIC THREAD NOTE

Appendix 17
ISO metric screw thread standard series (cont.)

Nominal Size Dia. (mm)			Pitches (mm)															Nominal Size Dia. (mm)
Column[a]			Series with Graded Pitches		Series with Constant Pitches													
1	2	3	Coarse	Fine	6	4	3	2	1.5	1.25	1	0.75	0.5	0.35	0.25	0.2	
	95		—	—	6	4	3	2	—	—	—	—	—	—	—	—	95
100			—	—	6	4	3	2	—	—	—	—	—	—	—	—	100
	105		—	—	6	4	3	2	—	—	—	—	—	—	—	—	105
110			—	—	6	4	3	2	—	—	—	—	—	—	—	—	110
	115		—	—	6	4	3	2	—	—	—	—	—	—	—	—	115
	120		—	—	6	4	3	2	—	—	—	—	—	—	—	—	120
125			—	—	6	4	3	2	—	—	—	—	—	—	—	—	125
	130		—	—	6	4	3	2	—	—	—	—	—	—	—	—	130
		135	—	—	6	4	3	2	—	—	—	—	—	—	—	—	135
140			—	—	6	4	3	2	—	—	—	—	—	—	—	—	140
		145	—	—	6	4	3	2	—	—	—	—	—	—	—	—	145
	150		—	—	6	4	3	2	—	—	—	—	—	—	—	—	150
		155	—	—	6	4	3	—	—	—	—	—	—	—	—	—	155
160			—	—	6	4	3	—	—	—	—	—	—	—	—	—	160
		165	—	—	6	4	3	—	—	—	—	—	—	—	—	—	165
	170		—	—	6	4	3	—	—	—	—	—	—	—	—	—	170
		175	—	—	6	4	3	—	—	—	—	—	—	—	—	—	175
180			—	—	6	4	3	—	—	—	—	—	—	—	—	—	180
		185	—	—	6	4	3	—	—	—	—	—	—	—	—	—	185
	190		—	—	6	4	3	—	—	—	—	—	—	—	—	—	190
		195	—	—	6	4	3	—	—	—	—	—	—	—	—	—	195
200			—	—	6	4	3	—	—	—	—	—	—	—	—	—	200
		205	—	—	6	4	3	—	—	—	—	—	—	—	—	—	205
	210		—	—	6	4	3	—	—	—	—	—	—	—	—	—	210
220			—	—	6	4	3	—	—	—	—	—	—	—	—	—	220
		225	—	—	6	4	3	—	—	—	—	—	—	—	—	—	225
		230	—	—	6	4	3	—	—	—	—	—	—	—	—	—	230
		235	—	—	6	4	3	—	—	—	—	—	—	—	—	—	235
	240		—	—	6	4	3	—	—	—	—	—	—	—	—	—	240
		245	—	—	6	4	3	—	—	—	—	—	—	—	—	—	245
250			—	—	6	4	3	—	—	—	—	—	—	—	—	—	250
		255	—	—	6	4	—	—	—	—	—	—	—	—	—	—	255
	260		—	—	6	4	—	—	—	—	—	—	—	—	—	—	260
		265	—	—	6	4	—	—	—	—	—	—	—	—	—	—	265
		270	—	—	6	4	—	—	—	—	—	—	—	—	—	—	270
		275	—	—	6	4	—	—	—	—	—	—	—	—	—	—	275
280			—	—	6	4	—	—	—	—	—	—	—	—	—	—	280
		285	—	—	6	4	—	—	—	—	—	—	—	—	—	—	285
		290	—	—	6	4	—	—	—	—	—	—	—	—	—	—	290
		295	—	—	6	4	—	—	—	—	—	—	—	—	—	—	295
	300		—	—	6	4	—	—	—	—	—	—	—	—	—	—	300

[1] Thread diameter should be selected from columns 1, 2, or 3; with preference being in that order.

Appendix 18
Square and acme threads

Size	Threads per Inch	Size	Threads per Inch
$\frac{3}{8}$	12	2	$2\frac{1}{2}$
$\frac{7}{16}$	10	$2\frac{1}{4}$	2
$\frac{1}{2}$	10	$2\frac{1}{2}$	2
$\frac{9}{16}$	8	$2\frac{3}{4}$	2
$\frac{5}{8}$	8	3	$1\frac{1}{2}$
$\frac{3}{4}$	6	$3\frac{1}{4}$	$1\frac{1}{2}$
$\frac{7}{8}$	5	$3\frac{1}{2}$	$1\frac{1}{3}$
1	5	$3\frac{3}{4}$	$1\frac{1}{3}$
$1\frac{1}{8}$	4	4	$1\frac{1}{3}$
$1\frac{1}{4}$	4	$4\frac{1}{4}$	$1\frac{1}{3}$
$1\frac{1}{2}$	3	$4\frac{1}{2}$	1
$1\frac{3}{4}$	$2\frac{1}{2}$	over $4\frac{1}{2}$	1

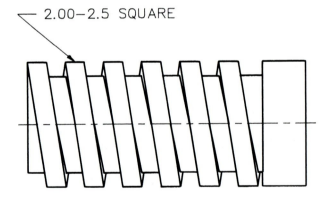

2.00–2.5 SQUARE

SQUARE THREAD NOTE

Appendix 19
American standard square bolts and nuts

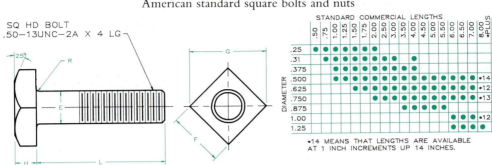

SQ HD BOLT
.50—13UNC—2A X 4 LG

*14 MEANS THAT LENGTHS ARE AVAILABLE
AT 1 INCH INCREMENTS UP 14 INCHES.

Dimensions of Square Bolts

Nominal Size or Basic Product Dia		Body Dia E	Width Across Flats F			Width Across Corners G		Height H			Radius of Fillet R
		Max	Basic	Max	Min	Max	Min	Basic	Max	Min	Max
1/4	0.2500	0.260	3/8	0.3750	0.362	0.530	0.498	11/64	0.188	0.156	0.031
5/16	0.3125	0.324	1/2	0.5000	0.484	0.707	0.665	13/64	0.220	0.186	0.031
3/8	0.3750	0.388	9/16	0.5625	0.544	0.795	0.747	1/4	0.268	0.232	0.031
7/16	0.4375	0.452	5/8	0.6250	0.603	0.884	0.828	19/64	0.316	0.278	0.031
1/2	0.5000	0.515	3/4	0.7500	0.725	1.061	0.995	21/64	0.348	0.308	0.031
5/8	0.6250	0.642	15/16	0.9375	0.906	1.326	1.244	27/64	0.444	0.400	0.062
3/4	0.7500	0.768	1 1/8	1.1250	1.088	1.591	1.494	1/2	0.524	0.476	0.062
7/8	0.8750	0.895	1 5/16	1.3125	1.269	1.856	1.742	19/32	0.620	0.568	0.062
1	1.0000	1.022	1 1/2	1.5000	1.450	2.121	1.991	21/32	0.684	0.628	0.093
1 1/8	1.1250	1.149	1 11/16	1.6875	1.631	2.386	2.239	3/4	0.780	0.720	0.093
1 1/4	1.2500	1.277	1 7/8	1.8750	1.812	2.652	2.489	27/32	0.876	0.812	0.093
1 3/8	1.3750	1.404	2 1/16	2.0625	1.994	2.917	2.738	29/32	0.940	0.872	0.093
1 1/2	1.5000	1.531	2 1/4	2.2500	2.175	3.182	2.986	1	1.036	0.964	0.093

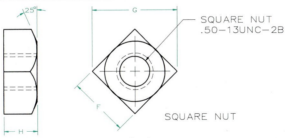

SQUARE NUT
.50—13UNC—2B

SQUARE NUT

Dimensions of Square Nuts

Nominal Size or Basic Major Dia of Thread		Width Across Flats F			Width Across Corners G		Thickness H		
		Basic	Max	Min	Max	Min	Basic	Max	Min
1/4	0.2500	7/16	0.4375	0.425	0.619	0.584	7/32	0.235	0.203
5/16	0.3125	9/16	0.5625	0.547	0.795	0.751	17/64	0.283	0.249
3/8	0.3750	5/8	0.6250	0.606	0.884	0.832	21/64	0.346	0.310
7/16	0.4375	3/4	0.7500	0.728	1.061	1.000	3/8	0.394	0.356
1/2	0.5000	13/16	0.8125	0.788	1.149	1.082	7/16	0.458	0.418
5/8	0.6250	1	1.0000	0.969	1.414	1.330	35/64	0.569	0.525
3/4	0.7500	1 1/8	1.1250	1.088	1.591	1.494	21/32	0.680	0.632
7/8	0.8750	1 5/16	1.3125	1.269	1.856	1.742	49/64	0.792	0.740
1	1.0000	1 1/2	1.5000	1.450	2.121	1.991	7/8	0.903	0.847
1 1/8	1.1250	1 11/16	1.6875	1.631	2.386	2.239	1	1.030	0.970
1 1/4	1.2500	1 7/8	1.8750	1.812	2.652	2.489	1 3/32	1.126	1.062
1 3/8	1.3750	2 1/16	2.0625	1.994	2.917	2.738	1 13/64	1.237	1.169
1 1/2	1.5000	2 1/4	2.2500	2.175	3.182	2.986	1 5/16	1.348	1.276

(Courtesy of ANSI; B18.2.1–1965 and ANSI; B18.2.2–1965.)

Appendix 20
American standard hexagon head bolts and nuts

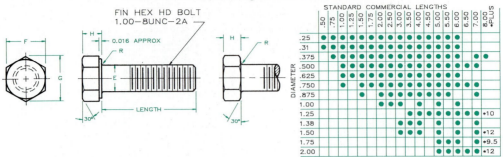

FIN HEX HD BOLT
1.00—8UNC—2A

*10 MEANS THAT LENGTHS ARE AVAILABLE AT 1 INCH INCREMENTS UP TO 10 INCES.

Dimensions of Hex Cap Screws (Finished Hex Bolts)

Nominal Size or Basic Product Dia		Body Dia E		Width Across Flats F			Width Across Corners G		Height H			Radius of Fillet R	
		Max	Min	Basic	Max	Min	Max	Min	Basic	Max	Min	Max	Min
1/4	0.2500	0.2500	0.2450	7/16	0.4375	0.428	0.505	0.488	5/32	0.163	0.150	0.025	0.015
5/16	0.3125	0.3125	0.3065	1/2	0.5000	0.489	0.577	0.557	13/64	0.211	0.195	0.025	0.015
3/8	0.3750	0.3750	0.3690	9/16	0.5625	0.551	0.650	0.628	15/64	0.243	0.226	0.025	0.015
7/16	0.4375	0.4375	0.4305	5/8	0.6250	0.612	0.722	0.698	9/32	0.291	0.272	0.025	0.015
1/2	0.5000	0.5000	0.4930	3/4	0.7500	0.736	0.866	0.840	5/16	0.323	0.302	0.025	0.015
9/16	0.5625	0.5625	0.5545	13/16	0.8125	0.798	0.938	0.910	23/64	0.371	0.348	0.045	0.020
5/8	0.6250	0.6250	0.6170	15/16	0.9375	0.922	1.083	1.051	25/64	0.403	0.378	0.045	0.020
3/4	0.7500	0.7500	0.7410	1 1/8	1.1250	1.100	1.299	1.254	15/32	0.483	0.455	0.045	0.020
7/8	0.8750	0.8750	0.8660	1 5/16	1.3125	1.285	1.516	1.465	35/64	0.563	0.531	0.065	0.040
1	1.0000	1.0000	0.9900	1 1/2	1.5000	1.469	1.732	1.675	39/64	0.627	0.591	0.095	0.060
1 1/8	1.1250	1.1250	1.1140	1 11/16	1.6875	1.631	1.949	1.859	11/16	0.718	0.658	0.095	0.060
1 1/4	1.2500	1.2500	1.2390	1 7/8	1.8750	1.812	2.165	2.066	25/32	0.813	0.749	0.095	0.060
1 3/8	1.3750	1.3750	1.3630	2 1/16	2.0625	1.994	2.382	2.273	27/32	0.878	0.810	0.095	0.060
1 1/2	1.5000	1.5000	1.4880	2 1/4	2.2500	2.175	2.598	2.480	15/16	0.974	0.902	0.095	0.060
1 3/4	1.7500	1.7500	1.7380	2 5/8	2.6250	2.538	3.031	2.893	1 3/32	1.134	1.054	0.095	0.060
2	2.0000	2.0000	1.9880	3	3.0000	2.900	3.464	3.306	1 7/32	1.263	1.175	0.095	0.060
2 1/4	2.2500	2.2500	2.2380	3 3/8	3.3750	3.262	3.897	3.719	1 3/8	1.423	1.327	0.095	0.060
2 1/2	2.5000	2.5000	2.4880	3 3/4	3.7500	3.625	4.330	4.133	1 17/32	1.583	1.479	0.095	0.060
2 3/4	2.7500	2.7500	2.7380	4 1/8	4.1250	3.988	4.763	4.546	1 11/16	1.744	1.632	0.095	0.060
3	3.0000	3.0000	2.9880	4 1/2	4.5000	4.350	5.196	4.959	1 7/8	1.935	1.815	0.095	0.060

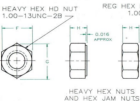

HEAVY HEX HD NUT
1.00—13UNC—2B

REG HEX HD JAM NUT
1.00—13UNC—2B

HEAVY HEX NUTS AND HEX JAM NUTS

REGULAR HEX NUTS AND HEX JAM NUTS

Dimensions of Hex Nuts and Hex Jam Nuts

Nominal Size or Basic Major Dia of Thread		Width Across Flats F			Width Across Corners G		Thickness Hex Nuts H			Thickness Hex Jam Nuts H		
		Basic	Max	Min	Max	Min	Basic	Max	Min	Basic	Max	Min
1/4	0.2500	7/16	0.4375	0.428	0.505	0.488	7/32	0.226	0.212	5/32	0.163	0.150
5/16	0.3125	1/2	0.5000	0.489	0.577	0.557	17/64	0.273	0.258	3/16	0.195	0.180
3/8	0.3750	9/16	0.5625	0.551	0.650	0.628	21/64	0.337	0.320	7/32	0.227	0.210
7/16	0.4375	11/16	0.6875	0.675	0.794	0.768	3/8	0.385	0.365	1/4	0.260	0.240
1/2	0.5000	3/4	0.7500	0.736	0.866	0.840	7/16	0.448	0.427	5/16	0.323	0.302
9/16	0.5625	7/8	0.8750	0.861	1.010	0.982	31/64	0.496	0.473	5/16	0.324	0.301
5/8	0.6250	15/16	0.9375	0.922	1.083	1.051	35/64	0.559	0.535	3/8	0.387	0.363
3/4	0.7500	1 1/8	1.1250	1.088	1.299	1.240	41/64	0.665	0.617	27/64	0.446	0.398
7/8	0.8750	1 5/16	1.3125	1.269	1.516	1.447	3/4	0.776	0.724	31/64	0.510	0.458
1	1.0000	1 1/2	1.5000	1.450	1.732	1.653	55/64	0.887	0.831	35/64	0.575	0.519
1 1/8	1.1250	1 11/16	1.6875	1.631	1.949	1.859	31/32	0.999	0.939	39/64	0.639	0.579
1 1/4	1.2500	1 7/8	1.8750	1.812	2.165	2.066	1 1/16	1.094	1.030	23/32	0.751	0.687
1 3/8	1.3750	2 1/16	2.0625	1.994	2.382	2.273	1 11/64	1.206	1.138	25/32	0.815	0.747
1 1/2	1.5000	2 1/4	2.2500	2.175	2.598	2.480	1 9/32	1.317	1.245	27/32	0.880	0.808

(Courtesy of ANSI; B18.2.1–1965 and ANSI; B18.2.2–1965.)

Appendix 21
Fillister head and round head cap screws

FILLISTER—HEAD CAP SCREW
.50—13UNC—2A X 2.00 LG

THREADS
LENGTH

*NOTE: OTHER LENGTHS ARE AVAILABLE. THESE LENGTHS ARE THE MOST STANDARD.

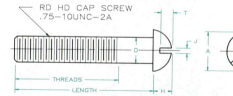

STANDARD LENGTHS

DIAMETER	.50	.75	1.00	1.25	1.50	1.75	2.00	2.50
.25	●	●	●	●	●	●	●	●
.313	●		●	●			●	
.375	●	●	●			●		
.500	●	●				●		
.625		●			●		●	
.750		●		●		●	●	
.875				●	●		●	
1.00			●	●	●	●	●	●

Nominal Size	D Body Diameter		A Head Diameter		H Height of Head		O Total Height of Head		J Width of Slot		T Depth of Slot	
	Max	Min	Max	Min	Max	Min	Max	Min	Max	Min	Max	Min
1/4	0.250	0.245	0.375	0.363	0.172	0.157	0.216	0.194	0.075	0.064	0.097	0.077
5/16	0.3125	0.307	0.437	0.424	0.203	0.186	0.253	0.230	0.084	0.072	0.115	0.090
3/8	0.375	0.369	0.562	0.547	0.250	0.229	0.314	0.284	0.094	0.081	0.142	0.112
7/16	0.4375	0.431	0.625	0.608	0.297	0.274	0.368	0.336	0.094	0.081	0.168	0.133
1/2	0.500	0.493	0.750	0.731	0.328	0.301	0.413	0.376	0.106	0.091	0.193	0.153
9/16	0.5625	0.555	0.812	0.792	0.375	0.346	0.467	0.427	0.118	0.102	0.213	0.168
5/8	0.625	0.617	0.875	0.853	0.422	0.391	0.521	0.478	0.133	0.116	0.239	0.189
3/4	0.750	0.742	1.000	0.976	0.500	0.466	0.612	0.566	0.149	0.131	0.283	0.223
7/8	0.875	0.866	1.125	1.098	0.594	0.556	0.720	0.668	0.167	0.147	0.334	0.264
1	1.000	0.990	1.312	1.282	0.656	0.612	0.803	0.743	0.188	0.166	0.371	0.291

All dimensions are given in inches.

The radius of the fillet at the base of the head:
For sizes 1/4 to 3/8 in. incl. is 0.016 min and 0.031 max,
7/16 to 9/16 in. incl. is 0.016 min and 0.047 max,
5/8 to 1 in. incl. is 0.031 min and 0.062 max.

RD HD CAP SCREW
.75—10UNC—2A

THREADS
LENGTH

FOR COMMERCIAL LENGTHS, REFER TO THE TABLE OF LENGTHS GIVEN WITH FILLISTER—HEAD SCREWS.

Nominal Size	D Body Diameter		A Head Diameter		H Height of Head		J Width of Slot		T Depth of Slot	
	Max	Min	Max	Min	Max	Min	Max	Min	Max	Min
1/4	0.250	0.245	0.437	0.418	0.191	0.175	0.075	0.064	0.117	0.097
5/16	0.3125	0.307	0.562	0.540	0.245	0.226	0.084	0.072	0.151	0.126
3/8	0.375	0.369	0.625	0.603	0.273	0.252	0.094	0.081	0.168	0.138
7/16	0.4375	0.431	0.750	0.725	0.328	0.302	0.094	0.081	0.202	0.167
1/2	0.500	0.493	0.812	0.786	0.354	0.327	0.106	0.091	0.218	0.178
9/16	0.5625	0.555	0.937	0.909	0.409	0.378	0.118	0.102	0.252	0.207
5/8	0.625	0.617	1.000	0.970	0.437	0.405	0.133	0.116	0.270	0.220
3/4	0.750	0.742	1.250	1.215	0.546	0.507	0.149	0.131	0.338	0.278

All dimensions are given in inches.

Radius of the fillet at the base of the head:
For sizes 1/4 to 3/8 in. incl. is 0.016 min and 0.031 max,
7/16 to 9/16 in. incl. is 0.016 min and 0.047 max,
5/8 to 1 in. incl. is 0.031 min and 0.062 max.

(Courtesy of ANSI; B18.6.2–1956.)

Appendix 22
Flat head cap screws

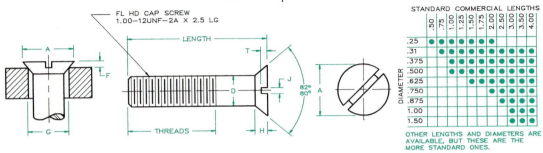

OTHER LENGTHS AND DIAMETERS ARE
AVAILABLE, BUT THESE ARE THE
MORE STANDARD ONES.

Nom-inal Size	D Body Diameter		A Head Diameter			G Gaging Diam-eter	H Height of Head	J Width of Slot		T Depth of Slot		F Protrusion Above Gaging Diameter	
	Max	Min	Max	Min	Absolute Min with Flat		Aver-age	Max	Min	Max	Min	Max	Min
1/4	0.250	0.245	0.500	0.477	0.452	0.4245	0.140	0.075	0.064	0.068	0.045	0.0452	0.0307
5/16	0.3125	0.307	0.625	0.598	0.567	0.5376	0.177	0.084	0.072	0.086	0.057	0.0523	0.0354
3/8	0.375	0.369	0.750	0.720	0.682	0.6507	0.210	0.094	0.081	0.103	0.068	0.0594	0.0401
7/16	0.4375	0.431	0.8125	0.780	0.736	0.7229	0.210	0.094	0.081	0.103	0.068	0.0649	0.0448
1/2	0.500	0.493	0.875	0.841	0.791	0.7560	0.210	0.106	0.091	0.103	0.068	0.0705	0.0495
9/16	0.5625	0.555	1.000	0.962	0.906	0.8691	0.244	0.118	0.102	0.120	0.080	0.0775	0.0542
5/8	0.625	0.617	1.125	1.083	1.020	0.9822	0.281	0.133	0.116	0.137	0.091	0.0846	0.0588
3/4	0.750	0.742	1.375	1.326	1.251	1.2085	0.352	0.149	0.131	0.171	0.115	0.0987	0.0682
7/8	0.875	0.866	1.625	1.568	1.480	1.4347	0.423	0.167	0.147	0.206	0.138	0.1128	0.0776
1	1.000	0.990	1.875	1.811	1.711	1.6610	0.494	0.188	0.166	0.240	0.162	0.1270	0.0870
1 1/8	1.125	1.114	2.062	1.992	1.880	1.8262	0.529	0.196	0.178	0.257	0.173	0.1401	0.0964
1 1/4	1.250	1.239	2.312	2.235	2.110	2.0525	0.600	0.211	0.193	0.291	0.197	0.1542	0.1056
1 3/8	1.375	1.363	2.562	2.477	2.340	2.2787	0.665	0.226	0.208	0.326	0.220	0.1684	0.1151
1 1/2	1.500	1.488	2.812	2.720	2.570	2.5050	0.742	0.258	0.240	0.360	0.244	0.1825	0.1245

All dimensions are given in inches.

The maximum and minimum head diameters, A, are extended to the theoretical sharp corners.

The radius of the fillet at the base of the head shall not exceed 0.4 Max. D.

*Edge of head may be flat as shown or slightly rounded.

(Courtesy of ANSI; B18.6.2–1956.)

Appendix 23
Machine screws

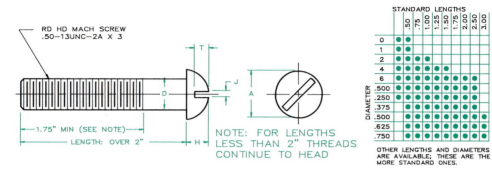

NOTE: FOR LENGTHS LESS THAN 2" THREADS CONTINUE TO HEAD

OTHER LENGTHS AND DIAMETERS ARE AVAILABLE; THESE ARE THE MORE STANDARD ONES.

Dimensions of Slotted Round Head Machine Screws

Nom-inal Size	D Diameter of Screw Basic	A Head Diameter Max	A Head Diameter Min	H Head Height Max	H Head Height Min	J Width of Slot Max	J Width of Slot Min	T Depth of Slot Max	T Depth of Slot Min
0	0.0600	0.113	0.099	0.053	0.043	0.023	0.016	0.039	0.029
1	0.0730	0.138	0.122	0.061	0.051	0.026	0.019	0.044	0.033
2	0.0860	0.162	0.146	0.069	0.059	0.031	0.023	0.048	0.037
3	0.0990	0.187	0.169	0.078	0.067	0.035	0.027	0.053	0.040
4	0.1120	0.211	0.193	0.086	0.075	0.039	0.031	0.058	0.044
5	0.1250	0.236	0.217	0.095	0.083	0.043	0.035	0.063	0.047
6	0.1380	0.260	0.240	0.103	0.091	0.048	0.039	0.068	0.051
8	0.1640	0.309	0.287	0.120	0.107	0.054	0.045	0.077	0.058
10	0.1900	0.359	0.334	0.137	0.123	0.060	0.050	0.087	0.065
12	0.2160	0.408	0.382	0.153	0.139	0.067	0.056	0.096	0.073
1/4	0.2500	0.472	0.443	0.175	0.160	0.075	0.064	0.109	0.082
5/16	0.3125	0.590	0.557	0.216	0.198	0.084	0.072	0.132	0.099
3/8	0.3750	0.708	0.670	0.256	0.237	0.094	0.081	0.155	0.117
7/16	0.4375	0.750	0.707	0.328	0.307	0.094	0.081	0.196	0.148
1/2	0.5000	0.813	0.766	0.355	0.332	0.106	0.091	0.211	0.159
9/16	0.5625	0.938	0.887	0.410	0.385	0.118	0.102	0.242	0.183
5/8	0.6250	1.000	0.944	0.438	0.411	0.133	0.116	0.258	0.195
3/4	0.7500	1.250	1.185	0.547	0.516	0.149	0.131	0.320	0.242

All dimensions are given in inches.

(Courtesy of ANSI; B18.6.3–1962.)

Appendix 24
American standard machine screws

(The proportions of the screws can be found by multiplying the major diameter, D, by the factors given below.)

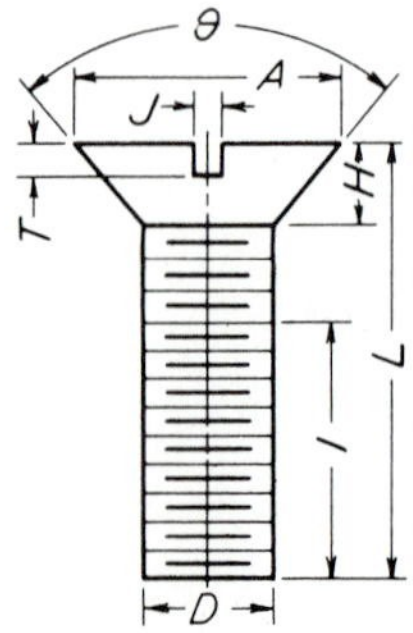

Flat Head

	Maximum	Minimum
A	2.04D + .003	1.84D
H	.619D −.002	.552D −.007
J	.182D +.020	.176 D +.010
T	.288D −.002	.192D − .002
θ	82°	80°

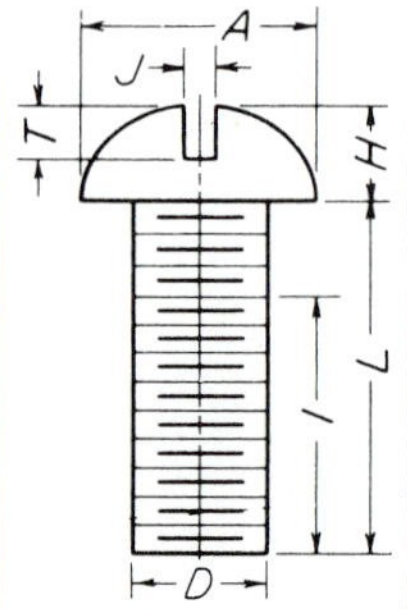

Round Head

	Maximum	Minimum
A	1.887D	1.813D − .010
H	.636D +.015	.624D +.005
J	.182D +.020	.176 D +.010
T	.362D +.017	.268D +.013

Profile of head is semi-elliptical

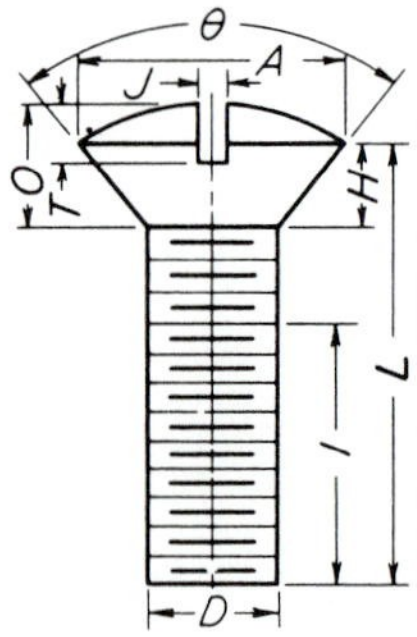

Oval Head

	Maximum	Minimum
A	2.04D +.003	1.84D
H	.619D −.002	.552D −.007
J	.182D +.020	.176 D +.010
O	.923D +.001	.820D −.008
T	.556D − .003	.460D −.003
θ	82°	80°

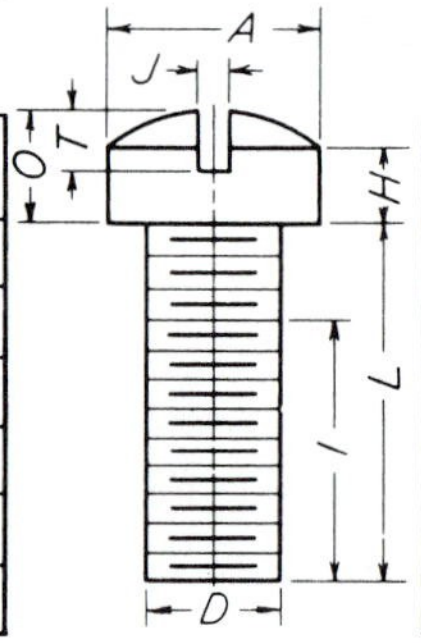

Fillister Head

	Maximum	Minimum
A	1.670D −.004	1.610D − .014
H	.620D +.010	.582D +.005
J	.182D +.020	.176 D +.010
O	.940D +.002	.820D −.008
T	.440D −.001	.374D −.011

Appendix 25
American standard machine tapers*

No. of Taper	Taper per Foot (Basic)	Origin of Series	No. of Taper	Taper per Foot (Basic)	Origin of Series	No. of Taper	Taper per Foot (Basic)	Origin of Series	No. of Taper	Taper per Foot (Basic)	Origin of Series
0.239	0.50200	Brown & Sharpe	*	0.62326	Morse	250	0.750	¾ in. per ft.	600	0.750	¾ in. per ft.
.299	.50200	Brown & Sharpe	4½	.62400	Morse	300	.750	¾ in. per ft.	800	0.750	¾ in. per ft.
.375	.50200	Brown & Sharpe	5	.63151	Morse	350	.750	¾ in. per ft.	1000	0.750	¾ in. per ft.
1	.59858	Morse	6	.62565	Morse	400	.750	¾ in. per ft.	1200	0.750	¾ in. per ft.
2	.59941	Morse	7	.62400	Morse	450	.750	¾ in. per ft.			
3	.60235	Morse	200	.750	¾ in. per ft.	500	.750	¾ in. per ft.			

All dimensions in inches.

* Extracted from American Standards, "Machine Tapers, Self-Holding and Steep Taper Series" (ASA B5,10-1960), with the permission of the publisher, The American Society of Mechanical Engineers.

Appendix 26
American national standard square head set screws (ANSI B18.6.2)

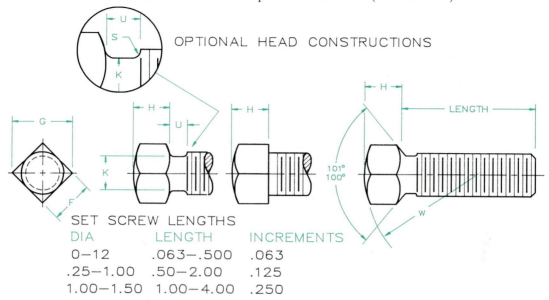

OPTIONAL HEAD CONSTRUCTIONS

SET SCREW LENGTHS

DIA	LENGTH	INCREMENTS
0–12	.063–.500	.063
.25–1.00	.50–2.00	.125
1.00–1.50	1.00–4.00	.250

Nominal Size[1] or Basic Screw Diameter		F Width Across Flats		G Width Across Corners		H Head Height		K Neck Relief Diameter		S Neck Relief Fillet Radius	U Neck Relief Width	W Head Radius
		Max	Min	Max	Min	Max	Min	Max	Min	Max	Min	Min
10	0.1900	0.188	0.180	0.265	0.247	0.148	0.134	0.145	0.140	0.027	0.083	0.48
1/4	0.2500	0.250	0.241	0.354	0.331	0.196	0.178	0.185	0.170	0.032	0.100	0.62
5/16	0.3125	0.312	0.302	0.442	0.415	0.245	0.224	0.240	0.225	0.036	0.111	0.78
3/8	0.3750	0.375	0.362	0.530	0.497	0.293	0.270	0.294	0.279	0.041	0.125	0.94
7/16	0.4375	0.438	0.423	0.619	0.581	0.341	0.315	0.345	0.330	0.046	0.143	1.09
1/2	0.5000	0.500	0.484	0.707	0.665	0.389	0.361	0.400	0.385	0.050	0.154	1.25
9/16	0.5625	0.562	0.545	0.795	0.748	0.437	0.407	0.454	0.439	0.054	0.167	1.41
5/8	0.6250	0.625	0.606	0.884	0.833	0.485	0.452	0.507	0.492	0.059	0.182	1.56
3/4	0.7500	0.750	0.729	1.060	1.001	0.582	0.544	0.620	0.605	0.065	0.200	1.88
7/8	0.8750	0.875	0.852	1.237	1.170	0.678	0.635	0.731	0.716	0.072	0.222	2.19
1	1.0000	1.000	0.974	1.414	1.337	0.774	0.726	0.838	0.823	0.081	0.250	2.50
1 1/8	1.1250	1.125	1.096	1.591	1.505	0.870	0.817	0.939	0.914	0.092	0.283	2.81
1 1/4	1.2500	1.250	1.219	1.768	1.674	0.966	0.908	1.064	1.039	0.092	0.283	3.12
1 3/8	1.3750	1.375	1.342	1.945	1.843	1.063	1.000	1.159	1.134	0.109	0.333	3.44
1 1/2	1.5000	1.500	1.464	2.121	2.010	1.159	1.091	1.284	1.259	0.109	0.333	3.75

[1] Where specifying nominal size in decimals, zeros preceding decimal and in the fourth decimal place shall be omitted.

Appendix 27
American national standard points for square head set screws (ANSI B18.6.2)

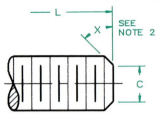

FLAT POINT

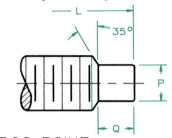

DOG POINT

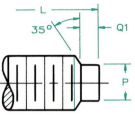

HALF DOG POINT

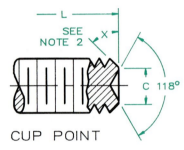

CUP POINT

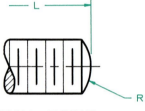

OVAL POINT

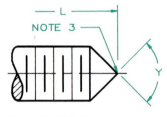

CONE POINT

Nominal Size[1] or Basic Screw Diameter		C Cup and Flat Point Diameters		P Dog and Half Dog Point Diameters		Q Point Length				R Oval Point Radius +0.031 -0.000	Y Cone Point Angle 90° ±2° For These Nominal Lengths or Longer; 118° ±2° For Shorter Screws
						Dog		Half Dog			
		Max	Min	Max	Min	Max	Min	Max	Min		
10	0.1900	0.102	0.088	0.127	0.120	0.095	0.085	0.050	0.040	0.142	1/4
1/4	0.2500	0.132	0.118	0.156	0.149	0.130	0.120	0.068	0.058	0.188	5/16
5/16	0.3125	0.172	0.156	0.203	0.195	0.161	0.151	0.083	0.073	0.234	3/8
3/8	0.3750	0.212	0.194	0.250	0.241	0.193	0.183	0.099	0.089	0.281	7/16
7/16	0.4375	0.252	0.232	0.297	0.287	0.224	0.214	0.114	0.104	0.328	1/2
1/2	0.5000	0.291	0.270	0.344	0.334	0.255	0.245	0.130	0.120	0.375	9/16
9/16	0.5625	0.332	0.309	0.391	0.379	0.287	0.275	0.146	0.134	0.422	5/8
5/8	0.6250	0.371	0.347	0.469	0.456	0.321	0.305	0.164	0.148	0.469	3/4
3/4	0.7500	0.450	0.425	0.562	0.549	0.383	0.367	0.196	0.180	0.562	7/8
7/8	0.8750	0.530	0.502	0.656	0.642	0.446	0.430	0.227	0.211	0.656	1
1	1.0000	0.609	0.579	0.750	0.734	0.510	0.490	0.260	0.240	0.750	1 1/8
1 1/8	1.1250	0.689	0.655	0.844	0.826	0.572	0.552	0.291	0.271	0.844	1 1/4
1 1/4	1.2500	0.767	0.733	0.938	0.920	0.635	0.615	0.323	0.303	0.938	1 1/2
1 3/8	1.3750	0.848	0.808	1.031	1.011	0.698	0.678	0.354	0.334	1.031	1 5/8
1 1/2	1.5000	0.926	0.886	1.125	1.105	0.760	0.740	0.385	0.365	1.125	1 3/4

[1] Where specifying nominal size in decimals, zeros preceding decimal and in the fourth decimal place shall be omitted.
[2] Point angle X shall be 45° plus 5°, minus 0°, for screws of nominal lengths equal to or longer than those listed in Column Y, and 30° minimum for screws of shorter nominal lengths.
[3] The extent of rounding or flat at apex of cone point shall not exceed an amount equivalent to 10 per cent of the basic screw diameter.

Appendix 28
American national standard slotted headless set screws (ANSI B18.6.2)

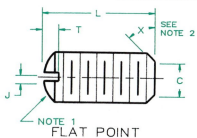

FLAT POINT

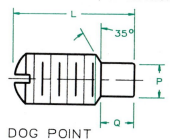

DOG POINT

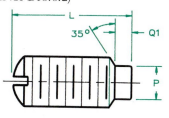

HALF DOG POINT

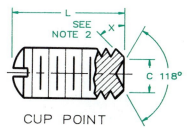

CUP POINT

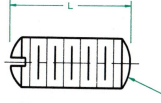

OVAL POINT

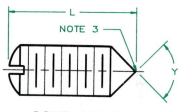

CONE POINT

Nominal Size[1] or Basic Screw Diameter		I^2 Crown Radius	J Slot Width		T Slot Depth		C Cup and Flat Point Diameters		P Dog Point Diameters		Q Point Length Dog		Q_1 Point Length Half Dog		R^2 Oval Point Radius	Y Cone Point Angle 90° ±2° For These Nominal Lengths or Longer; 118° ±2° For Shorter Screws
		Basic	Max	Min	Max	Min	Max	Min	Max	Min	Max	Min	Max	Min	Basic	
0	0.0600	0.060	0.014	0.010	0.020	0.016	0.033	0.027	0.040	0.037	0.032	0.028	0.017	0.013	0.045	5/64
1	0.0730	0.073	0.016	0.012	0.020	0.016	0.040	0.033	0.049	0.045	0.040	0.036	0.021	0.017	0.055	3/32
2	0.0860	0.086	0.018	0.014	0.025	0.019	0.047	0.039	0.057	0.053	0.046	0.042	0.024	0.020	0.064	7/64
3	0.0990	0.099	0.020	0.016	0.028	0.022	0.054	0.045	0.066	0.062	0.052	0.048	0.027	0.023	0.074	1/8
4	0.1120	0.112	0.024	0.018	0.031	0.025	0.061	0.051	0.075	0.070	0.058	0.054	0.030	0.026	0.084	5/32
5	0.1250	0.125	0.026	0.020	0.036	0.026	0.067	0.057	0.083	0.078	0.063	0.057	0.033	0.027	0.094	3/16
6	0.1380	0.138	0.028	0.022	0.040	0.030	0.074	0.064	0.092	0.087	0.073	0.067	0.038	0.032	0.104	3/16
8	0.1640	0.164	0.032	0.026	0.046	0.036	0.087	0.076	0.109	0.103	0.083	0.077	0.043	0.037	0.123	1/4
10	0.1900	0.190	0.035	0.029	0.053	0.043	0.102	0.088	0.127	0.120	0.095	0.085	0.050	0.040	0.142	1/4
12	0.2160	0.216	0.042	0.035	0.061	0.051	0.115	0.101	0.144	0.137	0.115	0.105	0.060	0.050	0.162	5/16
1/4	0.2500	0.250	0.049	0.041	0.068	0.058	0.132	0.118	0.156	0.149	0.130	0.120	0.068	0.058	0.188	5/16
5/16	0.3125	0.312	0.055	0.047	0.083	0.073	0.172	0.156	0.203	0.195	0.161	0.151	0.083	0.073	0.234	3/8
3/8	0.3750	0.375	0.068	0.060	0.099	0.089	0.212	0.194	0.250	0.241	0.193	0.183	0.099	0.089	0.281	7/16
7/16	0.4375	0.438	0.076	0.068	0.114	0.104	0.252	0.232	0.297	0.287	0.224	0.214	0.114	0.104	0.328	1/2
1/2	0.5000	0.500	0.086	0.076	0.130	0.120	0.291	0.270	0.344	0.334	0.255	0.245	0.130	0.120	0.375	9/16
9/16	0.5625	0.562	0.096	0.086	0.146	0.136	0.332	0.309	0.391	0.379	0.287	0.275	0.146	0.134	0.422	5/8
5/8	0.6250	0.625	0.107	0.097	0.161	0.151	0.371	0.347	0.469	0.456	0.321	0.305	0.164	0.148	0.469	3/4
3/4	0.7500	0.750	0.134	0.124	0.193	0.183	0.450	0.425	0.562	0.549	0.383	0.367	0.196	0.180	0.562	7/8

[1] Where specifying nominal size in decimals, zeros preceding decimal and in the fourth decimal place shall be omitted.

[2] Tolerance on radius for nominal sizes up to and including 5 (0.125 in.) shall be plus 0.015 in. and minus 0.000, and for larger sizes, plus 0.031 in. and minus 0.000. Slotted ends on screws may be flat at option of manufacturer.

[3] Point angle X shall be 45° plus 5°, minus 0°, for screws of nominal lengths equal to or longer than those listed in Column Y, and 30° minimum for screws of shorter nominal lengths.

[4] The extent of rounding or flat at apex of cone point shall not exceed an amount equivalent to 10 per cent of the basic screw diameter.

Appendix 29
Trist drill sizes

Number Size Drills

Size	Drill Diameter		Size	Drill Diameter		Size	Drill Diameter		Size	Drill Diameter	
	Inches	mm		Inches	mm		Inches	mm		Inches	mm
1	0.2280	5.7912	21	0.1590	4.0386	41	0.0960	2.4384	61	0.0390	0.9906
2	0.2210	5.6134	22	0.1570	3.9878	42	0.0935	2.3622	62	0.0380	0.9652
3	0.2130	5.4102	23	0.1540	3.9116	43	0.0890	2.2606	63	0.0370	0.9398
4	0.2090	5.3086	24	0.1520	3.8608	44	0.0860	2.1844	64	0.0360	0.9144
5	0.2055	5.2197	25	0.1495	3.7973	45	0.0820	2.0828	65	0.0350	0.8890
6	0.2040	5.1816	26	0.1470	3.7338	46	0.0810	2.0574	66	0.0330	0.8382
7	0.2010	5.1054	27	0.1440	3.6576	47	0.0785	1.9812	67	0.0320	0.8128
8	0.1990	5.0800	28	0.1405	3.5560	48	0.0760	1.9304	68	0.0310	0.7874
9	0.1960	4.9784	29	0.1360	3.4544	49	0.0730	1.8542	69	0.0292	0.7417
10	0.1935	4.9149	30	0.1285	3.2639	50	0.0700	1.7780	70	0.0280	0.7112
11	0.1910	4.8514	31	0.1200	3.0480	51	0.0670	1.7018	71	0.0260	0.6604
12	0.1890	4.8006	32	0.1160	2.9464	52	0.0635	1.6129	72	0.0250	0.6350
13	0.1850	4.6990	33	0.1130	2.8702	53	0.0595	1.5113	73	0.0240	0.6096
14	0.1820	4.6228	34	0.1110	2.8194	54	0.0550	1.3970	74	0.0225	0.5715
15	0.1800	4.5720	35	0.1100	2.7940	55	0.0520	1.3208	75	0.0210	0.5334
16	0.1770	4.4958	36	0.1065	2.7051	56	0.0465	1.1684	76	0.0200	0.5080
17	0.1730	4.3942	37	0.1040	2.6416	57	0.0430	1.0922	77	0.0180	0.4572
18	0.1695	4.3053	38	0.1015	2.5781	58	0.0420	1.0668	78	0.0160	0.4064
19	0.1660	4.2164	39	0.0995	2.5273	59	0.0410	1.0414	79	0.0145	0.3638
20	0.1610	4.0894	40	0.0980	2.4892	60	0.0400	1.0160	80	0.0135	0.3429

Metric Drill Sizes   Preferred sizes are in color type. Decimal-inch equivalents are for reference only.

Drill Diameter		Drill Diameter		Drill Diameter		Drill Diameter		Drill Diameter		Drill Diameter		Drill Diameter	
mm	in.	mm	in.	mm	in.	mm	in.	mm	in.	mm	in.	mm	in.
.40	.0157	1.03	.0406	2.20	.0866	5.00	.1969	10.00	.3937	21.50	.8465	48.00	1.8898
.42	.0165	1.05	.0413	2.30	.0906	5.20	.2047	10.30	.4055	22.00	.8661	50.00	1.9685
.45	.0177	1.08	.0425	2.40	.0945	5.30	.2087	10.50	.4134	23.00	.9055	51.50	2.0276
.48	.0189	1.10	.0433	2.50	.0984	5.40	.2126	10.80	.4252	24.00	.9449	53.00	2.0866
.50	.0197	1.15	.0453	2.60	.1024	5.60	.2205	11.00	.4331	25.00	.9843	54.00	2.1260
.52	.0205	1.20	.0472	2.70	.1063	5.80	.2283	11.50	.4528	26.00	1.0236	56.00	2.2047
.55	.0217	1.25	.0492	2.80	.1102	6.00	.2362	12.00	.4724	27.00	1.0630	58.00	2.2835
.58	.0228	1.30	.0512	2.90	.1142	6.20	.2441	12.50	.4921	28.00	1.1024	60.00	2.3622
.60	.0236	1.35	.0531	3.00	.1181	6.30	.2480	13.00	.5118	29.00	1.1417		
.62	.0244	1.40	.0551	3.10	.1220	6.50	.2559	13.50	.5315	30.00	1.1811		
.65	.0256	1.45	.0571	3.20	.1260	6.70	.2638	14.00	.5512	31.00	1.2205		
.68	.0268	1.50	.0591	3.30	.1299	6.80	.2677	14.50	.5709	32.00	1.2598		
.70	.0276	1.55	.0610	3.40	.1339	6.90	.2717	15.00	.5906	33.00	1.2992		
.72	.0283	1.60	.0630	3.50	.1378	7.10	.2795	15.50	.6102	34.00	1.3386		
.75	.0295	1.65	.0650	3.60	.1417	7.30	.2874	16.00	.6299	35.00	1.3780		
.78	.0307	1.70	.0669	3.70	.1457	7.50	.2953	16.50	.6496	36.00	1.4173		
.80	.0315	1.75	.0689	3.80	.1496	7.80	.3071	17.00	.6693	37.00	1.4567		
.82	.0323	1.80	.0709	3.90	.1535	8.00	.3150	17.50	.6890	38.00	1.4961		
.85	.0335	1.85	.0728	4.00	.1575	8.20	.3228	18.00	.7087	39.00	1.5354		
.88	.0346	1.90	.0748	4.10	.1614	8.50	.3346	18.50	.7283	40.00	1.5748		
.90	.0354	1.95	.0768	4.20	.1654	8.80	.3465	19.00	.7480	41.00	1.6142		
.92	.0362	2.00	.0787	4.40	.1732	9.00	.3543	19.50	.7677	42.00	1.6535		
.95	.0374	2.05	.0807	4.50	.1772	9.20	.3622	20.00	.7874	43.50	1.7126		
.98	.0386	2.10	.0827	4.60	.1811	9.50	.3740	20.50	.8071	45.00	1.7717		
1.00	.0394	2.15	.0846	4.80	.1890	9.80	.3858	21.00	.8268	46.50	1.8307		

Cont.

Appendix 29
Trist drill sizes (cont.)

Letter Size Drills

Size	Drill Diameter		Size	Drill Diameter		Size	Drill Diameter		Size	Drill Diameter	
	Inches	mm		Inches	mm		Inches	mm		Inches	mm
A	0.234	5.944	H	0.266	6.756	O	0.316	8.026	V	0.377	9.576
B	0.238	6.045	I	0.272	6.909	P	0.323	8.204	W	0.386	9.804
C	0.242	6.147	J	0.277	7.036	Q	0.332	8.433	X	0.397	10.084
D	0.246	6.248	K	0.281	7.137	R	0.339	8.611	Y	0.404	10.262
E	0.250	6.350	L	0.290	7.366	S	0.348	8.839	Z	0.413	10.490
F	0.257	6.528	M	0.295	7.493	T	0.358	9.093			
G	0.261	6.629	N	0.302	7.601	U	0.368	9.347			

(Courtesy of General Motors Corporation.)

Appendix 30
Straight pins

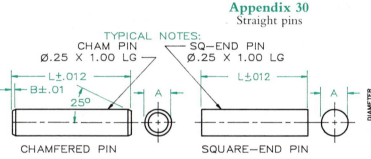

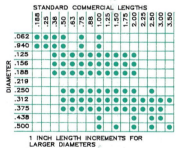

Nominal Diameter	Diameter A		Chamfer B
	Max	Min	
0.062	0.0625	0.0605	0.015
0.094	0.0937	0.0917	0.015
0.109	0.1094	0.1074	0.015
0.125	0.1250	0.1230	0.015
0.156	0.1562	0.1542	0.015
0.188	0.1875	0.1855	0.015
0.219	0.2187	0.2167	0.015
0.250	0.2500	0.2480	0.015
0.312	0.3125	0.3095	0.030
0.375	0.3750	0.3720	0.030
0.438	0.4375	0.4345	0.030
0.500	0.500	0.4970	0.030

All dimensions are given in inches.

These pins must be straight and free from burrs or any other defects that will affect their serviceability.

(Courtesy of ANSI; B5.20–1958.)

Appendix 31
Standard keys and keyways

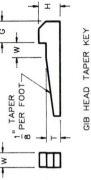

GIB HEAD TAPER KEY

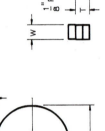

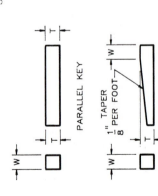

PARALLEL KEY

TAPER KEY

Sprocket Bore (= Shaft Diam.) Inches D	Keyway — For Square Key Width W	Depth T/2	Keyway — For Flat Key Width W	Depth T/2	Key — Square Width W	Height T	Key — Flat Width W	Height T	Tolerance on W and T (−)	Gib Head — Square Key H	G	Gib Head — Flat Key H	G	Key Tol W (−)	T (+)
1/2 — 9/16	1/8	1/16	1/8	3/64	1/8	1/8	1/8	3/32	0.002	1/4	7/32	3/16	1/8	0.002	0.002
5/8 — 7/8	3/16	3/32	3/16	1/16	3/16	3/16	3/16	1/8	0.002	5/16	9/32	1/4	3/16	0.002	0.002
13/16 — 1 1/4	1/4	1/8	1/4	3/32	1/4	1/4	1/4	3/16	0.002	7/16	11/32	5/16	1/4	0.002	0.002
15/16 — 1 3/8	5/16	5/32	5/16	1/8	5/16	5/16	5/16	1/4	0.002	9/16	13/32	3/8	5/16	0.002	0.002
1 7/16 — 1 3/4	3/8	3/16	3/8	1/8	3/8	3/8	3/8	1/4	0.002	11/16	15/32	7/16	3/8	0.002	0.002
1 13/16 — 2 1/4	1/2	1/4	1/2	3/16	1/2	1/2	1/2	3/8	0.0025	7/8	19/32	5/8	1/2	0.0025	0.0025
2 5/16 — 2 3/4	5/8	5/16	5/8	7/32	5/8	5/8	5/8	7/16	0.0025	1 1/16	23/32	3/4	5/8	0.0025	0.0025
2 7/8 — 3 1/4	3/4	3/8	3/4	1/4	3/4	3/4	3/4	1/2	0.0025	1 1/4	7/8	7/8	3/4	0.0025	0.0025
3 3/8 — 3 7/8	7/8	7/16	7/8	5/16	7/8	7/8	7/8	5/8	0.003	1 1/2	1	1 1/16	7/8	0.003	0.003
— 4 1/2	1	1/2	1	3/8	1	1	1	3/4	0.003	1 3/4	1 3/16	1 1/4	1	0.003	0.003
4 3/4 — 5 1/2	1 1/4	5/8	1 1/4	7/16	1 1/4	1 1/4	1 1/4	7/8	0.003	2	1 7/16	1 1/2	1 1/4	0.003	0.003
5 3/4 — 7 3/8	1 1/2	3/4	1 1/2	1/2	1 1/2	1 1/2	1 1/2	1	0.003	2 1/2	1 3/4	1 3/4	1 1/2	0.003	0.003
7 1/2 — 9 7/8	1 3/4	7/8	..	..	1 3/4	1 3/4	..	..	0.004	3	2	..	..	0.004	0.004
10 — 12 1/2	2	1	..	..	2	2	..	..	0.004	3 1/2	2 3/8	..	..	0.004	0.004

Standard Keyway Tolerances:　Straight Keyway — Width (W) + .005　Depth (T/2) + .010
　　　　　　　　　　　　　　　　　　　　　　　　　　　　− .000　　　　　　　　− .000

　　　　　　　　　　　　　　Taper Keyway — Width (W) + .005　Depth (T/2) + .000
　　　　　　　　　　　　　　　　　　　　　　　　　　　　− .000　　　　　　　　− .010

Appendix 32
Woodruff keys

FULL RADIUS TYPE   FLAT BOTTOM TYPE

WOODRUFF KEYS

BREAK CORNERS R 0.02 MAX

BREAK CORNERS R 0.02 MAX

Key No.	Nominal Key Size W × B	Actual Length F +0.000-0.010	Height of Key				Distance Below Center E
			C		D		
			Max	Min	Max	Min	
202	1/16 × 1/4	0.248	0.109	0.104	0.109	0.104	1/64
202.5	1/16 × 5/16	0.311	0.140	0.135	0.140	0.135	1/64
302.5	3/32 × 5/16	0.311	0.140	0.135	0.140	0.135	1/64
203	1/16 × 3/8	0.374	0.172	0.167	0.172	0.167	1/64
303	3/32 × 3/8	0.374	0.172	0.167	0.172	0.167	1/64
403	1/8 × 3/8	0.374	0.172	0.167	0.172	0.167	1/64
204	1/16 × 1/2	0.491	0.203	0.198	0.194	0.188	3/64
304	3/32 × 1/2	0.491	0.203	0.198	0.194	0.188	3/64
404	1/8 × 1/2	0.491	0.203	0.198	0.194	0.188	3/64
305	3/32 × 5/8	0.612	0.250	0.245	0.240	0.234	1/16
405	1/8 × 5/8	0.612	0.250	0.245	0.240	0.234	1/16
505	5/32 × 5/8	0.612	0.250	0.245	0.240	0.234	1/16
605	3/16 × 5/8	0.612	0.250	0.245	0.240	0.234	1/16
406	1/8 × 3/4	0.740	0.313	0.308	0.303	0.297	1/16
506	5/32 × 3/4	0.740	0.313	0.308	0.303	0.297	1/16
606	3/16 × 3/4	0.740	0.313	0.308	0.303	0.297	1/16
806	1/4 × 3/4	0.740	0.313	0.308	0.303	0.297	1/16
507	5/32 × 7/8	0.866	0.375	0.370	0.365	0.359	1/16
607	3/16 × 7/8	0.866	0.375	0.370	0.365	0.359	1/16
707	7/32 × 7/8	0.866	0.375	0.370	0.365	0.359	1/16
807	1/4 × 7/8	0.866	0.375	0.370	0.365	0.359	1/16
608	3/16 × 1	0.992	0.438	0.433	0.428	0.422	1/16
708	7/32 × 1	0.992	0.438	0.433	0.428	0.422	1/16
808	1/4 × 1	0.992	0.438	0.433	0.428	0.422	1/16
1008	5/16 × 1	0.992	0.438	0.433	0.428	0.422	1/16
1208	3/8 × 1	0.992	0.438	0.433	0.428	0.422	1/16
609	3/16 × 1 1/8	1.114	0.484	0.479	0.475	0.469	5/64
709	7/32 × 1 1/8	1.114	0.484	0.479	0.475	0.469	5/64
809	1/4 × 1 1/8	1.114	0.484	0.479	0.475	0.469	5/64
1009	5/16 × 1 1/8	1.114	0.484	0.479	0.475	0.469	5/64

(Courtesy of ANSI; B17.2-1967.)

Appendix 33
Woodruff keyseats

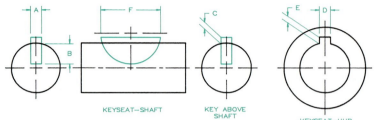

KEYSEAT—SHAFT   KEY ABOVE SHAFT   KEYSEAT—HUB

Keyseat Dimensions

Key Number	Nominal Size Key	Keyseat – Shaft					Key Above Shaft	Keyseat – Hub	
		Width A*		Depth B	Diameter F		Height C	Width D	Depth E
		Min	Max	+0.005 -0.000	Min	Max	+0.005 -0.005	+0.002 -0.000	+0.005 -0.000
202	1/16 × 1/4	0.0615	0.0630	0.0728	0.250	0.268	0.0312	0.0635	0.0372
202.5	1/16 × 5/16	0.0615	0.0630	0.1038	0.312	0.330	0.0312	0.0635	0.0372
302.5	3/32 × 5/16	0.0928	0.0943	0.0882	0.312	0.330	0.0469	0.0948	0.0529
203	1/16 × 3/8	0.0615	0.0630	0.1358	0.375	0.393	0.0312	0.0635	0.0372
303	3/32 × 3/8	0.0928	0.0943	0.1202	0.375	0.393	0.0469	0.0948	0.0529
403	1/8 × 3/8	0.1240	0.1255	0.1045	0.375	0.393	0.0625	0.1260	0.0685
204	1/16 × 1/2	0.0615	0.0630	0.1668	0.500	0.518	0.0312	0.0635	0.0372
304	3/32 × 1/2	0.0928	0.0943	0.1511	0.500	0.518	0.0469	0.0948	0.0529
404	1/8 × 1/2	0.1240	0.1255	0.1355	0.500	0.518	0.0625	0.1260	0.0685
305	3/32 × 5/8	0.0928	0.0943	0.1981	0.625	0.643	0.0469	0.0948	0.0529
405	1/8 × 5/8	0.1240	0.1255	0.1825	0.625	0.643	0.0625	0.1260	0.0685
505	5/32 × 5/8	0.1553	0.1568	0.1669	0.625	0.643	0.0781	0.1573	0.0841
605	3/16 × 5/8	0.1863	0.1880	0.1513	0.625	0.643	0.0937	0.1885	0.0997
406	1/8 × 3/4	0.1240	0.1255	0.2455	0.750	0.768	0.0625	0.1260	0.0685
506	5/32 × 3/4	0.1553	0.1568	0.2299	0.750	0.768	0.0781	0.1573	0.0841
606	3/16 × 3/4	0.1863	0.1880	0.2143	0.750	0.768	0.0937	0.1885	0.0997
806	1/4 × 3/4	0.2487	0.2505	0.1830	0.750	0.768	0.1250	0.2510	0.1310
507	5/32 × 7/8	0.1553	0.1568	0.2919	0.875	0.895	0.0781	0.1573	0.0841
607	3/16 × 7/8	0.1863	0.1880	0.2763	0.875	0.895	0.0937	0.1885	0.0997
707	7/32 × 7/8	0.2175	0.2193	0.2607	0.875	0.895	0.1093	0.2198	0.1153
807	1/4 × 7/8	0.2487	0.2505	0.2450	0.875	0.895	0.1250	0.2510	0.1310
608	3/16 × 1	0.1863	0.1880	0.3393	1.000	1.020	0.0937	0.1885	0.0997
708	7/32 × 1	0.2175	0.2193	0.3237	1.000	1.020	0.1093	0.2198	0.1153
808	1/4 × 1	0.2487	0.2505	0.3080	1.000	1.020	0.1250	0.2510	0.1310
1008	5/16 × 1	0.3111	0.3130	0.2768	1.000	1.020	0.1562	0.3135	0.1622
1208	3/8 × 1	0.3735	0.3755	0.2455	1.000	1.020	0.1875	0.3760	0.1935
609	3/16 × 1 1/8	0.1863	0.1880	0.3853	1.125	1.145	0.0937	0.1885	0.0997
709	7/32 × 1 1/8	0.2175	0.2193	0.3697	1.125	1.145	0.1093	0.2198	0.1153
809	1/4 × 1 1/8	0.2487	0.2505	0.3540	1.125	1.145	0.1250	0.2510	0.1310
1009	5/16 × 1 1/8	0.3111	0.3130	0.3228	1.125	1.145	0.1562	0.3135	0.1622

(Courtesy of ANSI; B17.2–1967.)

Appendix 34
Taper pins

TYPICAL NOTE:
— NO. 2 TAPER PIN—1.500 LG

Number	7/0	6/0	5/0	4/0	3/0	2/0	0	1	2	3	4	5	6	7	8	9	10
Size (large end)	0.0625	0.0780	0.0940	0.1090	0.1250	0.1410	0.1560	0.1720	0.1930	0.2190	0.2500	0.2890	0.3410	0.4090	0.4920	0.5910	0.7060
Length, L																	
0.375	X	X															
0.500	X	X	X	X													
0.625	X	X	X	X	X	X	X										
0.750		X	X	X	X	X	X	X									
0.875				X	X	X	X	X									
1.000			X		X	X	X	X	X								
1.250						X	X	X	X	X	X						
1.500						X	X	X	X	X	X	X					
1.750							X	X	X	X	X	X	X				
2.000								X	X	X	X	X	X	X			
2.250									X	X	X	X	X	X	X		
2.500									X	X	X	X	X	X	X		
2.750										X	X	X	X	X	X	X	
3.000										X	X	X	X	X	X	X	
3.250													X	X	X	X	
3.500													X	X	X	X	X
3.750															X	X	X
4.000															X	X	X
4.250															X	X	X
4.500															X	X	X
4.750															X	X	X
5.000															X	X	X
5.250																X	X
5.500																X	X
5.750																X	X
6.000																X	X

All dimensions are given in inches.

Standard reamers are available for pins given above the line.

Pins Nos. 11 (size 0.8600), 12 (size 1.032), 13 (size 1.241), and 14 (1.523) are special sizes—hence their lengths are special.

To find small diameter of pin, multiply the length by 0.02083 and subtract the result from the large diameter.

(Courtesy of ANSI; B5.20–1958.)

Appendix 35
Plain washers

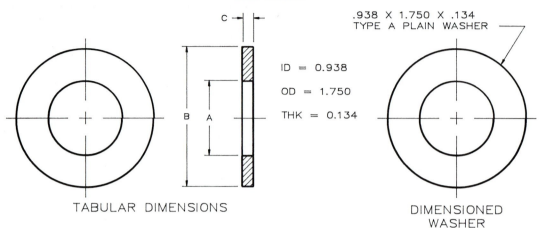

TABULAR DIMENSIONS

ID = 0.938

OD = 1.750

THK = 0.134

.938 X 1.750 X .134
TYPE A PLAIN WASHER

DIMENSIONED
WASHER

Dimensions of Preferred Sizes of Type A Plain Washers[a]

When specifying washers on drawings or in notes, give the inside diameter, outside diameter, and the thickness.
Example: 0.938 × 1.750 × 0.134 TYPE A PLAIN WASHER.

Nominal Washer Size[b]			Inside Diameter A			Outside Diameter B			Thickness C		
				Tolerance			Tolerance				
			Basic	Plus	Minus	Basic	Plus	Minus	Basic	Max	Min
—	—		0.078	0.000	0.005	0.188	0.000	0.005	0.020	0.025	0.016
—	—		0.094	0.000	0.005	0.250	0.000	0.005	0.020	0.025	0.016
—	—		0.125	0.008	0.005	0.312	0.008	0.005	0.032	0.040	0.025
No. 6	0.138		0.156	0.008	0.005	0.375	0.015	0.005	0.049	0.065	0.036
No. 8	0.164		0.188	0.008	0.005	0.438	0.015	0.005	0.049	0.065	0.036
No. 10	0.190		0.219	0.008	0.005	0.500	0.015	0.005	0.049	0.065	0.036
$\frac{3}{16}$	0.188		0.250	0.015	0.005	0.562	0.015	0.005	0.049	0.065	0.036
No. 12	0.216		0.250	0.015	0.005	0.562	0.015	0.005	0.065	0.080	0.051
$\frac{1}{4}$	0.250	N	0.281	0.015	0.005	0.625	0.015	0.005	0.065	0.080	0.051
$\frac{1}{4}$	0.250	W	0.312	0.015	0.005	0.734[c]	0.015	0.007	0.065	0.080	0.051
$\frac{5}{16}$	0.312	N	0.344	0.015	0.005	0.688	0.015	0.007	0.065	0.080	0.051
$\frac{5}{16}$	0.312	W	0.375	0.015	0.005	0.875	0.030	0.007	0.083	0.104	0.064
$\frac{3}{8}$	0.375	N	0.406	0.015	0.005	0.812	0.015	0.007	0.065	0.080	0.051
$\frac{3}{8}$	0.375	W	0.438	0.015	0.005	1.000	0.030	0.007	0.083	0.104	0.064
$\frac{7}{16}$	0.438	N	0.469	0.015	0.005	0.922	0.015	0.007	0.065	0.080	0.051
$\frac{7}{16}$	0.438	W	0.500	0.015	0.005	1.250	0.030	0.007	0.083	0.104	0.064
$\frac{1}{2}$	0.500	N	0.531	0.015	0.005	1.062	0.030	0.007	0.095	0.121	0.074
$\frac{1}{2}$	0.500	W	0.562	0.015	0.005	1.375	0.030	0.007	0.109	0.132	0.086

[a] Preferred sizes are for the most part from series previously designated "Standard Plate" and "SAE." Where common sizes existed in the two series, the SAE size is designated "N" (narrow) and the Standard Plate "W" (wide). These sizes as well as all other sizes of Type A Plain Washers are to be ordered by ID, OD, and thickness dimensions.

[b] Nominal washer sizes are intended for use with comparable nominal screw or bolt sizes.

[c] The 0.734 in., 1.156 in., and 1.469 in. outside diameters avoid washers which could be used in coin-operated devices.

Cont.

Appendix 35
Plain washers (cont.)

Nominal Washer Size[b]			Inside Diameter A			Outside Diameter B			Thickness C		
			Basic	Plus	Minus	Basic	Plus	Minus	Basic	Max	Min
				Tolerance			Tolerance				
$\frac{9}{16}$	0.562	N	0.594	0.015	0.005	1.156[c]	0.030	0.007	0.095	0.121	0.074
$\frac{9}{16}$	0.562	W	0.625	0.015	0.005	1.469[c]	0.030	0.007	0.109	0.132	0.086
$\frac{5}{8}$	0.625	N	0.656	0.030	0.007	1.312	0.030	0.007	0.095	0.121	0.074
$\frac{5}{8}$	0.625	W	0.688	0.030	0.007	1.750	0.030	0.007	0.134	0.160	0.108
$\frac{3}{4}$	0.750	N	0.812	0.030	0.007	1.469	0.030	0.007	0.134	0.160	0.108
$\frac{3}{4}$	0.750	W	0.812	0.030	0.007	2.000	0.030	0.007	0.148	0.177	0.122
$\frac{7}{8}$	0.875	N	0.938	0.030	0.007	1.750	0.030	0.007	0.134	0.160	0.108
$\frac{7}{8}$	0.875	W	0.938	0.030	0.007	2.250	0.030	0.007	0.165	0.192	0.136
1	1.000	N	1.062	0.030	0.007	2.000	0.030	0.007	0.134	0.160	0.108
1	1.000	W	1.062	0.030	0.007	2.500	0.030	0.007	0.165	0.192	0.136
$1\frac{1}{8}$	1.125	N	1.250	0.030	0.007	2.250	0.030	0.007	0.134	0.160	0.108
$1\frac{1}{8}$	1.125	W	1.250	0.030	0.007	2.750	0.030	0.007	0.165	0.192	0.136
$1\frac{1}{4}$	1.250	N	1.375	0.030	0.007	2.500	0.030	0.007	0.165	0.192	0.136
$1\frac{1}{4}$	1.250	W	1.375	0.030	0.007	3.000	0.030	0.007	0.165	0.192	0.136
$1\frac{3}{8}$	1.375	N	1.500	0.030	0.007	2.750	0.030	0.007	0.165	0.192	0.136
$1\frac{3}{8}$	1.375	W	1.500	0.045	0.010	3.250	0.045	0.010	0.180	0.213	0.153
$1\frac{1}{2}$	1.500	N	1.625	0.030	0.007	3.000	0.030	0.007	0.165	0.192	0.136
$1\frac{1}{2}$	1.500	W	1.625	0.045	0.010	3.500	0.045	0.010	0.180	0.213	0.153
$1\frac{5}{8}$	1.625		1.750	0.045	0.010	3.750	0.045	0.010	0.180	0.213	0.153
$1\frac{3}{4}$	1.750		1.875	0.045	0.010	4.000	0.045	0.010	0.180	0.213	0.153
$1\frac{7}{8}$	1.875		2.000	0.045	0.010	4.250	0.045	0.010	0.180	0.213	0.153
2	2.000		2.125	0.045	0.010	4.500	0.045	0.010	0.180	0.213	0.153
$2\frac{1}{4}$	2.250		2.375	0.045	0.010	4.750	0.045	0.010	0.220	0.248	0.193
$2\frac{1}{2}$	2.500		2.625	0.045	0.010	5.000	0.045	0.010	0.238	0.280	0.210
$2\frac{3}{4}$	2.750		2.875	0.065	0.010	5.250	0.065	0.010	0.259	0.310	0.228
3	3.000		3.125	0.065	0.010	5.500	0.065	0.010	0.284	0.327	0.249

(Courtesy of ANSI; B27.2–1965.)

Appendix 36
Lock washers (ANSI B27.1)

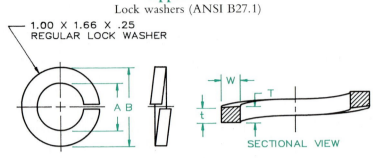

1.00 X 1.66 X .25
REGULAR LOCK WASHER

SECTIONAL VIEW

Dimensions of Regular* Helical Spring Lock Washers

Nominal Washer Size		Inside Diameter A		Outside Diameter B	Washer Section	
					Width W	Thickness $\dfrac{T+t}{2}$
		Min	Max	Max**	Min	Min
No. 2	0.086	0.088	0.094	0.172	0.035	0.020
No. 3	0.099	0.101	0.107	0.195	0.040	0.025
No. 4	0.112	0.115	0.121	0.209	0.040	0.025
No. 5	0.125	0.128	0.134	0.236	0.047	0.031
No. 6	0.138	0.141	0.148	0.250	0.047	0.031
No. 8	0.164	0.168	0.175	0.293	0.055	0.040
No. 10	0.190	0.194	0.202	0.334	0.062	0.047
No. 12	0.216	0.221	0.229	0.377	0.070	0.056
1/4	0.250	0.255	0.263	0.489	0.109	0.062
5/16	0.312	0.318	0.328	0.586	0.125	0.078
3/8	0.375	0.382	0.393	0.683	0.141	0.094
7/16	0.438	0.446	0.459	0.779	0.156	0.109
1/2	0.500	0.509	0.523	0.873	0.171	0.125
9/16	0.562	0.572	0.587	0.971	0.188	0.141
5/8	0.625	0.636	0.653	1.079	0.203	0.156
11/16	0.688	0.700	0.718	1.176	0.219	0.172
3/4	0.750	0.763	0.783	1.271	0.234	0.188
13/16	0.812	0.826	0.847	1.367	0.250	0.203
7/8	0.875	0.890	0.912	1.464	0.266	0.219
15/16	0.938	0.954	0.978	1.560	0.281	0.234
1	1.000	1.017	1.042	1.661	0.297	0.250
1 1/16	1.062	1.080	1.107	1.756	0.312	0.266
1 1/8	1.125	1.144	1.172	1.853	0.328	0.281
1 3/16	1.188	1.208	1.237	1.950	0.344	0.297
1 1/4	1.250	1.271	1.302	2.045	0.359	0.312
1 5/16	1.312	1.334	1.366	2.141	0.375	0.328
1 3/8	1.375	1.398	1.432	2.239	0.391	0.344
1 7/16	1.438	1.462	1.497	2.334	0.406	0.359
1 1/2	1.500	1.525	1.561	2.430	0.422	0.375

*Formerly designated Medium Helical Spring Lock Washers.

**The maximum outside diameters specified allow for the commercial tolerances on cold drawn wire.

Appendix 37
Cotter pins

STD COTTER PIN Ø.25 X 2.00 LG

B

L

A

STANDARD

NON—STANDARD NOTE:
BEVEL—POINT COTTET PIN Ø.25 X 2.00 LG

L

MITRE
END

L

EXTENDED
WIRE END

L

ROUND
SQUARE CUT

L

BEVEL
POINT

L

HAMMER
LOCK

L

CHISEL
POINT

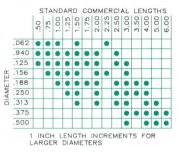

STANDARD COMMERCIAL LENGTHS

DIAMETER	.50	.75	1.00	1.25	1.50	1.75	2.00	2.25	2.50	3.00	3.50	4.00	5.00	6.00
.062	●	●	●		●									
.940	●	●	●	●	●	●		●						
.125	●	●	●	●	●	●	●		●					
.156		●	●	●	●	●	●		●	●	●			
.188			●	●	●	●	●		●	●	●	●		
.250				●	●	●		●	●	●	●	●		
.313						●			●	●	●	●	●	
.375											●	●	●	●
.500											●		●	●

1 INCH LENGTH INCREMENTS FOR LARGER DIAMETERS

Nominal Diameter	Diameter A		Outside Eye Diameter B Min	Hole Sizes Recommended
	Max	Min		
0.031	0.032	0.028	$\frac{1}{16}$	$\frac{3}{64}$
0.047	0.048	0.044	$\frac{3}{32}$	$\frac{1}{16}$
0.062	0.060	0.056	$\frac{1}{8}$	$\frac{5}{64}$
0.078	0.076	0.072	$\frac{5}{32}$	$\frac{3}{32}$
0.094	0.090	0.086	$\frac{3}{16}$	$\frac{7}{64}$
0.109	0.104	0.100	$\frac{7}{32}$	$\frac{1}{8}$
0.125	0.120	0.116	$\frac{1}{4}$	$\frac{9}{64}$
0.141	0.134	0.130	$\frac{9}{32}$	$\frac{5}{32}$
0.156	0.150	0.146	$\frac{5}{16}$	$\frac{11}{64}$
0.188	0.176	0.172	$\frac{3}{8}$	$\frac{13}{64}$
0.219	0.207	0.202	$\frac{7}{16}$	$\frac{15}{64}$
0.250	0.225	0.220	$\frac{1}{2}$	$\frac{17}{64}$
0.312	0.280	0.275	$\frac{5}{8}$	$\frac{5}{16}$
0.375	0.335	0.329	$\frac{3}{4}$	$\frac{3}{8}$
0.438	0.406	0.400	$\frac{7}{8}$	$\frac{7}{16}$
0.500	0.473	0.467	1	$\frac{1}{2}$
0.625	0.598	0.590	$1\frac{1}{4}$	$\frac{5}{8}$
0.750	0.723	0.715	$1\frac{1}{2}$	$\frac{3}{4}$

(Courtesy of ANSI; B5.20−1958.)

Appendix 38
American standard running and sliding fits (hole bases)

Limits are in thousandths of an inch.
Limits for hole and shaft are applied algebraically to the basic size to obtain the limits of size for the parts.
Data in bold face are in accordance with ABC agreements.
Symbols H5, g5, etc., are Hole and Shaft designations used in ABC System.

Nominal Size Range Inches (Over – To)	RC1 Limits of Clearance	RC1 Hole H5	RC1 Shaft g4	RC2 Limits of Clearance	RC2 Hole H6	RC2 Shaft g5	RC3 Limits of Clearance	RC3 Hole H7	RC3 Shaft f6	RC4 Limits of Clearance	RC4 Hole H8	RC4 Shaft f7
0 – 0.12	0.1 / 0.45	+0.2 / 0	−0.1 / −0.25	0.1 / 0.55	+0.25 / 0	−0.1 / −0.3	0.3 / 0.95	+0.4 / 0	−0.3 / −0.55	0.3 / 1.3	+0.6 / 0	−0.3 / −0.7
0.12 – 0.24	0.15 / 0.5	+0.2 / 0	−0.15 / −0.3	0.15 / 0.65	+0.3 / 0	−0.15 / −0.35	0.4 / 1.12	+0.5 / 0	−0.4 / −0.7	0.4 / 1.6	+0.7 / 0	−0.4 / −0.9
0.24 – 0.40	0.2 / 0.6	0.25 / 0	−0.2 / −0.35	0.2 / 0.85	+0.4 / 0	−0.2 / −0.45	0.5 / 1.5	+0.6 / 0	−0.5 / −0.9	0.5 / 2.0	+0.9 / 0	−0.5 / −1.1
0.40 – 0.71	0.25 / 0.75	+0.3 / 0	−0.25 / −0.45	0.25 / 0.95	+0.4 / 0	−0.25 / −0.55	0.6 / 1.7	+0.7 / 0	−0.6 / −1.0	0.6 / 2.3	+1.0 / 0	−0.6 / −1.3
0.71 – 1.19	0.3 / 0.95	+0.4 / 0	−0.3 / −0.55	0.3 / 1.2	+0.5 / 0	−0.3 / −0.7	0.8 / 2.1	+0.8 / 0	−0.8 / −1.3	0.8 / 2.8	+1.2 / 0	−0.8 / −1.6
1.19 – 1.97	0.4 / 1.1	+0.4 / 0	−0.4 / −0.7	0.4 / 1.4	+0.6 / 0	−0.4 / −0.8	1.0 / 2.6	+1.0 / 0	−1.0 / −1.6	1.0 / 3.6	+1.6 / 0	−1.0 / −2.0
1.97 – 3.15	0.4 / 1.2	+0.5 / 0	−0.4 / −0.7	0.4 / 1.6	+0.7 / 0	−0.4 / −0.9	1.2 / 3.1	+1.2 / 0	−1.2 / −1.9	1.2 / 4.2	+1.8 / 0	−1.2 / −2.4
3.15 – 4.73	0.5 / 1.5	+0.6 / 0	−0.5 / −0.9	0.5 / 2.0	+0.9 / 0	−0.5 / −1.1	1.4 / 3.7	+1.4 / 0	−1.4 / −2.3	1.4 / 5.0	+2.2 / 0	−1.4 / −2.8
4.73 – 7.09	0.6 / 1.8	+0.7 / 0	−0.6 / −1.1	0.6 / 2.3	+1.0 / 0	−0.6 / −1.3	1.6 / 4.2	+1.6 / 0	−1.6 / −2.6	1.6 / 5.7	+2.5 / 0	−1.6 / −3.2
7.09 – 9.85	0.6 / 2.0	+0.8 / 0	−0.6 / −1.2	0.6 / 2.6	+1.2 / 0	−0.6 / −1.4	2.0 / 5.0	+1.8 / 0	−2.0 / −3.2	2.0 / 6.6	+2.8 / 0	−2.0 / −3.8
9.85 – 12.41	0.8 / 2.3	+0.9 / 0	−0.8 / −1.4	0.8 / 2.9	+1.2 / 0	−0.8 / −1.7	2.5 / 5.7	+2.0 / 0	−2.5 / −3.7	2.5 / 7.5	+3.0 / 0	−2.5 / −4.5
12.41 – 15.75	1.0 / 2.7	+1.0 / 0	−1.0 / −1.7	1.0 / 3.4	+1.4 / 0	−1.0 / −2.0	3.0 / 6.6	+ / 0	−3.0 / −4.4	3.0 / 8.7	+3.5 / 0	−3.0 / −5.2
15.75 – 19.69	1.2 / 3.0	+1.0 / 0	−1.2 / −2.0	1.2 / 3.8	+1.6 / 0	−1.2 / −2.2	4.0 / 8.1	+1.6 / 0	−4.0 / −5.6	4.0 / 10.5	+4.0 / 0	−4.0 / −6.5
19.69 – 30.09	1.6 / 3.7	+1.2 / 0	−1.6 / −2.5	1.6 / 4.8	+2.0 / 0	−1.6 / −2.8	5.0 / 10.0	+3.0 / 0	−5.0 / −7.0	5.0 / 13.0	+5.0 / 0	−5.0 / −8.0
30.09 – 41.49	2.0 / 4.6	+1.6 / 0	−2.0 / −3.0	2.0 / 6.1	+2.5 / 0	−2.0 / −3.6	6.0 / 12.5	+4.0 / 0	−6.0 / −8.5	6.0 / 16.0	+6.0 / 0	−6.0 / −10.0
41.49 – 56.19	2.5 / 5.7	+2.0 / 0	−2.5 / −3.7	2.5 / 7.5	+3.0 / 0	−2.5 / −4.5	8.0 / 16.0	+5.0 / 0	−8.0 / −11.0	8.0 / 21.0	+8.0 / 0	−8.0 / −13.0
56.19 – 76.39	3.0 / 7.1	+2.5 / 0	−3.0 / −4.6	3.0 / 9.5	+4.0 / 0	−3.0 / −5.5	10.0 / 20.0	+6.0 / 0	−10.0 / −14.0	10.0 / 26.0	+10.0 / 0	−10.0 / −16.0
76.39 – 100.9	4.0 / 9.0	+3.0 / 0	−4.0 / −6.0	4.0 / 12.0	+5.0 / 0	−4.0 / −7.0	12.0 / 25.0	+8.0 / 0	−12.0 / −17.0	12.0 / 32.0	+12.0 / 0	−12.0 / −20.0
100.9 – 131.9	5.0 / 11.5	+4.0 / 0	−5.0 / −7.5	5.0 / 15.0	+6.0 / 0	−5.0 / −9.0	16.0 / 32.0	+10.0 / 0	−16.0 / −22.0	16.0 / 36.0	+16.0 / 0	−16.0 / −26.0
131.9 – 171.9	6.0 / 14.0	+5.0 / 0	−6.0 / −9.0	6.0 / 19.0	+8.0 / 0	−6.0 / −11.0	18.0 / 38.0	+8.0 / 0	−18.0 / −26.0	18.0 / 50.0	+20.0 / 0	−18.0 / −30.0
171.9 – 200	8.0 / 18.0	+6.0 / 0	−8.0 / −12.0	8.0 / 22.0	+10.0 / 0	−8.0 / −12.0	22.0 / 48.0	+16.0 / 0	−22.0 / −32.0	22.0 / 63.0	+25.0 / 0	−22.0 / −38.0

(Courtesy of USASI; B4.1–1955.)

Cont.

Appendix 38
American standard running and sliding fits (cont.)

Class RC 5			Class RC 6			Class RC 7			Class RC 8			Class RC 9			Nominal Size Range Inches	
Limits of Clearance	Standard Limits		Limits of Clearance	Standard Limits		Limits of Clearance	Standard Limits		Limits of Clearance	Standard Limits		Limits of Clearance	Standard Limits			
	Hole H8	Shaft e7		Hole H9	Shaft e8		Hole H9	Shaft d8		Hole H10	Shaft c9		Hole H11	Shaft	Over	To
0.6 / 1.6	+0.6 / −0	−0.6 / −1.0	0.6 / 2.2	+1.0 / −0	−0.6 / −1.2	1.0 / 2.6	+1.0 / 0	−1.0 / −1.6	2.5 / 5.1	+1.6 / 0	−2.5 / −3.5	4.0 / 8.1	+2.5 / 0	−4.0 / −5.6	0 − 0.12	
0.8 / 2.0	+0.7 / −0	−0.8 / −1.3	0.8 / 2.7	+1.2 / −0	−0.8 / −1.5	1.2 / 3.1	+1.2 / 0	−1.2 / −1.9	2.8 / 5.8	+1.8 / 0	−2.8 / −4.0	4.5 / 9.0	+3.0 / 0	−4.5 / −6.0	0.12 − 0.24	
1.0 / 2.5	+0.9 / −0	−1.0 / −1.6	1.0 / 3.3	+1.4 / −0	−1.0 / −1.9	1.6 / 3.9	+1.4 / 0	−1.6 / −2.5	3.0 / 6.6	+2.2 / 0	−3.0 / −4.4	5.0 / 10.7	+3.5 / 0	−5.0 / −7.2	0.24 − 0.40	
1.2 / 2.9	+1.0 / −0	−1.2 / −1.9	1.2 / 3.8	+1.6 / −0	−1.2 / −2.2	2.0 / 4.6	+1.6 / 0	−2.0 / −3.0	3.5 / 7.9	+2.8 / 0	−3.5 / −5.1	6.0 / 12.8	+4.0 / −	−6.0 / −8.8	0.40 − 0.71	
1.6 / 3.6	+1.2 / −0	−1.6 / −2.4	1.6 / 4.8	+2.0 / −0	−1.6 / −2.8	2.5 / 5.7	+2.0 / 0	−2.5 / −3.7	4.5 / 10.0	+3.5 / 0	−4.5 / −6.5	7.0 / 15.5	+5.0 / 0	−7.0 / −10.5	0.71 − 1.19	
2.0 / 4.6	+1.6 / −0	−2.0 / −3.0	2.0 / 6.1	+2.5 / −0	−2.0 / −3.6	3.0 / 7.1	+2.5 / 0	−3.0 / −4.6	5.0 / 11.5	+4.0 / 0	−5.0 / −7.5	8.0 / 18.0	+6.0 / 0	−8.0 / −12.0	1.19 − 1.97	
2.5 / 5.5	+1.8 / −0	−2.5 / −3.7	2.5 / 7.3	+3.0 / −0	−2.5 / −4.3	4.0 / 8.8	+3.0 / 0	−4.0 / −5.8	6.0 / 13.5	+4.5 / 0	−6.0 / −9.0	9.0 / 20.5	+7.0 / 0	−9.0 / −13.5	1.97 − 3.15	
3.0 / 6.6	+2.2 / −0	−3.0 / −4.4	3.0 / 8.7	+3.5 / −0	−3.0 / −5.2	5.0 / 10.7	+3.5 / 0	−5.0 / −7.2	7.0 / 15.5	+5.0 / 0	−7.0 / −10.5	10.0 / 24.0	+9.0 / 0	−10.0 / −15.0	3.15 − 4.73	
3.5 / 7.6	+2.5 / −0	−3.5 / −5.1	3.5 / 10.0	+4.0 / −0	−3.5 / −6.0	6.0 / 12.5	+4.0 / 0	−6.0 / −8.5	8.0 / 18.0	+6.0 / 0	−8.0 / −12.0	12.0 / 28.0	+10.0 / 0	−12.0 / −18.0	4.73 − 7.09	
4.0 / 8.6	+2.8 / −0	−4.0 / −5.8	4.0 / 11.3	+4.5 / 0	−4.0 / −6.8	7.0 / 14.3	+4.5 / 0	−7.0 / −9.8	10.0 / 21.5	+7.0 / 0	−10.0 / −14.5	15.0 / 34.0	+12.0 / 0	−15.0 / −22.0	7.09 − 9.85	
5.0 / 10.0	+3.0 / 0	−5.0 / −7.0	5.0 / 13.0	+5.0 / 0	−5.0 / −8.0	8.0 / 16.0	+5.0 / 0	−8.0 / −11.0	12.0 / 25.0	+8.0 / 0	−12.0 / −17.0	18.0 / 38.0	+12.0 / 0	−18.0 / −26.0	9.85 − 12.41	
6.0 / 11.7	+3.5 / 0	−6.0 / −8.2	6.0 / 15.5	+6.0 / 0	−6.0 / −9.5	10.0 / 19.5	+6.0 / 0	−10.0 / −13.5	14.0 / 29.0	+9.0 / 0	−14.0 / −20.0	22.0 / 45.0	+14.0 / 0	−22.0 / −31.0	12.41 − 15.75	
8.0 / 14.5	+4.0 / 0	−8.0 / −10.5	8.0 / 18.0	+6.0 / 0	−8.0 / −12.0	12.0 / 22.0	+6.0 / 0	−12.0 / −16.0	16.0 / 32.0	+10.0 / 0	−16.0 / −22.0	25.0 / 51.0	+16.0 / 0	−25.0 / −35.0	15.75 − 19.69	
10.0 / 18.0	+5.0 / 0	−10.0 / −13.0	10.0 / 23.0	+8.0 / 0	−10.0 / −15.0	16.0 / 29.0	+8.0 / 0	−16.0 / −21.0	20.0 / 40.0	+12.0 / 0	−20.0 / −28.0	30.0 / 62.0	+20.0 / 0	−30.0 / −42.0	19.69 − 30.09	
12.0 / 22.0	+6.0 / 0	−12.0 / −16.0	12.0 / 28.0	+10.0 / 0	−12.0 / −18.0	20.0 / 36.0	+10.0 / 0	−20.0 / −26.0	25.0 / 51.0	+16.0 / 0	−25.0 / −35.0	40.0 / 81.0	+25.0 / 0	−40.0 / −56.0	30.09 − 41.49	
16.0 / 29.0	+8.0 / 0	−16.0 / −21.0	16.0 / 36.0	+12.0 / 0	−16.0 / −24.0	25.0 / 45.0	+12.0 / 0	−25.0 / −33.0	30.0 / 62.0	+20.0 / 0	−30.0 / −42.0	50.0 / 100	+30.0 / 0	−50.0 / −70.0	41.49 − 56.19	
20.0 / 36.0	+10.0 / 0	−20.0 / −26.0	20.0 / 46.0	+16.0 / 0	−20.0 / −30.0	30.0 / 56.0	+16.0 / 0	−30.0 / −40.0	40.0 / 81.0	+25.0 / 0	−40.0 / −56.0	60.0 / 125	+40.0 / 0	−60.0 / −85.0	56.19 − 76.39	
25.0 / 45.0	+12.0 / 0	−25.0 / −33.0	25.0 / 57.0	+20.0 / 0	−25.0 / −37.0	40.0 / 72.0	+20.0 / 0	−40.0 / −52.0	50.0 / 100	+30.0 / 0	−50.0 / −70.0	80.0 / 160	+50.0 / 0	−80.0 / −110	76.39 − 100.9	
30.0 / 56.0	+16.0 / 0	−30.0 / −40.0	30.0 / 71.0	+25.0 / 0	−30.0 / −46.0	50.0 / 91.0	+25.0 / 0	−50.0 / −66.0	60.0 / 125	+40.0 / 0	−60.0 / −85.0	100 / 200	+60.0 / 0	−100 / −140	100.9 − 131.9	
35.0 / 67.0	+20.0 / 0	−35.0 / −47.0	35.0 / 85.0	+30.0 / 0	−35.0 / −55.0	60.0 / 110	+30.0 / 0	−60.0 / −80.0	80.0 / 160	+50.0 / 0	−80.0 / −110	130 / 260	+80.0 / 0	−130 / −180	131.9 − 171.9	
45.0 / 86.0	+25.0 / 0	−45.0 / −61.0	45.0 / 110.0	+40.0 / 0	−45.0 / −70.0	80.0 / 145.0	+40.0 / 0	−80.0 / −105.0	100 / 200	+60.0 / 0	−100 / −140	150 / 310	+100 / 0	−150 / −210	171.9 − 200	

(Courtesy of ANSI; B4.1–1955.)

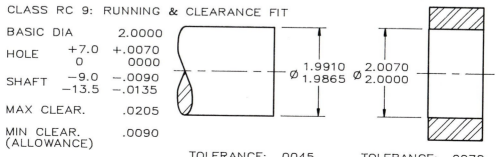

CLASS RC 9: RUNNING & CLEARANCE FIT

BASIC DIA	2.0000	
HOLE	+7.0 / 0	+.0070 / 0000
SHAFT	−9.0 / −13.5	−.0090 / −.0135
MAX CLEAR.	.0205	
MIN CLEAR. (ALLOWANCE)	.0090	

Ø 1.9910 / 1.9865   Ø 2.0070 / 2.0000

TOLERANCE: .0045   TOLERANCE: .0070

Appendix 39
American standard clearance locational fits (hole basis)

Limits are in thousandths of an inch.
Limits for hole and shaft are applied algebraically to the basic size to obtain the limits of size for the parts.
Data in bold face are in accordance with ABC agreements.
Symbols H9,f8, etc., are Hole and Shaft designations used in ABC System.

Nominal Size Range Inches Over	To	Class LC 1 Limits of Clearance	Class LC 1 Hole H6	Class LC 1 Shaft h5	Class LC 2 Limits of Clearance	Class LC 2 Hole H7	Class LC 2 Shaft h6	Class LC 3 Limits of Clearance	Class LC 3 Hole H8	Class LC 3 Shaft h7	Class LC 4 Limits of Clearance	Class LC 4 Hole H10	Class LC 4 Shaft h9	Class LC 5 Limits of Clearance	Class LC 5 Hole H7	Class LC 5 Shaft g6
0 —	0.12	0 / 0.45	+0.25 / −0	+0 / −0.2	0 / 0.65	+0.4 / −0	+0 / −0.25	0 / 1	+0.6 / −0	+0 / −0.4	0 / 2.6	+1.6 / −0	+0 / −1.0	0.1 / 0.75	+0.4 / −0	−0.1 / −0.35
0.12—	0.24	0 / 0.5	+0.3 / −0	+0 / −0.2	0 / 0.8	+0.5 / −0	+0 / −0.3	0 / 1.2	+0.7 / −0	+0 / −0.5	0 / 3.0	+1.8 / −0	+0 / −1.2	0.15 / 0.95	+0.5 / −0	−0.15 / −0.45
0.24—	0.40	0 / 0.65	+0.4 / −0	+0 / −0.25	0 / 1.0	+0.6 / −0	+0 / −0.4	0 / 1.5	+0.9 / −0	+0 / −0.6	0 / 3.6	+2.2 / −0	+0 / −1.4	0.2 / 1.2	+0.6 / −0	−0.2 / −0.6
0.40—	0.71	0 / 0.7	+0.4 / −0	+0 / −0.3	0 / 1.1	+0.7 / −0	+0 / −0.4	0 / 1.7	+1.0 / −0	+0 / −0.7	0 / 4.4	+2.8 / −0	+0 / −1.6	0.25 / 1.35	+0.7 / −0	−0.25 / −0.65
0.71—	1.19	0 / 0.9	+0.5 / −0	+0 / −0.4	0 / 1.3	+0.8 / −0	+0 / −0.5	0 / 2	+1.2 / −0	+0 / −0.8	0 / 5.5	+3.5 / −0	+0 / −2.0	0.3 / 1.6	+0.8 / −0	−0.3 / −0.8
1.19—	1.97	0 / 1.0	+0.6 / −0	+0 / −0.4	0 / 1.6	+1.0 / −0	+0 / −0.6	0 / 2.6	+1.6 / −0	+0 / −1	0 / 6.5	+4.0 / −0	+0 / −2.5	0.4 / 2.0	+1.0 / −0	−0.4 / −1.0
1.97—	3.15	0 / 1.2	+0.7 / −0	+0 / −0.5	0 / 1.9	+1.2 / −0	+0 / −0.7	0 / 3	+1.8 / −0	+0 / −1.2	0 / 7.5	+4.5 / −0	+0 / −3	0.4 / 2.3	+1.2 / −0	−0.4 / −1.1
3.15—	4.73	0 / 1.5	+0.9 / −0	+0 / −0.6	0 / 2.3	+1.4 / −0	+0 / −0.9	0 / 3.6	+2.2 / −0	+0 / −1.4	0 / 8.5	+5.0 / −0	+0 / −3.5	0.5 / 2.8	+1.4 / −0	−0.5 / −1.4
4.73—	7.09	0 / 1.7	+1.0 / −0	+0 / −0.7	0 / 2.6	+1.6 / −0	+0 / −1.0	0 / 4.1	+2.5 / −0	+0 / −1.6	0 / 10	+6.0 / −0	+0 / −4	0.6 / 3.2	+1.6 / −0	−0.6 / −1.6
7.09—	9.85	0 / 2.0	+1.2 / −0	+0 / −0.8	0 / 3.0	+1.8 / −0	+0 / −1.2	0 / 4.6	+2.8 / −0	+0 / −1.8	0 / 11.5	+7.0 / −0	+0 / −4.5	0.6 / 3.6	+1.8 / −0	−0.6 / −1.8
9.85—	12.41	0 / 2.1	+1.2 / −0	+0 / −0.9	0 / 3.2	+2.0 / −0	+0 / −1.2	0 / 5	+3.0 / −0	+0 / −2.0	0 / 13	+8.0 / −0	+0 / −5	0.7 / 3.9	+2.0 / −0	−0.7 / −1.9
12.41—	15.75	0 / 2.4	+1.4 / −0	+0 / −1.0	0 / 3.6	+2.2 / −0	+0 / −1.4	0 / 5.7	+3.5 / −0	+0 / −2.2	0 / 15	+9.0 / −0	+0 / −6	0.7 / 4.3	+2.2 / −0	−0.7 / −2.1
15.75—	19.69	0 / 2.6	+1.6 / −0	+0 / −1.0	0 / 4.1	+2.5 / −0	+0 / −1.6	0 / 6.5	+4 / −0	+0 / −2.5	0 / 16	+10.0 / −0	+0 / −6	0.8 / 4.9	+2.5 / −0	−0.8 / −2.4
19.69—	30.09	0 / 3.2	+2.0 / −0	+0 / −1.2	0 / 5.0	+3 / −0	+0 / −2	0 / 8	+5 / −0	+0 / −3	0 / 20	+12.0 / −0	+0 / −8	0.9 / 5.9	+3.0 / −0	−0.9 / −2.9
30.09—	41.49	0 / 4.1	+2.5 / −0	+0 / −1.6	0 / 6.5	+4 / −0	+0 / −2.5	0 / 10	+6 / −0	+0 / −4	0 / 26	+16.0 / −0	+0 / −10	1.0 / 7.5	+4.0 / −0	−1.0 / −3.5
41.49—	56.19	0 / 5.0	+3.0 / −0	+0 / −2.0	0 / 8.0	+5 / −0	+0 / −3	0 / 13	+8 / −0	+0 / −5	0 / 32	+20.0 / −0	+0 / −12	1.2 / 9.2	+5.0 / −0	−1.2 / −4.2
56.19—	76.39	0 / 6.5	+4.0 / −0	+0 / −2.5	0 / 10	+6 / −0	+0 / −4	0 / 16	+10 / −0	+0 / −6	0 / 41	+25.0 / −0	+0 / −16	1.2 / 11.2	+6.0 / −0	−1.2 / −5.2
76.39—	100.9	0 / 8.0	+5.0 / −0	+0 / −3.0	0 / 13	+8 / −0	+0 / −5	0 / 20	+12 / −0	+0 / −8	0 / 50	+30.0 / −0	+0 / −20	1.4 / 14.4	+8.0 / −0	−1.4 / −6.4
100.9 —	131.9	0 / 10.0	+6.0 / −0	+0 / −4.0	0 / 16	+10 / −0	+0 / −6	0 / 26	+16 / −0	+0 / −10	0 / 65	+40.0 / −0	+0 / −25	1.6 / 17.6	+10.0 / −0	−1.6 / −7.6
131.9 —	171.9	0 / 13.0	+8.0 / −0	+0 / −5.0	0 / 20	+12 / −0	+0 / −8	0 / 32	+20 / −0	+0 / −12	0 / 8	+50.0 / −0	+0 / −30	1.8 / 21.8	+12.0 / −0	−1.8 / −9.8
171.9 —	200	0 / 16.0	+10.0 / −0	+0 / −6.0	0 / 26	+16 / −0	+0 / −10	0 / 41	+25 / −0	+0 / −16	0 / 100	+60.0 / −0	+0 / −40	1.8 / 27.8	+16.0 / −0	−1.8 / −11.8

(Courtesy of USASI; B4.1–1955.)

Cont.

Appendix 39
American standard clearance locational fits (cont.)

Class LC 6			Class LC 7			Class LC 8			Class LC 9			Class LC 10			Class LC 11			Nominal Size Range Inches	
Limits of Clearance	Hole H9	Shaft f8	Limits of Clearance	Hole H10	Shaft e9	Limits of Clearance	Hole H10	Shaft d9	Limits of Clearance	Hole H11	Shaft c10	Limits of Clearance	Hole H12	Shaft	Limits of Clearance	Hole H13	Shaft	Over	To
0.3 / 1.9	+1.0 / 0	−0.3 / −0.9	0.6 / 3.2	+1.6 / 0	−0.6 / −1.6	1.0 / 3.6	+1.6 / −0	−1.0 / −2.0	2.5 / 6.6	+2.5 / −0	−2.5 / −4.1	4 / 12	+4 / −0	−4 / −8	5 / 17	+6 / −0	−5 / −11	0 − 0.12	
0.4 / 2.3	+1.2 / 0	−0.4 / −1.1	0.8 / 3.8	+1.8 / 0	−0.8 / −2.0	1.2 / 4.2	+1.8 / −0	−1.2 / −2.4	2.8 / 7.6	+3.0 / −0	−2.8 / −4.6	4.5 / 14.5	+5 / −0	−4.5 / −9.5	6 / 20	+7 / −0	−6 / −13	0.12 − 0.24	
0.5 / 2.8	+1.4 / 0	−0.5 / −1.4	1.0 / 4.6	+2.2 / 0	−1.0 / −2.4	1.6 / 5.2	+2.2 / −0	−1.6 / −3.0	3.0 / 8.7	+3.5 / −0	−3.0 / −5.2	5 / 17	+6 / −0	−5 / −11	7 / 25	+9 / −0	−7 / −16	0.24 − 0.40	
0.6 / 3.2	+1.6 / 0	−0.6 / −1.6	1.2 / 5.6	+2.8 / 0	−1.2 / −2.8	2.0 / 6.4	+2.8 / −0	−2.0 / −3.6	3.5 / 10.3	+4.0 / −0	−3.5 / −6.3	6 / 20	+7 / −0	−6 / −13	8 / 28	+10 / −0	−8 / −18	0.40 − 0.71	
0.8 / 4.0	+2.0 / 0	−0.8 / −2.0	1.6 / 7.1	+3.5 / 0	−1.6 / −3.6	2.5 / 8.0	+3.5 / −0	−2.5 / −4.5	4.5 / 13.0	+5.0 / −0	−4.5 / −8.0	7 / 23	+8 / −0	−7 / −15	10 / 34	+12 / −0	−10 / −22	0.71 − 1.19	
1.0 / 5.1	+2.5 / 0	−1.0 / −2.6	2.0 / 8.5	+4.0 / 0	−2.0 / −4.5	3.0 / 9.5	+4.0 / −0	−3.0 / −5.5	5 / 15	+6 / −0	−5 / −9	8 / 28	+10 / −0	−8 / −18	12 / 44	+16 / −0	−12 / −28	1.19 − 1.97	
1.2 / 6.0	+3.0 / 0	−1.2 / −3.0	2.5 / 10.0	+4.5 / 0	−2.5 / −5.5	4.0 / 11.5	+4.5 / −0	−4.0 / −7.0	6 / 17.5	+7 / −0	−6 / −10.5	10 / 34	+12 / −0	−10 / −22	14 / 50	+18 / −0	−14 / −32	1.97 − 3.15	
1.4 / 7.1	+3.5 / 0	−1.4 / −3.6	3.0 / 11.5	+5.0 / 0	−3.0 / −6.5	5.0 / 13.5	+5.0 / −0	−5.0 / −8.5	7 / 21	+9 / −0	−7 / −12	11 / 39	+14 / −0	−11 / −25	16 / 60	+22 / −0	−16 / −38	3.15 − 4.73	
1.6 / 8.1	+4.0 / 0	−1.6 / −4.1	3.5 / 13.5	+6.0 / 0	−3.5 / −7.5	6 / 16	+6 / −0	−6 / −10	8 / 24	+10 / −0	−8 / −14	12 / 44	+16 / −0	−12 / −28	18 / 68	+25 / −0	−18 / −43	4.73 − 7.09	
2.0 / 9.3	+4.5 / 0	−2.0 / −4.8	4.0 / 15.5	+7.0 / 0	−4.0 / −8.5	7 / 18.5	+7 / −0	−7 / −11.5	10 / 29	+12 / −0	−10 / −17	16 / 52	+18 / −0	−16 / −34	22 / 78	+28 / −0	−22 / −50	7.09 − 9.85	
2.2 / 10.2	+5.0 / 0	−2.2 / −5.2	4.5 / 17.5	+8.0 / 0	−4.5 / −9.5	7 / 20	+8 / −0	−7 / −12	12 / 32	+12 / −0	−12 / −20	20 / 60	+20 / −0	−20 / −40	28 / 88	+30 / −0	−28 / −58	9.85 − 12.41	
2.5 / 12.0	+6.0 / 0	−2.5 / −6.0	5.0 / 20.0	+9.0 / 0	−5 / −11	8 / 23	+9 / −0	−8 / −14	14 / 37	+14 / −0	−14 / −23	22 / 66	+22 / −0	−22 / −44	30 / 100	+35 / −0	−30 / −65	12.41 − 15.75	
2.8 / 12.8	+6.0 / 0	−2.8 / −6.8	5.0 / 21.0	+10.0 / 0	−5 / −11	9 / 25	+10 / −0	−9 / −15	16 / 42	+16 / −0	−16 / −26	25 / 75	+25 / −0	−25 / −50	35 / 115	+40 / −0	−35 / −75	15.75 − 19.69	
3.0 / 16.0	+8.0 / 0	−3.0 / −8.0	6.0 / 26.0	+12.0 / −0	−6 / −14	10 / 30	+12 / −0	−10 / −18	18 / 50	+20 / −0	−18 / −30	28 / 88	+30 / −0	−28 / −58	40 / 140	+50 / −0	−40 / −90	19.69 − 30.09	
3.5 / 19.5	+10.0 / 0	−3.5 / −9.5	7.0 / 33.0	+16.0 / −0	−7 / −17	12 / 38	+16 / −0	−12 / −22	20 / 61	+25 / −0	−20 / −36	30 / 110	+40 / −0	−30 / −70	45 / 165	+60 / −0	−45 / −105	30.09 − 41.49	
4.0 / 24.0	+12.0 / 0	−4.0 / −12.0	8.0 / 40.0	+20.0 / −0	−8 / −20	14 / 46	+20 / −0	−14 / −26	25 / 75	+30 / −0	−25 / −45	40 / 140	+50 / −0	−40 / −90	60 / 220	+80 / −0	−60 / −140	41.49 − 56.19	
4.5 / 30.5	+16.0 / 0	−4.5 / −14.5	9.0 / 50.0	+25.0 / −0	−9 / −25	16 / 57	+25 / −0	−16 / −32	30 / 95	+40 / −0	−30 / −55	50 / 170	+60 / −0	−50 / 110	70 / 270	+100 / −0	−70 / −170	56.19 − 76.39	
5.0 / 37.0	+20.0 / 0	−5 / −17	10.0 / 60.0	+30.0 / −0	−10 / −30	18 / 68	+30 / −0	−18 / −38	35 / 115	+50 / −0	−35 / −65	50 / 210	+80 / −0	−50 / −130	80 / 330	+125 / −0	−80 / −205	76.39 − 100.9	
6.0 / 47.0	+25.0 / 0	−6 / −22	12.0 / 67.0	+40.0 / −0	−12 / −27	20 / 85	+40 / −0	−20 / −45	40 / 140	+60 / −0	−40 / −80	60 / 260	+100 / −0	−60 / −160	90 / 410	+160 / −0	−90 / −250	100.9 − 131.9	
7.0 / 57.0	+30.0 / 0	−7 / −27	14.0 / 94.0	+50.0 / −0	−14 / −44	25 / 105	+50 / −0	−25 / −55	50 / 180	+80 / −0	−50 / −100	80 / 330	+125 / −0	−80 / −205	100 / 500	+200 / −0	−100 / −300	131.9 − 171.9	
7.0 / 72.0	+40.0 / 0	−7 / −32	14.0 / 114.0	+60.0 / −0	−14 / −54	25 / 125	+60 / −0	−25 / −65	50 / 210	+100 / −0	−50 / −110	90 / 410	+160 / −0	−90 / −250	125 / 625	+250 / −0	−125 / −375	171.9 − 200	

(Courtesy of ANSI; B4.1–1955.)

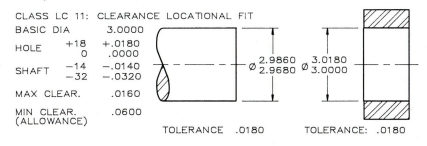

CLASS LC 11: CLEARANCE LOCATIONAL FIT		
BASIC DIA	3.0000	
HOLE	+18 / 0	+.0180 / .0000
SHAFT	−14 / −32	−.0140 / −.0320
MAX CLEAR.	.0160	
MIN CLEAR. (ALLOWANCE)	.0600	

Ø 2.9860 / 2.9680   Ø 3.0180 / 3.0000

TOLERANCE .0180   TOLERANCE: .0180

Appendix 40

American standard transition locational fits (hole basis)

Limits are in thousandths of an inch.

Limits for hole and shaft are applied algebraically to the basic size to obtain the limits of size for the mating parts.

Data in bold face are in accordance with ABC agreements.

"Fit" represents the maximum interference (minus values) and the maximum clearance (plus values).

Symbols H7, js6, etc., are Hole and Shaft designations used in ABC System.

Nominal Size Range Inches		Class LT 1			Class LT 2			Class LT 3			Class LT 4			Class LT 5			Class LT 6		
Over	To	Fit	Hole H7	Shaft js6	Fit	Hole H8	Shaft js7	Fit	Hole H7	Shaft k6	Fit	Hole H8	Shaft k7	Fit	Hole H7	Shaft n6	Fit	Hole H7	Shaft n7
0	0.12	-0.10 / +0.50	+0.4 / -0	+0.10 / -0.10	-0.2 / +0.8	+0.6 / -0	+0.2 / -0.2							-0.5 / +0.15	+0.4 / -0	+0.5 / +0.25	-0.65 / +0.15	+0.4 / -0	+0.65 / +0.25
0.12	0.24	-0.15 / +0.65	+0.5 / -0	+0.15 / -0.15	-0.25 / +0.95	+0.7 / -0	+0.25 / -0.25							-0.6 / +0.2	+0.5 / -0	+0.6 / +0.3	-0.8 / +0.2	+0.5 / -0	+0.8 / +0.3
0.24	0.40	-0.2 / +0.8	+0.6 / -0	+0.2 / -0.2	-0.3 / +1.2	+0.9 / -0	+0.3 / -0.3	-0.5 / +0.5	+0.6 / -0	+0.5 / +0.1	-0.7 / +0.8	+0.9 / -0	+0.7 / +0.1	-0.8 / +0.2	+0.6 / -0	+0.8 / +0.4	-1.0 / +0.2	+0.6 / -0	+1.0 / +0.4
0.40	0.71	-0.2 / +0.9	+0.7 / -0	+0.2 / -0.2	-0.35 / +1.35	+1.0 / -0	+0.35 / -0.35	-0.5 / +0.6	+0.7 / -0	+0.5 / +0.1	-0.8 / +0.9	+1.0 / -0	+0.8 / +0.1	-0.9 / +0.2	+0.7 / -0	+0.9 / +0.5	-1.2 / +0.2	+0.7 / -0	+1.2 / +0.5
0.71	1.19	-0.25 / +1.05	+0.8 / -0	+0.25 / -0.25	-0.4 / +1.6	+1.2 / -0	+0.4 / -0.4	-0.6 / +0.7	+0.8 / -0	+0.6 / +0.1	-0.9 / +1.1	+1.2 / -0	+0.9 / +0.1	-1.1 / +0.2	+0.8 / -0	+1.1 / +0.6	-1.4 / +0.2	+0.8 / -0	+1.4 / +0.6
1.19	1.97	-0.3 / +1.3	+1.0 / -0	+0.3 / -0.3	-0.5 / +2.1	+1.6 / -0	+0.5 / -0.5	-0.7 / +0.9	+1.0 / -0	+0.7 / +0.1	-1.1 / +1.5	+1.6 / -0	+1.1 / +0.1	-1.3 / +0.3	+1.0 / -0	+1.3 / +0.7	-1.7 / +0.3	+1.0 / -0	+1.7 / +0.7
1.97	3.15	-0.3 / +1.5	+1.2 / -0	+0.3 / -0.3	-0.6 / +2.4	+1.8 / -0	+0.6 / -0.6	-0.8 / +1.1	+1.2 / -0	+0.8 / +0.1	-1.3 / +1.7	+1.8 / -0	+1.3 / +0.1	-1.5 / +0.4	+1.2 / -0	+1.5 / +0.8	-2.0 / +0.4	+1.2 / -0	+2.0 / +0.8
3.15	4.73	-0.4 / +1.8	+1.4 / -0	+0.4 / -0.4	-0.7 / +2.9	+2.2 / -0	+0.7 / -0.7	-1.0 / +1.3	+1.4 / -0	+1.0 / +0.1	-1.5 / +2.1	+2.2 / -0	+1.5 / +0.1	-1.9 / +0.4	+1.4 / -0	+1.9 / +1.0	-2.4 / +0.4	+1.4 / -0	+2.4 / +1.0
4.73	7.09	-0.5 / +2.1	+1.6 / -0	+0.5 / -0.5	-0.8 / +3.3	+2.5 / -0	+0.8 / -0.8	-1.1 / +1.5	+1.6 / -0	+1.1 / +0.1	-1.7 / +2.4	+2.5 / -0	+1.7 / +0.1	-2.2 / +0.4	+1.6 / -0	+2.2 / +1.2	-2.8 / +0.4	+1.6 / -0	+2.8 / +1.2
7.09	9.85	-0.6 / +2.4	+1.8 / -0	+0.6 / -0.6	-0.9 / +3.7	+2.8 / -0	+0.9 / -0.9	-1.4 / +1.6	+1.8 / -0	+1.4 / +0.2	-2.0 / +2.6	+2.8 / -0	+2.0 / +0.2	-2.6 / +0.4	+1.8 / -0	+2.6 / +1.4	-3.2 / +0.4	+1.8 / -0	+3.2 / +1.4
9.85	12.41	-0.6 / +2.6	+2.0 / -0	+0.6 / -0.6	-1.0 / +4.0	+3.0 / -0	+1.0 / -1.0	-1.4 / +1.8	+2.0 / -0	+1.4 / +0.2	-2.2 / +2.8	+3.0 / -0	+2.2 / +0.2	-2.6 / +0.6	+2.0 / -0	+2.6 / +1.4	-3.4 / +0.6	+2.0 / -0	+3.4 / +1.4
12.41	15.75	-0.7 / +2.9	+2.2 / -0	+0.7 / -0.7	-1.0 / +4.5	+3.5 / -0	+1.0 / -1.0	-1.6 / +2.0	+2.2 / -0	+1.6 / +0.2	-2.4 / +3.3	+3.5 / -0	+2.4 / +0.2	-3.0 / +0.6	+2.2 / -0	+3.0 / +1.6	-3.8 / +0.6	+2.2 / -0	+3.8 / +1.6
15.75	19.69	-0.8 / +3.3	+2.5 / -0	+0.8 / -0.8	-1.2 / +5.2	+4.0 / -0	+1.2 / -1.2	-1.8 / +2.3	+2.5 / -0	+1.8 / +0.2	-2.7 / +3.8	+4.0 / -0	+2.7 / +0.2	-3.4 / +0.7	+2.5 / -0	+3.4 / +1.8	-4.3 / +0.7	+2.5 / -0	+4.3 / +1.8

(Courtesy of ANSI; B4.1–1955.)

Appendix 41
American standard interference locational fits (hole basis)

Limits are in thousandths of an inch.
Limits for hole and shaft are applied algebraically to the
basic size to obtain the limits of size for the parts.
Data in bold face are in accordance with ABC agreements,
Symbols H7, p6, etc., are Hole and Shaft designations
used in ABC System.

Nominal Size Range Inches (Over — To)	Class LN 1 Limits of Interference	Class LN 1 Standard Limits Hole H6	Class LN 1 Standard Limits Shaft n5	Class LN 2 Limits of Interference	Class LN 2 Standard Limits Hole H7	Class LN 2 Standard Limits Shaft p6	Class LN 3 Limits of Interference	Class LN 3 Standard Limits Hole H7	Class LN 3 Standard Limits Shaft r6
0 — 0.12	0 / 0.45	+ 0.25 / − 0	+0.45 / +0.25	0 / 0.65	+ 0.4 / − 0	+ 0.65 / + 0.4	0.1 / 0.75	+ 0.4 / − 0	+ 0.75 / + 0.5
0.12 — 0.24	0 / 0.5	+ 0.3 / − 0	+0.5 / +0.3	0 / 0.8	+ 0.5 / − 0	+ 0.8 / + 0.5	0.1 / 0.9	+ 0.5 / 0	+ 0.9 / + 0.6
0.24 — 0.40	0 / 0.65	+ 0.4 / − 0	+0.65 / +0.4	0 / 1.0	+ 0.6 / − 0	+ 1.0 / + 0.6	0.2 / 1.2	+ 0.6 / − 0	+ 1.2 / + 0.8
0.40 — 0.71	0 / 0.8	+ 0.4 / − 0	+0.8 / +0.4	0 / 1.1	+ 0.7 / − 0	+ 1.1 / + 0.7	0.3 / 1.4	+ 0.7 / − 0	+ 1.4 / + 1.0
0.71 — 1.19	0 / 1.0	+ 0.5 / − 0	+1.0 / +0.5	0 / 1.3	+ 0.8 / − 0	+ 1.3 / + 0.8	0.4 / 1.7	+ 0.8 / − 0	+ 1.7 / + 1.2
1.19 — 1.97	0 / 1.1	+ 0.6 / − 0	+1.1 / +0.6	0 / 1.6	+ 1.0 / − 0	+ 1.6 / + 1.0	0.4 / 2.0	+ 1.0 / − 0	+ 2.0 / + 1.4
1.97 — 3.15	0.1 / 1.3	+ 0.7 / − 0	+1.3 / +0.7	0.2 / 2.1	+ 1.2 / − 0	+ 2.1 / + 1.4	0.4 / 2.3	+ 1.2 / − 0	+ 2.3 / + 1.6
3.15 — 4.73	0.1 / 1.6	+ 0.9 / − 0	+1.6 / +1.0	0.2 / 2.5	+ 1.4 / − 0	+ 2.5 / + 1.6	0.6 / 2.9	+ 1.4 / − 0	+ 2.9 / + 2.0
4.73 — 7.09	0.2 / 1.9	+ 1.0 / − 0	+1.9 / +1.2	0.2 / 2.8	+ 1.6 / − 0	+ 2.8 / + 1.8	0.9 / 3.5	+ 1.6 / − 0	+ 3.5 / + 2.5
7.09 — 9.85	0.2 / 2.2	+ 1.2 / − 0	+2.2 / +1.4	0.2 / 3.2	+ 1.8 / − 0	+ 3.2 / + 2.0	1.2 / 4.2	+ 1.8 / − 0	+ 4.2 / + 3.0
9.85 — 12.41	0.2 / 2.3	+ 1.2 / − 0	+2.3 / +1.4	0.2 / 3.4	+ 2.0 / − 0	+ 3.4 / + 2.2	1.5 / 4.7	+ 2.0 / − 0	+ 4.7 / + 3.5
12.41 — 15.75	0.2 / 2.6	+ 1.4 / − 0	+2.6 / +1.6	0.3 / 3.9	+ 2.2 / − 0	+ 3.9 / + 2.5	2.3 / 5.9	+ 2.2 / − 0	+ 5.9 / + 4.5
15.75 — 19.69	0.2 / 2.8	+ 1.6 / − 0	+2.8 / +1.8	0.3 / 4.4	+ 2.5 / − 0	+ 4.4 / + 2.8	2.5 / 6.6	+ 2.5 / − 0	+ 6.6 / + 5.0
19.69 — 30.09		+ 2.0 / − 0		0.5 / 5.5	+ 3 / − 0	+ 5.5 / + 3.5	4 / 9	+ 3 / − 0	+ 9 / + 7
30.09 — 41.49		+ 2.5 / − 0		0.5 / 7.0	+ 4 / − 0	+ 7.0 / + 4.5	5 / 11.5	+ 4 / − 0	+11.5 / + 9
41.49 — 56.19		+ 3.0 / − 0		1 / 9	+ 5 / − 0	+ 9 / + 6	7 / 15	+ 5 / − 0	+15 / +12
56.19 — 76.39		+ 4.0 / − 0		1 / 11	+ 6 / − 0	+11 / + 7	10 / 20	+ 6 / − 0	+20 / +16
76.39 — 100.9		+ 5.0 / − 0		1 / 14	+ 8 / − 0	+14 / + 9	12 / 25	+ 8 / − 0	+25 / +20
100.9 — 131.9		+ 6.0 / − 0		2 / 18	+10 / − 0	+18 / +12	15 / 31	+10 / − 0	+31 / +25
131.9 — 171.9		+ 8.0 / − 0		4 / 24	+12 / − 0	+24 / +16	18 / 38	+12 / − 0	+38 / +30
171.9 — 200		+10.0 / − 0		4 / 30	+16 / − 0	+30 / +20	24 / 50	+16 / − 0	+50 / +40

(Courtesy of ANSI; B4.1–1955.)

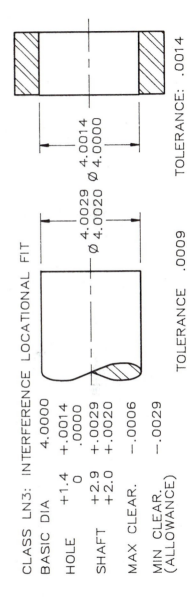

TOLERANCE: .0014

TOLERANCE .0009

Ø 4.0014 / Ø 4.0000

Ø 4.0029 / Ø 4.0020

CLASS LN3: INTERFERENCE LOCATIONAL FIT

BASIC DIA.	4.0000	
HOLE	+1.4 / 0	+.0014 / .0000
SHAFT	+2.9 / +2.0	+.0029 / +.0020
MAX CLEAR.		−.0006
MIN CLEAR. (ALLOWANCE)		−.0029

Appendix 42
American standard force and shrink fits (hole basis)

Limits are in thousandths of an inch.
Limits for hole and shaft are applied algebraically to the basic size to obtain the limits of size for the parts.
Data in bold face are in accordance with ABC agreements.
Symbols H7, s6, etc., are Hole and Shaft designations used in ABC System.

Nominal Size Range Inches Over — To	Class FN 1 Limits of Interference	FN1 Hole H6	FN1 Shaft	Class FN 2 Limits of Interference	FN2 Hole H7	FN2 Shaft s6	Class FN 3 Limits of Interference	FN3 Hole H7	FN3 Shaft t6	Class FN 4 Limits of Interference	FN4 Hole H7	FN4 Shaft u6	Class FN 5 Limits of Interference	FN5 Hole H8	FN5 Shaft x7
0 — 0.12	0.05 / 0.5	+0.25 / − 0	+ 0.5 / + 0.3	0.2 / 0.85	+0.4 / − 0	+ 0.85 / + 0.6				0.3 / 0.95	+0.4 / − 0	+ 0.95 / + 0.7	0.3 / 1.3	+0.6 / − 0	+ 1.3 / + 0.9
0.12 — 0.24	0.1 / 0.6	+0.3 / − 0	+ 0.6 / + 0.4	0.2 / 1.0	+0.5 / − 0	+ 1.0 / + 0.7				0.4 / 1.2	+0.5 / − 0	+ 1.2 / + 0.9	0.5 / 1.7	+ 0.7 / − 0	+ 1.7 / + 1.2
0.24 — 0.40	0.1 / 0.75	+0.4 / − 0	+ 0.75 / + 0.5	0.4 / 1.4	+0.6 / − 0	+ 1.4 / + 1.0				0.6 / 1.6	+0.6 / − 0	+ 1.6 / + 1.2	0.5 / 2.0	+ 0.9 / − 0	+ 2.0 / + 1.4
0.40 — 0.56	0.1 / 0.8	−0.4 / — 0	+ 0.8 / + 0.5	0.5 / 1.6	+0.7 / − 0	+ 1.6 / + 1.2				0.7 / 1.8	+0.7 / − 0	+ 1.8 / + 1.4	0.6 / 2.3	+ 1.0 / − 0	+ 2.3 / + 1.6
0.56 — 0.71	0:2 / 0.9	+0.4 / − 0	+ 0.9 / + 0.6	0.5 / 1.6	+0.7 / − 0	+ 1.6 / + 1.2				0.7 / 1.8	+0.7 / − 0	+ 1.8 / + 1.4	0.8 / 2.5	+ 1.0 / − 0	+ 2.5 / + 1.8
0.71 — 0.95	0.2 / 1.1	+0.5 / − 0	+ 1.1 / + 0.7	0.6 / 1.9	+0.8 / − 0	+ 1.9 / + 1.4				0.8 / 2.1	+0.8 / − 0	+ 2.1 / + 1.6	1.0 / 3.0	+ 1.2 / − 0	+ 3.0 / + 2.2
0.95 — 1.19	0.3 / 1.2	+0.5 / − 0	+ 1.2 / + 0.8	0.6 / 1.9	+0.8 / − 0	+ 1.9 / + 1.4	0.8 / 2.1	+0.8 / − 0	+ 2.1 / + 1.6	1.0 / 2.3	+0·8 / − 0	+ 2.3 / + 1.8	1.3 / 3.3	+ 1.2 / − 0	+ 3.3 / + 2.5
1.19 — 1.58	0.3 / 1.3	+0.6 / − 0	+ 1.3 / + 0.9	0.8 / 2.4	+1.0 / − 0	+ 2.4 / + 1.8	1.0 / 2.6	+1.0 / − 0	+ 2.6 / + 2.0	1.5 / 3.1	+1.0 / − 0	+ 3.1 / + 2.5	1.4 / 4.0	+ 1.6 / − 0	+ 4.0 / + 3.0
1.58 — 1.97	0.4 / 1.4	+0.6 / − 0	+ 1.4 / + 1.0	0.8 / 2.4	+1.0 / − 0	+ 2.4 / + 1.8	1.2 / 2.8	+1.0 / − 0	+ 2.8 / + 2.2	1.8 / 3.4	+1.0 / − 0	+ 3.4 / + 2.8	2.4 / 5.0	+ 1.6 / − 0	+ 5.0 / + 4.0
1.97 — 2.56	0.6 / 1.8	+0.7 / − 0	+ 1.8 / + 1.3	0.8 / 2.7	+1.2 / − 0	+ 2.7 / + 2.0	1.3 / 3.2	+1.2 / − 0	+ 3.2 / + 2.5	2.3 / 4.2	+1.2 / − 0	+ 4.2 / + 3.5	3.2 / 6.2	+ 1.8 / − 0	+ 6.2 / + 5.0
2.56 — 3.15	0.7 / 1.9	+0.7 / − 0	+ 1.9 / + 1.4	1.0 / 2.9	+1.2 / − 0	+ 2.9 / + 2.2	1.8 / 3.7	+1.2 / − 0	+ 3.7 / + 3.0	2.8 / 4.7	+1.2 / − 0	+ 4.7 / + 4.0	4.2 / 7.2	+ 1.8 / − 0	+ 7.2 / + 6.0
3.15 — 3.94	0.9 / 2.4	+0.9 / − 0	+ 2.4 / + 1.8	1.4 / 3.7	+1.4 / − 0	+ 3.7 / + 2.8	2.1 / 4.4	+1.4 / − 0	+ 4.4 / + 3.5	3.6 / 5.9	+1.4 / − 0	+ 5.9 / + 5.0	4.8 / 8.4	+ 2.2 / − 0	+ 8.4 / + 7.0
3.94 — 4.73	1.1 / 2.6	+0.9 / − 0	+ 2.6 / + 2.0	1.6 / 3.9	+1.4 / − 0	+ 3.9 / + 3.0	2.6 / 4.9	+1.4 / − 0	+ 4.9 / + 4.0	4.6 / 6.9	+1.4 / − 0	+ 6.9 / + 6.0	5.8 / 9.4	+ 2.2 / − 0	+ 9.4 / + 8.0
4.73 — 5.52	1.2 / 2.9	+1.0 / − 0	+ 2.9 / + 2.2	1.9 / 4.5	+1.6 / − 0	+ 4.5 / + 3.5	3.4 / 6.0	+1.6 / − 0	+ 6.0 / + 5.0	5.4 / 8.0	+1.6 / − 0	+ 8.0 / + 7.0	7.5 / 11.6	+ 2.5 / − 0	+11.6 / +10.0
5.52 — 6.30	1.5 / 3.2	+1.0 / − 0	+ 3.2 / + 2.5	2.4 / 5.0	+1.6 / − 0	+ 5.0 / + 4.0	3.4 / 6.0	+1.6 / − 0	+ 6.0 / + 5.0	5.4 / 8.0	+1.6 / − 0	+ 8.0 / + 7.0	9.5 / 13.6	+ 2.5 / − 0	+13.6 / +12.0
6.30 — 7.09	1.8 / 3.5	+1.0 / − 0	+ ·3.5 / + 2.8	2.9 / 5.5	+1.6 / − 0	+ 5.5 / + 4.5	4.4 / 7.0	+1.6 / − 0	+ 7.0 / + 6.0	6.4 / 9.0	+1.6 / − 0	+ 9.0 / + 8.0	9.5 / 13.6	+ 2.5 / − 0	+13.6 / +12.0
7.09 — 7.88	1.8 / 3.8	+1.2 / − 0	+ 3.8 / + 3.0	3.2 / 6.2	+1.8 / − 0	+ 6.2 / + 5.0	5.2 / 8.2	+1.8 / − 0	+ 8.2 / + 7.0	7.2 / 10.2	+1.8 / − 0	+10.2 / + 9.0	11.2 / 15.8	+ 2.8 / − 0	+15.8 / +14.0
7.88 — 8.86	2.3 / 4.3	+1.2 / − 0	+ 4.3 / + 3.5	3.2 / 6.2	+1.8 / − 0	+ 6.2 / + 5.0	5.2 / 8.2	+1.8 / − 0	+ 8.2 / + 7.0	8.2 / 11.2	+1.8 / − 0	+11.2 / +10.0	13.2 / 17.8	+ 2.8 / − 0	+17.8 / +16.0
8.86 — 9.85	2.3 / 4.3	+1.2 / − 0	+ 4.3 / + 3.5	4.2 / 7.2	+1.8 / − 0	+ 7.2 / + 6.0	6.2 / 9.2	+1.8 / − 0	+ 9.2 / + 8.0	10.2 / 13.2	+1.8 / − 0	+13.2 / +12.0	13.2 / 17.8	+ 2.8 / − 0	+17.8 / +16.0
9.85 — 11.03	2.8 / 4.9	+1.2 / − 0	+ 4.9 / + 4.0	4.0 / 7.2	+2.0 / − 0	+ 7.2 / + 6.0	7.0 / 10.2	+2.0 / − 0	+10.2 / + 9.0	10.0 / 13.2	+2.0 / − 0	+13.2 / +12.0	15.0 / 20.0	+ 3.0 / − 0	+20.0 / +18.0
11.03 — 12.41	2.8 / 4.9	+1.2 / − 0	+ 4.9 / + 4.0	5.0 / 8.2	+2.0 / − 0	+ 8.2 / + 7.0	7.0 / 10.2	+2.0 / − 0	+10.2 / + 9.0	12.0 / 15.2	+2.0 / − 0	+15.2 / +14.0	17.0 / 22.0	+ 3.0 / − 0	+22.0 / +20.0
12.41 — 13.98	3.1 / 5.5	+1.4 / − 0	+ 5.5 / + 4.5	5.8 / 9.4	+2.2 / − 0	+ 9.4 / + 8.0	7.8 / 11.4	+2.2 / − 0	+11.4 / +10.0	13.8 / 17.4	+2.2 / − 0	+17.4 / +16.0	18.5 / 24.2	+ 3.5 / + 0	+24.2 / +22.0
13.98 — 15.75	3.6 / 6.1	+1.4 / − 0	+ 6.1 / + 5.0	5.8 / 9.4	+2.2 / − 0	+ 9.4 / + 8.0	9.8 / 13.4	+2.2 / − 0	+13.4 / +12.0	15.8 / 19.4	+2.2 / − 0	+19.4 / +18.0	21.5 / 27.2	+ 3.5 / − 0	+27.2 / +25.0
15.75 — 17.72	4.4 / 7.0	+1.6 / − 0	+ 7.0 / + 6.0	6.5 / 10.6	+2.5 / − 0	+10.6 / + 9.0	9.5 / 13.6	+2.5 / − 0	+13.6 / +12.0	17.5 / 21.6	+2.5 / − 0	+21.6 / +20.0	24.0 / 30.5	+ 4.0 / − 0	+30.5 / +28.0
17.72 — 19.69	4.4 / 7.0	+1.6 / − 0	+ 7.0 / + 6.0	7.5 / 11.6	+2.5 / − 0	+11.6 / +10.0	11.5 / 15.6	+2.5 / − 0	+15.6 / +14.0	19.5 / 23.6	+2.5 / − 0	+23.6 / +22.0	26.0 / 32.5	+ 4.0 / − 0	+32.5 / +30.0

(Courtesy of ANSI; B4.1–1955.)

Appendix 43
The international tolerance grades (ANSI B4.2)

Dimensions are in mm.

Basic sizes Over	Up to and including	IT01	IT0	IT1	IT2	IT3	IT4	IT5	IT6	IT7	IT8	IT9	IT10	IT11	IT12	IT13	IT14	IT15	IT16
0	3	0.0003	0.0005	0.0008	0.0012	0.002	0.003	0.004	0.006	0.010	0.014	0.025	0.040	0.060	0.100	0.140	0.250	0.400	0.600
3	6	0.0004	0.0006	0.001	0.0015	0.0025	0.004	0.005	0.008	0.012	0.018	0.030	0.048	0.075	0.120	0.180	0.300	0.480	0.750
6	10	0.0004	0.0006	0.001	0.0015	0.0025	0.004	0.006	0.009	0.015	0.022	0.036	0.058	0.090	0.150	0.220	0.360	0.580	0.900
10	18	0.0005	0.0008	0.0012	0.002	0.003	0.005	0.008	0.011	0.018	0.027	0.043	0.070	0.110	0.180	0.270	0.430	0.700	1.100
18	30	0.0006	0.001	0.0015	0.0025	0.004	0.006	0.009	0.013	0.021	0.033	0.052	0.084	0.130	0.210	0.330	0.520	0.840	1.300
30	50	0.0006	0.001	0.0015	0.0025	0.004	0.007	0.011	0.016	0.025	0.039	0.062	0.100	0.160	0.250	0.390	0.620	1.000	1.600
50	80	0.0008	0.0012	0.002	0.003	0.005	0.008	0.013	0.019	0.030	0.046	0.074	0.120	0.190	0.300	0.460	0.740	1.200	1.900
80	120	0.001	0.0015	0.0025	0.004	0.006	0.010	0.015	0.022	0.035	0.054	0.087	0.140	0.220	0.350	0.540	0.870	1.400	2.200
120	180	0.0012	0.002	0.0035	0.005	0.008	0.012	0.018	0.025	0.040	0.063	0.100	0.160	0.250	0.400	0.630	1.000	1.600	2.500
180	250	0.002	0.003	0.0045	0.007	0.010	0.014	0.020	0.029	0.046	0.072	0.115	0.185	0.290	0.460	0.720	1.150	1.850	2.900
250	315	0.0025	0.004	0.006	0.008	0.012	0.016	0.023	0.032	0.052	0.081	0.130	0.210	0.320	0.520	0.810	1.300	2.100	3.200
315	400	0.003	0.005	0.007	0.009	0.013	0.018	0.025	0.036	0.057	0.089	0.140	0.230	0.360	0.570	0.890	1.400	2.300	3.600
400	500	0.004	0.006	0.008	0.010	0.015	0.020	0.027	0.040	0.063	0.097	0.155	0.250	0.400	0.630	0.970	1.550	2.500	4.000
500	630	0.0045	0.006	0.009	0.011	0.016	0.022	0.030	0.044	0.070	0.110	0.175	0.280	0.440	0.700	1.100	1.750	2.800	4.400
630	800	0.005	0.007	0.010	0.013	0.018	0.025	0.035	0.050	0.080	0.125	0.200	0.320	0.500	0.800	1.250	2.000	3.200	5.000
800	1000	0.0055	0.008	0.011	0.015	0.021	0.029	0.040	0.056	0.090	0.140	0.230	0.360	0.560	0.900	1.400	2.300	3.600	5.600
1000	1250	0.0065	0.009	0.013	0.018	0.024	0.034	0.046	0.066	0.105	0.165	0.260	0.420	0.660	1.050	1.650	2.600	4.200	6.600
1250	1600	0.008	0.011	0.015	0.021	0.029	0.040	0.054	0.078	0.125	0.195	0.310	0.500	0.780	1.250	1.950	3.100	5.000	7.800
1600	2000	0.009	0.013	0.018	0.025	0.035	0.048	0.065	0.092	0.150	0.230	0.370	0.600	0.920	1.500	2.300	3.700	6.000	9.200
2000	2500	0.011	0.015	0.022	0.030	0.041	0.057	0.077	0.110	0.175	0.280	0.440	0.700	1.100	1.750	2.800	4.400	7.000	11.000
2500	3150	0.013	0.018	0.026	0.036	0.050	0.069	0.093	0.135	0.210	0.330	0.540	0.860	1.350	2.100	3.300	5.400	8.600	13.500

Tolerance grades[3]

[3] IT Values for tolerance grades larger than IT16 can be calculated by using the following formulas:
IT17 = IT12 × 10; IT18 = IT13 × 10; etc.

Appendix 44
Preferred hole basis clearance fits—cylindrical fits (ANSI B4.2)

AMERICAN NATIONAL STANDARD
PREFERRED METRIC LIMITS AND FITS

ANSI B4.2-1978

Dimensions in mm.

BASIC SIZE		LOOSE RUNNING Hole H11	Shaft c11	Fit	FREE RUNNING Hole H9	Shaft d9	Fit	CLOSE RUNNING Hole H8	Shaft f7	Fit	SLIDING Hole H7	Shaft g6	Fit	LOCATIONAL CLEARANCE Hole H7	Shaft h6	Fit
1	MAX	1.060	0.940	0.180	1.025	0.980	0.070	1.014	0.994	0.030	1.010	0.998	0.018	1.010	1.000	0.016
	MIN	1.000	0.880	0.060	1.000	0.955	0.020	1.000	0.984	0.006	1.000	0.992	0.002	1.000	0.994	0.000
1.2	MAX	1.260	1.140	0.180	1.225	1.180	0.070	1.214	1.194	0.030	1.210	1.198	0.018	1.210	1.200	0.016
	MIN	1.200	1.080	0.060	1.200	1.155	0.020	1.200	1.184	0.006	1.200	1.192	0.002	1.200	1.194	0.000
1.6	MAX	1.660	1.540	0.180	1.625	1.580	0.070	1.614	1.594	0.030	1.610	1.598	0.018	1.610	1.600	0.016
	MIN	1.600	1.480	0.060	1.600	1.555	0.020	1.600	1.584	0.006	1.600	1.592	0.002	1.600	1.594	0.000
2	MAX	2.060	1.940	0.180	2.025	1.980	0.070	2.014	1.994	0.030	2.010	1.998	0.018	2.010	2.000	0.016
	MIN	2.000	1.880	0.060	2.000	1.955	0.020	2.000	1.984	0.006	2.000	1.992	0.002	2.000	1.994	0.000
2.5	MAX	2.560	2.440	0.180	2.525	2.480	0.070	2.514	2.494	0.030	2.510	2.498	0.018	2.510	2.500	0.016
	MIN	2.500	2.380	0.060	2.500	2.455	0.020	2.500	2.484	0.006	2.500	2.492	0.002	2.500	2.494	0.000
3	MAX	3.060	2.940	0.180	3.025	2.980	0.070	3.014	2.994	0.030	3.010	2.998	0.018	3.010	3.000	0.016
	MIN	3.000	2.880	0.060	3.000	2.955	0.020	3.000	2.984	0.006	3.000	2.992	0.002	3.000	2.994	0.000
4	MAX	4.075	3.930	0.220	4.030	3.970	0.090	4.018	3.990	0.040	4.012	3.996	0.024	4.012	4.000	0.020
	MIN	4.000	3.855	0.070	4.000	3.940	0.030	4.000	3.978	0.010	4.000	3.988	0.004	4.000	3.992	0.000
5	MAX	5.075	4.930	0.220	5.030	4.970	0.090	5.018	4.990	0.040	5.012	4.996	0.024	5.012	5.000	0.020
	MIN	5.000	4.855	0.070	5.000	4.940	0.030	5.000	4.978	0.010	5.000	4.988	0.004	5.000	4.992	0.000
6	MAX	6.075	5.930	0.220	6.030	5.970	0.090	6.018	5.990	0.040	6.012	5.996	0.024	6.012	6.000	0.020
	MIN	6.000	5.855	0.070	6.000	5.940	0.030	6.000	5.978	0.010	6.000	5.988	0.004	6.000	5.992	0.000
8	MAX	8.090	7.920	0.260	8.036	7.960	0.112	8.022	7.987	0.050	8.015	7.995	0.029	8.015	8.000	0.024
	MIN	8.000	7.830	0.080	8.000	7.924	0.040	8.000	7.972	0.013	8.000	7.986	0.005	8.000	7.991	0.000
10	MAX	10.090	9.920	0.260	10.036	9.960	0.112	10.022	9.987	0.050	10.015	9.995	0.029	10.015	10.000	0.024
	MIN	10.000	9.830	0.080	10.000	9.924	0.040	10.000	9.972	0.013	10.000	9.986	0.005	10.000	9.991	0.000
12	MAX	12.110	11.905	0.315	12.043	11.950	0.136	12.027	11.984	0.061	12.018	11.994	0.035	12.018	12.000	0.029
	MIN	12.000	11.795	0.095	12.000	11.907	0.050	12.000	11.966	0.016	12.000	11.983	0.006	12.000	11.989	0.000
16	MAX	16.110	15.905	0.315	16.043	15.950	0.136	16.027	15.984	0.061	16.018	15.994	0.035	16.018	16.000	0.029
	MIN	16.000	15.795	0.095	16.000	15.907	0.050	16.000	15.966	0.016	16.000	15.983	0.006	16.000	15.989	0.000
20	MAX	20.130	19.890	0.370	20.052	19.935	0.169	20.033	19.980	0.074	20.021	19.993	0.041	20.021	20.000	0.034
	MIN	20.000	19.760	0.110	20.000	19.883	0.065	20.000	19.959	0.020	20.000	19.980	0.007	20.000	19.987	0.000
25	MAX	25.130	24.890	0.370	25.052	24.935	0.169	25.033	24.980	0.074	25.021	24.993	0.041	25.021	25.000	0.034
	MIN	25.000	24.760	0.110	25.000	24.883	0.065	25.000	24.959	0.020	25.000	24.980	0.007	25.000	24.987	0.000
30	MAX	30.130	29.890	0.370	30.052	29.935	0.169	30.033	29.980	0.074	30.021	29.993	0.041	30.021	30.000	0.034
	MIN	30.000	29.760	0.110	30.000	29.883	0.065	30.000	29.959	0.020	30.000	29.980	0.007	30.000	29.987	0.000

Cont.

Appendix 44
Preferred hole basis clearance fits—cylindrical fits (cont.)

AMERICAN NATIONAL STANDARD
PREFERRED METRIC LIMITS AND FITS

ANSI B4.2-1978

Dimensions in mm.

BASIC SIZE		LOOSE RUNNING Hole H11	Shaft c11	Fit	FREE RUNNING Hole H9	Shaft d9	Fit	CLOSE RUNNING Hole H8	Shaft f7	Fit	SLIDING Hole H7	Shaft g6	Fit	LOCATIONAL CLEARANCE Hole H7	Shaft h6	Fit
40	MAX	40.160	39.880	0.440	40.062	39.920	0.204	40.039	39.975	0.089	40.025	39.991	0.050	40.025	40.000	0.041
	MIN	40.000	39.720	0.120	40.000	39.858	0.080	40.000	39.950	0.025	40.000	39.975	0.009	40.000	39.984	0.000
50	MAX	50.160	49.870	0.450	50.062	49.920	0.204	50.039	49.975	0.089	50.025	49.991	0.050	50.025	50.000	0.041
	MIN	50.000	49.710	0.130	50.000	49.858	0.080	50.000	49.950	0.025	50.000	49.975	0.009	50.000	49.984	0.000
60	MAX	60.190	59.860	0.520	60.074	59.900	0.248	60.046	59.970	0.106	60.030	59.990	0.059	60.030	60.000	0.049
	MIN	60.000	59.670	0.140	60.000	59.826	0.100	60.000	59.940	0.030	60.000	59.971	0.010	60.000	59.981	0.000
80	MAX	80.190	79.850	0.530	80.074	79.900	0.248	80.046	79.970	0.106	80.030	79.990	0.059	80.030	80.000	0.049
	MIN	80.000	79.660	0.150	80.000	79.826	0.100	80.000	79.940	0.030	80.000	79.971	0.010	80.000	79.981	0.000
100	MAX	100.220	99.830	0.610	100.087	99.880	0.294	100.054	99.964	0.125	100.035	99.988	0.069	100.035	100.000	0.057
	MIN	100.000	99.610	0.170	100.000	99.793	0.120	100.000	99.929	0.036	100.000	99.966	0.012	100.000	99.978	0.000
120	MAX	120.220	119.820	0.620	120.087	119.880	0.294	120.054	119.964	0.125	120.035	119.988	0.069	120.035	120.000	0.057
	MIN	120.000	119.600	0.180	120.000	119.793	0.120	120.000	119.929	0.036	120.000	119.966	0.012	120.000	119.978	0.000
160	MAX	160.250	159.790	0.710	160.100	159.855	0.345	160.063	159.957	0.146	160.040	159.986	0.079	160.040	160.000	0.065
	MIN	160.000	159.540	0.210	160.000	159.755	0.145	160.000	159.917	0.043	160.000	159.961	0.014	160.000	159.975	0.000
200	MAX	200.290	199.760	0.820	200.115	199.830	0.400	200.072	199.950	0.168	200.046	199.985	0.090	200.046	200.000	0.075
	MIN	200.000	199.470	0.240	200.000	199.715	0.170	200.000	199.904	0.050	200.000	199.956	0.015	200.000	199.971	0.000
250	MAX	250.290	249.720	0.860	250.115	249.830	0.400	250.072	249.950	0.168	250.046	249.985	0.090	250.046	250.000	0.075
	MIN	250.000	249.430	0.280	250.000	249.715	0.170	250.000	249.904	0.050	250.000	249.956	0.015	250.000	249.971	0.000
300	MAX	300.320	299.670	0.970	300.130	299.810	0.450	300.081	299.944	0.189	300.052	299.983	0.101	300.052	300.000	0.084
	MIN	300.000	299.350	0.330	300.000	299.680	0.190	300.000	299.892	0.056	300.000	299.951	0.017	300.000	299.968	0.000
400	MAX	400.360	399.600	1.120	400.140	399.790	0.490	400.089	399.938	0.208	400.057	399.982	0.111	400.057	400.000	0.093
	MIN	400.000	399.240	0.400	400.000	399.650	0.210	400.000	399.881	0.062	400.000	399.946	0.018	400.000	399.964	0.000
500	MAX	500.400	499.520	1.280	500.155	499.770	0.540	500.097	499.932	0.228	500.063	499.980	0.123	500.063	500.000	0.103
	MIN	500.000	499.120	0.480	500.000	499.615	0.230	500.000	499.869	0.068	500.000	499.940	0.020	500.000	499.960	0.000

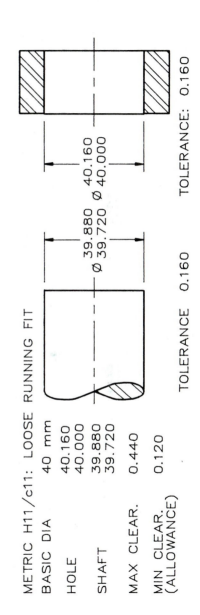

TOLERANCE: 0.160

Ø 39.880 Ø 40.160
Ø 39.720 Ø 40.000

TOLERANCE 0.160

METRIC H11/c11: LOOSE RUNNING FIT

BASIC DIA	40 mm
HOLE	40.160
	40.000
SHAFT	39.880
	39.720
MAX CLEAR.	0.440
MIN CLEAR. (ALLOWANCE)	0.120

TOLERANCE 0.160

Appendix 45
Preferred hole basis transition and interference fits—cylindrical fits (ANSI B4.2)

AMERICAN NATIONAL STANDARD
PREFERRED METRIC LIMITS AND FITS

ANSI B4.2-1978

Dimensions in mm.

BASIC SIZE		LOCATIONAL TRANSN. Hole H7	Shaft k6	Fit	LOCATIONAL TRANSN. Hole H7	Shaft n6	Fit	LOCATIONAL INTERF. Hole H7	Shaft p6	Fit	MEDIUM DRIVE Hole H7	Shaft s6	Fit	FORCE Hole H7	Shaft u6	Fit
1	MAX	1.010	1.006	0.010	1.010	1.010	0.006	1.010	1.012	0.004	1.010	1.020	-0.004	1.010	1.024	-0.008
	MIN	1.000	1.000	-0.006	1.000	1.004	-0.010	1.000	1.006	-0.012	1.000	1.014	-0.020	1.000	1.018	-0.024
1.2	MAX	1.210	1.206	0.010	1.210	1.210	0.006	1.210	1.212	0.004	1.210	1.220	-0.004	1.210	1.224	-0.008
	MIN	1.200	1.200	-0.006	1.200	1.204	-0.010	1.200	1.206	-0.012	1.200	1.214	-0.020	1.200	1.218	-0.024
1.6	MAX	1.610	1.606	0.010	1.610	1.610	0.006	1.610	1.612	0.004	1.610	1.620	-0.004	1.610	1.624	-0.008
	MIN	1.600	1.600	-0.006	1.600	1.604	-0.010	1.600	1.606	-0.012	1.600	1.614	-0.020	1.600	1.618	-0.024
2	MAX	2.010	2.006	0.010	2.010	2.010	0.006	2.010	2.012	0.004	2.010	2.020	-0.004	2.010	2.024	-0.008
	MIN	2.000	2.000	-0.006	2.000	2.004	-0.010	2.000	2.006	-0.012	2.000	2.014	-0.020	2.000	2.018	-0.024
2.5	MAX	2.510	2.506	0.010	2.510	2.510	0.006	2.510	2.512	0.004	2.510	2.520	-0.004	2.510	2.524	-0.008
	MIN	2.500	2.500	-0.006	2.500	2.504	-0.010	2.500	2.506	-0.012	2.500	2.514	-0.020	2.500	2.518	-0.024
3	MAX	3.010	3.006	0.010	3.010	3.010	0.006	3.010	3.012	0.004	3.010	3.020	-0.004	3.010	3.024	-0.008
	MIN	3.000	3.000	-0.006	3.000	3.004	-0.010	3.000	3.006	-0.012	3.000	3.014	-0.020	3.000	3.018	-0.024
4	MAX	4.012	4.009	0.011	4.012	4.016	0.004	4.012	4.020	0.000	4.012	4.027	-0.007	4.012	4.031	-0.011
	MIN	4.000	4.001	-0.009	4.000	4.008	-0.016	4.000	4.012	-0.020	4.000	4.019	-0.027	4.000	4.023	-0.031
5	MAX	5.012	5.009	0.011	5.012	5.016	0.004	5.012	5.020	0.000	5.012	5.027	-0.007	5.012	5.031	-0.011
	MIN	5.000	5.001	-0.009	5.000	5.008	-0.016	5.000	5.012	-0.020	5.000	5.019	-0.027	5.000	5.023	-0.031
6	MAX	6.012	6.009	0.011	6.012	6.016	0.004	6.012	6.020	0.000	6.012	6.027	-0.007	6.012	6.031	-0.011
	MIN	6.000	6.001	-0.009	6.000	6.008	-0.016	6.000	6.012	-0.020	6.000	6.019	-0.027	6.000	6.023	-0.031
8	MAX	8.015	8.010	0.014	8.015	8.019	0.005	8.015	8.024	0.000	8.015	8.032	-0.008	8.015	8.037	-0.013
	MIN	8.000	8.001	-0.010	8.000	8.010	-0.019	8.000	8.015	-0.024	8.000	8.023	-0.032	8.000	8.028	-0.037
10	MAX	10.015	10.010	0.014	10.015	10.019	0.005	10.015	10.024	0.000	10.015	10.032	-0.008	10.015	10.037	-0.013
	MIN	10.000	10.001	-0.010	10.000	10.010	-0.019	10.000	10.015	-0.024	10.000	10.023	-0.032	10.000	10.028	-0.037
12	MAX	12.018	12.012	0.017	12.018	12.023	0.006	12.018	12.029	0.000	12.018	12.039	-0.010	12.018	12.044	-0.015
	MIN	12.000	12.001	-0.012	12.000	12.012	-0.023	12.000	12.018	-0.029	12.000	12.028	-0.039	12.000	12.033	-0.044
16	MAX	16.018	16.012	0.017	16.018	16.023	0.006	16.018	16.029	0.000	16.018	16.039	-0.010	16.018	16.044	-0.015
	MIN	16.000	16.001	-0.012	16.000	16.012	-0.023	16.000	16.018	-0.029	16.000	16.028	-0.039	16.000	16.033	-0.044
20	MAX	20.021	20.015	0.019	20.021	20.028	0.006	20.021	20.035	-0.001	20.021	20.048	-0.014	20.021	20.054	-0.020
	MIN	20.000	20.002	-0.015	20.000	20.015	-0.028	20.000	20.022	-0.035	20.000	20.035	-0.048	20.000	20.041	-0.054
25	MAX	25.021	25.015	0.019	25.021	25.028	0.006	25.021	25.035	-0.001	25.021	25.048	-0.014	25.021	25.061	-0.027
	MIN	25.000	25.002	-0.015	25.000	25.015	-0.028	25.000	25.022	-0.035	25.000	25.035	-0.048	25.000	25.048	-0.061
30	MAX	30.021	30.015	0.019	30.021	30.028	0.006	30.021	30.035	-0.001	30.021	30.048	-0.014	30.021	30.061	-0.027
	MIN	30.000	30.002	-0.015	30.000	30.015	-0.028	30.000	30.022	-0.035	30.000	30.035	-0.048	30.000	30.048	-0.061

Cont.

AMERICAN NATIONAL STANDARD
PREFERRED METRIC LIMITS AND FITS

ANSI B4.2-1978

Dimensions in mm.

BASIC SIZE		LOCATIONAL TRANSN. Hole H7	Shaft k6	Fit	LOCATIONAL TRANSN. Hole H7	Shaft n6	Fit	LOCATIONAL INTERF. Hole H7	Shaft p6	Fit	MEDIUM DRIVE Hole H7	Shaft s6	Fit	FORCE Hole H7	Shaft u6	Fit
40	MAX	40.025	40.018	0.023	40.025	40.033	0.008	40.025	40.042	-0.001	40.025	40.059	-0.018	40.025	40.076	-0.035
	MIN	40.000	40.002	-0.018	40.000	40.017	-0.033	40.000	40.026	-0.042	40.000	40.043	-0.059	40.000	40.060	-0.076
50	MAX	50.025	50.018	0.023	50.025	50.033	0.008	50.025	50.042	-0.001	50.025	50.059	-0.018	50.025	50.086	-0.045
	MIN	50.000	50.002	-0.018	50.000	50.017	-0.033	50.000	50.026	-0.042	50.000	50.043	-0.059	50.000	50.070	-0.086
60	MAX	60.030	60.021	0.028	60.030	60.039	0.010	60.030	60.051	-0.002	60.030	60.072	-0.023	60.030	60.106	-0.057
	MIN	60.000	60.002	-0.021	60.000	60.020	-0.039	60.000	60.032	-0.051	60.000	60.053	-0.072	60.000	60.087	-0.106
80	MAX	80.030	80.021	0.028	80.030	80.039	0.010	80.030	80.051	-0.002	80.030	80.078	-0.029	80.030	80.121	-0.072
	MIN	80.000	80.002	-0.021	80.000	80.020	-0.039	80.000	80.032	-0.051	80.000	80.059	-0.078	80.000	80.102	-0.121
100	MAX	100.035	100.025	0.032	100.035	100.045	0.012	100.035	100.059	-0.002	100.035	100.093	-0.036	100.035	100.146	-0.089
	MIN	100.000	100.003	-0.025	100.000	100.023	-0.045	100.000	100.037	-0.059	100.000	100.071	-0.093	100.000	100.124	-0.146
120	MAX	120.035	120.025	0.032	120.035	120.045	0.012	120.035	120.059	-0.002	120.035	120.101	-0.044	120.035	120.166	-0.109
	MIN	120.000	120.003	-0.025	120.000	120.023	-0.045	120.000	120.037	-0.059	120.000	120.079	-0.101	120.000	120.144	-0.166
160	MAX	160.040	160.028	0.037	160.040	160.052	0.013	160.040	160.068	-0.003	160.040	160.125	-0.060	160.040	160.215	-0.150
	MIN	160.000	160.003	-0.028	160.000	160.027	-0.052	160.000	160.043	-0.068	160.000	160.100	-0.125	160.000	160.190	-0.215
200	MAX	200.046	200.033	0.042	200.046	200.060	0.015	200.046	200.079	-0.004	200.046	200.151	-0.076	200.046	200.265	-0.190
	MIN	200.000	200.004	-0.033	200.000	200.031	-0.060	200.000	200.050	-0.079	200.000	200.122	-0.151	200.000	200.236	-0.265
250	MAX	250.046	250.033	0.042	250.046	250.060	0.015	250.046	250.079	-0.004	250.046	250.169	-0.094	250.046	250.313	-0.238
	MIN	250.000	250.004	-0.033	250.000	250.031	-0.060	250.000	250.050	-0.079	250.000	250.140	-0.169	250.000	250.284	-0.313
300	MAX	300.052	300.036	0.048	300.052	300.066	0.018	300.052	300.088	-0.004	300.052	300.202	-0.118	300.052	300.382	-0.298
	MIN	300.000	300.004	-0.036	300.000	300.034	-0.066	300.000	300.056	-0.088	300.000	300.170	-0.202	300.000	300.350	-0.382
400	MAX	400.057	400.040	0.053	400.057	400.073	0.020	400.057	400.098	-0.005	400.057	400.244	-0.151	400.057	400.471	-0.378
	MIN	400.000	400.004	-0.040	400.000	400.037	-0.073	400.000	400.062	-0.098	400.000	400.208	-0.244	400.000	400.435	-0.471
500	MAX	500.063	500.045	0.058	500.063	500.080	0.023	500.063	500.108	-0.005	500.063	500.292	-0.189	500.063	500.580	-0.477
	MIN	500.000	500.005	-0.045	500.000	500.040	-0.080	500.000	500.068	-0.108	500.000	500.252	-0.292	500.000	500.540	-0.580

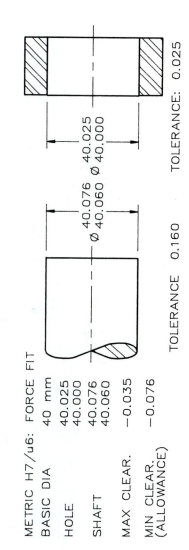

METRIC H7/u6: FORCE FIT

BASIC DIA	40 mm
HOLE	40.025 / 40.000
SHAFT	40.076 / 40.060
MAX CLEAR.	-0.035
MIN CLEAR. (ALLOWANCE)	-0.076

TOLERANCE 0.160

Ø 40.076 / Ø 40.025
Ø 40.060 / Ø 40.000

TOLERANCE: 0.025

Appendix 46
Preferred shaft basis clearance fits—cylindrical fits (ANSI B4.2)

AMERICAN NATIONAL STANDARD
PREFERRED METRIC LIMITS AND FITS

ANSI B4.2-1978

Dimensions in mm.

BASIC SIZE		LOOSE RUNNING Hole C11	Shaft h11	Fit	FREE RUNNING Hole D9	Shaft h9	Fit	CLOSE RUNNING Hole F8	Shaft h7	Fit	SLIDING Hole G7	Shaft h6	Fit	LOCATIONAL CLEARANCE Hole H7	Shaft h6	Fit
1	MAX	1.120	1.000	0.180	1.045	1.000	0.070	1.020	1.000	0.030	1.012	1.000	0.018	1.010	1.000	0.016
	MIN	1.060	0.940	0.060	1.020	0.975	0.020	1.006	0.990	0.006	1.002	0.994	0.002	1.000	0.994	0.000
1.2	MAX	1.320	1.200	0.180	1.245	1.200	0.070	1.220	1.200	0.030	1.212	1.200	0.018	1.210	1.200	0.016
	MIN	1.260	1.140	0.060	1.220	1.175	0.020	1.206	1.190	0.006	1.202	1.194	0.002	1.200	1.194	0.000
1.6	MAX	1.720	1.600	0.180	1.645	1.600	0.070	1.620	1.600	0.030	1.612	1.600	0.018	1.610	1.600	0.016
	MIN	1.660	1.540	0.060	1.620	1.575	0.020	1.606	1.590	0.006	1.602	1.594	0.002	1.600	1.594	0.000
2	MAX	2.120	2.000	0.180	2.045	2.000	0.070	2.020	2.000	0.030	2.012	2.000	0.018	2.010	2.000	0.016
	MIN	2.060	1.940	0.060	2.020	1.975	0.020	2.006	1.990	0.006	2.002	1.994	0.002	2.000	1.994	0.000
2.5	MAX	2.620	2.500	0.180	2.545	2.500	0.070	2.520	2.500	0.030	2.512	2.500	0.018	2.510	2.500	0.016
	MIN	2.560	2.440	0.060	2.520	2.475	0.020	2.506	2.490	0.006	2.502	2.494	0.002	2.500	2.494	0.000
3	MAX	3.120	3.000	0.180	3.045	3.000	0.070	3.020	3.000	0.030	3.012	3.000	0.018	3.010	3.000	0.016
	MIN	3.060	2.940	0.060	3.020	2.975	0.020	3.006	2.990	0.006	3.002	2.994	0.002	3.000	2.994	0.000
4	MAX	4.145	4.000	0.220	4.060	4.000	0.090	4.028	4.000	0.040	4.016	4.000	0.024	4.012	4.000	0.020
	MIN	4.070	3.925	0.070	4.030	3.970	0.030	4.010	3.988	0.010	4.004	3.992	0.004	4.000	3.992	0.000
5	MAX	5.145	5.000	0.220	5.060	5.000	0.090	5.028	5.000	0.040	5.016	5.000	0.024	5.012	5.000	0.020
	MIN	5.070	4.925	0.070	5.030	4.970	0.030	5.010	4.988	0.010	5.004	4.992	0.004	5.000	4.992	0.000
6	MAX	6.145	6.000	0.220	6.060	6.000	0.090	6.028	6.000	0.040	6.016	6.000	0.024	6.012	6.000	0.020
	MIN	6.070	5.925	0.070	6.030	5.970	0.030	6.010	5.988	0.010	6.004	5.992	0.004	6.000	5.992	0.000
8	MAX	8.170	8.000	0.260	8.076	8.000	0.112	8.035	8.000	0.050	8.020	8.000	0.029	8.015	8.000	0.024
	MIN	8.080	7.910	0.080	8.040	7.964	0.040	8.013	7.985	0.013	8.005	7.991	0.005	8.000	7.991	0.000
10	MAX	10.170	10.000	0.260	10.076	10.000	0.112	10.035	10.000	0.050	10.020	10.000	0.029	10.015	10.000	0.024
	MIN	10.080	9.910	0.080	10.040	9.964	0.040	10.013	9.985	0.013	10.005	9.991	0.005	10.000	9.991	0.000
12	MAX	12.205	12.000	0.315	12.093	12.000	0.136	12.043	12.000	0.061	12.024	12.000	0.035	12.018	12.000	0.029
	MIN	12.095	11.890	0.095	12.050	11.957	0.050	12.016	11.982	0.016	12.006	11.989	0.006	12.000	11.989	0.000
16	MAX	16.205	16.000	0.315	16.093	16.000	0.136	16.043	16.000	0.061	16.024	16.000	0.035	16.018	16.000	0.029
	MIN	16.095	15.890	0.095	16.050	15.957	0.050	16.016	15.982	0.016	16.006	15.989	0.006	16.000	15.989	0.000
20	MAX	20.240	20.000	0.370	20.117	20.000	0.169	20.053	20.000	0.074	20.028	20.000	0.041	20.021	20.000	0.034
	MIN	20.110	19.870	0.110	20.065	19.948	0.065	20.020	19.979	0.020	20.007	19.987	0.007	20.000	19.987	0.000
25	MAX	25.240	25.000	0.370	25.117	25.000	0.169	25.053	25.000	0.074	25.028	25.000	0.041	25.021	25.000	0.034
	MIN	25.110	24.870	0.110	25.065	24.948	0.065	25.020	24.979	0.020	25.007	24.987	0.007	25.000	24.987	0.000
30	MAX	30.240	30.000	0.370	30.117	30.000	0.169	30.053	30.000	0.074	30.028	30.000	0.041	30.021	30.000	0.034
	MIN	30.110	29.870	0.110	30.065	29.948	0.065	30.020	29.979	0.020	30.007	29.987	0.007	30.000	29.987	0.000

Cont.

Appendix 46
Preferred shaft basis clearance fits—cylindrical fits (cont.)

AMERICAN NATIONAL STANDARD
PREFERRED METRIC LIMITS AND FITS

ANSI B4.2–1978

Dimensions in mm.

BASIC SIZE		LOOSE RUNNING Hole C11	Shaft h11	Fit	FREE RUNNING Hole D9	Shaft h9	Fit	CLOSE RUNNING Hole F8	Shaft h7	Fit	SLIDING Hole G7	Shaft h6	Fit	LOCATIONAL CLEARANCE Hole H7	Shaft h6	Fit
40	MAX	40.280	40.000	0.440	40.142	40.000	0.204	40.064	40.000	0.089	40.034	40.000	0.050	40.025	40.000	0.041
	MIN	40.120	39.840	0.120	40.080	39.938	0.080	40.025	39.975	0.025	40.009	39.984	0.009	40.000	39.984	0.000
50	MAX	50.290	50.000	0.450	50.142	50.000	0.204	50.064	50.000	0.089	50.034	50.000	0.050	50.025	50.000	0.041
	MIN	50.130	49.840	0.130	50.080	49.938	0.080	50.025	49.975	0.025	50.009	49.984	0.009	50.000	49.984	0.000
60	MAX	60.330	60.000	0.520	60.174	60.000	0.248	60.076	60.000	0.106	60.040	60.000	0.059	60.030	60.000	0.049
	MIN	60.140	59.810	0.140	60.100	59.926	0.100	60.030	59.970	0.030	60.010	59.981	0.010	60.000	59.981	0.000
80	MAX	80.340	80.000	0.530	80.174	80.000	0.248	80.076	80.000	0.106	80.040	80.000	0.059	80.030	80.000	0.049
	MIN	80.150	79.810	0.150	80.100	79.926	0.100	80.030	79.970	0.030	80.010	79.981	0.010	80.000	79.981	0.000
100	MAX	100.390	100.000	0.610	100.207	100.000	0.294	100.090	100.000	0.125	100.047	100.000	0.069	100.035	100.000	0.057
	MIN	100.170	99.780	0.170	100.120	99.913	0.120	100.036	99.965	0.036	100.012	99.978	0.012	100.000	99.978	0.000
120	MAX	120.400	120.000	0.620	120.207	120.000	0.294	120.090	120.000	0.125	120.047	120.000	0.069	120.035	120.000	0.057
	MIN	120.180	119.780	0.180	120.120	119.913	0.120	120.036	119.965	0.036	120.012	119.978	0.012	120.000	119.978	0.000
160	MAX	160.460	160.000	0.710	160.245	160.000	0.345	160.106	160.000	0.146	160.054	160.000	0.079	160.040	160.000	0.065
	MIN	160.210	159.750	0.210	160.145	159.900	0.145	160.043	159.960	0.043	160.014	159.975	0.014	160.000	159.975	0.000
200	MAX	200.530	200.000	0.820	200.285	200.000	0.400	200.122	200.000	0.168	200.061	200.000	0.090	200.046	200.000	0.075
	MIN	200.240	199.710	0.240	200.170	199.885	0.170	200.050	199.954	0.050	200.015	199.971	0.015	200.000	199.971	0.000
250	MAX	250.570	250.000	0.860	250.285	250.000	0.400	250.122	250.000	0.168	250.061	250.000	0.090	250.046	250.000	0.075
	MIN	250.280	249.710	0.280	250.170	249.885	0.170	250.050	249.954	0.050	250.015	249.971	0.015	250.000	249.971	0.000
300	MAX	300.650	300.000	0.970	300.320	300.000	0.450	300.137	300.000	0.189	300.069	300.000	0.101	300.052	300.000	0.084
	MIN	300.330	299.680	0.330	300.190	299.870	0.190	300.056	299.948	0.056	300.017	299.968	0.017	300.000	299.968	0.000
400	MAX	400.760	400.000	1.120	400.350	400.000	0.490	400.151	400.000	0.208	400.075	400.000	0.111	400.057	400.000	0.093
	MIN	400.400	399.640	0.400	400.210	399.860	0.210	400.062	399.943	0.062	400.018	399.964	0.018	400.000	399.964	0.000
500	MAX	500.880	500.000	1.280	500.385	500.000	0.540	500.165	500.000	0.228	500.083	500.000	0.123	500.063	500.000	0.103
	MIN	500.480	499.600	0.480	500.230	499.845	0.230	500.068	499.937	0.068	500.020	499.960	0.020	500.000	499.960	0.000

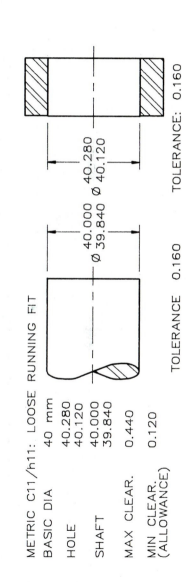

TOLERANCE: 0.160

TOLERANCE 0.160

METRIC C11/h11: LOOSE RUNNING FIT

BASIC DIA	40 mm
HOLE	40.280 / 40.120
SHAFT	40.000 / 39.840
MAX CLEAR.	0.440
MIN CLEAR. (ALLOWANCE)	0.120

TOLERANCE 0.160

Appendix 47

Preferred shaft basis transition and interference fits—cylindrical fits

**AMERICAN NATIONAL STANDARD
PREFERRED METRIC LIMITS AND FITS**

ANSI B4.2-1978

Dimensions in mm.

BASIC SIZE		LOCATIONAL TRANSN. Hole K7	Shaft h6	Fit	LOCATIONAL TRANSN. Hole N7	Shaft h6	Fit	LOCATIONAL INTERF. Hole P7	Shaft h6	Fit	MEDIUM DRIVE Hole S7	Shaft h6	Fit	FORCE Hole U7	Shaft h6	Fit
1	MAX	1.000	1.000	0.006	0.996	1.000	0.002	0.994	1.000	0.000	0.986	1.000	-0.008	0.982	1.000	-0.012
	MIN	0.990	0.994	-0.010	0.986	0.994	-0.014	0.984	0.994	-0.016	0.976	0.994	-0.024	0.972	0.994	-0.028
1.2	MAX	1.200	1.200	0.006	1.196	1.200	0.002	1.194	1.200	0.000	1.186	1.200	-0.008	1.182	1.200	-0.012
	MIN	1.190	1.194	-0.010	1.186	1.194	-0.014	1.184	1.194	-0.016	1.176	1.194	-0.024	1.172	1.194	-0.028
1.6	MAX	1.600	1.600	0.006	1.596	1.600	0.002	1.594	1.600	0.000	1.586	1.600	-0.008	1.582	1.600	-0.012
	MIN	1.590	1.594	-0.010	1.586	1.594	-0.014	1.584	1.594	-0.016	1.576	1.594	-0.024	1.572	1.594	-0.028
2	MAX	2.000	2.000	0.006	1.996	2.000	0.002	1.994	2.000	0.000	1.986	2.000	-0.008	1.982	2.000	-0.012
	MIN	1.990	1.994	-0.010	1.986	1.994	-0.014	1.984	1.994	-0.016	1.976	1.994	-0.024	1.972	1.994	-0.028
2.5	MAX	2.500	2.500	0.006	2.496	2.500	0.002	2.494	2.500	0.000	2.486	2.500	-0.008	2.482	2.500	-0.012
	MIN	2.490	2.494	-0.010	2.486	2.494	-0.014	2.484	2.494	-0.016	2.476	2.494	-0.024	2.472	2.494	-0.028
3	MAX	3.000	3.000	0.006	2.996	3.000	0.002	2.994	3.000	0.000	2.986	3.000	-0.008	2.982	3.000	-0.012
	MIN	2.990	2.994	-0.010	2.986	2.994	-0.014	2.984	2.994	-0.016	2.976	2.994	-0.024	2.972	2.994	-0.028
4	MAX	4.003	4.000	0.011	3.996	4.000	0.004	3.992	4.000	0.000	3.985	4.000	-0.007	3.981	4.000	-0.011
	MIN	3.991	3.992	-0.009	3.984	3.992	-0.016	3.980	3.992	-0.020	3.973	3.992	-0.027	3.969	3.992	-0.031
5	MAX	5.003	5.000	0.011	4.996	5.000	0.004	4.992	5.000	0.000	4.985	5.000	-0.007	4.981	5.000	-0.011
	MIN	4.991	4.992	-0.009	4.984	4.992	-0.016	4.980	4.992	-0.020	4.973	4.992	-0.027	4.969	4.992	-0.031
6	MAX	6.003	6.000	0.011	5.996	6.000	0.004	5.992	6.000	0.000	5.985	6.000	-0.007	5.981	6.000	-0.011
	MIN	5.991	5.992	-0.009	5.984	5.992	-0.016	5.980	5.992	-0.020	5.973	5.992	-0.027	5.969	5.992	-0.031
8	MAX	8.005	8.000	0.014	7.996	8.000	0.005	7.991	8.000	0.000	7.983	8.000	-0.008	7.978	8.000	-0.013
	MIN	7.990	7.991	-0.010	7.981	7.991	-0.019	7.976	7.991	-0.024	7.968	7.991	-0.032	7.963	7.991	-0.037
10	MAX	10.005	10.000	0.014	9.996	10.000	0.005	9.991	10.000	0.000	9.983	10.000	-0.008	9.978	10.000	-0.013
	MIN	9.991	9.991	-0.010	9.981	9.991	-0.019	9.976	9.991	-0.024	9.968	9.991	-0.032	9.963	9.991	-0.037
12	MAX	12.006	12.000	0.017	11.995	12.000	0.006	11.989	12.000	0.000	11.979	12.000	-0.010	11.974	12.000	-0.015
	MIN	11.988	11.989	-0.012	11.977	11.989	-0.023	11.971	11.989	-0.029	11.961	11.989	-0.039	11.956	11.989	-0.044
16	MAX	16.006	16.000	0.017	15.995	16.000	0.006	15.989	16.000	0.000	15.979	16.000	-0.010	15.974	16.000	-0.015
	MIN	15.988	15.989	-0.012	15.977	15.989	-0.023	15.971	15.989	-0.029	15.961	15.989	-0.039	15.956	15.989	-0.044
20	MAX	20.006	20.000	0.019	19.993	20.000	0.006	19.986	20.000	-0.001	19.973	20.000	-0.014	19.967	20.000	-0.020
	MIN	19.985	19.987	-0.015	19.972	19.987	-0.028	19.965	19.987	-0.035	19.952	19.987	-0.048	19.946	19.987	-0.054
25	MAX	25.006	25.000	0.019	24.993	25.000	0.006	24.986	25.000	-0.001	24.973	25.000	-0.014	24.960	25.000	-0.027
	MIN	24.985	24.987	-0.015	24.972	24.987	-0.028	24.965	24.987	-0.035	24.952	24.987	-0.048	24.939	24.987	-0.061
30	MAX	30.006	30.000	0.019	29.993	30.000	0.006	29.986	30.000	-0.001	29.973	30.000	-0.014	29.960	30.000	-0.027
	MIN	29.985	29.987	-0.015	29.972	29.987	-0.028	29.965	29.987	-0.035	29.952	29.987	-0.048	29.939	29.987	-0.061

Cont.

Appendix 47
Preferred shaft basis transition and interference fits—cylindrical fits (cont.)

AMERICAN NATIONAL STANDARD
PREFERRED METRIC LIMITS AND FITS

ANSI B4.2–1978

Dimensions in mm.

BASIC SIZE		LOCATIONAL TRANSN. Hole K7	Shaft h6	Fit	LOCATIONAL TRANSN. Hole N7	Shaft h6	Fit	LOCATIONAL INTERF. Hole P7	Shaft h6	Fit	MEDIUM DRIVE Hole S7	Shaft h6	Fit	FORCE Hole U7	Shaft h6	Fit
40	MAX	40.007	40.000	0.023	39.992	40.000	0.008	39.983	40.000	-0.001	39.966	40.000	-0.018	39.949	40.000	-0.035
	MIN	39.982	39.984	-0.018	39.967	39.984	-0.033	39.958	39.984	-0.042	39.941	39.984	-0.059	39.924	39.984	-0.076
50	MAX	50.007	50.000	0.023	49.992	50.000	0.008	49.983	50.000	-0.001	49.966	50.000	-0.018	49.939	50.000	-0.045
	MIN	49.982	49.984	-0.018	49.967	49.984	-0.033	49.958	49.984	-0.042	49.941	49.984	-0.059	49.914	49.984	-0.086
60	MAX	60.009	60.000	0.028	59.991	60.000	0.010	59.979	60.000	-0.002	59.958	60.000	-0.023	59.924	60.000	-0.087
	MIN	59.979	59.981	-0.021	59.961	59.981	-0.039	59.949	59.981	-0.051	59.928	59.981	-0.072	59.894	59.981	-0.106
80	MAX	80.009	80.000	0.028	79.991	80.000	0.010	79.979	80.000	-0.002	79.952	80.000	-0.029	79.909	80.000	-0.072
	MIN	79.979	79.981	-0.021	79.961	79.981	-0.039	79.949	79.981	-0.051	79.922	79.981	-0.078	79.879	79.981	-0.121
100	MAX	100.010	100.000	0.032	99.990	100.000	0.012	99.976	100.000	-0.002	99.942	100.000	-0.036	99.889	100.000	-0.089
	MIN	99.975	99.978	-0.025	99.955	99.978	-0.045	99.941	99.978	-0.059	99.907	99.978	-0.093	99.854	99.978	-0.146
120	MAX	120.010	120.000	0.032	119.990	120.000	0.012	119.976	120.000	-0.002	119.934	120.000	-0.044	119.869	120.000	-0.109
	MIN	119.975	119.978	-0.025	119.955	119.978	-0.045	119.941	119.978	-0.059	119.899	119.978	-0.101	119.834	119.978	-0.166
160	MAX	160.012	160.000	0.037	159.988	160.000	0.013	159.972	160.000	-0.003	159.915	160.000	-0.060	159.825	160.000	-0.150
	MIN	159.972	159.975	-0.028	159.948	159.975	-0.052	159.932	159.975	-0.068	159.875	159.975	-0.125	159.785	159.975	-0.215
200	MAX	200.013	200.000	0.042	199.986	200.000	0.015	199.967	200.000	-0.004	199.895	200.000	-0.076	199.781	200.000	-0.190
	MIN	199.967	199.971	-0.033	199.940	199.971	-0.060	199.921	199.971	-0.079	199.849	199.971	-0.151	199.735	199.971	-0.265
250	MAX	250.013	250.000	0.042	249.986	250.000	0.015	249.967	250.000	-0.004	249.877	250.000	-0.094	249.733	250.000	-0.238
	MIN	249.967	249.971	-0.033	249.940	249.971	-0.060	249.921	249.971	-0.079	249.831	249.971	-0.169	249.687	249.971	-0.313
300	MAX	300.016	300.000	0.048	299.986	300.000	0.018	299.964	300.000	-0.004	299.850	300.000	-0.118	299.670	300.000	-0.298
	MIN	299.964	299.968	-0.036	299.934	299.968	-0.066	299.912	299.968	-0.088	299.798	299.968	-0.202	299.618	299.968	-0.382
400	MAX	400.017	400.000	0.053	399.984	400.000	0.020	399.959	400.000	-0.005	399.813	400.000	-0.151	399.586	400.000	-0.378
	MIN	399.960	399.964	-0.040	399.927	399.964	-0.073	399.902	399.964	-0.098	399.756	399.964	-0.244	399.529	399.964	-0.471
500	MAX	500.018	500.000	0.058	499.983	500.000	0.023	499.955	500.000	-0.005	499.771	500.000	-0.189	499.483	500.000	-0.477
	MIN	499.955	499.960	-0.045	499.920	499.960	-0.080	499.892	499.960	-0.108	499.708	499.960	-0.292	499.420	499.960	-0.580

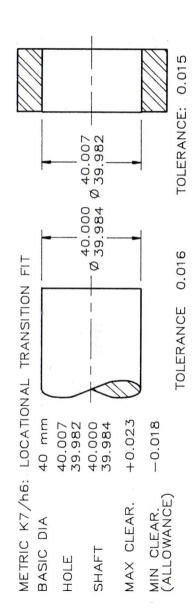

METRIC K7/h6: LOCATIONAL TRANSITION FIT

BASIC DIA   40 mm

HOLE   40.007 / 39.982

SHAFT   40.000 / 39.984

MAX CLEAR.   +0.023

MIN CLEAR.   -0.018
(ALLOWANCE)

TOLERANCE 0.016   TOLERANCE: 0.015

Ø 40.000 / Ø 39.984
Ø 40.007 / Ø 39.982

Appendix 48

Hole sizes for non-preferred diameters (dimensions are in millimeters)

Basic Size		C11	D9	F8	G7	H7	H8	H9	H11	K7	N7	P7	S7	U7
OVER 0 / TO 3		+0.120 +0.060	+0.045 +0.020	+0.020 +0.006	+0.012 +0.002	+0.010 0.000	+0.014 0.000	+0.025 0.000	+0.060 0.000	0.000 −0.010	−0.004 −0.014	−0.006 −0.016	−0.014 −0.024	−0.018 −0.028
OVER 3 / TO 6		+0.145 +0.070	+0.060 +0.030	+0.028 +0.010	+0.016 +0.004	+0.012 0.000	+0.018 0.000	+0.030 0.000	+0.075 0.000	+0.003 −0.009	−0.004 −0.016	−0.008 −0.020	−0.015 −0.027	−0.019 −0.031
OVER 6 / TO 10		+0.170 +0.080	+0.076 +0.040	+0.035 +0.013	+0.020 +0.005	+0.015 0.000	+0.022 0.000	+0.036 0.000	+0.090 0.000	+0.005 −0.010	−0.004 −0.019	−0.009 −0.024	−0.017 −0.032	−0.022 −0.037
OVER 10 / TO 14		+0.205 +0.095	+0.093 +0.050	+0.043 +0.016	+0.024 +0.006	+0.018 0.000	+0.027 0.000	+0.043 0.000	+0.110 0.000	+0.006 −0.012	−0.005 −0.023	−0.011 −0.029	−0.021 −0.039	−0.026 −0.044
OVER 14 / TO 18		+0.205 +0.095	+0.093 +0.050	+0.043 +0.016	+0.024 +0.006	+0.018 0.000	+0.027 0.000	+0.043 0.000	+0.110 0.000	+0.006 −0.012	−0.005 −0.023	−0.011 −0.029	−0.021 −0.039	−0.026 −0.044
OVER 18 / TO 24		+0.240 +0.110	+0.117 +0.065	+0.053 +0.020	+0.028 +0.007	+0.021 0.000	+0.033 0.000	+0.052 0.000	+0.130 0.000	+0.006 −0.015	−0.007 −0.028	−0.014 −0.035	−0.027 −0.048	−0.033 −0.054
OVER 24 / TO 30		+0.240 +0.110	+0.117 +0.065	+0.053 +0.020	+0.028 +0.007	+0.021 0.000	+0.033 0.000	+0.052 0.000	+0.130 0.000	+0.006 −0.015	−0.007 −0.028	−0.014 −0.035	−0.027 −0.048	−0.040 −0.061
OVER 30 / TO 40		+0.280 +0.120	+0.142 +0.080	+0.064 +0.025	+0.034 +0.009	+0.025 0.000	+0.039 0.000	+0.062 0.000	+0.160 0.000	+0.007 −0.018	−0.008 −0.033	−0.017 −0.042	−0.034 −0.059	−0.051 −0.076
OVER 40 / TO 50		+0.290 +0.130	+0.142 +0.080	+0.064 +0.025	+0.034 +0.009	+0.025 0.000	+0.039 0.000	+0.062 0.000	+0.160 0.000	+0.007 −0.018	−0.008 −0.033	−0.017 −0.042	−0.034 −0.059	−0.061 −0.086
OVER 50 / TO 65		+0.330 +0.140	+0.174 +0.100	+0.076 +0.030	+0.040 +0.010	+0.030 0.000	+0.046 0.000	+0.074 0.000	+0.190 0.000	+0.009 −0.021	−0.009 −0.039	−0.021 −0.051	−0.042 −0.072	−0.076 −0.106
OVER 65 / TO 80		+0.340 +0.150	+0.174 +0.100	+0.076 +0.030	+0.040 +0.010	+0.030 0.000	+0.046 0.000	+0.074 0.000	+0.190 0.000	+0.009 −0.021	−0.009 −0.039	−0.021 −0.051	−0.048 −0.078	−0.091 −0.121
OVER 80 / TO 100		+0.390 +0.170	+0.207 +0.120	+0.090 +0.036	+0.047 +0.012	+0.035 0.000	+0.054 0.000	+0.087 0.000	+0.220 0.000	+0.010 −0.025	−0.010 −0.045	−0.024 −0.059	−0.058 −0.093	−0.111 −0.146

Cont.

Appendix 48

Hole sizes for non-preferred diameters (dimensions are in millimeters) (cont.)

Basic Size	c11	d9	f8	g7	h7	h8	h9	h11	k7	n7	p7	s7	u7
OVER 100 TO 120	+0.400 +0.180	+0.207 +0.120	+0.090 +0.036	+0.047 +0.012	+0.035 0.000	+0.054 0.000	+0.087 0.000	+0.220 0.000	+0.010 −0.025	−0.010 −0.045	−0.024 −0.059	−0.066 −0.101	−0.131 −0.166
OVER 120 TO 140	+0.450 +0.200	+0.245 +0.145	+0.106 +0.043	+0.054 +0.014	+0.040 0.000	+0.063 0.000	+0.100 0.000	+0.250 0.000	+0.012 −0.028	−0.012 −0.052	−0.028 −0.068	−0.077 −0.117	−0.155 −0.195
OVER 140 TO 160	+0.460 +0.210	+0.245 +0.145	+0.106 +0.043	+0.054 +0.014	+0.040 0.000	+0.063 0.000	+0.100 0.000	+0.250 0.000	+0.012 −0.028	−0.012 −0.052	−0.028 −0.068	−0.085 −0.125	−0.175 −0.215
OVER 160 TO 180	+0.480 +0.230	+0.245 +0.145	+0.106 +0.043	+0.054 +0.014	+0.040 0.000	+0.063 0.000	+0.100 0.000	+0.250 0.000	+0.012 −0.028	−0.012 −0.052	−0.028 −0.068	−0.093 −0.133	−0.195 −0.235
OVER 180 TO 200	+0.530 +0.240	+0.285 +0.170	+0.122 +0.050	+0.061 +0.015	+0.046 0.000	+0.072 0.000	+0.115 0.000	+0.290 0.000	+0.013 −0.033	−0.014 −0.060	−0.033 −0.079	−0.105 −0.151	−0.219 −0.265
OVER 200 TO 225	+0.550 +0.260	+0.285 +0.170	+0.122 +0.050	+0.061 +0.015	+0.046 0.000	+0.072 0.000	+0.115 0.000	+0.290 0.000	+0.013 −0.033	−0.014 −0.060	−0.033 −0.079	−0.113 −0.159	−0.241 −0.287
OVER 225 TO 250	+0.570 +0.280	+0.285 +0.170	+0.122 +0.050	+0.061 +0.015	+0.046 0.000	+0.072 0.000	+0.115 0.000	+0.290 0.000	+0.013 −0.033	−0.014 −0.060	−0.033 −0.079	−0.123 −0.169	−0.267 −0.313
OVER 250 TO 280	+0.620 +0.300	+0.320 +0.190	+0.137 +0.056	+0.069 +0.017	+0.052 0.000	+0.081 0.000	+0.130 0.000	+0.320 0.000	+0.016 −0.036	−0.014 −0.066	−0.036 −0.088	−0.138 −0.190	−0.295 −0.347
OVER 280 TO 315	+0.650 +0.330	+0.320 +0.190	+0.137 +0.056	+0.069 0.017	+0.052 0.000	+0.081 0.000	+0.130 0.000	+0.320 0.000	+0.016 −0.036	−0.014 −0.066	−0.036 −0.088	−0.150 −0.202	−0.330 −0.382
OVER 315 TO 355	+0.720 +0.360	+0.350 +0.210	+0.151 +0.062	+0.075 +0.018	+0.057 0.000	+0.089 0.000	+0.140 0.000	+0.360 0.000	+0.017 −0.040	−0.016 −0.073	−0.041 −0.098	−0.169 −0.226	−0.369 −0.426
OVER 355 TO 400	+0.760 +0.400	+0.350 +0.210	+0.151 +0.062	+0.075 +0.018	+0.057 0.000	+0.089 0.000	+0.140 0.000	+0.360 0.000	+0.017 −0.040	−0.016 −0.073	−0.041 −0.098	−0.187 −0.244	−0.414 −0.471
OVER 400 TO 450	+0.840 +0.440	+0.385 +0.230	+0.165 +0.068	+0.083 +0.020	+0.063 0.000	+0.097 0.000	+0.155 0.000	+0.400 0.000	+0.018 −0.045	−0.017 −0.080	−0.045 −0.108	−0.209 −0.272	−0.467 −0.530
OVER 450 TO 500	+0.880 +0.480	+0.385 +0.230	+0.165 +0.068	+0.083 +0.020	+0.063 0.000	+0.097 0.000	+0.155 0.000	+0.400 0.000	+0.018 −0.045	−0.017 −0.080	−0.045 −0.108	−0.229 −0.292	−0.517 −0.580

Appendix 49
Shaft sizes for non-preferred diameters (dimensions are in millimeters)

Basic Size	c11	d9	f7	g6	h6	h7	h9	h11	k6	n6	p6	s6	u6
OVER 0 TO 3	−0.060 / −0.120	−0.020 / −0.045	−0.006 / −0.016	−0.002 / −0.008	0.000 / −0.006	0.000 / −0.010	0.000 / −0.025	0.000 / −0.060	+0.006 / 0.000	+0.010 / +0.004	+0.012 / +0.006	+0.020 / +0.014	+0.024 / +0.018
OVER 3 TO 6	−0.070 / −0.145	−0.030 / −0.060	−0.010 / −0.022	−0.004 / −0.012	0.000 / −0.008	0.000 / −0.012	0.000 / −0.030	0.000 / −0.075	+0.009 / +0.001	+0.016 / +0.008	+0.020 / +0.012	+0.027 / +0.019	+0.031 / +0.023
OVER 6 TO 10	−0.080 / −0.170	−0.040 / −0.076	−0.013 / −0.028	−0.005 / −0.014	0.000 / −0.009	0.000 / −0.015	0.000 / −0.036	0.000 / −0.090	+0.010 / +0.001	+0.019 / +0.010	+0.024 / +0.015	+0.032 / +0.023	+0.037 / +0.028
OVER 10 TO 14	−0.095 / −0.205	−0.050 / −0.093	−0.016 / −0.034	−0.006 / −0.017	0.000 / −0.011	0.000 / −0.018	0.000 / −0.043	0.000 / −0.110	+0.012 / +0.001	+0.023 / +0.012	+0.029 / +0.018	+0.039 / +0.028	+0.044 / +0.033
OVER 14 TO 18	−0.095 / −0.205	−0.050 / −0.093	−0.016 / −0.034	−0.006 / −0.017	0.000 / −0.011	0.000 / −0.018	0.000 / −0.043	0.000 / −0.110	+0.012 / +0.001	+0.023 / +0.012	+0.029 / +0.018	+0.039 / +0.028	+0.044 / +0.033
OVER 18 TO 24	−0.110 / −0.240	−0.065 / −0.117	−0.020 / −0.041	−0.007 / −0.020	0.000 / −0.013	0.000 / −0.021	0.000 / −0.052	0.000 / −0.130	+0.015 / +0.002	+0.028 / +0.015	+0.035 / +0.022	+0.048 / +0.035	+0.054 / +0.041
OVER 24 TO 30	−0.110 / −0.240	−0.065 / −0.117	−0.020 / −0.041	−0.007 / −0.020	0.000 / −0.013	0.000 / −0.021	0.000 / −0.052	0.000 / −0.130	+0.015 / +0.002	+0.028 / +0.015	+0.035 / +0.022	+0.048 / +0.035	+0.061 / +0.048
OVER 30 TO 40	−0.120 / −0.280	−0.080 / −0.142	−0.025 / −0.050	−0.009 / −0.025	0.000 / −0.016	0.000 / −0.025	0.000 / −0.062	0.000 / −0.160	+0.018 / +0.002	+0.033 / +0.017	+0.042 / +0.026	+0.059 / +0.043	+0.076 / +0.060
OVER 40 TO 50	−0.130 / −0.290	−0.080 / −0.142	−0.025 / −0.050	−0.009 / −0.025	0.000 / −0.016	0.000 / −0.025	0.000 / −0.062	0.000 / −0.160	+0.018 / +0.002	+0.033 / +0.017	+0.042 / +0.026	+0.059 / +0.043	+0.086 / +0.070
OVER 50 TO 65	−0.140 / −0.330	−0.100 / −0.174	−0.030 / −0.060	−0.010 / −0.029	0.000 / −0.019	0.000 / −0.030	0.000 / −0.074	0.000 / −0.190	+0.021 / +0.002	+0.039 / +0.020	+0.051 / −0.032	+0.072 / +0.053	+0.106 / +0.087
OVER 65 TO 80	−0.150 / −0.340	−0.100 / −0.174	−0.030 / −0.060	−0.010 / −0.029	0.000 / −0.019	0.000 / −0.030	0.000 / −0.074	0.000 / −0.190	+0.021 / +0.002	+0.039 / +0.020	+0.051 / +0.032	+0.078 / +0.059	+0.121 / +0.102
OVER 80 TO 100	−0.170 / −0.390	−0.120 / −0.207	−0.036 / −0.071	−0.012 / −0.034	0.000 / −0.022	0.000 / −0.035	0.000 / −0.087	0.000 / −0.220	+0.025 / +0.003	+0.045 / +0.023	+0.059 / +0.037	+0.093 / +0.071	+0.146 / +0.124

Cont.

Appendix 49

Shaft sizes for non-preferred diameters (dimensions are in millimeters) (cont.)

Basic Size	c11	d9	f7	g6	h6	h7	h9	h11	k6	n6	p6	s6	u6
OVER 100 TO 120	−0.180 −0.400	−0.120 −0.207	−0.036 −0.071	−0.012 −0.034	0.000 −0.022	0.000 −0.035	0.000 −0.087	0.000 −0.220	+0.025 +0.003	+0.045 +0.023	+0.059 +0.037	+0.101 +0.079	+0.166 +0.144
OVER 120 TO 140	−0.200 −0.450	−0.145 −0.245	−0.043 −0.083	−0.014 −0.039	0.000 −0.025	0.000 −0.040	0.000 −0.100	0.000 −0.250	+0.028 +0.003	+0.052 +0.027	+0.068 +0.043	+0.117 +0.092	+0.195 +0.170
OVER 140 TO 160	−0.210 −0.460	−0.145 −0.245	−0.043 −0.083	−0.014 −0.039	0.000 −0.025	0.000 −0.040	0.000 −0.100	0.000 −0.250	+0.028 +0.003	+0.052 +0.027	+0.068 +0.043	+0.125 +0.100	+0.215 +0.190
OVER 160 TO 180	−0.230 −0.480	−0.145 −0.245	−0.043 −0.083	−0.014 −0.039	0.000 −0.025	0.000 −0.040	0.000 −0.100	0.000 −0.250	+0.028 +0.003	+0.052 +0.027	+0.068 +0.043	+0.133 +0.108	+0.235 +0.210
OVER 180 TO 200	−0.240 −0.530	−0.170 −0.285	−0.050 −0.096	−0.015 −0.044	0.000 −0.029	0.000 −0.046	0.000 −0.115	0.000 −0.290	+0.033 +0.004	+0.060 +0.031	+0.079 +0.050	+0.151 +0.122	+0.265 +0.236
OVER 200 TO 225	−0.260 −0.550	−0.170 −0.285	−0.050 −0.096	−0.015 −0.044	0.000 −0.029	0.000 −0.046	0.000 −0.115	0.000 −0.290	+0.033 +0.004	+0.060 +0.031	+0.079 +0.050	+0.159 +0.130	+0.287 +0.258
OVER 225 TO 250	−0.280 −0.570	−0.170 −0.285	−0.050 −0.096	−0.015 −0.044	0.000 −0.029	0.000 −0.046	0.000 −0.115	0.000 −0.290	+0.033 +0.004	+0.060 +0.031	+0.079 +0.050	+0.169 +0.140	+0.313 +0.284
OVER 250 TO 280	−0.300 −0.620	−0.190 −0.320	−0.056 −0.108	−0.017 −0.049	0.000 −0.032	0.000 −0.052	0.000 −0.130	0.000 −0.320	+0.036 +0.004	+0.066 +0.034	+0.088 +0.056	+0.190 +0.158	+0.347 +0.315
OVER 280 TO 315	−0.330 −0.650	−0.190 −0.320	−0.056 −0.108	−0.017 −0.049	0.000 −0.032	0.000 −0.052	0.000 −0.130	0.000 −0.320	+0.036 +0.004	+0.066 +0.034	+0.088 +0.056	+0.202 +0.170	+0.382 +0.350
OVER 315 TO 355	−0.360 −0.720	−0.210 −0.350	−0.062 −0.119	−0.018 −0.054	0.000 −0.036	0.000 −0.057	0.000 −0.140	0.000 −0.360	+0.040 +0.004	+0.073 +0.037	+0.098 +0.062	+0.226 +0.190	+0.426 +0.390
OVER 355 TO 400	−0.400 −0.760	−0.210 −0.350	−0.062 −0.119	−0.018 −0.054	0.000 −0.036	0.000 −0.057	0.000 −0.140	0.000 −0.360	+0.040 +0.004	+0.073 +0.037	+0.098 +0.062	+0.244 +0.208	+0.471 +0.435
OVER 400 TO 450	−0.440 −0.840	−0.230 −0.385	−0.068 −0.131	−0.020 −0.060	0.000 −0.040	0.000 −0.063	0.000 −0.155	0.000 −0.400	+0.045 +0.005	+0.080 +0.040	+0.108 +0.068	+0.272 +0.232	+0.530 +0.490
OVER 450 TO 500	−0.480 −0.880	−0.230 −0.385	−0.068 −0.131	−0.020 −0.060	0.000 −0.040	0.000 −0.063	0.000 −0.155	0.000 −0.400	+0.045 +0.005	+0.080 +0.040	+0.108 +0.068	+0.292 +0.252	+0.580 +0.540

Appendix 50
Engineering formulas

Motion

S = distance (inches, feet, miles)
t = time (seconds, minutes, hours)
v = average velocity (feet per second, miles per hour, etc.
v_1 = initial velocity
v_2 = final velocity
a = acceleration (feet per second per second)

(1) $S = vt$

(2) $V(\text{avg.}) = \dfrac{v_2 - v_1}{2}$

(3) $S = \left(\dfrac{v_1 + v_2}{2}\right)\dfrac{\text{ft}}{\text{sec}}(t \text{ sec})$

(4) $a = \dfrac{v_2 - v_1}{t}$

(5) $S = v_1 t + \frac{1}{2}at^2$

Angular Motion

V = linear velocity
N = number of revolutions per min
θ = angular distance in radians
1 radian = $360°/2\pi = 57.3°$
ω(omega) = average angular velocity
        = θ/t (rad per sec, rev per min)
ω_1 = initial velocity
ω_2 = final velocity
α(alpha) = angular acceleration = rad per sec^2
S = length of arc
r = radius of arc
D = diameter

(6) $\theta = (\text{avg. } \omega)t$

(7) $\omega(\text{avg.}) = \dfrac{\omega_2 + \omega_1}{2}$

(8) $\alpha = \dfrac{\omega_2 - \omega_1}{t}$

(9) $\omega_2 = \omega_1 + \alpha t$

(10) $\theta = \omega_1 t + \dfrac{\alpha t^2}{2}$

(11) $V = \pi DN$ or $V = r\omega$ (ft per sec, ft per min, etc.)

(12) $\omega = \dfrac{2\pi rn}{r} = 2\pi N$

Force and Acceleration

F = force (pounds)
M = mass
a = acceleration (ft per sec^2)
g = gravitational acceleration = 32.2 ft/sec^2
W = weight (pounds)
$M = F/a = W/g$ (units of mass in slugs)

Work

W = work (ft · lb)
F = force (lb)
d = distance
$W = Fd$

Power

W = work
t = time

1 horsepower = $550 \dfrac{\text{ft} \cdot \text{lb}}{\text{sec}}$

Avg. power = $\dfrac{W}{t} = \dfrac{\text{ft} \cdot \text{lb}}{\text{sec}}$ or $\dfrac{\text{ft} \cdot \text{lb}}{\text{min}}$ etc.

Kinetic Energy

W = weight
V = velocity
g = 32.2 ft per sec^2
K.E. = $WV^2/2g$

Appendix 51
Grading graph

This graph can be used to determine the individual grades of members of a team and to compute grade averages for those who do extra assignments.

The percent participation of each team member should be determined by the team as a whole (see Chapter 2 problems).

Example: written or oral report grades

Overall team grade: 82

Team members N=5	Contribution C=%	F=CN	Grade (graph)
J. Doe	20%	100	82.0
H. Brown	16%	80	76.4
L. Smith	24%	120	86.0
R. Black	20%	100	82.0
T. Jones	20%	100	82.0
	100%		

Example: quiz or problem sheet grades

Number assigned: 30
Number extra: 6

Total 36

Average grade for total (36): 82

$$F = \frac{\text{No. completed} \times 100}{\text{No. assigned}} = \frac{36 \times 100}{30} = 120$$

Final grade (from graph): 86.0

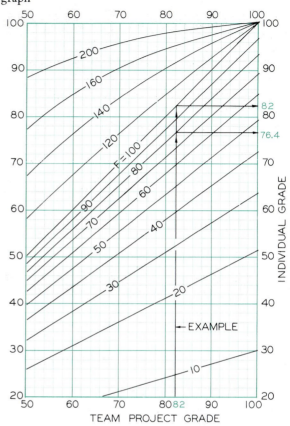

Fig. A51.−1. Grading graph.

Appendix 52
Lisp programs

The following programs were written by Professor Leendert Kersten of the University of Nebraska at Lincoln and they have been reproduced here with his permission. These programs were introduced in Chapter 20, in which the principles of descriptive geometry are covered. These very valuable programs can be duplicated and added as supplements to your AutoCAD software.

```
)
(defun C:PARALLEL ( )
  (setvar "aperture" 5)
    (setq sp (getpoint "\nSelect START point of parallel line:"))
    (setq ep (getpoint "\nSelect END point of parallel line:"))
    (setvar "osmode" 1)
    (setq sl (getpoint "\nSelect 1st point on line for parallelism:"))
    (setq el (getpoint "\nSelect 2nd point on line for parallelism:"))
    (setvar "osmode" 0)
    (setq pa (angle sl el))
    (setq la (angle sp ep))
    (setq ll (distance sp ep))
    (setq m -1)
    (setq d 0)
    (if (> pa d) (setq m 1))
    (if (< la d) (setq d 1))
    (if (/= m d) (setq pa (+ pa 3.141593)))
    (setq ep (polar sp pa ll))
    (setvar "cmdecho" 0)
    (command "line" sp ep"")
    (restore)
)

(defun C:PERPLINE ( )
  (setvar "aperture" 5) (setvar "cmdecho" 0)
    (setq sp (getpoint "\nSelect START point of perpendicular line:"))
    (setvar "osmode" 128)
    (setq cc (getpoint sp "\nSelect ANY point on line to which perp'lr:"))
    (setq beta (angle sp cc)) (setvar "osmode" 0)
(setq ep
(getpoint "\nSelect END point of desired perpendicular (for length only): "))
    (setq length (distance sp ep))
    (setq ep (polar sp beta length))
    (command "line" sp ep "")
    (restore)
)
```

Appendix 52
Lisp programs (cont.)

```
(defun C: TRANSFER ( )
    (setvar "aperture" 5) (setvar "cmdecho" 0)
    (setvar "osmode" 1)
    (setq aa (getpoint "\nSelect start of transfer distance:"))
    (setvar "osmode" 128)
    (setq bb (getpoint aa "\nSelect the reference plane:"))
    (setq length (distance aa bb))
    (setvar "osmode" 1)
    (setq cc (getpoint "\nSelect point to be projected:"))
    (setvar "osmode" 128)
    (setq dd (getpoint cc "\nSelect other reference plane:"))
(setvar "osmode" 0)
    (setq alpha (angle cc dd))
    (setq ep (polar dd alpha length))
    (COMMAND "CIRCLE" EP 0.05)
    (restore)
)

(defun RESTORE ( )
    (setvar "aperture" 10)
    (setvar "cmdecho" 1)
    (setvar "osmode" 0)
(defun command: ( )
    (defun *error* (st)
        (setvar "osmode" 0)
    (princ))
    (quit))
)

(defun C:COPYDIST ( )
    (setvar "aperture"5)
    (setvar "cmdecho" 0)
(setvar "osmode" 1)
    (setq p1 (getpoint "\nSelect start point of line distance to be copied:"))
    (setq p2 (getpoint "\nEnd point?:"))
(setvar "osmode" 0)
    (setq dist (distance p1 p2))
    (setq p1 (getpoint "\nStart point of new distance location:"))
    (setq ang (getangle p1 "\nWhich direction?:"))
    (setq p2 (polar p1 ang dist))
(setvar "osmode" 0) (command "circle" p2 0.05)
    (restore)
```

Appendix 52
Lisp programs (cont.)

```
(defun C:BISECT ( )
  (setvar "aperture" 5)
  (setvar "osmode" 32)
  (setq sp (getpoint "\nSelect Corner of angle:"))
  (setvar "osmode" 2)
  (setq aa (getpoint "\nSelect first side (remember CCW):"))
  (setq alpha (angle sp aa))
  (setq bb (getpoint "\nSelect other side:"))
  (setvar "osmode" 0)
  (setq beta (angle sp bb))
  (setq m (/ ( + alpha beta) 2))
  (if (> alpha beta) (setq ang ( + pi m)) (setq ang m))
  (setq ep
    (getpoint "\nSelect endpoint of bisecting line (for length only):"))
  (setq length (distance sp ep))
  (setq ep (polar sp ang length))
  (setvar "cmdecho" 0)
  (command "line" sp ep "")
  (restore)
```

Index

(Primary Screen Menu Hierarchy)

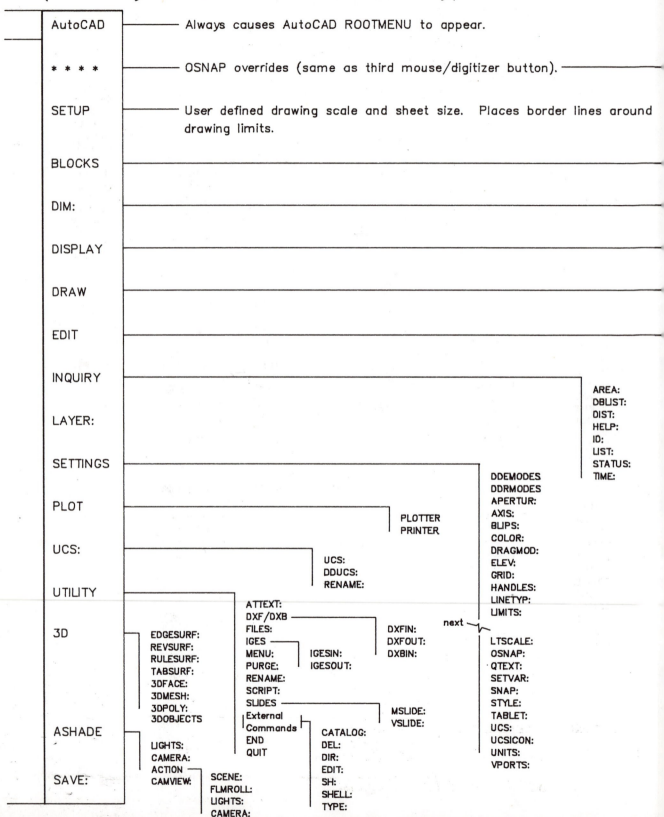

AutoCAD —— Always causes AutoCAD ROOTMENU to appear.

* * * * —— OSNAP overrides (same as third mouse/digitizer button). ——

SETUP —— User defined drawing scale and sheet size. Places border lines around drawing limits.

BLOCKS

DIM:

DISPLAY

DRAW

EDIT

INQUIRY

AREA:
DBLIST:
DIST:
HELP:
ID:
LIST:
STATUS:
TIME:

LAYER:

SETTINGS

DDEMODES
DDRMODES
APERTUR:
AXIS:
BLIPS:
COLOR:
DRAGMOD:
ELEV:
GRID:
HANDLES:
LINETYP:
LIMITS:

PLOT

PLOTTER
PRINTER

UCS:

UCS:
DDUCS:
RENAME:

UTILITY

ATTEXT:
DXF/DXB ———— DXFIN:
FILES: DXFOUT:
IGES ——   DXBIN:
MENU:
PURGE: IGESIN:
RENAME: IGESOUT:
SCRIPT:
SLIDES ——
External
Commands
END
QUIT

next

LTSCALE:
OSNAP:
QTEXT:
SETVAR:
SNAP:
STYLE:
TABLET:
UCS:
UCSICON:
UNITS:
VPORTS:

MSLIDE:
VSLIDE:

CATALOG:
DEL:
DIR:
EDIT:
SH:
SHELL:
TYPE:

3D

EDGESURF:
REVSURF:
RULESURF:
TABSURF:
3DFACE:
3DMESH:
3DPOLY:
3DOBJECTS

ASHADE

LIGHTS:
CAMERA:
ACTION
CAMVIEW:

SCENE:
FLMROLL:
LIGHTS:
CAMERA:

SAVE: